RESEARCH DECISIONS AND ESTIMATION WITH CONFIDENCE AND POWER

Up To Date Statistics Reference and Text From Classical and Recent Literature

L.E. MacCarter

Copyright © 2021 by L.E.MacCarter
ISBN-13: 978-1532721076
Library of Congress Control Number: 2021913804
Morro Bay, California

Dedicated to, and in appreciation of,
Dr. Dale H. Habeck and Dr. O.C. Ruelke

INTRODUCTION

This book is about research with an emphasis on inference, and on confidence intervals, and a rational approach to power. As is implied by the extensive bibliography, most of the material presented here has been around long enough to be verified and widely accepted by theoreticians and many applied statisticians.

Cohen (1962), who is justifiably credited with calling our attention to sample size and power, published a set of magic numbers, "small", "medium", and "large", as a set of normative effect sizes (Section 5.9). He later halfheartedly backed away from such standardized constants (1988, p. 25). Unfortunately, these magic numbers are likely to be misleading, but are persistently associated in literature with power analysis, and sample sizes. It turns out that small, medium and large are not the same for major and minor league sports, let alone different medical research groups, etc.

There is history based confusion as to how parametric statistical tests work; most authors started with an assumption that the data will be normally distributed. Normally distributed raw data is extremely rare, a fact that should almost eliminate the use of any statistical procedure that depends on the assumption of normally distributed data. And, lo and behold, the very common and widespread uses of the ANOVA are dependent on this usually invalid presumption. Even though many texts elaborate on the Central Limit Theorem, CLT, they do not make it clear that tests like Student's t are sufficiently justified for large samples by the CLT whether Student and the other authors realized it or not. The fact that, as per the CLT, the means are normally distributed with a standard error seen in the denominator of the t test saves the day. However, the fact that the CLT is not often effective for samples less than large renders Student's t less good as a small sample test than Student and many texts expected, but it is great as a large sample test that can be totally derived from the CLT. That leaves nonparametric tests which work well for many small samples and which are the only resort for samples which are not on an equal interval scale (such as IQ, and Likert scales such as survey scores).

Students need to know that the reasonable minimum difference and the sample size necessary to achieve the power (sensitivity) to detect it are all considerations that must be made before undertaking any experiment or set of observations. Confidence interval estimation, a superior way to display results, is given the attention it deserves.

> **NOTE TO GRADERS: The accuracy of the source of Z or t available to students, along with computer rounding, can make a difference in the fourth or fifth figure.** This is unavoidable, especially in a teaching environment involving homework, and in testing situations. Likewise, large sample size estimates might vary by one or two.

Although most researchers use a computer for statistical calculations, simply learning how to use computer programs, and even a vocabulary compliant with the more popular commercial software, is not going to provide a firm understanding of what to ask the computer, why to ask, how to collect the data or what the results will mean. Those questions aside, few, if any, computer packages are going to meet all the researchers needs without a lot of informed thought regarding what is eventually going to be input, and also regarding what will be subsequently output. Once you understand the material presented here, you should be able to use many of the multitude of statistical computer packages intelligently, instead of blindly risking your reputation by basing it on black box voodoo.

> This book is produced by a print on demand process, so please **notify me of any errors** and I will correct them sooner than is possible in more traditionally published books (mac@terranrobotics.com).

Acknowledgments

Mason Celli and J.K. MacCarter assisted with the editing and provided suggestions for this book.

The setting of the type and the layout of this book used L$_Y$X, a WYSIWYM word processor, that can be freely downloaded from the internet. I had to install TexLive before L$_Y$X to provide the Latex library for L$_Y$X. It was necessary to reset the Page Layout to the page size to be used in the published book. That led to the necessity to adjust the Page Margins. I have also overridden many of its features, so my typographical and other abominations are no one's fault other than my own. Please notify me of errors and I will correct them sooner than is possible in more traditionally published books. I downloaded many free programs and statistical or mathematical packages to produce this book. So, I am grateful to the L$_Y$X people and to the people who produced all the open source software used in this project. Some of the initial citations and illustrations, for instance Figure 11.3, which was almost indispensable, were found via Wikipedia. **Only** Figure 11.1, and figures that are clearly attributed in their captions to Creative Commons sources a are available for copying, the latter under only their Creative Commons limitations, that includes, but is not always limited to, citing original sources. Although not always accurate, Wikipedia is a great place from which to start a literature survey on any of a vast number of subjects. Many thanks to them, to their contributing authors, and to their financial supporters. Products mentioned in this book are the property of their respective owners and the author makes no claims regarding them.

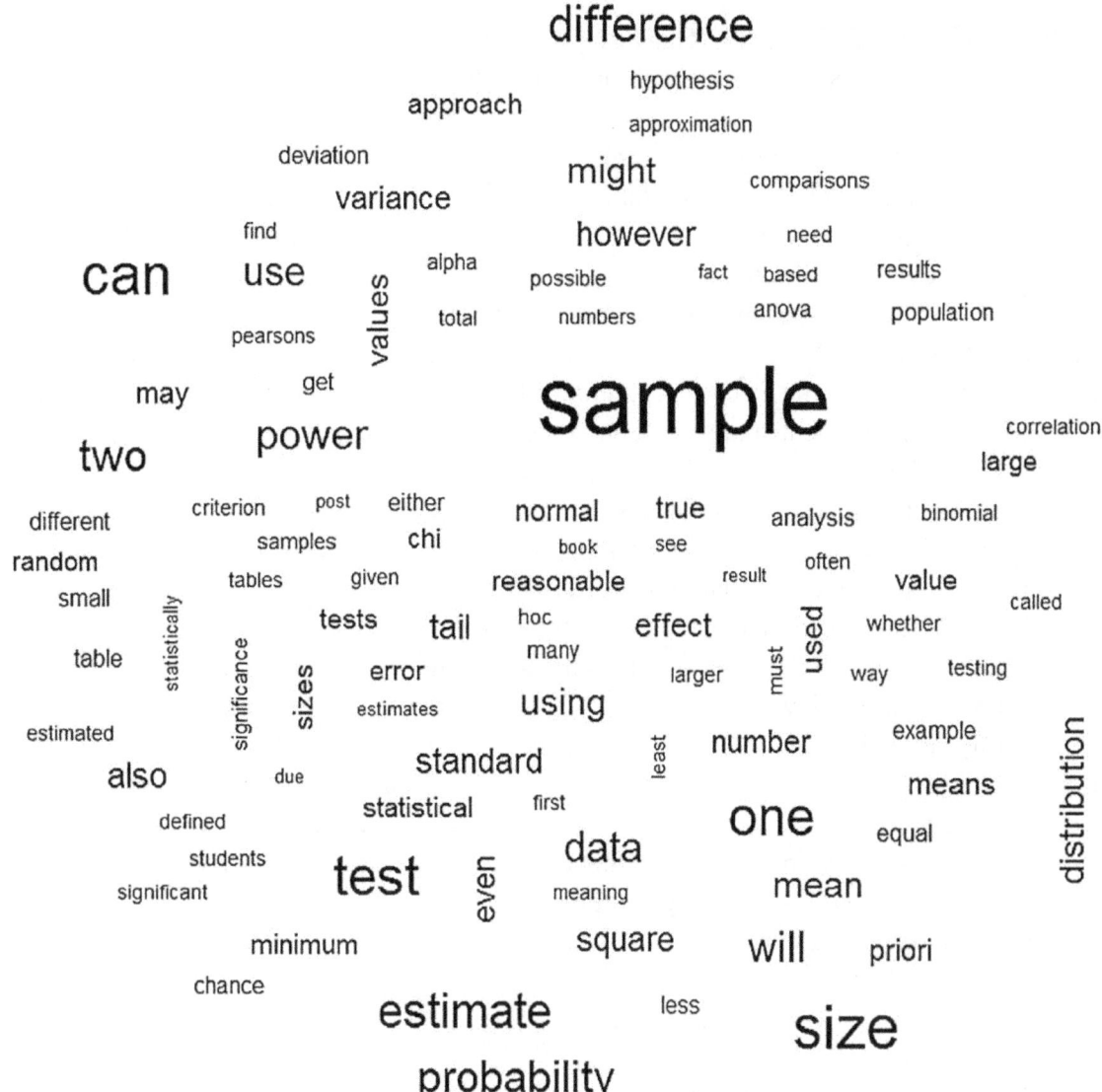

The sizes of the words are alleged to be* "proportional" to their frequency, and standard stop words (very common words with little import to the analysis such as 'the') are left out. However, the spacial arrangement is roughly random in this instance. Whether art or statistics, word clouds (also called tag clouds) may be handy for writers who want to contemplate their word choices. This one was created by RStudio using the `tm`, `wordcloud`, and `ggplot` packages. Somewhat more organized versions that arrange the words in concentric rings from frequent to less frequent are also possible.

* Zipf's Law says that in any text the second most common word is used half as much as the first, the third, one third as much, etc. It works for English and most phonetically written language texts (Auerbach F. 1913. Das Gesetz der Bevölkerungskonzentration. Petermann's Geographische Mitteilungen 59, 74–76; Zipf George K. 1935. The psychology of language, Houghton-Mifflin.). If the stop words were left in, and size was really proportional, the area covered by the smallest word here would be 1/100 as large as the largest. The letters in the smallest word here are only about 1/36 the size of those in the largest word. So, the words are merely roughly sized in order of their occurrence, but it is still a very handy visual device. Incidentally 'the' is the most common word in most texts, but it is a stop word for obvious reasons, it is just too obvious.

Contents

Contents

Contents

Contents

1 SUFFERING the FICKLE FINGER of FATE

This book is about the science of dealing with uncertainty. It is intended for researchers and future researchers. Specifically, it is about the dicey, often daunting necessity to estimate or detect effects, or at least differences. I, like Einstein (no resemblance other than unruly hair), used to think that, from a scientific view, that chance was entirely due to cause and effect. Carried to an extreme, this leads to *determinism*, the idea that there is no chance in our experience. The determinist view is that chance doesn't really exist, but is an illusion created by complexity beyond our unraveling. If that is true, then even human thought is no more or less than whatever preceded it and thus the final outcome is predetermined. It turns out that although the complexity of existence plays a major part, there is more to chance than can be explained by determinism. Most physicists since Einstein disagree with him on the subject of chance and accept that 'chance exists'. For instance, there even seems to be evidence that the firing of individual neurons in our brains has a truly random, that is, non-cause-effect, quantum component (e.g. Jedlicka 2017). However, with stimulation, the chance of firing is increased, so the mass effect of the hundreds of millions of neurons involved usually sums up to result in the appropriate response or perception. Nonetheless, if a researcher is only able to examine a sample of limited size, the researcher's conclusions are themselves subject to chance.

In applied mathematics, *chaos*, meaning sudden and extreme changes diverging from a trend, may pop up in situations potentially described by reasonable, sometimes even simple, equations or models.[1] The *sensitivity to initial conditions theorem* says that any difference in initial conditions, **no matter how small**, will result in different long term behavior. This phenomenon was first noticed by the weather modeler Edward Lorenz, and in the case of weather, initial conditions have the potential to lead to chaos. Such phenomena can be a source of *outliers*, values that differ greatly from most of the other values in a sample.

1.0.1 Surely not; science has limitations *! ! ! ?*

> "As human beings, we tend to love information, but we hate uncertainty – especially when we need to make decisions. Information and uncertainty, however, are actually two sides of the same coin." (Meng, Xiao-Li 2009)

Science is restricted to phenomena which are both **repeatable** and **observable**. Thus science cannot deal with one specific time only, or even with events that are too rare. In the fields using statistics, you, as a researcher, must make multiple observations in order to conclude anything, and even then, chance might invalidate your conclusion. We frequently hear in the news the results of a research study that appears to contradict the results of a study published just a few years earlier. Although there is occasional deliberate falsification, most of the problem comes from lack of understanding of statistical techniques, their proper use, and their limitations (Smith 2012).

---

[1] In some instances data read by a computer's game port can be startlingly chaotic. Tiny differences in boat propellers can have great effects on when cavitation sets in. Gleick (1988), Pickover (1980), and several other authors have popularized the concept of chaos.

Statistics can be viewed as a tool to help you cope with the combined effects of both real chance and the apparent chance due to complex cause-effect. Chance interferes with attempts to put an exact number on such things as the number of lampreys in Lake Superior, the maximum combined weight of elevator passengers who might crowd into a certain size elevator, the swimming speed of copepod larvae with low *Reynold's numbers*,[2] or whether substance X is more fungicidal than substance Y. Statistics considers populations via samples. A ***population*** is all the possible people, things, or outcomes under consideration. A ***sample*** can be defined as a subset of a population. In each case you must randomly take, or observe, a random sample out of the whole population.

1.1 Defining Statistics

Statistics is the summarizing, presentation, and use, often by testing, of data in general. It started as the amassing and processing of government data. But statistics has expanded into almost all areas of research, even such things as who wrote a particular manuscript, or how many reproductively mature albacore are in the Pacific Ocean. Perhaps statistics is operationally defined by what leading statisticians say about it. Paul F. Velleman (2008) has stated, "One of the major goals of Statistics—in fact, the principal goal of Statistics—is facilitating the discovery, understanding, quantification, modeling, and communication of facts about the world." So, it was reasonable for William G. Hunter (1981) to state, "At the outset the most important question for the statistician to ask is: What is the objective of this investigation?" Statistics involves a lot of probability and other mathematics (especially algebra), and these subjects take up a lot of the space in this book. However, how the data is obtained and how it is interpreted is more important. If it is not done mindfully, needed research may fall victim to the information scientist's dreaded nemesis, 'garbage in, garbage out'.

1.1.1 The limitations of the examples

Although this book is mostly about scientific decision making, including how to collect and process data, it is often impractical to illustrate the techniques using data sets that are large enough. Thus many of the examples involve practical applications applied to inadequate and fictional data for purposes of clarification and also to sometimes demonstrate common misuses of statistical techniques.

You can use calculators, tables, or the R language for your sources of important values. The **accuracy of your source** of such values (e.g. Z) can make a difference of up to a little more or less than one in the fourth or fifth figure. This will also affect sample size estimates.

1.1.2 Great (and small) expectations

When you publish your results the reader must very clearly understand 'what' or 'who' was studied and exactly how and under what conditions the study was performed. Then sample size is an additional and inescapable issue that recurs throughout this book.

A *distribution* is a description of the frequency of each distinct result (thing, happening, or outcome) occurring in the population being considered. Each member is distinguished by a measure or count (a score). The distribution could be the top speed of family cars; it might have the engineers'

---

[2]Reynold's number was originally an aerodynamic term that has been extended to both the microscopic as well as the macroscopic and to fluids in addition to air. It relates to inertial force and the viscosity of the fluid in which the object is moving or which is moving past it. However, at small enough sizes the effect of viscosity greatly increases due to laminar layers. Very tiny organisms have very low *Reynold's numbers* and live in a very different sort of universe than do we. In that universe, surface tension, laminar layers, adhesion, and even Brownian motion may have considerable effects that may far exceed the effects of inertia or gravity (see Purcell, E.M. 1977).

intended speeds of car models as the data points or it could have the top speeds of individual cars as the data points. These speeds could be measured on a continuous scale, conceptually having an infinite number of possible values (assuming you could measure the speeds to an infinite number of decimal places). Or the data might only have two possible values such as flips of a coin, and the distribution would be of only two categories, heads and tails. Distributions can be very usefully defined by equations, but we humans often benefit by representing them by graphs such as a binomial distribution of some raw data, or a normal curve (so called bell shaped) distribution of the means of sufficiently large samples (Section 4.2).

A *statistic* is an estimate and is subject to chance, whereas a *parameter* is the true value and is not subject to chance, but it is often unknown and unknowable.

The parameter

$$\mu = \frac{\sum_{i=1}^{N} x_i}{N},$$

is the *true mean*, the true average of all the N things in the population, a numerical midpoint of the population. The *summation operator* is symbolized by capital sigma, Σ, and is commonly used to tell you to add things together. Humans can rarely measure everything in a population. So, instead of measuring all N things in the population, you usually only take a sample of some practical size n.

The *expected value*, $E_{(x)}$ is defined by, and defines, the *mean* of all the individual, say, x values in population X,

$$E_{(X)} = \mu,$$

where μ is the Greek letter *mu*. It is also called the *true mean*. In practice the mean is often assumed to be due to the major factor, the central cause of the distribution.[3]

In application, you have to settle for an estimate of the true mean, the **sample mean**, the average of a sample. It is calculated using summation of all the scores (ie. x values) observed in a sample of n things, and is then divided by n,

$$\bar{x} = \frac{\sum_{i=1}^{n} x_i}{n},$$

and it is only a statistic, an estimate. It can be calculated from samples of any size in the range $1 < n < N$, where N symbolizes the total number of whatever in the whole population. This statistic, $\bar{x}$, only estimates the $E_{(x)}$, the expected, or true, mean. Statisticians consider a sample of only one from a whole population to be meaningless.

1.1.2.1 Randomness is required

Random means that every member of a population has an equal chance of being selected.

A sample mean can only be taken seriously as an estimate of μ if the sample is taken in a truly random fashion. Both the true mean, μ, and its estimate, the statistic $\bar{x}$, the sample mean, are often referred to as 'the mean'. Look out for context, it is sneaky!

Unfortunately, sample means are just estimates and are subject to chance, meaning that they too have a distribution, although less variable than the original data.

---

[3]This $E_{(X)}$ notation and jargon is a little redundant from the viewpoint of someone simply interested in application. Subsection 1.1.2 is not the whole story; the expected value of X can be written u, $E_{(X)}$ or $E(X)$. The latter can also be seen as an *operator* (something that does something), $E(x) = \sum_{i=1}^{i=N} x_i p(x_i)$, where N is the size of the entire population. So this defines the *expectation operator*. You simply have to realize that, in a theoretical context, the 'expected' value of the mean is the true mean.

1.1.3 Law of large numbers

The *Law of Large Numbers* says that for **increasingly large sample sizes**, n, the sample mean approaches ever closer to the population mean, μ (Bernoulli 1713, Mlodinow 2008). It is often, perhaps more accurately, called the *Law of Averages*. So, if you perform the same experiment (trial) a large number of times, the average of the results obtained should be close to the expected value (the true mean, μ), and will **tend to become closer as more trials[4] are performed**. This is a simple theorem, but is actually one of the most important facts in this book. Some authors incorporate the Law of Large Numbers into the Central Limit Theorem (Subsection 4.3). However, especially on the internet, it is mistakenly given as the Central Limit Theorem. This latter is a serious error and the source of considerable confusion. The Law of Large Numbers is **not the same** as the Central Limit Theorem (Subsection 4.3).

1.1.4 Central measures

A mean (average) is a *central measure*, one of the values that, in some contexts, are viewed as the center of the population or sample.[5] However, here are another two central measures that have their own common uses. One is the *median*, $\widetilde{\mu}$, which would be the value in the middle if you could line up all the possible values by size; its estimate is written $\widetilde{x}$. Another is the *mode*, which could be said to be the fashion because it is defined as the most common outcome (most frequently occurring).

1.1.4.1 Don't lose your marbles

To better illustrate the above central measures let's count the wealth of seven children handily named A, B, C, D, E, F, and G by the Unnameable Orphanage. Child A has 10 marbles, B has 10, C has 9, D has 7, E has 6, F has 11, and child G has 3 marbles. The total is all the marbles added together, $10 + 10 + 9 + 7 + 6 + 11 + 3$, which equals 56 marbles. The mean is the number of marbles per child, $56/7 = 8$, if they were distributed equally. The median is the middle value. If we rearrange the number of marbles in order, 3, 6, 7, 9, 10, 10, 11, the median is 9, it has an equal number of values on each side of it. Both the mean and the median are sorts of centers, called central measures, but although the mode is classified as a central measure and is sometimes in the middle, as when marble collectors are striving to be like each other, it is not as likely to really be at any sort of center. The mode is the most common thing (in this case the most common number of marbles), also seen as the most popular. In this case the mode is 10, because this number of marbles occurs more often than the other counts, which each occur only once. Fashion designers are very interested in the mode and the mode is what is commonly called fashionable. Social workers tend to be interested in the median. Tax collectors might be most interested in the mean.

1.1.5 Relative frequency and probability defined

Relative frequency, $\hat{f}_B$, is the estimated fraction of a population that has characteristic B. When you are sampling, the *relative frequency* or *sample relative frequency* is the percentage (or ratio) of times,

---

[4]A *trial* is a procedure that can be repeated infinitely many times, and has a well-defined set of possible outcomes, known as the *sample space*. Technically, in statistics, *experiment* is a synonym for a trial.

[5]Another mean is the geometric mean, $\sqrt[n]{\prod_1^n x}$, a central measure that is sometimes less than the mean itself. The $\prod$ symbol (capitol pi) represents the product operator. It means multiply the following together. The geometric mean only turns up occasionally in applied statistics of the sort covered here.

n_B, out of an overall sample of total size **n**, in which the event $_B$ occurs,

$$\hat{f}_B = n_B/\mathbf{n}\,. \tag{1.1}$$

So, when sampling, the ***sample relative frequency***, which is often written $\hat{f} = r/n$, estimates the ***proportion*** of successes (what you want to count). ***An event*** could be an explosion or an accident, but it could also be a color, or catching a fish weighing over 5 kg. Under the frequentist approach used in most of this book, to get a meaningful sample you will have to obtain a random sample. Getting a random sample can be tricky, as discussed in Subsection **1.1.2.1**.

The ***absolute frequency***, N_B of B, is the usually unknowable number of times event $_B$ (fish weighing over 5 kg, for instance) occurs in the **whole population** of size N (the total number of fish in the sea). The ***probability*** of event $_B$, the probability of the particular event (big fish for instance) being observed **at random**, is

$$p_B = \frac{N_B}{N} = \frac{absolute\,frequency\,of\,B}{number\,in\,whole\,population}. \tag{1.2}$$

This harks back to the definition of random in which every member of the population has an equal probability of being included in the sample. But be aware that estimation, although the best you can manage to attain, is not precise certainty. Trying to estimate probabilities by sampling is called the ***frequentist approach*** or ***frequentist school*** and is discussed in Subsection 1.3.1. So, the frequentist viewpoint is that the **distribution** is the whole collection of all the possible samples (how many of each size) from a defined group of measures such as human heights. The **sample relative frequency (the proportion)**, $\hat{f}_B = n_B/\mathbf{n}$, of event $_B$, assuming randomness, is an <u>**estimate**</u>, $\widehat{p}_B$, **of the probability**.

> **NOTE for f and p:** When frequentist refer to probability they are assuming randomness, meaning that every member of the population has an equal probability of being selected at random. Thus, references to probability infer a numerical equivalence to frequency, f, and, unless stated otherwise, references to frequency infer equivalence to probability, p.

Assume for now that, as commonly alleged, adult human heights tend to follow the normal, or 'bell shaped', curve fairly well.[6] The normal curve theoretically goes on forever in either direction, but the distribution of human heights is a ***truncated,*** meaning cut off, normal curve. There are no potential sample outcomes that go to infinity in either direction. As of this writing (as far as the information we have globally), the low end is cut off at the height of Mr. He Pingping's 76 centimeters, in China, and the high end is cut off at the height of Mr. Bao Xishun, 2.36 meters high in Mongolia.

1.1.5.1 Example: Frequency

The Smallest Adult Person in China A way to look at the **probability** is to look at it as the proportion (the fraction or %) of a whole population that includes some particular outcome. When this book was being written there was not known to be any adult person anywhere who

---

[6]Herrnstein and Murry's (1994) book, The Bell Curve, examines the normal curve in some detail. However, they messed up a good job by coming to genetic (hence racial) conclusions based on a so called IQ test that required a minimal level of education not available to everyone. Likewise, these conclusions did not take into account that malnutrition, especially in early childhood, affects performance on such tests. Although, when properly tested, IQ may relate to academic potential, Arthur R. Jensen (1970) concluded that actual scholarly performance tends to be more related to environment than to genetics.

was smaller than the 76 cm tall Mr. He Pingping[7] in China, $N_A = 1$. Assume $N = 1,321,852,000$ total people in China (OK, it's rounded a little because the exact number fluctuates). If an alien abductor who only had enough budget to capture one adult human specimen from China and used an ***unbiased*** (meaning not influenced by anything outside what it intended to sample) random way of selecting and abducting a sample of one adult human from China, the alien would have had 1 chance in $1,321,852,000$ chances of getting someone that small. That is a probability (also a frequency) of $p = 1/1,321,852,000$ (or $p = 0.000000000756514$ (which is the same thing) of getting someone as small as 76 cm tall. Notice in this case that the aliens are somewhat limited in their potential to estimate probabilities because their budget is so limited.

If these aliens get a larger budget, they might more typically sample, say, $n = 40,000$ Chinese adults to count how many, let's say 40, so $n_{small} = 40$, are less than a meter (100 cm) tall. That would give them a sample relative frequency (a proportion), $\hat{f}_{small} = n_{small}/n$, or $\hat{f}_{small} = 40/40,000$, which would also (only) be an **estimate**, $\hat{p}_{small} = 1/1000$, of their probability[8] of randomly abducting a single adult person less than a meter tall. Frequentists assume that the true probability, p, of any outcome can be estimated by its relative frequency in random samples. Although there is another school of thought, most applied statisticians use such a frequentist approach to make decisions and many kinds of estimates are based on samples. **If a sample is not random**, a frequentist based statistical test or estimate is **meaningless** (Section 1.3.1). Note that the word **estimate does not imply** accuracy or certainty, it only implies an attempt to approach a reasonable precision and probability.

For additional understanding, you can also try Question 9, which is answered at the back of the book.

1.2 Don't Look Now!

Suppose equal numbers of each gender walk past your neighbor's window, but your neighbor claims to be able to predict the gender *a priori*, meaning before the fact (in this case before observing the gender of the next person). If every time you visit this allegedly prescient person, they a priori make a correct prediction of the next person's gender, that would be remarkable. However, if this neighbor simply said, ***post hoc***, meaning after the fact (in this case after observing the gender of the next person), that they had known what gender that person was going to be, that would probably just be eccentric. The difference between a priori actions and post hoc actions and conclusions is one of the most important concepts in statistics. You can attempt to hypothesize, to predict, an outcome that contains an element of chance and then procure data to see if you may have been wrong, but as soon as you examine that data, you lose any opportunity to test further predictions by using that data.

Some rules for avoiding bias while obtaining the best possible answer to yes/no questions when dealing with chance are:

- Always be certain the data you use to form a proposed explanation for something (a hypothesis) **is not the same data** you are going to use for the test. Doing otherwise biases, and thus invalidates, the results.

- **Never examine** the data you intend to test **before you test it**. Finding something is not a chance event if you have already found it. Some authors and YouTube lecturers carelessly say that you

---

[7]He Pingping, a very popular person wherever he went, had primordial dwarfism, meaning that he was unusually small even as a fetus. He died visiting Rome, while this book was still being written, on 13 March 2010, of heart congestion. He was there to take part in a television series. He apparently was also the smallest adult in the world.
[8]The hat, ^ on $\hat{p}_{small}$ is to indicate that it is an estimate.

should graph the data which you are going to use in order to see its mathematical characteristics before testing. This is peeking and will lead you into the post hoc error. If you must check the mathematical qualities before the test, you must **first obtain a separate random subset of data** that **cannot** subsequently enter into the actual test.

Even if you were to merge all the data so that you could not tell A from B for graphing before the test, there is the possibility that the graph would have, or maybe lack-having, two humps (or visible clusters) suggesting something about the hypothesis, suggesting an adjustment of the experimental design, or suggesting a new hypothesis. Whatever happens, you cannot test a hypothesis which is based on, or even verified to be suitable for, a particular test, with the data which provided that hypothesis. Statistical inference is all about an outcome being too unlikely to suspect that it is merely due to chance.

Data can be *prospective*, meaning that you are planning to collect it, for instance when you run field trials for a new way to grow rice. Or it can be *retrospective*, meaning that you intend to sample data you have not already examined, from preexisting data such as national surveys or from accumulated hospital records. You can examine subsets of data from the source you intend to analyze, but they cannot subsequently enter into the actual analysis. The inferential procedure is invalid if you are in any way influenced a priori by the sample data you intend to analyze, before you analyze it.

It may be necessary to check for outliers and for misleading coding practices ahead of time, but, if essential, the examination had best be done in a double blind manner, as by a second person, or maybe by a machine, without a priori feedback. The person must understand a little about the data, but had best not know exactly what you are asking. For instance, some data might code 0 for missing, and this may need to be recoded to something like NA or NULL (in the R language) so it doesn't get treated as a value and wreck your analysis. Some data might be scanned for extreme outliers using MAD (Section 15.6.2) or a similar mechanized approach. Remember that if you check any data in enough detail, you will probably encounter anomalies which are purely coincidental. The best approach, if possible, is to examine data from the same population (a so-called *training set*), but which will not be included in the analysis (in the *test set*). Deviation from this approach creates post hoc outcomes which are invalid for inference.

1.3 Randomness and the Struggle to Achieve Disorder

To produce valid research results you have to deal with trying to obtain order out of complexity. Ironically, it is necessary to use disorder dependent on chance to even approach perceiving order. *Random* means that every member of a population has an equal probability of being selected. You need to strive for randomness by striving to achieve a lack of pattern or lack of predictability in procedures such as sampling and observation. Sometimes you will hear the word *stochastic* applied to a process. This is a bit of jargon that means something that is a result of a mixture of both random processes and cause-effect. For instance, several years ago there was a group at Caltech experimenting with stochastic computer **algorithms** (sets of instructions) to simulate evolution. The probability of sampling a particular value, characteristic, or kind of thing, is equal to the proportion of it in the population, *iff* [9] the sample is drawn from the population in a truly **random fashion**. But in evolution, organisms inherit their characteristics at random, then cause-effect such as climate, etc., favors the survival to reproductive age of the best adapted organisms from this randomly generated population.

---

[9]*iff* means if and only if.

1.3.1 The frequentist approach and randomness

The *frequentist* school of statistical thought has provided most of the techniques presented in this book. Most, but not all, of the pioneers in developing the methods presented in this book were frequentists. They accept that any given experiment can be considered as one of an infinite sequence of possible repetitions of the same experiment, each capable of producing statistically independent results (Everitt 2003). They accept that random means that everything in a population has an equal chance of being sampled and would say that if 12 in every 1000 wolves are solid black, then the probability of one randomly selected wolf being black is 1.2%.

There is a strong emphasis on obtaining yes/no answers and estimates that, if they meet certain generally accepted standards of probability, are accepted as fact for the time being. Under this approach, the first condition for any research that involves sampling is that the sampling must be **truly random**. Getting a random sample (Subsection 1.3.2.1) is not always (ever?) easy. Nonetheless, it is up to the researcher to consider, before the fact, whether the researcher's experimental design will guarantee that each result or observation can really be considered to be one of all possible events, each of which is no more or less likely to have been sampled than any other event. What the frequentist researcher intends is to discover reality, despite the fact that this school also accepts a small probability that the fickle finger of fate may provide an incorrect answer no matter how well designed and executed the collection of the data happens to be.

1.3.2 The Bayesian school

There is another school of statistical thought called *Bayesian statistics*, that was inspired by, and often uses, Bayes' theorem (Subsection 1.8.1). It has a strong focus on the fact that we are unlikely to ever know the answer with perfect certainty for questions for which a statistical procedure might be necessary. Bayesian statistics starts with some degree of belief in a ***prior probability***, $P(\mathbf{H})$, (the probability of a hypothesis) that is assumed to have some evidence based support, and then looks for new evidence (data), $\mathbf{E}$, not used in computing $P(\mathsf{H})$, to adjust that degree of belief. The result is a '***posterior probability***, $P(\mathbf{E}|\mathbf{H})$, and can be accepted as being a potential prior probability for adjustment for a succeeding posterior probability. These sometimes successive levels of belief are also called 'prior beliefs' and 'posterior beliefs'. Bayesians frequently warn one another to be careful about selecting their priors, because they do not always agree on what is a suitable initial prior for a given question. However, with Bayesian statistics, even hypothetical probabilities can sometimes be used to prove, regardless of true values, that some outcomes are impossible, or some approaches are impractical. So, whereas the frequentist approach tends to look for yes/no answers, the Bayesian approach is more oriented towards estimates and general principles and better maybe's by exploration of what is most probable.[10]

It is often possible to create Bayesian models which explain processes, and which can sometimes also set certain undeniable limits on what is possible. In some huge data analyses, Bayesian approaches might even be preferable for inference. For instance, huge data sets can sometimes provide an excellent foundation for making Bayesian predictions and adjustments of belief. Indeed a characteristic of most Bayesian techniques is that they are very computer intensive. Even relatively simple Bayesian techniques along with sufficient literature review can sometimes be especially useful for deciding whether frequentist experiments, such as medical trials, should even be initiated. This

---

[10]Some frequentists see the Bayesians as heretics and some Bayesians see frequentists as just plain wrong. However, frequentist statistics have brought us a long way, and the Bayesians have produced a great variety of useful applications, not all of them using (though often implying) the Bayes' theorem. One might wonder whether some of the Bayesians most consequential approaches are modeling or not, or if Bayesians are 'data scientists' or 'statisticians'. Though hotly debated, both 'schools' are useful.

could lead to tremendous savings in costs, and, through reallocation of resources, speed research in general.

Fields where data is scarce or randomness is unobtainable have been the source of much criticism of the frequentist approach. Unfortunately, as the frequentist approach has been classically applied, these criticisms are often valid and reform in their application is needed. For instance, the Bayesians are always aware of the possibility of a 'black swan', a completely unexpected discovery or event,[11] and never accept a probability estimate that equals 0 or 1.

Some frequentists see Bayesian approaches as being more susceptible to bias, whether conscious or unconscious, when substituted for inferential frequentist techniques. Hopefully, this book, although almost totally frequentist, will aid in adjusting or establishing your approach to research.

This book is mostly about specific decisions, inference, and inference-based estimates, all drawn directly from raw data, and it is largely about frequentist approaches. As this book reveals, some kind of prior information (generally about variability) turns out to be necessary for the efficient application of even frequentist approaches. Maybe the future will see a merging of frequentist and Bayesian approaches.

Students might wish to take at least an introductory course in Bayesian techniques after obtaining a firm grounding in the proper application of frequentist techniques. Because Bayesian techniques so often include intensive calculation, a course in computer programming or the R statistical language (originally the proprietary S language) might be a reasonable prerequisite.

1.3.2.1 How to randomize

Even though randomization is the key to estimating frequencies, many researchers mistakenly and disastrously think that randomization is easy. So, they behave as though it is not terribly important. In order to prevent selection bias in randomized trials, it is probable that the NIH and many peer reviewed journals will be requiring increasingly more detailed descriptions of researchers randomization techniques in the future. Randomization and the literature search are usually the two most important procedures in any research project. Random numbers generated on a PC or Mac will **not** dependably do the job, and there are numerous pitfalls to various methods of drawing numbers, as out of a hat. In general it is safest to use a well authenticated table of random numbers (often labeled ***pseudo random*** because of being pre-checked),[12] whether in a book or on a CD. Such sources tend to meet the requirements of what you need, but beware how you use them. Some students have to be warned to look out for row numbers (which are not the random numbers themselves and which are ordered, if present) and to use the others. Also, each time you get a **sequence** (a list or series) of numbers out of a table, **mark or record the position** of those numbers and **never use them again**.

---

[11]The European world believed all swans to be white, but then the black species of swan was discovered in far away Australia in the late eighteenth century. Taleb (2007) discusses black swan events and their impact.

[12]In randomization you come up against some of the many realities that make research difficult in unsuspected ways. There are a number of ways to generate true random numbers (to draw numbers at random). For instance, measuring the time between Geiger counter clicks is among the most popular. If you start with a table, meaning a list, of millions of such randomly drawn numbers in the order they were drawn, you would observe regions of apparent randomness and areas in which chance provided sequences of numbers in some sort of order. For instance, it would not be impossible to draw 1045, 1046, ..., 1049, each after the other, nor would it be impossible to draw 1045, 1055, ..., 1095. Most random number generators are quite capable of generating 1045, 1045, ..., 1045. A long enough list of truly random numbers would provide you with all the possible sequences of some kind of order as well as all the sequences of disorder. However, statisticians have made many theoretical and simulation studies and accepted that by computer screening they can eliminate such **runs** (sequences) of patterned numbers without damaging the effective (or sufficiently apparent) randomness. At best, the results are, in reality, **locally random numbers**, also called pseudo random numbers. An example is the classic table by The Rand Corporation, A Million Random Digits with 100,000 Normal Deviates (The Free Press 1955).

Otherwise you or your successor will be going over the same pattern and this can cause severe bias. Often the tables of random numbers will have columns that are very wide, say six to ten or more digit numbers. But if you are only randomizing less than ten things you will use the rightmost single digits. If you consider the circled numerals in the simple table shown in Figure 1.1 you will see we are only using the rightmost meaningful integer because we are only interested in values less than ten. If we had to randomize, say, 420 items we would use a table with wider rows and only use the rightmost three integers.

For simplicity we will consider a randomization problem in sports where the number allotted to a player is a potential source of contention. Suppose you are in charge of allotting six numbers, let's call them #1 through #6, to a middle school basketball team. You must always choose a part of the table you have never used before, and use numbers as they come. Let's say you are using the simple two digit randomization table in Figure 1.1 and the first number on the right that you come to is a 6, so you assign it to the first player in line; the next number is a 5 so assign it to the next player; then comes a 4 so assign it to the next; the next is a 1 so assign it accordingly; however the next is a 6 which is a **duplicate** so you skip it; the next is 8 which is **out of range** so you skip it too, and so on until you come to the final number you need which is a 2 for the remaining player. It is important to **mark that spot** so you never use that part of the table again.

Table 1.1: Selecting From Table of Random Numbers

96	54	40	7
35	83	58	30
54	50	70	21
21	1	93	89
16	19	82	67
58	80	90	18
26	55	94	59
36	74	65	38
93	2	61	81
57	97	26	18
79	2	60	50
9	68	81	48
17	95	42	20
2	8	22	28
84	8	51	82
59	62	40	53

Selection of random integers 1 through 6 from a pseudo random number table. Because only numbers less than 10 were needed, the last numeral in each random number was used, giving 6, 5, 4, 1, 3, 2. The gaps are simply due to skipping repeats and numbers that are out of range (such as 8). The line you mark after the last selection will remind you that you should start any future selection beyond this group. As the sample grows larger, any effects of short runs, like the first four numbers here, will tend to average out. If you are willing to accept computer generated, or maybe stored random numbers, the starting place is called a **seed**, and you must select a different seed each time.

In taking samples for statistical decision making, the harmful effects of a failed attempt at randomization may lead you to conclusions that have nothing to do with the question you are researching. In a real life situation, even if you do everything correctly, you might be unable to avoid unknowingly rejecting a wonderful new cancer drug, or mislabeling a delicious food as potentially causing cancer due to something analogous to the spin of a roulette wheel. However, the chances are more strongly against you if you use a defective source of 'random' numbers or use those numbers carelessly. Imperfect randomization is as though the roulette wheel is off balance and you cannot even be certain in which direction it is biased.[13]

Although not the only challenge, human bias is one of the researchers greatest challenges. Our minds are designed for being hunter-gatherers and they can fool us when we are searching for truth rather than game or fruit. How often have you been convinced you were observing something that you later found out was something else? Have you ever discovered that you had been defending a falsehood because you wanted it to be true, or have you missed the sought after truth because you wanted to be 'fair'? Ideally you should have a **double blind** approach to research; different people do different parts of the process, the first step being applied according to a predefined randomization and then arranging that the data collector never knows what has gone before. However, double blind is not always possible. A surgeon performing the operation knows whether the patient is receiving a radical mastectomy or a partial one. So, in such cases you must use **allocation concealment**, meaning that at least the allocation of treatments is inaccessible to the person mandated to apply the treatment.

But, how would you feel about experimenting on human beings? In medicine, clinicians tend to find various ways to circumvent randomization based on what they would do without the treatment being allocated at random for some research program. For instance, it is common for a clinician who feels a certain patient is in need of a certain treatment, rather than getting a placebo or another treatment, to (without the PI's knowledge) leave that patient out of the study.[14] This may have such effects as leaving the most extreme patients out of the study, or it may leave the least extreme patients out of the study. Then the outcomes are usually very misleading. As a result the outcome of the study is biased by leaving certain kinds of patients out of the study, but it is still published as though the results apply to a random (that includes with respect to symptoms as well as other effects) selection of patients (Vickers 2006). With subsequent publication in a peer reviewed journal the lie generated by trying to force doctors to ignore their personal standards is reported to the world. Such published outcomes may lead to the suffering of tens or hundreds of thousands of future patients who get the wrong treatment as a result.[15]

If, instead of randomizing, you use a systematic experimental design, then in addition to mocking the basic idea behind statistical tests, you have no way of knowing that some mechanical process (or accounting process, etc.) has not already created a corresponding regular pattern into which your pattern might dovetail to generate erroneous results. For example, if you give the first patient of the day the first treatment, the second patient of the day the second treatment, and so on each day, it is very possible that the effect in early risers and late risers alone, let alone biological clock and other related effects, will create a strong bias even if the treatments had identical effects. Systematizing may tend to produce lower variability, often making it possible to falsely conclude there is an ef-

---

[13] Incidentally, roulette wheels are always a tiny bit biased even though it would be to the casino's advantage to avoid bias. So, they change roulette wheels after a certain amount of play and send them to get rebuilt with new bearings, etc. They may do this several times a night if the betting is heavy. Otherwise someone might discover the bias and could win big at the expense of the casino.

[14] Some rules for randomization are, more or less, specific to the field. Perhaps we live in an economically desperate, or simply immoral society, where doctors are forced, or at least coerced, to take part in studies that they would refuse on what they personally see as moral grounds if they had a choice. Many doctors have been caught using bright lights to see through sealed envelopes to find out the next patient's treatment so that they can decide who that patient will (or at least, will not) be. As a result, aluminum foil is now routinely included in the envelopes to thwart this behavior. There is reason to doubt that even this is adequate when job tenure and desperation is the only incentive to people unwillingly performing a front line part of the research.

[15] Such forcing does not have to be overt or even conscious. If doctors love, or are dedicated to, the usual responsibilities of their jobs and then they are suddenly asked to take part in a study, they are acutely aware of becoming a 'problem', even a 'liability' upon openly refusing.

fect that does not exist. The untruth produced can be especially deceptive, with the erroneous results seeming to have very strong reported probability estimates supporting them.

Extreme effects due to doctors thwarting randomization involving surgery, etc. are shocking and thus memorable if you read them in a book. However, in prosaic agricultural on-farm research, there is a constant problem with the farmer deciding what appears to be obvious and quietly interfering. If the farmer sees that your experimental row of corn is being chewed to ribbons, the farmer may spray it because, after all, if you are getting such bad results, you must just be fooling around anyway.

Especially when doing preliminary experiments, researchers sometimes attempt to randomize out of their head. A problem with you attempting to choose what is random out of your head is that brains tend toward systematic patterns. You might not think so, but it is true of us all. Humans **cannot just pull randomness** out of their heads. The truth is that what goes on in our brains starts out with considerable randomness. Unstimulated neurons seem to fire in a quantum (true random) fashion. There is a low but not tiny probability of any single unstimulated neuron firing at a given time and there are millions of neurons in a brain. Then even when a potential stimulus is present, neurons simply take on a higher probability, not a certainty, of firing. However, each neuron is attached to a number of other neurons that when stimulated in turn, influence the chance that the neuron of interest will fire. So, the brain generates a lot of random mistakes to start with, but it compensates for most of this by having a lot of built in mechanisms for systematization and completion (finding, even creating, pattern in the apparent anarchy). Therefore, your brain is fighting the randomness you are trying to achieve, without you being conscious of the battle.

In any research involving experimenting, sampling or surveys, you, the researcher, are trying to minimize the probability that your results are false by trying to eliminate irrelevant, potentially confusing patterns. Randomization is the first step, maybe the most important step, towards extracting statistically valid information from the complex and dynamic real world.

1.3.2.2 Some places where you can obtain random numbers:

As of this writing, some sources of random values are:

http://www.random.org   Provides true random numbers. It also has a variety of other related and interesting links, all within its overall site. If advanced users with certain special needs check out the links on that site, they can also find a normally (Gaussian) distributed random number (Section 4.2) generator.

http://www.rand.org/pubs/monograph_reports/MR1418   Is possibly the most useful in that it can help provide selected groups of properly screened, locally random numbers, also called pseudo random numbers. If you examine the site carefully you will find that they have provided a million such numbers for free as a zip file. It also has an option for random normally distributed numbers.

http://www.fourmilab.ch/hotbits   HotBits: True random numbers generated by radioactive decay. It is a very handy site for programmers, researchers, etc.

The R language, conveniently handled by using RStudio[16] or various python packages, provides for generating computer random numbers, and this is sometimes the only practical way to select huge random samples from even **huger** data sets. But, the other above sources are generally preferred when using them is practical.

---

[16]RStudio can be freely downloaded from the internet.

1.4 Let's Get Legalistic About Chance

In games like blackjack, it is possible to make money by memorizing which high value cards have already been dealt out of the large combined deck from which they are being drawn. You are betting against the other players and paying a small rake-off to the house. The memorizer then bets against tourists and other naive gamblers who do not follow what is happening and have created a situation in which the *odds* (defined in Section 1.5) are strongly in a memorizer's favor. Thus you might think that you could learn *mnemonics* and make a living that way.[17] Some people have made pin money that way. However, the State of Nevada does not see these games as games of skill, but as games of chance. So, 'counting cards' is illegal there. For instance, in Reno, if your lips are moving silently and you tend to win a lot, or you are almost always winning more than a couple of bucks, you will be detained and photographed and no casino anywhere in Reno will let you in ever again. But that is not the sort of law upon which this chapter focuses.

1.4.1 The whole is the sum of the parts

You need to know that $\geq$ **means greater than or equal to**, and that $\leq$ **means less than or equal to**, whereas $>$ and $<$ **mean greater than** and **less than**. Some laws of chance (as opposed to the laws of humans) seem obvious, such as

$$0 \leq p \leq 1, \tag{1.3}$$

meaning that the probability, p, of anything is between zero percent and one hundred percent, **inclusive**. A Bayesian would say that $0 < p < 1$, meaning that the probability, p, of anything is between zero percent and one hundred percent, **not inclusive**. (It's a Bayesian thing; you'll have to look it up elsewhere.)

1.4.2 Independent probability, and a booby trap

Two or more events are *independent* if the occurrence (or value) of one event in a random sample does not affect the probability of the other. If the events are not independent, they are dependent (Section 1.6); the occurrence of one affects the probability of the other.

Slot machine **trials** (individual bets) are independent. Suppose you lived in Reno, Nevada where there are still a few casinos that offer meals that are a bargain. On the way to your favorite restaurants within these glitzy places you would pass hundreds of slot machines, often manned by frail players who are hooked up to their little wheeled oxygen carts while smoking. When they have to go to the bathroom, these slot machine jockeys glamorously signal for an attendant who ropes off the machine upon which they have been losing. When they finally decide to go home or run out of money, other gamblers who have seen them losing for hours at that machine rush to take their place. All this place holding and going to losing machines is because these folk believe the *gambler's fallacy* that somehow, after enough losses, the player must get a win. Some even think that the more the losses, the bigger the guaranteed win.

This simply isn't so; **it's a trap** for people who wish to believe it. The above definition of independent probability, also called the *law of* **independent** *probability*, shows that if the bets are independent, you have the **same probability** of losing every new *bet* (trial) you make. A now deceased relative of mine who was a street child during the Great Depression, was taken in by an oddly kindly minor gangster. As a result, this relative had been both a slot machine mechanic and had briefly owned a bunch of the things. He agreed with mathematical statisticians (the old gangsters call them

---

[17]*Mnemonics* are techniques of associating things with each other so as to make them easier to memorize.

lay off men) that this law of chance regarding (as well as defining) independent probability holds up in practice. His machines tended to keep ten cents for every dollar put into them, when averaged across a few thousand dollars (in those days put in as quarters so that there were four trials per dollar). If one machine paid a lot of jackpots one month, another might pay nothing, but chance worked out so that most of them just egged the customer along with frequent small paybacks, not quite covering the sum of their accumulating losses. The machine that paid lots of jackpots this month might not pay at all next month or might give lots of jackpots, even though the owner did nothing but put in or take out the difference and maintain the machines. All the machines together were certain to produce a profit in the long run for the owner, the 'house', **not** the player.[18] Modern slot machines may take as much as fifteen percent from players, but the law of independent probability has not changed. Incidentally, the machines rarely cheat because, where legal, they are inspected, and especially because there is no need to cheat. The machines that are paying off at the moment are simply generating random outcomes; they are not different from the ones that aren't paying off at the moment.

The same law of independent probability applies if someone wants a baby of a given gender; they only have a roughly 50% probability of conceiving the one they want **each time** they try.[19] They can end up with a lot of extra babies that way.[20]

1.4.3 Multiplication law, independent *and* together

In statistics two occurrences are **independent** if the occurrence of one does not affect the probability of the other happening. The probability of slipping and falling on an icy sidewalk is independent of the probability of winning the lottery that same day. Independence is the focal point of the multiplication law; it applies to the probability of two independent occurrences happening together by chance alone. The word together in this context can be seen many ways; for instance, it can be instantaneously together as in time, such as being born the same day and hour as your future spouse, or it can be conceptually together, such as being innocent but being accused of a murder that happened today and having searched poisons on your computer for a college essay a couple of days ago. This law is one of the **most important** in statistics.

So, to define the ***multiplication law***, also called the ***multiplication rule***, or the ***sampling with replacement rule***, if two randomly drawn outcomes, A and B, are **independent**, then their probabilities, P_A and P_B, are independent, and the probability of both of them occurring together is the product of these probabilities (Figure 1.1). Technically the independence has to be maintained by sampling with replacement (like catch-release fishing).

A 'math nerd' might point out that the multiplication law applies **only to independent events**, often called A and B, and write it as

$$P_{A\cap B} = P_A \times P_B. \tag{1.4}$$

---

[18]For a casino or a large scale slot machine operator, a huge portion of the gambling world is the 'house'. When a large casino has a 'bad run of luck', it simply sends runners out for a temporary loan from one or more other large casinos. But a small gambler, given enough time, will eventually have a long enough 'bad run of luck' to run out of money even if the odds are somehow in the small gambler's favor. This is because even the unlikely will eventually happen. When the inevitable unlikely happens, such as a run of losses for a small player, it will consume all the reserves the small player would need to continue playing until a win occurs.

[19]Note that I wrote "roughly". The sperm which would create a boy are a little more active than the sperm which would create a girl. Even though there are other small effects that affect gender ratios, the world wide ratio is close to $50:50$.

[20]OK, this assumes this is in a country where medical intervention in gender selection or infanticide is not widespread. It is common practice to assume that the number of boys born equals the number of girls; in reality there is a tiny effect in nature. Slightly more boys are born and slightly more girls survive infancy.

The *cap* symbol ∩, also called **hat**, is read as 'and', and is also called the **intersection operator**. It is merely a symbol to indicate that two things are together, no matter in what way, whether related or not. So it is used to symbolize **togetherness**. In statistics ∩ is often associated with randomness, hence probabilities.

1.4.3.1 Example

Lake health You can often get an estimate of the health of a body of fresh water by considering the predominant kinds of insect larvae that live in it. Assume you are an ecologist and you are lowering a tool and taking random grab samples of identical quantities of mud from a lake bottom. Such random samples are independent, and let's imagine that you only have 1 chance in 4 of 'event A', getting at least 1 insect larva in a single grab when sampling this lake, $p = 1/4$. (You can think of this $1/4$ as the proportion of the number of grabs that would have larvae if you could take an infinite number of possible grab samples.) No matter how many grabs you have already made, you still have only 1 chance in 4 of getting at least 1 larva, $p = 1/4$. For perfect independence, assume you put the larvae back into the lake, a procedure called **sampling with replacement**). Figure 1.1 below shows the probabilities for successive independent successes, getting at least one insect larva again and again:

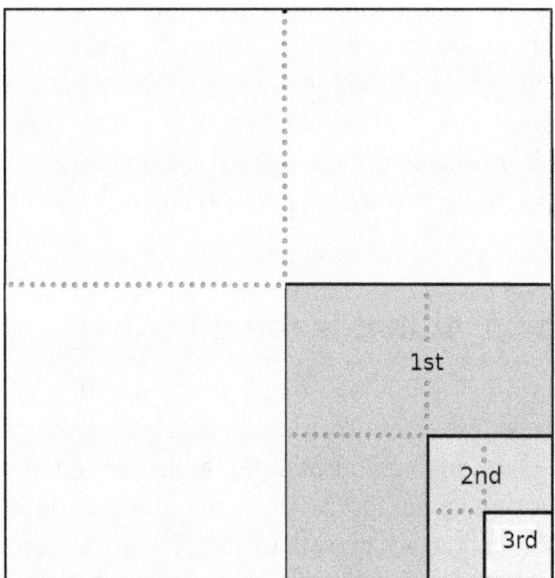

Figure 1.1: Multiplication Law: Probabilities for Successive Independent Successes with $p = 1/4$
The whole box represents the population of all the possible outcomes if you were to sample every spot in a lake. The whole one fourth square marked off in the lower right corner (1st) represents just the proportion, one fourth, of the samples that would contain at least one insect larva in this particular lake. If you took just one random sample you would have a $p_A = 1/4$ chance of a **success** (event A) of getting insect larvae in one sample. The next smaller box (2nd) is a fourth of that, and is the proportion of times you would get larvae in the first sample (A_1) **and** also in the second sample (A_2), if you randomly sampled at the lake just twice. That is, according to the multiplication law, under random selection, the probability of two things happening together, $A_1 \cap A_2$ (for example as in succession), is the product of their independent probabilities (corresponding to their proportions), $P_{A_1 \cap A_2} = 1/4 \times 1/4 = 1/16$. Then the probability that three successive random samples **all** will have larvae is $P_{(A_1 \cap A_2) \cap A_3} = (1/4 \times 1/4) \times 1/4 = 1/64$. That is, only one time in sixty four visits to the lake would you get insects $(A_1 \cap A_2) \cap A_3$ in all the first three random draws. And so on and so on for foursomes, etc. (And because they are independent they would really not obligingly all crowd down into one corner for you to find a bonanza of successes.)

Two different kinds of independent event This lake also has fresh water clams and let's imagine you have 1 chance in 5 of event B bringing up at least one fresh water clam. Now assuming A and B are really independent (they are not attracted or repelled by each other), you still bring up at least one larva 1 time in 4, and you also bring up at least 1 clam one time in 5, whether larvae are present or not. Let's determine the probability of at least 1 larva *and* at least 1 clam **together** in a grab sample. Just one grab sample will have larvae 1 time in 4 and of those times $1/5$ of them will also have at least 1 clam. That is, the chance of a larva *and* a clam **together** in a grab sample would be $P_{A \cap B} = 1/4 \times 1/5 = 1/20$. Notice I used the togetherness symbol cap, $\cap$, symbol for *and* because it is easier to write.

In reality an ecologist needs the insect larvae identified by an entomologist and is unlikely to throw the insects back; in the lake sampling example, taking a few out of the millions of larvae in the lake would only change the probability a tiny bit, such as maybe in the eighth or ninth decimal place. You can assume the effect will be lost in rounding anyway when taking a political poll in India, or when sampling some characteristic of sardines (pilchards) in the sea. But, when populations are not vastly larger than your samples, then replacement, whether in theory or in reality, is so important to this rule that the multiplication rule is often referred to as the ***sampling with replacement rule***.

1.4.3.2 Losing to the multiplication law

To get an intuitive picture of the effect of the sampling with replacement rule on something measurable, imagine you had $\$9,000$ and you gambled by putting it all, one dollar at a time, through a slot machine that was designed to, on the average, only give you two thirds of your money back, $p = 2/3$. If you were crazy enough to gamble this way, the first time you did this you would be expected to find you just had **about** $\frac{2}{3}$ of your money left, or about $\$6,000$. If you then put all the remainder through, you would be expected to find you only had about $\frac{2}{3}$ of that left, or about $\$4,000$. That would be $4/9$ of what you started with ($\frac{2}{3} \times \frac{2}{3} = \frac{4}{9}$). (Generally, slot machines in the US take up to 15%.)

1.4.4 To be A, or ! to be A; that is the question

The logic symbol for **not** is ! (A or !A); it is also called the ***negation operator***. So let's go back to grab samples from a lake bottom. Getting insect larvae is A, and getting clams is B. So, not getting any clams is !B. Then, assuming that your random grabs are independent, P_B is the probability of getting clams regardless of whether larvae are in the grab or not, and $1 - P_B = P_{!B}$ is the probability of no clams whether larvae are present or not. Then according to the multiplication law, the probability of both having larvae, A, and having no clams, !B, occurring in the same grab is $P_{A \cap !B} = P_A \times P_{!B}$. Think of the cap as only covering categories or kinds of things that overlap between two groups, the larvae set, and the empty clam set, only the grabs that include larvae *and* include *no* clams.

1.4.4.1 Example: Not !

Lake health (discussion continued) Recall that a probability, $p = 1$, means there is a 100% probability. For instance, the probability of it not raining plus the probability of it raining leaves nothing out, so that these (or any) two mutually exclusive probabilities add to 100%. Given that the probability of getting at least 1 larva in a grab (also confusingly called a grab 'sample') from the lake is $P_A = 1/4$ and the probability that you would **not** get a single larva in a grab is $P_{!A} = 1 - P_A = 3/4$, where $!A$ reads **not** A. So, after two grabs, your probability of not getting any larvae at all would be $P_{!A} \cap P_{!A} = 3/4 \times 3/4 = 9/16$, which is more than 50%.

1.4.5 Addition law

We must discuss counts and their associated probabilities in order to understand the addition law which enables us to correctly add probabilities; before we add we must count. To calculate the probability that a randomly drawn student, from a certain school with population size N, is on at least one team, whether tennis **or** soccer, we must count. We can call the tennis team category **A**, and the soccer team category **B**. They are mutually exclusive; a specific team cannot be both tennis and soccer, but some students might be members of both **A**, the tennis team, and **B**, the soccer team. To calculate P_A, the probability of a student being in category **A**, we get the count A, the number of students on the tennis team. Then the probability of a randomly drawn student being on the tennis team is $P_A = A/N$. Likewise $P_B = B/N$, the count B (soccer students) divided by the population size.

Figure 1.2: The Challenge of What Counts

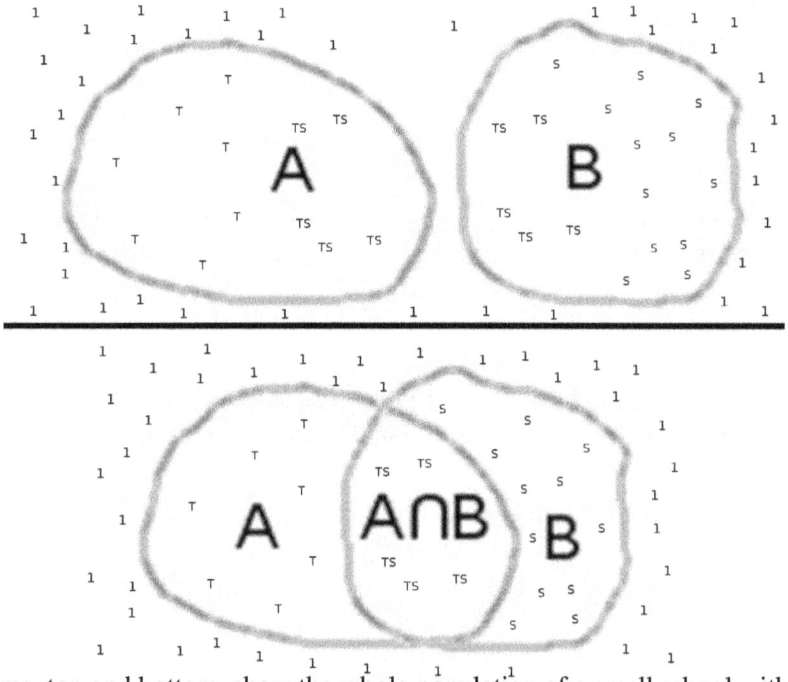

These Venn diagrams, top and bottom, show the whole population of a small school with respect to membership in two sports teams. The Ts are members of the tennis team (cloud **A**), the Ss are members of the soccer team (cloud **B**), the TSs are on both teams, and the 1s are students that are on neither team. When we count the number of Ts we get the number of tennis team members, A, even though we counted both Ts and TSs. Likewise when we count the number of Ss we get the number of soccer team members, B, even though we counted the same TSs again; team membership is disjoint (upper figure). But when we are counting to see how many seats we need for a combined sports dinner, team membership is not disjoint (lower figure). If we just add the number of team members counted in the top, we have counted the TS students twice. Thus for counting togetherness, there is extra overlap, $A \cap B$. We must subtract the extra overlap as in the addition rule, $A \cup B = A + B - A \cap B$.

If there were no students in both teams, no overlap, then $A \cap B = 0$, so $A \cup B = A + B$. The count of 1s would not come into the team counts or dinner count, but are part of the whole school's population size, N, needed to estimate the probability of a randomly chosen student having a certain single or combined affiliation.

Consider the counts and the associated probabilities that we will need for the addition law, which is used to find the $P_{A\,\text{or}\,B}$ (also written $P_{A\cup B}$):

- A, the number of students on the tennis team, probability P_A,

- B, the number of students on the soccer team, probability P_B,

- $A \cap B$, the number of students on both teams, probability $P_{A \text{ and } B}$, also written $P_{A \cap B}$ (defined in, Section 1.4.3).

The **union operator**, $\cup$, is called **cup** and is sometimes read out loud as 'or'; the math phrase 'A *or* B' is equivalent to the English phrase 'A or B or both A and B together'.[21] In logic '**or**' means at least one is true; they can both be true. A subject could belong to any one of the 'three sets', A or B or the 'third set', the intersection of A and B. The count of all athletes, whether on either team or both teams, is written $A \cup B$. Count A includes all the students on the tennis team, which includes $A \cap B$, the students that are **also** on the soccer team. Likewise count B includes all the students on the soccer team, which includes $A \cap B$, all the students **also** on the tennis team. We are **counting** $A \cap B$ **twice**. So to estimate the probability of a randomly chosen student being on a sports team by adding A to B to get $A \cup B$, we must **subtract the extra** $A \cap B$ **once**. (That means we must also keep a count $A \cap B$ of the number of times A and B occur together.) This describes the **addition law**, sometimes called the **addition rule**,

$$P_{A \cup B} = P_A + P_B - P_{A \cap B}. \tag{1.5}$$

where $P_{A \cup B}$ can also be written $P_{A \text{ or } B}$.

There are sometimes situations in which $P_{A \cup B} = 0$, that is, it is impossible for A and B to happen together. This is sometimes seen as its own law, the **disjoint addition law**,[22]

$$P_{A \cup B} = P_A + P_B. \tag{1.6}$$

The counting challenge affecting both the disjoint multiplication law (Equation 1.4) and the **nondisjoint multiplication law** (Equation 1.6) are visually represented in Figure 1.2 (top and bottom respectively).

Of course these probabilities are found by dividing each count by the school's student population size N,

$$P_{A \cup B} = \frac{A}{N} + \frac{B}{N} - \frac{A \cap B}{N}$$
$$= \frac{(tennis + both)}{N} + \frac{(soccer + both)}{N} - \frac{both}{N}.$$

If I were an entomologist studying ants by trapping them with a tin can pitfall in a desert, one of the first things I might notice is that some are red and some are black; there are other colors too, but I am interested in the proportion of ants that have red or black or both, regardless of what other colors they might or might not have. So, I decide to call the number of ants that have a reasonable amount (meaning not just a few little lines or speckles) of **red** on them A, and I call the number of ants with a reasonable amount of **black** on them B.

Suppose my two technicians (the people who go out there and get sunburned) find that the trap has $n = 611$ total ants. But they figure that they can divide the work of counting red ants and black ants by having one of them count just the ants with red on them, and that comes to $A = 400$, and the other later counts just the ants with black on them, and that comes to $B = 200$. After they come back with the ants preserved in alcohol, I count the ants which have enough red or black on them to be in $A \cup B$, and they $= 500$. My sample count of all the ants that have red **or** black **or** both on them only come out to 500 ants, **not** $400 + 200 = 600$ ants like the technicians counted. One

---

[21] In math there is also a word, **xor**, which means only one thing or the other, but **not** both.
[22] **Disjoint** means that the categories cannot occur together (not jointly).

technician had counted all the ants in the sample that had reasonable amounts of red **and** the ones that happened to include $A \cap B$, the number of ants that had red and **also** had reasonable amounts of black. Then while that guy was busy trying to cool off under a thorn bush but getting stung by a tarantula hawk, the other technician counted all the ants in the sample that had reasonable amounts of black **and** the ones that happened to include $A \cap B$, the number of ants that had black and **also** had reasonable amounts of red. These $A \cap B$ ants **had been counted twice**. So, I need to count the **overlap**, $A \cap B = 100$, that have the two characteristics together, and then subtract the extra $A \cap B$ count to get $A \cup B = A + B - A \cap B$,giving $A \cup B = 400 + 200 - 100$, which comes out to $A \cup B = 500$ ants.

As frequentists, we try to sample as **randomly** as possible in order to get an accurate as possible estimate of the relative frequency, which, given randomness, we accept as an estimate of the probability. We will calculate the probabilities by dividing the counts in each category by the total sample size, $n = 611$. So, it is customary to estimate the probability (as the relative frequency) of the combined set, $A \cup B$, by substituting the counts into Equation 1.5 and dividing each term by the total sample size,

$$P_{(A \cup B)} = A/\boldsymbol{n} + B/\boldsymbol{n} - (A \cap B)/\boldsymbol{n}$$

to get $P_{(A \cup B)} = (400/611) + (200/611) - (100/611)$ to estimate that $P_{red \cup black} = 0.8183306$ or about 82%.

Of course this is just a made up story. In terms of my hope of measuring the proportion of ants of each color type, this sampling process is **biased**. Experience shows that ants of different species **behave differently** in ways that **affect** how likely it is that they will fall into a pitfall trap and therefore will affect the proportion of each species that falls in, which also affects the proportion of colors. The fact that I used pitfalls defines my experiment as much as the colors of the ant populations. That is, the sampling technique did not meet the definition of randomness. The result could be roughly said to represent only the population of ants that would fall into a tin can pitfall in that desert. This sort of sampling challenge is not restricted to the broad area of biological studies; educationists, political scientists, sociologists, health workers, etc. also have to frequently face the question of how to get a random sample that represents what they want to study.

Trapping ants with pitfalls is an example of the pitfalls that we must struggle not to fall into even if we are not in the ant counting business. **Data represents the experiment**, rather than reality; we must struggle to get data that approaches representing reality. Physicists say you cannot observe subatomic processes without changing them. Well, this tends to be true for most things we measure or observe. If I were an anthropologist observing aunts rather than an entomologist trapping ants, my presence would almost certainly affect their behavior.

So, the ant example was analogous to comparing sidewalk political polls taken in Brooklyn on successive Thursdays to determine the effect of the local campaign for Congress. That could not show how many votes there are, or even the proportion for each candidate, but repeating it and applying a more advanced statistical technique might give a hint as to what effect the campaign strategies are having (especially regarding chatty people on Brooklyn sidewalks on Thursday).

You can try the addition law yourself via Question 7, which is answered at the back of this book.

1.4.6 Notation, P and p

Unfortunately the use of large P and small p is rather arbitrary, but mainly used to help you follow what is where in the equations. One exception is when considering the binomial distribution; lowercase p is always the probability of a single success.

1.5 Odds

Odds is a common term used in gambling. ***The odds*** is a ratio, the probability of something happening divided by the probability of something not happening, and can be given in two different ways: 'odds in favor' and 'odds against',

$$\mathcal{O} = \frac{p}{1-p} = \frac{probability\ of\ \textbf{\textit{something}}\ happening}{(probability\ of\ \textbf{\textit{that\ something}}\ not\ happening)},$$

where ***something*** can be chosen to represent either winning or losing.

Odds are handy in gambling, where if you win you get your money back, plus the winnings. Thus if you are rolling a fair six sided die and want to know the odds of getting a 1, then your probability of getting a one is $p = \frac{1}{6}$ and of getting any other outcome than a one is $q = 1 - p = \frac{6}{6} - \frac{1}{6} = \frac{5}{6}$. When dividing fractions, recall that you invert the divisor and multiply. So, the odds of winning are $\mathcal{O} = \frac{p}{1-p}$, so

$$\left(\frac{1}{6}\right) / \left(1 - \frac{1}{6}\right) = \left(\frac{1}{6}\right) / \left(\frac{5}{6}\right) = \frac{1}{6} \times \frac{6}{5} = 1/5,$$

which is usually written $1 : 5$ (one to five). However, informal games use odds of losing as a ***payoff***; that is, if the odds are $5 : 1$ (five to one) against rolling a one, the other guy will pay the winner, who rolled a one, 5 yen for each 1 yen that was bet. So, if a casino included such a simple game, it might tantalizingly report the odds as four point seventy five to one, written $4.75 : 1$, called the payoff; the 0.25 goes to the house cut (casino profits and costs).

To avoid confusion, odds are **never** expressed as a percent in print, but probability is often expressed as a percent. For future reference, you can find the corresponding probability for odds by using

$$p = \frac{\mathcal{O}}{\mathcal{O}+1}. \tag{1.7}$$

Odds are popular in medicine as a measure of the probability that the patient gets well, divided by the probability that the patient does not, or conversely that the patient dies or the patient does not. Also popular in medicine is the *odds ratio*, the ratio of the odds after treatment divided by the odds before treatment or vice versa (Section 12.14).

1.6 B Given A

If the fact that some event, A, has happened affects the probability of another event B happening, then the probability of B happening when A has happened is written $P_{B|A}$. This is referred to as the probability of B, *given* A. Such relations are called **dependent** because the probability of one has that kind of relation to the probability of the other. For instance, $P_{A|B}$ could be the probability of A, your laptop breaking down, given that B, the warranty has very recently expired (companies hire actuaries to determine how long warranties should be so as to minimize profit loss). This probability could also be called a conditional probability (Subsection 1.7). Although $P_{A|B}$ refers to the order of happenings, A and B, the ordering can simply be in logical order; it does not have to be in temporal order (ie., given that A, a woody plant, is in the family Rosaceae, what is the probability, $P_{B|A}$, of B, that it is an apple tree?).

1.6.1 Sampling without replacement

A common form of dependence becomes obvious in a small population if you *sample without replacement*,

$$P = 1/(N - n),$$

where N is the size of the whole population and (given random sampling) n is the size of the sample. If there are only four unmarked dollar grab bags and you are hoping for the one with the rubber duck, you have a $p_1 = 1/4$ probability of getting it on the first random draw. Then if after you buy one and find it is not the one you desire, if you do not put it back, you have a $p_2 = P_{success|one\,failure} = 1/3$ probability of getting it on the next draw.

1.6.2 Probabilities involving complements

Registered voters are called for jury duty at random, but it is generally easy to cop out. Designate a registered voter being willing and able to serve if called, A, and designate any registered voter being called, B. Then the probability of a registered voter being called at random is P_B, and the probability of a registered voter being willing and able to serve is P_A. So, the probability of a registered voter both being called at random **and** also being willing and able to serve is probability $P_{B \cap A}$, the probability of A *and* B (also written $P_{A \cap B}$).

Not being in group A is written $!A$ (not-A), and it is called the **complement** of A. The probability of a randomly drawn registered voter not being willing and able to serve if called is $P_{!A}$. So, the probability of a voter both being called **and** also not being willing and able is $P_{B \cap !A}$. For a registered voter to be both, $A \cap !A$, is not possible ($A \cap !A = 0$). Something cannot be both A and $!A$; there is absolutely no possible overlap; A and $!A$ are said to be **mutually exclusive**. The **law of complements** says that $P_A + P_{!A} = 1$ (100%). (A variant on the law of complements is $P_{!A} = 1 - P_A$.) Although complementary outcomes A, $!A$ are mutually exclusive, they can define subsets of another group, such as B, that has only two possible subgroups which are $B \cap A$ and $B \cap !A$, and they are obviously mutually exclusive, $(B \cap A) \cap (B \cap !A) = 0$.

For this example we'll use C and D, but variables in general can refer to individual things, and they can refer to compound, or conceptual things, so that we can say $C = (B \cap A)$ and $D = (B \cap !A)$. The addition law says that the probability of either or both of two things happening is $P_{C \cup D} = P_C + P_D - P_{C \cap D}$, but $P_{C \cap D} = 0$, because $C = (B \cap A)$ and $D = (B \cap !A)$ are mutually exclusive and cannot happen together. So, in situations like this with two mutually exclusive complementary subsets within an overall group such as B, then $P_D = P_{B \cap A} + P_{B \cap !A} - 0$; and simply stated, is

$$P_B = P_{B \cap A} + P_{B \cap !A}. \tag{1.8}$$

When dealing with seemingly complicated situations, remember to look for and recognize a mutually exclusive subset within the whole, say, B. In some instances a set, B, can be composed of n different parts, where each part is called $A_i = A_1, A_2, ... A_n$, that is, $1 \leq i \leq n$, that together make up the whole, so that, provided they do not overlap, B is the sum of all its, B's, parts: $B = \Sigma_{i=1}^{i=n} A_i$. Pretend you are learning statistics from Yoda.[23]

---

[23] Although not essential, in this book, other than for completion (throwing in the kitchen sink), De Morgan's law says that $!(X \cap Y)$ is the same as $!X \cup !Y$; and $!(X \cup Y)$ is the same as $!X \cap !Y$. Note the symmetry.

1.7 Conditional Probability

Let's use P_A for the probability of a randomly selected person being drunk (A, for *Alcohol*), and use P_B for the probability that someone will be a driver in an auto accident (B for *Bang*). Then the conditional event, $B|A$ (B given A) is the probability that a person will be a driver in an auto accident, given that the person is drunk. Assume that P_A, the probability of a person being drunk, and $P_{B|A}$, the probability of a person being a driver in an auto accident, given that the person is drunk, are independent. Then by the multiplication law, you can calculate the probability, $P_{A \cap B}$, of **any randomly selected person** both being drunk and a driver in an auto accident,

$$P_{A \cap B} = P_A P_{B|A}. \tag{1.9}$$

Then if you want to figure the ***conditional probability***, $P_{B|A}$, that a person will be a driver in an auto accident, given that the person is drunk, then Equation 1.9 (above) can be algebraically rearranged to get the conditional probability,

$$P_{B|A} = \frac{P_{A \cap B}}{P_A} \tag{1.10}$$

the probability of B occurring (or being true), given that A occurs (or is true). Likewise $P_{A|B} = \frac{P_{A \cap B}}{P_B}$.

1.7.0.1 Example: Conditional Probability

D given A There are various ways to classify people as either A, abused in childhood, or to classify them as D, having an eating disorder. Suppose that your team classifies $15\% = 0.15$ of a study population as having been abused in childhood, A. Therefore

$$0.15 = P_A.$$

Assume that a national survey reports $0.3\% = 0.003$ of this same **whole** study population as suffering from eating disorders **and** also as having been abused as children, $A \cap D = D \cap A$, so

$$0.003 = P_{A \cap D}, \text{ which also equals } P_{D \cap A}.$$

Patient confidentiality regulations and bureaucracy prevented your team from getting more information about how the national survey came to this conclusion.

Can you estimate the conditional probability, $\widehat{P}_{D|A}$, that a person in this study population will have an eating disorder, **given** that the person was abused as a child? (By the way, $\widehat{P}_{D|A}$ usually does **not** equal $\widehat{P}_{A|D}$.) Assuming these frequencies (percentages) are from adequately random huge samples, the resulting probabilities facilitate the estimation of the conditional probability,

$$\widehat{P}_{D|A} = \frac{P_{D \cap A}}{P_A} = \frac{0.003}{0.15} = 0.02 \,,$$

where the hat, $\widehat{}$, is simply included to emphasize that this is an estimate. So, in this (fictitious) study population, you would estimate the conditional probability, given that a person has been abused as a child, that the person will also have an eating disorder, is 2%. If this data is true, the probability of eating disorders would be much larger among those who were abused than for the overall population.

1.8 Inverse Probability

Inverse probability is a surprising and useful area of applied probability that may save us from getting into some very scary misunderstandings and government programs. (Understanding the preceding section might be crucial for understanding what follows.)

1.8.1 From the grave: Bayes awakens the medical establishment

> "All instruction given or received by way of argument proceeds from pre-existent knowledge. This becomes evident upon a survey of all the species of such instruction."

> *Posterior Analytics.* Aristotle 350 BC (Translated by G. R. G. Mure, edited by W. D. Ross, 1925.)

Bayes' theorem, also called ***Bayes' law***, also called ***Bayes' rule***, involves what Bayes (published posthumously in 1763) called "a problem of ***inverse probability***":

$$P_{A|B} = \frac{P_A P_{B|A}}{P_B},\qquad(1.11)$$

which can be found by substituting $P_A P_{B|A}$ for $P_{A\cap B}$ in the conditional probability $P_{A|B} = \frac{P_{A\cap B}}{P_B}$ (Subsection 1.7).

It is relatively easy to find statistics for estimating the probability $P_{B\cap A}$ of people being in auto accidents and also being drunk, likewise for the probability P_B of being a driver in an auto accident. As we have seen, we can then calculate the conditional probability, $P_{B|A} = P_{B\cap A}/P_A$, that a person will be a driver in an auto accident, given that they are drunk. Then, assuming that we also knew the probability P_A of a randomly selected person being drunk, we can use Bayes' theorem to find the probability $P_{A|B}$ that a driver was drunk given that the driver was in an automobile accident.

Bayes' theorem is accepted and used by all statisticians whether Bayesian or frequentist (refer to Subsection 1.3.1). For instance, it leads to some remarkable conclusions regarding mass health screening and certain vaccinations (Subsection 1.8.3).

Although this discussion refers to law enforcement and medical practice, there are many other potential uses. Bayes' theorem is a wonderful development of the early eighteenth century that should be applied more in medicine, public health and many other areas today. It is great for 'what if modeling'. The following subsections have a few handy details for applying it for medical, public health or other issues.

Getting the show together

The probability that event A **really has** happened when B, a medical test, reads positive, is written $P_{A|B}$ (probability of A *given* B). So, let's start with the simplest form of Bayes' theorem, the conditional probability Equation 1.11,

$$P_{A|B} = \frac{P_{B|A} P_A}{P_B},$$

and referring to Equation 1.8 substitute into the denominator (the bottom), $P_{B\cap A} + P_{B\cap !A}$ for P_B to get

$$P_{A|B} = \frac{P_{B|A} P_A}{P_{B\cap A} + P_{B\cap !A}}.$$

Recalling that $P_{!A} = 1 - P_A$, hence $P_A + P_{!A} = 1$, it is possible to derive the most popular form of Bayes' theorem by substituting complementary versions of the Equation 1.9 $P_{B \cap A} = P_{B|A}P_A$ and $P_{B \cap !A} = P_{B|A}P_{!A}$ in place of their sum, P_B,

$$P_{A|B} = \frac{P_{B|A}P_A}{P_{B|A}P_A + P_{B||A}P_{!A}} . \tag{1.12}$$

In Bayesian statistics, the input values are what they call **priors**, and the result is what Bayesians call the **posterior**, which is an updated estimate. Bayesians do not see the posterior, or probabilities in general, as fixed point estimates, they do not see them as fixed in stone. Although this book is primarily frequentist and not Bayesian, either viewpoint, frequentist or Bayesian, finds Bayes' law acceptable in application.

The denominator of Equation 1.12 could even be a sum of a still larger number of complementary probabilities (as a summed group) if more than two priors are involved. However, this book does use this very useful form, Equation 1.12, that contains only two mutually exclusive (non overlapping) complementary probabilities.

If you need to commit this to memory, notice that **it always helps to look for symmetry** in learning equations. One underlying pattern here might crudely be remembered as having the

$$P_{something}/(P_{something} + P_{the\,rest})$$

frequentist pattern of finding 'inverse' probabilities[24].

Think in terms of testing for a contagious disease, or for drug use. Bayes' law is often about testing to see whether A has actually occurred when B, the test, appears positive. The probability that a test is truly detecting anything when it shows positive results is equal to the probability of the test actually being positive when true, times the probability that a random person from the population is positive, divided by the total probability of a positive test whether it is correct or not.

Question 11 allows you to try working an example of a common application of Bayes' theorem (answer at the rear of this book).

1.8.2 Mass screening

Drug testing is often required for getting jobs, continuing parole, keeping jobs, and even child custody in the case of parents who have had difficulty with substance addiction. A false positive may lead to not getting the job without the applicant knowing the cause. Accusing someone, or even demanding another test, is awkward for the employer to say the least. The already over burdened courts have an additional case load involving prolonged legal battles over drug test results that are suspicious or even false. However, some schools have considered routinely drug testing all the children over age 12. As is shown in the following example, mass testing for substance abuse will turn up more false positives than true positives. A false positive at school would cause suspension while the authorities attempt to confirm the test, if they even do bother to confirm. That would result in traumatic confrontations between parents and innocent youngsters, and other complications. Repeatedly testing the same person, as in a school program or court mandate of monthly testing, would greatly increase the hazard of false positives for that individual. If tested often enough, an eventual false positive is almost guaranteed.

---

[24] The denominator of Equation 1.12 could even be written, $\Sigma_{all\,A} P_{B|A}P_A$.

1.8.2.1 Example: Bayes' Theorem and Drug Enforcement

WARNING, APPROXIMATION: The values used below are a very crude compromise from various 2008 internet sites where the values varied over a wide range. They might be acceptable **for the last month**, because that is the maximum time range an individual test is likely to cover and because many users do not use continuously. However, with up-to-date verifiable accurate data, the outcome would be similar, and that could serve as a model that would be acceptable in court.

Drug enforcement: This example demonstrates how to apply ballpark estimates of values for multi-substance (except alcohol) drug testing, approximated as of 2009. Let A represent the fact that the subject has used a targeted substance in the last month, and let B represent a positive test result.[25]

By the accepted frequentist approach, and assuming true randomness, the probability of the subject having used such a substance is

$$P_A = f_A,$$

where f_A is the absolute frequency representing the proportion of the population that is using such a substance. We will call the probability of a positive test result, whether it is a true positive or a false positive, P_B.

The following involves applying the test a single time to a single person:

$P_{B|!A}$ is the probability of a false positive: $\approx 3\%$ in a trustworthy laboratory (not all are trustworthy).

$P_{!B|A}$ is the probability of a false negative: maybe as high as $\approx 1\%$ (lower or higher depending on the substance).

$P_A = N_A/N$ is the probability of the person having used illegal drugs within the last month $\approx 7\%$ for everyone in the US over age 12.

$P_{!A} = 100\% - P_A \approx 93\%$ is the probability of not having used within the last month.

$P_{B|A} = 100\% - P_{!B|A} \approx 99\%$ is the probability of a positive when A is true, meaning that there really was drug use.

To get a more complete understanding, start by using Bayes' theorem to calculate the probability, when testing everyone over 12 years old in the whole country, that a **positive test result is a true positive**. The probability that the subject actually has used such a substance, given a positive test result, can be found by:

$$P_{A|B} = \frac{P_{B|A}P_A}{P_{B|A}P_A + P_{B|!A}P_{!A}} = \frac{P_{B|A}P_A}{P_B} \qquad \text{(Bayes' theorem, Equation 1.12)}$$

$$P_{A|B} = \frac{0.99 \times 0.07}{0.99 \times 0.07 + 0.03 \times 0.93} \qquad \text{(substituting values given above)}$$

$$P_{A|B} \approx 0.713 \qquad \text{(probability that any positive test is correct)}$$

$P_{!A|B} \approx 1 - 0.713 \approx 0.287$ (by subtraction: probability **that any positive test is false**)

A decision tree is another way to look at the process used in this example:

---

[25]In the example, Subsection 1.8.2.1, a purist might assert that you are using systematic sampling and assuming randomness. Alternatively, it could be argued that the false positives occur at random, and there is also some randomness in who gets the 'opportunity' to be a participant in such programs.

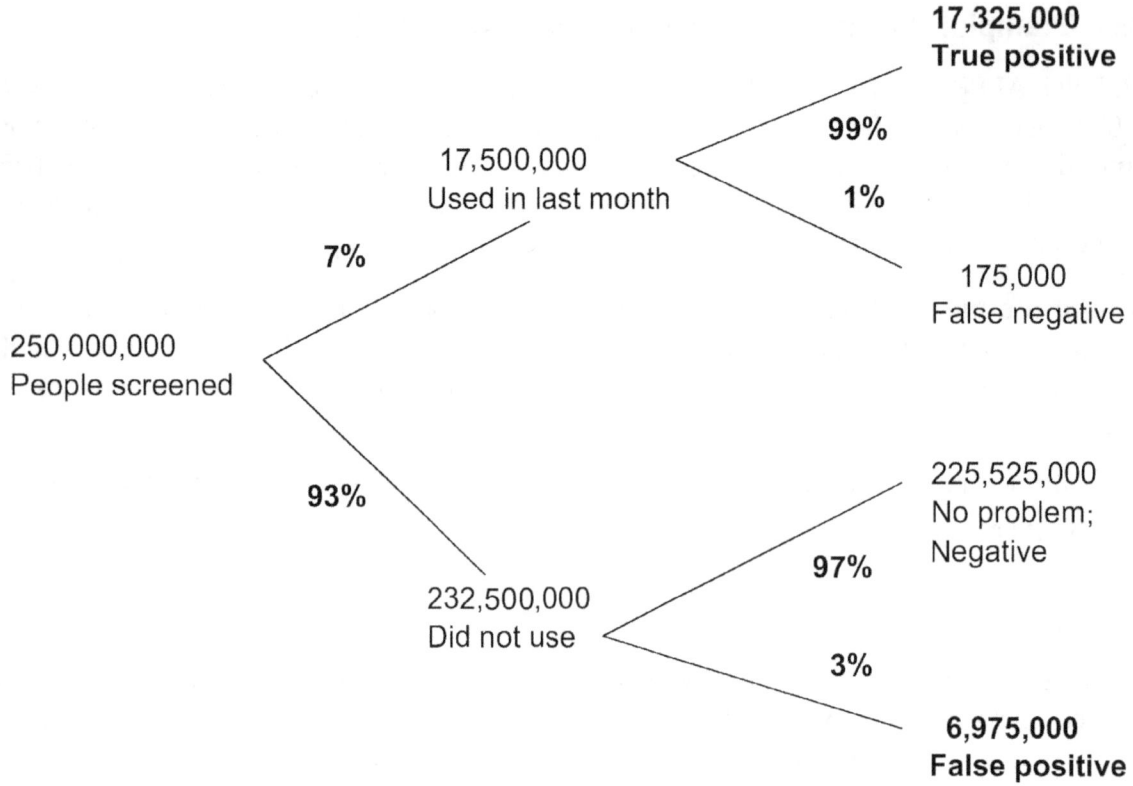

Discussion: This seemingly non-intuitive outcome, 6975000/(17325000 + 6975000), almost 29% of the positives being **false positives** when testing the whole country, is because the proportion of people taking the test who were non-users during the last month is so huge (93%), the number of false positives is uncomfortably large. On the other hand, although numerically very large, the proportion of the population who really should read positive are almost all detected, but they are only part of the whole who test positive. Let's look at the numbers involved here; N is the population over age 12 in the US; $N \approx 250,000,000$ in 2009 (assume all are tested just once) and $N_A = P_A \times N = 17,500,000$ who have used illegal substances in the last month. Of the $P_{B|A} = 0.99$ of them, $17,325,000$ will according to this estimate be detected by true positive tests. However, a much larger number, $N - N_A = 232,667,000$ of non-users that month took the test and about $P_{B|!A} = 0.03$ of them will get false positive tests. So, 3% of $232,667,000$ **non-users** equals $6,975,000$ false positive results. That means that almost seven million **innocent people** would be mistakenly accused of using drugs.

Think of the misery of being a 13 year old and being confronted with a false positive drug test at school. Think of being permanently laid off because your employer does not want the legal hassle of possibly being sued due to an accusation based on a potentially false positive drug test. **Testing too widely** for rarer abuses or conditions, as is threatening to become common practice, even with very sensitive tests, **can produce so many false positives** that they overwhelm the true positives. If you were mass testing for only one uncommon drug, even with a very sensitive test, you would find that you were getting far more false positives than true positives!

No doubt some readers will discount this result as 'trying to go easy on drug users'. Many websites are by people who think screwing up your mind is merely recreation. I emphatically disagree with them concerning the 'safety' of mind altering substances and I did not draw my information from the propaganda or self deception of the so-called drug community.

I think the above result is nearly correct (approaches accuracy), and it demonstrates the futility, and even danger, of routine or monthly drug testing of people. Even if I have mis-estimated one of

Figure 1.3: Home Test Kits for Family Unity?

Home test kits marketed for about $1 that say that they are "97% accurate" (top) and "98% accurate" (bottom). If you really use one of these, any person with a positive home test should be referred immediately and without fuss (on your part) to a certified laboratory for verification. What will happen if you use one of these on your child once a week for years on end?

the more controversial values by quite a bit, they, when updated, would still show that extreme safeguards are needed before coming to any conclusion concerning guilt, or even the slightest suspicion, based on mass screening, routine screening, or unprovoked home testing. Monthly screening by employers, schools, or court order would add up to huge numbers of tests over time and produce huge numbers of false positives. If applied to the general workforce or the public in general, the false positives would create a legal and management nightmare. This demonstrates that without the pre-testing requirement of a reasonable reason for suspicion such as changes in behavior, etc., testing can cause more problems than it can solve.

This example did not take into account the fact that many labs produce worse results than a mere 1% of false positives, or that some labs claim 99.9% detection rates. The former means that this issue is more of a problem than the example shows, and the latter, if true, would provide only a small improvement. Updating the example with more recent and accurate estimates based on peer reviewed citations would make a start on a good student term paper for any of a variety of subject areas.

1.8.3 Screening and health care

Debauchery involving drugs is not a rare 'disease' as diseases go, but the more rare (the less probable) a disease is, the greater the probability that a positive test result is false. So, nationwide testing for less common diseases will often produce more false positives than true positives. A 2009 article in the Los Angles Times is typical of many like it, not only in the media, but also in professional journals. It reported results from a number of reputable sources, including the *Journal of the American Medical Association* and interviews at outstanding medical facilities, expressing alarm about results of mass cancer screening, that they were producing too many false positives. Among other recommended reductions in screening practices, these sources recommended drastic reductions in screening young women using pap smears and mammograms, and many sources expressed concern regarding the prostate-specific antigen test to screen for prostate cancer in men. False positives for these tests can lead to unnecessary surgery.

Bayes' theorem is also essential when considering universal vaccinations. Unlike mass health screening through tests, the appropriate procedures have been applied to be certain that the vaccinations are vastly more likely to help you than harm you. However, vaccinations do have a tiny probability of causing adverse events, or even more rarely, deaths. So, in light of the fact that many people might never be exposed to the disease, and the fact that some of those actually getting the disease will be little harmed, the public health community has had to consider possible injury and deaths caused by vaccination versus how many will be seriously sick or die if they are not vaccinated. Mass vaccination of prepubescent females to avoid the terrible effects of rubella in pregnancy is presently judged to be prudent; whereas mass vaccination for extinct or almost extinct smallpox is not. Likewise, serious effects from the rubella vaccine is very rare, whereas the vaccine for smallpox is comparatively risky. When the author was a child, the UN vaccinated the entire world for smallpox, some of it (myself included) twice. However, I have personally known several people killed by crossing the street. Do you know anyone killed by a vaccine? One of my readers, somewhere, might say yes, but I doubt that any of my readers, even professors older than me, can say they know several.

1.8.3.1 Example: Bayes' Theorem, Public Health

Public health screening This totally fictitious example demonstrates how things might come out if you have widespread public health screening for a relatively uncommon contagious disease with a unusually sensitive test that had an unusually low false positive rate. Let A represent that the person is infected, and let B represent that the test is positive.

$P_{B|!A} \approx 1\%$ is the probability of a false positive,
($P_{!B|!A} = 100\% - P_{B|A} \approx 99\%$ is the **specificity**, the probability that a test is negative given that the subject does not have the condition, not used directly in this example.)

$P_A = N_A/N \approx 0.1\%$ is the probability of the person having the condition (the **prevalence**).

$P_{!A} = 100\% - P_A \approx 99.9\%$ is the probability of not having the condition.

$P_{!B|A} \approx 0.001\%$ is the probability of a false negative, given that the person has the condition.

$P_{B|A} = 100\% - P_{!B|A} \approx 99.999\%$ is medical **sensitivity**, the probability that a test is positive when the condition exists.

To find the effect of mass screening with an unusually sensitive test of individuals in the population at random without any prescreening for specific symptoms, **start by using Bayes' theorem** to calculate the probability that a positive test really indicates the presence of the condition. The probability that a subject actually has the targeted condition, given a positive test result, is:

$$P_{A|B} = \frac{P_{B|A}P_A}{P_{B|A}P_A + P_{B|!A}P_{!A}} = \frac{P_{B|A}P_A}{P_B}$$ (**Bayes *factor***, equation expressing the Bayes' theorem)

$$P_{A|B} \approx \frac{0.99999 \times 0.001}{0.99999 \times 0.001 + 0.01 \times 0.999}$$ (substituting values given above)

$$P_{A|B} \approx 0.090991$$ (probability that any positive test is correct, **so**)

$P_{!A|B} \approx 1 - P_{A|B} \approx 0.909009$ (by subtraction is the probability that any positive test is false)
in other words, a little over 90% of all the positive test results will be false positives.

Discussion: (Why didn't I use lowercase p in the examples? It was a purely arbitrary choice to match certain of the preceding notations.) Notice that the last two probabilities are with respect to the persons getting a positive test (whether true positive or false positive respectively).

If you used this test to mass screen $250,000,000$ people, the test would correctly find almost all, $B|A \approx P_{B|A} \times 250,000000 = 2,499,800$, out of the $A = 2,500,000$ **people with the disease**, but you would have **about ten times as many false positives**, $!A|B \approx 24,973,200$. So you can treat about 2.5 million sufferers, if you can sort out the almost 25 million false positives. The input values for this example are hypothetical, and yet, it is not uncommon in such mass screenings to get as many as ten times more false positives than negatives when you screen everyone with some medical tests, even when they are as extremely sensitive as this hypothetical test. This outcome is because you are screening such a multitude to find a rare instance of the malady. The false positives add up because a huge proportion of the people that would be tested would not have the condition, as opposed to the tiny proportion of the people tested that would have the condition.

Diagnostic tests for uncommon conditions, even when extremely good at detection, are much more appropriate for situations where they can be used to choose between potential alternative diagnoses than for screening apparently healthy patients.

During many months of the covid-19 outbreak the US suffered a shortage of material to test for the virus, and one of the first available tests was reported to produce 30% false positives. Since then, extremely sensitive tests have become available, but their false positive rate is not well established (as of August 2020). Even if the tests had only had a half percent false positive rate, and **one tenth of one percent** of the population was infected in that period, mass testing the whole population of the US then would have resulted in additional chaos, because the flood of false positives would have made medical services less available to the people that really had the virus. What was really needed was for physicians and hospitals to have access to enough testing material **to verify** that people who had possible covid-19 symptoms actually had COVID, and to test the contacts of those who indeed had it. If there were enough compliance with social distancing, mask wearing, and testing temperatures at the entrances of schools and other crowded places such as theaters, transit stations, etc. we might have avoided the collapse of so many businesses and kept the schools open. Taiwan, for instance, was actually able to attain this goal[26].

Bayes' theorem is used by both frequentist and Bayesian statisticians. There is no evidence whether its author, Bayes, was a Bayesian or not. Bayesian statistics is so named because it focuses on prior knowledge (including opinions or assumptions) and hence tends to use Bayes' theorem more often than does frequentist statistics. In his 1814 second edition *Essai philosophique,* Pierre Simon Laplace set out a mathematical system of inductive reasoning that could be seen as a refinement of 'Bayesian statistics'. So, maybe Bayesian statistics should be called 'Laplacian statistics'.

In a sense, both of the above examples tend to be more Bayesian than frequentist because they could be seen as based on the author's estimation of the probabilities, what might be seen as my prior belief. Such estimates are sometimes based on what the researcher deems to be prior knowledge, but with the understanding that (for the Bayesian) we never know anything exactly. They decisively demonstrate the pitfalls associated with mass screenings. A more conservative estimate based on frequencies or means of carefully controlled sampling, maybe data from peer reviewed journals, could nail down a more exact but similar conclusion. Improving the accuracy of the values in the examples in Sub-subsection 1.8.2.1, and in Sub subsection 1.8.3.1 will not change the conclusion as to the difficulties associated with mass screenings.[27]

In many situations, especially when time or funding is a substantial constraint, obtaining sufficient sample based information is not practical, if even possible. In such situations, a Bayesian approach can be useful because of its potential for exploring the question at hand and for discovering a wider range of not precisely quantified possibilities. Bayesian techniques along with a sufficient literature review (a source of prior information) can be especially useful for deciding whether experiments, such as medical trials or screenings, should even be initiated. They can also be an excellent source of hypotheses for planning frequentist research.

---

[26]https://www.worldometers.info/coronavirus/country/taiwan

[27]Bayesian statisticians use Bayes' theorem (Section 1.8.1) in a number of situations involving *Markov chains*, stochastic models describing a sequence of possible events in which the probability of each event depends only on the state attained in the previous event. They would phrase Bayes' theorem, as applied in such a manner, as "the posterior probability is proportional to prior probability times likelihood." Their use of Markov chain calculations often requires extreme computing capacity, but the calculations tend to converge to meaningful outcomes. This approach is especially useful when the amount and complexity of the data would otherwise be overwhelming.

However, assume you were trying to assess whether a given fish is being overfished, but you do not have the funds to get meaningful results. Is it reasonable to shut off a major portion of a nation's food supply or the traditional income of thousands of families because, in your opinion, something might be overfished? We, as the peoples of the earth, must expend adequate resources so that researchers can at least get adequate data for such critical questions. Whether a frequentist approach or a Bayesian approach is to be used to make the crucial decision, your preliminary approach to budgeting for obtaining such data might benefit greatly by using an appropriate Bayesian estimate of the logistics for which you must budget. Various approaches to exploratory and descriptive statistics will help you to find strategies to design valid experiments to actually obtain the real data that is needed. A popular and delightful book on data exploration using a variety of approaches is John W. Tukey's classic, Exploratory Data Analysis (1977).

The context of this section, and this book in general, is mostly about sample based sources of data. With vast databases, the preexisting information providing the information to adjust the researcher's 'degree of belief' can be quite substantial, and there is a question of whether a frequentist or a Bayesian approach might be best, even for inference. Expect to see a near future where large data analysts, 'data scientists', routinely apply both approaches, frequentist and Bayesian.

1.8.4 Prosecutor's fallacy

The above discussion (Section 1.8.1) is similar to a legal risk to the innocent that is called the ***prosecutor's fallacy***. Suppose that the government has a DNA sample from a crime. Most of these crime scene samples are only partial (the DNA at the crime scene is often degraded so that the code cannot be read in its entirety, or fingerprints are only partial). If the prosecutor had a reason to compare the partial sample to someone already suspected, a false match might be extremely unlikely. However, the DNA database for a large country like the US can contain hundreds of thousands of samples. Several innocent person's samples could match the partial sample, while the perpetrator's sample could possibly not even be in the database.

The partial sample might match one in many thousands if they were chosen at random. If it also matched someone who was already otherwise a likely suspect, that would be strong evidence. However, suppose the police have found a match in an enormous database and the match is their sole reason for the prosecution. The larger the database, the more probable a false match becomes.

In order for statistical evidence to be of value in court, there has to be some sort of *a priori* (meaning before the fact) reason for resorting to a statistical approach. In referring to this sort of problem, a Bayesian would say, "Keep your priors straight." 'Unlikely' means it is only likely to have occurred once in a very large sample, but a database might be a huge sample and might easily contain one or more such false positives. Put simply, the very unusual occurrence will often be included if your sample is large enough.

Assume a suspect was apprehended while hiding in the park when, and where, the crime occurred. Then if such a partial DNA match had 1 chance in 10, 000 of occurring by chance alone it would be very good evidence in combination with evidence that the person was there in that place and time. But what if the **only** evidence was that the match was found in a data base of 22, 000 persons; there was no other evidence to definitely associate them with the crime? Then it should not be able to convict by itself, because we expect that there will be about 2 matches, based on chance alone, in that size of data base, which would correspond to two different people. Even then the actual perpetrator's data might not be in the data base, because the majority of the population isn't even in the data base. This is not an unusual courtroom problem.

Another problem involves sudden infant death syndrome. One natural SIDS child death in a family is only a bit rare. However, mothers have been prosecuted because there was later a second death and under the multiplication law that is very rare. Well, there are billions of mothers in the world and the extremely rare has to happen sometime, to some family; it, in itself, is not sufficient evidence of wrongdoing. A tragic example of this may be found by looking up Sally Clark in Wikipedia.

A frequentist approach, especially using the Neyman-Pearson model, Subsection 4.7.2 and looking for statistical significance using an established, a priori, absolute cutoff can greatly reduce the probability of bias based (such as prejudice based) decisions, but nothing can totally eliminate the chance of false significance (or false conclusions).

Questions for Ch 1

Write the equations where appropriate and fill in the values. **Odd numbered questions** have worked answers at the back of the book and can be studied as additional examples. The data tends to be fictitious.

Always show your work and explain.

1. Sue lives in a tiny rural village, and her dog that she let run loose is missing. Then she discovers that it has displayed **certain kinds of uncharacteristic behavior** and **bitten someone**, and even more amazing, the pound has actually chopped its head off and shipped its head away to some laboratory (one of those places with mad scientists in it). Does this validly confirm that rumor that you need to hide your dog whenever it bites someone or the authorities will fiendishly murder it?

2. Joe and his pregnant wife are sharecroppers in an impoverished area and they are aware that there are lots of miscarriages; at least, among the people they meet in town, miscarriages are very frequent. However, when Joe is plowing the east forty, he encounters his closest neighbor, a widower who tells him that even though the FBI will arrest anyone caught selling it, if his wife eats Mississippi mud it will make sure that she has a fine strong baby. Joe obtained some of this mud and his wife ate quite a bit of it. She produces a fine healthy baby. Is that sufficient evidence that this natural substance really prevents miscarriage?

3. Your company's board of directors believe that they can fit the construction of a new headquarters into their five year development plan, but they wish you to verify and maybe improve their estimate of how long it would take. So you take the proposed plans and site details to several experts on each essential stage such as earthworks, electrical, mechanical, concrete, glazing, and so forth and get them to estimate time ranges based on personal experience for each phase of construction. You then use a recommended procedure to make a combined overall estimate of the time required. Would you associate such an approach with frequentists or Bayesians?

4. Your company is considering putting soft serve machines in all their thousands of fast food shops. However, they wonder if climate might be involved in such a way as to determine which shops get such machines. You sample soft serve sales and corresponding weather records of 65 shops that already have such machines so that you can compare the returns with their local weather reports. Is your approach here more frequentist or more Bayesian?

5. You do not know it, but there are 620 sea lions in your local area that have a certain parasite out of an unknown total of 2000 animals. You randomly sample 50 sea lions with only one visit

to the emergency room and a few stitches, and you find 10 animals that have the parasite. Because it is evident that the number of sea lions is extremely large, the result would be like sampling with replacement. How would Allseeing Mother Nature, who knows the true numbers of healthy and parasitized animals, calculate (in a human statistical way) the **absolute frequency**, f, and the true **probability**, P, of a randomly examined sea lion in your area having this parasite? How would limited human you, who only sees an extremely large number of sea lions and has taken a random sample of 50 and found 10 that have the parasite, estimate the sample relative frequency, $\hat{f}$, and the estimated probability, $\hat{p}$? Are $\hat{p}$ and P the same? Why?

6. Suppose you have a 0.02 probability of losing money on a forestry planting and a 0.035 probability of losing money in producing soybeans. You decide to invest in 1 of each. What is the probability that you will lose money in both if you invest in both? What assumption is involved?

7. A geothermal company is drilling for natural steam to run power plants at the foot of active volcanoes. Assume 20% of the wells produce adequate steam, 51% cost less than was budgeted, and 13% of the wells both produce adequate steam and cost less than was budgeted. Estimate the combined percent of the wells drilled that cost less than was budgeted or produced adequate steam or both.

8. Cloud forests may occur high on tropical mountains at about the same altitude for centuries. Over longer periods they move up and down a little but they may occur on the same mountain for hundreds of thousands of years. If you visited one you would find yourself in a cloud (a fog bank) and organisms there would include many things unique to cloud forests. Because mountain peaks are usually quite effectively separated from each other, cloud forests are like islands with respect to the organisms specialized to live in them. Some organisms may be unique, only occurring in one particular cloud forest. Suppose you only have a 47% probability of collecting a slime mold that produces a potentially useful new antibiotic when intensively sampling a cloud forest and you have a 62% probability of collecting a bacteria that produces a useful antibiotic. Assuming independence, what is a reasonable estimate of the combined probability of collecting either or both in a single randomly chosen cloud forest?

9. Plant explorers have provided a great many new introductions from their travels throughout the world. H.Y. Pothetical, MAg, is going to randomly screen a large chicory collection for varieties that are A, resistant to the chicksucker insect. He decides to call any **positive** test B whether it is a true positive or a false positive. Past experience has shown that the probability of the test detecting a plant that is truly resistant is $P_{B|A} = 80\%$ and that the probability of a false positive is $P_{B|!A} = 5\%$. Past experience also indicates that about $P_A = 1\%$ of newly collected chicory varieties are resistant. What is the **probability**, $P_{A|B}$, that a variety that **tests positive** (result B) will **really** be resistant (have property A)?

10. What is the probability of a person dying on their birthday?

11. X-Tra-Spensiv Drug Company has analyzed previous data using multivariate approaches (not included in this book)[28] that are similar to the way marketers target customers and new markets, to find the most promising places to collect marine sediments that might contain new

---

[28]The multivariate approaches to selecting soil sampling sites might include principal component analysis, or factor analysis, neither of which are treated here.

antibiotic producing micro organisms.[29] Using automated machinery, they are able to screen thousands of field collected samples a dozen ways every week, but their biggest challenge is the high cost of field collection. Their experience, since this targeted exploration began, indicates that there is a $P_A = 0.0050$ probability of a single randomly taken specimen from the targeted sites containing a new antibiotic producing organism that they can hope to culture for further study. Altogether their dozen **in vitro** (Latin, *in glass*) tests are $P_{B|A} = 0.90$ likely to detect a suitable organism, if present, that produces antibiotic effects against any of the dozens of pathogens that they are targeting. The probability of a false positive is $P_{B|!A} = 0.08$. What is the probability that a sample that produces a positive test is actually a true positive?

FAT KATZ

"Are you nuts? I have a reputation to protect...
I wouldn't dare forecast the stock market."

[29]**Antibiotic** literally means anti-living. So even after finding one, the drug company will have to study not only its efficacy against pathogens, but its potential toxicity to humans. Also recall that these numbers are made up even though such screenings do exist and are typically automated.

2 YES, NO and BEING DISCRETE

Measurements produce continuous numbers; you can keep measuring more and more accurately; the more accurately you measure, the more decimal places there are.[1] **Continuous** numbers have continuous numbers between them that, of course, have continuous numbers between them. **Discrete** refers to numbers that have distinct values rather than being continuous from one number to the next. For example, integers ($\pm$ counting numbers and 0) are discrete. Each integer has a distinct value; it is not continuous, it does not vary smoothly from one number to the next, as do the real (the measuring) numbers. Discrete values are countable.[2]

Sometimes we are counting each of only two possible outcomes, such as 0 and 1, yes and no, or other mutually exclusive pairs (Section 2.4).

2.1 The Cost of Achieving Order

A lady told a true story of going to a race track and buying a $2 ticket that included choosing which horses would win or place throughout the day. She bet at random and won $2000. The probability of doing so was very low, but the very improbable occasionally happens. She fell victim to gambling madness and during the next few weeks she lost the $2000, sold everything she had, went into debt, and did more serious things that could have ruined her entire life. She has since recovered and is living happily ever after, but by her own account it was a very unpleasant experience.

The factorial operator, $!$, merely says that you count down while multiplying. Thus, $6! = 6 \times 5 \times 4 \times 3 \times 2 \times 1$ is six **factorial**.[3]

As in the above example, the number of ways you can arrange n things one after another is **n factorial**, $n!$; it is the number of ways you can arrange n things without using any of them twice.[4] The last term could be written $n - (n - 1) = 1$, which, at first glance might seem a little confusing[5] and sometimes the next to last term confuses more than it helps unless you realize that, $n - (n - 2) = 2$. The $!$ (**after** something) is called the **factorial operator** and thus $n!$ is called n factorial.

2.1.1 Order of outcomes

Randomly getting n outcomes in a desired order has a probability of:

$$P_{(n\,\textbf{\textit{in order}}\,out\,of\,n)} = 1/n!$$

---

[1] The above definition of continuous is crude. For those of you who are familiar with such things, the real number system is continuous.

[2] The square roots of integers are almost all either other integers, or they are irrational numbers (unending numbers) represented by real numbers. However, unlike the real numbers themselves, the square roots of integers are discrete because they do not flow smoothly from one to the next. They are countable, hence discrete (have gaps between them).

[3] If you are sufficiently math oriented you could formally write $n! = \prod_{i=1}^{to\,i=n} i = 1 \times 2 \times 3 \times \ldots \times (n-1) \times n$ where $\prod$ is the **product operator,** which means multiply everything indicated together.

[4] If you use the R language factorial(x) returns $x!$.

[5] Recall that subtraction can be seen as a form of addition so that, $4 - 2 = 2$ can just as easily, and sometimes more usefully be thought of as $4 + (-2) = 2$ because, in many contexts, -2 can be a number in itself.

2.1.1.1 Example: Racing

The Nags Suppose you go to the race track without any knowledge of the horses. So, you get the feature race's $2 special and bet on the order in which six horses in a given race will arrive at the finish line. You have 6 ways to choose the first horse, 'the win bet', and you have 5 remaining ways to choose the second, the place bet, and so on. The total number of ways you can *order* (arrange, or sequentially choose) six horses is $6 \times 5 \times 4 \times 3 \times 2 \times 1$ or 720. So your probability, p, of randomly guessing the exact order in which the horses will arrive out of $n = 6$ is $p = 1/720$.

Discussion: So, when $n = 6$, $n! = 720$ and the probability, p, of choosing the outcome (the order) at random is $p = 1/n! = 1/720$. Computers round huge numbers at every step of a calculation, and n does not have to be very large before $n!$ will be so huge that all that rounding would spoil the accuracy[6]. There are many other uses of factorials besides gambling. If you have scheduled the interviews of $n = 5$ people for a study, there are $5! = 120$ **ways to order** (arrange) the 5 different interviews. For an acoustics expert, or musician, there are $4! = 24$ ways to arrange the instruments in a row for a quartet. Regardless of your field, knowing about the possible ways of arranging things can be useful. It is especially useful in understanding much of this book.

Question 17 might provide good practice relating to ordered outcomes (answer in back of book).

2.2 Permutations and the Parimutuel

A *permutation* is the number of ways you can arrange n things **taken r at a time**.[7]
    The formula for the total number of *permutations* is

$$^n\mathcal{P}_r = \frac{n!}{(n-r)!}.$$

(2.1)

I have used a calligraphic $\mathcal{P}$ to warn you that this letter refers to discrete values (integers) and does **not refer to a probability**. Permutations apply to situations in which **order is important** and you are taking n things r at a time. Sometimes it is necessary to know that $0! = 1$ because there is one

---

[6]If you win based on odds, you get your money back plus the amount specified by the 'odds' shown at the horse room or track. But, assuming the race track used simple probabilities to calculate the odds and the probability, $p = 1/720$, then from a gambler's view the **odds** would be

$$\mathcal{O} = \frac{p}{1-p} = \frac{1/720}{1 - 1/720} = \frac{1/720}{719/720} = \frac{1}{719},$$

or 1 chance of winning to 719 chances of losing (Section 1.5). In real life the parimutuel (the bet pooling service) calculates the odds based on the amount expected to be bet on each horse, after the track and taxes have taken a house cut. However, odds are only of use in a few applications covered in this book. Also note that the simple odds calculated here are based on randomness and therefore would not be valid if you had any real knowledge about the individual horses.

[7]It does not require a very large number before the factorial operator will return a number bigger than your scientific calculator or your computer can present exactly. The numerator and sometimes the denominator will in some instances be so huge, using common software, that the computer will not be able to represent it accurately. However, you are dividing an exponential by an exponential, that is, by the same size as the remaining terms in the denominator, and this has the effect of eliminating those remaining terms. So, if you do your own computer programming you have to be careful to avoid disastrous rounding when n is large. You will need to use Equation 2.2 to reduce the tendency of the numerator to grow so huge as to be unavoidably vulnerable to rounding.

way to arrange 0 things. An approach for hand calculation and for computer programming that minimizes rounding uses canceling out (illustrated here using arrows that point to a small copy of what was canceled) to get

$$^{n}P_{r} = \frac{n \times (n-1) \times \ldots \times ((n-r)+1)\cancel{(n-r)!}^{\;(n-r)!}}{\cancel{(n-r)!}^{\;(n-r)!}}.$$

This might be easier to remember than its technically simplified version,

$$^{n}P_{r} = n \times (n-1) \times \ldots \times ((n-r)+1). \tag{2.2}$$

An example to help you understand permutation follows.

2.2.0.1 Example: Permutations to Arrange

Vehicular permutations Suppose that you are in charge of the automobile show at the county fair. You have $n = 12$ very rare cars available and you have only one row of $r = 4$ display places (positions). What is the probability, if you assign their spaces at random, that you will get the one best ordered arrangement of the best four cars for patron satisfaction?

You need to calculate how many different **ordered** arrangements (lineups, or queues) of only $r = 4$ cars you can get out of 12 cars. There are 12 cars from which to choose the one to put in first place, then that leaves 11 cars from which to choose one to put in the second place, and so on, but after you do the fourth place in the row you are out of display places and you are finished. So, you have $12 \times 11 \times 10 \times 9 = n \times (n-1) \times \ldots \times (n-r+1) = \frac{n!}{(n-r)!}$. Notice that $(n-r)$ would be the next term if you had another space, but you do not. Dividing by $(n-r)!$ is like canceling the remaining multiplications, including $n-r$, all the way down to 1.

In the example, $^{12}P_{4} = 11,880$. So, you have a frequency, hence a probability, of $p = 1/^{n}P_{r} = \frac{1}{11,880}$, that you will get the one perfect arrangement of cars by chance alone. Note that you cannot put the same car in more than one place, meaning you can use it only once in each permutation.[8]

2.2.0.2 Example: Permutations and Probability

We will stick to simple probabilities in this section even though horse racing is mentioned.

Guessing? You go to the races at the state fair. There are 9 horses in a particular race and your roommate, who has never seen a horse before, claims the spirits will reveal the first 3 horses; win, place, and show. What is the probability of your roommate choosing the first 3 horses **and** their order by chance alone? Is which horse wins a race really a random event for the experienced bettor?

Solution: This is a question of how to get the one perfect way to arrange 3 things out of 9 things by chance alone. So you must find the inverse of the number of permutations of 3 things from 9

---

[8]If you take the permutations of n things n at a time you will find that

$$^{n}P_{r=n} = \frac{n!}{(n-n)!} = \frac{n!}{0!} = n!$$

because 0! turns out to equal 1. If $n = 0$ is one possible outcome, it must be able to take 1 place. It confirms the previous statement that n things can be arranged $n!$ ways.

things,

$$1/\,^9\mathcal{P}_3 = 1/\left[\frac{9!}{(9-3)!}\right] = \frac{1}{9 \times 8 \times 7} = 1/504\,,$$

or 0.001984. Another way to do this is by counting; there are 9 ways to choose the first, leaving $9 - 1 = 8$ ways to choose the second (one has already been chosen), and then only 7 ways to choose the third horse. Thus, applying the definition of relative frequency and the multiplication law,

$$\frac{1}{9} \times \frac{1}{8} \times \frac{1}{7} = \frac{1}{9 \times 8 \times 7} = \frac{1}{504}\,,$$

the same outcome, 0.001984, or about 0.2%. Several other variations for finding this answer are possible, but the answer should be the same.

For the experienced horse player or for racing professionals, the probability of a particular horse winning a race is not random. Race horses are athletes, and although there is a strong element of chance, each participant in a race has a different probability of being the winner. This is one of the reasons that the parimutuel betting system used at race tracks does not assume randomness.

Working Question 21 which is answered at the rear of the book may help you get an understanding of this sort of problem.

2.3 When Order is Unimportant: Combinations

Combinations are the result of taking n things r at a time **without** any concern about **how they are ordered** (arranged). You are only interested in selections that are **not rearrangements** of another selection. ABC, CBA, BCA, BAC, CAB, and ACB are all the **same combination** because combinations do not care about order. It is as though you can keep any one permutation and ignore all the others. So, because you can order the r things $r!$ ways, the number of combinations, $^n\mathcal{C}_r$, is simply the **number of permutations divided by** $r!$,

$$^n\mathcal{C}_r = \frac{n!}{r!(n-r)!} = \binom{n}{r}\left(=\,^n\mathcal{P}_r/r!\right), \tag{2.3}$$

where $\binom{n}{r}$ is just another way of writing $^n\mathcal{C}_r$. It is important to remember that $\binom{n}{r}$ and $^n\mathcal{C}_r$ represent the same thing. The latter is sometimes, without adequate context, confusingly written C_r^n. Learning common notations helps you look things up when you forget the details.[9] When doing homework, programming, or hand calculation you might look for opportunities to cancel $r!$ or $(n-r)!$ (as in Section 2.2).

Suppose you are selling surplus flowers to supplement the funding for your plant genetics research. You decide to sell bouquets of $r = 4$ different kinds of flowers out of $n = 13$ different kinds of flowers. The reason you insist on mixing the kinds is partly to help ensure that all the kinds of flowers get sold, rather than just two or three favorite kinds. Each bouquet will have $r = 4$ different kinds of flowers. The number of possible *combinations without replacement* are $^n\mathcal{C}_r = \frac{n!}{r!(n-r)!}$, the number of ways to do this. This is $^{13}\mathcal{C}_4 = (13 \times 12 \times 11 \times 10)/(4 \times 3 \times 2 \times 1) = 715$. The probability that one of your customers will pick the perfect one at random to take home to his wife is $p = 1/\binom{13}{4} = 1/715$.

---

[9]If you use R language, factorial(x) returns $x!$. So, $\binom{8}{6}$ could be found using factorial(8)/(factorial(6)×factorial(8 − 6)).

2.3.0.1 Example: Combinations and Probability

A friend from North Carolina observed what he described as "elephant snot hanging from telephone lines in the early morning." Upon repeated observation he found out that large yellow slugs, *Aromas spp*, also called banana slugs, mate while hanging and slowly descending from a high place such as a telephone line by a string of mucus.[10] The mucus is allegedly the greatest of natural glues and supposedly science has not yet been able to synthesize it. And of course, that made me think, not of chucking my lunch, but rather of a statistics related example.

Glue Science Suppose you are a banana slug geneticist working for a glue factory.

Methods and Materials: You know that banana slugs are hermaphrodites so that they can mate with any other banana slug.[11] You guess six pairs should be a good starting number, so you collect $n = 12$ of the immature (so as to be unmated) banana slugs from the wild and rear them to adulthood. Now, excluding self-mating, how many different ways can you pair them (not using any slug in more than one mating)?

Analysis: Because you want pairs (twos) and gender does not come into it, $r = 2$. In addition to the divisor (the ***denominator***) used for permutations, you divide by the $r! = 2!$ ways to order (arrange) them to estimate the number of combinations. Because their order, $r!$, is not important you get it out of there by division. So, you calculate the combination, using the Equation 2.3,

$$^{12}C_2 = \frac{12!}{2!(10!)} = \frac{12 \times 11}{1 \times 2} = \frac{6 \times 11}{1} = 66 \,.$$

To avoid rounding using a computer or calculator, or for plain convenience, it is easiest to work it out using

$$^nC_r = [n \times (n-1) \times \ldots \times ((n-r)+1)] / r! = \frac{12 \times 11}{2 \times 1} = 132/2 = 66 \,,$$

where the innermost parenthesis includes the rest of the denominator after the rth term. If the example asked the **probability** of getting some **uniquely ideal** group of size $r = 2$, the probability of choosing that ideal group out of all the possible pairs at random from $n = 12$ things would be $p = 1/^{12}C_2 = 1/66 = 0.015152$ or about 1.5%.

So, if order does not matter, as in a handful of mixed candies, it is a combination. Note that combinations are smaller than permutations because $^nC_r =^n P_r/r! = \binom{n}{r}$; the first and last terms mean the same thing, the number of combinations of r things possible out of n things. Incidentally, you may have noted that $2! = 2$, and that is not remarkable, $1!$ also equals 1. But, $0! = 1$ seems surprising until you realize there is 1, and only 1, way to select zero things.

2.3.0.2 Example: Combinations and Fortune Telling

Cards Assume a deck with 52 cards has been shuffled sufficiently that drawing from it should produce random results. Your roommate, who claims to be regularly visited by spirits, claims to

---

[10] Maybe if human biology were like that of banana slugs it would have eliminated a certain typical unnecessary scene from a quarter of the movies made in the last decade.

[11] Banana slugs can also mate with themselves, but they prefer not to do so. If you can get them to do it more or less on command, as by solitary confinement, you can get intense inbreeding when you need it.

be able to predict which cards will be the first 3 draws (not necessarily in order).[12] What is the probability that your roommate will get it correct by chance alone?

Solution: The probability of randomly getting 1 of 3 particular cards from a whole pack of 52 cards by chance alone would be $3/52$. But if that happened there will be only 2 chosen cards left in the remaining 51 cards, so the probability of drawing another of them would be $2/51$. Next, assuming the other 2 events happened, only 1 of the chosen cards would be left in the remaining 50 cards and the probability of getting it would be $1/50$. Having found this by counting, then applying the multiplication law twice, the probability that your roommate would guess all 3 randomly drawn cards would be

$$p_{3\,correct} = \frac{3}{52} \times \frac{2}{51} \times \frac{1}{50} = 0.00004525.$$

There is another way to approach this problem. You can see that this is just the probability of selecting $r = 3$ things out of $n = 52$ things in **one way** without regard to order,

$$1/{}^nC_r = 1/\left(\frac{52!}{3!\,(52-3)!}\right) = 1/(\frac{52 \times 51 \times 50}{3 \times 2 \times 1}) = 6/132600 = 0.00004525,$$

so you get the same answer using combinations.

2.4 Briefly Binomial

When describing populations, the word *distribution* describes the frequencies of distinct results, often called outcomes, such as different things or different measurements. If you recorded all the heights of third grade children in Iowa to the nearest foot, you might get one event in which a child was six feet tall to the nearest foot and there might be one in which one or more would be counted as 2 feet tall, but there would be thousands that were recorded as being 4 feet tall. The collection of counts of all the events in some category, such as the number of each possible nearest foot height measurement, in a population is a distribution. We are now going to consider the distribution of events with only two outcomes, such as butter up and butter down. We generally choose one outcome, such as butter up, as a *success*, and the other outcome, such as butter down, as a *failure*.

So, the number of successes, r, in n independent random *trials* with only two mutually exclusive possible outcomes, + and -, is distributed as a **binomial distribution**. Each of the two mutually exclusive outcomes has a probability $0 \le p \le 1$ of a success and a probability of a failure, $q = 1 - p$. But, p is the true probability of only a single trial, and you are generally interested in the results of a collection of many trials. Each trial is sometimes called a **binomial trial**, or a **Bernoulli trial**.[13]

The **binomial distribution function**,

$$P_r = \binom{n}{r} p^r q^{(n-r)}, \tag{2.4}$$

gives the probability of getting r successes out of n (coin flip-like, *yes-no*) random trials. The superscripts are the powers, the number of times p or q has to be multiplied times itself. This is part of the equation because of the law of probability, that says that the probability of two independent random

---

[12]You cannot use binomial probabilities in the above sample, because you are drawing from a small population without replacement and changing the probability with each draw.

[13]A Bernoulli trial is sometimes called a *Bernoulli experiment*, but such a terminology can be confusing in applied statistics because the word *experiment*, used alone in the more usual context, can be a number of samples each consisting of n trials. Always be aware of context; the language of mathematics is not always as precise as it pretends to be.

events is the product of their outcomes. Thus the probability of r things each with probability p, is p^r, but if p happens r times then q must happen the remainder of the times, $n - r$ times, and that has a probability of $q^{(n-r)}$. These, p^r and $q^{(n-r)}$, must also be multiplied together. Then the leading multiplier is $\binom{n}{r}$, the number of possible combinations (Equation 2.3, $\binom{n}{r} = \frac{n!}{r!(n-r)!}$). **Coefficient** is a somewhat vaguely defined word in mathematics that can be generalized to mean a multiplicative factor in some term (part) of an expression. Thus $\binom{n}{r}$ in the binomial formula is considered its coefficient. So combinations, $\binom{n}{r}$, also written ^nC_r, are often called **binomial coefficients** even in other contexts.

Because r and n are counting numbers, they are discrete, and the resulting binomial coefficients are **rational numbers**, meaning they can be represented by fractions with integers in both the top and the bottom. Though the rational numbers, as a group, have an infinite collection of numbers between numbers they do not include all the real (continuous) numbers like π, which is an irrational number. That means that, although they may not appear so, the rational numbers are discrete.[14]

For testing with a sample size that is less than about $n = 60$, the gaps between the resulting binomial probabilities for a given p and q may sometimes need to be considered when using the binomial distribution (Figure 3.2).

Example: Too Erratic for Application in Decision Making?

Background: When we take random samples the results often represent what we might call more or less the truth, but out of all possible samples rare coincidences may deceive us.

Methods and materials: Suppose I believe a coin honest (my hypothesis is that the coin will come up heads $1/2$ of the time), but my buddy says that it is cleverly weighted so it usually comes up heads. Well, I cannot say that the flips would somehow affect each other, so I can use the multiplicative law to figure out probabilities. If you have gone on much further in this chapter you might see that I should only accept the results if there is a 5% or less chance of what happens in my test of this coin. I decide that I will give up believing it to be honest if flipping it produces a result that has a probability of happening 5% or less of the time. In that case, I will reject my hypothesis (my belief) and accept that the outcome is too unlikely for my hypothesis to be true. I choose in advance that $n = 5$ flips should be adequate.

Results and analysis I flip the coin $n = 5$ times and get all heads, $r = 5$. A fair coin should come up heads $1/2$ of the time, so the probability of a random sample of $n = 5$ flips all coming up heads is

$$1/2 \times 1/2 \times 1/2 \times 1/2 \times 1/2 = (1/2)^5 = 0.03125$$

To be conservative I round it all the way up to 4%. It is still less than 5% so I decide this outcome is too unlikely to be true.

---

[14]The ancient Greeks thought all numbers were **rational**, meaning equal to the ratio of two integers. However, in disagreement with the ancients, there are gaps, including an infinite number of irrational numbers, between any two adjacent rational numbers. Together all the numbers, rational and irrational, are the real numbers, which means that they have no breaks and are therefore called continuous. This mistake by the Greeks treating all numbers as discrete was a stumbling block until the renaissance. Numbers with gaps between them such as the integers and the rational numbers are discrete. Frequentist probabilities are indeed rational because they are ratios of counts, and therefore also discrete. Researchers sometimes treat observed binomial probabilities as though they include all the real numbers even though they are really rational. This works fine in many applications. But, when attempting to use continuous approximations of the binomial for small size samples, this reality, concerning the fact that rational numbers are discrete, may become a serious stumbling block (e.g. Figure 3.2).

Discussion: The testing procedure outlined in Part I will be formalized as you proceed through this chapter. However, you can already see that this reliance on 5% as the threshold of too unlikely to be true is just a wee bit too liberal. As a child you no doubt saw that almost any honest coin if flipped enough times will have runs of 5 heads or even 5 tails. But for such a run to come along **immediately after hypothesizing** that the coin is fair does seem to be hard to call mere coincidence. So, conceptually the method looks acceptable.

What if I had just flipped the coin $n = 4$ times? Well, $(1/2)^4 = 0.0625$, or about 6%, that is twice as likely with only a one flip difference and, if you think about it, there is no way that $n = 4$ flips are able to prove anything at a 5% level.

However: I still wonder whether the outcome could have gotten 5% or at least a bit closer to it. So, I ask what if I had gotten $r = 4$ heads out of my original $n = 5$ flips (binomial experiments). Oh, that is a little more complicated to figure out and requires Equation 2.4,

$$P_r = \binom{n}{r} p^r q^{(n-r)};$$

$$P_r = \binom{5}{4} (1/2)^4 (1/2)^1,$$

$$P_r = \frac{5!}{4!\,(5-4)!} (1/2)^4 (1/2)^1,$$

canceling, gives,

$$P_r = \frac{5 \times 4!}{4!\,(1!)} (1/2)^4 (1/2)^1 = \frac{5}{(1)!} (1/2)^4 (1/2)^1,$$

$$P_r = 0.15625$$

Well 15.6% certainly is a long way from 5% whether we are looking at possible outcomes or actually challenging an idea. This hopping around of probabilities is not conducive to our comfortably using the binomial theorem for testing. Since we are at it let's see what would happen if we took $n = 6$ samples and got $r = 5$ heads.

$$P_r = \binom{6}{5} (1/2)^5 (1/2)^1$$

$$P_r = \frac{6!}{5!\,(6-5)!} (1/2)^5 (1/2)^1$$

$$P_r = \frac{6 \times 5!}{5!\,(1!)} (1/2)^5 (1/2)^1 = \frac{6}{(1!)} (1/2)^5 (1/2)^1$$

$$P_r = 0.0975\,.$$

Well, we have pretty much covered ways to approach 5% in our coin flipping and its outcomes. It is like the corny old saw, "You cannot get there from here." As we shall see later, less jumping would make a variety of other useful calculations work better. Large jumps for a single change in n in a binomial calculation is a problem for most sample sizes up to almost $n = 60$. There is a method of using a normal approximation to approximate binomial outcomes (Section 5.6), but it would only make the gaps between outcomes with small samples less obvious. So, using the binomial theorem to deal with counts and proportions works best with larger than large samples, but with small samples it would only mask this problem (e.g. Figure 3.1).

When doing such calculations, it will sometimes help to be aware that the number of possible r values is $n + 1$ because the number of successes can be any integer between 0 and n inclusive.[15] As you may recall, $q = 1 - p$ is also between 0 and 1. Because the distribution is different for each probability of a success and each sample size (number of trials), there is an infinite number of possible binomial distributions, but they are lumped together as a category, '**the binomial distribution**'.

2.4.1 It only takes two

It should now be obvious that you only need to know just two numbers, the sample size, n, and the probability of a success, p, to completely *define* (describe) any particular binomial distribution by using

$$P_r = \binom{n}{r} p^r (1 - p)^{(n-r)},$$

which gives the probability of any outcome, r (where $r = 0, \ldots, n$).

2.4.1.1 Example: $p \neq q$ Simple Inheritance

Heredity Suppose you are interested in a genetics experiment with flowers that may have either one or both of two alleles (genetic instructions regarding alternative occurrences) for a color gene pair. Assume that the allele for white flowers is *recessive*, meaning that the flowers are white only if all genetic information inherited for that gene, one allele from each of that plant's parents, is for white. Also assume that the allele for orange is *dominant* over the allele for white, meaning that if either or both alleles are orange, then the flowers will be orange. Now if you cross plants that are *homozygous* (only have one kind of allele) for recessive white flowers with plants homozygous for dominant orange flowers, you will get a mixed '*F1 generation* that is all orange. The pure orange flower parent, OO, would provide an orange allele, O, and the pure white flower parent WW would provide a white allele, W. By inbreeding the orange appearing offspring (all of which would be OW) of the *F1 generation* (from 'brother-sister' pollination), you would get what geneticists call the *F2 generation*, with the offspring receiving one allele from each parent. In theory, meaning in situations where there are not other genetic or cytological effects interfering, it would be random whether the offspring received an orange or a white allele. That would work out so that there is

Parents	O	W
O	OO	OW
W	OW	WW

only a $p = 1/4$ probability that any one seed from this cross will produce a plant with white flowers, WW.

This is like using a biased coin such that the probability of getting heads is $p = 1/4$ and getting tails is $q = (1 - p) = (1 - 1/4) = 3/4$. A landscaping company might ask what is the probability of observing exactly $r = 3$ white flowers out of $n = 10$ *F2* offspring from this cross. To get the answer, apply the binomial theorem:

$$P_{(r=3, n=10)} = \binom{10}{3} (\tfrac{1}{4})^3 (1 - \tfrac{1}{4})^{(10-3)} = \frac{10!}{3!(7)!} (\tfrac{1}{4})^3 (\tfrac{3}{4})^7 = \frac{10 \times 9 \times 8}{3 \times 2 \times 1} (\tfrac{1}{4})^3 (\tfrac{3}{4})^7$$

$$= 5 \times 3 \times 8 (\tfrac{1}{64})(\tfrac{2187}{16384}) \approx 0.250282 .$$

---

[15]The awkwardness of using continuous approximations of the binomial was indirectly brought to my attention by Dr. Calvin Burgoyne, more recently at UG Athens.

It is about 25%, but not exactly. There is no exact rule as how much to round in this case because the inputs were all fractions (rational numbers) and integers, which are exact.

In application, basic genetic theory does not always work as well as it appeared to for Gregor Mendel. Often in genetics, other genes than the pair in which you are interested modify the outcome to some degree. Another difficulty, that is associated with multiple pairs of genes, is that groups of genes are stuck together in strings called chromosomes and the redistribution of genes from the same chromosome is not independent. The distance between genes on the same chromosome affects the probability of their being inherited separately.

Another example related to Mendel's model, in which all the genes were on different chromosomes, is Question 23 (answered at rear of this book).

2.4.2 p and q unequal (or not)

Remember that p and q can be anywhere between 0 and 1, so long as $p + q = 1$; equivalently, $p\% + q\% = 100\%$, as in the following example.

The top set of column graphs in Figure 2.1, also shown on the back cover of this book, include the distribution of successes, r, for various numbers of trials, n, with probability $p = 1/2$. The bottom of the same Figure 2.1 shows another set of binomial distributions with probability $p = 1/4$ and the same sample sizes. Notice that this bottom set of distributions is not symmetrical (is skewed) because $p \neq 1/2$. As the sample size, n, increases (regardless of the probability), the symmetry tends to increase and the high point (highest frequency) of a distribution tends to move to the right away from $p = 0$.

2.4.3 Equal p and q

The classical idea of tossing a *fair coin*, meaning $p = 1/2$ exactly, is perhaps the best illustration of equal p and q because $q = 1 - p = 1/2$. Figure 2.1, top, demonstrates the binomial theorem to calculate the expected probability of getting a certain outcome when the probability, p, of a success is known to be $1/2$.

2.4.3.1 Example: Coin Flips

Heads or Tails What is the probability of getting heads exactly $r = 3$ times, by chance alone, if you flip a fair coin $n = 10$ times? The answer is

$$P_{(r=3)} = \binom{10}{3} p^3 q^{(10-3)}$$

$$= \frac{10!}{3!(7)!}\left(\frac{1}{2}\right)^3\left(\frac{1}{2}\right)^7 = \frac{10 \times 9 \times 8}{3 \times 2 \times 1}\left(\frac{1}{2}\right)^{10} = 5 \times 3 \times 8\left(\frac{1}{1024}\right) \approx 0.11719$$

a little less than 12% probability of getting heads exactly 3 times by chance alone out of $n = 10$ trials.[16] The top part of the next to the last step is obtained by division of 10! by 7! that can be done by marking off that part of the multiplication. This is because $7 \times 6 \times 5 \times 4 \times 3 \times 2 \times 1 = 7!$ is also included in $10! = 10 \times 9 \times 8 \times (7!)$. The next step is just algebra, simplification by division, dividing 10 by 2 and 9 by 3.

---

[16]The Examples in Subsection 2.3.0.1 and Sub-subsection 2.4.3.1, are not a statistical test of anything. They can be seen as simply probability estimates to help you understand the distribution.

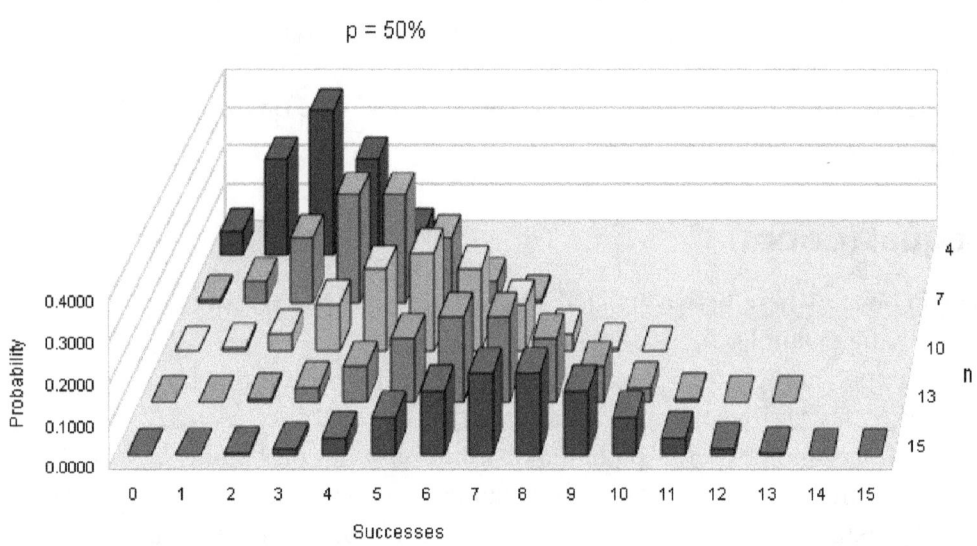

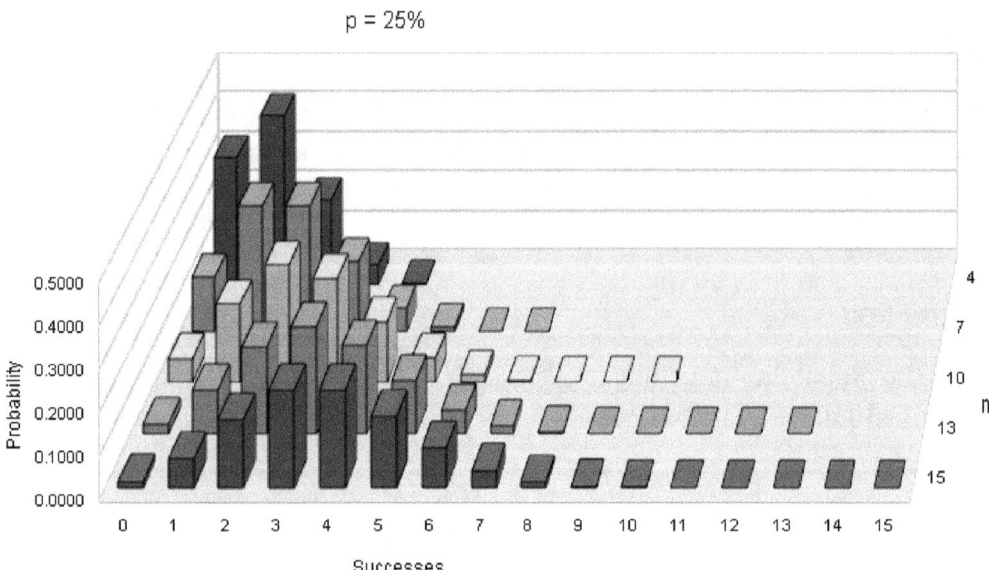

Figure 2.1: Binomial Distributions of Successes, r, shown at
two values of p with various numbers of trials n.

So, the probability of flipping a perfect coin $r = 3$ times (3 trials) and getting heads each time is $1/2 \times 1/2 \times 1/2 = (1/2)^3 = p^3$, and the probability of flipping it $r = 7$ times and getting tails (the alternative to heads) each time has a probability of $(1/2)^7 = q^7$. Because the outcomes are two independent things happening at the same time, you can combine their probabilities by the multiplication law, and because they are equal you can add their exponents in the process $\left(A^c A^d = A^{c+d}\right)$ to get $p^3 q^{(10-3)} = (1/2)^3 (1/2)^7 = (1/2)^{10}$ which is the probability of exactly 3 heads **in one particular order** out of $n = 10$ flips. However, there are $\binom{10}{3} = 720/6 = 120$ different ways to order (arrange) these outcomes and still get this same result, 7 heads and 3 tails.

The **simplified formula** starts off with the usual explanation of the cumulative binomial. The multiplication law of probability says that if two independent occurrences (such as heads and tails) have probabilities p and q, respectively, then the probability of them both happening in two trials is pq. Extending this, the probability of p happening n times in n independent trials is p^n. The probability of r successes (say heads), and of its alternative, $n - r$, failures happening in n trials is $p^r q^{(n-r)}$ times the number of ways it can happen, $\binom{n}{r}$. However, there is a difference, because if you multiply values that are powers of the same constant together, the result is that constant to the sum of its powers, $C^r C^{n-r} = C^n$. **When the probabilities are equal**, $p = 1/2 = 1 - p$, the constant is $C = 1/2$. Thus, when $p = q = 1/2$, the binomial formula can be simplified to

$$P_{(r,n\,p=0.5)} = \binom{n}{r} \left(\frac{1}{2}\right)^n. \tag{2.5}$$

2.4.3.2 Example: Simplified Coin Flips

Equality made simple The probability of getting a particular outcome, $r = 3$ heads out of $n = 10$ trials, when $p = 1/2$, could have been calculated by

$$P_{(r=3)} = \binom{10}{3} (1/2)^{10} = \frac{10!}{3!(7)!}\left(\frac{1}{2}\right)^{10} = \frac{10 \times 9 \times 8}{3 \times 2 \times 1}\left(\frac{1}{2}\right)^{10} = 5 \times 3 \times 8\left(\frac{1}{1024}\right) \approx 0.11718.$$

You may wish to try finding a single binomial probability, given $p = q$, via Question 5 and Question 25, both of which have answers in the back of the book.

2.4.4 You may approach the Throne of Symmetry

As you shall see, symmetrical distributions are easier to work with. Larger sample sizes, n, tend to produce more symmetrical binomial distributions, but are never perfectly symmetrical unless $p = 1/2$. So, the $p = 1/4$, $n = 7$ graph at the bottom of Figure 2.1 is less symmetrical than the $p = 1/4$, $n = 15$ graph. However, as p approaches $1/2$, the resulting distributions also come closer and closer to symmetrical. At large sample sizes, definitely at $n \geq 30$ and close to $p = 1/2$, the binomial distribution, although not continuous, definitely approaches a bell shape distribution. This leads to the normal approximation (Section 5.6), which is especially useful for estimating probabilities and frequencies involving counting data.

2.4.5 An accumulation of successes: Summed binomial probabilities

Sometimes you need to know the probability of a group of n binomial outcomes between and including two numbers. For this you can use the formula for a cumulative binomial distribution.

$$B(n,\ p,\ first,\ last) = \sum_{i=first\,r}^{i=last\,r} \binom{n}{i} p^i q^{(n-i)}. \tag{2.6}$$

Note that B is an ordinary Latin letter and stands for binomial. It is assumed that the n trials are independent and random. You can also think of this as the area under some part of a graph such as the ones in Figure 2.1. This can also be viewed as an application of the *binomial theorem*, $(a+b)^n = \sum_{k=0}^{k=n} \binom{n}{k} a^n b^{(n-k)}$, to probability.

If you needed to know the probability of getting **any number between** $r_1 = 3$ and $r_3 = 5$, inclusive, out of $n = 7$ random trials with $p = 0.4$, you could calculate the cumulative probability,

$$B(7,\ 0.4,\ 3,\ 5) = \sum_{i=3}^{i=5} \binom{7}{i} 0.4^i (1-0.4)^{(7-i)}.$$

The hand calculations for this might look something like

$$B(7,\ 0.4,\ 3,\ 5) = \binom{7}{3} 0.4^3 0.6^4 + \binom{7}{4} 0.4^4 0.6^3 + \binom{7}{5} 0.4^5 0.6^2$$

$$= \frac{7!}{3!4!} 0.4^3 0.6^4 + \frac{7!}{4!3!} 0.4^4 0.6^3 + \frac{7!}{5!2!} 0.4^5 0.6^2$$

$$= \frac{7 \times 6 \times 5}{3 \times 2} 0.4^3 0.6^4 + \frac{7 \times 6 \times 5!}{3 \times 2} 0.4^4 0.6^3 + \frac{7 \times 6}{2} 0.4^5 0.6^2$$

$$= \frac{210}{6} (0.06400)(0.12960) + \frac{210}{6} (0.02560)(0.21600) + \frac{42}{2} (0.01024)(0.36000)$$

$$= 35 (0.0082944) + 35 (0.0055296) + 21 (0.0036864) = 0.5612544,$$

giving a probability of about 56.1%.

2.4.5.1 Equality is nice

Recall that success and failure are common designations for the two outcomes of a binomial trial, and p is the probability of a success, whereas q is the probability of a failure. When $p = 1/2$, then $q = 1 - p = 1/2$. When p and q are equal like this, it greatly simplifies the evaluation of the cumulative *number of successes* observed, r, using

$$P_r = \left(\frac{1}{2}\right)^n \sum_{i=r}^{i=n} \binom{n}{i}, \tag{2.7}$$

where n is the sample size. Equation 2.5 is a simplification (of Equation 2.6) that can be used when $p = 1/2$. It finds the probability of a result as great as or greater than r; it can be used to test whether one outcome is more common than the other outcome, because you can choose either of the two possible outcomes to represent the probability of a success.

You may also see the above written as $P_r = \sum_{i=r}^{i=n} \binom{n}{i} \left(\frac{1}{2}\right)^n$, there is no difference; it is just a rearrangement that depends on the fact that $(1/2)^n$ is a constant.

As will be clarified in the next section, this is handy for challenging the *null hypothesis*:

H0: TWO TREATMENTS ARE THE SAME,

and if **H0** fails this test, then you accept the *alternative hypothesis*:

HA: TREATMENT A IS GREATER (MORE NUMEROUS IN THE POPULATION) THAN TREATMENT B.

Note that the **0** in **H0** is **not** a capital letter, it is the **number zero**. **H0** is pronounced 'h-nought'. The small capital **A** in **HA** stands for alternative.

Under this null hypothesis, **H0**, $p = q = 1/2$, and, with random sampling, you expect to overshoot the true probability no more than you expect to undershoot it. So, when p and q are equal, this simplification of Equation 2.6 can be used instead of doing the same calculation using the longer version (which does not require equality).

As you shall see it is easy to adapt this equation to challenge **H0** in a two directional hypothesis pair,

H0: THERE IS NO DIFFERENCE,

HA: THERE IS A DIFFERENCE.

Using Equation 2.7, you can start someplace other than $i = r$ or finish someplace other than $i = n$, if that meets your need for some application other than the common tests that follow.[17]

2.5 A Quick Look at Testing

This book tends to focus on testing to find **meaningful** <u>**differences**</u> between things. The general method is that you make a statement, **H0**, about something, a *hypothesis*, and then set out to disprove it.

If two categories are **equally frequent**, the probability under random sampling of a success is $p = 50\%$, in other words, the probability of either is $p = 1/2$. The following shows how to use the binomial theorem to challenge the hypothesis that there are equal numbers in the population.

2.5.1 Focus on frequency: the 50% test

You provisionally provide a ***null hypothesis*** (or ***working hypothesis***), **H0**, that there is no meaningful difference, a null difference, between two things, which you suspect to be possible, and per scientific tradition you challenge, to see if it can be proved wrong **beyond a reasonable doubt**, in which case you reject it. That process of trying to see if **H0**, your null hypothesis, can be proved wrong beyond a reasonable doubt, consists of demonstrating that the probability of **H0** being true is equal to or less than α (Greek letter, ***alpha***), a criterion, a limit, that you choose before examining the test data.

Your chosen α is the limit at and below which the probability of **H0** being true must be rejected. Alpha, α, is commonly defined as 'the probability of getting a Type I error' because we choose it to limit how much of a chance we are willing to take of rejecting H0 when it is in fact true. To put it another way, alpha is the probability of incorrectly rejecting a true null hypothesis, **H0** (probability of getting a Type I error). It is commonly accepted that α should be equal to or less than 5%, because we are looking to see if the probability of **H0** being true is too tiny to be accepted as true, which would justify rejecting the hypothesis.

The ***50% test*** is a binomial test involving a sample of n things (*factors*) that can be classified into two ***categories*** (classifications) that might be called X and Y. These categories can be chosen on the

---

[17] Another cumulative extension of Equation 2.5 is $P_{<r} = \left(\frac{1}{2}\right)^n \sum_{i=0}^{i=r-1} \binom{n}{i}$; it finds the probability of a result being smaller than r.

basis of any of many different properties, male and female, toast landing butter up and butter down, species X and species Y, fans who think Elvis's spiritual music was his best and and fans who do not, etc.

You know that because this is only a sample, and thus subject to chance, you cannot be sure it has the same proportion of Xs and Ys as does the whole population. For the moment you are going to use what is called the classical Neyman-Pearson approach (Subsection 4.7.2) to challenge

H0: THEY ARE EQUALLY LIKELY (meaning that the population has 50% of each),

versus

HA: THEY ARE NOT (some other ratio).

You are going to challenge the null hypothesis that the absolute frequency, N_X, is equal to another absolute frequency, N_Y. It is important to know that the term ***Type I error*** means erroneously rejecting **H0** when it is in fact true (in this case, equally likely, meaning no effect). The criterion, α, is sometimes inaccurately called the ***Type I error rate***. The p_value is the estimated likelihood of the difference found by random sampling being as large as observed due to chance alone (a kind of after the test probability). The p_value can be seen in various ways, for instance as the probability of a coincidence or similarly the probability that the researcher is just observing an effect of variability, like vibration, insufficient accuracy, etc. Under this common approach, ***statistical significance testing***, you make a before-the-fact commitment that if the p_value is greater than α, there is **not a sufficient reason to doubt H0**, and you cannot reject it. If there is sufficient reason to doubt **H0** (too small a likelihood that the observed outcome is due to chance alone) based on (smaller than) the fixed probability criterion, α, you declare ***statistical significance*** and reject **H0**. You can use this process, statistical significance testing, to avoid bias such as preconceptions and wishful thinking, while making a try at determining whether your outcome was too unlikely for **H0** to be true. Journals and granting agencies tend to demand that conclusions in papers be supported by statistical significance testing.

Given a sample of a certain size, n, you may count the number of successes, r, which in this case equals the number of outcomes in category Y, to see whether they support the alternative hypothesis, **HA** (research hypothesis). Notice that you are working with frequencies and because this is a frequentist approach, you must rely on random sampling to enable you to estimate probabilities.

Because your null hypothesis, **H0**, states that the frequencies of X and Y are the same, you calculate the summed binomial probability of observing a particular number of successes, r or more, on the basis of the assumption that **H0** is correct, meaning $p = 1/2$. Hence, this test is called the 50% test.

As already shown, the necessary calculation can be somewhat simplified when $p = 1/2$. So, because you are hypothesizing that $p = 1/2$, the 50% test uses Equation 2.5, $P = \left(\frac{1}{2}\right)^n \sum_{i=r}^{i=n} \binom{n}{i}$, the simplified version of Equation 2.6, to estimate the probability of getting a Type I error with this value of r. Notice that P is the test statistic and serves as the p_value. If it is too small, the result is deemed **too unlikely to be true**, and therefore P is deemed statistically significant. That is, you are checking to see if your sample produced a result that would be ridiculous if p really equaled $1/2$, that is, 50%.

The p_value is the probability of obtaining a test statistic that is **at least as much different from the hypothesized value** as the one that you actually observed by random sampling, assuming that the null hypothesis, **H0**, is exactly true. It is a cumulative probability, including the estimated probability of the observed outcome, plus any outcome that was less likely in the same direction (+ or -). In the **50% test**, P is used as the estimate of the p_value and is then evaluated against the criterion, α, that **must have been chosen before** the test. **H0** is rejected if the $p\_value \leq \alpha$. Philosophical difficulties regarding the exact nature of the results (the probability of what, under what assumptions) aside, this provides you with some control of the number of Type I errors that you will experience during your career. **Type I errors are inevitable**, all you can do is **attempt to reduce** how often they happen to you.

Such a test can conclude that Y is more common than X (more successes), or it could conclude that Y is less common than X (more failures). These are ***one tail versions of the 50% test*** (meaning one direction, + or -). A ***tail*** is an extreme of a distribution in either direction from its center. The tests can also have two tails challenging likeness (equality) which is not tied to a specific direction, and these are called ***two tail tests***.[18]

The 50% test relates to a variety of things that apply to situations in which you are evaluating successes (evergreen versus deciduous trees, crows present vs no crows present, etc.) when, under **H0**, THE PROPORTION OF SUCCESSES EQUALS THE PROPORTION OF FAILURES (50% and 50% respectively). Assuming your design is appropriate, you can run a one tail test in either (one) direction depending **on how** you phrase your alternative hypothesis, **HA**.

2.5.1.1 Example: 50% Test, Hamburger

A one tail 50% test applies Equation 2.5, $p\left(r \geq \cdots\right) = \left(\frac{1}{2}\right)^n \sum_{i=r}^{i=n} \binom{n}{i}$. Suppose I wished to estimate the probability of getting $r = 20$ successes **or more** out of $n = 30$ trials,

$$p\left(r \geq 20\right) = \left(\frac{1}{2}\right)^{30} \sum_{i=20}^{i=30} \binom{30}{20},$$

where $p\left(r \geq 20\right)$ is the *p_value*. An online calculator returns the probability, $p = 0.04936857$, or about 0.049%. This also illustrates the equation used for calculating the values in Appendix 1.

Fast Food Madness I thought that there were about equal numbers of men and women using the local fast food place. But, my neighbor thought there might be more women using it than men. So, let's suppose that I was somewhat statistically inexperienced, and I decided to try to test this. Before obtaining the data, I had to choose a *criterion*, α, the probability at or below which I would say that the observed outcome is too extremely small to be the result of chance alone, so it must be rejected. I selected $\alpha = 5\%$. (Unfortunately this approach claims to test whether there is absolutely no difference between the number of men and women at all. We will eventually discuss how to remedy this focus on absolute zero.) Then I created a pair of hypotheses with a criterion, and went forth to test this:

H0: THERE IS NO GENDER EFFECT IN THE NUMBER OF CUSTOMERS.

HA: THERE ARE MORE FEMALE CUSTOMERS THAN MALE (meaning that $p > 1/2$ if true). This **HA** makes it a one tail hypothesis pair. The hypothesis pair is **always chosen before the data is tested**.

Criterion: $\alpha = 0.05$ (Reject **H0** and accept **HA** *iff* the *p_value* **equals or is less than** 0.05.)

Methods: I designated the number of female customers r. The criterion is always chosen before the data is obtained. Then I went and counted the customers on that particular summer day at about 12:30 pm. After taking this sample I started the analysis (I did say I was being statistically ignorant) by noting that the number of adult female customers did seem to be greater than the males. If the number had been smaller or equal the test would have been finished right

---

[18]If you reject the null hypothesis for a two tail test, you are saying $B \neq A$. But if you reject the null hypothesis for a one tail test, you are, **depending on the tail you chose**, saying either that $B > A$, or that $B < A$. The reason I still write **H0:** $B = A$ is because 0 difference is the numerical reference point upon which the probability (the *p_value*) of a false positive is being calculated, **regardless of whether the alternative is about one tail or two tails**.

there without having any reason to reject **H0**. However, since there was a majority of female customers, I needed to determine if there was enough of a majority of female customers to conclude that **H0** was sufficiently improbable to be rejected. (If I were being less ignorant, I would have been visiting at random on a suitable number of days and times and each time designating the first person to make a transaction as being whichever gender.)

Results: There were $n = 28$ people, of which more than half, $r = 16$, were female.

Analysis: Inspecting the data **after** all the previous steps was part of the analysis, and because there were more females, I had to go on to the next step, using Equation 2.7,

$$p\_value = \left(\frac{1}{2}\right)^n \sum_{i=r}^{i=n} \binom{n}{i} = \left(\frac{1}{2}\right)^{28} \sum_{i=16}^{i=28} \binom{n}{i} = 0.2858\,.$$

This estimates that this outcome would have a $p\_value \approx 0.29$, meaning that it would happen about 29% of the times I sampled at random if **H0** were true. This was much more likely than $\alpha = 5\%$, my already agreed upon criterion for which anything as small or smaller would demonstrate beyond a reasonable doubt that **H0** is false. So, because $p\_value > \alpha$, my challenge must concede that the test did not find sufficient evidence to reject **H0**. (Not rejecting **H0** does not necessarily mean that **H0** is true.)

Conclusion: The observed difference is **not statistically significant** because there is inadequate evidence to reject **H0** using $\alpha = 5\%$ (also because there are many inadequacies in the experimental design). It did not prove beyond a reasonable doubt that $p_r > 1/2$. But, it did **not show** beyond a reasonable (or any) doubt that $p_r = 1/2$ either. The hypothesis, **H0**, was merely provisional while I was running the test. **H0** only provides a what-if, used for calculation to see if it is too unlikely for you to accept. The result did not reject **H0**, so I could not accept the alternative, **HA**. So I gained no new information as to the difference, or lack thereof. (This example will be continued later.)

Discussion: There were no persons of visually unidentifiable gender that day so I did not have to reduce the sample size, n. Classically some researchers might instead have used Fisher's approach (see Chapter 3) for this extremely simple example, in which case I would not have created an alternative hypothesis, **HA**. Even though under that approach the data would have been directly judged on its $p\_value \approx 0.29$, it would not have supported any conclusion that the observed difference was meaningful, no matter what my personal feelings were about the outcome. Otherwise it was a pretty silly experiment applying to what? A single place at a single hour of a single day? Where was the randomization? This example is about calculation, but its experimental technique would be **unacceptable in practice**.

I did not do this calculation in my head. I used an online calculator and checked with a package in the R language. There is a theoretical difficulty with the fact that I seemed to be challenging a true zero difference. In effect, if you do not target a particular non-zero difference, **your sample size and the options** you choose will do it for you. In a sense, statistical tests can only provide evidence against something, not evidence for it.

You can also try Question 7, a similar 50% test (answer is at back of book).

2.5.2 Two tail 50% testing is OK too

For a **two tail test** you need only change the alternate hypothesis to be more general, **HA**: THERE IS A DIFFERENCE. You then start the analysis, designating whichever **outcome is largest** (for convenient calculation) **as r successes**, then you apply the same formula as in the above example, except you will compare the results (the p_value to a cutoff that is half of the criterion) and you can reject **H0**: THERE IS NO DIFFERENCE if the $|p\_value| \leq \alpha/2$, where the $| \,|$ operator means the absolute value (treat it as positive even if negative). You are testing both tails, so you must a priori split the criterion, α, between the tails. So you shall use an $\alpha/2$ *cutoff* for each tail. Remember that you must choose both $\alpha/2$ and the hypothesis pairs **before** inspecting the data. You may wish to try Question 9.

For your convenience, Appendix 1 is a table of r values for one tail statistical significance of the 50% test, given sample sizes $n < 51$ and α (or $\alpha/2$) equal to any of 0.005, 0.01, 0.025, or 0.05. These statistical significance criteria are the ones most commonly used and they are the most expected in peer reviewed journals. Smaller values of α (or $\alpha/2$) are sometimes appropriate and occasionally essential in applied statistics, but they are really only practical with much larger sample sizes, and cost generally grows with sample size. Larger values of α (or $\alpha/2$) are not likely to be well received for publication. (Actually there is a trend in some fields to use $\alpha = 10\%$ for two tail tests, in order to have $\alpha/2 = 5\%$ in each tail, but that is really a very risky practice).

> JARGON: Rather than calling both α and $\alpha/2$ 'criterion', this book uses 'criterion' to only refer to α and uses the word ***cutoff*** to refer to the criterion, α, or half of it, $\alpha/2$, depending on the context (1 tail or 2). This convention assures you that you can be certain that the word 'criterion' applies to α, the whole α, and nothing but the α.

2.6 The Sign Test

If you have paired data, meaning that every data point consists of two values, one from each of two different categories, then you can use the ***sign test*** to challenge a hypothesis that there is no difference between the two categories. This is achieved by choosing one category to be a 'success' so that the other is a 'failure'. Then you give a plus, $+$, to each pair in which the success category is the larger, and you give a minus, $-$, to all the pairs in which the failure category value is larger. You do not count the ties, and you reduce the sample size, n, by one for each tie. This test converts continuous data into discrete (binomial) data. The calculations then resemble a 50% test. Here is an example:

2.6.0.1 Example: At Sea With the Sign Test

April 19-27, 2009 Percent Cloud Cover Assume two fishing boat captains, one from Los Angeles (LA), and one from Santa Barbara (SB) were about to go to lunch together. One claims that there is more cloud cover over the Santa Barbara fishing region than L.A., from April 19 to 27, and the other says 'ah, baloney' (LA speak for that's ridiculous–not to be confused with abalone) and they bet on it. The loser is to pay for lunch. They agree that a criterion of $\alpha = 5\%$ is sufficiently improbable to justify declaring the difference statistically significant. And they also agree to the hypothesis pair:

H0: THEY ARE THE SAME.

HA: THERE WAS MORE CLOUD COVER OFF SB THAN OFF LA.

Criterion: $\alpha = 0.05$ (*Reject* **H0** *if r corresponds to a $p\_value \leq 0.05$*)

Method and Materials They consulted the NOAA website at http://www.wrh.noaa.gov to get the data from two NOAA buoys, one each from their respective areas. They paired the data by the recorded times. Even though this data appears rather difficult to interpret, it did provide simultaneously paired weather readings taken by the two buoys, each near where each respective captain takes clients to fish. Because they anticipated a set of naturally paired observations from two populations with continuous distributions, they decided in advance to use a sign test. After all, only the cost of lunch was at stake. They agreed in advance that each measurement pair in which there was more cloud cover over Santa Barbara is a success, +. So, as shown, **they subtracted** the Los Angeles pair members from the Santa Barbara pair members.

Buoy Sta B.	Buoy L.A.	Diff.	Sign
43	39	4	+
43	40	3	+
43	39	4	+
43	40	3	+
10	10	0	(tie)
5	95	−90	-
11	15	−4	-
43	40	3	+
25	10	15	+
10	10	0	+
43	40	3	+
11	15	−4	-
84	10	74	+
9	17	−8	-
43	40	3	+
25	10	15	+
9	17	−8	-
10	10	0	(tie)

Results and Analysis: There were 2 ties, so the sample size for the analysis was, in effect, reduced to $n = 16$ pairs (or rather, $18 - 2$ ties) and $r = 11$ + signs, successes.

They could have calculated the binomial probability that $r \geq 11$ when $n = 16$ to directly find the p_value and then compared that with α. However, they referred r and n to the same table, as for a (one tail) 50% test in Appendix 1, where it had already been done for them. Because a $p\_value \leq \alpha/2$ was required that $r \geq 12$ for $n = 16$, to reject at $\alpha/2 = 5\%$, they **did not reject H0**. Because $p\_value > \alpha$, there is not a sufficient reason to doubt based on their particular r and final sample size, less ties, n, with $p_0 = 1/2$. So, because there was not enough evidence to support her claim, the Santa Barbara captain paid for their lunch.

Discussion: Notice that the data was cloud cover, a continuous measure, but in pairing and scoring + or - they created a discrete (binomial) distribution. This data required pairing because of the way it was collected at the same two times each day for the two locations. **Part of the test** was observing that there were more successes; if not the test would be finished right there without rejecting **H0** and before any calculation. The data populations sampled by the buoys may only pertain to their own small immediate areas. The data was also only collected in daylight hours, but that is appropriate for charter fishing boats needs. The decision to use a sign test is reasonable because (at least with this small sample) the sizes of the differences is almost chaotic. The sign test avoids much of the confusion because it only works with +, meaning above the hypothesized, **H0**, zero (a tie) and −, meaning below zero.[19] However, the reader might reasonably ask whether the sample was sufficiently large. What was its power (its sensitivity)? There

---

[19]There is a group of tests at the end of this book (last but certainly not least) called nonparametric tests which might have a little more chance of discovering a meaningful difference between the two buoys (Chapter 14).

might also be some question regarding randomization, but the general approach is otherwise consistent with the classical practice of only considering statistical significance or statistical nonsignificance based solely on the probability of your data representing a Type I error **if H0** were perfectly true. You reject if the test finds that **H0** is too unlikely to be accepted as true.

The choice of which hypothesis to test was arbitrary since one captain believed **H0** and the other believed its alternative, **HA**. But, this readily available test happens to be handy for challenging equality (no effect of location). So, as in many instances where the choice is arbitrary, the captains chose to challenge equality.

Note that this was a **one tail version** of the sign test. The fact that they did not reject **H0** does not provide any evidence supporting **H0**; it simply failed to contradict. Absence of evidence is not evidence of absence (Altman and Bland 1995).

You can try Question 11 to be certain you understand this test.

2.6.1 Beware the tie

Because ties do not count, you want to use the data to the greatest possible precision. That way you avoid rounding before comparing pair members. Some statisticians would classify the sign test as comparing medians, but other statisticians who value more precise statements would find fault with that and point out that it merely compares distributions, the same versus not the same.

2.6.2 Only one tail!? Why not two?

Hopefully you have gotten used to the idea that a tail of a distribution is the diminishing area of a distribution that extends away (in either direction) from the conceptual center of that distribution. The captains in the example Subsection 2.6.0.1 attempted a small sample inferential test based on the binomial theorem. It considers only one tail, referring to one direction away from equality. Greater than (+) relates to one tail and less than (-) relates to the other. So far you have been introduced to a one tail test that attempts to reject **H0**: THEY ARE THE SAME, versus **HA**: Y IS LARGER OR MORE FREQUENT THAN X.

Now let's examine **two tail tests,** which ask the more general question of whether X and Y are meaningfully different. Both one tail tests and two tail tests will use the cumulative binomial formula to compare population X with population Y. So suppose the lady captain had simply said that there was a difference in terms of cloud cover due to location.

2.6.2.1 Example: Two tail sign test (at sea about)

At sea About Weather (cont. for two tails) Sign Test; April 19-27, 2009 Percent cloud cover.

H0: CLOUD COVER OFF SB IS THE SAME AS OFF LA.

HA: THEY ARE NOT THE SAME.

Criterion: $\alpha = 0.05$ (Reject **H0** if r corresponds to a $p\_vlaue \leq \alpha/2 = 0.025$.) Note the **adaptation** for two tails.

Methods: Data was derived from NOAH on the internet.

Results: $r = 11$ successes (**+**); $n = 18 - 2$ ties= 16 pairs;

Referring r and $18 - 2 = n = 16$ to Appendix 1, the $p\_value > \alpha/2 = 0.025$ because the required $r \geq 13$ was necessary to reject at $\alpha/2 = 0.025$ and $n = 16$. So, they **did not reject H0**.

Discussion: The two **changes** (from the one tail test to a two tail test) **were the wording of HA, and the use of $\alpha/2$** for the decision as to whether there was statistical significance. **The greater of either + or - was counted** as r, the number of successes. This process of choosing which direction has the larger count (if either) was part of the actual analysis. This choosing of which direction was counted as successes divided the criterion equally between the 2 tails and therefore required the use of $\alpha/2$ as a probability cutoff. Notice that because they had to use $\alpha/2 < \alpha$, a larger number of successes is usually required to reject **H0**. In this case it would have required $r \geq 13$ (rather than 12) to reject **H0** and declare statistical significance. Of course the method of analysis must be **selected before** the actual data is examined, so the test could not **validly** be done one way and then the other.

2.6.2.2 50% test, only a basis for understanding

The sample sizes that are commonly seen in popular use of the sign test are often so small that they have little hope of finding statistical significance even when a meaningful difference really exists. So the captain that said there was no difference had an unfair advantage. Both binomial tests, the sign test and the 50% test, are popular but weak. They are so weak that some mathematical proofs assume that if a test fails to find statistical significance and the sign test gets significance, the test that failed is wrong.

> **BEWARE:** such simple testing approaches as the sign test and the 50% test are inadequate. They are only part of the story; there are still questions as to **what degree two things are different**, and as to **how sensitive** your test really is. Without answering these questions, your results, especially a failure to reject, will be ambiguous to say the least.

2.6.3 For potential spreadsheet users:

Be careful of the 0 on hand calculators which sometimes looks like an 8, but it is best to avoid using spreadsheets for statistical **calculation**, in favor of scientific hand calculators (which have great accuracy), or languages such as R or Python. Excel is a source of pretty graphics, but spreadsheets, especially Excel, will **not** provide trustworthy calculation results, particularly when small probabilities are involved. Although spreadsheets are not ideal for professional statistical analysis, Gnumeric is at least more accurate than Excel for almost any other calculation.[20] The R or python languages are the most suitable for those who regularly do a lot of statistical calculations, or who work with large data sets. Google Docs has been reported to have deficiencies and maybe should not be used for scientific statistical analysis (Keeling and Pavur 2012). Excel tends to have many shortcomings for statistical calculations (McCullough and Wilson 2005, Yalta,2008).

---

[20]The Gnumeric spreadsheet, is the most accurate of all spreadsheets and it is freely downloadable from the internet (http://www.gnumeric.org/) for Linux, but for Windows you must search for alternative sources). It is often helpful to export Gnumeric spreadsheets as Excel files (.xls) in order to use the excellent Excel graphics. The lack of attractiveness of the Gnumeric spreadsheet may have to do with what you used before you switched to Gnumeric. You could use Open Office CALC, or Excel for graphics, but Gnumeric is almost satisfactory for accuracy in statistical calculation.

Spreadsheets are great for entering data and storing it like little data bases. If you intend to analyze data from a Gnumeric or Excel spreadsheet in the R language, then export them as tab, space separated, or comma separated text files. In Excel tab delimited text file export is available under save as. Be careful; the Excel files can have unexpected variations. Tab delimited are fine for most uses and can usually be treated as space delimited ("") or tab delimited ("\t") in R. However, for either Python with Pandas or for R, comma separated files (.csv) are especially handy.

2.6.3.1 Ranking on a spreadsheet

You can do relatively large scale rankings on a spreadsheet. Enter the data you wish to rank with a parallel column of numbers showing the order in which you entered the data (this will enable you to separate any combined data later). Then use the spreadsheet's sort command to order these two columns together using the data column as the key. Next number a parallel column from 1 to n and have the spreadsheet order the three columns together using the entry order column as the key. This will give the original two columns with the order (rank) values in the third column. For ranking both ways you can form another column where you subtract the ranks in the original direction from $(n + 1)$. A little experimentation will help you understand all this.

2.7 Getting Graphic Adds Up

Figure 2.2 (also on the front cover) illustrates some basic ideas using two discrete populations, binomial distributions represented by overlapping column graphs of the probabilities of various numbers of successes, r. In this case you are only interested in the cumulative probabilities of each population. It is easy to calculate cumulative probabilities for discrete populations because all you have to do is sum them. This can be visualized as adding up the heights of corresponding columns in a column graph of the population from the probability of some value, $r = a$, to some other value, $r = b$, inclusive. (For continuous curves this could be seen as measuring selected areas under a curve.)

The heights of the leftmost bars (darker) in Figure 2.2 represent individual probabilities in a population that is completely described mathematically by the information that is generated by taking binomial samples of size $n = 30$ when the true probability $p = 1/2$. The number of successes, r, is shown on the horizontal axis. In a real testing situation, the probability, p, is unknown, but suppose you have hypothesized **H0**: $p = p_0 = 1/2$, that half the outcomes are successes. The use of p_0, pronounced **p-nought**, here represents the probability hypothesized by **H0** (h-nought).

Then suppose you challenged **H0** with an alternative hypothesis, **HA**: THERE ARE MORE SUCCESSES THAN FAILURES, meaning that $p > p_0$. Under this one tail **HA**, the true probability could be **any value greater than** $p_0 = 1/2$; that is, that $p > 1/2$. The initial assumption of equality, or sameness, means that you proceed as though **H0** is true until proved otherwise beyond a reasonable doubt, making the commonly used statistical tests easier to work with than some less used approaches.[21] Figure 2.2 will eventually lead to an understanding of how the Neyman-Pearson approach works (Subsection 4.7.2). The next chapter will expand upon how all this also leads to the process of sub-

---

[21] Starting as though **H0**: $p_0 = 1/2$ is true could be called a ***central approach***. ***Noncentral approaches*** start with the assumption that **H0**: $p \neq 1/2$ is true. Working with them is more complex than central approaches and usually involves more computing. They introduce at least one more variable, a ***non-centrality factor***, a measure of how different from **H0** the truth is assumed to be (often denoted by the Greek letter *lambda*, λ). Noncentral approaches often include more than just one more variable. Tabulating them becomes complicated and computer programs are usually required. Although they are sometimes used to estimate sample sizes, a major theme in this book, they will not be covered as such, because they are not suited to the targeted audience of this book and any savings in cost (the sample sizes required) is often minimal.

sequently calculating the sensitivity, the power of such a test (with respect to a particular alternative probability, say, $p_A = 60\%$).

Assume you have chosen $\alpha = 0.05$ and a one tail alternative, in the r increasing direction, in advance of obtaining the data for Figure 2.2. The vertical line cuts through the leftmost distribution, the null distribution (represented by the dark bars) at the point on the horizontal r scale beyond which (as r increases), **if HO**: $p = 1/2$ **were true**, the remaining dark bars would add up to 5% of the total probability. That is, it would cut off a total of 5% of all the (dark) bars in the leftmost distribution. This represents the probability, 0.05 or 5%, that you would get a value of r **that large or larger** if you took a random sample of size $n = 30$ and the probability of a success really was exactly $p = 1/2$. Keep in mind that you are considering the one tail model, and you have chosen this much, $\alpha = 0.05$, probability of rejecting **HO** and incorrectly declaring a statistically significant difference by chance alone, *iff* **HO** was in fact true (getting a Type I error). By your selection of α in advance, you have agreed that you will not accept any greater probability of incorrectly rejecting **HO** and declaring statistical significance if **HO** were in fact true, and your data were collected in a truly random fashion. If **HO** were true you could sample an infinite number of times and 5% of these times you would incorrectly reject **HO** due to chance alone.

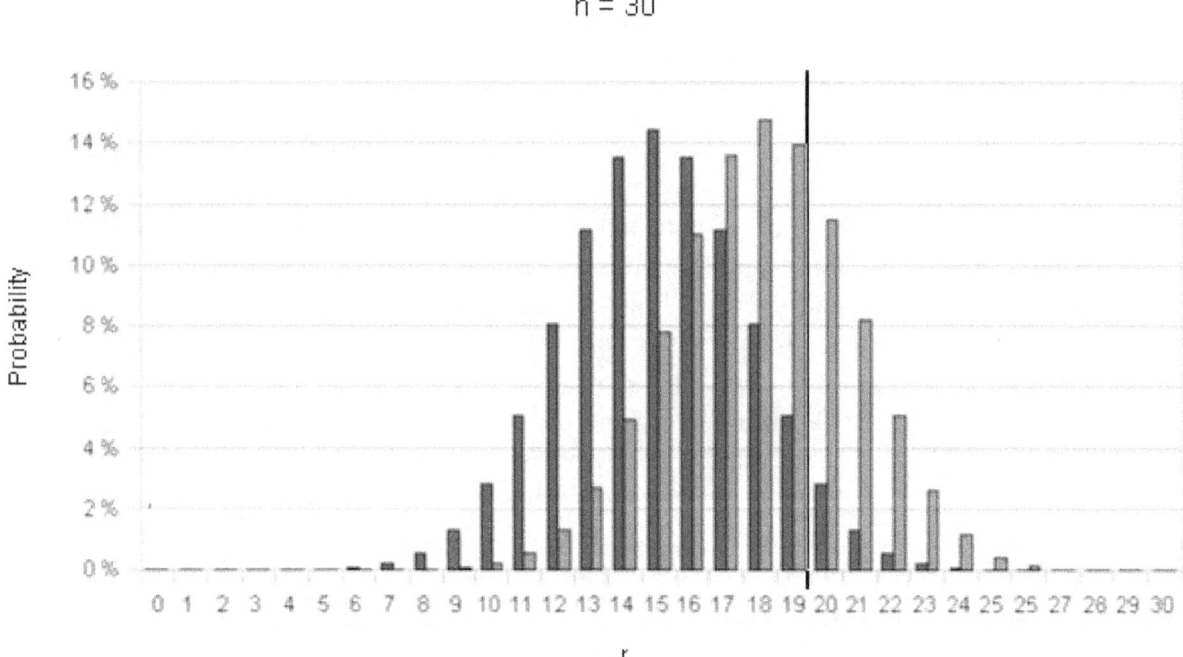

Figure 2.2: Overlapping Binomial Distributions, $n = 30$

Individual probabilities at $n = 30$:

Leftmost dist. $p_0 = 50\%$ (the darker bars; most common **HO**, null distribution).

Rightmost dist. $p_A = 60\%$ (the lighter bars; one of all possible realities).

Vertical line: is just above 19 successes

Right of line, $\alpha \leq 5\%$ *significance, at $p_0 = 50\%$* (leftmost dist.).

Also right of line, ***Power*** $(1 - \beta) \approx 29\%\,at,\ p_A = 60\%$ (rightmost dist.), where β (beta) is the probability of failing to reject **HO** when it is false.

(These values can be found by summing successive bars.)

Now assume that the truth (reality) is that the probability of a success is really 60%, so that the rightmost set of bars (lighter) also represent a binomial distribution with $n = 30$, but the truth is that $p_A = 0.60$. This is **only one of an infinite number** of possible alternative values of $p \neq 1/2$, and each probability, p_A, completely defines a whole binomial distribution for even just one sample size, $n = 30$. In this instance the same vertical cutoff line cuts off 29% of the alternative distribution, with $p_A = 0.60$. So, with $\alpha = 5\%$, the **power** of a one tail 50% test to detect that $p \neq 0.50$, when the truth is that p exactly equals 0.60, is just 29% at this sample size, $n = 30$. Although power is calculated for only one possibility out of an infinite number of possibilities, it still tells you that the sign test at this sample size, which was classically called large, is not particularly sensitive. In many political opinion surveys, for instance, this difference between p_0 and p_A would be unusually large; some would say ridiculously large.

This chapter should leave you with some idea of how classical statistical testing works, and a hint of how it might be improved if you contrive to include a measure of sensitivity (power).

So how could I calculate this power?

To estimate a power, the chance you will detect a particular difference, of either a 50% test or a sign test, it is often necessary to first calculate beta (β), which is the chance of getting a Type II error, of not detecting a particular difference, $\mathcal{D}_{min}$, or rather the cumulative probability of getting $r - 1$ number of successes, $p_r = p_0 + \mathcal{D}$, **when** the observed difference, $\mathcal{D}$, is at least as much as $\mathcal{D}_{min}$. In the graph above (Figure 2.2), the true probability is $p_A = 0.60$, which is $\mathcal{D} = 0.10$ more than the hypothesized $p_0 = 0.50$. In a real life situation you or your experimental technique (whether you realize it or not) will choose the actual value of $\mathcal{D}_{min}$ (even then your estimate of $\mathcal{D}$ will be somewhat obscured by noise due to chance). So let's consider the alternative distribution (the lighter bars, representing the true distribution) in Figure 2.2.

Start by finding an appropriate number of successes, r, needed for statistical significance, given your chosen number of tails, sample size, and criterion, α. For a one tail situation like this, Appendix 1 says that with $p_0 = 0.50$ and $n = 30$, then $r = 20$ **or more** has a cumulative probability equal to, or less than, $\alpha = 5\%$ (given **H0**, the sum of the probabilities, darker bars, **for r greater than** $r = 19$). To find the power, a probability equal to $1 - \beta$, start by finding beta, β, the probability of getting a Type II error, of not detecting that $p_A = 60\%$. Given your chosen alternative (**HA**) distribution with exactly $p_A = 0.60$, this β is equal to the cumulative sum of all the probabilities, the lighter bars, of all r values up to and including $r - 1 = 19$. Then the power is the probability, $1 - \beta$, of actually detecting an $r > 19$, that is detecting a $\mathcal{D}_{min} \geq 0.10$.

You calculate β by substituting the appropriate terms into Equation 2.7 to get,[22]

$$\beta = B(n,\, p,\, first,\, last) = \sum_{i=0}^{i=r-1} \binom{n}{i} p_A^i (1-p)_A^{(n-i)},$$

which is presented later in more detail as Equation 3.1. Then substituting actual values, you get

$$B(30,\, 0.60,\, 0,\, 19) = \sum_{i=0}^{i=19} \binom{30}{i} 0.60^i 0.30^{30-i} \quad where\, n = 30.$$

which gives $\beta = 0.7085281$, as can be found in RStudio, etc. So, the power, the sensitivity, is $1 - \beta = 0.2914719$, or about 29%. Using this classically popular test size and criterion, your power to detect a difference of ten percent more than half, at this popular sample size and criterion, is less than

---

[22]The Gnumeric spreadsheet's, cumulative probability function for a binomial range is $bimom.dist.range$
$(n,\, p, first,\, last)$. You get to it via the $f(x)$ button. Be cautious such operations in spreadsheets are not always sufficiently accurate.

one chance in three. And remember that you are also taking a one chance in twenty (5%) of falsely declaring statistical significance if no difference actually exists. If you took a one in twenty chance of an accident at every intersection when driving, you would not get very far. There definitely seems to be a hint here that the application of statistics to research needs greater care than was classically taken.

An alternative to the Neyman-Pearson approach, which uses α for testing, is Fisher's approach, which does not use α (Subsection 4.7.1). In Fisher's approach you a priori decide to report the p_value as a mere piece of evidence regarding your experiment (Subsection 4.7.1), instead of making a definitive decision on the basis of a criterion, α. You must choose one approach or the other before even collecting the data, and they cannot be combined. There are good arguments for either approach, but you need to understand both approaches before getting into which you might choose as best, **and for what**. Be sure to check with your journal policy to see if Fisher's approach will be accepted.

Question 13 and its answer at the back of the book further clarifies these ideas. This chapter barely touches on the inescapable subject of power; the following chapters **will expand on this material**.

Questions for Ch 2

Write the equations where appropriate and fill in the values. **Odd numbered questions** have worked answers at the back of the book and can be studied as additional examples. The data tends to be fictitious.

1. Anthropologists camping near nomads to observe them leave the camp for a couple of hours, having deliberately left something of moderate value behind, to see whether the nomads being studied will take it. When they return, nothing remains; it is even hard to tell where the camp had been. Was this an experiment, an observation or an abomination? Does it represent the social conditions among the nomads?

2. You count the number of adult musk ox per herd just before calving season and just after it by flying high over them as to not disturb them and taking aerial photos. Is this an experiment or an observation? Are the results purely the number of musk ox in herds at these seasons? Does this give a hint of why the word experiment is used rather generally?

3. One theory of criminal law is that such law exists in order to limit revenge. In a US court room you must settle the matter once and for all by declaring the defendant innocent if there is any reasonable doubt. Is it the same in science? Is the initial hypothesis really accepted if there is not enough evidence to reject it?

4. What does double blind mean? Would it create a problem to tell each of the patients in a pharmaceutical trial which of the treatments they were getting? Explain.

5. What is the probability of randomly getting no more or less than $r = 4$ barrels of wine that have turned to vinegar if you buy $n = 12$ barrels at an auction from a large population of barrels, half of which have turned to vinegar?

6. If 70% of a shipment of bolts of cotton cloth are defective, what is the probability of getting exactly 3 bad bolts when you buy 5 bolts? (Assume randomness.)

7. Suppose you are working with an introduced plant (a weed of foreign origin) that has two genders, male and female, but experts disagree whether there are more females or equal numbers

of each gender. You decide to use a hypothesis pair, **H0**: THERE ARE THE SAME NUMBER OF EACH GENDER and **HA**: THERE ARE MORE FEMALES, with an $\alpha = 0.05$ criterion. Then you divide a map into hundreds of one hectare squares and number the ones that are mostly, or all, pasture. Then you use a random number table to select $n = 50$ of these *quadrats* (sampled squares), and sample the plant nearest the southeast corner of each selected quadrat in that random order. You find $r = 35$ females and the rest are males. Use the 50% test to ascertain whether this is statistically significant at the given criterion so that you may reject **H0** and accept **HA**. Explain how you got an answer.

8. You are investigating an eating disorder called 'mega-snacker's syndrome' and your coworkers say only about half of the victims have some sort of detectable emotional comorbidity. You use a hypothesis pair, **H0**: THERE ARE 50% WITH COMORBIDITY and **HA**: THERE ARE MORE THAN 50% WITH COMORBIDITY and you decide to use an $\alpha = 0.05$ criterion. You cannot test all of the 1406 of the already cataloged cases, so you use a table of random numbers to select $n = 49$ mega-snacker victims from this database. Then you arrange for someone else to examine them, again at random, for comorbidity without telling the examiner why you are looking for emotional illness, or even that they are mega-snackers. This independent evaluation finds $r = 30$ cases of comorbidity. Use an appropriate test for statistical significance (ignore the question of test sensitivity for the moment). Show your work, including writing out the hypothesis and criterion, and state your conclusion. Why did you use an independent evaluator (someone 'not in the know') rather than doing it yourself? Are your results with respect to the question at hand inconclusive (lacking a definite conclusion)?

9. Referring to the samples taken in Question 7, suppose you sampled $n = 50$, but used a hypothesis pair that included a different alternative hypothesis, **HA**: THE NUMBER OF FEMALES IS NOT THE SAME AS THE NUMBER OF MALES. You got the same result, $r = 35$. Disregard the issue of how you define different and use the 50% test with $\alpha = 0.05$ criterion to ascertain whether this is statistically significant at the given criterion so that you may reject **H0** and accept **HA**. Explain what you did.

10. Referring to Question 8 you still have $n = 49$ (for the moment, disregard the issue of how much unlike). Assume you never tested this data before and assume you use a different hypothesis pair, **H0**: THERE ARE 50% WITH COMORBIDITY, and **HA**: THE AMOUNT OF COMORBIDITY IS NOT 50%, and still use an overall $\alpha = 0.05$ criterion. Explain what you did and how it is different from Question 8.

11. The values that are **paired** vertically below, are daily results for the concentration of an industrial waste at two different stream monitoring stations, A and B.

A : 5.5 7.3 2.7 6.2 4.2 0.1 3.8 5.4 2.9 7.1 1.1 5.2 5.7 2.3 3.5 *ppm*
B : 2.8 2.2 9.4 8.2 1.2 7.1 3.6 5.8 5.2 4.3 7.7 0.8 9.5 0.9 1.1 *ppm*

Use the appropriate test based on the binomial at $\alpha = 0.05$ to test **H0**: THE OVERALL CONCENTRATIONS ARE THE SAME AT THE TWO SITES versus **HA**: THE OVERALL CONCENTRATION AT SITE A IS GREATER THAN B. How did you do it, what was the sample size, and how many successes were observed? What did you conclude? (Note this is only for greater than.)

12. Two treatments are vertically **paired** daily measurements of the concentration of a pollutant at two different harbor monitoring stations, A and B.

A : 6.6 6.5 3.3 9.0 4.0 6.8 0.0 7.9 6.7 4.5 0.2 2.5 2.8 5.1 6.4 *ppm.*
B : 3.9 8.4 1.5 9.3 9.9 5.0 2.0 7.8 6.0 5.5 7.9 9.7 8.3 5.1 6.2 *ppm*

Use the appropriate test based on the binomial with a criterion, $\alpha = 5\%$, to test **H0**: THE OVERALL CONCENTRATIONS ARE THE SAME AT THE TWO SITES versus **HA**: THE OVER-ALL CONCENTRATIONS ARE NOT THE SAME AT THE TWO SITES. How did you do it? What is the sample size, **and** what is the final value of n used for your test? Why? So, how many successes were observed and what did you conclude? How did the method of evaluating this question differ from Question 11?

13. The table in Appendix 1 shows that $r = 5$ is statistically significant at $\alpha = 0.05$ for a **one tail** binomial test (either a 50% test or a sign test) when the sample size (excluding non-tied pairs in the sign test) is $n = 5$. Your hypothesis pair is equivalent to **H0**: THE NUMBERS OF SUC-CESSES AND FAILURES ARE EQUAL versus **HA**: THERE ARE MORE SUCCESSES. Using this $r_a = 5$ and Equation 3.1, $\beta = \sum_{i=0}^{i=r_a-1} \binom{n}{i} (p_A)^i (q_A)^{n-i}$ estimates the probability of getting a Type II error for a targeted p_A. (A scientific calculator might save some pain.) Can you explain why the formula uses $r_a - 1$? **What is the power** of a sign test or a 50% test to detect a difference this large or larger with a sample size of $n = 5$ and a targeted alternative probability of $p_A = 0.80$? Recall that β is the chance of getting a Type II error, and is something you must find in order to estimate the power.

14. Find the power for a binomial test of the hypothesis pair, **H0**: THE NUMBER OF SUCCESSES AND FAILURES IN THE POPULATION IS THE SAME, versus **HA**: THEY ARE NOT THE SAME, with only $n = 7$ samples, a targeted $p_A = 90\%$, using either a sign test or a 50% test at the $\alpha = 5\%$ level. Because this hypothesis pair indicates that an outcome in either direction would be statistically significant, you will use $\alpha/2 = 0.050/2 = 0.025$ in each tail. The table in Appendix 1 shows that for a two tail binomial test with $n = 7$, and a criterion, $\alpha = 0.05$, statistical significance may be declared when $r \geq (r_{\alpha/2=0.025} = 7)$ successes. So, because this is a two tail test, you will examine the data for the most extreme outcome (furthest out in its tail away from the middle). The largest count, whether A or B, left or right, yes or no, is r. However, in order to estimate the power, $1 - \beta$, you must estimate β, the probability of getting a Type II error. It is the sum of all the $n - 1$ probabilities which would have less than $n = 7$ successes, given that your targeted $p_A = 90\%$ is the true probability, p. Use Equation 3.1 to find $\beta = \sum_{i=0}^{i=r_{\alpha/2}-1} \binom{n}{i} (p_A)^i (q_A)^{n-i}$, where $q_A = 1 - p_A$. **Also explain** why 1 is subtracted from r_{cutoff}, in this case $r_{\alpha/2} - 1$, of the summation, $\sum_{i=0}^{i=r_{\alpha/2}-1}$.

15. Assume a common card deck with 52 cards has been shuffled sufficiently that drawing from it should produce random results. Your roommate who claims to be regularly visited by spirits claims to be able to draw a queen on each of the next 4 draws. What is the probability that your roommate does this if the cards are drawn at random? (Note that there are four queens in a pack and your roommate is drawing from a small population without replacement).

16. Unlike the example in Subsection 2.2.0.2, suppose you and your roommate are on military furlough and go to the races at the state fair. There are 9 horses in a particular race and, **unlike** the example mentioned, your roommate, who has never seen a horse before, only claims the spirits will reveal the first 3 horses, but not their order. What is the probability of your roommate choosing the first 3 horses by chance alone? Is which horse wins a race really a random event for an experienced bettor?

17. You are a UN Peace Keeper and the leaders of 7 different factions want to individually discuss partial disarmament with you. The order in which you meet with them may affect the outcome of your negotiations. In how many different ways can you arrange the interviews? Assuming

there is only one best way, what is the probability that you would get it right if you interview them at random?

18. Due to logistics and bureaucratic craziness, a UN Peace Keeper needs to meet with a group of 5 out of 7 leaders; the choice of which 5 is critical. How many possible groups of 5 can be selected? What is the probability of the best possible outcome if the choice of which 5 is made at random?

19. You are a tomb raider (not a particularly common profession) and your Nazi occult enemies have sealed you in with a bomb. There are 5 wires and you must cut no more or less than 3 of them **in the correct order** to avoid a sudden ending of your career. How many different ways can you try to do this? Assuming complete randomness, what is your probability of survival?

20. The early plug and play computers were very sensitive to the order in which you plugged the cards into the slots. Thus they were also called 'plug and pray'. Suppose you have 3 cards with 3 slots. In how many ways could you order them in the slots? Assuming your choice is truly random and that there is only one correct way, what is the probability of doing it correctly on the first try?

21. You are deploying 9 medical teams, but 12 countries have requested emergency medical aid. Each country can only get one team. How many different ways of allotting 9 teams to 12 countries are there? Because of politics, you're not given the information to triage which countries get teams and you are required to proceed **randomly**. Assuming the teams are identical, what is the probability of getting the one best result?

22. You are a stir fry cook and you have 9 different ingredients available, but the boss will only allow you to bring 7 out of the cold storage at a time. How many different ways can you do this? Assuming that your assistant does this at random before you get to work, what is the probability of you arriving to find your one favorite assortment?

23. What is the probability of observing exactly $r = 5$ plants with white flowers out of $n = 10$ of the F2 offspring from inbreeding F1 parents that are crosses of dominant orange flowering plants and recessive white flowering plants? (Referring to the example in Subsection 2.4.1.1, recall that $p_{white} = 1/4$ in this F2 generation.)

24. Assume that the right rear tire of a certain model of car has a 56% probability of being the one of the two rear tires that blows out. Out of $n = 7$ rear tire blowouts on this model, what is the probability that $r = 5$ of them will be this tire?

25. You and your spouse want to have $n = 4$ children with $r = 2$ girls and $n - r = 2$ boys. What is the probability you can get 2 girls and 2 boys out of 4 births? (Assume no medical manipulations and that the population-wide rates of male and female births are equal.)

26. A hotel in Manila was served by **equal numbers** of air conditioned and non-air conditioned taxis. The latter involved a higher security risk. Assuming randomness, what is the probability that out of $n = 5$ unsuspecting tourists, exactly $r = 3$ of them got the safer 'aircon' taxi?

27. Due mostly to waterborne disease, there are a number of places in the world where there is only a $p = 3/4$ probability that a baby will survive from birth to four years old[23]. After that, the survival rate is a great deal higher. Suppose a particular couple is fated to have $n = 7$ live

---

[23]In some of these places, malaria competes with water borne diseases to cause even more infant deaths.

2 YES, NO and BEING DISCRETE

births under such conditions. What is the probability that no less than $j = 2$ or more than $j = 4$ of these children will survive to age four?

28. Micronesian islands are close to the equator and are near where typhoons form and usually move away without hitting a populated island. Suppose your island has a $p = 20\%$ probability of actually being hit directly by any locally generated typhoon before it moves away. If you are teaching on the island through a long period that includes the generation of $n = 5$ typhoons, what is the chance of you experiencing between $j = 1$ and $j = 3$ (inclusive) direct hits? Show the formula with integers substituted in before resorting to software or built in calculator functions to get an answer.

3 WHERE NO TWO THINGS are ALIKE

There are two common conceptual approaches classically used by frequentists when testing for statistical significance. They have their supporters, their enemies, and those who would invalidly try to use both of them simultaneously. The bad news is that the error of using both together is commonly found in published research. It involves the misuse of the p_value. It is necessary to do a quick review and a little discussion of using the p_value alone, vs with the criterion, α, along with a review of some recently learned concepts and definitions.

Using the earliest approach, Fisher's approach, the p_value can be seen as the probability of your results having turned out the way they did by chance alone; it can be taken to stand by itself as **evidence** against **H0**. **Or** using the other, the Neyman-Pearson approach, the p_value may be taken to be a step in the **decision making** process of discovering whether the probability of your sample turning out the way it did was less than or equal to some probability, the criterion α, chosen a priori by you to demonstrate **H0** to be too unlikely to be true. But the p_value **cannot play both roles** at once. That is, **before** obtaining the data, you must choose whether you are going to report a p_value, or whether you are going to report that the p_value was, or was not, less than or equal to an a priori criterion, α. Each approach has its advantages and its disadvantages. Using only the p_value only allows you to reject the null hypothesis, **H0**, based on the p_value combined with whatever additional evidence you consider relevant. However, using a precise criterion, α, produces more reliable evidence by eliminating some of the effect of the variability of the p_value itself and allows you to accept an alternative hypothesis, **HA**, as well as rejecting the null hypothesis, **H0**. Using both approaches at the same time, especially the common practice of using α and then post hoc justifying or strengthening your conclusion by referring back to the p_value, gives an incorrect assessment of the results. Even when appropriately separating these two approaches to testing, be certain to check with intended editors before choosing not to use an α.

Be forewarned that many researchers use *"alpha"* or even the letter itself, the criterion, α, incorrectly for the p_value. This practice is also common in verbal usage, as in seminars. In such cases you must rely on context in order to follow what is being reported or discussed. "What was your alpha?" might, depending on context, unfortunately be used to mean, "What was the p_value you compared with your criterion, α?" or it might correctly be used to mean the criterion α itself.

3.0.0.1 The nill-null kerfuffle

Both approaches are (often derogatorily, especially when α is involved) referred to as *NHST*, nil hypothesis significance testing, but it can also mean null hypothesis significance testing. Null means having no **binding** force, whereas *nil* means a zero score, or no effect. If you could **really test** to see if a difference in some measurement equaled **exactly zero difference** between things, you would **always reject** the nil hypothesis, because no two things (maybe even at the subatomic level), especially no two samples, are perfectly alike. A possible exception is where conditions are such that both 'treatment' and 'control' turn out to be exactly the same thing.

In this sense the critics perhaps misunderstand the workings in the application of statistical tests because the models (or at least their operation in application) have been inadequately described. The descriptions most commonly leave out the fact that although the models have a **zero reference point**, the tests **become less sensitive** as the difference from zero becomes small, and that sensitivity (power to detect a difference) fades to meaninglessly small even while at a potentially measurable, and sometimes consequential distance from zero. Hence the word null applies operationally while the word nil only mathematically (conceptually) refers to a point of reference.

To be certain of zero true difference, assuming it exists, you would have to sample the whole population. However that is rarely possible, and by definition, a sample is not the whole population. A sample is **only a part**, a surrogate for what is being tested. The result of sampling is blurred by variation around the central effect. However, recall that the *law of large numbers* says that if you perform the same experiment a large number of times, the average of the results obtained should be close to the expected value (the true mean), and will tend to become closer as more trials are performed. When considering sample size you must understand that the **largeness** of the sample, or equivalently, the **number of observations**, is a **major factor** in determining the sensitivity (**power**) with regard to a minimum difference that can **never** operationally be zero, and which, in almost all contexts, cannot be zero in reality.

Many researchers who have gone before were not conscious of the fact that their calculations implicitly targeted some minimum difference largely dependent on the sample size they chose to use. As with aiming arrows, there are a lot of outside factors involved, like gusts of wind, even when a great archer shoots. A badly shot arrow might even hit the bullseye because of a gust of wind, but the worse the archer, the less probable that the bullseye will be hit. Although other details of planning and execution are also critical to successful research, the **larger the sample size**, the greater the probability of the estimate being close to the true central measure or other parameter of interest.

3.0.0.2 Randomness is required

Recall that random means that every member of a population has an equal chance of being selected. A sample mean can only be taken seriously as an estimate of the true mean, μ, if the sample is taken in a truly random fashion. Both μ and its estimate, the statistic $\bar{x}$, the sample mean, are often referred to as 'the mean'. Look out for context, it is sneaky!

Sample means are just estimates and subject to chance, meaning that they too have a distribution, although less variable than the original data. A great deal of modern statistics and of the material in this book is about the distributions of means themselves. For instance, the famed normal distribution is the expected distribution of the sample means. Thus the distributions of the means of random samples is a major theme in most of what follows in this book.

3.1 The α and the β of Inference

These two values, α and β, (discussed below) are exact values; you cannot round them, because they are used to define an exact point. This is especially important when you compare another probability to α. You cannot round a probability, say, 0.050000000001, and say it equals $\alpha = 0.05$ (5%). **It is larger**, no matter how infinitesimal the difference. Especially in the context in which α is most often used, $\alpha = 0.05$, it is almost too liberal, as you shall see.

3.1.1 α and what is too unlikely to be true

Historically, looking for the answers to yes-no questions with statistics followed a pattern[1] that attempted to answer the question: Is the probability that two things (populations included) are alike ('the same') **too small to be accepted as true**? An early approach (Subsection 4.7.1) was to test to

---

[1]Gigerenzer, Gerd et al. 1989. The Empire of Chance: How Probability Changed Science and Everyday Life (Ideas in Context), is intended for social science students, but can be useful for wider applications.

see whether an outcome was unlikely with no specific minimum probability; that was left to the investigator. A later approach (Subsection 4.7.2) was generally based on an a priori, meaning before the fact, *criterion*, a probability below or at which the researcher has decided the outcome would be **too unlikely** to be true. This criterion is almost always written as the Greek letter alpha, α. It can interchangeably be presented as either a percent, usually $0.0\% < \alpha \ll 50\%$, or as a pure decimal, $0.0000 < \alpha \ll 0.5000$, where $\ll$ means much less than.[2] In application, the common values of α are 1% and 5%.

It may surprise those who struggle with statistical tests that they are just formalized attempts to determine whether some outcome is too unlikely to be accepted as true. **H0** is the *null hypothesis*, the hypothesis that there is no real difference other than a possible difference in the sample that is merely due to chance. (Unfortunately this classical definition should say 'no meaningful difference'; as we will eventually cover; the tests really test for a particular non-zero meaningful difference that you can usually control. If you do not take charge, the procedure will choose its own target difference.)

The p_value is the probability of obtaining a test statistic that is **at least** as much different from the hypothesized value as the one that you actually observed, assuming that the null hypothesis, **H0**, is exactly true. Another, likewise valid, way to look at the p_value is to see it as an **estimate of the probability** that whatever results you get are merely due to chance. In the most common approach, the Neyman-Pearson method, the p_value is compared with a criterion, α, a **predetermined** probability of **getting a Type I error**. If the p_value is less than or equal to α, we say that the test result is statistically significant, and therefore we reject **H0** and accept **HA** (more formal and applied detail in Section 3.3.1).

3.1.1.1 Formally cutting off the cutoff

So far we have cut off the border between what is statistically significant and what is not, at a probability, the criterion, α. We have a priori accepted that below or equal to α, the probability of our sampling being from a population like what our null hypothesis, **H0**, described, would be so extremely small that **H0** would be **too unlikely to be true**.

So, in this book we define the *cutoff* as being the probability below which, or equal to, **H0** would be too unlikely to be true. The cutoff can be either of two different probabilities, α or $\alpha/2$. That is, in situations where you are testing for **only** too large a difference, the cutoff equals α, and testing for **only** too small a difference, the cutoff still equals the criterion α, but for **either** too large or too small, the cutoff equals $\alpha/2$.

Do you really want to know if there is exactly no difference between two random samples of children under different treatment regimes, or between two harvests? Reality rarely, if ever, contains zero difference between any two things. In application, **zero difference is only a conceptual reference point** to facilitate the process of estimation or of inference.

3.1.2 When is a difference a difference?

If a Ford costs one cent less than a Honda the cost difference is of no importance to the buyer. For statistical inference you are testing for a reasonable meaningful difference between two outcomes, such as a treatment and a control in a clinical trial, or the cost of living in one city as opposed to another. In application, differences near absolutely no difference have no use and the smaller the dif-

---

[2]Either notation, percent or decimal, can be written to whatever number of figures is appropriate. In publications α is often only reported to two or three figure accuracy for the criterion, but you may need four or more figure accuracy in the tables or other source for some calculations. Recall that the $\%$ symbol simply means move the decimal point two places to the right; $0.12345 = 12.345\%$, or the other way around (equal signs do not imply direction).

ference the greater the cost to detect it. How big the difference is relates to the power, the probability that it will be detected. The power also relates to how big your sample is, the variability of the measurement you are using, and your chosen α and β. So, in application, you must first decide how small a difference you think is reasonable and you will usually use it to choose an appropriate sample size. So the **reasonable minimum difference** is defined as the smallest difference that would be of any real use in the actual situation.

3.2 Isn't Any Unusual Occurrence Significant?

If we had never seen a real Chevrolet automobile before and we went to the US and saw 6 red Chevrolets before we saw any others, could we conclude that most Chevrolets are red?[3] Well, if we hypothesized that more than half of all Chevrolets were red, we might think that there was a less than $(0.5)^6 = 0.0156$ or a less than 2% probability that we would be wrong. **But we would be wrong**. If enough people count colors of the first 6 Chevrolets they ever see, some will see only one color, and that is no basis for saying most of the rest must be that color. If we make inferences **after** observing in this manner, we may find all sorts of things erroneously 'statistically significant'. However, if we had previously decided on a sample size of $n = 6$ and hypothesized that the cars would tend to be red, with an $\alpha = 0.05$ criterion, and **then** the first 6 Chevrolets we see are red, that would be a surprising, thus statistically significant, outcome. Statistical significance is really a frequentist standard for deciding that something is too unlikely to be accepted as a coincidence. That is, it would be a rare coincidence to see 6 red cars after we decided on a hypothesis and set the criterion. To produce a situation in which we can consider the probability of a coincidence, we must design our experiments **a priori** and make our hypotheses **a priori** (before the fact).

Although these are frequentist examples, there may be Bayesian analogies. Bayesian statisticians are interested in questions such as how often do you have to observe something before you can accept that it is true. It is doubtful that any Bayesian statistician would agree that you could decide that all Chevrolets, or even a majority of Chevrolets were red, because without any previous conjecture, your first observation included six Chevrolets and they were all red. However, a Bayesian might point out that in the absence of any other information, you might have a better probability of survival if your life depended on predicting the color of the next one to be red. For instance, maybe the local gangster prefers to use red cars for carrying out assassinations.

3.2.1 Is your career a quality control process?

Even in scientific publication, single experiments are overemphasized. "A recent study of randomized, controlled trials reported in major medical journals showed that very few referred to the body of previous evidence from such trials in the same field." (Clarke 1998). A great deal of the media 'news' contains statements like 'scientists agree that ... has no effect'. Such statements can have very strong political implications and result in public opinions leading to governmental acts that can result in great hardship for individuals, or damage the environment, or both. You must be careful what you publish or report to regulatory agencies, or even to your institution's press officer!

Each time you make a set of observations, perform an experiment or do a survey, you are taking a chance of being wrong. Both the Bayesian and frequentist approaches make it obvious that you are only going to get the correct answer some of the time. Despite the attempt of Fisher's approach to be

---

[3]Bayesians would never assume that because they just happened to see 6 red Chevrolets, that most Chevrolets were red. However, they tend, perhaps unfairly in most cases, to accuse the frequentists of doing just that. The important thing to keep in mind is that we cannot use frequentist statistics as a black box, something that thinks for us.

about nature, the truth is that your data is indeed largely defined by how you got it. A research career is a little like a screening process in which you attempt to balance the cost of doing experiments against the proportion of your experiments which will lead to wrong decisions, due to chance. This could be seen as one of several facts justifying the use of the Neyman-Pearson approach, which focuses on populations of experiments or observations (Subsection 4.7.2).

A valid justification for the use of a fixed a priori cutoff based on a criterion, α, is that chance is involved and you are trying to minimize the chance of coincidence being the reason why your test came out statistically significantly different. When using the Neyman-Pearson approach, the p_value should never be used as a measure of the strength of some difference, it is only a step on the way to finding if the probability of your results are less than or equal to α.

To control the number of errors chance inflicts upon you, use α to determine if an a priori chosen reasonable minimum difference has been reached or exceeded; it is more likely to avoid incorrect outcomes than will a freely subjective examination of the p_value. In addition to reducing the bias that is unavoidable in subjective mental processes (a person's judgment), the use of a criterion, α, which you a priori select, is also a necessary part of taking control of the sought after sensitivity, the power.

3.3 Not to be Hyper- or Hypo- Sensitive, Let's Further Examine Type II error

A starting point, zero, 0, is essential to the measurement of something, as with a measuring tape or a timer. In most fields it would be difficult indeed to specify a difference in units of measure without reference to zero difference. However, the very existence of zero can be a subject of fierce debate in some fields of endeavor. The application of frequentist statistics, particularly the Neyman-Pearson approach (Subsection 4.7.2), with its α and β both dependent on zero for a point of reference for their very existence, is a point of contention.

Because it is only possible to define β with respect to a particular difference point and it is also customary in innumerable texts to formally, but vaguely, define β as 'the probability of not rejecting **H0** when it is in fact false due to chance alone', we do seem to have a wee difficulty! Let's be practical and accept that **in application**, a Type II error is failure, due to chance alone, to detect a particular minimum difference. This definition changes the focus, but still implies not rejecting **H0** due to chance alone when it is in fact false (but to a specified minimum degree of falseness). Hopefully, you will designate a reasonable minimum difference and then make it the difference you are looking for, rather than let it choose itself, as has previously been the custom. Reasonable minimum difference is a value often required in grant applications because of its relation to β, which, in turn, relates to the sensitivity of your experiment (its power, $1 - \beta$).

Star Trek's Mr. Data may rattle off time, distance, and percentages to fourteen figure accuracy, but if he were to strive for perfect accuracy he would go on with more and more decimal points for the rest of his existence. The criterion, α, and even the p_value, are in reference to no difference from the point called 0, meaning zero difference from whatever **H0** hypothesizes, even though **no two real things can be exactly alike** in terms of continuous units of measure.

In general there cannot be a zero difference, however, there are situations in which there is zero difference. For instance, there may only be one population, even though you are testing to see if there may be two. That is, there is **zero difference** between some one thing and itself. Similarly, you may be testing for an effect, and it is possible for something to have zero effect, but it is **impossible** for you or I to exactly **measure** a **zero effect**. So, the criterion, α, relates to random variation about a hypothesized zero value whereas β relates to random variation about your pre-chosen reasonable minimum difference, which is a priori chosen by you as a distance from the conceptual zero difference.

Statistical inference is a tool, not a magical black box. It is up to the research community to set standards for its use, and it is necessary for the researcher to use judgment in applying the standards.

3.3.1 How different is a *reasonable* difference?

If a Type II error is extremely probable, an experiment may not be worth the investment of time and resources. It is counterproductive to spend your time and resources doing experiments that are not sensitive enough to almost always detect all true differences that are big enough to be important to you. For that matter, it is often too expensive to look for tiny differences, especially differences which are too tiny to have any meaning in application. Unfortunately, much of what has gotten published is research regarding situations in which the true difference was huge, plus a very few improbable detections of a smaller true difference and some false positives (e.g. Subsection 7.1.1). This widespread use of insensitive experimental designs tends to screen out a lot of moderate true differences that may have been of interest to the researcher and available at an acceptable cost. So, let's reflect upon what has been covered so far, with the addition of a couple of details.

Recall the law of probability which says the sum of the probability of something happening and of it not happening is always one (100%). The probability of getting a Type II error is β, so because all the possibilities regarding a particular event add to one (100%), then *power,* the probability of **not** getting a Type II error, equals $1 - \beta$. In application, **power** is the sensitivity of an experiment (or observation), the probability of detecting the reasonable minimum difference in which you are interested.

It is essential that you should **specify** what you consider a **reasonable minimum difference** (along with α and the power $1 - \beta$) before you begin the expensive process of sampling or observing. If at all possible, you should target a true difference that is neither too big nor too tiny, based on the context in which you intend to apply your results. Aiming specifically for either extreme may lead to disappointment. For various reasons, you should be able to document your reasonable minimum difference and be able to justify why you chose it. For instance, if a fertilizer application is going to cost y dollars per acre, then the increased crop value should be at least y dollars. The a priori consideration of β (or of its complement $1 - \beta$) is also intrinsic to estimating the cost (in terms of the sample size n) of having an acceptable probability of detecting a reasonable minimum difference (e.g. Chapter 7). However, do not panic yet; read on for a few more chapters and then panic (or maybe not).

3.4 Try to Keep It Simple:

Keep your hypothesis simple and direct in order to avoid going beyond the evidence. Some studies a few decades ago concluded that because bilingual children (who were mostly impoverished immigrants and first generation immigrants) were tending to do poorly in school, parents should not teach their children more than one language. Assume your test did show that the dual language children were, on the average, not doing as well in school as the single language children. That would be likely to be a valid outcome, but you would not be certain whether it was neurological (the brain filling up with too much stuff?), or if it was environmental (poverty?, disrupted home environments?, both parents with two jobs?, school location?). It would not confirm that it could or did show that having two languages was the cause. One of the pioneers in statistics is justifiably accused of having been an eugenicist monster and encouraging what became the Nazi holocaust.

So, how do we avoid falling into similar traps? The **safest hypothesis** pairs are of a very general pattern, little more than **H0**: THERE IS NO DIFFERENCE versus either **HA**: THERE IS A MEANINGFUL DIFFERENCE, or versus **HA**: A IS MEANINGFULLY MORE THAN B. Beware of putting a subjective definition on the word 'meaningful'; that could lead to major pitfalls. Then explain what you intended in designing your experiment or observation plan and what you concluded separately. At least you will be able to provide something useful even if you were not the first to see the point of the outcome. And, as implied in Section 3.3.1, the rejection of **H0**, and thus the meaning of DIFFERENT (or MORE THAN), will not be independent of the power (sensitivity) of the experiment. This way **HA** is largely defined as equaling or exceeding a reference point, the reasonable minimum difference.

3.5 Power Analysis

Power analysis is estimating the power of an experiment that is already finished, or estimating the sample size required for an experiment to have the power to find the answer you want. Classically, power has been neglected in the training of researchers, but funding agencies and editors have increasingly required that the power be specified. This will almost always include steps not, until recently, regularly included in research procedures. To meet this requirement you will have to make some very clear decisions and be able to demonstrate that they were indeed a priori (before the fact). If you are intending to use a test that assumes normality, you also need a measure of variability from data from the population to be tested. This information must be from a random subsample that will not be used for the actual testing. Sometimes it can be obtained from already existing records, or it might be necessary to do a complete pilot study. Occasionally the information you need might be available from the literature.

If possible, an a priori power analysis to find the sample size, n, can save you a lot of misery. For your power estimate or sample size estimate, you will need to have a priori chosen a reasonable minimum difference, $\mathcal{D}_{min}$, the smallest difference that would be worthy of your notice, and a criterion, α or half of it, $\alpha/2$, before proceeding. For many testing approaches, a valid power analysis will also require a trustworthy a priori measure of variability, such as the variance, which is explained in Section 4.1. Unfortunately, finding such a measure of variability is not always easy, but not considering power when choosing a sample size is disastrous, and funding agencies often demand a proposed power in your proposal. As you shall see, this requirement is often simpler for binomial tests because of the fact that hypothesized p_0, along with n, determines the variance of the hypothesized distribution. For actual sample size estimation you will also need to a priori specify either β or the power, $1 - \beta$.

When you cannot achieve the chosen sample size, or when you have no control of the sample size, you can, assuming you provided the appropriate a priori inputs, use your achieved sample size, n, to validly ask," Given my **before the fact choices**, what is the power?" But you must have a priori targeted (at least selected) a specific reasonable minimum difference **before obtaining the data**. For the binomial, which relates to ratios (percents), this book uses a calligraphic $\mathcal{D}_{min}$ to symbolize a sample based binomial difference and some percentages or proportions involving counts.[4] For continuous data the book will use D_{min} (not calligraphic). When calculating the power you can only use the **achieved sample size**, n, with a priori inputs, α and $\mathcal{D}_{min}$ (or D_{min}), that you **choose before** the inferential test. Either $\mathcal{D}_{min}$ (or D_{min}) is simply called 'Dee - min'. If the sample size cannot be controlled, or if it changes during the experiment, you will need them for an after the fact power analysis of your test. Unfortunately these two challenges are somewhat common. So, even if you have **no control** of n, you must have a priori chosen α and D_{min}. It be would reasonable for your colleagues to accept that your a priori α was no less than 5% whether you stated it or not, because that is the minimum that is accepted in the research community, but for any other value of α, lacking a preplanned value is unacceptable. Likewise, a preexisting $\mathcal{D}_{min}$ might (but very often will not) be

---

[4]Because they are ratios, percentages do not always behave the same as measurement data.

evident from a previous statement of your results. However, lacking an a priori analysis to see what sample size is needed to achieve your goal will leave you at a disadvantage. Especially when using small samples, if you have to get the variance (the measure of variability) from the test itself instead of beforehand, your estimate of power will be weakened.

It is common for a researcher's sample size to be too small so that the researcher will fall into the trap of missing a meaningful true difference because the experiment did not have the power, $1 - \beta$, to dependably detect the reasonable minimum difference of interest. Although not done in the examples in this book, it is good practice to provide a **few more** measurements or counts than you have estimated if dropouts or other excusable mishaps are expected. Because hypersensitivity is usually impossibly expensive, you are usually, but not always, safer with the largest samples you can afford than you are with smaller samples. However, there are pitfalls such as hyper sensitivity associated with having vastly larger sample sizes than your a priori estimate of n.

Recall Figure 2.2 (also shown in color on the cover), which shows a typical inferential situation, but with a somewhat larger reasonable minimum difference of $\mathcal{D} = 10\%$. However, if the true (unknown) distribution were really this particular alternative distribution defined by $p_A = 60\%$ and $n = 30$, then the lighter rightmost bars represent it. Of those rightmost bars (red), the ones that fall below the cutoff sum up to $\beta = 0.71$. That is, if you took a random sample of size $n = 30$, there will be a $\beta = 71\%$ probability of getting a Type II error, failing to reject **H0** when it is false **if** this distribution, defined by $p_A = 0.60$ and $n = 30$, define the real distribution. You would then only have a **power** (a probability) of $1 - \beta = 29\%$ of rejecting the false **H0**, if the true probability, p, were equal to this $p_A = p_0 + 10\%$. Note that this probability, β, of getting a Type II error is defined in terms of the particular reasonable minimum difference, in this case $\mathcal{D}_{min} = 10\%$.

So, only 29% of the true **HA** distribution (lighter bars) is below the criterion (the vertical cutoff line). This is not what the older texts may have gotten around to telling you. That seems pretty wretched, and yet many texts specifically define a *large* sample size as $n = 30$. They do not consider how weak (lacking in power) your experimental design would be even if a big, $\mathcal{D} = 10\%$, difference existed, because the true probability is $p = 60\%$. Even a power, $1 - \beta$, approaching detection of a true difference almost $1/3$ of the time is better than nothing, but non-detection can be both unnecessarily disappointing, wasteful, and very misleading. There are mathematical advantages to using sample sizes $n \geq 30$ (Subsection 4.3), but $n = 30$ by no means justifies the term 'large' with respect to the power to detect small but meaningful differences.

So, before any inferential test, even before obtaining the data, you must choose the reasonable minimum difference, herein called $\mathcal{D}_{min}$ (dee-min), just as you a priori choose the minimum criterion, α. They are up to you, but you must irrevocably choose them in the very earliest planning stages of any experiment or observation. Then when you report your outcome you should report the value of $\mathcal{D}_{min}$ that you targeted and the associated power.[5] The minimum reasonable difference, D_{min}, must clearly be chosen a priori even if you do not have any control of the sample size, or you will have no basis for even a *subsequent* (meaning after the initial calculation) power analysis. Unless you somehow target a specific true difference **not equal to 0**, you will fall into a trap because of the fact that no two things are perfectly the same. If your test were able to detect the slightest nonzero true difference, you would always get statistical significance whether there is a meaningful true difference or not. And, because you do not know the real difference, you report your estimate of the power, $1 - \beta$, on the basis of $\mathcal{D}_{min}$, your chosen minimum difference (**also to be reported**). So, you

---

[5]Notation (symbolism) is a math curse. Some heavy math books use three alphabets and run out of letters. This simple book used two alphabets, Latin and Greek, and could not use many letters because the readers might recall them in another familiar, but confusing, context. For between the means this book will use the Greek letter δ for the true unknown effect and a D_{min} for the targeted reasonable minimum difference. For the binomial distribution it might have been more consistent to have used something like $D\%$ for the true unknown binomial rather than the simpler calligraphic $\mathcal{D}$, but maybe the $\%$ would have just confused the readers more.

must a priori (before obtaining the data) choose α and $\mathcal{D}_{min}$. If you can control sample size, you can do an a priori power analysis and that will require you to also a priori choose either β (or the power, $1 - \beta$). However, even if you will not be able to control the ultimate sample size, a targeted power is still needed to estimate the sample size, or better yet the minimum sample size, you are targeting

The following is important because it is critical theory for understanding, and because it can be useful in some small sample instances. Because of the limitations of paper and pencil calculations, and even of rounding in computers, I really do not suggest you apply these actual equations for samples greater than $n = 60$ trials. For samples of size $n > 60$, I suggest you wait and use the approximations suggested in Section 7.10. These approximations are quite acceptable at binomial sample sizes as small as $n = 30$ trials, but even at less than $n = 61$ trials binomial tests tend to have so little power as to usually be useless.[6]

The sign test (Section 2.6) and the 50% test (Subsection 2.5.1) are ***exact tests***, meaning that you can (at least in theory) calculate the exact probability of an outcome. For these two tests, at small sample sizes it is relatively easy to make an after analysis calculation of the power of the test, if you could not control the sample size. Assume you actually had defined a reasonable minimum difference, $\mathcal{D}_{min}$, the smallest difference from p_0 that would be worthy of your notice, and a criterion, before getting the data. To find the power, $1 - \beta$, after the initial calculation, you first need to estimate β, the probability of getting a Type II error, with respect to this reasonable minimum difference, $\mathcal{D}_{min}$, given an a priori criterion, α. You also need a before the fact, or hypothetical, probability, p_0, which for these tests is always $p_0 = 1/2$, from the fact that that is what these particular tests, a 50% test or a sign test, are for. You are looking for a probability, or ratio, as extreme or more extreme than $p_A = \pm(p_0 + \mathcal{D}_{min})$. Remember that for a publishable result, only **n** and (given a lack of control of n) the resultant power, $1 - \beta$, can be different from what you planned.

In the next subsection, either **n** or n refer to the sample size, but when there are two or more samples, the **bold** letter, **n**, refers to the total size of all the included samples. In that case, n, with or without various subscripts, refers to a particular or single sample size.

3.5.1 Subsequent binomial power (n out of control)

Assume there is no way to control sample size, such as when counting the gender of drivers who become obviously confused about when to yield in a traffic circle, over a standard 40 hr work week period of observation. Males are called 'successes' and the a priori formal hypothesis pair is

  H0: $p_0 = 1/2$
  and
  HA: $p > 1/2$

where p is the true, unknown and unknowable probability, and the $cutoff$ can be α or $\alpha/2$ for one tail (**HA**:$p > 1/2$) or two tail (**HA**:$p \neq 1/2$) tests respectively. As one chooses smaller and smaller probabilities, α, the cutoff (number of successes), r_{cutoff}, gets larger.

3.5.1.1 Subsequent one tail binomial power

Again refer to Figure *2.2* (or the cover), where the rightmost distribution (the lighter bars) would be true if the effect really was exactly $\mathcal{D}_{min} = 10\%$. Then the true distribution really would be defined by $p_A = p_0 + \mathcal{D}_{min}$ (and n of course). That means that r_{cutoff} cuts off a different area (probability) that would be, under the Neyman-Pearson approach, counted as statistically significant. So, given the

---

[6]Unless p is rather distant from $p = 1/2$, the fact that approximations are valid with binomial $n = 30$ has both to do with the symmetry of large binomial samples and with a principle called the 'Central Limit Theorem' (Subsection 4.3). However, I have noticed that binomial distributions tend to be more suitable for application when $n > 50$.

population was really defined by the same n, and $p_A = p_0 + \mathcal{D}_{min}$, then the resulting β, the probability of getting a Type II error, given $\mathcal{D}_{min}$, could be found by summing the probabilities represented by the rightmost (lighter, or red) set of bars that are **below** r_{cutoff}. This is done by using a cumulative binomial equation as follows:[7]

$$\beta = \sum_{i=0}^{i=\left(r_{cutoff}-1\right)} \binom{n}{i} \left(p_A\right)^i \left(q_A\right)^{n-i}, \tag{3.1}$$

and the power is $1 - \beta$. Of course, $q_A = 1 - p_A$. Values of r_{cutoff} are available from Appendix 1, or from a number of online sources. This equation corresponds to Equation 2.6 with the appropriate values filled in. So, this equation sums all the probabilities for a binomial equation defined by n and p_A, all the way from the probability of $r = 0$ successes up to the probability of $r = r_{cutoff} - 1$, the number of successes necessary to have a probability **just falling short of the criterion**. The power would be $1 - \beta$. Notice that the summing started with the smallest possible outcome, 0 (**not** 1).

Is, say, 60% really all that far from 50%? Yes, $\mathcal{D} = 10\%$ would have been a huge difference for many US presidential races, so pollsters must use huge samples to detect these tiny differences. I constructed Figure *2.2* as an example of such a difference, with $n = 30$, and **H0**, as usual, hypothesized to be $p_0 = 50\%$ (the leftmost set of bars). The vertical line cuts off $\alpha = 5\%$ of this **H0** distribution. However, the rightmost (lighter bars) distribution is a possible alternative **HA** distribution that has probability $p_A = 60\%$ and overlaps the **H0** distribution. The vertical line is in the same place with respect to the null hypothesis, but it cuts the alternative distribution at a different place. The sum of bars that are **above** the criterion (the vertical line) on this rightmost (darker distribution) equal $1 - \beta$, the power, which is 29% (assuming the unknown and unknowable true difference, $\delta = p_A - p$).

3.5.1.2 Example: Power and Simple Binomial Tests

To demonstrate the concepts, if not best practice, let's return to my very crude (for the sake of simplicity) survey observation of the genders of fast food customers. From the very first I had my one tail hypothesis pair, in which **H0** claims that the probability was $p_0 = 1/2$, and I had selected $\alpha = 0.05$ as the criterion relative to the probability of getting a Type I error, given that **H0** is true, and therefore $p_0 = 1/2$. The usual number of customers at that time of day seemed to range from $n = 2$ to $n = 50$, unless an army reserve truck, or a swarm of Brazilian motorcyclists stopped by. Either of the latter two events would not have fitted my intent and would have aborted the experiment due to a violation of the design premises anyway. **Due to the omission** of an a priori reasonable minimum difference, I could not even do a subsequent power analysis. But assume I had had the good sense to provide an a priori minimum difference, $\mathcal{D}_{min}$.

Fast Food Madness (continued from Subsection 2.5.1.1): This is only about the power of my experiment *iff* I had really taken a proper **random** sample[8] and *iff* I had first chosen an a priori reasonable minimum difference $\mathcal{D}_{min} = 0.20$. Recall that the criterion was $\alpha = 0.05$, the initial sample size was $n = 28$, and that a binomial distribution can be completely defined by any two values, n and p. However, I did not know the sample size, n, in advance, so I had to do a subsequent power analysis. The null hypothesis, **H0**: THERE IS NO TRUE DIFFERENCE BETWEEN THE NUMBERS OF FEMALE AND MALE CUSTOMERS, corresponds to $p_0 = 0.50$,

---

[7]We will eventually get to a handier more versatile way to approximate binomial β etc.

[8]Randomization can be tricky and it is difficult to see how you could achieve randomness in a grab sample anything like Subsubsection 2.5.1.1. For now we will assume that the sample was random. But there is no way you could even discuss this sample without prefacing it with "if this sample had been random...". Even so, it is not worthy of valid research reporting. But it is handy for illustrating concepts and procedures.

versus **HA**: THERE ARE **MORE FEMALE CUSTOMERS** THAN MALE CUSTOMERS, which corresponds to any alternative probability, $p_A > 0.50$, of a success.[9] For my **subsequent** power analysis, estimating $1 - \beta$, I first need to estimate β, the probability of getting a Type II error associated with some targeted reasonable minimum difference, $\mathcal{D}_{min}$. The reporting of the power, 1-β, is now required by many editors. The customarily targeted power to avoid getting a Type II error, at the time of this writing, was $1 - \beta = 80\%$. For planning, this is commonly viewed as an acceptable minimum.

Let's assume the fast food marketing division had previously indicated to me that they would be interested in changing their advertising program **if there were a gender effect** resulting in 20% **more** female customers. This also gives an a priori justification for choosing $\mathcal{D}_{min} = 20\%$ with $p_0 = 50\%$. So, then I would have targeted $p_A = 50\% + 20\% = 70\%$, meaning that $q_A = 30\%$. Some authors would have chosen from a set of values called 'small', 'medium', or 'large', but these have not turned out to be universally applicable and should **not** be used. The **larger r is, the smaller the corresponding** p_value will be. To start with, I will have calculated r_α, or obtained it from a table such as Appendix 1. So, an adequate number of successes, r_α, is required to reject **H0**: $p_0 = 1/2$ for a particular sample size and criterion; in this case $n = 28$ and criterion $\alpha = 5\%$. In Appendix 1. I found $r_{\alpha=0.05} = 19$ when $n = 28$ and $\alpha = 5\%$. Because r_α is an integer and probabilities are continuous values, the exact probability of an r equaling or exceeding r_α may often be somewhat less than $\alpha = 0.0500$, especially with smallish sample sizes. In this case, the closest value was $p_{r,\alpha} = 0.0436$ for $r_\alpha = 19$. This unavoidable inexactness of r_α is a bit discouraging, but it also means that any declaration of statistical significance would be conservative. Being conservative gives me a little insurance against being wrong even though it also costs a little.

After obtaining $r_{\alpha=5\%} = 19$, I was able to calculate the probability, β, of not finding a p_value greater than α *iff* $p_A = 70\%$. I started by subtracting one to get the next lower integer, $r_\alpha - 1 = 18$, the largest number of successes that were **not** statistically significant. Then using my alleged sponsor's stated minimum alternative probability of interest, $p_A = 0.70$, I just needed to sum all the probabilities of all the possible successes, $i = 0$ to $i = r_\alpha - 1$, with $n = 28$, $p_A = 0.7000$, and $q_A = 1 - p_A = 0.3000$. So by using Equation 3.1,

$$\beta = \sum_{i=0}^{i=(19-1)} \binom{28}{i} (0.7000)^i (0.3000)^{28-i} = 0.31751,$$

or about 32%. So, the estimated **power**, $1 - \beta = 68\%$.

Discussion: At this sample size, I **did not** achieve the customarily minimum acceptable power, $1 - \beta = 80\%$ probability of specifically detecting the reasonable minimum difference; in this instance, $\mathcal{D}_{min} = 0.20$. The power $1 - \beta = 68\%$ probability of rejecting **H0** if $p_A = 0.70$ with a sample size of $n = 28$ was also only for a huge $\mathcal{D}_{min} = 0.20$. Techniques for large sample sizes involving approximation are presented in the Section 7.10. Large samples, hence approximate approaches, are usually preferable in real life binomial inference. Although you would **never re-analyze** the data using a new hypothesis in practice, there will be a two tail version of this example for demonstration purposes.

If, in this example, I had simply added up the probabilities in the rightmost distribution corresponding to numbers of successes, r, that are **less than** the criterion (the vertical line), I would have gotten

---

[9]Calculating β in the Fast Food Madness example does not relate to the experimental results, r, but it does relate to the sample size, n. It is about the power I had to detect the effect, $\mathcal{D}_{min} = 20\%$, between $p_0 = 50\%$ and in this instance, $p_A = p_0 + \mathcal{D}_{min} = 70\%$, given the sample size attained (in this case $n = 28$).

$\beta = 32\%$. This is the same thing that I would have been doing using Equation 3.1.

> **CAUTION:** Suddenly we have an ambiguity that flim-flam artists adore, the fact that percent more can be interpreted in many ways. If the request had been for 20% more female than male customers and we called the percent male customers M, then the percent of female customers would be $M + 20\%$, so the total would be $2M + 20\%$. But that equals the whole population, so $2M + 20\% = 100\%$ and $M = 40\%$. The request seemed to be to test for $p_A = 40\% + 20\% = 60\%$ and if so, $q_A = 40\%$. But **beware when folks throw percents around**. In this case, the advertising folks assumed that there 'should be' 50% of each gender as in the population outside the store. At least that agreed with the **H0** that $p_0 = 50\%$ and that therefore $p_A = 50\% + \mathcal{D}_{min}$.

The probabilities of the outcome **would not be** declared statistically significant, given $\mathcal{D} = 10\%$. The probabilities would add up to a $\beta = 71\%$ probability of getting a Type II error. So the power is only $100\% - \beta = 29\%$, whichever direction you count from. As your chosen $\mathcal{D}_{min}$ approaches 0, its probability of detection rapidly decreases because the medians (the middle values) of the two distributions move closer together.

Note that in this case all the critical inputs were **a priori determined**, that is, before the analysis.

3.5.1.3 Subsequent two tail binomial power

For two tails you have to use $cutoff = \alpha/2$, instead of α, for finding the cutoff number of successes; $r_{cutoff} = r_{\alpha/2}$, equal to the minimum r needed to reject **H0**: THERE IS NO TRUE DIFFERENCE. As part of the analysis for **two tails**, you designate the outcome (like heads or tails, female or male, etc.) that got the **highest number** of successes to be r in the test. That is, if the toast lands butter side up more often than butter side down, you use the number of times it landed butter up as r, not the number of times it lands butter down. The alternative distribution has the same sample size, n, but is defined by using that n **with** p_A, where p_A depends on your a priori chosen reasonable minimum difference, $\mathcal{D}_{min}$, $p_A = 1/2 + \mathcal{D}_{min}$. So, you sum all the probabilities of $r_A \leq (r_{\alpha/2} - 1)$ successes in that alternative distribution, to get β under the null hypothesis (that implies $p_0 = 1/2$). (And, yes, you could get the power, $1 - \beta$, directly if you were to somehow sum all the probabilities, $r_{\alpha/2} \leq r \leq n$.)

If you picture this as a pair of overlapping graphs like Figure **2.2**, they would look the same, except for two cutoff lines, one in each tail, with respect to your hypothesized $p_0 = 1/2$ distribution. However, whether the alternative distribution was on the right or left depends on the sampling outcome that leads the analyst to choose a particular tail, right or left, as the success direction.

3.5.1.4 Example: Sketchy Binomial Test

This is continued from Subsection:2.5.1.1, but as though I had chosen a different null hypothesis.

Fast Food Madness (continued as though two tailed): Suppose the hypothesis pair is

H0: THERE IS NO GENDER EFFECT IN THE NUMBER OF FEMALE AND MALE CUSTOMERS
This corresponds to $p_0 = 0.50$, as in the original example, but now it is paired with

HA: THE NUMBER OF **MALE AND FEMALE CUSTOMERS** IS DIFFERENT,
making it two tailed. I was still interested in 1-β, the power, after the initial calculation, with a reasonable minimum difference , $\mathcal{D}_{min}$, that had been a priori selected by the desires of my company. Finding the power meant finding its complement, β, the probability of getting a Type II error.

Calculation: I had a priori chosen to be able to detect if at least $\mathcal{D}_{min} = 20\%$ more than half of the customers were one or the other of the genders, giving a $p_A = 70\%$. The reasonable minimum difference, $\mathcal{D}_{min}$, is not an exact minimum because there is a small probability, that rapidly drops towards none, of detecting differences a little less than any targeted D_{min}. However, because this is assumed to be a minimal effect, I, as is generally accepted, calculated **as though** D_{min} were an exact cutoff, below which there was no chance involved, other than the chance associated with α, of declaring significance.

Because the a priori chosen α was split in either direction, it had to be divided between the two tails as the cutoff, $\alpha/2$. After collecting the data, I was able to use n to calculate the value of $r_{a/2}$, which is always dependent on the sample size, n, that I achieved, and on my chosen cutoff, $\alpha/2$. Recall that as r gets larger, the p_value (the value that is to be compared with $\alpha/2$) will tend to get smaller.

I had chosen a criterion, $\alpha = 0.050$, which resulted in an $\alpha/2 = 0.025\ cutoff$ probability in each tail. I found $r_{\alpha/2}$ by referring to the table in Appendix 1, choosing the $n = 28$ row and the column corresponding to 2.5% in one tail (meaning 5% total in the two tails together). A success was a member of the most numerous gender in this two tail test. With $n = 28$ samples, it would have required $r_{\alpha/2} \geq 20$ successes (at least 20 of the most numerous, whether males or females) to reject **H0** and declare statistical significance. I used $r_{\alpha/2} - 1 = 19$ and the other values given here by substitution into the formula $\beta = \sum_{i=0}^{i=r_\alpha - 1} \binom{n}{i} (p_A)^i (q_A)^{n-i}$, to find

$$\beta = \sum_{i=0}^{i=20-1} \binom{28}{i} (0.7)^i (0.3)^{28-i} = 0.47248048035317 \,.$$

Again notice that the summing began with the smallest possible outcome, 0 (**not** 1).

Discussion: $\beta \approx 42.48\%$, and the **power** is $1 - \beta \approx 57.52\%$ chance of detecting a $\mathcal{D} = 20\%$ true difference from the hypothesized $p_0 = 50\%$. The power tends to drop off rapidly as the true difference falls increasingly below the reasonable minimum difference, $\mathcal{D}_{min}$. I would have been able to detect a true difference this big a bit more than half of the time. But that isn't very good, is it? Of course the unknown true power would be higher if the true probability of a success, p, was actually higher than 70% (meaning $\mathcal{D}_{min} > 20\%\ greater\ than\ equality$). However, this reasonable minimum difference, $\mathcal{D}_{min} = 0.20$, is already a huge difference.

Note that, in this instance, it takes as many or more successes, r, to achieve the same power for a two tail test as for a one tail test. For most other tests two tails will require more successes than one tail. Also note the reduced power, as compared to the preceding one tail example.

3.5.2 A priori sample size, n, for binomial tests deferred until later

A sample size estimate to achieve a given power is called an ***a priori power analysis***. As implied, it is made before you even take a sample. If at all possible, it is essential to do an a priori power analysis. Note that this process is only estimating a practical sample size, not testing anything. Like the subsequent power analysis, it uses a p_0, based on your null hypothesis, **H0**.

For an exact method you could try out various sample sizes, n, and substitute values from a table of r_α, maybe Appendix 1, into Equation 3.1 again and again until you get a satisfactory answer. However, such an iteration approach would be tedious, and rounding difficulties would set in somewhere between $30 \leq n \leq 60$. So, a direct a priori power analysis for either the 50% test or for the sign test (or their equivalents) is most easily done by using approximation methods in Subsection 7.10.1 and Subsection 7.10.2. Such approximations work well in both theory and reality for sample sizes

$n \geq 30$. Discussion of large sample size determination will be deferred until you get to those sections.

Your ability to provide an a priori power analysis can be limited in situations such as wildlife management studies, where you cannot be certain of achieving a sample size equal to n before drawing the sample. However, you would be well advised to attempt an a priori analysis with a guess of what your minimum sample size could be to see whether it is worth trying, or if you need more resources. Whether you can do an a priori power analysis or not, you should always provide an a priori reasonable minimum difference, and a p_0 for any binomial.

3.5.3 Binomial n, small sample 80% power $\alpha = 5\%$ table

If you consider Appendix 2, you will realize that for small samples, even at the popular choices of $\alpha = 5\%$ and $1 - \beta = 80\%$, the 50% test and sign test are rather weak in the sense of only being able to detect almost uselessly large true differences.

IMPORTANT: Your colleagues are unlikely to accept a criterion $\alpha \quad > \quad 5\%$ or a power, $(1 - \beta) \quad < \quad 80\%$. These values are extremely liberal because they are very risky: 1 chance in 20 of false statistical significance, and 1 chance in 5 of not detecting the smallest difference you would deem meaningful.

Whether you are interested in one tail or two tails, if you are targeting very large true differences with the popular $\alpha = 0.05$ and $\beta = 0.20$, Appendix 2 will tell you how large the samples need to be for either the sign test or the 50% test ($p_0 = 1/2$), given that you choose a minimum true difference, $\mathcal{D}_{min} \geq 20\%$. This table might be handy for classroom exercises, as in laboratory exercises under conditions where larger sample sizes would not be possible in the time provided, and for quick checks in situations that may lead to more consequential studies. You might say, "I doubt there are huge true differences, but let's do a small sample binomial test to see whether there is any evidence of there being a **big** true difference."

In order to achieve the desired $\beta \leq 0.20$ with $D_{min} = 20\%$, targeting $p = 50\% + 20\% = 70\%$ and $q = 30\%$ in the example Sub-subsection 3.5.1.2, I could have gone to the table Appendix 2 (calculated by an exact approach) **before the experiment** (the observation) and seen that I would need random sample sizes of at least $n = 42$ for a one tail test or $n = 51$ for a two tail test. When possible, target a reasonable minimum difference, $\mathcal{D}_{min}$, that requires a sample size, n, that is no larger than you need in order to avoid hypersensitivity, but that is large enough to attain sufficient sensitivity (power). For samples larger than $n = 51$, you should consider using an approach involving normal approximation, to be covered later (see: Subsection 7.10.1).

3.6 Sign and 50% Tests: Exact But Not Exactly Strong?

The reason for so much discussion of these two binomial tests is that they illustrate the **basic principles** upon which almost all other tests are founded. Even the traditionally defined word *large*, meaning sample sizes $n \geq 30$, are quite small with respect to power. This is especially true for these lowest power tests. Occasionally these tests are unavoidable. However, they are particularly disappointing for samples less than, say, $n = 52$, partly because of reasons such as those demonstrated by the scatter plots in Figure 3.1

and in Figure 3.2, and their lack of power for anything but very large true differences.

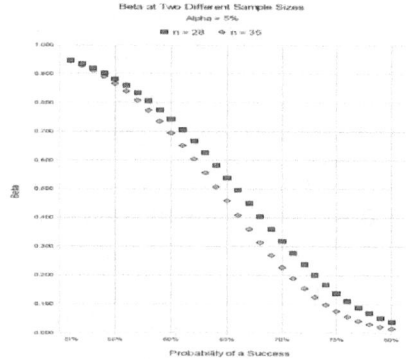

Figure 3.1: β for 50% Test or Sign Test at Two Different Sample Sizes with $\alpha = 5\%$.
The probability of not detecting the difference from $p_0 = 1/2$, given the true probability of a success (horizontal axis). The number of successes necessary to achieve statistical significance at $\alpha = 5\%$ and $p_0 = 1/2$ is $r_\alpha = 19$ at $n = 28$, and is $r_\alpha = 23$ at $n = 35$.

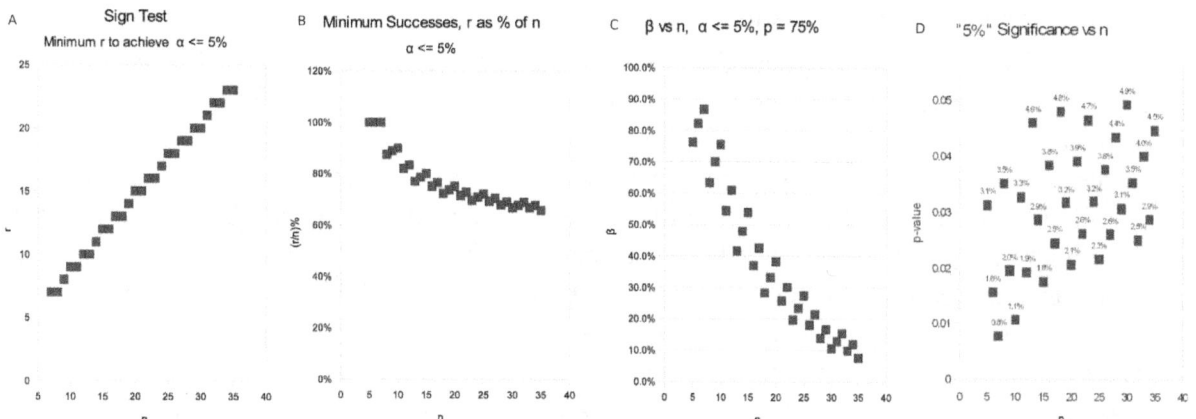

Figure 3.2: Small Sample Single Tail Sign Test or 50% $\mathcal{D}_{min} = 25\%$ Test, $\alpha \leq 5\%$:

A Contrary to intuition, slightly larger n might not always mean you need more successes, r, to declare statistical significance in a binomial test.

B Successes, r, needed to reject as a % of n also shows the above effect, but this tendency becomes less marked as n becomes larger.

C The probability of getting a Type II error, β, tends to get smaller as n gets larger, and there is less relative dispersion of β. $\mathcal{D}_{min} = 25\%$.

D The true achievable probability', the p_value' **below and nearest to** $\alpha = 5\%$ associated with the minimum value of r required for statistical significance. It tends to grow with n, and it will approach y converging to near 5%, but only for huge samples.

As you increase sample size from there, the power that is gained suffers diminishing returns and you will need huge sample sizes for tiny true differences, such as the margins commonly seen in election statistics. To some extent, all statistical tests are subject to a problem of diminishing rates of increase of power with increasing sample sizes. However, increasing sample size is, as you shall see, the most used and generally required strategy to improving the power (sensitivity, Chapter 7) of your tests and estimates.

As mentioned at the beginning of Chapter 2, measurements, unlike counts, produce continuous numbers; the more accurately you measure, the more possible measurements there are. That is, continuous measurements are like straight lines with no breaks no matter how much you magnify them. That is, you can name two continuous values such as 5.0005 and 5.0006 and if they are continuous, 5.00055 will be between them along with an uncountable number of other numbers such as 5.000551 or 5.00055101. Newton was challenged by such continuous values, because he could not exactly sum them to find the area under a curve. Oh yes, he could write $\sum_{i=0}^{i=\infty} x_i$, but where could working with that crazy thing, ∞, get him? Well, he started with chopping the line into a very large number of small pieces and then summed them. This technique is called Newtonian integration, and although not perfectly accurate, it is still used in some computer processes. But eventually this led him and Leibniz to integration, which can be **roughly** explained (in a cart before the horse way) as involving $\int x\, dx$, where the dx refers to the size of each of what approaches an infinite number of infinitesimal pieces of x that are being integrated (summed). In passing, I might point out that the rules for summation, denoted $\sum$, tend to apply for integration, denoted $\int$. Well, this is not a calculus book, and we will not be going into the details of calculus except to say that it often makes problems involving continuous data easier to work with.

Counts of two things, such as prefer Trump vs prefer Putin are very common, and combinations involving large numbers are a challenge, often a defeat, for even very powerful computers. We have ways of approximating binomial results for large samples by treating them as though continuous and distributed like a bell shaped curve (Section 5.6). However, when used for small samples, these approximation techniques merely mask the problems we now discuss.

The otherwise very useful Neyman-Pearson approach of trying to shoot down **H0** at a fixed α (such as $\alpha = 5\%$), using small sample size binomial tests, wastes power erratically and excessively. Because the binomial theorem is about discrete counts (of successes), it produces a distribution of probabilities that is likewise discrete, in the sense that each r is a separate number with a gap between it and the next number, and neither has to fall exactly on a given percent. The resulting confusion, along with the size of some of the disparities between the nearest obtainable $p\_value < \alpha$ and α itself, is demonstrated in the lower right corner of Figure 3.2. Yes, the very center of each of those little square dots is as close as you are going to come, in this case when attempting to approach as closely as possible to $\alpha = 5\%$ without exceeding it, but it may not be as large as α.

Other awkward characteristics include a stair-step effect for either the sign test or the 50% test with respect to the minimum number of successes, r_α, needed to attain statistical significance at $\alpha = 5\%$, at various smallish sample sizes, as can be seen in the top left of the scatter plots in Figure 3.2. This stair-step effect is because the probabilities of different binomial outcomes, r, for a given sample size are discrete (have gaps between them), but probability itself is continuous (has all possible values between outcomes). Additional unfortunate characteristics of these binomial tests at low sample sizes is revealed by examining the critical $(r_\alpha/n)\,\%$ necessary for statistical significance at the $\alpha \leq 5\%$ level at various sample sizes, n, as seen in the scatter diagram (upper right in Figure 3.2). However, that scatter diagram also suggests **some stabilizing** of this effect in application, when viewed from a proportional viewpoint $(\%)$, at large sample sizes somewhere well above $n = 40$. The graph on the lower left of Figure 3.2 also shows some stabilizing with respect to β, hence with power, $1 - \beta$, at larger sample sizes. Beware, this scatter diagram is for an **extravagant effect of** $\mathcal{D}_{min} = 25\%$. Figure 3.1 also shows a stepping for two sample sizes, one on either side of 'large', when β is compared with achievable p_values targeting 5%. So, stepping is unavoidable when using any but rather large samples with the binomial. These small sample difficulties are **masked, but not avoided**, by using continuous approximations (Section 5.6) of binomial tests.

Thus, if you are faced with a situation in which the data is binomial, you should consider much larger sample sizes (often samples of size $n \gg 50$, where $\gg$ means much greater than) even if you are trying to detect somewhat large true differences. Because they are interested in small true differences of a few percent from equality ($p_0 = 50\%$), political pollsters generally work with samples greater than $n = 1000$. Likewise, even though informal, a major online opinion straw vote does not report the differences (the effect estimates from the sample) until they have collected somewhat above $n = 1000$ answers.

Questions for Ch 3

Write the equations where appropriate and fill in the values. **Odd numbered questions** have answers at the back of the book and can be studied as additional examples. The data tends to be fictitious.

1. The critics of hypothesis testing point out the nonexistence of exactly zero difference. What are they (and many other users) misunderstanding about the role of zero in hypothesis testing?

2. What does the law of large numbers say?

3. Describe and contrast the p_value and α as they are commonly applied.

4. What is the reasonable minimum difference? What important probability can only be defined in terms of some reasonable minimum difference?

5. Contrast the meanings of the two words that are used for the N in NHST?

6. What causes the stair-step like effect in graphs of the binomial probabilities?

7. β is the probability of failure to detect a particular minimum difference by chance alone. So, what is $1 - \beta$ called, and what probability does it represent?

8. What is the complement of β, the probability of getting a Type II error? That is, what is it called and what does it measure?

9. What are δ, $\mathcal{D}$, and $\mathcal{D}_{min}$?

10. Name three things you must specify before obtaining your data, when testing with a hypothesis pair **H0** and **HA**.

11. What is the effect of different sizes of the true difference, δ, on the power you can achieve?

12. What would be the effect of different values of the reasonable minimum difference on required sample size to attain a particular power?

13. What is a good way to select a reasonable minimum difference that is actually reasonable?

14. What would be the value of p_0 for a binomial test challenging **H0:** THE TWO THINGS ARE EQUAL ?

15. In application, what kind of data would be suitable for a sign test, **and** which kind of data would be suitable for a 50% test?

16. What in addition to α would you have to **also report** to make your report of $1 - \beta$, the power, meaningful?

17. Can you ever accept **H0**?

18. Why must you learn about the Neyman-Pearson approach to testing which uses a rigid cutoff, α, even if you prefer and your peers would allow you to use Fisher's approach, which leaves acceptance or rejection up to the researcher?

19. When would it be appropriate to do a subsequent power analysis and when would you do an a priori power analysis?

20. Power analysis can be either of two things; what are they? What might happen if your sample size was a **lot** bigger than a valid estimate, $\hat{n}$, of what it should be?

21. What target do you always need before obtaining the data and analyzing it?

22. In addition to α, and β or $1 - \beta$, what should you choose for a binomial test before obtaining the data? Hint: this is not the sample size.

23. What input values do you need to make an after the fact (subsequent)) binomial power analysis for a 50% test or for a sign test?

24. Why does this book suggest using a notary or two witnesses to sign and date a statement to verify the date of your research plans, including such things as the planned reasonable minimum difference? Most importantly, what should you always be able to prove?

25. Why do exact binomial tests such as the sign test and the 50% test produce stairstep effects?

26. For a subsequent power analysis this book uses $\beta = \sum_{i=0}^{i=r_{cutoff}-1} \binom{n}{i} (p_A)^i (q_A)^{n-i}$ to find the probability of getting a Type II error for either a sign test or a 50% test, and then that is used to find $1 - \beta$. Why must you use $r_{cutoff} - 1$ in the summation?

4 APPROACHING NORMALITY

This chapter relates to ***parameters***, population characteristics, such as the mean, $\mu = \left(\sum x \right) / N$; and various measures of variability. Because these parameters are so important to the concept of '*normality*', which means being normally distributed, or close enough to it, this subject and the related processes are called ***parametric statistics***.

4.1 'Deviance' (Variance and Standard deviation)

ATTENTION: This section and its subsections are about a **subject that is relevant to all** the rest of the book. Knowing and understanding the definitions and display equations here **will serve you well**. You may even wish to come back to review this section and its subsections occasionally as you proceed.

When counting, 1, 2, 3, ... each number is as precise and potentially as accurate as things can get. Of course there might be questions about classification; is this plant to be counted as a sapling or as a tree? But assuming correct classification, a correct count is perfectly precise.

However, measurement is not like counting, and our measurements cannot be perfect. In measurement you can use various devices to get closer and closer, even close enough for your purpose, but never reach a perfectly exact measurement. Measurement is an activity where the rule that no two things (in this case measurements) are alike is inescapable. The outcomes of measuring all the things in the same population would all be different, even if you could measure perfectly. Measurements are ***continuous***, meaning that if you graph their possible values, there would be no break in the line regardless of how much you magnify it. Because of expansion and contraction, and swelling and shrinking due to humidity changes, you need to take a measurement at a point in time that is instantaneous. Oops, time, a measure related to the passage of events, is never really measured perfectly either. Then there is the problem of atomic orbits, is the outermost electron in the furthermost atom going to be on the outside or on the inside when you take this measurement? Similar difficulties relate to timing a runner over a $10,000$ meters course, even if you could measure the distance itself perfectly.

Even discrete processes such as counting can be tricky. Dynamic populations such as the number of rats in New Your vary from day to day and place to place (yes, New York is wonderful, but rats love it too). The number of rats is variable because it is changing through birth and death associated with food supply, weather, and chance, right while you are counting. Is that rat lying there alive (dying) or dead (finished dying)?

In this book the word ***point*** implies a set of exact coordinates, a precise location in space, an infinitely small dot. If it happens to be defined by an actual count or counts, it can easily be written exactly, but a lot of the points we will consider involve *continuous* values, measurements. It may be represented by a single number coordinate on a straight line, 1D; two number coordinates for an impossibly tiny dot on a piece of paper, 2D; an equally tiny exact location in space, 3D; a point in space-time, 4D (and even higher dimensions in economics and quantum mechanics). Even if you could measure to the last atomic orbital, or had futuristic bureau of standards measuring equipment, imagine working with the enormous numbers of figures needed to approach exactness.[1] You cannot achieve perfection, so data tends to be rounded and to have variability that, in some contexts, masks information. Even though estimates are inexact by definition, you must attempt to make a ***point estimate***, a best estimate, of such things as the variance of a population. However, attempting a point estimate of variability, such as the variance, although some might call it a 'best guess', can be very useful and informative in itself.

There is a term, ***error***, used throughout most of statistics, which refers to the **estimated differences** between things. It can be the differences between individual scores and their means, or it can even be (and often is) the differences between sample means and their grand mean. In such contexts, **it is about differences and estimates of differences**, not about mistakes. For students this is testable information because it is so important.

There are two very common measurements of variability, the amount by which things in the same population tend to differ from their mean. One is the ***variance***,

$$\sigma^2 = \frac{\sum_{i=1}^{i=N} (x_i - \mu)^2}{N}.$$

It is a **parameter**, the average of the **squares of** all the $i = 1, 2, \ldots, N$ possible differences between individual measurements, x_i, and the mean, μ. In most cases, a probable estimate, a *statistic*, is as much as you can hope to know. The Greek letter, σ, is called *sigma*. So, the symbol for the variance can be called **sigma squared**, σ^2. The estimate of the variance is the statistic, S^2, the *sample variance*. It can be calculated a number of different ways, including the popular

$$S^2 = \frac{\sum_{i=1}^{i=n} (x_i - \bar{x})^2}{n-1}, \tag{4.1}$$

where n is the size of a sample. Note that the numerator here, $\sum_{i=1}^{i=n} (x - \bar{x})^2$, is called the ***sum of squares***, ***sum of squared errors*** or ***sum of squared deviations*** and is sometimes written as SSX or SSE.

The other measure of variability, the ***standard deviation***, σ, is the **square root of the variance**. It can be seen as an average variability. Its estimate is often handier than the estimate of the variance because **it is in the original units of measure** (rather than squared measures),

$$S = \sqrt{S^2} = \sqrt{SSX/(n-1)}, \tag{4.2}$$

the square root of the estimated variance. As the variability of our data (the variance, or standard deviation) grows we need larger samples to have the same degree of confidence in our results.

In many situations the words 'variance' and 'standard deviation' are used almost interchangeably because if you know one, you know (can exactly calculate) the other. It is also true that, in context, 'variance' and 'standard deviation', which are words referring to parameters are used to refer to the estimated variance and estimated standard deviation which, of course are mere statistics (estimates). Both estimates of variation will be essential to remember as you read on.

Some say that 'the truth shall set you free'. Maybe that relates to the fact that the divisor, $n - 1$, which can be seen as representing available information, is called the ***degrees of freedom*** and abbreviated *df*'. More exactly, the number of independent pieces of information that go into the estimate of a parameter is called the ***degrees of freedom***. The denominator, unlike the usual divisor, in finding

---

[1]Irrational numbers cannot be represented by a fraction with integers in both the numerator and denominator. Our decimal system of writing numbers can be thought of as involving rational numbers with powers of ten in the denominator.

a mean, seemingly lost one piece of independent data. You have had to estimate one piece of information because you did not know one parameter, μ, the exact mean, and therefore replaced it by an estimate, the statistic, $\bar{x}$. This statistic is part of the estimate, but its calculation negates the independence of one piece of information ('uses up a degree of freedom'). So, in some calculations, you will have to **make more subtractions** when more things have to be estimated. The resulting reduction, by one, in the divisor, increases the estimate of the variance. Both the variance and its square root, the standard deviation, can be seen as **measures of uncertainty** (or 'spread').

Try Question 3 if you wish practice calculating these key values.

4.1.0.1 The standard deviation, the mean, or both?

A public health investigator might be most interested in the mean age of mothers first giving birth in Iowa as contrasted to the ones giving birth in Texas. However, there are situations in which the variability is the information that you want. For instance, a sociologist might be interested in the variability of the age of mothers first giving birth in Iowa as contrasted to Texas, rather than the means of these ages.

But you will generally be working with both together. Even if you are primarily interested in the difference between the means, you will commonly need a measure of variability to help determine the strength of evidence concerning any observed difference between them. Some whole distributions can be defined in terms of only these two parameters (the mean, and the standard deviation or variance).

4.1.1 Pooled estimates of variance, $S_{1,2}^2$

Many statistical tests (generally challenges of likeness) require an estimate of variability, and it is common to start off assuming that two things being compared, such as, say, the effect of aspirin versus buffered aspirin, before versus after, or whatever, have the same variance, σ^2. This assumption of equal dispersion seems reasonable because the experimenter tries to hold everything other than the treatment constant. It is likewise implicit in any null hypothesis analogous to **H0**: THE POPULATIONS ARE THE SAME. However, when you are comparing two sample means, chance will, wherever possible, provide you with two different estimates of a single underlying variance, σ^2. This subsection shows how to cope with that by a combined, or rather, *pooled* estimate of σ^2, using

$$S_{1,2}^2 = \frac{\sum_{i=1}^{i=n_1} \left(x_{1,i} - \bar{x}_1\right)^2 + \sum_{j=1}^{j=n_2} \left(x_{2,j} - \bar{x}_2\right)^2}{n_1 + n_2 - 2}, \tag{4.3}$$

where $\bar{x}...$ is an estimate of the mean from the sample ... and the two populations here are designated with the subscripts $_1$ and $_2$. This is called the ***pooled variance*** for two categories. It is commonly, but not exclusively, used in tests to compare means. Notice that on the way to the answer, you have to estimate two additional parameters, μ_1 and μ_2, to calculate this variance statistic. That relates to why you need to use $n_1 + n_2 - 2$ in the denominator; you have used up 2 of the degrees of freedom. For practice you may wish to try Question 17, which has an answer at the end of the book.

The pooled variance can also be found from separate estimates of σ^2, using

$$S_{1,2}^2 = \frac{(n_1 - 1) S_1^2 + (n_2 - 1) S_2^2}{n_1 + n_2 - 2}, \tag{4.4}$$

where S_1^2 and S_2^2 are the sample variances of the two data sets (samples) being compared. This is really just Equation 4.1.1 in disguise.

When using the pooled variance you **assume** (under **H0**) that the samples are from the **same population with variance** σ^2, but that the **chance variations** in sampling causes the individual samples to have their **own variances** that approach the expected variance, σ^2. You will soon be using the pooled variance to compare the estimated means, $\bar{x}_1$ and $\bar{x}_2$, of two populations. If, as is often the case, the assumption of equal true variances is true, then you get a better approximation of σ^2 by pooling S_1^2 and S_2^2 than you would get by using either alone or even by pooling all the data before attempting to estimate the variance. Avoid rounding as much as possible until the final step of calculating any S^2.

4.1.2 Pooled standard deviation estimate

Sometimes you have to estimate the standard deviation from a pooled variance, $S_{1,2}^2$, so

$$S_{1,2} = \sqrt{S_{1,2}^2}, \tag{4.5}$$

is the ***pooled sample standard deviation***, or more exactly, the ***pooled standard deviation estimate***. If possible, avoid rounding $S_{1,2}^2$ before taking the square root.

4.1.3 Rounding

The following are all to four significant figures: $1,023,000$, 10.23, 1.023, 0.01023, etc. ***Significant figures*** are a series of numerals including zeros which can contribute to the accuracy of any calculations in which they are used. However zeros before the first nonzero numeral do not count as significant figures. The inclusion of zeros after the last non zero numeral should indicate additional significant figures, but might simply be careless, and might not really be significant figures. So, the set of data figures, 0.989, 0.020, 0.140, and 0.001, are all to three figure accuracy, and 2.00 is, at least implied, to be to three figure accuracy.

While calculating a variance estimate S^2 or $S_{1,2}^2$, try to maintain as many figures in the numbers being added and subtracted as you can. You cannot subtract 1.96 from 2.0 and get a meaningful answer, however $2.00 - 1.96 = 0.04$ is at least correct even though the number of significant figures is reduced. So, if you are going to estimate σ by using the sample standard deviation, $S_{1,2}$ (or any S), avoid rounding as much as possible until after you take the square root of the variance, $S_{1,2}^2$. Bluman (2005) sets down some rules for rounding for reporting in statistics. One is to **include one more digit** than is used in the data when reporting your estimates of **means, variances**, and **standard deviations**. (In numbers like 0.0122222, zeros before first non-zero numerals do not count as digits.)

Do **not round** until you must. Although they are often inappropriate for final reporting, unrounded values are useful when there is a need to break ties or if there is to be a subtraction before taking a square root.

In traditional engineering applications, some students would regularly round during ongoing calculation to the same number of figures as the input data. This definitely will not work for calculations estimating the variance, $\widehat{\sigma^2} = S^2$. For instance, assume the data was reported to an accuracy of one digit; then if $S^2 = 0.25$, an engineering student would supposedly round to $S^2 = 0.2$. However, the similarly rounded square root of $S = \sqrt{0.2} = 0.4$, whereas the variance estimate when properly calculated would be $S = \sqrt{0.25} = 0.5$. The introduced error would be one fifth of the estimate. Never be so careless for squared values. Some estimates, that are covered in this book, involve subtraction. Suppose your raw data is accurate to two significant figures, and you have a step in the calculation where you need to subtract two sums of squared numbers, say $70.026 - 70.012 = 0.014$. If you round to even three figure accuracy before subtracting $70.0 - 70.0$, you are going to come to a miserably incorrect conclusion.

At first it may seem peculiar, but whole number results need to be appropriately rounded for reporting. If your mean was 33333333.00 and your data was accurate to only four significant figures, the correct value to report would be $\bar{x} = 33333000$ or, less ambiguously, 3.3333×10^7 in scientific notation. Incidentally, an integer, such as a counting number, is exact, and is accurate to an infinite number of decimal places. When you use true integers ($\pm$counting numbers) in your calculations, the integers do not, in themselves, limit the accuracy. Likewise, there is no rule for rounding the unending figures after the decimal place resulting from evaluating $\sqrt{2}$ or $1/\pi$. You should report your results rounded to no more than one more figure than the least number of meaningful figures that you use anywhere in the calculation of the number being reported. Otherwise, the decision is up to your good judgment. The final accuracy is dependent on the lowest accuracy of your inputs. If you get even one value in your calculation from a three decimal place table, the result, say estimated sample size, might require more rounding than if you had used a five decimal place table.

4.2 Normality at Last!

Distribution curves are mathematically ideal graphs representing, and often being treated as though equivalent to, *distributions* that are tabulations of all the frequencies of each individual outcome. If you knew N, the true number of fish in a pond, and could count the scales on each of all N of them, you could tabulate how many fish there were for each scale count and that would be their scale count distribution. However, many distributions, like coin flips, have no defined population size, and are conceptual in the sense that they are about what would happen if you repeated an experiment an infinite number of times (ie. forever).

4.2.0.1 What is normal anyway?

Suppose you put 36 coins in a big enough cup and shake them up and cast them on a table 1000 times and tally, on a list, the number of heads you observe at each throw. If you graph the results as a column graph of frequencies, and then if you draw a smooth line through the top center of each of the 37 bars of the column graph, you would, unless you experienced a rare example of the effects of chance, observe that the line forms a symmetrical smooth continuous curve, except possibly at the extremes (0 and 36). It would approximate the shape of what is called a normal curve. Figure 4.1 shows the graph of the frequencies of heads that you would observe if you could throw groups of 36 perfectly unbiased (meaning $p = 1/2$) coins an infinite number of times. Among other things, the resulting binomial distribution (Equation 2.4) in this example does not go on forever in either direction. If you were to compare this binomial curve to the smooth line, its normal approximation (described in Section 5.6), the binomial distribution, is cut off (*truncated*), in this instance at 0 and 36, but the smooth line for a normal curve would go on forever. A single throw of a group of $n = 36$ coins can only have $n + 1 = 37$ possible outcomes.[2] A common way to define an *experiment* is that it is all the replica-

---

[2]In statistics a *replication* is a random repetition (duplicate) of an experiment or experimental condition for the sake of estimating such parameters as the mean and the variance that you would find if you could repeat the experiment an infinite number of times. When doing groups of things like throwing 30 coins at a time, or repeating an experiment on several different farms, the word replication seems reasonable for each throw or farm and we can call the number of events we include in each replicate the *sample size*. However in simpler situations where we are just throwing one coin at a time, then the number of replicates is the sample size. Because ambiguity encroaches when things start becoming the slightest bit complicated, I simply refer to the sample size, the number of observations or measurements, n, whenever possible.

tions, or in the simple binomial instance, all the *trials*.[3] The experiment described in this paragraph included all of 1000 replications of sampling $n = 36$ coin tosses for a grand total of $\mathbf{n} = 36,000$.

Figure 4.1: Binomial Distribution with $n = 36\ p = 0.5$ with the Top Center of Each Column Joined Smoothly

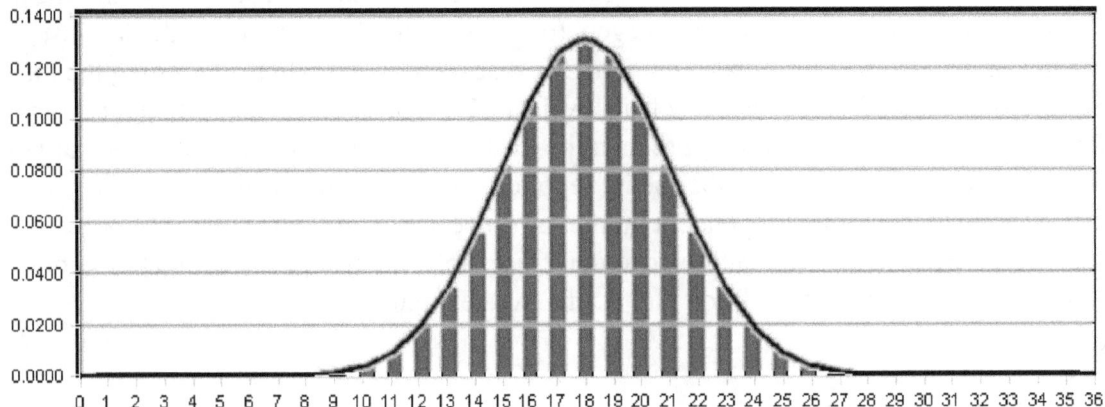

A real normal curve would go on forever in either direction because it is like a graph of the results of throwing an infinite number of ideal, $p = 1/2$, coins an infinite number of times (Figure 4.2). That sounds impossible, but it is an idealization that has many applications.

   Normality is a word describing the character of a population distribution that conforms to a mathematical ideal, 'the normal probability distribution'. It can be represented by the frequency function,

$$f_{(x)} = \frac{1}{\sigma\sqrt{2\pi}} e^{-\frac{1}{2}\left((x-u)/\sigma\right)^2},$$

Where e is an unending (*irrational*) number, $2.71828\ldots$, that like pi, turns up throughout mathematics, physics, statistics, etc[4]. This function describes the probability of the occurrence of a single point, x, in the infinite number of possible points. It can be difficult to evaluate by the techniques of higher mathematics. This is **not** an equation you need to memorize, but it does suggest, correctly, that a given normal distribution is described by the mean, μ, and the standard deviation, σ. The mean, median, and mode of the normal distribution are all equal (Subsection 1.1.4).

   A normal distribution is usually visualized as the smooth curve in Figure 4.2, a ***normal curve***. A normal curve is generally drawn with x on the horizontal axis and frequency, $f_{(x)}$, on the vertical axis. It is mound shaped, outlined by a smooth line somewhat like the smooth line in Figure 4.1, but the line goes forever in both the plus and the minus *directions* from the mean. So x goes from $-\infty$ to $+\infty$, whereas its frequency, $f_{(x)}$ is highest at the mean and approaches the straight base line, $f(x) = 0$ very closely after a relatively short distance on either side of its center, the mean. It is *symmetrical* about (the same shape on either side of) the mean, and asymptotic to $f_{(x)} = 0$.

4.2.0.2 Empirical rule

The ***empirical rule***, also known as the three-sigma rule of thumb or the 68-95-99.7 rule, provides a quick estimate of the spread of data in a normal distribution, given the mean and standard deviation (Figure 4.2). Specifically, the empirical rule states that for a normal distribution,

---

[3] Some authors inconsistently use the word 'experiment' when referring to a binomial situation such as a single Bernoulli trial (a coin flip), but this book attempts to achieve a more logical consistency. So, here an experiment is a whole process rather than just a part of a larger process.

[4] One mathematical definition of e is $e = \sum n_{n=0}^{n=\infty} \frac{1}{n!}$.

Figure 4.2: Normal Curve

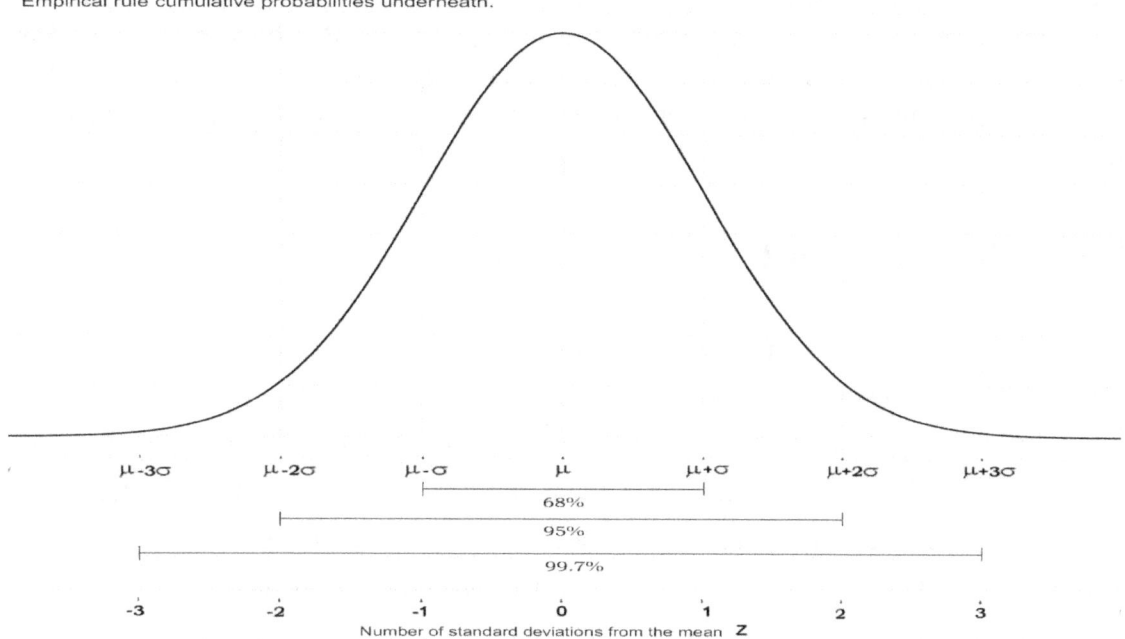

- about 68% of the data will fall within 1 standard deviation of the mean.

- about 95% of the data will fall within 2 standard deviations of the mean.

- about 99.7% of the data will fall within 3 standard deviations of the mean.

The latter, the 99.7%, is **close to** all of a normally distributed population and, in application, some people think of it ($3*\sigma$) as all of a normal distribution. The empirical rule can be used as a rough gauge of normality. When a number of data points fall outside the three standard deviation range, it could indicate a non-normal distribution.

4.2.0.3 Fame and infamy

The normal curve is famous because it is a model that works for properly applied estimation, and it is infamous for its misapplication. The practice of grading students on the normal curve is open to criticism, in part because, despite the involvement of summed scores, the grades are being summed from a population of 1 student, not a random sample of n students in the class (Herrnstein and Murray 1994). The grades of individuals tend to have different variances, each being in its own population of that individual's means. They are not the class-wide means which might meet the requirements of the CLT for that class. However, the means of large random samples of test scores of different students would tend toward normally distributed means. That is, normality can be useful, but its usefulness is usually about the distribution of the means of **random** large samples. There are also some difficulties associated with comparing samples that are drastically different sizes.

4.2.0.4 The standardized normal distribution

The total area under the normal curve, or under any probability curve is 1 (meaning all the probability, or you may say 100% of the probabilities). Statistical significance testing often focuses on probabilities equaling summed frequencies (seen as combined probabilities of ranges of outcomes), which can be represented as chosen areas of interest under their graphed curve. Values of the normal distribution are difficult to calculate (not nice to figure by hand, and not amenable to integration by calculus). So, you will want to use tables (or software with built in tables, or approximations) based on the **standard normal curve** that represents a particular normal distribution. By standardization (Subsubsection 4.9), this commonly tabulated **standard normal distribution** has mean $\mu = 0$ and standard deviation $\sigma = 1$. Appendix 4 contains one version of such a table that will be explained and used as you proceed.[5]

4.3 Enter the Central Limit Theorem

The normal distribution is about a central measure of populations of means, not necessarily about particular individuals. Even when the distribution of the individual data values is not normal, the **Central Limit Theorem** (**CLT**) often comes to the rescue. It is one of the most important things you can learn. It states: Given that a distribution has a finite (as in fixed) mean, μ, and finite standard deviation, σ, then as the sample size, n, **becomes larger**, the **distribution of the sample means** will approach a normal distribution with mean $\mu_{\bar{x}} = \mu$ and **standard error of the mean**,

$$\sigma_{\bar{x}} = \sigma/\sqrt{n},$$

even when the underlying frequency distribution is not normal (de Moivre, 1733; Laplace 1812).[6] The common phrase 'sigma sub bar x' refers to $\sigma_{\bar{x}}$ which is used in equations to signify the **standard deviation of the mean**, which is also commonly called the **standard error of the mean**.[7] The **Central Limit Theorem**, the **concept of normality**, and the **Law of Large Numbers** which says that as n gets bigger, $\bar{x}$ approaches the true mean, μ (Section 1.1.2), become particularly useful and important as a tool for human understanding of our world.

Some internet sites give the Law of Large Numbers instead of the CLT, mistaking one for the

---

[5]Common tables of the normal distribution are in terms of a standardized value Z, where one Z unit is one standard deviation (usually one standard deviation of a mean) in either direction away from the center, $Z = 0$. Most tables tabulate the range between Z and ∞ (or $-\infty$). The positive values in tables of the standard normal distribution only account for half of the symmetrical curve; the other half represents the probabilities of the negative values. That is, a range in the table, say $Z > 2$, has a probability of about 2.5%, and $Z < -2$ also has a probability of about 2.5%.

[6]It is difficult to be absolutely certain that the variance is finite. Despite a great deal of data, there are justifiable controversies as to whether economic measures relating to the stock market are finite or not. Likewise, there are controversies about whether certain long range weather data have finite variances. However, the many sorts of application shown in this book (except a few such as the charter fishing example that demonstrates a minimal concept of testing) are generally suitable for applying the CLT. For instance, the CLT might apply where the variation is about numbers of small effects creating variations around the expected value, such as the mean. However, the stock market can be influenced by the mood of the public, creating tens of millions of small effects, but a large combined industry suddenly selling their holdings or a single insurance company buying or selling heavily can have an effect that equals millions or tens of millions of small investors. Complicating such models is the fact that over short terms the variance may be affected by small effects, but the longer the measure (or the larger the sample), the more probability of large effects perturbating the variance. Except for relatively pure physics, some weather measures, complex economic indexes, and measures which could contain one large central cause one time and another another time, do not be overly fearful about these in inferential or very simple contexts. Nonparametric tests (Chapter 14) might be safer if you suspect non-finite variance, but they do not always avoid reflecting the oddities of some of the possible underlying data distributions. Where possible designing your experiments (sampling approaches) to answer clear simple questions will help to avoid nasty surprises of many kinds.

[7]Even variances which have their own *chi square distribution* (Chapter 8) can be seen as means, and in application, for sample sizes greater than or equal to $n = 25$, the chi square distribution tends to approach an acceptable approximation of the normal distribution (Section 8.9).

other. The Law of Large Numbers is **not the same** as the Central Limit Theorem (Section 4.3). Some authors combine the Law of Large Numbers and the Central Limit Theorem into one definition. This approach seems to be presented rather chaotically on page 130 of Sokal and Rohlf's 1966 edition of their classic textbook, Biometry.

So the one sample **estimate** of the standard deviation of the mean, $\sigma_{\bar{x}}$, is the *standard error of the mean,*

$$S_{\bar{x}} = \sqrt{S^2/n},\tag{4.6}$$

where n is the number of things in the sample. There are a number of ways to rearrange this equation, including

$$S_{\bar{x}} = S/\sqrt{n}.$$

So, the CLT is about the population of means of random samples. The standard error of the mean estimates a theoretical standard deviation of the means, $\sigma_{\bar{x}}$. It is dependent on sample size and is **different for each sample size**. It is not about a population parameter other than a population of means of random samples of size n from the population being sampled. Question 9 and Question 15 provide opportunities to try estimating $\sigma_{\bar{x}}$.

Most raw data populations are not normally distributed. The normal distribution generally develops as a consequence of the relation of the means of random samples from non normal distributions via the Central Limit Theorem. The normal distribution is the **limiting distribution of the means** as the sample size n becomes large. In practice, you are unlikely to be able to be certain that any values on a continuous scale other than means based on 'large' ($n \geq 30$) sample sizes will follow a normal distribution. There are measures of the amount that distributions deviate from normal, called skew and kurtosis, which have formal definitions in Subsection 15.2.1) and (Equation 15.3). For now, think of skew as meaning the graph of the distribution is lopsided, or lacks symmetry,

A Skewed Distribution

. Think of kurtosis as making a graph too sharply peaked, or too flat.

DO NOT PANIC (YET): Almost daily I encounter instances of folks using normal statistics who have made a graph of the raw data and are worrying because it is **not** normally distributed. But except for **sample means**, normally distributed data is an extremely rare item. This confusion is not surprising because many of the classical papers start off with the convenient premise that the data is normally distributed. However, in almost every case the authors would have come to almost the same conclusions if they had started off by assuming that, **with adequate sample size**, the means obtained in the calculation of these procedures would adequately approach a normally distributed population of means. That is, at least in application for large samples, these common statistical procedures depend on the Central Limit Theorem, and given reasonable planning and execution plus **sufficient sample sizes**, they almost always work because of the effect of the CLT as the sample size increases. As will be discussed in later chapters, samples of $n \geq 30$ tend to be just right to compensate for almost all kurtosis, and for most skew. Perhaps that is why it is traditional to refer to $n \geq 30$ as "large".

Detecting skew and kurtosis is difficult when using **small samples**, $n < 30$. If the sample is small, the raw data might fail with respect to skew or kurtosis; in that case, statistics based on the normal distribution are inadequate, and nonparametric tests (Chapter 14) or transformations must be considered.

However, some **discrete** data can readily approximate a normal distribution for applied testing and estimation. Data consisting of counts such as binomial data are truncated at 0 and n, meaning there are no counts below 0 or above n. So, what about the fact that the true normal distribution goes on to minus infinity and to plus infinity?[8] The vast majority of the normal distribution is quite close to its center. So, in application, the distant tails are minimal, and the means of sufficiently large samples consisting of counts, especially binomial distributed counts with probabilities, p, that are not too far from $1/2$, are usually sufficiently good approximations of the areas of interest under the normal curve to meet the need of statistical applications. Similar phenomena occur with some other discrete data such as the multinomial distribution, which is like an extension of the binomial distribution. The CLT acting on the means of random samples to approach a normal distribution may reduce the risk associated with your investment in your research, but be careful to achieve randomness and do not under-capitalize in terms of sample size, n.

Skew and kurtosis are not the only things causing data to deviate from normal. If you lump all the students taking a particular class into one data set, you might find a somewhat confounded population (*confounding* is the inseparable mixing of sources of variation). That is, each student can be seen as including hopelessly mixed components of individuality, and, in addition, one or more major factors which that student shares with the rest of the class. When I graphed the grades in my math and statistics classes, the result usually appeared as two mounds. Such distributions suggest the confounding of at least two different populations with very different means. Although female Micronesian students tended to do better than male Micronesian students in my classes, the two mounds could not be confirmed to be associated with this gender difference.

As previously stated, the law of large numbers says that if you perform the same experiment a large number of times, the average of the results obtained should be close to the expected value (the mean), and will tend to become closer as more trials are performed (Bernoulli,1713, Mlodinow 2008).[9] Researchers often have a vague idea that nature tends to form a symmetrical mound shaped distribution, the normal distribution, due to small chance effects acting to distort the larger results of a central effect that tends to be represented by the mean. That is, they assume they are looking a main effect that they assume creates the mean, around which they design their approach to random sampling. If there were no other effects, it seems reasonable that all the values would be the same, but there are usually a multitude of tiny random effects that create the variation observed in the data. The ones on each side of the main effect can add up for each direction. These small effects are assumed to average out in very large random samples, creating about the same effect in each direction, with the overall outcome somewhat resembling a normal distribution. This long accepted idea describes a process that is believed to have the potential to result in a normal distribution or something close to a normal distribution. But is this true in application? What if you have a few minor random effects that are meaningfully large, perhaps contributing, say, 5% to 20% of the observed response? How are we to know that these few not-so-tiny effects could not happen to be in one direction or the other and each have their own group of tiny modifying effects?

4.3.0.1 Normality of continuous raw data in nature?

A few human-made devices are generally assumed (but not proved) to produce something very close to normally distributed data. Some survey instruments and micrometers seem to have only one strong main effect and no distracting not-so-tiny lesser effects. Astronomers of the nineteenth century tended to believe that the chance of getting a telescope to return to the same exact orientation as the day before was normally distributed through the accumulation, in effect the summation, of a huge num-

---

[8]*Infinity*, ∞, can be described as the biggest of all numbers, but in pure mathematics infinity is said to be *undefined* because, depending on your viewpoint, $\infty + 1 = \infty$. $\infty \times 5 = \infty$ but then, $\infty - 1 = ?$ However, in R, which is a statistical language, infinity is seen as a number, 'inf", defined by $1/0 = \infty$ and $1/\infty = 0$, implying that $x/\infty = 0$ and $x/0 = \infty$, for any $x \neq 0$.

[9]Some texts mistake Law of Large Numbers for the Central Limit Theorem, which it is not (Subsection 4.3).

ber of tiny, but somehow similar, effects. I am still wondering if I have personally observed such a thing. At one point I thought the answer might be yes. In 1978, I went to Heron Island in the Great Barrier Reef and used a surveying instrument, a theodolite equipped for stadia based distance measurements, on a plane table, to sight on the locations of a reasonably large number of tagged sea cucumbers (*Holothuria* spp), to observe whether such organisms might move according to some sort of random walk model.

My wife waded around in the lagoon among the coral bommies with a calibrated staff, a specialized stadia rod, and held it by each of the many sea cucumber specimens and shouted its tag number. This repetition, day after day for much of a lunar month, was science at its thrilling best (sarcasm). Each day our measurements gave a slightly different position for each of the tagged black sea cucumbers (*Holothuria atra*). Then for a few days at the spring tide, we observed a large movement for almost all of them, but we also observed that the current of the outgoing tide had been so strong that the lagoon floor showed erosion. So, I drew graphs of the measurements on the days with less extremes of current, that is, all but the last few days of that lunar cycle that was creating the strong tides. The graph of the data for the apparent movement of each *H. atra* on such days had a distribution showing movements as close to a normal curve as I could hope to measure in the field with that technique.

This suspiciously precise curve, and, more so, the fact that the supposed movements were very small, made us unable to accept that the data was generated by organismic movement. I believe that the *H. atra* had not been moving or had rarely moved of their own volition. They had just sat there out on the sand processing the surface sand as the gentler tides changed it for them little by little, and as photosynthesis in resident algae accumulated nutrients in the surrounding and slowly shifting surface perturbation of the sand. Then the final movement after almost a lunar month had been created by currents sweeping the animals to new positions. If the apparent movement on the calm days had been larger, I might have concluded the distribution was created by a random walk. However, instead, I provisionally concluded that the approximate normal curve had almost certainly been generated by my use of the surveying instrument distances, with tiny effects due to rod placement, my coordination, eyesight, and other factors directly affecting the instrument itself. The central effect, the true movement of the organisms themselves, seems to have been almost 0. A smaller sample of tagged similar black sea cucumbers, but with pink undersides (*Holothuria edulis*), mostly in or near the coral bommies, moved quite a distance in the same area right from the first, were hard to find after just a few days and were soon lost. So, one species (*H. edulis*) moved too much for us to use the techniques I had available, and I hypothesize that the other (*H. atra*) was almost stationary except when moved by extreme tidal currents, and thus not suitable for random walk studies. Unfortunately we ran out of time and never got back to Heron Island to follow up on this hypothesis.[10]

Assuming what was, in effect, no movement, we were measuring exactly the same thing, no movement, over and over again each day. If a normal distribution existed it was almost certainly not nature, but rather the distribution of our ability to measure.[11] If I were to get out my notes from that expedition and test that data to see if it was significantly different from a normal distribution, I suspect it wouldn't be. That would prove nothing. However, R.A. Fisher stated, "I have never found

---

[10]Math note: The distance estimates involved very small angles. Accepting a little allowance for the difference, the sine of very small angles is almost linearly proportional to the angular measure to which it applies. However, measuring so many very small angles under such conditions could theoretically lead to the summing, perhaps averaging, of a multitude of small errors. But to add to the confusion, part of a sine wave might come out symmetrical and mound shaped. Thus, in this past conclusion, I am confronted with the controversial debate of the early development of normal statistics: 'Do telescopic aiming errors really sum in such a way as to be normally distributed?'

[11]Note that bivariate normal distributions which should graph as ovals are a popular theoretical model for random walks. But, perhaps due to goal oriented movements of organisms (tropisms, etc.), clear-cut bivariate distributions are rarely if ever observed for organismic movement in nature.

a normal curve to fit anything if there are enough observations. The astronomical data provided to prove that errors of observation follow normal curves are pitiably scanty and if proper tests are applied usually show that they do not! The fact is that all these descriptions by mathematical curves in no case represent 'natural laws'. They have nothing in this sense to do with 'hypothesis' or 'reverse of hypothesis'. They are merely graduation curves, mathematics constructs to describe more or less accurately what we have observed." (1935, also quoted in Inman, 1994). Likewise, Micceri (1986, 1989) has produced papers that also deny the widespread occurrence of normality. Although Fisher's comment may have been part of the numerous heated exchanges in which he was involved, I am strongly tempted to accept it as valid with respect to my tentative conclusions regarding our sea cucumber observations. Nonetheless, I would point out that the normal curve is a valuable mathematical fact, and in application, a basis for tool building. Due to the widely observable effects of the results of the Central Limit Theorem (Subsection 4.3), it is indeed something more than a "mere graduation curve".

Independent and identically distributed, written as *iid*, is a clear way of saying 'from the same distribution'. Given a large enough random iid[12] sample, the **means** from almost any distribution tend to be normally distributed as per the Central Limit Theorem (Subsection 4.3). Although this normal distribution of the means is more easily attained when the distributions are closer to normal, the normal distribution is the mathematical result of forming means.

4.3.1 Pooled standard error of the mean

If you are interested in the causes of two mounds in a distribution, and you can find a reasonable hypothesis for the existence of the two mounds, you could devise your experimental design using double blind new random data so as to verify whether two different populations exist. For a small sample, it would be advisable to use a nonparametric test (Chapter 14), but then, due to small sample size, your test might lack sufficient sensitivity to give you a fair chance of detecting a meaningful difference.

It is important to know that, for two samples, the standard error of the mean can be estimated by the ***pooled standard deviation of the mean***, (Subsection) also called the ***pooled standard error of the mean***,

$$S_{\bar{x}_1, \bar{x}_2} = S_{1,2}\sqrt{1/n_1 + 1/n_2},\tag{4.7}$$

where it is irrelevant whether you write the subscript as $\bar{x}_1, \bar{x}_2$ or, $\bar{x}_{1,2}$; they are equivalent. This calculation may also use the pooled variance under the radical sign and be represented as

$$S_{\bar{x}_1, \bar{x}_2} = \sqrt{S_{1,2}^2 \left(1/n_1 + 1/n_2\right)}.$$

4.4 Z: Test or Standardized Unit?

For creating certain useful tables we can use standard deviations of something as a unit of measure (z for transformations, or Z for tests). For testing with very large sample sizes, or some tests involving counts, Z is often used as such a measure; it is the number of **standard deviations of the mean** between one sample mean and another mean (from another sample mean or from a hypothesized mean). A mathematician would say that it is a measure of the **distance** between the means.

---

[12]Sufficiently large random samples that are iid are almost certain to have normally distributed means under the Central Limit Theorem. I have some doubts that the cumulative variations in optical instruments are, in themselves, truly iid, in that they would not necessarily all be independent of each other, and I doubt that they are all identically distributed. The question regarding such instrumentation remains open.

Then, assuming randomness, and given that the sample size is such that we can assume a normal distribution of the means, Z can be converted, usually using a table, to find the probability that your sample means were that different by chance alone. If the probability was too small to be accepted as true, it is customary to reject **H0** (as for the tests in the preceding chapters). For such a normal distribution based test, the difference can also be visualized as the area under a normal curve like

this, , where the whole area under the curve represents all the probable outcomes (100%); larger values of Z (bottom axis) cut off increasingly smaller and smaller areas of the curve (shaded). That is, as Z increases, these little areas represent smaller and smaller percents of the distribution, meaning smaller and smaller probabilities that the differences between the means are merely due to chance alone.

Some folks say that there is no specific 'Z test', but this is a technical quibble. There are a number of situations in which we use or approximate Equation 4.8 in very useful applications. So, for easier understanding, the name 'Z test' **as used in this book** refers to a category of tests using huge samples, or to normal approximations (to follow in Section 5.6), and to any tests for which the standard deviation does not have to be separately estimated.[13] (For historical reasons, a **two tail** 'Z test' is properly classified as a 'Student's t test'. But I will not follow that confusing convention here because a somewhat different two tail test described in Section 5.2 shares that name.) In application the Z test assumes, as per the CLT, that the means of very large samples are sufficiently close to normally distributed. It involves standardizing the difference between the sample means by using the **standard error** (the standard deviation **of the mean**). For huge samples it enables us to estimate the likelihood of the difference having occurred by chance alone, which can be read from relatively simple tables. It can be calculated by using

$$Z = \frac{D}{S_{\bar{x}}} \text{ or } Z = \frac{D}{S_{\bar{x}_1, \bar{x}_2}}. \tag{4.8}$$

Both the numerator and denominator are in original units, which cancel out. The first (leftmost) of the two versions, $Z = D/\left(S/\sqrt{n}\right)$, refers to a one sample situation, $D = \bar{x} - \mu_0$ (where μ_0 is often hypothesized to equal 0). The second (rightmost) version, $Z = D/\left(S\sqrt{\frac{1}{n_1} + \frac{1}{n_2}}\right)$, is for the two sample situation where $D = \bar{x}_2 - \bar{x}_1$ refers to the observed difference between two sample means.

You may also write a number of variations on these two equations,

$$Z = \frac{D}{\sqrt{S_{\bar{x}}^2}} \text{ or } Z = \frac{D}{\sqrt{S_{\bar{x}_1, \bar{x}_2}^2}}.$$

It is common to write the equation with the D expanded, often as

$$Z = \frac{\bar{x} - \mu_0}{\sqrt{S_{\bar{x}}^2}} \text{ or } Z = \frac{\bar{x}_2 - \bar{x}_1}{\sqrt{S_{\bar{x}_1, \bar{x}_2}^2}},$$

where μ_0 can even be zero ($D = \bar{x} - 0$, often simplified to $\bar{x}$). So, despite being calculated in a similar fashion, uppercase Z is about marking off cumulative areas under curves as part of a test to see if two samples are different, but it can also be used to refer to an approximation of the binomial distri-

---

[13]Sometimes I will put Z test 'in quotes', to remind you that it is not standard use.

bution called *the normal approximation* (Section 5.6).[14] Lowercase z usually refers to the result of a transformation that standardizes data values to have a mean of zero and a standard deviation of one. (It uses the standard deviation of the raw data to do this standardizing, **not** the standard error of the means.) Such a standardization is useful for looking at how measures, say, x and y, vary with respect **to each other** (if at all), but it is not suitable for testing whether x and y are from different populations. However, the same symbol, z, may also be used as a placeholder when the value of Z, or a corrected value of Z called t, is being sought (Section 5.2).[15] In all these instances, Z or z, whether capitalized or not, can be somehow related to the normal curve (Figure 4.2), in theory if not always in application.

D is an estimate of δ (delta), the real unknowable true difference, $\delta = \mu_2 - \mu_1$. δ is an ideal, so $\delta = 0$, as hypothesized under **H0**, **if there is no true difference**. In the one sample instance, the estimate can be $D = \bar{x} - \mu_0$, where μ_0 is hypothetical, or it can be $D = \bar{x}_2 - \bar{x}_1$, where $\bar{x}_1$ and $\bar{x}_2$ are means from random samples. Which is sample$_1$ and which is sample$_2$ can be somewhat arbitrary, but it can also be used to designate before and after, or treated and untreated. Also in the two sample instance, D can be assumed to **approach** 0 when the samples turn out to be from **only one population** or when the true mean **equals** the hypothesized mean, $\mu = \mu_0$. So, whether only one or two input sample means are involved, under the provisionally assumed null hypothesis, $D = 0$ and the test value Z can be seen as a standardization of your sample from the distribution of estimates of the one true mean. Then, even though the mean and standard deviation are different for different experiments, for huge samples, a single standardized Z table can, as explained below, provide you with a probability of a difference as great or greater than observed, assuming **H0** is true for that particular test. (Keep in mind that we often use a lowercase z_i to designate a standardized value that is not necessarily appropriate for an inferential test value.)

So, uppercase Z refers to a difference between means divided by the pooled standard error of the means. Because the units of measure cancel out, Z can be thought of as having no units of measurement other than itself; however, that self is in terms of standard errors of the mean. But, it is especially important to know that Z refers to a measure, the number of standard errors of the mean (or **standard deviations of the mean**) away from where $Z = 0$. Under the null hypothesis, the true mean is where $\delta = 0$. Be aware that D is on the **same scale as the data**, whereas Z is standardized with regard to the standard error of the means.

Z values are commonly used as points on the abscissa, meaning the horizontal axis of the standardized normal curve. The ordinate, meaning the vertical axis, represents the relative frequencies. The *cumulative* (meaning total) area under the standard normal curve **between two Z values** represents a **summed frequency** equal to an **estimated probability**. For the whole distribution, 100% is between $Z = -\infty$ and $Z = \infty$ with $Z = 0$ being at 50%.

Z values can either represent hypothetical means, or they can represent sample based estimated means from a distribution or distributions from which random samples were collected. Most Z values considered in this book are used to find cumulative probabilities between $+Z$ from sampling and ∞ (right or positive tail), or $-Z$ from sampling and $-\infty$ (left tail). Do not be concerned; you will not have to work with the undefined values $\pm\infty$, these are about the 'tail' areas (probabilities) cutoff from the total (=100%). The little sketch of the normal curve that follows further on shows a darkened right tail.

For a 'Z test' challenging **H0**: THERE IS NO TRUE DIFFERENCE versus **HA**: TREATMENT 2 IS DIFFERENT, you do not know the truth. So, **during the calculation stage**, you usually assume **H0**

---

[14]The normal approximation almost always uses capital Z for testing, but it could, in theory, be used to derive something else.

[15]Some folks might say that the Z test, if it exists, is that to which the t test (Section 5.2) defaults when $n \to \infty$ (becomes huge).

to be *tentatively true*, meaning until proven otherwise beyond a reasonable doubt. This calculation is also made on the assumption that the two means (one of which could be $\mu_0 = 0$) come from the same, or at least from identically shaped distributions. That is, **you assume the variances are the same**, **regardless** of whether you ultimately observe a large enough difference to justify rejecting H0 or not.

Recall that we are testing to see if the null hypothesis is too unlikely (p too small) to be accepted as true. The corresponding Z would have to be **large enough** that the area cut off in the corresponding tail is too small to be accepted as likely.

First consider just one tail at a time, as do many statistical tests. Such tests use the same **H0**, but they word the alternative hypothesis a little differently, such as **HA**: THE DIFFERENCE BETWEEN THE MEANS IS GREATER THAN ZERO. Notice that the GREATER THAN refers to the right tail; an alternative **HA** could say LESS THAN, which would refer to the left tail. Due to the CLT, you can accept that if the true difference is zero ($\delta = \mu_2 - \mu_1 = 0$), then the probability of getting a result that is Z_A standard errors of the mean away from the middle ($Z = 0$) can be found from the corresponding area under a normal curve, if the **sample size is very large**. That is, corresponding areas under the normal curve represent cumulative probabilities when the whole of the normal curve is taken to represent 1 (100%).

Directly estimating areas under such a continuous curve is not handy. So, tables of areas under the normal curve such as Appendix 4 (or computer approximations) are made widely available. They usually report a cumulative probability such as is represented by a darkened area under this normal

curve sketch, . Indeed tables should include such curves to show you what each table is about. This common darkened and widely tabulated area is called the right tail, and for normal distributions it extends to the right all the way to $+\infty$. In this case the tabulated Z values would be growing larger as the probability (dark area) grows smaller. That is, as we move further from the true mean (the center), the area beyond becomes smaller, and theoretically the probability of that happening by chance alone becomes smaller. In some tests you will be interested in whether the difference is in the negative direction. A negative value of Z can be read from the table to find the same probability as though it were positive because the normal curve is perfectly symmetrical. That is, for cumulative probabilities you approach the table with the absolute value of Z.

> **CAUTION:** some tables and calculators give you the white area rather than the dark area. When confronted by a new table or computer calculator of Z, you simply have to be careful and examine a few values to see what is happening to the cumulative probabilities as the Z values grow larger.

Z can sometimes be more comfortably viewed as a distance (sometimes signed) from $Z = 0$, the center, the mean, of the distribution tentatively assumed under **H0**. All this will become clearer as you progress through this book.

Lots of authors, the author of this book included, often find it convenient to write the generalized 'Z test' as

$$Z = \frac{\bar{x}_2 - \bar{x}_1}{S/\sqrt{n}}.$$

4.4.0.1 Example: Z test for Two Samples

Check your bulbs Suppose I was trained in the old school of statistics where we rarely considered power and I am comparing the lifespan of two kinds of light bulbs. I happen to have the resources to test a rather large sample size. So, I feel justified in using a 'Z test'. Recall that Z is the number of standard deviations of the mean between one sample mean and another mean. My criterion is $\alpha = 1\%$ and my hypothesis pair is:

H0: THEY HAVE THE SAME LIFESPAN

HA: BRAND 2 HAS A GREATER LIFESPAN

Procedure: I then draw random samples of $n_1 = 600$ Brand 1 light bulbs and $n_2 = 600$ Brand 2 light bulbs. After all, that seems handy. Brand 1 produced a sample mean lifespan of $\bar{x}_1 = 1206$ hours and a sample variance of $S_1^2 = 1651$ and Brand 2 had a sample mean lifespan of $\bar{x}_2 = 1250$ hours and a sample variance of $S_2^2 = 1598$. Notice that variances are in terms of units square, but authors often neglect to write out such things as 'hours squared'. I have the variances, so I estimate the pooled variance by using Equation 4.4,

$$ S_{1,2}^2 = \frac{(n_1 - 1)\,S_1^2 + (n_2 - 1)\,S_2^2}{n_1 + n_2 - 2}\,, $$

and with $600 + 600 - 2 = 1198\ df$, this gives

$$ S_{1,2}^2 = \frac{(599 \times 1598) + (599 \times 1651)}{600 + 600 - 2} $$

$$ = 1624.5\,. $$

However, I consider the standard deviation easier to work with, so I take the square root of the variance to estimate it, $S_{1,2} = \sqrt{S_{1,2}^2} = 40.3051$ hours. Now for the actual test. I use as many decimals as my old comptometer provides and compare the sample means by using Equation 4.8,

$$ Z = \frac{\bar{x}_2 - \bar{x}_1}{S_{\bar{x}}} = \frac{\bar{x}_2 - \bar{x}_1}{S_{1,2}\sqrt{1/n_1 + 1/n_2}} $$

$$ = \frac{1250 - 1206}{40.3051\sqrt{1/600 + 1/600}} = \frac{44}{2.32702} = 18.9083\,. $$

I go to a table of Z and find that the value of Z necessary to reject **H0** at the 1% level for one tail is $Z_{\alpha=0.01} = 2.3263$. Because my **result is larger** than that, I reject **H0** and accept **HA**. A popular way to say this is 'I found the difference to be **statistically significant**' (statistically too unlikely to be true).

Discussion: Of course, as we are now becoming aware these days, I should have first sought a standard deviation estimate for light bulbs through industry journals, or even by contacting manufacturers and then considered sample sizes.

There is another question, to be covered later (even the term *Z test* is not perfectly correct), as to whether this particular table should be used for this sample size. Ideally I would use a slightly different approach, but this illustrates the basic concept of comparing differences between means.

It may sometimes be useful to recall, from the common way to add fractions with different denominators, that

$$ 1/n_1 + 1/n_2 = \frac{n_1 + n_2}{n_1 n_2}\,. $$

A rule you may have learned verbally in school is to multiply the top and bottom of each of the two fractions by the denominator (the bottom) of the other to *create a common denominator* (convert them into the same units) so they can be added.

4.4.0.2 Nothing in one of your samples?

So far, two sample tests have been discussed, but **one sample tests** in which the sample is compared with a hypothesized value, μ_0, are also important. The hypothesized value can have any value, including 0. So, for one sample tests you use $D = \bar{x} - \mu_0$. So,

$$Z = \frac{\bar{x} - \mu_0}{S_{\bar{x}}},$$

and when $\mu_0 = 0$,

$$Z = \frac{\bar{x} - 0}{S_{\bar{x}}} = \frac{\bar{x}}{S_{\bar{x}}}.$$

Question 11 provides an example for you to try this calculation yourself.

Variations on this theme occur over and over again. Students should be aware that the stats in this short chapter is likely to somehow bite you on the bottom at the final. Notation changes to fit the situation, but it will still make sense if you make an effort to understand its initial presentation.. Such basics and elementary algebra can carry you through this book, and if ever needed, through a world of statistics.[16]

4.5 Effect Size, ES

If you decide two things are different, as when you declare statistical significance, it is only reasonable to report how much different. The ***effect size***, symbolized ES, is the observed difference between two samples, or between a sample and a hypothesized value. Because the effect size is only found by sampling, **it only <u>estimates</u> the true difference**, such as the difference, δ (delta), between two means, which equals $\mu_2 - \mu_1$. It is almost never perfectly accurate because you cannot observe, let alone manipulate, nature without including additional effects due to your technique. Although ES is usually an observed difference with respect to subtraction, such as $\bar{x}_2 - \bar{x}_1$ (the difference between sample means), it is sometimes an observed ratio, such as $variability_2/variability_1$. Like the common English word 'difference', its meaning depends on context, the context being what you are trying to evaluate. The ideal effect size is expressed in the original units of measure.

When you choose to target a reasonable minimum difference, D_{min}, you are targeting a particular true difference, δ, if it exists. However, despite what some folks confusingly write, the reasonable minimum difference, D_{min}, is not an effect size, ES. If your experiment is sufficiently well designed and there is a **moderate** true difference, then as you **increase your sample size**, the ES will tend to **approach a true difference**. The ES only tends to approach the true difference; it rarely if ever actually reaches the exact difference.

---

[16]Of course, some calculus might expand what you can do with statistical concepts, but simple algebra is particularly useful for understanding most statistical processes and it is almost all you need here. However, to continue much further than this book, linear algebra would be a big help.

Unless specifically stated, 'effect size' as used in this book does not refer to standardized effect sizes mentioned in Section 5.9.

4.5.0.1 Misuse of effect size

John M. Hoening and Dennis M. Henisey (2001) describe and make a valid case against the alarmingly common misuse of 'effect sizes' and power estimates to erroneously attribute some validity to **H0** because **HA** was not proved due to lack of evidence supporting it. Although they seem to have been unaware of the extreme value of effect size estimates and power determinations (estimating the sensitivity of your experiments) when properly applied, their paper is well worth reading.

4.6 Difference Between Means, δ

For lack of a standard notation with respect to **means**, this text symbolizes the **sample difference between** the **means** as D, and the **reasonable minimum difference** with respect to the means as D_{min} (pronounced *d-min*)[17]. The reasonable minimum difference, D_{min}, is that point value that **you choose** to target as being the reasonable minimum difference between means, even though you will never know the true $\delta = \mu_2 - \mu_1$. So when testing, the estimate of δ is

$$D = \bar{x}_2 - \bar{x}_1 \text{ for one tail, or for two tails, } D = |\bar{x}_2 - \bar{x}_1| ,$$

the difference between the sample means, where $|\ |$ means *the absolute value of* (treat as though positive even if not positive). For comparing two means, the notation could be $\bar{x}_2 - \bar{x}_1$, but for comparing a hypothesized mean under H0, the notation could be $\bar{x}_1 - \bar{x}_0$. You can determine which tail you are examining by selecting which mean you call μ_1 and which you call μ_2. As you shall see, choosing a D_{min} will not only facilitate estimates and inferences, but will also enable you to attain power. Note that D_{min} (no standard notation available) is a fixed target, the reasonable minimum difference, not an estimate.

4.6.1 Crudely confident (a preview of confidence intervals)

Suppose someone was explaining their latest scheme to their spouse and they asserted that they were 95% confident that they would make an average of between fourteen million and sixteen million dollars next year and each year thereafter. Such an estimate of the **range** of between $14,000,000$ and $16,000,000$ when presented along with an estimate of its probability, is a ***confidence interval***, abbreviated ***CI*** (Neyman 1937).

Recall the law of probability which says the sum of the probability of something happening and of it not happening is always one (100%). This **confidence interval** of between $14,000,000$ and $16,000,000$ is asserted to have a $1 - \alpha = 0.95$ probability of **including** the true statistic, in this case, the true mean, μ. The probability, $1 - \alpha = 0.95$, can be called the ***confidence*** and it is numerically the complement of α; that is, $(\alpha + (1 - \alpha) = 1)$.

However, α can be used in different contexts. In testing, α refers to an exact probability point that marks the upper probability limit of what is too unlikely to be true, but with respect to CIs, α refers to the whole summed probability of not including the sought after result. This estimated confidence interval implies that there is only an $\alpha = 0.05$ probability that it does not include the estimated result of implementing the scheme.

Notice that a confidence interval has a **lower bound** and an **upper bound** ($14 million and $16 million respectively) as well as the alleged probability (95% in the above instance), the confidence,

---

[17]Some authors use $\bar{x}_d$ and μ_d rather than using D and δ.

that it **includes** the true parameter. We can also create confidence intervals that are intended to include variances or effect sizes. However, confidence intervals for the mean are the most common, which is what is meant by saying "the confidence interval," unless specified otherwise. The confidence interval is such a useful tool that a whole chapter is dedicated to it (Chapter 6).

4.7 Ongoing History

4.7.1 R.A. Fisher's approach, focus on p_value

R.A. Fisher's approach is the oldest of the two approaches (inferential testing models) in this book. Fisher (1935, 1953) devised his model with the intent of providing evidence against a hypothesis, **H0**, about the actual population in nature. It assumes that sometimes the researcher can not reject **H0** because the risk of being wrong is too large. There is no alternative hypothesis **HA** in Fisher's model. Fisher's approach potentially provides the researcher the freedom to choose to *reject* **H0** on the basis of the researcher's assessment of a probability, called the **p_value**, as a measure of the strength of evidence (the smaller the p_value the stronger the evidence against **H0**). In this sense, the p_value can be seen as a part of a collection of evidence. Under the assumption (as in **H0**) of no true difference or no effect, the p_value can be defined as the estimated probability of having obtained a result equal to what was actually observed just by chance. (Fisher's model is a little like the Bayesian model in that it can be seen to be intended to adjust the researcher's degree of belief.)

According to Fisher, the p_value is estimated from an experiment designed so as to be as representative of nature as is humanly possible. It is an **estimate of the probability** of the observed outcome having occurred by **chance alone**, given that **H0** is true. This approach is entirely about rejecting **H0**, if, in the researcher's opinion, it is too unlikely to be true. Although Fisher originated the concept of a cutoff α (including the popular $\alpha = 5\%$) he did **not provide for accepting an alternative hypothesis** and later turned against strict cutoffs except in applications like quality control.

The lack of a rigid preordained criterion, α, is especially attractive to researchers in fields where the data is expensive and where uncertainty often tempts the researcher to be subjective. Fisher's approach, in itself, cannot provide you with any control over the size of the true difference you are targeting. It is best to consider Fisher's approach and the Neyman-Pearson approach as separate and distinct to avoid the pitfalls of inappropriately mixing them. However, it does not preclude reporting some sort of *interval estimate* such as a boxplot (Subsection 15.1.1), or even a Neyman-Pearson **based** (Subsection 4.7.2) confidence interval (Chapter 6).

Fisher's approach does not, in itself, define the degree of the true difference necessary to reject **H0** at any particular p_value, and it can not provide a method of estimating a suitable sample size for doing so. But that does not necessarily mean that you cannot choose a sample size that targets a particular true difference by a **completely separate** a priori estimate that resorts to a method dependent on the Neyman-Pearson approach (Subsection 4.7.2, below next).

4.7.2 The Neyman-Pearson approach, providing an alternative hypothesis

Modern statisticians may see Fisher's approach as preliminary, leading to an evolutionary emergence of the Neyman-Pearson approach, which, although less flexible, is less likely to result in a Type I error. That is, in application, the Neyman-Pearson not only demands that the result be unlikely, but also that it be unlikely to have been equal to or beyond a cutoff predetermined by your a priori chosen criterion, α. The *Neyman-Pearson approach* (1933) is a method (an inferential model) for choosing between hypotheses, called *a hypothesis test*. The underlying concept visualizes a population of experiments that would result if you exactly repeated the one experiment you actually performed again and again. This conforms with the approach used by Student (1908, Section 5.2). Under it, the researcher can only reject **H0** based on the data and **a rigid preselected a priori criterion**, α. In addition to a priori selecting α, the researcher must a priori (before the fact) specify an alternative hypothesis, **HA**, to be accepted if **H0** is rejected. A single test of a sample of n things is considered to be representative of the whole population of potential tests.[18] Declaring *statistical significance* means rejecting **H0** and is dependent on comparing the p_value to some a priori chosen cutoff, α, or to $\alpha/2$. You can **only** reject **H0** if the p_value is less than or equal to the $cutoff$. Common $cutoff$s are (or represent) the criterion, α itself, or half of it, $\alpha/2$, if the criterion is divided between two tails (greater than and less than).

Under this approach, no matter **how tiny the amount** by which the p_value is more than the $cutoff$, it is necessary to rigidly adhere to the $cutoff$. You, the researcher, have a priori chosen your α, and you must live with it. This approach is only concerned with a yes-no decision about whether the p_value is above, versus the same or less than the $cutoff$. This rigid restriction is seen as essential, in part due to the fact that the p_value is a cumulative probability, implying an inexactness (an inseparable variability) concerning its true location within the variation about that cumulative probability (also see Subsection 4.7.3). Another advantage of using a rigid criterion, α, is that it strengthens the evidence that the rejection of **H0** was not merely a coincidence.

This, to some, seemingly outrageous restriction, the requirement that you set an a priori maximum probability criterion, α, is essential to enable you to estimate sample sizes and to target reasonable minimum differences with a power, $1 - \beta$, that is a priori selected by you. Modern approaches drawing upon the Neyman-Pearson approach can be adapted to avoid both hyper-sensitivity and hypo-sensitivity, and they allow you to make estimates of research costs as is required by granting agencies, as well as by good sense. The use of exact yes-no cutoffs also eliminates confounding associated with the fact that the p_value is itself an estimate and has its own tendency to be variable. Subjective interpretation of a p_value cannot be used for anything other than evidence partially supporting the rejection of **H0**. It cannot be used with **HA**, and Fisher's approach cannot be chosen as a post hoc approach. Unfortunately, many textbooks now in print still appear to allow this, but **most peer reviewers and journal editors reject it**. Please remember that **Fisher's approach and the Neyman-Pearson approach cannot be mixed**, especially not post hoc (Goodman 1999), even though that is still mistakenly common at the time of this writing.

4.7.3 That confounded p_value?

A very good definition of p_value is provided by Wikipedia citing Aschwanden (2015), and Wasserstein and Lazar (2016), that "In statistical hypothesis testing, the p-value or probability value is the probability of obtaining test results at least as extreme as the results actually observed during the test, assuming that the null hypothesis, **H0**, is correct.".

A small p_value can result from a very large sample and a tiny true difference, or it may result from a small sample and a very large true difference (Lewicki and Hill 2007). The results of estimates such as estimates from samples are variable. Variability is the challenger here; statistics attempts to sort out the truth despite it. **Confounding** occurs when the individual true sources of variability, particularly of variance, of one or more factors cannot be separated from each other.

---

[18]Referring to Pearson's 1941 work, Harry F. Inman (1994) states that random sampling provided Pearson the chance framework for computing an "objective" probability with a corresponding frequency interpretation. This probability, in turn, approximated the degree of belief appropriate for outcomes of future observations like those in the sample. In effect, the observed sample data were the sequence of past sense impressions on which predictions of the future were to be based, and the observed objective probability was taken "as the basis of belief as to the future".

Personality characteristics, such as a tendency to get into fights, tend to be both a result of nature and nurture, and they can be seen as confounded in such a way that a judge has no hope of precisely separating them when considering the sentencing of an offender. Thus, monsters who should not be on the streets often get minimal parole and people who are eager to turn their lives around and who have potential to do so, get maximum incarcerations for the same crime.

The p_value is a single estimate of probability on a continuous scale, rather than being something capable of being separated into variances. It is often called confounded in the sense that it combines both the strength of effect and the precision of the estimate in an inseparable manner. Reporting the effect size and the confidence interval provides the user with separate outcomes arising from these two factors (Cumming 2012). But is it truly confounded if we can at least separate the underlying characteristics by using an effect size and a confidence interval? In any case, to do a dichotomous test, it is necessary to find a p_value as a step in the analysis.

If you reject **H0**, based on a p_value without the inflexible constraint of a criterion, you are using **Fisher's approach, under which any** p_value **can be used as evidence regarding whether you can reject a hypothesis**, **H0**. It assumes that if the probability of **H0** is sufficiently small, in your opinion, then that is **evidence against H0**. There will be no alternative hypothesis, **HA**; any decision to reject **H0** will be made by the researcher using this evidence along with any other available evidence. This approach leaves a lot up to you, the researcher, and after all, it is your project, and you might even be the only expert in the exact niche you fill in your field. That is, it is your project unless the sources of income, the journal editors, and the tenure committee feel that you have allowed yourself too much opportunity for bias (whether conscious or unconscious). Note that you cannot change horses in midstream; the decision to use Fisher's approach is only valid if it is made **in advance** of obtaining the data (a priori).

Neyman and Pearson designed their approach so as to automate the decision about whether your experimental outcome is **so improbable as to be declared false**, in the sense of not being a member of a hypothesized population of similar experiments for which **H0** is true. As already mentioned, the rejection of **H0** under this model is called *statistical significance*.

Using this approach, the p_value is subject to the sample size and you will tend to get a smaller p_value with larger sample sizes, and a larger p_value with smaller sample sizes. Even when using a cutoff that is based on a criterion, α, very large sample sizes can lead to hypersensitivity, getting statistical significance for no more reason than that different things are different even if the difference is so tiny as to be meaningless in application. On the other hand, small sample sizes may miss truly meaningful differences. So, the **proper management of sample size avoids these pitfalls** to the degree that chance will allow us to avoid them. However, appropriate sample size management will require the use of the Neyman-Pearson approach, at least for initially getting an appropriate sample size estimate. Under the Neyman-Pearson approach, it is definitely incorrect to start off with an $\alpha = 0.05$ and then say that your result was "stronger" because you got a $p\_value = 0.0001$. The p_value, being a confounded value, has a variance of its own and thus cannot be accepted as a fixed point.

It used to be popular to initially state a hypothesis pair, **H0** and **HA**, an $\alpha = 5\%$ and then post hoc use one * for statistically significant at 5% and ** for a calculated p_value that would equal statistical significance at 1%, and *** for 0.1%. This is part and parcel with the hybridization of Fisher's approach with the Neyman-Pearson approach and most editors no longer accept this nonsense. However, some computer packages, including R 3.0.2, still include functions which produce asterisks in the listed results to provide you with the relevant tabular criterion values. Hopefully, these are tolerated because a machine cannot know what your a priori criterion was; do not transcribe such to your reports. That is, if you started with $\alpha = 0.05$ and get a $p\_value = 0.003$, you cannot opt to include ** in your reported work.

Because the p_value cannot be used in a manner suitable for assessing either **HA** or power (sensitivity), using a fixed point cutoff is a minimally adequate decision making strategy for controlling the number of Type I errors over the long run. **Without** a strict a priori criterion, α, it is completely **outside any valid approach to combine an alternative hypothesis**, **HA**, with the rejection of **H0**, based on a post hoc p_value.

Your objective is to maximize the likelihood that you are not just observing a meaningless coincidence. To report an outcome you did not originally clearly define is post hoc, and thus invalid. So, the Neyman-Pearson approach requires you to specify an inflexible a priori criterion, α, as a fixed and non-confounded decision point. Setting out to only reject if the p_value falls below a criterion, α, that you have **chosen in advance** (not after) minimizes subjectivity and potential bias.[19]

4.7.4 These approaches to decision making cannot be mixed; so which is 'best'?

Many researchers moan about 'the tyranny of α', because they are dismayed at not being allowed the same free interpretation of the p_value under the Neyman-Pearson approach as was allowed under Fisher's approach. These same people have also included some of the leaders in the great strides that have been made in determining sample size for controlling power. Ironically, these improvements are dependent on the Neyman-Pearson approach. This is the present state of frustration about the application of inferential statistics to research. It comes from the many decades of teaching an improperly mixed approach in which the researchers exploited some of the advantages of the Neyman-Pearson approach (although neglecting others), and then threw in a totally incorrect post hoc interpretation of the p_value. But the p_value under the Neyman-Pearson approach is **nothing more than a computational step**. A difficulty subtly underlying the controversy, one which emerges throughout this book, is that larger sample sizes, and thus more hard work and investment, is needed in order for the researchers to obtain what they have been deceived into thinking they were already getting. However, among the several other difficulties which haunt the subject is the acrimony, much of it ugly and personal, between Karl Pearson and R. Fisher (Inman 1949).

Under the Neyman-Pearson approach the phrase '**null hypothesis**' refers to a hypothesis that implies that the reasonable minimum difference, D_{min}, that has been, a priori, stated by you, has **not been equaled or exceeded**, so your test challenges that hypothesis. Even when you clearly target a reasonable minimum difference, true differences on either side of the reasonable minimum difference are sometimes detected.

Recall that a larger absolute true difference (e.g. $|\mu_2 - \mu_1|$) favors a smaller p_value, and vice versa (see Table 4.1). That is, the further the true difference is above the reasonable minimum difference, the more likely the p_value is to be less than the pre-chosen $cutoff$ (statistical significance), **and** the further the true difference is below the reasonable minimum difference, the less likely a p_value is to be less than the given $cutoff$ (not statistical significance).

> **NOTE:** If you do not find statistical significance, consider at least publishing a paper regarding the distribution, standard deviation, and other observed characteristics of the population from which you sampled. It would be great if this would become common practice because it would facilitate meta studies and also be of great assistance to future researchers attempting to estimate sample sizes.

---

[19]The a priori declaration of a probability, α, above which **H0** cannot be rejected minimizes the chance of rejecting when the evidence is inadequate to validate the rejection. It leaves us no second chance to change bets after the race. Some lecturers would go so far as to say that the purpose of statistical testing is to avoid [even unconscious] bias.

4.7.4.1 Only for quality control?

Opponents of the Neyman-Pearson approach (primarily R.A. Fisher) have declared that its application should be restricted to quality control, or similar applications. Supporters of the Neyman-Pearson approach might claim that it is objective and that Fisher's approach of allowing the researcher to evaluate the meaning of the p_value is too subjective.

4.8 Summarizing this Foundation Material

The case for the Neyman-Pearson approach to statistical testing is rather strong, so let's consider Table 4.1, which summarizes it:

Table 4.1: Neyman-Pearson Testing Process

START by tentatively assuming H0 to be true		Unknown truth (whether RMD, D_{min}, exceeded)	
		H0 false	**H0** true
	$p\_value \le \alpha$ Reject **H0**	True **HA** accepted as true Statistical Significance	False **HA** accepted as true Type I error
	$p\_value > \alpha$ Don't Reject **H0**	**H0** returns to **unknown** Type II error	**H0** returns to **unknown** No evidence against **H0**

A Neyman-Pearson test challenges a null hypothesis **H0** that there is no difference between two populations of values, or between a hypothetical population and a real population, or even that we are really observing the same population. However, in reality it can only provide statistical evidence of a difference, called the 'reasonable minimum difference' (RMD) and when comparing populations it can never quite reach zero. The starting assumption that **H0** is true is purely for calculational purposes such as providing a zero difference reference point. If the test fails to reject H0, it gives no evidence of anything, especially **not** evidence supporting **H0**.

4.8.0.1 Have a plan, and sample size matters

As you shall become increasingly aware as we proceed, the selection of an **appropriate sample size** is your handiest tool for controlling the sensitivity of your research and the acceptability of its outcome. You are, by your choice of sample size and the intrinsic sensitivity of the test, targeting a reasonable minimum difference at which your test will reach a particular power. Many journals now require you to report a power whenever you report statistical significance (Section 7.3). In order for you to report a power, the α used to evaluate or obtain that final power would have to be the same as used for the test. An interesting lack in some of the instructions to authors is that some journals fail to require a reasonable minimum difference along with the power, $1 - \beta$ (or its complement, β). This is a peculiar omission because **neither β nor $1 - \beta$, the power, is meaningfully defined without a reasonable minimum difference**, as well as α. It is likely that that omission will soon change.

However, accepting **H0**: THERE IS NO DIFFERENCE as true, as opposed to simply 'not discredited' **cannot** be justified (Gigerenzer, et al. 1989). You are only challenging **H0**. Even the failure of a very powerful experiment to detect a difference from **H0** would not justify reporting that it was true, that there was no difference. Such a report would not be the sort of thing a responsible researcher would ever release to the media without further tests that challenged the consequences of **H0**. The result only means that there is no evidence of a difference, or better yet that **there is no evidence of the targeted reasonable minimum difference**.

There is an irony associated with scrapping the Neyman-Pearson approach for making decisions in favor of Fisher's approach; you can only usefully and validly rely on Fisher's approach to decision making in lieu of the Neyman-Pearson use of an α criterion for decisions by relying on the Neyman-Pearson model, including a tentative α to estimate appropriate sample sizes, and where appropriate, to both control and evaluate power. Even if you scrap statistical significance testing based on an α criterion, you will still have to understand the ideas behind the Neyman-Pearson approach in order to use the spin-off techniques resulting from it. But let's assume your editor says that it is OK to simply report a p_value, or a confidence interval (a sort of formal probability based range estimate, Chapter 6). The latter, a confidence interval, is not particularly open to bias and is especially informative. There are some excellent arguments for scrapping the use of α as a decision point and leaving more of the process of making decisions up to the reader. Scrapping α as a decision maker may seem especially reasonable in the presence of a confidence interval such as "I am $1 - \alpha = 95\%$ confident that the mean weight of the adults of these fish is between 4 grams and 5 grams". However, estimating a reasonable sample size to have a confidence interval which is unlikely to be ridiculous requires using the Neyman-Pearson approach, which includes choosing an at least tentative α, as well as a β and a reasonable minimum difference, D_{min}. You will need the same general Neyman-Pearson approach in order to estimate a sample size needed to have the power for a reasonable p_value.

If your peers are going to allow you to validly avoid using α for your actual decision you are going to be jumping from Neyman-Pearson based estimations for sample size and power to reporting a p_value. That is, you are going to give up the option of declaring statistical significance and of accepting a formal alternative hypothesis, **HA**. **If you are allowed to use this Fisher's approach**, simply follow the explanations and samples in the following chapters, but stop with the p_value, and report it instead of comparing it to α. You must make it clear that you were not testing for statistical significance if you present a p_value and then a confidence interval. Otherwise you might be seen as fudging to combine Fisher's approach and the Neyman-Pearson approach (boxplots as in Section 15.1.1 giving an idea of the distribution of the differences might be OK). You must a priori choose which approach (Neyman-Pearson's or Fisher's) you intend to use for the final decision, **before you obtain or inspect the data**.[20]

You may a priori obtain sample sizes using methods based on the Neyman-Pearson approach, but if you then run a test using Fisher's approach, that must be all Fisher's without Neyman-Pearson contamination. The Neyman-Pearson's approach gives you so much more information than Fisher's approach alone. Many medical practitioners are not very receptive to articles which only give them confidence intervals, but do not provide more decisive information. Final users in other fields probably have similar attitudes to reporting only confidence intervals or p_values. Because of the present controversy concerning statistical approaches, you may want to see if the instructions to authors take a stand on what approach to use.

Having read all this, you, as a researcher or consultant, might still prefer to access papers which report only p_values **or** confidence intervals, provided their sample sizes are sufficient. Such papers, especially those with only CIs, would have a number of advantages, including being more readily acceptable for publication when the results did not strongly argue against **H0**. Favoring Fisher's approach might reduce the publication bias in which only statistically significant outcomes are being reported. If journals also required the publication of information such as measures of variability, it

---

[20]To R.A. Fisher, what is herein called the 'p_value' merely served as a "well-defined measure of reluctance to the acceptance" (Fisher1956, p. 44). Some critics may say that you should report a p_value and then leave the decision up to the reader. This is not exactly what Fisher intended. Researchers reporting p_values under Fisher's model are likely to also report a conclusion that is a blend of their experience and opinion as well as the p_value as Fisher intended. There is a good deal of sense in this if the researcher is an expert in that specialized area. However, opponents of this approach may point out that the researcher has a vested interest in seeming to make progress, having something to publish, etc. Such feelings and biases are often not even conscious. In any case it is cheating to pretend that a decision is from an a priori fixed criterion, α, when it really is based on a p_value the researcher post hoc considers adequate for rejecting **H0**.

would provide material for the design of further experiments. The answer to which is best, a purely Neyman-Pearson approach or Fisher's approach supported by Neyman-Pearson based techniques, remains open at the moment. However, it appears that α, at least as a reference point, is very likely to survive. (Verifiable, as in notarized, dated laboratory records should ideally be available to prove an a priori decision to use Fisher's approach, because changing during the analytical process is a deliberate introduction of bias. Some would call it cheating.)

4.9 Standardization and Normalization

When studying how things (such as height vs protein intake) vary with respect to each other, it is often useful to transform the data to a different scale. If you were to take any data set and subtract its mean from each data score, the results would have a mean equal to zero. If you then divided each difference by its sample standard deviation, S, where S can be simple or pooled, the standard deviation (and the variance) would be 1. Doing both, as in the equation

$$z_i = \frac{x_i - \bar{x}}{S},\qquad(4.9)$$

is **standardizing** the data. Standardization transforms the data from x values that may have any mean and sample standard deviation to (lowercase) z 'units' (sample standard deviation units) that have mean $\bar{z} = 0$ and standard deviation $\sigma_z = 1$. This transformation does not change the shape of the graph of the distribution, but the subtraction shifts it to the right or left, and the division either magnifies or shrinks it. An important use of standardizing values is as an internal part (at least a calculation step) in tests that you will soon explore; it allows the use of simple standardized tables for the evaluation of the tests themselves.

You can **transform standardized data back into its original form** by multiplying the transformed values by the original standard deviation and then adding the original mean,

$$S \times z_i + \bar{x} = x_i.$$

If you are careful to take outliers (extreme anomalous outcomes) into consideration, standardization is very useful in various processes such as correlation (Section 11.3) or building models or trying to find if things vary together, but should not be used as a data conversion before simple inferential tests that compare treatments. Internally, that is within, various inferential calculations, we reach a point at which we can standardize, not the data, but the difference between treatments, which lets us use a reasonable instead of an impossibly large number of tables to interpret the test results.

Do not be misled by claims that imply that by standardizing such things as the total scores of individuals so that $x_i \to z_i$, you can compare effect sizes from different experiments, or assign valid percentile values to, say, the average 'intelligence' of people that have been tested using different instruments (testing procedures, often questionnaires) designed by different persons. For instance, instrument B might include some spatial perception questions and have a bias for mechanical ability, and instrument C might emphasize scholarly ability and have a bias for literary ability and vocabulary. Each of these 'intelligence measures' might exceed the other for evaluating the suitability of job applicants for particular occupations, but however you label what each is testing, the likeness is not going to be complete.[21] One could be seen as relating to blue collar potential and the other as relating

---

[21] Attempts at standardizing effect sizes such as $D = \bar{x}_2 - \bar{x}_1$, based on different subjective scales or procedures, by using such approaches as $d = (\bar{x}_2 - \bar{x}_1)/\sigma$, are still somewhat popular, but do not produce valid results in practice (Section 5.9). The resulting effect size is like a broken t value with the $\sqrt{n}$ left out, and it is a difference between means measured in terms of the standard deviation of the raw data rather than of the means; it is, among other issues,

to white collar potential. No amount of mumbo-jumbo would put them on the same scale for inferential comparisons. A lower (or higher) score on one would have a somewhat different meaning from a lower (or higher) score on the other.

However, standardized data from different populations might be usefully examined by correlation (Section 11.3) to see how, and how much, they vary together. That would be more often useful for examining complex non-inferential relationships. A similar process is what is often meant when articles refer to '*normalization*',[22] where

$$x' = (x - \min(x)) / (\max(x) - \min(x)).$$ (4.10)

Normalization can be especially sensitive to outliers, so be careful. Both standardization and normalization produce values with a range of 0 to 1, and this has the effect of putting them on similar scales, facilitating some multivariate (in the sense of more than two variable) processes.

Note that standardizing the service of fast food places in a chain of franchises is not a statistical procedure. However, the effect is in a sense similar in that it makes things, whether numbers or other things, more alike. So, be careful of the **context** of the word '**standardize**'.

4.10 Question of Evidence vs 'Statistical Significance'

There has been growing concern over the credibility of claims of new discoveries based on "statistically significant" results. A few years ago it was suggested that we should always say 'statistical significance' rather than simply 'significance'; now in 2020 there is a flap decrying even the phrase statistical significance. Similarly, for decades there has been a constant flow of papers, mostly by folks in fields where the availability of data is expensive or simply difficult to obtain, that complained that the 5% minimum criterion is too constraining, indeed that it is a tyranny.

In reality the prevailing minimum $\alpha = 5\%$ means taking an occurrence that has a $1/20$ probability of something happening by chance alone as scientific proof. That a $1/20$ probability is proof of anything is **absurd wishful thinking** and yet it is the basis upon which we researchers supposedly must base a good deal of our careers. However the number of published results that cannot be reproduced appears to **far exceed** even one in twenty. For instance Nosek, Aarts, Anderson and Kappes conducted replications of 100 experimental and correlational studies published in three psychology journals and only 39% of effects were subjectively rated to have replicated the original result (2015). Referring to a wider variety of fields, John P. A. Ioannidis said, "There is increasing concern that most current published research findings are false." (2005). As he implies, there are many more papers and editorial comments that have expressed or even demonstrated similar conclusions.

In response to this crisis, Benjamin et al (2020, et al being 72 researchers) proposed that we should change "... the default P-value threshold for statistical significance for claims of new discoveries from 0.05 to 0.005." It should be noted that they seem to be qualifying their well justified petition to target this suggestion of 0.005 to apply to fields presently using the deplorable 0.05, and they make it clear that within reason different fields may have different, more restrictive standards. However, in their published statement they propose,

---

awkwardly dependent on sample sizes.

[22]The normalization from Equation 4.10 does not make a data distribution normal, but another process that can be seen as true normalization is *probit conversion*, in which the data is ranked, arranged into percentiles and then given values equal to 5 + their corresponding Z score from the normal distribution. The results can be fed into processes that would otherwise assume that the raw data is normally distributed. This creates an effect similar to rank tests (Section 14.1). This can be appropriately used in big data applications if there is justification for assuming that the population has a sigmoid-like cumulative distribution. It is also used by toxicologists (Finney 1990).

"Results that would currently be called 'significant' but do not meet the new threshold should instead be called 'suggestive'. The 'new threshold' referring to '$\alpha < 0.005$' and the 'do not meet the new threshold' refers to '$\alpha < 0.05$'."

Unfortunately this use of two criteria would continue the lamentable practice of post hoc reassessing the p_value after running a Neyman-Pearson test. Such a misuse of the p_value as a measure of the strength of the results would mix the Neyman-Pearson approach with the Fisher's approach, weakening the results, and encouraging a variety of misunderstandings.

Even though we should have never been using 0.05, a 0.005 criterion would lead to an $\alpha/2 = 0.0025\%$ cutoff for a two tail test and that is so small as to make many projects totally impractical in terms of sample size. However, **many researchers would be willing to agree** to some adjustment that involved existing standard practices. For instance, **I would strongly support** a reform such that the $\alpha = 5\%$ minimum criterion should at least be restricted to overall omnibus tests (tests comparing more than two things at once) **when subsequent pair by pair comparisons** are planned. **Otherwise** for comparing two things, or for subsequent pair by pair comparisons an $\alpha \leq 1\%$ criterion ($\alpha/2 \leq 0.005$ for two tail tests) would be a familiar criterion to conscientious researchers and might be accepted as minimal. In practice it would routinely be adjusted lower if adjustment for more than $k = 5$ multiple comparisons following an omnibus test were needed (Section 10.1).

4.10.0.1 Infinitesimals vanish into nothingness

Many statistics texts say the minimum probability for what is commonly called 'statistical significance' should be **less than** 5% and this book says that the probability should be 5% **or less**. How dare it say that! Well **probability is on a continuous scale**; less than or equal to 5% is the range 0 to 5%, and less than 5% is the range from 0 to

$$4.99999999999999999999999999999999999999999999999999999999 \ldots \% \, ;$$

the 9s run on forever. In practice and **without allowing any downward rounding** (other than some undefined infinitesimal) the two ranges are functionally identical. Otherwise if we insist on $\alpha < 0.01$ rather than $\alpha \leq 0.01$ the less mathematical readers are belabored with the meaningless question of how much less than 1% qualifies for rejecting **H0**, and that would also encourage p-hacking (fudging).

4.10.1 Are the statistics in this book obsolete?

It is likely that in some cases big data, also called data science, will appropriately displace the techniques covered in this book when huge amounts of data are available. However, the huge data sets necessary for such wonderful approaches as random forests or neural nets are not possible in a great many research fields due to costs and often due to the impracticality of obtaining them. The tests outlined in this book are in reality medium data tests mostly suitable for sample sizes, $30 \leq n < (\approx 2500)$. Please consider the examples using $\alpha \leq 0.05$ in this book as demonstrations of the process of statistical testing, and **not the author's** personal approval of the prevailing practice of using an excessively **reckless high value like** $\alpha = 5\%$ to declare that the researcher has made an original discovery.

4.10.2 What if there is no evidence?

Already there is a gathering storm based on complaints that lowering the minimum criterion, α, would reduce the publication rate and slow science. But, as is pointed out in this book, there is a need to publish the not covered characteristics of the population resulting from well described methods and materials regardless of whether comparisons within them were significant or not. What if the criterion is not met? Well, if there is no evidence against **H0**, there is no evidence for it. If a brutal wife beater's wife goes missing and a jury cannot find him guilty, we must behave as though his innocence is true, But do we believe it? How will the justice system behave if his next marriage also ends with his wife vanishing? Once again, 'lack of proof is not proof of a lack'. However, you as a researcher are better off than you would be as a prosecutor; you still have the data and **you can and should publish** the variance and other characteristics of either the combined data or of the control, including at least one CI (either of the combined data or of a control) or a boxplot along with a reproducible methods and materials. Such publications with adequate methods and materials will be of **great assistance to future researchers and to meta studies** (the latter an essential need to facilitate more research). Such publications will even justifiably add to the publication record that might be essential to your own survival.

Questions for Ch 4

Write the equations where appropriate and fill in the values. **Odd numbered questions** have worked answers at the back of the book and can be studied as additional examples. The data tends to be fictitious.

1. Ideally, when should you round when using statistical tests?

2. Are there any kinds of numbers which are exact in application as well as theory?

3. Hand calculate a variance estimate and a standard deviation estimate for the following sample of 6 measurements in cubic centimeters:
   4.5cc 1.2cc 3.7cc 1.3cc 4.3cc 1.8cc
   You need an estimate of the mean to do this. Instead of doing it in a row like in a book, it would be more convenient to put the additions, etc. **in columns** beside each other. That might include a column to get the various differences and another next to it for their squares.

4. Bluman (2005) suggests that when reporting your results you round means, variances, and standard deviations to how many digits? You should conform to this for means, but why might it be useful to keep (not necessarily report) your records to a greater number of digits for variances and standard deviations?

5. Why did you use 5 degrees of freedom in Question 3?

6. Hand calculate the **pooled variance** for the following two sets of numbers. **Also** calculate the **degrees of freedom** for your result.
   3.5 1.3 2.7 2.3 1.8
   4.0 3.1 2.0 1.1 1.5 1.6
   (Note that these would be far too few values to have much hope of detecting anything but an enormous difference.)

7. You should be able to state the Central Limit Theorem, (CLT) from memory as soon as possible. What is it?

8. Can you describe the weakness of standardizing effect sizes rather than using original units of measure?

9. What is the standard deviation of the mean (the standard error of the mean) for the numbers
   4.5cc 1.2cc 3.7cc 1.3cc 4.3cc 1.8cc
   from Question 3?

10. What is the (pooled) standard error of the **mean** for the data
    3.5 1.3 2.7 2.3 1.8
    4.0 3.1 2.0 1.1 1.5 1.6
    from Question 6?

11. Hand calculate (other than say, any square roots) a one tail 'Z test' for
    H0: THERE IS NO DIFFERENCE
    HA: THE TRUE MEAN μ IS LESS THAN $\mu_0 = 5.0$ METERS, with criterion $\alpha = 5\%$,
    given the following data in meters:
    5.5m 5.9m 5.0m 5.1m 5.3m 5.4m 5.1m 5.5m 5.8m 5.4m
    Do not worry about β for this exercise, even though not considering it, or the power, would be
    a dreadful mistake in a real research situation. This is just to demonstrate the calculation; the
    sample size, $n = 10$, is too small to have much power. It is only to demonstrate the calculation
    process; **never** use Z in this way for such a **little** sample size.

12. Assume you have taken a random sample and you have estimated $S^2 = 144$, $\bar{x} = 5.20$ and
    $n = 1000$. Given that a final accuracy of three figures is sufficient, use this to test for **H0**:
    THERE IS NO DIFFERENCE vs **HA**: μ IS GREATER THAN $\mu_0 = 5.0$, with $\alpha = 5\%$. Although
    not good practice in application, you need not estimate power or β for this purely academic
    exercise.

13. What is the difference between a standardized z_i value and a Z value? (Hint: how are they cal-
    culated?)

14. Standardize the numbers
    4.5cc 1.2cc 3.7cc 1.3cc 4.3cc 1.8cc
    from Question 3 (Hint: You can also use the unrounded $\bar{x}$ and S from Question 9). As a check,
    add all the standardized values together. They should have a mean of 0, hence the sum should
    also be very close to 0 (a small inaccuracy due to rounding may occur in the last decimal place.).

15. Any set of numbers can be seen as a population, and has a variance. Random numbers can be
    in any order. So assume you drew a random sample of $n = 5$ integers that actually happened to
    come out in the order 1, 2, 3, 4, 5. Estimate the **mean** and then estimate the **variance** by hand;
    what is the other, very important, name for the variance? What is the divisor called? Then esti-
    mate the **standard deviation** (also called the estimated standard deviation or the sample stan-
    dard deviation). A calculator can be used for square roots and long division.

16. Estimate both the variance and the standard deviation for a sample with values 1, 4, 6, 4, 1 as
    though they were a **random sample** of size $n = 5$. State the equations, and substitute numbers
    into them. Incidentally, you will not often find yourself calculating measures of dispersion for
    such a small sample, nor for a systematic series of numbers.

17. Use the first version of the pooled variance Equation 4.3 to estimate the pooled variance and
    the standard deviation of the following as though they were two different treatments with the
    same underlying variance, σ^2. Include the sum of squares when showing your work.
    Treatment $_1$: 5.5, 7.3, 2.7, 6.2, 4.2, 0.1, 3.8, 5.4, 2.9, 7.1, 1.1, 5.2, 5.7, 2.3, 3.5 Kg.
    Treatment $_2$: 2.8, 2.2, 9.4, 8.2, 1.2, 7.1, 3.6, 5.8, 5.2, 4.3, 7.7, 0.8, 9.5, 0.9, 1.1 Kg.

Show all the equations, as well as how you found $S_{1,2}^2$. What are the degrees of freedom, df? Then estimate the pooled standard deviation, $S_{1,2}$.

18. Estimate the individual variances for A and B as though they were two independent sets of observations. What is the df for each? Then estimate the pooled variance using Equation 4.4. You do need include the individual sum of squares when showing your work. What is its df? Also give a pooled estimate of the standard deviation, for the following samples, in degrees Centigrade:
Observ. A: 6.6, 6.5, 3.3, 9.0, 4.0, 6.8, 0.0, 7.9, 6.7, 4.5, 0.2, 2.5, 2.8, 5.1, 6.4 °C.
Observ. B: 3.9, 8.4, 1.5, 9.3, 9.9, 5.0, 2.0, 7.8, 6.0, 5.5, 7.9, 9.7 °C.
Show all the equations for estimating the single variances with and without the numbers. Explain how you got the df for each variance and the combined variance. Is there a problem with the sample size?

19. Does the term standard error refer to the raw data population?

20. What is the sum of 1/5 plus 1/6 ? Show how to do it without a calculator and leave the answer as a rational number (the ratio of two integers). The reason for this question is that the process is very important to understanding many simple derivations.

21. How does the Neyman-Pearson approach use the p_value differently from Fisher's approach?

22. Although often disastrously mixed with Fisher's approach (model), the Neyman-Pearson approach is the most popular. The Neyman-Pearson approach is cognizant that your experiments and observations define your outcome and it estimates the outcome that you should expect if you repeat the experiment or observation an unlimited number of times. Explain what the p_value is not used for in the Neyman-Pearson approach.

23. R.A. Fisher is a name that will recur in this book, and even in instructions to authors from some journals because he did so much pioneering work. What did he intend for the researcher to do with the p_value? His approach assumes that sometimes the researcher can not reject **H0** because . . .?

24. The p_value is said to be confounded by the advocates of the Neyman-Pearson approach. What does confounded mean?

25. Which approach, Neyman-Pearson's or Fisher's, might be best for quality control?

26. What is an effect size, ES? Is an ES different from a reasonable minimum difference?

27. What is a confidence interval?

5 STUDENT SPILLS BOSS'S BEER!

Many great developments that occurred far back in the mists of time are sufficiently complete that we can only polish and extend them. So, on our way to yet another glimpse of the true, but elusive, basis of power (Subsection 7.1.1 and Section 7.3), let's retire to our favorite shadowy den of iniquitous chance to discuss the test published by someone who, back then could only be known as 'Student'.

5.1 Why Student Had to Do It

William Sealy Gosset was employed by the Guinness beer company. Beer fermented in small vats does not come out the same as in large commercial vats. So he, allegedly was having to pour out and waste n huge vats of Guinness beer for each experiment. This was enormously expensive, and it was also very upsetting to the guys at the pub. So he heroically fired up his brain cells to devise a test which at least adjusts for sample size (number of vats), so that fewer numbers of huge vats of beer needed to be sacrificed. Everyone knew what company employed Gosset, so publishing the new test under his name would have given away a trade secret. Allegedly after some deliberation, he was required to use the pseudonym *Student*. So, **Student's t test** was intended to be a small sample test (Section 5.2). Well, this was the apocryphal story told to me as a student decades ago, but modern less romantic biographers now say, that although frantic for a small sample test, Student was only testing smallish samples of the barley and other ingredients going into the beer.

Traditionally, 'small' (as opposed to 'large') meant sample sizes of $n < 30$. Student's 1909 paper assumed **randomness**, and that the underlying distribution was **close enough to normal** as to make no difference (Student 1909). Unfortunately the normality assumption is rarely met by raw data, and now, students are often told that Students t works because as the sample size increases, the distribution of the **means** increasingly approaches normality. Sample means are usually, though not always, sufficiently close to normally distributed when the sample size $n \geq 30$. Therefore, statisticians have recently become increasingly aware that student's t test excels as a 'large' sample test.

5.2 What He Did and How He Did It.

Student published his paper, "The probable error of a mean" (1908), because of the need to compensate for the fact that S (which is like a mean sample variability and estimates the true standard deviation, σ), **and** D, the difference between the sample means, are each an estimate, are each **independently** subject to chance, and each have their own variances. The effect of the fact that S is only an estimate of the standard deviation, σ, which is independent of the estimation of the mean, is particularly severe for smallish samples. So, for small to relatively big sample sizes, Student's t test uses the **t distribution** which adapts for the fact that you are **additionally** <u>estimating</u> the value of an independent **standard deviation**, $\sigma = \sqrt{\sigma^2}$, by correcting the normal distribution of the means for the distribution of the sample standard deviation, S, which has a chi-square distribution, and this correction requires different tables for different sample sizes. The t distribution only approaches the standardized normal distribution represented by Z as the sample size becomes larger.

This book often uses a single variable, D, for $D = \bar{x} - \mu_0$, where μ_0 is the hypothesized mean, or $D = \bar{x}_2 - \bar{x}_1$, where there are two means, such as $_1$ for **before**, and $_2$ for **after**. Recall that the degrees of freedom is generally the denominator of the variance estimate, S^2. **Student's t test** is calculated like the 'Z test' ($Z = \frac{D \text{ or } \mathcal{D}}{S/\sqrt{n}}$), except that the resulting t values have a different set of probabilities for **each number of degrees of freedom** associated with estimating S. Its two generalized forms are

$$t = \frac{\bar{x} - \mu_0}{S/\sqrt{n}} \text{ and } t = \frac{\bar{x}_2 - \bar{x}_1}{S_{1,2}\sqrt{1/n_1 + 1/n_2}}, \tag{5.1}$$

where $n_{...}$ is the sample size. An obvious difference between Z and t tests is the dependence of the probabilities of t on the degrees of freedom, so for Student's t test you evaluate the results **using different tables** such as appendix 3 to find if you have statistical significance at the attained degrees of freedom, df, and your preselected level, the *cutoff* values of α for one tail, and of $\alpha/2$ for two tails. Note that for the first (leftmost) version of Equation 5.1, the **single sample** version, the degrees of freedom equal $n - 1$, and for the second (rightmost), the **two sample** version, the degrees of freedom equal $= n - 2$. Either denominator $S/\sqrt{n}$ or $S_{1,2}\sqrt{1/n_1 + 1/n_2}$ in Equation 5.1 above is the **standard error of the mean**, and is often written $S_{\bar{x}}$, or sometimes when appropriate, $S_{\bar{x}_2 - \bar{x}_1}$. It estimates an ideal, the standard deviation of the means, $\sigma/\sqrt{n}$, that is guaranteed by the Central Limit Theorem (Subsection 4.3), and is written $\sigma_{\bar{x}}$. Earlier authors were aware of the t distribution, but Student was the first to realize that it must be interpreted using the degrees of freedom, such as $df = n - 1$ or $df = 2n - 2$ rather than just the sample size, n. Hence both the test and the layout of the tables are due to Student.

Unfortunately, we now know that small samples, even samples quite a bit bigger than 'large', $n \geq 30$, are **not sufficiently sensitive** to have a useful chance of detecting any differences which are not huge. Sensitivity aside, for simple tests with $30 \leq n \leq 60$, it is essential to use t tables instead of tables of Z.

For good reasons the tables of t go up to very large sizes of df, and thus very large sample sizes. This relates to its many uses, including sample size estimation (Section 7.5). t can give useful accuracy of such calculations as sample size estimates when $n \leq 1000$ and $\alpha \geq 0.025$ for three figure accuracy (and for four figure accuracy when $n \leq 4500$ and α even smaller than $\alpha = 0.010$). These sample sizes are crude rules of thumb found by examining the way in which the tables change. The estimated standard deviation can be either S, the plain standard deviation estimate for one sample, or $S_{1,2}$, the pooled standard deviation estimate for two samples.

When used for two samples, Student's t test assumes that there is **just one** underlying standard deviation, σ, and that this can be estimated as a pooled standard deviation. This homogeneity of variances assumption is rarely true in real life, and cannot be taken for granted when performing a statistical test (Erceg-Hurn and Mirosevich, 2008; Zumbo and Coulombe, 1997). This can become a serious problem when the sample sizes are different. So, although you should be aware of the principle of how the now almost universally applied two sample Student's t test works, even when the sample sizes are unequal, in practice it is best that you choose the **Welch's t test** (Section 5.3) whenever the **sample sizes are not equal** (Delacre, Lakens, and Leys 2017; this is worth reading).[1]

Recall that $\sigma_{\bar{x}} = \sigma/\sqrt{n}$ and is estimated by $S_{\bar{x}} = S/\sqrt{n}$. Z is theoretically equal to $D/\sigma_{\bar{x}}$, and, for continuous data, it theoretically applies only for an infinitely large sample size, which is impossible, or for the whole population, which, if you could measure it, would not require testing. However, $t = D/S_{\bar{x}} = D/(S/\sqrt{n})$, which is evaluated by using tables of t rather than of Z, is usually very good for testing samples of continuous data in application. But for continuous data, Z, to a three figure accuracy, can sometimes be used in a starting step in sample size estimation. In situations where

---

[1] Minitab and the base package of the R language default to Welch's t test unless programmed otherwise.

the underlying distribution is discrete, such as the binomial, Z is appropriate, because the mean and the standard deviation are then not independent of each other. The correct values of t for the appropriate degrees of freedom are available in tables such as Appendix 3, and a table of Z values is in Appendix 4. Either t or Z values are also available from many sources, including online.

It is important to remember that the calculation of $S_{\bar{x}}$ can be somewhat different depending on whether you are referring to the mean of one sample or of two. So, there are two basic versions of Student's t test, but they are simply called the ***t test***.

> **MUST KNOW:** Be aware that t is a set of distributions corrected from Z for when the difference between the means and the variance are independent. **Each t distribution is dependent on the degrees of freedom** used to estimate the particular variance of interest. Therefore, the degrees of freedom with which you enter a table of t is dependent on the size of the sample upon which the variance is estimated, and the degrees of freedom in turn chooses the distribution of t. This point will prove to be very important later when you are estimating sample sizes, and also for multiple comparisons (Chapter 10) following certain tests which compare more than two treatments (categories).

A **one sample t test** starts with a hypothesized mean, μ_0, and estimates only one sample mean, $\bar{x}$, to challenge

H0: THERE IS NO TRUE DIFFERENCE, $\mu = \mu_0$

versus

HA: THERE IS A DIFFERENCE.

So, for one sample versus a known or hypothetical μ_0,

$$t = \frac{\bar{x} - \mu_0}{S/\sqrt{n}} \quad df = n - 1. \tag{5.2}$$

The denominator is $S/\sqrt{n}$, an estimate of the standard error of the mean, where the lowercase italic n is the size of the single sample. The hypothesized mean, μ_0, can be any value you hypothesize, including $\mu_0 = 0$, in which case you can write $t = \frac{\bar{x}}{S/\sqrt{n}}$ or $t = \frac{\bar{x}}{\sqrt{S^2/n}}$, where $D = \bar{x} - 0$, and $\bar{x}$ is the sample based estimate of the true mean, μ. Estimating $\bar{x}$ for one mean uses up a degree of freedom, $1\,df$, so your remaining degrees of freedom are $df = n - 1$.

A **two sample t test** challenges whether there is only one population versus two populations when you don't know the true means for either. You need to create a null hypothesis, something like **H0**: THEY ARE THE SAME, and under the Neyman-Pearson model this would be versus **HA**: THEY ARE NOT THE SAME (possibly indicating a direction). Thus, it requires two samples and makes two estimates of the means, $\bar{x}_1$ and $\bar{x}_2$, each estimate using up a degree of freedom. For historical reasons, the two sample t test should technically **not be called** a 'Z test', even when your sample size is so big that you could use a table of Z.

In big sample testing, Student's t test can be treated as a '*Z test*' **if** the total sample size, **n**, is sufficiently larger than large ($n \gg 30$ as distinguished from merely large, $n \geq 30$) which could be more than a thousand, or even thousands depending on your choice of α. Most statisticians consider $\alpha/2 = 2.5\%$ to be a minimum cutoff for any reasonable two tail test, and for that cutoff the approximation $Z_{0.025} \approx 2$ can be used conservatively for any $n > 60$, although, say, $n = 61$ gives very minimal power and it is only accurate to three figures. A much greater accuracy is necessary for power analysis calculations, or for smaller values of α. So, when planning for power, you will often need to use t values for $df < 1001$ and $cutoff < 2.5\%$ (and for $df < 4500$ if $cutoff \ll 2.5\%$).

For pitting two samples against each other, it is customary to estimate the single common $\sigma_{\bar{x}}$ by a pooled standard error of the population of sample means, also called the *pooled standard error of the mean*, Equation 4.7. This pooled standard error ($\sqrt{S_{1,2}^2 \left(\frac{1}{n_1} + \frac{1}{n_2}\right)}$) becomes the denominator of the two sample Student's t test,

$$t = \frac{\bar{x}_2 - \bar{x}_1}{\sqrt{S_{1,2}^2 \left(\frac{1}{n_1} + \frac{1}{n_2}\right)}} \qquad df = n_1 + n_2 - 2. \qquad (5.3)$$

However, the pooled standard error of the mean can also be estimated by rearranging $\sqrt{S_{1,2}^2 \left(\frac{1}{n_1} + \frac{1}{n_2}\right)}$

which is shown in the divisor (*denominator*) in any of several ways, such as $S_{\bar{x}_2 - \bar{x}_1} = S_{1,2}\sqrt{\left(\frac{1}{n_1} + \frac{1}{n_2}\right)}$

$= \sqrt{S_{1,2}^2 \left(\frac{n_2 + n_1}{n_1 n_2}\right)}$, etc. If $n_1 = n_2 = n$ then you can rewrite the calculation for a two sample t test as

$$t = \frac{\bar{x}_2 - \bar{x}_1}{S_{1,2}\sqrt{2/n}} \qquad df = 2n - 2. \qquad (5.4)$$

Any of the above denominators are assumed to relate to a single (or common) standard deviation parameter σ. That is, it is assumed that that standard deviation is the same even though there are two samples and that the only difference, if any, is due to the effect of the treatment or treatments on the means. This seems reasonable because this process could be seen as challenging something equivalent to **H0**: THE RESULTS REALLY REPRESENT THE SAME POPULATION OF MEANS, to discover if there is sufficient evidence to justify rejecting it. The calculations are based on the provisional assumption that **H0** is true. You proceed in that manner until you have either reached a conclusion or, as sometimes happens, you are left without any conclusion (due to lack of statistical significance).

In its **one tail** form the t test can be used to challenge **H0**: THE MEANS ARE THE SAME, which can be seen as saying they are in the same population. When applying the Neyman-Pearson approach, generally used here and by custom, you use a pre -chosen criterion, α, and test some hypothesized pair, say the above **H0** against **HA**: MEAN μ_B IS LARGER THAN MEAN μ_A. However, because the direction is pre-chosen by the researcher, this one tail test could, a priori, be **alternately** applied by using **HA**:$\mu_A < \mu_B$.

Student's t test for a **two tail** test using the Neyman-Pearson approach would use some form of **H0**: THEY ARE THE SAME, but challenge them with **HA**:THE MEANS ARE NOT THE SAME, or less specifically, **HA**: THE POPULATIONS ARE NOT THE SAME. One tail tests use a pre-chosen criterion, α, **but two tail tests** must **divide the criterion by 2** and assign a cutoff of $\alpha/2$ to each tail. This is in accord with the law of probability that says that a total probability is the sum of the probabilities contributing to it.

Although it is convenient to think of μ_1 and μ_2 as before and after, it is merely an arbitrary decision as to which mean is represented by which. Often the idea of before and after does not apply, even though there is a possibility that there is just one population with one mean. The convention of using μ_0 to represent some purely hypothetical mean and μ_1 for a possibly unlike mean is only slightly more consistent. Sometimes it is handy to simply use μ to signify a true but unknowable mean.

5.2.0.1 Example: One Tail, One Sample t Test

Fructose Suppose a public health student suspected that the average of the weights of male students at Little U College that use soft drinks with high fructose sweetening is more than the national average for males in the same age group. To be certain she was using the right approach,

she made a sketch , and given that her available degrees of freedom (due to limited size samples) would be small, she used a t scale rather than an approximation using a Z scale. The sketch reminded her that given that she believed she was testing the idea that high fructose drinks increased the student's weight, she needed to put all her criterion probability, represented by the dark area, out there in one tail, the positive tail. She formed a hypothesis pair:

H0: THERE IS NO TRUE DIFFERENCE BETWEEN THE AVERAGE OF THE WEIGHTS OF MALE STUDENTS DRINKING HIGH FRUCTOSE SWEETENED SOFT DRINKS AT LITTLE U AND THE CURRENT NATIONAL AVERAGE FOR COLLEGE AGE MALES.

HA: THE AVERAGE OF THE WEIGHTS OF MALES DRINKING HIGH FRUCTOSE SWEETENED SOFT DRINKS AT LITTLE U IS GREATER THAN THE NATIONAL AVERAGE FOR THEIR AGE GROUP. (Notice that this alternative hypothesis is customarily about the one direction in which she, the researcher, was interested as well as about the fact that she was comparing an estimate, the sample mean $\bar{x}$, to what she considers to be a known value, μ.)

Criterion: She selected a popular criterion, $\alpha = 5\%$. She suspected that $\alpha = 5\%$ was a bit liberal since it is the largest criterion that is accepted by peer reviewed journals, but she did not expect to be able to get a very large sample. **All decisions to this point were a priori**; that is, they were made **before** the experiment, in this case before collecting and processing the data.

Methods: In the time available she was only able to collect a random sample of size $n = 13$ male students who drank high fructose soft drinks. Then from a national database for that year, she found that officially, $\mu_0 = 162.70$ pounds for the appropriate age group. This value was reported to five figure accuracy and limited the sample mean, $\bar{x}$, to five figure accuracy in the subtraction that constituted the numerator of her calculation of t. Otherwise it might have been a little better to round less until evaluating and actually reporting. She used a medical scale, technically a balance, that was easy to read to ± 0.1 lb, to weigh the $n = 13$ students.

Results: Her raw data in pounds was:
$x_i =$189.4 191.6 126.4 197.2 153.8 224.6 106.8 241.8 156.7 151.7 104.1 197.4 183.8

Analysis: Because adverse rounding effects are magnified in estimating σ, she avoided rounding as much as was reasonable and possible. She then calculated that $\bar{x} = 171.18$ lb and $S = 42.480$ lb. The table below is presented as though the student had not rounded for reporting yet; it shows the mid-calculation values to fourteen figures.

Estimate mean	Differences fr Mean	Differences Sq.
189.4	18.22307692308	332.08053254438
191.6	20.42307692308	417.10207100592
126.4	−44.77692307692	2004.97284023669
197.2	26.02307692308	677.20053254438
153.8	−17.37692307692	301.95745562130
224.6	53.42307692308	2854.02514792899
106.8	−64.37692307692	4144.38822485207
241.8	70.62307692308	4987.61899408284
156.7	−14.47692307692	209.58130177515
151.7	−19.47692307692	379.35053254438
104.1	−67.07692307692	4499.31360946746
197.4	26.22307692308	687.64976331361
183.8	12.62307692308	159.34207100592
2225.3 T	0.00000000000 Sum dif.	21654.58307692308 SS
171.176923076923 T/n		1804.54858974359 MS
		42.4799786928335 S

As you can see, the sum of the differences, $\sum_{i=1}^{i=n} (x_i - \bar{x}) = 0$; classically this used to be a check on hand calculations. Also note that SS is the sum of squares, and the mean squares, $MS = \frac{SS}{df} = \sum_{i=1}^{i=n} (x_i - \bar{x})^2 / (n-1)$, so this $MS = S^2$, and the sample standard deviation, $S = \sqrt{S^2} = 42.480$. So,

$$t = \frac{(171.18 - 162.70) \text{ lb.}}{42.480 \text{ lb.}/\sqrt{12}} = 0.6915156,$$

with $df = n-1 = 12$. The accuracy here was restricted by the limited accuracy of the available national mean, μ_o, but if possible, while calculating t, it is a good idea to use even more than one more figure than the number of figures in the data she collected. Finally, everything calculated directly from the data should be reported to no more than one more figure than the data. The pounds over pounds canceled out, so, as always, t is unitless, but it could also be thought of as being in units which could be called 'standard errors of the sample mean'.

(There are many ways to write summation notation and they must be taken in context, especially in the literature and in other books. The above could have simply used $\sum_1^n$, or even more simply, just $\sum$.)

Results: Referring to a table of t (Appendix 3) and finding $\alpha = 0.05$ with $df = 13 - 1 = 12$, she found that the one tail value of $t_{0.05,12}$ required to reject **H0** is 1.782. So because she only found $t = 0.6915$ she could **not** reject **H0**.

Discussion: Assuming other experimental details were handled with sufficient care, does this mean she had enough evidence to even suspect **H0** was true? And, what did she really mean by 'greater than'? Would a true difference of 0.10 lb have been meaningful to her? Laxative advertisements aside, their weight could easily change by 0.75 lb or maybe more when they defecate (Hosseini 2000, carefully and usefully studied this). Was a t test appropriate, considering that she had a $df < 30$? Well, I'll leave y'all to discuss it.

The Number of tails and samples is up to you

Such variables as daily variation in weight are part of the chance variation with which a medical researcher has to cope. The student researcher could **not accept H0** because the results were **incon-**

clusive. Randomization helps with uncontrollable variation, but maybe weighing humans to four figure accuracy is not very useful, except for meeting the tie breaking requirements of some other tests, such as the sign test (Section 2.6). This was a one sample one tail test, but she could have designed her research to have a one sample test with two tails. She could also have a two sample test with one tail. That is, the number of tails and the number of samples are up to the researcher, so long as such choices are a priori.

5.2.0.2 Example:Two Tail Two Sample t Test

Pretend the previous example **never happened**: Suppose our student collected a sample of control data from the other students and had actually decided to use a two tail ('not the same', usually meaning larger or smaller), two sample ($\bar{x}_1$ vs $\bar{x}_2$) test.

Fructose (another way) In this scenario our public health student had suspected that the average of the weights of male students at Little U College using soft drinks with high fructose sweetening is **unlike** the average for those Little U male students who were not using soft drinks with high fructose sweeteners. To help herself consider what she was going to do, she made a

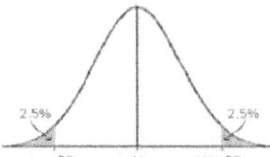

rough sketch where the shaded areas share the criterion, α, so each tail (the + and the - directions) will get $\alpha/2$. She formulated a hypothesis pair:

H0: THERE IS NO TRUE DIFFERENCE BETWEEN THE AVERAGE OF THE WEIGHTS OF MALE STUDENTS AT LITTLE U DRINKING HIGH FRUCTOSE SWEETENED SOFT DRINKS AND THE AVERAGE OF THE WEIGHTS OF MALE STUDENTS AT LITTLE U WHO DO NOT.

HA: THERE IS A DIFFERENCE.

Criterion: She selected the popular $\alpha = 5\%$ criterion, so because the difference can be either way, the cutoff $\alpha/2 = 2.5\%$ for statistical significance in either direction.

Methods: Assume that in the time available, she was able to randomly collect a sample of $n_1 = 15$ students who drank soft drinks sweetened with high fructose and $n_2 = 13$ control students who did not. Her roommate, who did not even know that there were two categories, used a medical scale, technically a balance, that was easily read to ± 0.1 lb, to weigh the male students, presented in a random order. Having her roommate help was to get a double blind experiment. She, herself, knew which were in each of the two groups, X_1 non-users and X_2 high fructose soft drink users, but her roommate took the data without knowing which was which.

Results: Her raw data in pounds was:
$x_{1,j}$ = 125.6 209.0 147.5 173.9 195.9 122.2 87.6 173.6 174.1 133.4 157.4 176.2 122.2 223.3 110.1
$x_{2,i}$ = 189.4 191.6 126.4 197.2 153.8 224.6 106.8 241.8 156.7 151.7 104.1 197.4 183.8
(Note that she used i and j in the subscripts rather than i for the individuals in both 'treatments', 1 and 2, because she wished to emphasize that they were not paired in any way. For instance, $\sum_{i=1}^{i=n_i} x_{1,i}/n_1 = \bar{x}_1$ and $\sum_{j=1}^{j=n_2} x_{2,j}/n_2 = \bar{x}_2$. The choice of notation is up to you, but try to keep your context obvious.).

Analysis: She then calculated $\bar{x}_1 = 155.47$ lb, $S_1^2 = 1490.80$ lb, and $\bar{x}_2 = 171.18$ lb, $S_2^2 = 1804.55$ lb.

Because she assumed that both sample standard deviations really estimated the same standard deviation and that any differences were due to chance alone, she decided to estimate a pooled variance using one of several possible computational approaches. So she decided to take advantage of the fact that the sums of squares are additive and used

$$S_{1,2}^2 = \frac{\sum_{i=1}^{i=13} (x_{1,i} - \bar{x}_1)^2 + \sum_{j=1}^{j=15} (x_{2,j} - \bar{x}_2)^2}{n_1 + n_2 - 2}$$

to get the pooled standard deviation,

$$= \frac{(20871.23 + 21654.58)}{(15 + 13 - 2)} = \frac{42525.82}{26} = 1635.608$$

She then calculated standard error of the means,

$$S_{1,2} = \sqrt{S_{1,2}^2} = \sqrt{1635.608 \, \text{lb}^2} = 40.44265 \, \text{lb}.$$

Then,

$$t = \frac{\bar{x}_2 - \bar{x}_1}{S_{1,2}\sqrt{\frac{n_1+n_2}{n_2 n_1}}} = \frac{(171.18 - 155.47) \, \text{lb}}{40.44265 \, \text{lb}\sqrt{\frac{15+13}{(13)(15)}}} = \frac{15.71 \, \text{lb}}{40.44265 \, (0.37893) \, \text{lb}}.$$

(Note that, as in the denominator, multiplication with a value less than one can be seen as having an effect equivalent to division.)

The pounds cancel out, so, $t = \frac{\bar{x}_2 - \bar{x}_1}{S_{\bar{x}_1 \bar{x}_2}} = \frac{15.710}{15.325} = 1.02512$ with $df = n_1 + n_2 - 2 = 26$.

Results: Referring to a table of t (Appendix 3), she found that for $\alpha/2 = 0.025$ and $df = 28 - 2 = 26$, the cutoff value required to reject **H0** is $t_{0.025,26} = 2.0553$. So, she could **not** reject **H0** and the results are **inconclusive**.

Discussion: In using Student's t test with such a small sample, she had assumed that the distribution of human weights for subjects in the same age range, community, and gender would be sufficiently near normal to use Student's t test. Because she had obviously already calculated the individual variances, she could have used the simpler $S_{1,2}^2 = \frac{S_1^2(n_1-1)+S_2^2(n_2-1)}{n_1+n_2-2}$ instead of the pooled standard deviation equation that she used. The result would have been the same because $S^2(n-1) = \sum_{i=1}^{i=n} (x_i - \bar{x})^2$.

In any case, editors should **require**, and be **encouraged** to require, the reporting of either the **standard deviation**, S, as an estimate of σ, **or the sample variance**, S^2, as an estimate of σ^2. If the methods and materials are sufficiently reproducible, **variability estimates are valuable data** that future researchers may use to design their research.

The data was not real for either this or the preceding sample. For this example, I sampled at random from a larger truly random sample manipulated so as to produce a near normal random distribution, and both the control means were in the same ballpark, as such a survey of weights might find. However, I do not know what the real means might have been. The standard deviations of the contrived sample populations were equal. I built a small effect into my manipulated data, but the 'public health student' had little hope of detecting it at such low sample sizes.

Once again, the concept of 'unlike' is vague. She would have had to define how much of a difference should be considered 'unlike' in order to determine what size her sample should have been. Even if her sample size had been a lot larger, there is at least a theoretical possibility that the observed difference for positive results was too small to have meaning. The t distribution is continuous, so its graph is a solid line at any magnification. However, sample sizes are discrete. So, there is a different t distribution for each sample size.

5.2.1 Robustness of Student's t Test

Remember that t tables **do not compensate** for a lack of population normality; they simply adjust for your having to estimate the standard deviation. So, you will still need a sufficiently large sample in order for the CLT to adjust for a lack of normality. Over more than a century of its use, it has been observed that **t tends to be sufficiently robust** when $n \geq 30$ and it is almost always very robust when $n > 80$. This robustness is because of the Central Limit Theorem (CLT, Subsection 4.3), which helps meet the assumption of normality. **Below this sample size of** $n = 30$, a nonparametric equivalent such as a Wilcoxon-Mann-Whitney summed rank test might be preferable (Subsection 14.1.1).

5.2.1.1 Sample size and skew

An exception to the adequacy of sample sizes of $n = 30$ is extreme skew (Subsection 15.2.1), meaning that the distribution of the data is very asymmetrical, tending to be steep on one side and gradually sloping down to a long tail on the other. This might occasionally be severe enough to require sample sizes as large as $n = 80$ (Ratcliffe 1968), or in rare cases, up to 500 (Thomas, Diehr, Emerson, and Chen 2002). Ratcliffe (1968) examined sample sizes of up to 80 and concluded that "extreme non normality can as **much as double** the value of t at the 2.5% (one tail) probability level for small samples, but increasing the sample sizes to 80, 50, 30, and 15 will for practical purposes remove the effect of extreme skewness, moderate skewness, extreme flatness, and moderate flatness, respectively." Looking at these references, it would appear that kurtosis, at least in the form of flatness, is not a problem at sample sizes as large as $n = 30$ (Section 15.2).

Skew is commonly measured as third moment skew,[2]

$$\widehat{skew} = \frac{1}{n-1} \sum_{i=1}^{i=n} \left(\frac{x_i - \bar{x}}{S} \right)^3.$$

For example, Lovric (2010) asserts that a sample of 50 observations should be plenty even if the skew is as large as 2 or –2.

Equal sample sizes are best for Student's t tests, **otherwise** consider Welch's t (Subsection 5.3). Welch's t test is always best in application when sample sizes are **unequal**. However, to understand both testing and sample size estimation, we need to carefully consider simpler classical t tests, even for sample sizes that are not equal.

5.2.1.2 Example: Larger Two Sample Two Tail t Test

Let's look at a larger sample size Student's t test where both the robustness and the power will be greater.

Fructose, $\bar{x}_1$ vs $\bar{x}_2$ Suppose our public health student, who suspected that the average of the weights of male students at Little U College using soft drinks with high fructose sweetening is more than the average for those male students who were not using soft drinks with high fructose

---

[2]The Wilcoxon tests (Section 14.1.1) may be better and even more powerful than Student's t tests even with quite large samples when skew is extreme. This text covers the larger sample version of the Wilcoxon test for when at least one sample is greater than 20, and you do not feel comfortable with a Student's t test. Simple tables are available for smaller sample size Wilcoxon tests (with both samples ≤ 20); for instance in Langley 1970, which is still in print.

sweeteners, had been able to use **larger samples**, with $n_1 = n_2 = 200$. She made a sketch like the drawing in the example in Subsection 5.2.0.2. In this two tail situation, she had to place half, $\alpha/2$, of her criterion, $\alpha = 5\%$ in each tail (the + and the - directions). Because the sample seemed large enough, she at first decided to use $Z_{\alpha/2} = 1.960$ as her Z cutoff for statistical significance. However, referring $df = 400 - 2 = 398$ to a table of t (Appendix 3), she found $t_{\alpha/2=0.025,\,df=298} = 1.966$. That is enough greater than $Z_{\alpha/2=0.05} = 1.960$ to affect the outcome in some situations. So she chose the t value instead of the Z value (also avoiding possible criticism). She then formulated a hypothesis pair:

H0: THERE IS NO TRUE DIFFERENCE BETWEEN THE MEAN WEIGHT OF MALE STUDENTS AT LITTLE U DRINKING HIGH FRUCTOSE SWEETENED SOFT DRINKS AND THE MEAN WEIGHT OF THE ONES WHO DO NOT.

HA: THERE IS A DIFFERENCE.

Criterion: She had selected a 5% criterion, meaning that the cutoff was $\alpha/2 = 2.5\%$ for statistical significance in either direction.

Methods: She was able to collect a sample of $n_1 = 200$ students who drank high fructose soft drinks and $n_2 = 200$ control students, who did not. She weighed these male students in random order to the nearest tenth of a pound.

Analysis: She calculated $\bar{x}_1 = 161.28$ lb and $S_1^2 = 719.62$ lb^2; $\bar{x}_2 = 167.12$ lb and $S_2^2 = 848.58$ lb^2, and $n_1 = n_2 = n = 200$.

Because she assumed that the true standard deviations were really the same and that any differences were due to chance alone, she pooled the variances using one of several possible equivalent computational approaches. She decided to take advantage of the fact that the sums of squares (found by using $(n-1)\,S^2$) are additive, so she used $S_{1,2}^2 = \frac{(n_1-1)S_1^2+(n_2-1)S_2^2}{n_1+n_2-2}$,

$S_{1,2}^2 = \frac{(200-1)719.62+(200-1)848.58}{200+200-2} = \frac{199(719.62+848.58)}{398} = 783.9$

So, $S_{1,2} = \sqrt{S_{1,2}^2} = \sqrt{783.9} = 28.00$ lb. Technically this would have been called a t test because it has two samples, even if the sample size had exceeded 4500 (where the tables would have defaulted to Z).
She found

$$t = \frac{\bar{x}_2 - \bar{x}_1}{S_{1,2}\sqrt{\frac{1}{n_1} + \frac{1}{n_2}}} = \frac{\bar{x}_2 - \bar{x}_1}{S_{1,2}\sqrt{\frac{2}{200}}} = \frac{(167.12 - 161.28)\text{ lb}}{28.00\text{ lb} \times 0.1} = \frac{5.84\text{lb}}{2.800\text{ lb}},$$

so,

$$t = 2.086.$$

Subtraction only produced an apparent three figure result despite the fact that she estimated the means to a five figure precision. So, she could only report three figure precision for the effect size of $\hat{\delta} = 5.84$. This tendency to lose apparent accuracy is common when subtracting and is why we try to avoid rounding before subtracting. Thus the number she compared with the t cutoff value, corresponding to her criterion, was really only precise to three figures, whereas the table is precise to four figures, as was the justifiable reporting accuracy. Because the difference, $\bar{x}_2 - \bar{x}_1$ was sufficient, that did not harm her in this situation, but it could have misled her if her results had been closer to each other (if the effect size had been smaller).

Results: All went well and she was able to randomly collect n $= 400$ gross weights, the net weights only being accurate to about a pound, with $n_1 = n_2 = 200$ students in each of the 'treatment' categories. Listing all the data point values was not necessary for publication, but she would be reporting the means and the estimated standard deviations from the analysis. Referring to a table of t (Appendix 3), she confirmed that $t_{\alpha/2=2.5\%, df=398} = 1.966$ when $df = 398$. Because her result exceeded the critical value for t at that size, **she rejected H0** at the $\alpha = 5\%$ criterion and **accepted H$_A$**. Note that the rejection was for $\frac{\alpha}{2} + \frac{\alpha}{2} = 5\%$ because she was testing for either greater **or** smaller than, and chance could have gone against her either way.

Discussion: If you use this example as a model, note that although the approach is correct, **she rounded too early**. This example leaves several questions unanswered in addition to the question of why she rounded too early, **potentially losing** some of the information contained in the data. Another shortcoming was that this example does not say if she made any effort to achieve a double blind effect, but it is not impossible that it was simply left out of her description. You might wish to consider how a student with limited time and resources could achieve that.

However, the data was not real. It was generated from a truly random sample manipulated so as to produce a near normal distribution, and both the control means and standard deviations were in the same ballpark, as such a survey might find. Thus I happen to know the unknowable, that the real effect for the whole population, $\delta = \mu_1 - \mu_2$, is 3 lb. That means that chance appears to have caused her results to be off by about one standard deviation of the mean ($\hat{\delta} - \delta = (5.8 - 3.0)$ lb $= 2.8$ lb and $S_{\bar{x}} = 2.8$ lb). Inaccuracy is to be expected when the research is subject to chance. In a normal distribution, about 32% of all samples will be more than one standard deviation from the true mean. Clearly, the results could have been different due to chance and these samples could have undershot due to chance alone, in which case our make-believe student would have been unable to reject. Yep, statistics are dicey. She also has left out all consideration of β, the probability of getting a Type II error, so stay tuned for eventual continuations of this example.

If her sample size had been large enough to result in, say, $df > 4500$ (or $df > 1000$ for three figure accuracy), she might have simply used a Z table to find and work with $Z_{\alpha/2=2.5\%} = 1.960$ (see Figure 5.1). However, in this situation with any $df > 61$, she could have rounded $t_{\alpha/2=2.5\%, df=398}$ to $= 1.966 \approx 2 \approx Z$ and used 2 as a Z cutoff. That would have cost her a tiny bit of power, but would also be more conservative.

Figure 5.1: Two Tail $\alpha = 5\%$

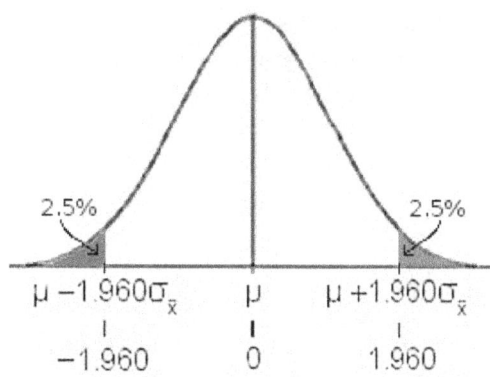

Uses normal curve based measure, Z, because of sample size.
Upper horizontal scale in original units.
Lower horizontal scale in standard error of the mean units, Z.

Question 3 is handy as additional practice of a t test with $n_1 = n_2$, and, like all odd numbered questions, it is answered at the end of this book.

> **CAUTION:** It would have been more correct in Example 5.2.1.2 to have carried the calculation of the means and standard deviations to a greater precision **until the last subtraction**, even though the data was only collected to four figure accuracy. Subtraction is often **a stumbling block** in such situations, if the researcher rounds prematurely.

5.3 Maybe $\sigma_1 \neq \sigma_2$, So If $n_1 \neq n_2$, Default to Welch's t

For data that is not paired, if you suspect that the population variances could be unequal, $\sigma_1 \neq \sigma_2$ (as is usually true), **then if** $n_1 \neq n_2$, you should estimate the pooled standard error using the **Satterthwaite approximation**,

$$S_S = \sqrt{\frac{S_1^2}{n_1} + \frac{S_2^2}{n_2}}. \tag{5.5}$$

S_S gives a slightly different estimate of the standard error of the mean than does S_{x_1,x_2}, when the sample sizes are different from each other, $n_1 \neq n_2$. Also be aware that **the degrees of freedom are different** (Equation 5.7). However, the decision to use it must be a priori (Moser and Stevens 1992, Hayes and Cai 2007)[3].

The t test is not always robust for the assumption of equal population variances (Ramsey 1980). The assumption of equal variances is rarely true in real life, and cannot be taken for granted when performing a statistical test (Erceg-Hurn and Mirosevich, 2008; Ruxton, 2006; Zumbo and Coulombe, 1997). So, especially when $n_1 \neq n_2$, you get more certain results using the Satterthwaite approximation in place of the standard error of the means in the t test to get

$$t = \frac{\bar{x}_2 - \bar{x}_1}{\sqrt{\frac{S_1^2}{n_1} + \frac{S_2^2}{n_2}}}, \tag{5.6}$$

which is a version of the t test called the **Welch's t test**. The denominator (lower term) is technically not a pooled standard deviation, but it is still useful in place of one when appropriate. **The** *df* **for the Welch's t test** and **for the Satterthwaite approximation** is found by using

$$df = \frac{\left(\frac{S_1^2}{n_1} + \frac{S_2^2}{n_2}\right)^2}{\left[\left(\frac{S_1^2}{n_1}\right)^2 / (n_1 - 1)\right] + \left[\left(\frac{S_2^2}{n_2}\right)^2 / (n_2 - 1)\right]}. \tag{5.7}$$

Then you simply find your p_value from a table of t such as Appendix 3 and compare it with your a priori chosen α to see if you should declare statistical significance. This approach is from the work of

---

[3]To check that two variances are equal to each other some researchers use an F test (Section 8.3) or Levene's test, both of which are available on popular software suites, but they are not recommended because they are only valid under the assumption of normality, and they collapse as soon as one variable deviates even slightly from the normal distribution (Hayes and Cai, 2007; Rakotomalala, 2008; Rasch, Kubinger, and Moder 2011).

Welch (1947) and Satterthwaite (1946)[4].

5.3.0.1 Example: Large $n_1 \neq n_2$ (variation on example above)

Methods: Suppose in all this horde of 400 guys, the public health student found, through some error of classification, that she only weighed $n_1 = 185$ non users and ended up weighing $n_2 = 215$ high fructose users. However, she bore up under the shame and went forward with the process.

Analysis: For ease of illustration of Welch's test, we'll just assume the means and standard deviations were the same as in the above example: $\bar{x}_2 = 171.18$ lb, $S_2^2 = 1804.55$ lb, and $\bar{x}_1 = 155.47$ lb, and $S_1^2 = 1490.80$ lb.

Because she has both unequal sample sizes **and** suspiciously different estimates of the variance, she realizes that a Welch's t test is best,

$$t = \frac{167.12 - 161.28}{\sqrt{\frac{848.58}{185} + \frac{719.62}{215}}} = 2.0733 \,,$$

and the degrees of freedom needs calculation,

$$df = \frac{\left(\frac{719.62}{185} + \frac{848.58}{215}\right)^2}{\left[\left(\frac{719.62}{185}\right)^2 / 184\right] + \left[\left(\frac{848.58}{215}\right)^2 / 214\right]} = 396 \,.$$

Results: She looked up the value $t_{\alpha/2=2.5\%,\, df=396} = 1.966$ and found the results statistically significant. So, with both large sample sizes similar and the effect size, $\bar{x}_2 - \bar{x}_1$, relatively small, the effect of the change in calculation methods was so small as to have no difference in application. However, this was the **best t test** for analyzing such unbalanced data.

5.3.0.2 Small $n_1 \neq n_2$ (resurrected)

Methods: Harking back to the scenario where she could only get a couple of small unequal samples in the time available; $n_1 = 13$ and $n_2 = 15$.

Analysis: She decided the best way to do the above small sample test was to use Welch's test. She had calculated $\bar{x}_2 = 171.18$ lb, $S_2^2 = 1804.55$ lb, and $\bar{x}_1 = 155.47$ lb, $S_1^2 = 1490.80$ lb. Because she has unequal sample sizes **and** suspiciously different estimates of the variance, she decides to use a Welch's t test,

$$t = \frac{171.18 - 155.47}{\sqrt{\frac{1490.80}{13} + \frac{1804.55}{15}}} = 1.02485 \,.$$

The denominator is a tiny bit **less** than the pooled variance approach. However, remember that

---

[4]The df for Welch's test may alternately be found using a rearrangement of Equation 5.7,

$$df = \frac{\left(\frac{S_1^2}{n_1} + \frac{S_2^2}{n_2}\right)^2}{\left[\frac{\sum(x_{i,1} - \bar{x}_1)^2}{n_1}\right] + \left[\frac{\sum(x_{j,2} - \bar{x}_2)^2}{n_2}\right]} \,.$$

the degrees of freedom also needs some calculation,

$$df_S = \frac{\left(\frac{1490.80}{13} + \frac{1804.55}{15}\right)^2}{\left[\left(\frac{1490.80}{13}\right)^2 / 12\right] + \left[\left(\frac{1804.55}{15}\right)^2 / 14\right]} = 26 .$$

Results: Once again the result is tiny, but the point is that it can be much larger and even make the difference between statistically significant and not statistically significant. When the sample sizes are **different**, **Welch's t test** is the best choice. Notice in the above two examples that the apparent loss of power (sensitivity implied by a smaller t value) by Welch's t test is so small as to be meaningless in this situation.

If you use Welch's t test followed by a confidence interval for the effect size (say, the difference between the means) it would be reasonable to use the degrees of freedom, df_S, from the Welch's and the Satterthwaite standard deviation, S_S.

Some computer packages default to Welch's t test if the sample sizes are unequal, in case the variances are really unequal.[5] However, unless the sample sizes are different, and especially if the variances are the same, there should be little loss of power. Even though Welch's test does cost you a little power, you cannot validly, before analysis, go poking around to assess whether the variances are equal in the data you are actually going to analyze.[6] So, it is prudent (more conservative) to use Welch's t test for simple testing between two means if the sample sizes are **different**, and it is essential if you suspect that the variances are also different. The required sample sizes can be assumed to be about the same as needed for Student's t test, but never smaller. As we shall see, **for small samples**, nonparametric Wilcoxon tests are to be preferred over either Student's or Welch's t tests (Section 14.1.1). Welch's t tests is **usually** an **acceptable** alternative to Student's t test.

5.4 Paired t Test

Suppose a company is testing a new nail polish against its established product, Blotto, both of which have a color changing feature to show when they are fully cured. The researchers measure the length of time it takes to apply and cure on each hand. Each of these products are in a special class that have a color change indicating when they are done. They are a little slow, but a lot of customers like the color change feature. Because each woman has two hands and the ladies' nail sizes and conditions, the researchers realize they can use one selected hand of each of n_d ladies for each of the two treatments. This would surely result in a substantially lower variability than if a different lady was used for each comparison. As always, researchers have to randomize all the details including which hand, right or left, gets which product and which treatment is applied first on each lady.

To test pair i of products H and K, they treat the $\mathbf{d}_i = x_{K,i} - x_{H,i}$ values as the variable of interest and proceed with a one sample Student's t test with the hypothesized $\mu_{\mathbf{d}_0} = 0$. (Using various forms of the letter d, this one being lowercase bold, for difference in various contexts is unavoidable. You might actually consider reading any $\mathbf{d}$ as 'difference', but in this case it is the *paired difference*.)

---

[5]In the R language, $t.test(x, y)$ defaults to Welch's t test. If the sample sizes are equal or if you are certain (is that possible?) the variances are equal, use $t.test(x, y, var.equal = TRUE)$ in the R language. Minitab similarly defaults to Welch's test.

[6]Do not use the popular two-step approach for a two sample test, where the assumption of equal variances is first tested using Levene's test, and then a choice between Student's t test or Welch's t test is made, based on the outcome of Levene's test (Section 5.3). Because the statistical power for Levene's test is often low, researchers will inappropriately choose Student's t test instead of the more robust alternative, Welch's test (Delacre, Lakens, and Leys 2017; this citation is worth reading).

They collected the data (in a double blind manner) on n_{d} pairs (of ladies' hands in this case). This is formalized as the ***paired t test***,

$$t = \frac{\overline{d}}{S_{\mathrm{d}}/\sqrt{n_{\mathrm{d}}}} \qquad df = n_{\mathrm{d}} - 1\,, \tag{5.8}$$

where $S_{\overline{\mathrm{d}}} = S_{\mathrm{d}}/\sqrt{n_{\mathrm{d}}}$. It assumes the sample mean difference, $\overline{\mathrm{d}} = \sum_{i=1}^{i=n_{\mathrm{d}}} \mathrm{d}_i/n_{\mathrm{d}}$, is normally distributed and this assumption could be met by the CLT, if the sample is large enough (Larsen and Marx 1986). Notice that once they have the total, n_{d}, of the individual differences (each d_i), the process of calculation is the same as for the single sample t test; they are just using the per lady differences instead of simple x_i values. Thus they can choose to evaluate this using either $t_{\alpha, n_{\mathrm{d}}-1}$ or $t_{\alpha/2, n_{\mathrm{d}}-1}$, depending on whether they have a one tail or a two tail hypothesis, respectively. This number, $n_{\mathrm{d}} - 1$ degrees of freedom may look small, however, the estimate of the standard deviation is often very greatly improved by pairing (Rubin 1973). Such would definitely be the case in this hypothetical situation and for many before-and-after treatment situations, or if each of a trial of a pair of treatments was run on a different farm.[7] The resulting test is equivalent to a complete block ANOVA (an RCBD, Subsection 9.4) with only two treatments. Pairing is a very important design in itself and this text will eventually reveal additional applications.[8]

5.4.0.1 Example: Pairing

Pairing on one subject per pair is common in many fields.

Folks with two hands Assume a group of researchers really did the nail polish experiment as follows. They formed a hypothesis pair

H0: THERE IS A DIFFERENCE

HA: THERE IS NO DIFFERENCE,
and an $\alpha = 5\%$ criterion. Calculations resembling an ordinary two tail t test with an $\alpha/2 = 2.5\%$ cutoff would follow.

Method: As described above, which of the same lady's hands was to receive either product H or product K would be at random. The product was applied in a double blind manner to a total of $n_{\mathrm{d}} = 20$ pairs of hands. Even the time keeper didn't know which brand went to which hand.

Analysis: They chose to create a table of times, $x_{i,j}$, and calculations:

---

[7]Note that the sign test, although based on a rather weak binomial model, is another example of pairing.

[8]The *covariance* of X_1 and X_2, $\mathrm{Cov}(X_1, X_2)$ is discussed later, but can be seen as a measure of how much of the variance estimate is really due to X_1 and X_2 varying together. $S_{\overline{\mathrm{d}}}^2 = S_{\mathrm{d}}^2/n_{\mathrm{d}}$ is smaller or larger because it lacks the confounding influence of the variables varying together. Thus it is a better estimate than the pooled variance of the means, $S_{\overline{x}_{1,2}}^2$. This is because when X_1 and X_2 are unpaired, the observed variance is $\sigma_{\mathbf{X_2}-\mathbf{X_1}}^2 = \sigma_{X_1}^2 + \sigma_{X_2}^2 + 2\mathrm{Cov}(X_1, X_2)$, and will be an over (or under) estimate if $\mathrm{Cov}(\overline{x}_2, \overline{x}_2) \neq 0$ (Snedecor and Cochran 1967, p 106). That is, $\sigma_{\mathrm{d}}^2 = \sigma_{\mathbf{X_2}-\mathbf{X_1}}^2 - 2\mathrm{Cov}(X_1, X_2)$.

x_H	x_K	$d = x_K - x_H$	$d - \bar{d}$		$(d - \bar{d})^2$
19.52	19.88	0.36	2.478		6.140484
18.55	21.95	3.40	5.518		30.448324
22.37	10.53	-11.84	-9.722		94.517284
19.37	10.53	-8.84	-6.722		45.185284
22.55	19.95	-2.60	-0.482		0.232324
17.31	19.39	2.08	4.198		17.623204
12.43	13.67	1.24	3.358		11.276164
10.19	10.11	-0.08	2.038		4.153444
10.40	12.60	2.20	4.318		18.645124
21.22	15.18	-6.04	-3.922		15.382084
23.37	9.53	-13.84	-11.722		137.405284
19.34	18.46	-0.88	1.238		1.532644
13.58	13.02	-0.56	1.558		2.427364
11.31	9.39	-1.92	0.198		0.039204
10.19	11.11	0.92	3.038		9.229444
22.34	10.46	-11.88	-9.762		95.296644
13.49	10.81	-2.68	-0.562		0.315844
12.25	22.25	10.00	12.118		146.845924
21.46	10.74	-10.72	-8.602		73.994404
13.49	22.81	9.32	11.438		130.827844
		$\bar{d}$ =-2.1180	Σ =0.0000	SS	SS_d =841.5183

They used the first columns to find the $i = 1$ to 20 differences, d_i, and then they used these differences in the equivalent of a t test

(This approach allows easy visualization and calculation of the sum of squares and their precursors. It is the classical way of doing hand calculations, and it is useful to show what is happening. It is not particularly different from how the preliminary calculations could be done on a spreadsheet. Ideally, a brief program could be written to do this in any of many computer languages.) The last column totaled to the sum of squares, $\sum(d - \bar{d})^2 = 841.518320 \, \text{min}^2$. Therefore the variance was $S_d^2 = \sum(d - \bar{d})^2/n - 1 = 44.290438$, where $n_d = 20$ is the number of pairs. So, the standard deviation was $S_d = \sqrt{S_d^2} = 6.15511$ min. (This could be called 'the standard error of the difference', but Student's paired t test usually designates it 'the **standard error of the mean difference**', $S_{\bar{d}} = S_d/\sqrt{n}$.)

But they used an alternative t equation that uses the variance $S_d^2 = 841.5183/19$ of the difference. (Either approach can be used to fill in $S_{\bar{d}}$.) They found Student's t for the paired differences, using

$$t = \frac{\bar{d}}{\sqrt{S_d^2/n_d}} = \frac{-2.1180 \, \text{min}}{\sqrt{(44.2904 \, \text{min}^2)/20}} = -1.42327,$$

which rounds to a four significant figure value of t, (as is commonly tabled), in this case, $t = -1.4233$. Because the t distribution is symmetrical, the direction, whether + (favoring K) or - (favoring H), is irrelevant to the question of statistical significance. For that, they were only interested in the absolute value (as though positive) of t. Their result was so much smaller (meaning **more** likely to be due to chance alone) than the t-cutoff, $t_{\alpha/2=0.025, \, 19df} = \pm 2.0930$, that they could **not** reject **H0**. Therefore they did not know anything new, other than the estimates of the variance and the standard deviation that could be useful for planning later research.

Discussion: The Student's t test does not correct for non normality; it corrects for having to estimate the variance with a less than huge sample size that is **big enough that the CLT** (the Central Limit Theorem) **is able to compensate for any lack of normality**. The sample size, $n_d = 20$ pairs, might be too small to depend on the CLT, and in any case, small samples usually lack the sensitivity (power) to detect the difference you have chosen. Except in cases of data that is known to have extreme skew, the author generally trusts that $n \geq 30$ **individual** (unpaired) sample values is large enough to assure that the distribution of the means will converge to normal. But there remains the question of whether $n_d = 20$ **pairs** of sample values is large enough to assure that the distribution of the means converges towards normality as much as does the distribution of the means of $n = 40$ **individual** sample values or even $n = 30$ **individual** sample values. One way to look at it is that subtraction is simply adding a negative number to another number so that the result is a weighted sum, as is a mean. **If this is true in application**, then $n_d \geq 15$ **pairs** of sample values should usually produce a distribution of means that is sufficiently close to normal for a Student's paired t test, but I am unable to find a study demonstrating that to be true. So, I suggest using the Wilcoxon paired sample test (Subsection 14.2.0.1) for $n_d < 30$ pairs of sample values because it is not dependent on an assumption of normality.

Yes, I contrived the data, starting with random numbers that had the same mean and variance. Then I added a small random difference which had a normal distribution (for H) and subtracted a similar small random difference (for K). As for other details, the real population difference should have been about $\mu_d = 2.5$ minutes. Apparently the sample size was not sufficient to detect this difference, given the variance of this particular experiment. However, the real life experimenters would not have the advantage of knowing the unknowable true mean difference.

5.5 Some Measurement Scales Do Not Work for t or ANOVA

Tests based on the normal curve require equal interval scales. *Equal interval* means that the size of a unit (or range of variables) is the same no matter where it occurs on the scale. *Likert scales* (for example, satisfied, very satisfied, no opinion, dissatisfied, very dissatisfied, even rating on a 1 to 10 scale) do **not** provide uniform distances between units, so they **can't** be used for parametric tests such as the t test and its relatives.

> **CAUTION:** Likert scales and other non-equal interval scales are common in psychology and the social sciences, but using them with inferential tests based on the normal distribution **cannot** produce valid results. Likewise their usefulness in regression and correlation is limited by the fact that parametric tests of slope will be invalid.

You cannot really add the values from such scales and get anything like a meaningful sum. So, how can you really feel comfortable calculating a mean or a standard deviation, both of which involve summation, using Likert scales or non-equal interval scales? Yes, it is common for researchers, the media, and government agencies to do this. However, this distorts the distribution and can even have an effect on randomization. Maybe people prefer 1, 5, 7 and 10 above some of the other values regardless of what the question might be. IQ scores may have similar problems; the lower IQs, 70 and below, are often linked to illnesses or terrible environmental deprivation, whereas the highest IQs, above say 135, have been relegated to completely different tests. Two times an IQ of 75 does not make a genius!

    This is a warning that instead of using Student's t test and its relatives, all of which involve means and standard deviations, for Likert scales and non-equal interval scales, you should use tests which **do not require** the means and variances of the data. These include nonparametric tests and Bayesian approaches. Nonparametric tests are primarily **dependent on ordinal position** and are suitable for a lot of social science and psychological data. (Nonparametric tests have been relegated, as per custom, and to facilitate easy understanding, to the latter part of this book, Chapter 14; they are not particularly complex or difficult to use. Bayesian tests seem more alien to the uninitiated and often require a lot of computer resources, so they are not included in this book.)

5.6 Life Without Student?! Normal Approximation of the Binomial

Before going on, consider that calculating exact binomial probabilities of some number, r, of successes, such as coin flips, involves combinations, $\binom{n}{r}$, the so called binomial coefficients (Section 2.4 and Section 2.3). These often require calculating factorial numbers such as $n!$. It turns out that even the factorial of a barely 'large' sample size, $n! = 30!$, can be too large for convenient hand calculation. Even worse is the fact that such huge numbers often lead to an unfortunate degree of rounding by computers. This makes the direct calculation of binomial tests awkward and impractical.

However, we can use an approximation of the normal distribution for an inferential test for discrete data that can be treated as binomial, such as counts, percentages, and discrete (as opposed to continuous) probabilities. Generally, the phrase 'normal approximation' refers to using this approximation in an inferential binomial test.

Recall that a binomial distribution can be completely described with just two values, the sample size, n, and the probability, p, of a success, given a single trial. Thus $\hat{p} = r/n$ is a mean, and it also estimates the probability as well as the relative frequency, $f = \mathbf{r}/N$, of successes in the whole population (bold $\mathbf{r}$ is the number of successes in the whole population). The CLT, the Central Limit Theorem, applies to means and thus to $\hat{p}$. So, with larger and larger sample sizes, n, the distribution of $\hat{p}$ moves closer and closer to a normal distribution. A more formal algebraic definition is the **mean** of the proportion of the successes is

$$\mu = p,$$

and to approximate a normal distribution, its estimate is

$$\hat{\mu} = \hat{p}.$$

The corresponding formal **variance** is defined as,

$$\sigma^2 = p\left(1 - p\right),$$

the approximation of which is estimated by

$$\hat{\sigma}^2 = \hat{p}\left(1 - \hat{p}\right) = S^2.$$

The use of these estimated approximations for inference, as though working with the normal distribution, is called the ***normal approximation***.

Graphs of continuous probability distributions can be thought of as having infinitely many bars, each with its own relative frequency (height), resulting in smooth graphs (Figure 4.1). So another way to think of continuous distributions is to visualize the smooth graph as a line joining the tops of an infinite number of bars. As sample sizes grow larger, the distribution of means of samples drawn from a discrete distribution such as the binomial distribution tends to approach a smooth distribution.

The more symmetrical the true population distribution, the more quickly the distribution of the means will approach normality. Despite the fact that the binomial distribution is truncated at $r = 0$ and at $r = n$, the CLT starts to be satisfactorily applicable at not quite large sample sizes (say $n = 25$), when $p \approx 1/2$, the probability at which the binomial distribution is perfectly symmetrical. This is obvious in Figure 2.1 and Figure 2.2, where $p = 1/2$ results in a symmetrical curve-like pattern approximating the graph of the normal distribution (Section 4.2). This approximation is further justified by the fact that most of the normal curve is within a few standard deviations of the mean.

It is handy to define q as the probability of a failure, so q equals $1 - p$. You can either have a success or a failure, and nothing else, so $p + q = 1$. The probability estimate, $\hat{p}$, is assumed to be variable and to be random in the samples, whereas n, the sample size that you choose, is a constant, and together, these two values determine **both** the sample mean and the sample standard deviation. Note that n and p are all that is needed to define a binomial distribution. Because n and p define the whole distribution, the sample variance is not estimated independently of the sample mean. So, because the mean and variance only require the estimate of one value, p, the normal approximation follows the normal distribution and uses the table of Z; it has **no need for the many tables** of t which are corrected for the separate estimation of the variance as was necessary to create Student's t. Of course such an approximation based test needs a standard error equivalent to $S_{\bar{x}} = \sqrt{S^2/n}$ in the denominator, and substitution gives

$$S_p = \sqrt{\hat{p}\hat{q}/n} = \sqrt{\hat{p}(1-\hat{p})/n} = \hat{\sigma}_p, \tag{5.9}$$

the estimated standard error of the mean, $\hat{\mu} = \hat{p}$. (Question 9 is an example of calculating the above.) A hypothetical probability is often written p_0. This and the above lead us to a t-test-like test (**without** t) for a difference between the hypothetical probability (or hypothetical frequency), p_0, and the sample probability, $\hat{p}$ (which equals the estimated relative frequency), giving

$$Z = \frac{\hat{p} - p_0}{\sqrt{p_0 q_0/n}}, \tag{5.10}$$

where $q_0 = 1 - p_0$. The denominator is the hypothesized standard error.

From a frequentist point of view, probabilities represent frequencies (proportions or percentages of successes in the population). Recall that random sampling means that every member of a population has an equal chance of being sampled. So this single sample normal approximation test uses the proportion of the successes in a random sample to estimate the relative frequency (such as a percentage or proportion), $\hat{f} = r/n$, of successes in the population, where r is the number of successes observed in a sample of size n. Since $\hat{p} = r/n$ is equivalent to an estimated relative frequency, $\hat{f} = r/n$, the normal approximation can be seen as comparing either probabilities or proportions. For further understanding, some of the implications of Equation 5.10 can also be written as

$$Z = \frac{r/n - p_0}{\sqrt{p_0 (1 - p_0)/n}}, \tag{5.11}$$

and this leads to using the binomial approximation when working with counts (below).

5.6.0.1 A view of the approximation that counts

You can also apply the appropriate above approximation to calculate the theoretical numeric mean, μ, and standard deviation σ from a sample of counts of a particular **size**, n. Notice that the mean can be found by multiplying p by n:

$$\mu_r = np,$$

which is estimated by the observed $r = n\hat{p}$, the number of successes in a sample of size n. The standard error of a particular binomial probability distribution can be seen as a mean of the deviations (like vibrations) around the mean probability, so likewise, multiplying $\sqrt{p_0 q_0 / n}$ by n to get the standard error of the counts, r, gives

$$\hat{\sigma}_r = \sqrt{np(1-p)}.$$

Therefore, multiplying the top and bottom of Equation 5.11, $Z = \frac{r/n - p_0}{\sqrt{p_0(1-p_0)/n}}$, by n, you can obtain

$$Z = \frac{r - np_0}{\sqrt{np_0 q_0}}, \tag{5.12}$$

where $q = 1 - p$.

The sign of Z merely indicates the tail, negative being in the lower tail and positive being in the upper tail. For any value of Z, $+Z$ and $-Z$ have the same probability because the normal curve is symmetrical. This is also called the ***normal approximation*** of the **binomial distribution**; the binomial is two valued, success and failure, but Equation 5.12 refers to the actual observed number of successes. Be aware of context. This or variations of it can be used for testing to see if there really is a difference between the number of successes and a hypothesized number of successes..

Do not use t for the normal approximation because the mean and standard deviation are not independent (whereas the assumption underlying Student's t test is that they are independent). The normal approximation works well with large samples using Z because both the mean and variance depend on the estimate of p, so they are not independent.

In any case, you will usually find that in real applications you need large sample sizes to use this approximation (Section 7.24), or for that matter the binomial itself in most situations. Unless p is near $1/2$, and $n > 1000$, which is a minimum design sample size in political opinion polls, the normal approximation for either testing or estimation might not produce the precision you want. However, in some applications, where the reasonable minimum difference is relatively large, $0.25 < p < 0.75$ might be close enough to $1/2$ to be useful in application with less extreme sample sizes.

So far we have been looking at a special one sample case of the normal approximation associated with the ***Bernoulli distribution*** of binomial probabilities, which can be seen as the probability of one particular outcome out of two possible outcomes (such as win rather than lose). Another version of the normal approximation can be extended to two samples in order to compare two different proportions of outcomes (such as more male customers rather than more female customers).

5.6.0.2 Two sample normal approximation, p

For the two sample 'Z test' challenging equal probability (or proportions), you can use the normal approximation,

$$Z = \frac{\hat{p}_2 - \hat{p}_1}{\sqrt{\hat{p}(1-\hat{p})\left(\frac{1}{n_1} + \frac{1}{n_2}\right)}}, \quad \hat{p} = \frac{r_1 + r_2}{n_1 + n_2}. \tag{5.13}$$

It compares the independent probabilities where $\hat{p}_1 = r_1/n_1$, $\hat{p}_2 = r_2/n_2$, and the unsubscripted $\hat{p}$ is the pooled probability $\frac{r_1 + r_2}{n_1 + n_2}$. The numbers of successes, r_1 and r_2, are independent and must be randomly obtained (Snedecor and Cochran 1967). This equation uses a pooled sample standard deviation because the simple null hypothesis is **H0**: THE DISTRIBUTIONS ARE ALIKE. As with most tests, you should strive for equal sample sizes where possible. So, recall that if $n_1 = n_2 = n$, then $2/n$ can replace $\left(\frac{1}{n_1} + \frac{1}{n_2}\right)$ in the denominator.

However, under some circumstances you may wish to avoid such explicit two stage pooling by using

$$Z = \frac{\hat{p}_2 - \hat{p}_1}{\sqrt{(\hat{p}_1 \hat{q}_1 / n_1) + (\hat{p}_2 \hat{q}_2 / n_2)}}. \tag{5.14}$$

This equation is essentially the same as Equation 5.13, but it can be handier for some derivations.

5.6.0.3 Example: Two Sample Two Tail Normal Approximation

Diabetes Suppose you were contrasting two obese traditional populations, A and B, with respect to the probability of a given population member having diabetes. (This may also be seen as contrasting the proportion of diabetes victims in each population.) You use the Neyman-Pearson hypothesis model, which requires a hypothesis pair. In this instance you hypothesize:

H0: THERE IS NO TRUE DIFFERENCE,

HA: THERE IS A TRUE DIFFERENCE.
You decide to take two random samples, $n_A = 600$ and $n_B = 600$, and you choose a cutoff, $\alpha/2 = 0.5\%$ for each tail (reject **H0** if $|Z| > 2.5756$) for an overall $\alpha = 1\%$, whether + or -, for the two tails.

Analysis: You count 'successes' (diabetes victims) and $r_A = 500$ persons that have diabetes, so your estimated probability that a member of population A will have diabetes is $\hat{p}_A = 500/600 = 0.833333$, and for population B, the number of successes is $r_B = 240$ that have diabetes, so $\hat{p}_B = 240/600 = 0.400000$. Referring to Equation 5.13, you find the pooled probability estimate,

$$\hat{p} = \frac{r_A + r_B}{n_A + n_B} = \frac{740}{1200} = 0.6166667$$

$$Z = \frac{\hat{p}_B - \hat{p}_A}{\sqrt{\hat{p}(1-\hat{p})\left(\frac{1}{n_A} + \frac{1}{n_B}\right)}} = \frac{0.400000 - 0.833333}{\sqrt{0.616667\,(0.383333)\,(2/600)}} = \frac{-0.433333}{0.0280707} = -15.4372,$$

the absolute value of which exceeds $Z_{\alpha/2=0.005} = 2.5756$. Therefore you reject **H0** at the $\alpha = 1\%$ overall (two tail) level and you accept **HA**, that there is a difference.

Discussion: The very high absolute $Z\,value$ (which means the observed p_value is very tiny) may seem interesting, **but** that has nothing to do with the model behind the Neyman-Pearson approach, because under it, p_values are appropriately considered confounded in that they have their own inseparable sources of variation. So, you can only claim significance with your a priori $\alpha = 1\%$. This strict demand that you stick with your initial assumptions and standards adds strength to your conclusions.

The difference between the two samples in this example was huge, so it is not surprising that you got statistical significance. Tests that can detect huge differences will not necessarily be able to detect small differences. This essential question of the power of a test to detect different sizes of differences will be covered in future chapters.

Testing with the normal approximation of the binomial distribution is essentially like any other test. However, as discussed and demonstrated in Section 3.6, there are difficulties with **smallish samples that might be masked, but not necessarily corrected** by using this continuous approximation of a discrete distribution. These difficulties with smallish samples, $n < 60$, will be made worse if the probability is very large or very small ($p < 0.25$ or $p > 0.75$). Another possible source of imprecision associated with using this normal approximation for two sample normal approximations is that parametric tests such as the Z based test assume that the variances remain the same even if the means may differ. The means and standard deviations of binomial populations vary together. Some authors suggest corrective transformations,[9] but these approaches have their own weaknesses. However, actual testing is often handled more conveniently by using 2×2 chi square tests (Section 12.4.1).

5.6.1 Rules limiting the normal approximation

Be sure to remember that binomial sampling cannot approximate a t distribution because the mean and the variance are not independent. You can use the normal approximation by using Z instead of t.

Unfortunately, not all uses of the normal approximation involve $p_0 = 1/2$, so as p_0 moves away from $1/2$, in either direction, you will need larger and larger samples to make the approximation valid. An almost universally accepted rule is that at the **very least**, $np \geq 5$ **and** $nq \geq 5$. **But,** some authors say that it is essential that $np \geq 10$ **and** $nq \geq 10$, and I would agree with this more conservative rule, **if** $p < 0.25$ or $p > 0.75$. Note that you will still have to **consider the power** to determine what sample size, n, will be **large enough** to detect the **minimum difference** that would be meaningful to you.

Different authors have additional validity rules of their own. Box, Hunter and Hunter (1978 p. 130) require $\left| \frac{1}{\sqrt{n}} \left(\sqrt{q/p} - \sqrt{p/q} \right) \right| < 0.3$. Recall that $| \ |$ indicates the absolute value, meaning that whatever the sign, treat the value within the vertical delimiters as positive. Ramsey and Ramsey (1988) imply that if you are using $\alpha = 0.01$ rather than $\alpha = 0.05$, you should require $np \geq 35$. At the very least, if your planned sample size, n, and your hypothesized p_0 fall short of any of these rules, you may wish to increase n, but you should **not** use these validity rules to **limit** the maximum size of n.

Probable Alcoholism, One Tail Test Suppose a highway patrol chief wants to know if it is true that only 40% of truck drivers have drinking problems.[10] The patrol officers (without regard to driving behavior) randomly checked $n = 1200$ truck drivers for both high blood alcohol and for large concentrations of alcohol metabolites, such as certain aldehydes (which involved either blood or urine samples), before 10:30 in the morning. The hypothesis pair is:

H0: THERE ARE $p_0 = 0.40$, OR 40% OF TRUCK DRIVERS THAT ARE DRINKING EXCESSIVELY.

HA: p IS LARGER ($p > 0.40$)
and the criterion is $\alpha = 5\%$, which means (referring to Appendix 4, or whatever) that the one tail $Z_\alpha = 1.640$.
Using the definition created by the testing procedure, there were $r = 500$ drivers who turned out to have been recently drinking too much..

Analysis: Checking the above rules to see if there is any hope of even continuing with the normal approximation shows that obviously $1200 \times 0.40 > 5$, and $\left| (1/\sqrt{n})(\sqrt{(1-p)/p} - \sqrt{p/(1-p)}) \right| = 1\sqrt{1200} \left| \sqrt{0.6/0.4} - \sqrt{0.4/0.6} \right|$, which equals $0.0117 < 0.3$ as per two rules for using the

---

[9] A *transformation* can be seen as putting your data through a function that replaces it with something derived from it, such as its cosign. However, the word transformation could also be used as a synonym for a function.

[10] A CBS news report, March 1, 2010, quoted a National Institute on Alcohol Abuse and Alcoholism study as saying that about 42% of American men and about 19% of American women reported a history of either alcohol abuse or alcoholism during their lives. Unfortunately, the reality might be higher because many people never accept that they have a problem, despite the resulting divorces and estranged family members and arrests and property damage and lawsuits and news photos.

normal approximation (Subsection 5.6.1). So this passes both those checks of whether the normal approximation is acceptable. The estimation $\hat{p} = r/n = 500/1200 = 0.4166667$ is substituted into Equation 5.11 $Z = \frac{r/n - p_0}{\sqrt{p_0 q_0/n}}$, producing

$$Z = \frac{0.4166667 - 0.4000000}{\sqrt{0.4000000 \times 0.6000000/1200}} = 1.178511 \,.$$

Results: Because $1.1785 < Z_{\alpha=0.05}$, **H0** cannot be rejected.

Discussion: OK, ideally, for calculation lots of meaningful figures should be used (actually a scientific calculator handles a lot of those unending sixes).

Not only is the sample not statistically significant, there is no conclusion as to what is true because **H0** is only provisional, under the classical interpretation of the Neyman-Pearson model. Tisk, they went to so much trouble, and now the attorney general is fussing about something.

Question 11 provides an example of this test for you to work (answer at back of book).

5.6.1.1 Example (continued): Probable Alcoholism using counts

Let's just do the calculation, emphasizing counts rather than probabilities:

Analysis: Now, validly using the above Equation 5.12 ($Z = \frac{r - np_0}{\sqrt{np_0 q_0}}$),

$$Z = \frac{500 - 1200 \times 0.40}{\sqrt{1200 \times 0.40 \times 0.60}} = 1.17851 \,.$$

Results: Because $1.1785 < Z_{\alpha=0.05}$, **H0** cannot be rejected.

Discussion: Either approach, comparing probabilities as in the previous version of this example, or counts as done here, comes out with the same conclusion (even the same $Z\ value$): not statistically significant.

You could use the same approach to compare two counts, r_1 and r_2, out of the same sample, or out of two equal samples of the same size. However, actual comparison testing of counts is often handled more conveniently by using 2×2 chi square tests (Section 12.4.1).

5.6.2 Continuity correction

When testing to see if r is significantly different from your hypothesized mean, the stair stepping illustrated in Figure 3.1 and Figure 3.2 results in gaps that may have an effect as though their bounds had equal probabilities, that is, **as though** $P(r) = P(r+1)$. So, it is popular to compensate by adding $\pm 1/2$ to your observed number of successes, r, and to test if $r + 1/2 > np_0$ or if $r - 1/2 < np_0$,. This is a conservative approach in which you **add** $1/2$ to correct your estimate of the **lower tail** probability or **subtract** $1/2$ to correct the **upper tail** probability. That is, this adjustment reduces your chance of false statistical significance that is due to the stepping artifact created by evaluating discrete data using a continuous scale. That is, if $r = 7$, the continuity correction stretches the number 7 between 6.5 and 7.5 and then uses the upper or lower value, whichever is most conservative (least likely to achieve false statistical significance). Many authors support this ***continuity correction*** of adding **or** subtracting $1/2$, but many others disagree. In simple testing, the continuity correction can be appropriate, especially when small sample sizes are involved. As samples grow large, $n > 30$, its effects tend to become negligible.

5.7 Finite Population Correction

We have been discussing sampling with replacement. With it, you can consider the population as being infinitely large, that is, as though you replace each sample before randomly taking another. Under that assumption, you can conceptually treat the results as though they represented one of an infinite number of repeated results. You could take a random sample of orphans in the world, take their blood counts, and then ship them to adoptive mothers without actually replacing them, but the number of orphans in the world is so enormous that you would not measurably bias the estimate with regard to the proportion not examined. So, you can treat samples that are only an infinitesimal proportion of the population as though they were sampled with replacement, even though you do not really replace them.

But what if your sample is half of the population, wouldn't that distort such things as your estimate of variability? *Finite* means having definite limits; it is the opposite of infinite. Merely knowing the total population size, N, does not mean you can completely survey it. Even sampling $> 5\%$ of a population may require corrections for population size. In addition to adjusting for the estimation of standard error of the mean (below), you would have to adjust your sample size estimate, $\hat{n}$, as in Subsection 7.12.

5.7.1 Finite population, correction for $S_{\bar{x}}$

Sampling from a finite population actually **reduces** what the estimated **standard error of the mean** will be, regardless of whether you are sampling discrete or continuous data. Suppose you are studying the progress of children in a small school district with only, say, $N = 400$ total students. However, you only have the resources to measure the social integration of some of the children, say a random sample of $n = 100$. You cannot put the already studied children back in to be exposed to random selection again; that would cause a lot of children to be sampled twice, or even more times. So, you must sample at random without replacement. This sample provides an estimate of the mean, $\bar{x} = \sum x/n$, that has its own standard deviation, $\sigma/\sqrt{n}$, the standard error (or S.D.) of the mean. If you could sample all the children, $n = N$, there would be only one possible estimate of the mean and its variation; because there can only be one such sample, the standard error of the mean would be 0. That is, as your sample gets larger, its estimate of the sample mean has less and less standard error of the mean (it is less variable). But whenever you did such an experiment, sampling $n = 100$ out of $N = 400$, you would be leaving out about $1 - {}^n/_N = 3/4$ of the population of potential sample means and you would have about $1 - (100/400) = 3/4$ of the variability of the possible outcomes left over. More specialized books go into sampling finite populations, using more complex designs (e.g. Valliant, Dorfman, and Royall 2000), but for uncomplicated random sampling you only have to apply a simple correction to your estimate, the standard error of the mean.

5.7.1.1 Finite correction for continuous distributions

If the sample, n, is **greater than or equal to** one twentieth (5%) of the whole population size, N, you should use the corrected estimate of the standard error of the mean, which is adjusted for a finite population,

$$S'_{\bar{x}} = S_{\bar{x}} \sqrt{\frac{N-n}{N-1}}, \tag{5.15}$$

where $\sqrt{(N-n)/(N-1)}$ is the *finite population correction* (Isserlis 1918).[11]

---

[11]The -1 in $\sqrt{(N-n)/(N-1)}$ can be seen as caused by the effect of estimation. However, with a reasonably large N, it is such a small effect that some authors leave out the -1 (e.g. Snedecor and Cochran 1967).

The finite approximation correction can be adapted to other standard error estimates (other parameters), especially when referring to the standard error of a sum of counting data, sometimes symbolized by the letters FPC instead of $S'_{\bar{x}}$.

5.7.1.2 S'_p, finite population correction for normal approx.

The finite population corrected standard error of the probability (the proportion of successes) with random sampling is

$$S'_p = \sqrt{\frac{\hat{p}\,(1-\hat{p})}{n}} \left(\sqrt{\frac{N-n}{N}} \right) \qquad (5.16)$$

when applying the normal approximation for proportions, regardless of whether $p = p_0$, as in testing, or if $p = r/n$, as in considering numbers of observed successes. This could be written $S'_p = S_p \left(\sqrt{\frac{N-n}{N}} \right)$, where S_p can be either $\sqrt{\hat{p}\hat{q}/n}$ or $\sqrt{\hat{p}\hat{q}\left(\frac{1}{n_1}+\frac{1}{n_2}\right)}$, for one sample or two samples respectively. The estimated proportion, $\hat{p}$, is also the sample relative frequency of successes (the probability estimate under the frequentist approach) for the population[12].

5.8 Failure to Reject

Suppose a researcher *does not get statistical significance*, meaning that the researcher cannot reject a null hypothesis, **H0**. Often the researcher has been **incorrectly told** to claim H0 is true just because it is not rejected. The absurdity of attempting to accept **H0** is summarized by the title of a paper by D.G. Altman and J.M. Bland (1995), "**Absence of evidence is not evidence of absence**". Some professors and graduate advisers correctly tell their students that **even negative results** should be **reported** so that everyone does not go on repeating possibly useless experiments. However, the question might remain as to whether the experiment had enough power (sensitivity) to be worth reporting. When a researcher publicly reports failure to reject, whether in print or in an official report, it must not be taken as implying meaningful evidence refuting the alternative hypothesis, **HA**, or supporting the null hypothesis, **H0**. An a priori power analysis will certainly give more credence to the need to report your particular statistically nonsignificant results.

It is commonly said that every experiment gets redone again and again so that you needn't worry about reporting an occasional false outcome. This is often **far from true** if the experiments were very expensive or if they involved difficult human subject arrangements, sacrificed animals, were dependent on rare opportunities, etc. Also consider how your negative results might be misrepresented in the press, which is usually more interested in shock and entertainment than in accuracy. Such distorted reports can create a bias such that a chance false statistical nonsignificance might unfortunately be taken seriously. That can have harmful effects on people's behavior and on political decisions, even though statistical nonsignificance is meaningless.

Due to the effects of chance, not all these difficulties are completely avoidable. However, sample sizes that are adequate without being ridiculously large will reduce the number of false negatives. Although they do not provide conclusions regarding the hypothesis pair itself, it would be very useful if there were provision for a brief description of all properly obtained negative outcomes to be available

---

[12]For the normal approximation, $\sqrt{\hat{p}\hat{q}/n}$ or $\sqrt{\hat{p}\hat{q}\left(\frac{1}{n_1}+\frac{1}{n_2}\right)}$, $\hat{p}$ and $\bar{x} = n\hat{p}$ are jointly dependent on the estimation of p, thus reducing any need for the -1 in Equation 5.7.1.2. So, Equation 5.16 uses the simpler Snedecor and Cochran (1967) version of Equation 5.15. This correction, $\sqrt{\frac{N-n}{N}}$, would also be appropriate for the multinomial distribution.

from literature. These should include experimental technique, sample sizes, means, standard deviations or variances, achieved power, an estimate of skew where available, and a description of any tendency for outliers or loss of subjects. In some instances a single CI would be useful, unless they are CIs of effect sizes.

5.9 Cohen's d, etc.

Cohen was a hero in that he called the attention of researchers to how important a priori sample size estimation is (1962). Unfortunately almost all his indices and standards were proved to be dead ends decades ago and yet they are common in textbooks and the internet to this day.

He suggested a standardized effect size created by dividing the original difference between means, $D = (\bar{x}_2 - \bar{x}_1)$, by the population standard deviation,[13]

$$d = \frac{\bar{x}_2 - \bar{x}_1}{\sigma}.$$

He thought his d could also be used for sample size estimation. His work in this area has led to a group of variations on this theme that are generically referred to as **Cohen's** d. Cohen's d is unitless because the top and bottom units cancel out. It implies the use of the unknown and unknowable standard deviation, σ, as though it were a universal unit of measure. Figure 7.1 happens to have a Cohen's $d = 2$, known because it is an example with a contrived σ for illustration. In application, it is **impossible to calculate Cohen's** d as defined because you never know the exact value of σ, so you must use **some estimate** of the standard deviation. Depending on the estimate of σ you use, there are a number of different names for these standardizations. Each variant gives a slightly different value. Of course you could view Student's t test as $t = d\sqrt{n/2}$ and see that d is part of the calculation of various results. But d alone leaves out the $\sqrt{n/2}$ (or $\sqrt{1/(1/n_1 + 1/n_2)}$), and loses such advantages as Student's use of t to adjust for variance, and even if we tried to relate it to Z, the **denominator no longer** refers to the distribution of the means or the differences between them. It certainly is not a reasonable standardization for discerning differences; for inference, you cannot standardize something (the means) with the standard deviation of something else (the raw data).

The way Cohen's d is used, and is intended to be used, is as though different sets of data may be put on the same scale whether they are in the same universe or not. In almost all attempts to use it in application, careful reexamination suggests that the sets of data were indeed not in the same universe. Aside from that, everyone, myself included, has also, at first, been excited by his promising looking unit-free approach by using what appears to be standardization. Imagine, a standardized way to compare effect sizes from tests of differences in happiness with effect sizes of tests of contentment, and being able to do so even if they are measured on different, even semi-arbitrary scales! Well, that does not exaggerate some folks intent in using Cohen's d.

Note that the standard deviation is independent of the mean and both of them are subject to chance. Such an estimate would need additional adjustment for the involvement of chance in the **estimation of** σ, which is what Student did by using t, which deviates considerably from $Z = \frac{\bar{x}_2 - \bar{x}_1}{\sigma/\sqrt{n}}$, one of the details that Cohen audaciously skipped. This certainly relates to reports of its lack of robustness (e.g. Guillaume, Rousselet and Garstats 2018 https://garstats.wordpress.com/2016/05/02/robust-effect-sizes-for-2-independent-groups/). Although not completely rejecting every possible use of Cohen's d, Wilcox (2006) reviews many faults of it, including that a tiny difference in one tail or a single out-

---

[13] The absolute value of Cohen's d is a manipulation of an older measure of distance (from the mean), Mahalanobis distance, usefully applied in large data involving many variables at a time, such as factor analysis (Mahalanobis, 1936).

lier can create a huge difference in Cohen's d. Estimates of d tend to be confounded (Lenth 2001), meaning they have inseparable variance components. Also, because the standard deviation is never known and must be estimated, it is affected by sample size (in effect the df). Barnette and Mclean (1999, 2001) actually ran simulations that demonstrated that Cohen's d, and standardized effect size (SES) in general, are **strongly affected by sample size** (also see Barnette and Mclean 2002, and Owuegbuzie and Lavine 2003). Therefore, Cohen's d **cannot serve** the needs of a reasonable minimum difference for sample size estimation. **Cohen's d, and similar 'standardized' effect sizes** are a disaster, an ongoing train wreck.

Cohen published a set of values of d for "small effect", "medium effect", and "large effect" (Cohen 1977). These magic numbers are still widely published on the internet and in texts, even though Cohen, who worked primarily in behavioral science, has since stated:

> "The terms "small," "medium," and "large" are relative, not only to each other, but to the area of behavioral science or even more particularly to the specific content and research method being employed in any given investigation. In the face of this relativity, there is a certain risk inherent in offering conventional operational definitions for these terms for use in power analysis in as diverse a field of inquiry as behavioral science. This risk is nevertheless accepted in the belief that more is to be gained than lost by supplying a common conventional frame of reference which is recommended for use only when no better basis for estimating the ES index is available." (1988, p. 25).[14]

This admission understates the problem; it resembles the blithe contention of some instructors that the importance of statistical tests is to avoid bias. If these folks and Cohen are correct we might as well forget about statistical testing in general.

The use of 'small', 'medium', and 'large' even with attempts to recalibrate them for various fields have proved to be misleading at best (Barnette and McLean 2002, 2006; Lenth 2001, Paterson, et al. 2016). It has pervaded the social sciences with a concept of 'small', 'medium', and 'large' reasonable minimum differences which are harmfully in use and common on the internet, in texts, and even in publication and some grant requirements at the time of this writing, 2016. They even crop up in fields entirely outside the social sciences, such as biology, medicine, and business. These erroneous values are commonly used to estimate sample sizes for research projects **even though we have known** for decades that such effect size standards are arbitrary, as are many, if not all, of the alleged effect size measures from which they originate. The frequent inappropriate use of 'small', 'medium' and 'large' as reasonable minimum differences for power analysis has messed up a lot of research and it has also given an undeserved bad name to power analysis.

McCartney and Rosenthal (2000) showed that in intervention areas that involve hard to change low base rate outcomes, such as the incidence of heart attacks, the most impressively large effect sizes found to date fall well below the 0.20 that Cohen characterized as small. Those 'small' effects correspond to reducing the incidence of heart attacks by about half, an effect of enormous practical significance (cited from Lipsey et al. 2012).

Now there are a variety of standardized effect size measures (e.g. Section 9.5), and some of them may relate to differences between parameters other than the means (e.g. Subsection 12.5.1.1). However, different standardized effect size measures are not as easily interconverted as was once thought (e.g. Barnette and McLean 2006).[15]

---

[14]Cohen (1988) also stated that "The meaning of any given ES is, in the final analysis, a function of the context in which it is embedded". Certainly this is especially true of most unitless effect sizes.

[15]Cohen published an assortment of other effect size indexes for various kinds of statistical testing, most of which were erroneously supposed to be interconvertible, and almost all of which have widely published "small", "medium", and "large" values too. These effect indexes almost all suffer from various difficulties, and their published values of

If reporting in original units of measure is impossible, consider using some sort of potentially meaningful ratio such as a percent or probability.[16]

Questions for Ch 5

Write the equations where appropriate and fill in the values. **Odd numbered questions** have worked answers at the back of the book and can be studied as additional examples. The data tends to be fictitious.

1. Student's t test (as compared to a 'Z test') adjusts for what?

2. Operationally, what is the difference between a Student's t test and a 'Z test'?

3. Given **H0**: THE TREATMENTS HAVE THE SAME EFFECT vs **HA**: TREATMENT $_2$ HAS MORE EFFECT THAN TREATMENT $_1$: You need to find out if the result of Treatment $_2$ is really **greater than** the result of Treatment $_1$ at $\alpha = 1\%$ and you get two samples of equal size, $n_1 = n_2 = 15$. Assume the total sample size is large enough that you can be reasonably certain that the means are from a normally distributed population of means. In the initial calculations of a Student's t test, you get means $\bar{x}_1 = 4.200000$ kg, $\bar{x}_2 = 4.653333$ kg, and the pooled sample variance is $S_{1,2}^2 = 7.23847619$ kg^2. Finish the t test.

4. Two observers are each using a very high precision temperature measuring technique. You hope to find if the results from Observer A represents a **different** population from the population represented by Observer B at $\alpha = 0.05$. Test **H0**: THERE IS NO TRUE DIFFERENCE vs **HA**: THE POPULATION OBSERVED BY A IS UNLIKE THAT OBSERVED BY B, given $\bar{x}_A = 8.39174242424243 \,°K$, $\bar{x}_B = 7.15028571428571 \,°K$, and $S_{A,B} = 2.77426146328472 \, K°$. You are only able to obtain $n_A = 15$ observations from Observer A and $n_B = 12$ observations from Observer B. Show formulas and fill in the numbers; what do you conclude? What if t turns out negative? Is there a problem with the sample size?

5. A paired t test reduces the degrees of freedom. Does that make it harder to detect a meaningful difference?

6. Does Student's t test correct for the parent distribution being non normal?

7. The normal approximation does not work well for extreme values of p or q. In application I would avoid situations in which $p > 0.75$ or $p < 0.25$. If I must use a p outside this range, I minimally insist that $np \geq 10$ and $nq \geq 10$ (some authors insist on this minimum in any case). However, **almost all** authors would agree that the normal approximation would be unacceptable when $np \cdots$?

8. Write the one sample normal approximation of the 'Z test' using the $q_0 = (p_0 - 1)$ notation.

"small"," medium", and "large" are additional booby traps for the unwary. Do not under any circumstances use them for power analysis.

[16]Beware of how you interpret the ratio-like value called correlation (the product moment correlation, Section 11.3) with respect to inference. As intended by early workers, it mostly relates to how well the effect fits a straight line relationship. That can be quite misleading because your subject matter can involve curvilinear relationships. If you are only interested in roughly straight lines, correlation is OK, but in many applications it may leave out some information (see Figure 11.3). Do not use the popular interconversions between various effect size measures and correlation for the same reason. They appear to work algebraically, but they are usually converting between unrelated things.

9. What is the **standard error of the proportion** (the probability or relative frequency) of a binomial distribution defined by $n = 72$ and $p = 0.42$? Show an equation and fill in numbers; using a calculator is OK.

10. Estimate the standard error of the successes, r, given $r = 220$ and $n = 1000$. Show the equation and plug in the values (you will have to divide and subtract a little), and then you may use a calculator for the final step. (Recall that $q = (1 - p)$.)

11. Schistosomiasis, or snail fever, is a chronic illness of humans and other mammals caused by a parasite which lives part of its life in certain species of snails. The snails are common in the rice paddies that are critical to the food supply of tropical Asia. It is second behind malaria in terms of socioeconomic and public health importance in tropical and subtropical areas. Assume you have received a grant to control the host snails, hence the disease, in rice paddies. The granting agency has based the funding on the assumption that the snails are common in 45% of the rice paddy watersheds within your very large developing world tropical rice growing region. You suspect that the real percentage is larger. So you form a hypothesis pair and choose a 1% criterion. Your crews randomly sample 360 rice paddy watersheds and find the snails are prevalent in 252 of them. Test whether this is sufficient evidence to make a case for more funding.

12. As of this writing, little is known about the statistics of 911 emergency calls, especially those which are not real emergencies. The percentage that are phantom wireless calls are thought to account for between 25 and 70% of all 911 calls alone. Then there are lonely people who call, drunks who think the fast food place has cheated them, etc. Newspaper workers have told me that, in general, nothing happens on Thursdays. However, both subjects need more detailed study. For instance, are more people going to get drunk or lonely and frivolously call 911 because they are bored on Thursdays? Suppose that a preliminary study shows that out of a random sample of 1000 calls on Thursdays, 620 of them are not emergencies and out of a random sample of 1000 calls on Mondays, 560 are not emergencies. Well, this data cannot completely answer the question because all the days of the week could differ from one another. However, at least pretend that you have a priori produced a hypothesis pair for this preliminary study, **H0**: THERE IS NO DIFFERENCE vs **HA**: THERE IS A HIGHER PROPORTION OF NON-EMERGENCIES ON THURSDAYS VS MONDAYS with a criterion of $\alpha = 5\%$. Complete the test.

13. Do you need to correct for continuity for very large samples, say $n > 60$?

14. Assume you have randomly sampled $n = 50$ from a radiation exposed population of $N = 256$ soldiers and get a standard error of the mean for a prohibitively expensive test, $S_{\bar{x}} = 28.2$. What is the finite population corrected standard error of the mean, $S'_{\bar{x}}$?

15. What is Cohen's d, and how could it be useful?

16. What widely published values regarding standardizing effect sizes have led many authors and researchers into incorrect sample size choices?

Youngster Casting Net

6 CONFIDENCE INTERVALS

In Saipan, my wife and I used to see an old man, accompanied by his grandchild, throwing a net so that it spread out in a circle and fell into the waves. This fisherman could give you an estimate of the probability that one throw would catch a fish. Even if he was 80% confident that a fish would be somewhere in there, he didn't really know if any fish would be anywhere in the net. Using confidence intervals (CIs) is like casting that net, because sometimes what you are so confident of including is not there at all. But, statistical sampling is worse than casting a net; you are not ever perfectly certain that what you want to 'catch' has been included. Even after you finish, you do not know if you have caught what you are after or not, you only know what is the probability of having caught it. All you can report is what part of the real number line the CI covers, and the probability that the parameter really is in there. Because confidence intervals include a range of possibilities and a measure of how confident you are that the value of interest is in there, they might have an advantage over inference alone, which simply puts a limit on how likely you are of having a false statistical significance. Confidence intervals are used for a variety of parameters, including the standard deviation, σ (Equation 8.8), the difference between the means, δ, or whatever. However, most of this section focuses on the CI, which attempts to enclose the mean, μ.

Recall the preliminary introduction to confidence intervals back in Section 4.6.1; we supposed that someone was explaining their latest scheme to their spouse and they asserted that they were 95% confident that they would make between $\$14,000,000$ and $\$16,000,000$ next year. Aside from its lack of an explanation of how they got it, that is a pretty good example of a statistical confidence interval. If properly researched and calculated, this CI would be an interval centering at some sample mean.

Here is a little review and background with respect to the confidence interval for the mean. Consider this sketch of an ideally normal distribution of means:

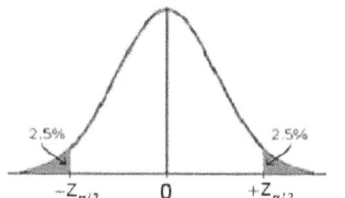

The uppercase letter Z stands for a distance in standard errors of the mean as though they were units. That is, 'one standard error of the mean', $Z = 1$, represents $\sigma_{\bar{x}} = \sigma/\sqrt{n}$ in the **plus or minus direction** from the mean, $Z = 0$. The subscripts $\alpha/2$ in $\pm Z_{\alpha/2}$ in the sketch pertain to the context of what is being measured. The symbol Z is most often used in terms of standard deviations of a population of means assumed to be normal as per the CLT. And, yes the standard error of a population of means is the standard deviation of the means, but it is **not the standard deviation of the population** from which they were drawn. The heights of points on this curve represent the relative frequencies for individual values after *standardization via the standard error of the mean* (conversion to Z values from their original units). From a frequentist approach, these heights also reflect the probabilities of exact values, given random sampling and normality. A horizontal Z value can be seen as cutting off various areas representing summed (integrated) probabilities. In the above sketch, the values $Z_{\alpha/2} = 1.960$ and $-Z_{\alpha/2} = -1.960$; both refer to $\alpha = 5\%$, so each cuts off $\alpha/2 = 2.5\%$ of the probability in each tail (extremity). All values which are more likely than those in the shaded tails are included in between the tails; the probability of a given value being between them is $1 - \alpha$, which in this case equals 95%.

Confidence intervals and tests that assume the normal distribution of the sample means use the same data, but be careful, because the distribution for a CI is centered at a sample mean, which is at a different place than the otherwise similar distribution for testing which is centered at a hypothesized mean under **H0** (the little bell shape curve in the diagram above is located in a different place on the real number line for estimating a CI than for testing).

So, a confidence interval designates $1 - \alpha$ as the probability of including (you could say **trapping**) a value in **original units of measure**. It should not be seen as some sort of opposite of statistical significance testing (Wolfe and Hanley 2002).

Confidence intervals have estimated endpoints, the ***confidence limits*** (or *bounds*), L_1 and L_2, that are values in original units of measure between which the unknown mean (or parameter of interest) has a $1 - \alpha$ probability of really being located. This means we say I am $1 - \alpha$ confident that the mean is between L_1 and L_2. That is, $1 - \alpha$ is called the ***confidence level***, or ***confidence coefficient***.

The confidence interval is part of the parameter space (all the possible outcomes). It is generally reported in original units of measure. Mathematical statisticians hold that the confidence level $(1 - \alpha)$ should **not** be thought of as the complementary probability of α when α is being used **as the level of statistical significance**; confidence intervals and tests are not in the same context. That is, statistical significance refers to the probability of something happening when **H0** is true, so it is based on the distance from the **hypothesized mean**. But the bounds of the confidence interval are measured from either side of the **sample mean**.

For confidence intervals for normal approximation (e.g. using percents or counts) or for very big samples, the confidence limits are

$$L_1 = \bar{x} - Z_{\alpha/2}S_{\bar{x}} \text{ and } L_2 = \bar{x} + Z_{\alpha/2}S_{\bar{x}},$$

or, for continuous data,

$$L_1 = \bar{x} - t_{\alpha/2,\,df}S_{\bar{x}} \text{ and } L_2 = \bar{x} + t_{\alpha/2,\,df}S_{\bar{x}},$$

if the degrees of freedom are not large enough to use Z (which needs $df > 60$ for a 95% CI, and much larger df for larger percentage CI estimates). In either case there is an $(\alpha/2) + (\alpha/2) = \alpha$ probability that the mean will fall outside this darkened area

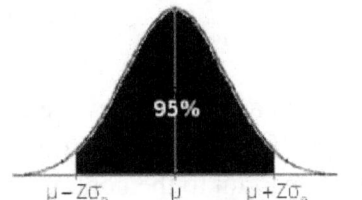

The CIs shown here assume a normal distribution of the means, and random sampling. As for t and 'Z tests', sample size has to be at least $n = 30$ in order for the CLT to usually assure that the means are normally distributed. For reporting a CI for a mean with $n < 30$, you are going to have to justify assuming normality, and that might not be easy. At the 95% confidence level for means with greater than 60 degrees of freedom, you can use $Z_{\alpha=0.05} \approx 2 \leq t_{0.025,\, df>60}$ as a conservative input. That is, for $df > 60$,

$$\bar{x} - 2S_{\bar{x}} \leq \mu \leq \bar{x} + 2S_{\bar{x}}$$

captures the true mean at least 95% of the time. This approximation is only reportable to a maximum of three meaningful figure accuracy for CIs, but this accuracy is only permissible because the approximation of $t_{0.95,df>60}$ is conservative.

The confidence interval for a parameter is **not an acceptance region** of some test for a given parameter, as is sometimes thought. When you are referring to a CI, you are describing a probability of including the true parameter in a given range. But when performing a statistical test, you are considering the probability of falsely rejecting an idea. You could select some total probability, α, other than the common $\alpha = 5\%$ criterion, but it is the one most commonly used for CIs. Using $\alpha = 1\%$ for a confidence interval with probability $1 - \alpha = 99\%$ of including the mean is an extremely conservative alternative that might be suitable if your samples are enormous. If $n \geq 30$ (or $n \geq 50$ with skewness), you are generally safe in assuming the distribution of the means will approach normality sufficiently to assure that using t is acceptable.

Even though testing and CI estimation are in different spaces (contexts), if you use a $1 - \alpha$ CI following a test, then it is less confusing to base it on the same α as you used for the sample size determination. OK, some folks use a 90% confidence interval (reflecting the probability of trapping the true mean), to accompany a significant **one tail** test with $\alpha = 5\%$. This can be seen as a CI leaving off $\alpha = 5\%$ in each tail, and results in a wider, riskier CI.

To summarize, with equations:

For a 95% probability of capturing the value of the true mean between the confidence limits, L_1 and L_2 (Subsection), where $df > 60$, or for the normal approximation, it is reasonable, and conservative, and handier, to round to the traditional $Z_{2.5\%} = 2$. Then you can calculate your 95% CI to roughly three figures as

$$\bar{x} - 2S_{\bar{x}} \leq \mu \leq \bar{x} + 2S_{\bar{x}} \quad df > 60\,. \tag{6.1}$$

However, for greater precision you must use $Z_{\alpha/2}$ for any level of confidence other than 95%:

$$\bar{x} - Z_{\alpha/2}S_{\bar{x}\,or\,p} \leq \mu \leq \bar{x} + Z_{\alpha/2}S_{\bar{x}\,or\,p}\,. \tag{6.2}$$

For $df < 61$ with $(1 - \alpha \leq 95\%)$, or for $1 - \alpha\% > 95$ and $n \leq 1001$, use t for continuous data. So, in terms of confidence limits,

$$L_1 = \bar{x} - t_{\alpha/2, df} S/\sqrt{n}; \quad L_2 = \bar{x} + t_{\alpha/2, df} S/\sqrt{n}. \tag{6.3}$$

Remember to use Z if your CI involved an estimate based on the normal approximation.

So, if you could repeat the experiment forever, the CIs would include the mean, $1 - \alpha = 95\%$ of the time (nineteen times out of twenty), That is,

$$P(L_1 \leq \mu \leq L_2) = 1 - \alpha.$$

At best, most scientists and researchers can only discover, not the truth, but only the **probable** truth about the results of their experiments and observations. Confidence intervals are often presented graphically, as in Figure 6.1. Be certain not to round your data until you are finished calculating. Then round to a reasonable precision in consideration of the precision of the values that you actually obtained (as a rule, to one more figure than the data). **Note**: in application, the difference between $<$ and $\leq$ is meaningless on a continuous scale but $<$ and $\leq$ (or $>$ and $\geq$) might be different (often by one) on a discrete scale, such as a counting scale. When in doubt, use $\leq$ (or $\geq$).

There remains the jarring reality that CIs are often used with non-normal distributions, and in that role, they can be asymmetrical (skewed as after a back-transform of L_1 and L_2). I would prefer to call the corresponding values in this general category the ***upper whisker length***, and the ***lower whisker length***, or, if they are symmetrical, then simply the 'margin of error' (Subsection 7.13, below). When asymmetry (unequal whiskers) is being described, boxplots might be worth considering; see Subsection 15.1.1.

6.1 CI, the Most Useful Approach?

Confidence intervals are the least controversial of all statistical procedures, but through no fault of their own, they are not popular with all authors. Recall that statistical significance may be obtained by chance alone, and that chance sometimes leads us to fail to find differences when they exist. The entire August 2001 issue of *Educational and Psychological Measurement* was devoted to articles about the use of confidence intervals for effect sizes. The a priori choice of confidence intervals without tests might produce very wide CIs when you have no choice but to use small samples, but they might produce less ambiguous results than tests. Jeff Gill (1999) and others have long supported the use of CIs instead of inferential tests; such an approach with suitable sample sizes could be very informative in application if your readers are able to appreciate the ramifications of the CIs. The widespread acceptance of such a stance seems to be a good idea but it also seems unlikely in the near future. However, it could greatly reduce ***non-publication bias***, the phenomenon in which inaccuracy is introduced into the literature through the fact that almost nothing except statistically significant tests are published.

After testing and having observed statistical nonsignificance, you **cannot** in any way suggest that there is or is not a difference between treatments, so comparative CIs, or CIs of effect sizes should not be used because that would tend to mislead the reader that you had somehow found a meaningful difference despite your a priori agreement to adhere to a cutoff in the probability space.

However, a lack of significance does not invalidate your data. If possible you should still publish estimates of means, standard deviations or other parameters, and where appropriate their CIs. In any case, following a test and in the absence of statistical significance, you can only publish two or more CIs (or estimates) if you make it very clear that they are **not presented for purposes of comparison**.

Figure 6.1: Two Ways to Show Confidence intervals

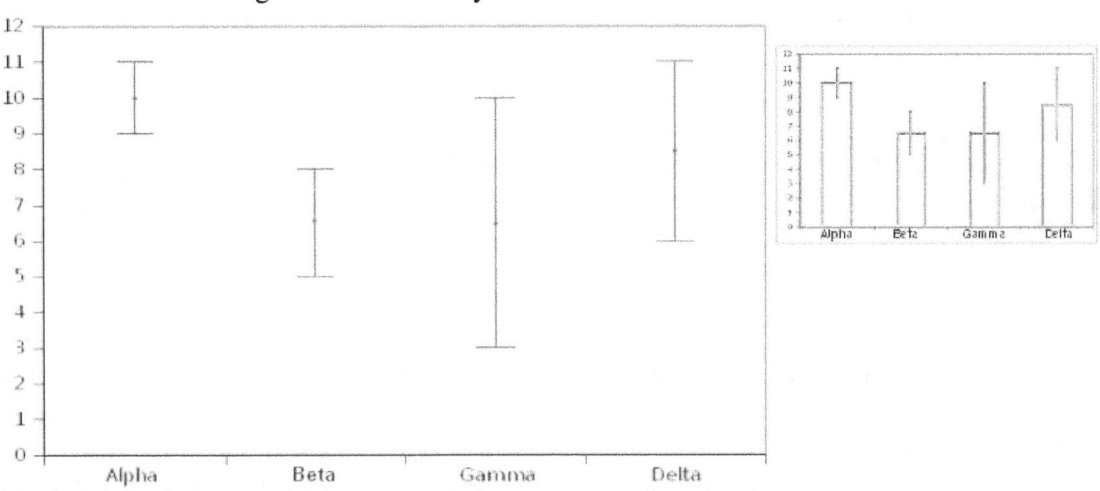

The larger figure is the classical tie fighter representation that is what most older researchers immediately think of when a confidence interval is mentioned. The dot in the center of each CI is the location of the point estimate (e.g. $\bar{x}$). The smaller figure shows the increasingly more popular representation as column graphs in which the height of the bars may represent estimated means. The **whiskers** in either direction from the middle (e.g. the sample mean) run out to the confidence limits. (The names and values are arbitrary and fictitious.)

You should always think about the eventual communication, hence the description of your results. Descriptive statistics such as visual representations are usually as valuable as, often more informative than, numerical information and inference. However, they often work well together, and diagrams such as Figure 6.1 should be accompanied by actual numerical values rounded to no more than one figure more than the input data. Recall that counts are infinitely accurate and can be represented to any number of places; in such cases you should always use good sense.

Confidence intervals are not more commonly reported because they can be embarrassing. Is that science? Authors occasionally leave out confidence intervals because if they are too wide, it suggests that the researcher did not invest enough effort, and if they are too small, it may suggest the results are somehow trivial (Lewicki and Hill 2007). Even though limited resources are rarely the fault of the researcher, failing to report a confidence interval because it is too large is misleading. To not report the CI because it is too small is foolish because the information could be very useful, no matter how seemingly 'trivial'. The probability of having to face either of these situations can be reduced by using the techniques of Section 7.3 to select an initial sample size, n. If you have used an a priori sample size estimation, your confidence interval is clearly justified, correctly centered, and not open to criticism.

Some researchers who hate inferential tests may find CIs (rather than reporting a p_value) worth considering. On the other hand, the techniques in this book are not pure NHST testing, in that this book's orientation is not to discover any true difference no matter how infinitesimal, but rather to **specify what true difference is meaningful and then target it**. Likewise your objective should almost always be to detect a reasonable minimum difference caused by the treatment that is large enough to be of interest to you. Some social science journals are strongly encouraging the use of CIs, particularly because of the following quote from page 22 of the 2001 Publication manual of the American Psychological Association (5th ed.):

> "The reporting of confidence intervals (for estimates of parameters, for functions of parameters such as effects regarding means, and for effect sizes) can be an extremely effective way of reporting results. Because confidence intervals combine information on location and precision and can often be directly used to infer statistical significance lev-

els [sic], they are, in general, the best reporting strategy. The use of confidence intervals is therefore strongly recommended. As a rule, it is best to use a single confidence interval size (e.g. a 95% or 99% confidence interval) throughout the course of the paper."

Most of this is good advice in that it encourages an improvement in prevailing practice. However, it is **unfortunate** the quote included the part about "Because confidence intervals . . . can often be directly used to infer statistical significance levels . . ." because there are **many pitfalls** associated with using CIs for inferring statistical significance. If multiple comparisons are involved, their CIs generally do not correspond to the per-comparison cutoffs for statistical significance, and a huge gaping pitfall for using CIs for simple two sample tests is discussed in Julious (2004). I strongly recommend against risking the use of CIs for inference (Subsection 6.6).

6.2 Chebyshev's Theorem

What if there is no way your data is going to fit a normal distribution? What, then, is the ultimate worst case scenario? ***Chebyshev's theorem***, also called the Bienaymé-Tchebichef theorem (Bienaymé 1853, Tchebichef 1867) applies to distributions of populations or random samples thereof. It gives a bound on the probability of deviation of a given random variable from its mathematical expectation (the mean) in terms of the variance.[1] For any number, $k > 1$, the proportion of the data values that lie within k standard deviations of the mean is at least

$$f = 1 - 1/k^2. \tag{6.4}$$

The bounds implied by the word 'within' are equidistant from the mean. From a frequentist viewpoint, f estimates p, the probability, often expressed as a percent. You can even make estimates based on a hypothetical mean, μ_0, with your randomly sampled results using $k = (\bar{x} - u_0)/S$ as an estimator of k and then estimate the most extreme probability of your sampling outcome being between your two extremes, $\pm kS$, as $p = 1 - 1/k^2$. Of course the probability of $\bar{x}$ being outside your estimate, given μ is true, would be estimated as $1 - p$, because all the probabilities must add up to 1, meaning 100%.

This approach would produce very wide confidence intervals. But because of its simplicity and universality, Chebyshev's theorem has been useful in a number of proofs of the laws of probability, such as the law of large numbers (Subsection 4.3).

6.3 Caution When Planning and Sampling

6.3.0.1 Never graphically explore the same data you are going to test before you test it

Remember that statistical testing is about finding occurrences that are too unusual to occur by chance alone. This does not mean combing through the whole world until you come across something unusual; that has nothing to do with chance. Graphs and other graphics including boxplots, stem and leaf diagrams, etc. are often great for forming hypotheses, but you **cannot** form your hypothesis from the data you are going to use for testing the hypothesis. If, as seemingly implied in some texts, and occasionally on YouTube, you use graphs to examine the sample that is to be tested before doing the inferential testing on it, you are biasing the result.

---

[1] We English speakers are perhaps too free with the spelling of names originating in Cyrillic. Hence 'Chebyshev' and Tchebichef are the same.

6.3.0.2 But, for parametric tests you may need extra data

If you wish to estimate a sample size for a test that assumes normality, you must get a standard deviation or a variance from extra data (also called *training data* in data science). It must have been obtained in a manner that is, except maybe for sample size, obtained **exactly in the same way** as for the tests you intend to run for inference. In some cases this might require running a smaller test without necessarily intending to actually finish calculating test values such as t, Z, or F.

To get a truly representative result, it is best to use the same approaches for finding suitable sample sizes for CIs as for tests. Although smaller sample sizes are technically acceptable for CIs, it is misleading to report a CI for a difference, the existence of which you have not even shown to be likely (statistically significant).

If you have an immense data set, such as from the BRFSS, the Behavioral Risk Factor Surveillance System, you can get a standard deviation, S, from a random sample of the main data base and then run your inferential test on a subsample from what is left over.

So, you can get the standard deviation and you can create graphs to get an idea of the distribution, perhaps including the skew of the raw data, by using data from previous studies or a subsample, but that data can not be included in the test. You can also get a variance estimate, S^2, or its square root, the standard deviation, S, for sample size estimation (e.g. Section 11.8.0.1).

Remember that in addition to other approaches for getting sample size estimates using the standard deviation, S, you should try to get a total sample size, **n**, that is at least 25 times the third moment skew (Section 15.2.1). However, checking that is not always practical because quite large amounts of data are required to really estimate skew. If you have no alternative, assume this requirement is met if you have a total sample size of $\mathbf{n} = 30$ or more and you have no reason to think that there is a great deal of skew (if highly skewed, it would require sample sizes up to at least $\mathbf{n} = 80$).

Nonparametric procedures, Chapter 14, and tests of binomial data sometimes have less demanding ways to find sample sizes and can avoid a big outlay for extra data that cannot be reused.

6.3.0.3 But you might discover something

You will find all sorts of oddities by searching data after the fact, but you must obtain new unexamined random samples to test the hypothesis arising from your exploration. However, you can a priori do all these things **using a random subsample**. If you can obtain sufficient data, you definitely should do so. But, the subset examined **must not** be included in any subsequent inference or prediction, or used for CIs.

6.3.0.4 Why so restrictive?

By now some readers are going to be annoyed that this book is more restrictive than what they have been doing. However, this material arises from peer reviewed journal literature, most of which has been out there and confirmed for years, sometimes decades, or more. Do explore and graph your **preliminary** data in detail, but do **not** include it or its results in your analysis for publication. Exploratory graphs are informative and appropriate after the analysis of the data for tests or CIs is complete.[2]

---

[2]Spreadsheets, especially Excel, are poor for accuracy in statistical analysis (Pavur 2012, McCullough and Wilson 2005, Almiron et al. 2010, Yalta,2008). However, Mélard 2014 (slightly revised 2015) reports a slight, but insufficient improvement in Excel 2010. Excel also produces a selection of attractive graphics which need to be carefully checked against the data due to inaccuracies such as incorrect intercepts. The R and Python languages are more accurate and provide a limitless variety of graphics possibilities; unfortunately, learning R or Python is a bit challenging and time consuming. Python is more versatile than R and is starting to include most of the advantages of R, except that RStudio outshines Python's available editors. Tableau (not a spreadsheet) is good, and more varied,

If we think about the result of a long career in which some researcher based all hypotheses on what was observed in a priori graphical examination of the very data that was to be inferentially tested, or modeled, we might come to expect the outcome of that researcher's career to be retrogression rather than advancement in the researcher's field. Likewise, it is reasonable to expect that researchers who did not pay adequate attention to power have contributed less to their fields than they would have using a priori power analysis. Statistics is not some irrelevant ritual inflicted upon us by some sort of mathematician in need of a job. It is a group of techniques, mostly developed by our fellow researchers, for intelligently coping with reality, the unfortunate pervasiveness of chance effects.

6.3.0.5 CI, error bar or what?

The results of an experiment are sometimes presented in a form like 5.1 ± 0.39, which usually means $\bar{x} \pm Z_{0.025}S_{\bar{x}}$, but in some journals it means $\bar{x} \pm S_{\bar{x}}$, which is not at all the same thing. Fortunately, the latter is becoming less common in modern journals. However, some authors still draw unexplained **error bars** that look like graphical presentations of CIs (left side Figure 6.1), but which may represent + or - the standard deviation, or maybe + or - the standard error of the mean. Even when including standard errors of the mean, these are not confidence intervals unless you pretend you are specifying an $\alpha = 15.9\%$, in which case you might be able to call it a 68.2% CI ($Z = 1$). Always state whether you are presenting **confidence intervals or** some sort of **error bars** and describe the meaning of the latter. Unfortunately each journal has its own house style, and you cannot be certain what is being reported with this $\pm$ notation, especially in older publications, from journals which may have changed their styles one or more times in the past, or may have allowed each author to do it her or his own way (*aaaarrrgh*). The safest and most informative way is to clearly present and label confidence intervals. Sometimes you might even present bars for the CIs in an illustration, but make that clear to the reader if you do it.

6.3.1 Single sample CI for μ or $\mu - \mu_0$

The most common CIs are simply about the **mean**, μ, as shown here:

$$\bar{x} - Z_{\alpha/2}S_{\bar{x}} \leq \mu \leq \bar{x} + Z_{\alpha/2}S_{\bar{x}} .$$

However, **CIs that target the effect size** $\delta = \mu - \mu_0$ using $D = \bar{x} - \mu_0$ can be very informative, but they are only acceptable **when there is statistical significance** (evidence that there is an effect). They are estimated by

$$|D| - Z_{\alpha/2}S_{\bar{x}} \leq \delta \leq |D| + Z_{\alpha/2}S_{\bar{x}}, \tag{6.5}$$

where $S_{\bar{x}} = S/\sqrt{n}$, $D = \bar{x} - \mu_0$. Note that $\delta = \mu - \mu_0$, and that μ_0 is the hypothesized mean, and μ is the unknown true mean. This CI uses the difference between the sample mean and the hypothesized mean $D = \bar{x} - \mu_0$ to try to trap the true difference, the **effect size**, $\delta = \mu - \mu_0$. However, your hypothetical value could be $\mu_0 = 0$, in which case the effect size is estimated by the true mean, $D = \bar{x}$.

Adding or subtracting the same constant to every member of a population, X, will not affect the variance or the standard deviation of the mean upon which the length of the whiskers depends. So, it can be shown that these two, the CI for the mean and the CI for the effect size, will have the same

---

for data visualization and is not excessively expensive. If you want to learn it before purchase, get Tableau public at https://public.tableau.com/en-us/s/ and use the tutorials at https://public.tableau.com/en-us/s/resources. Be aware that if you save your work on Tableau public, it goes **online as freely available**.

width if $S_{\bar{x}}$ and α are the same. However, they will have **different centers**, $\bar{x}$ and $D = \bar{x} - \mu_0$, respectively.

As you shall see in later chapters, a similar approach can be adapted to obtain a CI for almost any parameter. We generally refer to $D = \bar{x} - \mu_0$ as 'the effect size' even though it is only an estimate of the effect size.

Question 7 provides a chance to try this and is worked out in the answers at the end.

6.3.2 CI for a two sample effect size $\delta = \mu_2 - \mu_1$

Figure 6.2: Two independent samples: From Example 5.2.0.2

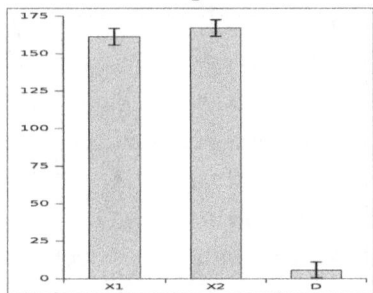

Xs are for μ_1 and μ_2; D is for δ between them; the tie fighters at the tops represent the CIs for the effect sizes of the respective bars; the CI for $D = \bar{x}_2 - \bar{x}_1$ does **not quite** reach $D = 0$.

The CI estimation in this section follows a pattern analogous to the t test, that assumes both means are associated with the same variance, hence the same standard deviation. The true difference, δ, can be treated as a mean (Lane 2011). When reporting outcomes for a statistically significant two sample t test, you can also estimate a CI for each **independent** $\bar{x}$, but you can alternatively estimate a CI using $D = \bar{x}_2 - \bar{x}_1$, which estimates $\delta = \mu_2 - \mu_1$, the **effect size** (e.g. D in Figure 6.2). An effect size is only meaningful if it is associated with a statistically significant test. It is hard to imagine a better way of reporting effect size than using confidence intervals.

Given a sample size that is large enough, the Central Limit Theorem allows you to assume that D is normally distributed. Because you are not estimating each μ_i individually as $\bar{x}_i$, for the true difference between the means, you must use a pooled population standard deviation such as $S_{1,2}$ as the estimate of σ.

The defaulting of t to Z at high degrees of freedom, $df > 1000$, is justified because t approaches Z so closely at sufficient degrees of freedom that rounding will lose the difference anyway. Even at $cutoff = 2.5\%$, if your $df \leq 1000$ (or $df < 4500$ for four figure accuracy), you get more accurate limits by using t, rather than Z, from a table such as Appendix 3. However, for 95% CIs when $df > 60$, you may conservatively use $Z_{\alpha/2=2.5\%} \approx 2$ (1.96 rounded up) for reporting CIs, but with an accuracy of only three figures.[3] For 99% CIs with $df > 4500$, you can use $Z = 2.577$, but that rounded to three numerals, 2.580, as per many tables, would only be close enough for reporting if your data is only to three figures or less.

Using t all the way up to 1000 df (or better yet, up to 4500 df for more accuracy, or if $cutoff < 0.025$) will give your confidence intervals greater precision than classical textbook values that **mistakenly** default to Z when $df = 30$. You can estimate the **limits** of the confidence interval that has a probability equal to $1 - \alpha$ of enclosing $\delta = \mu_2 - \mu_1$, using the degrees of freedom, $df = n_1 + n_2 - 2$, as though for a two tail t test, by using

$$(\bar{x}_2 - \bar{x}_1) - t_{\alpha/2, n_1+n_2-2} S_{1,2} \sqrt{1/n_1 + 1/n_2} \leq \delta \leq (\bar{x}_2 - \bar{x}_1) + t_{\alpha/2, n_1+n_2-2} S_{1,2} \sqrt{1/n_1 + 1/n_2}, \quad (6.6)$$

where $\bar{x}_1$ and $\bar{x}_2$ and $S_{1,2}$ are rounded as little as possible until the last step. For $df > 1000$ you can

---

[3]Using $Z \approx 2$ is conservative because as the degrees of freedom exceed 60 for $\alpha/2 = 0.05$ it will make the CI a little too large in the second or third meaningful figure, but it will never inaccurately reduce the CI (a harmfully misleading outcome) unless $df < 61$.

use

$$(\bar{x}_2 - \bar{x}_1) - Z_{\alpha/2}S_{1,2}\sqrt{(1/n_1 + 1/n_2)} \le \delta \le (\bar{x}_2 - \bar{x}_1) + Z_{\alpha/2}S_{1,2}\sqrt{(1/n_1 + 1/n_2)}. \qquad (6.7)$$

But, most researchers find that their 95% CIs, or confidence limits, are satisfactory to **three places** if they use 2 in place of $t_{\alpha/2=2.5\%,\,n_1+n_2-2}$ at $n > 60$.

6.3.2.1 Simplified 2 sample 95% CIs, $n_1 + n_2 - 2 > 60$

So for 95% confidence intervals, **with** $df > 60$, for two samples assumed to have the same standard deviation, you can be conservative and estimate the CI for the **difference** between the means by

$$|D| - Z_{0.025}\sqrt{S_{1,2}^2\,(n_1 + n_2)\,/(n_1 n_2)} \le \delta \le |D| + Z_{0.025}\sqrt{S_{1,2}^2\,(n_1 + n_2)\,/(n_1 n_2)}, \qquad (6.8)$$

where $\sqrt{S_{1,2}^2\,(n_1 + n_2)\,/(n_1 n_2)}$ is just another way to write $S_{1,2}\sqrt{(1/n_1 + 1/n_2)}$. This approach is conservative, and might even be standard for some journals.[4] Of course, if $n_1 = n_2 = n$, then

$$|D| - Z_{0.025}\sqrt{S_{1,2}^2\,(2/n)} \le \delta \le |D| + Z_{0.025}\sqrt{S_{1,2}^2(2/n)}.$$

You may round $Z_{0.025} = 1.960$ to 2 if you wish, but **beware** of the results from some older papers that may have rounded to "$Z = 2$" even when $df < 61$. Because, at such smallish sample sizes, $t_{0.025,\,df\le 60} > 2$, this is rounding in the wrong direction, resulting in false and misleading reported 'CIs'. The potential for a false interpretation, incorrectly involving statistical significance, is only one of several examples of why using CIs **as tests should be avoided** when looking for probable yes-no answers (e.g. Wolfe and Hanley 2002).

6.3.3 CI for effect size of paired t test

The confidence interval for a paired t test is much the same as for a usual t test (first part of Chapter 6), but with appropriate substitutions, including $|\bar{d}|$, the average of the difference between the members of the pairs, instead of $\bar{x}$, the mean of the individual values. For instance, for $(n_d - 1) > 60$ pairs, you could conservatively report a 95% CI with limits

$$L_1 = |\bar{d}| - 2S_d/\sqrt{n_d} \text{ and } L_2 = |\bar{d}| + 2S_d/\sqrt{n_d}, \qquad (6.9)$$

$$\text{giving } |\bar{d}| - 2S_d/\sqrt{n_d} \le \delta_d \le |\bar{d}| + 2S_d/\sqrt{n_d},$$

with about three digit accuracy. You will have to use t values from tables such as in Appendix 3; when $30 \le (n_d - 1) \le 60$, use $t_{\alpha/2=0.025,\,n_d-1}$. However, you can use Z for $df \le 1000$ for three figure accuracy and $\alpha = 5\%$. For greater accuracy, or $\alpha < 5\%$, use t as from Appendix 3 for up to $df = 4500$. You could also use $Z_{0.005} = 2.576$ from a table such as Appendix 4 for a 99% CI, if $df > 4500$.

It would be nice to know the third moment skew (Section 15.2) and then compensate by being certain that your chosen sample size is **at least** $n > 25 \times skew$. However, enough extra data to a priori estimate the population skew after pairing is often not available. A box plot might help you decide if the skew is excessive (Subsection 15.2.1).

---

[4]The absolute difference notation in Equation 6.8 is to make the equation more readable. It simply means 'put the smaller on the left and the larger on the right'.

6.4 CI for Proportion, p, Using the Normal Approximation

The proportion or relative frequency, $\hat{f}$, of successes estimates the probability, p, of a success on a single random draw. Thus the single sample estimated likelihood (probability) is

$$\hat{p} = r/n,$$

where r is the number of successes observed in a sample and n is the sample size, equivalently the number of trials. The normal approximation is particularly useful when you wish to produce a confidence interval for **probabilities or binomial proportions**. Recall that the normal approximation of the **estimated probability** is $\hat{p} = r/n$ and its variance estimate is

$$S_p^2 = \left(\hat{p}\left(1 - \hat{p}\right)n\right)/n^2 = \hat{p}\hat{q}/n.$$

So the square root of that is

$$S_p = \sqrt{\hat{p}\left(1 - \hat{p}\right)/n} = \sqrt{\hat{p}\hat{q}/n},$$

the *standard error of the probability*, or, what is the same thing (Section 5.6),

$$S_p = \sqrt{(r/n)\left(1 - r/n\right)/n}.$$

Then, a CI of the approximation, using the **normal approximation** of the normal distribution from a binomial sample, is

$$\hat{p} - Z_{\alpha/2}\sqrt{\hat{p}\hat{q}/n} \le p \le \hat{p} + Z_{\alpha/2}\sqrt{\hat{p}\hat{q}/n}, \tag{6.10}$$

where $\hat{q} = 1 - \hat{p}$. The accuracy of the $Z_{\alpha/2}$ value can limit the number of meaningful figures in your result, but r and n are integers and therefore exact.

A more conservative 95% CI, but only for three figure accuracy when $df > 60$ for p is

$$\hat{p} - 2\sqrt{\hat{p}\hat{q}/n} \le p \le \hat{p} + 2\sqrt{\hat{p}\hat{q}/n},$$

where the 2 represents $Z_{\alpha/2=0.025} = 1.960$, rounded up a little. For greater accuracy, or $30 \le n \le 60$, or for confidences other than 95%, you would have to refer to a table such as Appendix 4 and substitute some other $Z_{\alpha/2}$ value in Equation 6.10. Question 9 is an example of this as well as an exercise for this.

The CI from Equation 6.10 is called the *Wald confidence interval* and requires a rather large sample size. Agresti and Coull (1998) provide a more complete discussion and favor their more complicated and less immediately obvious confidence interval, especially for small and not very large sample sizes.

Start with their estimate of $\widetilde{p} = \left(r + Z_{1-\alpha}^2/2\right)/\left(n + Z_{1-\alpha/2}^2\right)$; then their CI estimate is

$$\widetilde{p} - Z_{1-\alpha/2}\sqrt{(\widetilde{p}/n)\left(1 - \widetilde{p}\right)} \le p \le \widetilde{p} + Z_{1-\alpha/2}\sqrt{(\widetilde{p}/n)\left(1 - \widetilde{p}\right)}. \tag{6.11}$$

This *Agresti and Coull CI* might be best at most sample sizes.

6.4.0.1 Example: Normal Approximation CI, One Sample

Normal approximation CI, alcoholism Let's use information from the example in Subsubsection 5.11 to create a $1 - \alpha = 95\%$ confidence interval for the probability, p, that a randomly chosen job applicant will have a drinking problem. To be conservative and for convenience let's use $Z \approx 2$ vs $Z_{0.025} = 1.960$. That is, assume you took a random sample of $n = 1200$

applicants and found $r = 500$ with drinking problems. You calculate $\hat{p} = r/n$, which is $500/1200 = 0.416667$. Then, use

$$\hat{p} - 2\sqrt{\hat{p}\hat{q}/n} \leq p \leq \hat{p} + 2\sqrt{\hat{p}\hat{q}/n}\,,$$

$$.416667 - 2\sqrt{.41667 \times .58333/1200} \leq p \leq .416667 + 2\sqrt{.41667 \times .58333/1200},$$

$$.416667 - 2 \times .014232 \leq p \leq .416667 + 2 \times .014232,$$

$$0.3882 \leq p \leq 0.4451\,.$$

I left this excessively unrounded to consider the question of what if you defaulted to $Z_{\alpha/2=0.25}$, which equals 1.9600, and you wished to be a little more accurate? Using Equation 6.10, this improved estimate gives

$$.41667 - 1.960 \times .01423 \leq p \leq .41667 + 1.960 \times .01423,$$

$$0.3888 \leq p \leq 0.4446,$$

a potentially meaningful change if you are trying to squeeze four (maybe even five) figure accuracy out of your data. That might be reasonable, given the sizes of your n and r.

Discussion: $Z_{\alpha/2=0.025} = 1.959964$, which handily rounds to four decimal places as 1.960. However, even assuming your tables are really that accurate, which is unusual, rounding the result to less than five figures is unnecessary. Z is used rather than t because t does not apply to the normal approximation. The first estimate of the CI was roughly accurate to three figures because it used $Z_{0.024} \approx 2$. So, for publication, it should be reported as

$$0.389 \leq p \leq 0.445\,,$$

which rounds to be conservatively a tiny bit wider (third decimal place). Always round wider, because in statistics you should always give the worst case the benefit of the doubt.

If using a frequentist approach, by equating p to the frequency, you might see this process as estimating a **CI for the frequency**,

$$\hat{p} - 2\sqrt{\hat{p}\hat{q}/n} \leq f \leq \hat{p} + 2\sqrt{\hat{p}\hat{q}/n}\,.$$

This may also be seen as a **CI for the proportion** of the population that represents successes.

6.5 CIs for Difference

More and more journals are requiring that you report your estimate of the **effect size,**, ES, the **difference between the treatments** (observations, etc.), or between your result and a hypothesized ideal. ES can refer to differences in terms of almost any parameter estimates (between two values of the same statistic). Unfortunately, the phrase 'effect size' has become associated with standardized effect sizes, a group of indexes such as Cohen's d, which have long ago been abandoned in theory but not necessarily in practice (Subsection 5.9). If you **conclude** there is a **difference**, as by finding statistical significance, it is only good sense to report an estimate of **how big** a difference you observe, in original units of measure (or as a ratio such as a percent).

> **CAUTION:** Effect size should, where possible, be expressed in the **most meaningful** units of measure, usually the original units of measure.

Although they do not seem to agree with this book's stance that null hypothesis testing is often the best approach we have, Barnette and McLean are brilliant at validly pointing out and even suggesting corrections for the extremely common misuses and abuses of prevailing statistical practice. In a particularly useful presentation, they point out, "Confidence intervals are usually not included in categorizations of effect sizes, but because they are also ways of **describing the magnitude** of effects, they probably should be **included**" (2006, bolding added). The most common effect size in original units is the difference between the means, $D = \bar{x}_2 - \bar{x}_1$, or $D = \bar{x} - \mu_0$. It can be either positive or negative. The symbol μ_0 (mu nought) in $\bar{x} - \mu_0$ is a hypothesized ideal mean provided by you, so μ_0 is not, itself, an estimate. What could be more informative than reporting the ES of D, along with its associated CI?

6.5.0.1 Example: CI for Differences

Plaster pills Assume the food and drug enforcement agency has intercepted an enormous shipment of plaster tablets that externally appear to be genuine standard name brand anti-malarial tablets.[5] The agents are curious about the fact that these plaster fakes are so well made; do they even weigh the same as the real thing?

Procedure: They **a priori** decide to check this out and include a 95% confidence interval in their evidence description document. The technician was double blinded, in the sense of not being told which batch is which. The technician weighed random samples of size $n = 1000$ of each of the real tablets, X, and the plaster tablets, Y, reporting the weights to the nearest tenth milligram and reported the means for the real tablets as $\bar{x} = 104.1$ mg and for the plaster imitations as $\bar{y} = 106.8$ mg, with standard deviations $S_x^2 = 40.1$ and $S_y^2 = 35.9$. The technician had rounded the means reasonably for reporting, but too early for the subtraction process, which produced $D = 106.8 - 104.1 = 2.7$ mg. That is, despite a very large sample, information had been lost due to rounding too early. So the assistant investigator went back to the weigh tickets and found a less rounded $\bar{x} = 104.83$ mg for the sample of real medicine and $\bar{y} = 106.13$ mg for the plaster imitations. This provided a more acceptable accuracy of $D = 106.83 - 104.13 = 2.70$ mg, rather than cutting the accuracy of the final subtraction to only two figures. Notice that, in appropriate **context**, a trailing zero is indeed a significant figure.

Analysis: The difference of $D = 106.83 - 104.13 = 2.70$ mg is better than the first D, but still does not reveal all the information in the original data. However, the boss was demanding a report right away. The assistant investigator chose to calculate the pooled variance using Equation 4.4 combined with Equation 4.5 to get

$$S_{x,y} = \sqrt{\frac{S_x^2 (n_x - 1) + S_Y^2 (n_y - 1)}{n_x + n_y - 2}} = \sqrt{\frac{40.1 \times 999 + 35.9 \times 999}{1000 + 1000 - 2}} = 6.16441 \,.$$

This value is more or less unrounded because the assistant wasn't finished calculating. Then,

---

[5] Very convincing counterfeit medications are common in S.E. Asia, where even well intended honest druggists are often deceived, along with the customer. Counterfeit antimalarials and diluted antimalarials are helping sustain the prevalence of malaria in the developing world. However, a number of other potentially life saving medicines are also being counterfeited and their resemblance to the real products, down to the fine details of the packaging, is surprising. Ironically, dilution is even worse than the fakes. The diluted medications help the disease causing organisms to develop resistance more rapidly than would happen if the medications were full strength.

given $n = n_x = n_y$, the standard error of the effect size (the difference between the means) is

$$S_{\bar{x},\bar{y}} = S_{x,y}\sqrt{2/n} = 6.16441/\sqrt{500} = 0.27568\,,$$

which, at this stage, can be rounded to five figures for convenient use of four decimal place tables of Z (five figures for much of the table). Classically the assistant investigator might have seen $df = 1998$ as large enough to default from t to $Z_{0.025} = 1.9600$. However, comparing values in a table such as Appendix 4 suggests that it could be more accurate to use t all the way up to at least $df = 4500$, so that Z would have been a little small. The result uses $t_{0.025,1998} = 1.961$ to get

$$D - t_{\alpha/2,1998} \times S_{\bar{x},\bar{y}} \leq \delta \leq D + t_{\alpha/2,1998} \times S_{\bar{x},\bar{y}}\,, \tag{6.12}$$

which works out to

$$2.70 - 1.961 \times 0.27568 \leq \delta \leq 2.70 + 1.961 \times 0.27568\,.$$

The calculations produced

$$2.159392\,\text{mg} \leq \delta \leq 3.240608\,\text{mg}.$$

So, rounding the result (**up to the right** and **down to the left**) to only one more digit than the lowest accuracy input, D, (above) gives

$$2.15\,\text{mg} \leq \delta \leq 3.25\,\text{mg}.$$

Discussion: OK, the assistant investigator did not really need to be concerned with five figure accuracy for t; four would have been more than adequate, because even the final difference between the means had been mishandled in such a way as to enforce a maximum final accuracy of three figures. That is, each of the means in the second pair are prematurely rounded. Adding 1000 values, some of which exceed 99, and finding the average will give you a lot of figures with which to work. Even with the improved estimates of the means, premature rounding sacrificed information and lost accuracy when subtracting one rounded mean from the other. The original data contained enough information to justify the reported L_1 and L_2 to four figures. Caution dictates that neither the means nor the variance (nor the standard deviation) be rounded to their final accuracy until the very last step.

In this example the reported accuracy was so low and the df so large that sticking with t had no effect after rounding. However, that is not always true, even with large df, especially if $df < 4500$ or $\alpha/2 < 0.025$. Although you cannot report to an accuracy exceeding recommended practice, you can report to a lesser accuracy if you feel it is more convenient.

You cannot produce a valid CI without **a priori intent**, including randomness, double blind procedure, and careful thought. Contrary to common procedure, CIs following statistical tests should only be reported comparatively if the tests were statistically significant. Otherwise only the CI for the control is justifiable. This may seem a technicality, but it is not. You might avoid this restriction if you honestly have not tested, but you would do so with a loss of information that might have supported your alternative hypothesis. There is also the question of whether a test would be required for the publication of that part of your research.

6.6 CIs and Ambiguous Statistical Significance

It is widely accepted that if a CI for differences between means does not include zero, then it is statistically significant (Harlow, Mulaik and Steiger 1997), but is using this assumption in application good practice? The related practice of looking at a collection of two or more single sample CIs and accepting that the ones that do not overlap are statistically significant was even encouraged by the 2001 Publication Manual of the American Psychological Association (5th ed.), but it is **far from best practice**, and often misleading. Even worse is the common assumption that if the non overlap indicates statistical significance, then the overlap indicates a lack of statistical significance. Both of these mistaken notions will be considered here. For simplicity, when considering the following, it might help to recall that confidence intervals have whiskers on either side of the sample mean and use Z for normal approximations or for large sample normally distributed samples.

The most common CIs are for means, and these are usually symmetrical, **each whisker** having a length of $t_{\alpha/2,df}\sqrt{S_{\bar{x}}^2}$ or $Z_{\alpha/2}\sqrt{S_{\bar{x}}^2}$. The calculation and hence the size of $S_{\bar{x}}^2$ and its effect on the length of the CI differs considerably, depending on whether one or two samples are involved. The CI for the difference between means uses calculations like those for a two sample test, but CIs for individual means use calculations like those for a one sample test. Other contributing factors such as degrees of freedom and number of CIs involved may further confuse the results of using CIs for testing; their effects on the length of the CI differs considerably, depending on whether one or two samples are involved.

6.6.1 Risky practice of inference from two CIs

To understand the effect of having to calculate the standard error, $S_{\bar{x}}^2$, differently for two sample consideration versus for individual means, assume a situation that is sufficiently uncomplicated that we can trust the analogy that for huge samples, the length of either whisker of a CI for difference is numerically equal to the distance between zero difference and significance in a test as in the following Figure 6.3:

Figure 6.3: CIs Are Not Tests

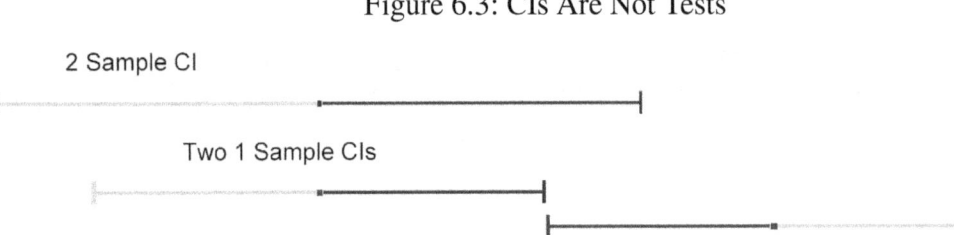

2 Sample CI

Two 1 Sample CIs

These confidence intervals, all drawn on the same original units of measure scale, have the same underlying standard deviation, σ, sample size, n, and level of confidence, $1 - \alpha$. The top is a 2 sample CI which is also called a CI for the effect size or a CI for the difference. The remaining two are single sample CIs intended to trap the true means of individually sampled populations and have the same two means as are being compared for effect size in the top CI. The CIs are centered on the dots in the middle, either the observed effect size (top), or the sample means (below it). Assume that the sample means of the lower two happened to be such that their CIs just barely reach each other but neither overlap nor have a gap.

In the top confidence interval, which attempts to trap the difference between two samples (the effect size), the dark whisker has a length equal to the difference between two sample means that would be numerically equal to the distance required to be statistically significant at the α level. Like all such two sample CIs for $\mu_2 - \mu_1$, its length is determined using units that are each equal, $S_{\bar{x}} =$

$\sqrt{S^2\,(2/n)}$ ($S_{\bar{x}}$ is calculated the same way as for a two sample test to see if there is a difference between the means). Its actual length is $Z_{\alpha/2}\sqrt{S^2\,(2/n_1)}$.

The pair of CIs at the bottom both have darkened whiskers, each of which is $Z_{\alpha/2}\sqrt{S^2\,(1/n_1)}$ long, that is numerically equal to the difference that would be barely statistically significant at the same α level for a one tail test. Like all such one sample CIs, their whisker lengths are determined by using units that are each equal, $S_{\bar{x}} = \sqrt{S^2\,(1/n)}$ (this $S_{\bar{x}}$ is calculated the same way for all one sample tests to see if there is a difference between the mean and zero, or between the mean and a hypothesized mean). Their sample means (center dots) just happen to be different enough that these two CIs are neither overlapping nor do they have a gap between them, and if lack of overlap of them could be used to rule out statistical difference, their two black whiskers together would be the same length as the dark whisker of the two sample CI above them and their means would be only on the verge of being statistically significantly different. But the means of the two bottom CIs are even **further apart** than the distance associated with a two sample test, $Z_{\alpha/2}\sqrt{S^2\,(2/n_1)}$, and are really statistically significantly different from each other.

Comparing the whisker length that is numerically analogous to non significance of the actual two sample test (what might be called the 'confidence'), with the summed whisker length of two one-sample CIs, we find that

$$Z_{\alpha/2}\sqrt{S^2\,(2/n_1)} \neq Z_{\alpha/2} \times 2\sqrt{S^2/n_1}.$$

This non equality can be verified by simplifying to eliminate the like terms on either side of the $\neq$ to get $\sqrt{2} \neq 2$. The combined whisker length of two CIs is $\sqrt{2} = 1.414\ldots$ times longer than the whisker length corresponding to the test comparing them. As Wolfe and Hanley (2002) point out, it takes less difference to attain statistical significance under the t test than it does to guarantee that the CIs will not overlap. So, the sample means could come closer and closer to each other with their CIs overlapping more and more for a considerable distance ($\sqrt{2}$ times the length of the dark whisker of the 2 sample CI) before they would stop being statistically different.

If more than 2 CIs occur in the same figure, the need to treat such comparisons as multiple comparisons will usually reverse the situation and the needed gap to be associated with statistical significance will **increase** as more CIs are involved. This fact, that statistical significance is more difficult to infer as the number of samples involved increases, will become apparent after you get to Subsection 10.1 . Even if confidence intervals were somehow equivalent to (or even complementary to) tests, you could **not graphically add** the equivalent of two one sample tests **to get a two sample test**.

6.6.1.1 Risky inference using just a CI for difference

As mathematical statisticians will tell you, 'confidence intervals and tests are not in the same realm' (not in the same context). A test looks at an observed **apparent** effect size to see if it is so large as to be unlikely to really be zero, but a confidence interval for an effect size is just a calculated set of limits that have a given **probability of having trapped** the effect size within. Because the calculations are similar, some folks mistakenly use the fact that a particular confidence interval for the effect size (for instance the difference between sample means) does not include zero as a test to infer statistical difference. And it is true that I just used the length of a whisker of such a CI as analogous to a similarly calculated test distance to reach a point theoretically equaling statistical significance (above). However, such a provisional procedure was handy for demonstrating that the overlap of CIs even under simplified and idealized conditions could not be a valid test. From yet another viewpoint and in application, making the declaration of statistical significance dependent on not overlapping zero, assuming such a practice approached validity, would be nil hypothesis testing; an actual test in application will judge a difference to be statistically significant even if the observed difference **falls short** of zero by a distance equal to your reasonable minimum difference.

> **BEWARE:** Confidence intervals are not suited to be used as statistical tests and attempting to interpret them as demonstrating statistical significance involves pitfalls and theoretical contradictions.

Question 3 and Question 5 provide CI calculations with answers at the end of this book.

Questions for Ch 6

Write the equations where appropriate and fill in the values. **Odd numbered questions** have worked answers at the back of the book and can be studied as additional examples. The data tends to be fictitious.

1. Is the true mean, μ (or whatever other parameter of interest) at the center of its $1 - \alpha = 99\%$ confidence interval? If not, where is it?

2. Why is it reasonable to approximate $Z_{0.025}$, or $t_{0.025, df > 60}$ using $Z \approx 2$ or $t \approx 2$, for estimating a 95% CI? (You may wish to check some tables.)

3. Your sample of size $n = 100$ estimates the variance, $S^2 = 35.5$, and the mean, $\bar{x} = 5.20$. What is the 95% confidence interval for this result? What is the length of the right whisker? What is the length of the left whisker? Use the **conservative approximation** of $Z_{\alpha/2 = 2.5\%}$.

4. What is the margin of error; what does it have to do with a CI?

5. Even though the sample size is really too small to assume its mean would be from a normally distributed population, let's assume a normal distribution in order to practice calculating a 95% confidence interval for the following random sample: 5.5, 7.3, 2.7, 6.2, 4.2, 0.1, 3.8, 5.4, 2.9, 7.1, 1.1, 5.2, 5.7, 2.3, 3.5 Kg. Report to the appropriate number of figures. (This data happens to be Treatment $_1$ in Question 7.)

6. Calculate the 99% CI for the mean of the following sample: 2.8, 2.2, 9.4, 8.2, 1.2, 7.1, 3.6, 5.8, 5.2, 4.3, 7.7, 0.8, 9.5, 0.9, 1.1 Kg (Assume normality even though unlikely at such a sample size. At least show the variance or standard deviation estimate and the sample mean to some extra figures on the way.) Finally report your results to the appropriate number of figures. (Question 7 calls this data Treatment $_2$.)

7. Find the 99% confidence interval for the difference, δ, between the two treatments:
   Treatment $_1$: 5.5, 7.3, 2.7, 6.2, 4.2, 0.1, 3.8, 5.4, 2.9, 7.1, 1.1, 5.2, 5.7, 2.3, 3.5 Kg.
   Treatment $_2$: 2.8, 2.2, 9.4, 8.2, 1.2, 7.1, 3.6, 5.8, 5.2, 4.3, 7.7, 0.8, 9.5, 0.9, 1.1 Kg.
   Because these random sample sizes are equal, use the same calculations as when the underlying variance is equal. Can you infer anything about statistical significance from this CI? (You could use a calculator.)

8. Given the following observer measurements,
   Obsvr A: 6.6, 6.5, 3.3, 9.0, 4.0, 6.8, 0.0, 7.9, 6.7, 4.5, 0.2, 2.5, 2.8, 5.1, 6.4 degrees C,
   Obsvr B: 3.9, 8.4, 1.5, 9.3, 9.9, 5.0, 2.0, 7.8, 6.0, 5.5, 7.9, 9.7 degrees C,
   find the 95% confidence limits for the difference $\delta = \mu_A - \mu_B$. (You can estimate the pooled variance using Equation 4.4.) What is the df used for the calculation of the CI? Show some of the steps and calculations.

9. You take a sample of $n = 250$ salmon and find $r = 155$ have sea lice, Lepeophtheirus salmonis (Pacific external parasitic copepod). What is the 99% confidence interval for the proportion of this fish population that have sea lice?

10. It is the year 2070 and you are an alien who has abducted a random sample of $n = 45$ humans, Homo sapiens, at random from all over Planet Earth for the purpose of discovering the effect of the widespread practice of using a sonogram to detect and then abort female fetuses. You find only $r = 11$ females. What was your 95% **continuity corrected** CI for the proportion of females, using $Z_{0.05/2} = 1.9600$? Why not use $t_{0.025,43}$?

11. In your country's army the companies are modeled after those of some WW1 armies and have 250 soldiers each. A war is starting, and you take a random sample of $n = 250$ out of the tens of thousands of new inductees, and your very accurate but expensive test indicates that $r = 88$ inductees in this sample have never been exposed to chickenpox, Herpes zoster. Estimate a 95% CI for the number of soldiers per company which have never been exposed to this common viral infection and which might be inconveniently susceptible. ($Z_{0.025} \approx 2$ is OK. No continuity correction is needed.)

12. Why not use the overlap or non overlap of pairs of CIs as a test?

13. If **more** than two CIs are involved, you cannot compare the CIs to get a clue to statistical significance, but if you were to only have two, what would be the primary difficulty in comparing them to get a clue?

14. Assume that, as in Sub-subsection 6.5.0.1, the manufacturer sold the real medicine as $\mu_0 = 106.00$ mg tablets and that the assistant investigator had a priori asked for a confidence interval for δ_x, the difference between this alleged mean and the sample mean, $\bar{x} = 104.13$ mg of the batch of real tablets from which the sample was drawn, given that the variance was $S_x^2 = 40.10$mg^2. Calculate a 95% CI for δ_x. Recall that $n = 1000$. Show the equations and fill them in to get this CI for the effect size, the difference between the manufacturer's claim and the sample mean. (The available information has limitations and there is an inconvenient loss due to rounding. But, a final three figure accuracy is OK. Don't use a continuity correction).

7 FOR POWER SEEKERS

Classically, there has been an unfortunate tendency to accept any sample size of $n \geq 30$ measurements or observations as being sufficient, but the resulting sensitivity, power, is usually disastrously low. Power is a long neglected, misunderstood tool without which research can be meaningless. This neglect has slowed the progress of many fields of study. A major misunderstanding of power analysis is due to its inappropriate association with targeting standardized 'small', 'medium', and 'large' power indexes (Section 5.9).

What issue do researchers most often ask statisticians about? When I worked as a consulting and support statistician, I was most often asked for statistical advice about sample size, meaning *n*, the number of data items. Either it was how small could their experiment be or it was how big should their experiment be? These two are actually the same question, differing only to a small degree in implied motivation. It is a question that relates to, or perhaps exposes, the very nature (clay feet?) of a good deal of the relatively recent history of applied statistics.

The truth of the matter is that the statisticians and our clients, the researchers, were, and many still are, operating in an imperfect, underfunded, behavioral type A world. Prevailing practice did not, and often still does not consider the power, or rather the lack of power, of these experimental designs with respect to their subject matter. Ironically, the custom that underestimated the need to consider power was a large part of the reason why the funds were so limited. 'Everyone' was using 6 treatments in each of 4 field plots, or in 4 hospital wards. 'Everyone' was using $n = 15$ pairs. Often the effects were so huge that these recommended sample sizes actually got meaningful results. So, it was assumed that the funding that only provided for such wretchedly insensitive experimental designs was satisfactory.

The crucial question of what sample size you need will come up again and again as a central feature, not only of this book, but also of your career. It is your career and you need to understand the issue of cost versus the power to discover and make the greatest possible gains. When choosing a sample size, in order to understand what you are up against in terms of time, space, and money, you must consider how big an effect, D_{min}, a **reasonable minimum difference**, has to be before it is of interest to you.[1] You, the researcher, need to become able to make your own decisions when it comes to the minimum needs of your research. Once you have a strategy worked out and on paper, then find a professional statistician to either help you polish your plan, or to show you where you might be going wrong. But **do it before collecting the data**!

7.1 Be Power Hungry!

Power is the sensitivity of an experiment (or observation), the probability of detecting the reasonable minimum difference in which you are interested. With respect to 'Z tests' or Student's t tests, you are

---

[1]Epidemiologists and drug testing teams sometimes use 'equivalence testing'. They start with a ***threshold***, Δ (Greek capital delta), the ***interval of tolerable difference***', a difference that represents a divergence that is barely not large enough to have any clinical implications (Barker, Luman, McCauley, and Chu 2002, Shtaynberger, and Bar 2013). Then they construct a hypothesis pair in the form, **H0**: $|u_2 - \mu_1| < \Delta$ and **HA**: $|u_2 - \mu_1| \geq \Delta$, which threatens to get them into the awkward world of non-centric distributions.

interested in a reasonable minimum difference, D_{min}, which you choose, but you never know the actual true difference, $\delta = \mu_2 - \mu_1$. Power is the probability of **not** getting a Type II error (as redefined in Section 3.3). The probability of getting a Type II error is β, so because all the possibilities regarding a particular event add to one (100%), then *power,* the probability of not getting a Type II error, equals $1 - \beta$.

As you may recall, power, $1 - \beta$, measures the sensitivity of an experiment to discover a **targeted** true difference, $\delta = D_{min}$, of a certain size, if it exists. If a larger true difference exists, the ability of your experiment to detect it is even greater than the power you targeted for your given D_{min}. If you chose $D_{min} = 3\,K°$, you are even more likely to find statistical significance if the true difference is $\mu_2 - \mu_1 = 4\,K°$. Twentieth century style research reports that conclude that something is, or is not, statistically significant, but which include **neither** a power analysis **nor** a confidence interval are **inadequate**, and more and more editors will refuse them.

Unfortunately, agricultural research is problematic because the variance might vary from year to year. But a university might have local data for something which could be used as a control. All the better if it covers several years. (eg, Hales Best used to be a common control for cantaloupe research.) Nonparametric tests such as the Kruskal-Wallis (Section 14.3), or Friedman's complete block test (Section 14.5) are somewhat more robust at the cost of a little power. In the absence of pre-existing information, nonparametric sample sizes can be easier to find.

> **CAUTION:** Many research projects using **parametric** tests with more than two variables, and thus using an ANOVA based design (Chapter 9), will **need a pilot study** that is like a full dress rehearsal with even variance partitioning to estimate the underlying variance. This can be a real stumbling block if you are on a tight schedule. For instance, an agriculture masters degree student may intend to complete the course requirements one year, and to plant a set of experimental plots involving an annual crop the next, thus finishing in 2 years. However, lacking a good estimate of the underlying standard deviation or its square, the variance, the student might not have a basis for estimating a sample size for the usual parametric tests.

To estimate a sample size for tests that assume normality, you will have to obtain or determine the standard deviation, S, or the variance, S^2, by using some data from the same source as the test or by obtaining it from some preexisting source, such as hospital or laboratory records if your experiment can duplicate the conditions and the manner in which the data was collected. In most cases this data for sample size estimation will represent the control, and the sample size estimate may not seem to be perfectly representative of the complete comparison. However, this might be justified by a common model for parametric inferential tests that only tries to detect differences between the means; other parameters, such as the standard deviation, are treated by the model as being unchanged. That would imply that the variance of the control is a suitable representative of the overall variance. Ideally, you should look for a priori information such as the expected third moment **skew** (Equation 15.1), because your sample size should be no less than **twenty five** times it.

7.1.0.1 Improving on, even avoiding, failure

Researchers had better be prepared for challenges when trying to get peer review and acceptance of their research reports by editors. The solution is to **design your research** so as **to avoid**, as much as possible, getting Type II as well as Type I errors. Then be sure to put a number on the probability of each of them, α and β (or $1 - \beta$), along with how big a reasonable minimum difference you targeted. You must report the observed difference, the effect size, ES, in original units of measure. With respect to Student's t test this is $D = \bar{x}_2 - \bar{x}_1$.

You do not want **hypersensitivity**, a tendency to detect meaningless tiny true differences, that are not even relative to the factor or the treatment that you are considering. However, hypersensitivity is often so expensive that you might not have the opportunity, let alone the need, to guard against it. Without necessarily realizing it, my clients generally approached this question from both directions, trying to get enough power and avoiding too much. They sought an answer to what they may have seen as a different question, "How large should the sample be?" A suitably chosen sample size, n, targets a specific difference, the reasonable minimum difference, D_{min}. It prevents you from wasting your time and money on an insufficiently sensitive procedure, or from wasting your investment by paying for the added cost of misleading hypersensitivity.

7.1.1 Limping along bereft of power

To illustrate the wretched condition of much of modern research, consider some of the following examples in the field of orthopedic medicine. Out of 90 manuscripts that **did not** report a statistically significant difference, Tornetta, Locher, and Bhandar (2000) found only 5.6% referred to the question of power in the manuscript. They concluded that the vast majority of randomized trials in orthopedic trauma are not of sufficient sample size to draw accurate conclusions. To quote them, "Thus, when a surgeon reads papers that conclude that there is no difference between treatment methods, he or she must be extremely suspicious to avoid incorrect conclusions." They expressed the opinion that "**it is incumbent upon the journal reviewers to insist that power analysis is included for all studies demonstrating no difference in treatment groups.**"(bold highlighting added). Notice the word "demonstrating" describes the impression that these statistical reviewers thought that doctors would draw from the many meaningless statistically nonsignificant studies that have been reported. Not rejecting **H0** demonstrates nothing. It is your responsibility as a researcher (even more than the reader's) to be certain that you have not given erroneous impressions.

Freedman, Back, and Bernstein did a similar study of literature with similar results, also in orthopedics. They came across some studies chancing to get positive results despite being pitifully lacking in power (2001). The literature has a multitude of such examples covering many fields in addition to medicine and social sciences, most with extensive examples (e.g. Calin-Jageman, Robert J., and Geoff Cumming 2019; Dumas-Mallet, Button, Boraud, Gonon, and Munafò 2017; Reinhart 2015; Tsang, Colley, and Lynd 2009; Vankov, and Munafò 2014). The researchers who got positive results were just lucky not to have gotten falsely negative results. In addition to the points made in the preceding paragraph, this suggests that many researchers are taking rather extreme risks of not finding anything important even if it exists.

It is difficult to conceive what experiments like these surgical procedures and the associated follow ups might cost in monetary terms. Perhaps they cost hundreds of thousands of dollars per project beyond the medical costs involved in ordinary clinical treatment without research. Imagine going to your boss after all that expense and the usual hassle with the human subjects folks, and then saying, "I found no statistical significance, but I cannot publish anything because, as the reviewers found, my study did not have enough power to detect whether there was any meaningful result." In the next section you will start to learn how to minimize such awkwardness.

The above studies imply, correctly, that the sample sizes in terms of number of patients and investment necessary to have a reasonable probability of producing something that the researchers should report, whether positive or negative, would have to be much larger.

Similar reports in another area of medicine include Fox and Mathers (1997). In agricultural research, Roush and Tozer (2004) found lots of published research based on experiments which are woefully lacking in power.

7.2 Sources of Estimates of Variance (for μ_2 vs μ_1)

An a priori, meaning **before the fact**, power analysis is a part of the experimental design that enables you to attempt to **avoid the pitfalls** of either investing too much or too little resources into obtaining the actual and final test data. It must be performed before the data is examined, and even before it is collected, if possible. In application, either too large a sample or too small a sample can adversely affect your outcome. An a priori power analysis for a parametric test requires an adequate estimate, S, of the standard deviation, σ.

> **CAUTION: Beware of rules of thumb** for a priori sample sizes, as they tend to underestimate sample sizes and thus get you started without much hope of detecting a meaningful real difference! The ideal starting point for power analysis, preceding a parametric test, would be a full dress rehearsal of data collection, ideally with sample sizes of $n \geq 30$ (including variance or standard deviation estimation, at least with respect to a control). The result can be used for sample size estimation, and maybe for estimating other parameters, but, because it is pre-examined, and its sample **size is so small**, it must **not** be used as part of the actual test data.

For the actual test, as opposed to the pilot data examination, you are trying to have an acceptable probability of detecting a reasonable minimum difference, D_{min}. To do that, you must refer to your choice of the maximum probability, β, of getting a Type II error (not detecting your chosen reasonable minimum difference). To justify rejection of **H0** you must also refer to your choice of α, the maximum probability of getting a Type I error (rejecting hypothetical 0 difference) when there **really is no difference big enough to be detected**. These cutoffs, α and β (or $1 - \beta$), along with a reasonable minimum difference, D_{min}, should be stated in your methods and materials.

However, to achieve this goal, you must know what sample size, n, to use. So, you use an a priori power analysis to estimate the sample size that targets your desired power, $1 - \beta$ (usually reported as a %), for a true difference as small as your chosen reasonable minimum difference. Even if the journals were to change over to requiring confidence intervals (Chapter 6) instead of reports about statistical significance, a similar (or identical) a priori analysis would be necessary to have a reasonable chance of obtaining a confidence interval that is neither uselessly wide nor ridiculously narrow, and in particular, would provide the power to correctly **center** it (Subsection 7.13).

But, you must have at least an estimate, S or S^2, of the underlying standard deviation, σ, or its square, the variance σ^2, in order to do a power analysis for tests that assume a normal distribution, or more reasonably, tests which assume the means are normally distributed via the CLT. Provided you can get the necessary trustworthy estimate of σ, an a priori power analysis will enable you to select a practical sample size. This sample size will, within limits imposed by the accuracy of your estimate of the true standard deviation, σ (or variance σ^2), target your a priori chosen power, $1 - \beta$, of detecting a reasonable minimum difference, D_{min}, as small as you feel it is important to detect. Your tests will still have some sensitivity to smaller true differences, $\delta = \mu_2 - \mu_1$, but this sensitivity will rapidly drop as the true difference gets smaller and smaller with respect to your chosen reasonable minimum difference, D_{min}.[2]

Unfortunately, for tests that assume a normal distribution such as the t test and ANOVA (Chapter 9), obtaining an estimate of σ is an essential step on the way to estimating the necessary sample size, and it is not always easy to obtain. If you do not have information about variability, such as from your team's similar experiments, or from literature, or from sources such as patient records, then it is necessary for you to do some sort of preliminary data collection to find the standard deviation, S. (If desperate see: Subsection 7.2.3, or maybe better, consider a nonparametric approach,

---

[2]sample size estimate, source of variances

Chapter 14). However, for complex designs such as an ANOVA that estimates an underlying variance by partitioning, you may need to run a ***pilot study***, using the procedure you **intend to apply** and most of the actual calculation, especially partitioning, to get an estimate of the variance. The idea is to find the best available estimate of the variability of yield of field plots of corn, of patient outcomes, of fish catches, or whatever.

Data collection for power analysis can most usefully be a rehearsal of the larger experiment to follow, but it may not always need to include more than a single suitably chosen control category (or treatment). Lancaster, Dodd, and Williamson (2004) discuss pilot studies in clinical situations. Festing (2010) referring, in part, to Mead (1988), suggests that due to diminishing returns in the estimation of σ, sample sizes of $n = 10$ to $n = 20$ may suffice for pilot studies. These extremely small sizes were intended for an initial estimate of S when something drastic such as animal sacrifice is involved.

Such small samples tend to give standard deviation estimates that might be sufficient to get your sample size into an almost acceptable range (best inflate the resulting estimate, $\widehat{n}$, a little in such cases). However, if you were to take such small random samples again and again from a large data set that was not intended for research, as you could do **experimentally** on a Jupyter notebook or a spreadsheet, you might be alarmed at the variety of differences between means that you would get. Small samples seem to be adequate in many instances if only used to get a variance to estimate sample size. Small samples ($n < 30$) will not be adequate for actual tests. **If at all practical**, pilot studies for **estimating sample sizes** for subsequent testing should minimally use random subsample sizes of $n \geq 25$ to estimate S or S^2.

Whatever your source of the variance or standard deviation, the **accuracy** of your estimate will be **restricted** by the number of degrees of freedom in your estimator, S^2 or S, from records, or pilot samples, pilot studies, or whatever. If the number of degrees of freedom in your estimate of S or S^2, is too small, the estimated sample size, $\hat{n}$, will be larger than necessary.

Much greater samples might be necessary if you wish to estimate skew. If a sufficient sample size is available, it would be advisable to estimate the third moment skew, so that you can make sure that your final sample will **not be less than** $25 \times skew$. If you cannot get an estimate of skew, but have reason to suspect it might be a problem, a total sample size of 60 or 80 will almost always avoid problems resulting from relatively unusually large skew, etc, for the t test[3]. As compared to the t test which can only involve one, or at most, two populations, the ANOVA (Chapter 9) involves three or more treatments, so the effects of skew might be more serious.

The estimation of σ or σ^2 must be from data collected using the exact same procedure as the experiment or observation that you intend to do. When your analysis will involve large amounts of **already existing (*retrospective*) data**, you can take a random subsample, what modelers might call a 'training sample', to not only estimate the standard deviation, but to graph and observe its distribution. If your data source is huge, such as the BRFSS health behavioral survey data, you should sample a nice big random subset, say a couple of hundred. But such training subsets **must not be recombined with the test data** intended for analysis.

When planning an ANOVA (Chapter 9), you might be able to avoid a formal pilot study by using historical data and some partitioning calculations. You must be able to closely duplicate the conditions under which the data was recorded. For instance, the orthopedic surgeons in the preceding discussion could have gotten the information for an a priori power analysis by first monitoring and recording the results of whatever usual procedures they intended to use as a control. This data might already exist and be found in the hospital or the practitioner's own records, without any pilot experimentation. Check to see if hospital data is risk adjusted, and if so, report that you used it, along with the standard by which it was adjusted. For thoracic surgery risk, adjusted data from the STS National

---

[3]n and skew

Database is available.[4] Local medical data accumulated under the same conditions as the research is going to be conducted would be best. Hospitals that are interested in **getting research grants** would do well to keep suitable data on all standard procedures and have it available, with stringent individual patient confidentiality safeguards, for statistical background when needed for research design. When they have enough data, it is then possible to estimate the standard deviation, as S, to be used in an a priori power analysis. That would assist both them and the granting agency to determine what sample size they need, to have an acceptable probability of detecting a reasonable minimum difference. Large data accumulations or databases could even enable the estimation of third moment skew to assure that the sample size at least equals or exceeds twenty five times the skew (Equation 15.1).

Whatever your source of S (or S^2), your research technique **must conform** to the conditions under which it was obtained or derived.

If your pilot study or other data source has extreme outliers (crazy numbers above or below the rest), you should consider Subsection 15.6.2 for data cleaning before this initial estimate of the variability for power analysis. However, if you clean your data for sample size estimation, your sample size estimate should be for experimental data that has been cleaned the same way, and that would have to be done **in a double blind manner** (without pre-examination) before it is used. Using MAD is not always possible, but it might be possible when your research data is to be obtained by random sampling from preexisting big data collections.

7.2.1 Is this project practical?

You could use a variety of ways to estimate σ to see if your study might be too costly in terms of sample size. Although it is unlikely that you can perfectly duplicate someone else's experimental techniques to obtain a trustworthy sample size estimate, you can sometimes get an estimate, S, of σ, or S^2 of σ^2, from literature or in various other ways. Stating the estimated standard deviation or variance, S or S^2, in your own publication is a very good idea, especially if you want to facilitate further research in your field. However, as is implicit in true science anyway, your methods and materials must be presented in sufficient detail to be repeatable. The quality of such estimates is dependent on sample size; larger sample sizes tend to produce results that are more precise and more accurate. Reports of t tests should ideally include the mean, the effect size ($\hat{\delta} = \bar{x}_2 - \bar{x}_1$ or $\bar{x}_0 - \bar{x}_1$), and an estimate of the standard deviation,[5] **all in original units**, to facilitate further research.[6] If your sources of your estimate, S, do not agree, then you should try very hard to select the source most like the experiment and analysis you intend to perform. You can combine two or more estimates of the variance, using

$$\widehat{\sigma^2} = \sum_{all\ i}(S_i^2 \times df_i)/\sum_{all\ i}(df_i),\qquad(7.1)$$

---

[4]The STS National Database, which has more than 5.4 million patient records, uses risk adjustment to provide more accurate patient outcomes. If not risk adjusted, the records of surgeons who perform operations on higher-risk patients such as the elderly would be unfairly evaluated. If you use a variance or standard deviation from risk adjusted data, you may have to consider how it relates to your research subjects.

[5]If you have the standard error of the mean, $S/\sqrt{n}$, you can multiply that by $\sqrt{n}$ to get S. Similarly, with a little thought, if you have a t value, and the effect size, and the sample size (or df), you can unravel the t test to estimate S.

[6]A more complicated group of tests, the ANOVAs, also yield such information as the mean square error, $MSE = MS_W$, to get your estimate of S to use in subsequent t tests of the data actually in that ANOVA (Chapter 9). However, to **plan** the sample size for a t test of a **future** experiment, you might want to be more cautious and use $S^2 = TSS/(n-1)$ from a previous ANOVA as your variance estimate, where n is the total sample size for the whole ANOVA. That is, a small error leading to a larger sample size is far better than the small error of even a very small underestimate leading to a too small sample size.

but the sources of the variance estimates S_i must be equivalent. That is, the variance estimates must estimate the variance of the same thing, including identical experimental or observational techniques. If after everything else you are still at a loss for a standard deviation, then consider Subsection 7.2.3.

For your own research, you might even write to authors who have done research resembling yours and ask them about their estimates of S, (and maybe even *skew*) along with the sample sizes involved (be prepared for shrieks of dismay). But, if you wish to duplicate their techniques sufficiently to use sample sizes (hence costs) based on their estimates of σ in research grants, or in the final plan, you should consider spending time with them, actually taking part in their research. You can only be confident of the same distribution and variance if you can very closely duplicate their research.

7.2.2 Binomial and an a priori S

For using the normal approximation (Section 5.6) for a binomial distribution (yes-no, red-white, success-failure), you often need the a priori probability, p, of success to approximate $S = \sqrt{np(1-p)}$ (See Section 7.10). You need an estimate of only one parameter, the probability p, assuming that **H0** is true. For **one sample tests** you will be using your hypothesized probability p_0 **without a pilot study**. For instance, if you hypothesize that there is no difference, or two things are equally likely, then $p_0 = 0.5$. But, sometimes the required value of p might have to be obtained from a pilot study, or from preexisting records obtained under the exact conditions of your planned measurements or observations. If this is impractical, $p = 0.5$ is likely to be adequate, but may overestimate the required sample size.

7.2.3 Using guesstimates to obtain a tentative S:

For research, the approach discussed here is an act of desperation, primarily used to see if the proposed research can be fitted into your budget. Ideally it involves as many sources of guesstimates from suitable experienced persons as possible. Unfortunately, in application, that is usually not a large number. In the field of **project management** it is not unusual to use

$$S \approx (greatest - smallest)/6,$$

where *greatest* is an outcome (an expected x value, for instance how long to complete) that the managers (experienced 'experts', experienced 'experimenters', or ...) estimate of what they consider the largest possible. This x could be days to completion, length in microns, weight in metric tons, etc. Then the *smallest* is the smallest estimate of x that they consider possible. For lack of a better answer as to the degrees of freedom used for this guesstimate of S to estimate the sample size of a subsequent study, I would use $df = 6$ or $df = n - 1$ (where $n =$ number of contributors), whichever is smaller. This might seem a silly approach, but such guesstimates are definitely better than nothing when considering whether to proceed at all. As an estimate of σ for sample size estimation, it is useful in **estimating the potential cost** when considering whether or not to proceed to even a pilot study. Favor caution whenever you use such techniques, especially if your decision involves major expenditures.

Especially for small to moderately large samples, the need to already know S might be avoided by choosing to use nonparametric tests (Chapter 14).

7.3 n, the Costly Road to Power

At a sufficiently large sample size, the Central Limit Theorem (Subsection 4.3) tells us that means approach being normally distributed despite varying degrees of non normality of the raw data that goes into them. This behavior of the means of samples is the only justification for the widespread use of most of the statistical testing procedures presently in use. The mean of means is simply the population mean itself, but samples of means have their own distributions that are strongly dependent on sample size. So the **standard error of the mean**, $S/\sqrt{n}$, is particularly important to your research. Obviously this variable, $S_{\bar{x}} = S/\sqrt{n}$, is also the **estimated standard deviation** of the mean. It gets smaller as the sample size, the number of replicates, n, gets larger, giving different values for different sample sizes. For t (and Z) tests, you are working with means and standardizing from the original units of measure to a new scale, Z, that is based on the normal distribution of **the means**, and it puts the initial reference mean at $Z = 0$ (Section 4.9).

Student exploited the Central Limit Theorem and the normal distribution for testing. Student's 1908 publication was intended to be for small sample tests and has been used for that for a long time now despite the fact that in application it is extremely useful for, and only appropriate for, large sample tests with $n \geq 30$. (For small samples, nonparametric tests are preferable.) Because sample sizes are counting numbers, and not continuous, you need a large, but not infinite, table of t values. The various t distributions, that are each based on a sample size (by way of degrees of freedom), are adjusted values of the standard normal, or Z, distribution, and as the sample size grows huge, the t distributions become more and more like the Z distribution. These eventually become so close to Z that you default to a table of Z at maybe $n > 4500$ (or $n > 1000$ for three figure accuracy and $\alpha \geq 2.5\%$).

7.4 The Great Separation

Recall that you can standardize individual raw data values, x_i, using $z_i = (x_i - \bar{x})/S$ so that they have mean 0 and standard deviation 1. For testing we do something like this to **means** with their standard deviation, the standard error of the **mean**, $S_{\bar{x}} = S/\sqrt{n}$, to get such standardizations as $(\bar{x} - \mu_0)/S_{\bar{x}}$. This provides us with Z and t values that have enabled the creation of manageable tables. This scaling by the standard error of the mean also fits in nicely with the Central Limit Theorem. Modern applied statisticians extend the achievements of Student's era to facilitate the sensible application of these tests by first estimating the sample size that will give them the power to detect what they looking for, if it is there. Sample size considerations are also (through similar relationships) a key part of estimating useful confidence intervals.

As discussed in (Subsection 4.3) the Central Limit Theorem applies despite the shape of the curve (the graph) that results when you plot the frequency of the various measures (or observations) making up any population of raw data with a finite mean, μ, and standard deviation, σ. It is hard to imagine having to work with a population which does not qualify, but, although not common in most research, they do exist. However, the CLT is not about the values of the raw data, it is about the values of the means of random samples. In reality there is a different distribution, and therefore a different plot for each sample size, n. But it is sometimes easier to visualize them as though they were all one dynamic (like living) population of means that changes as the sample gets larger. As n gets larger, the mean of the means is the original mean, but the graph of their distribution gets more and more symmetrical and mound shaped, approaching an ideal, the normal curve. Even if there were not any normal curves out there in nature for you to work with, the means of random samples from any of the infinite variety of qualifying populations will produce a population of means that is close to normally distributed if you take a large enough number of measurements or observations per mean. YES! The normal curve lives!

Figure 7.1: The Great Separation

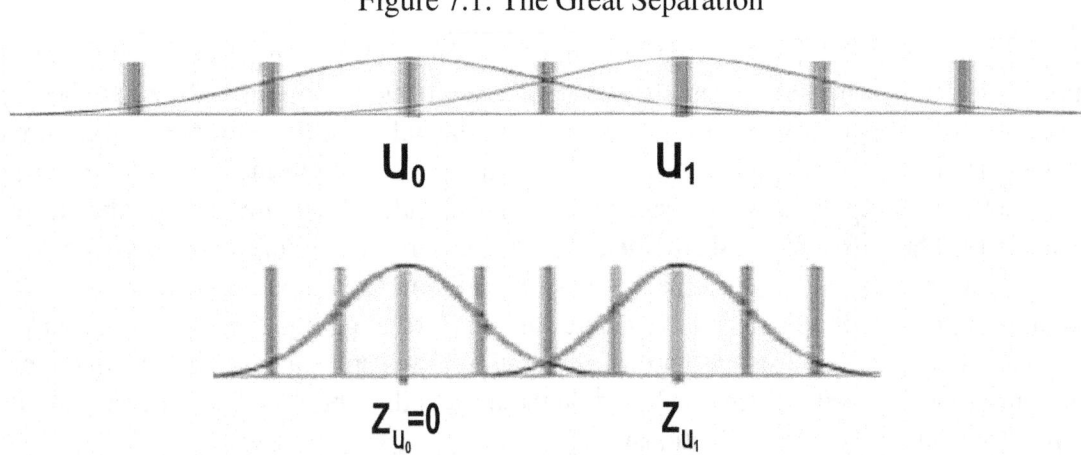

Top: An idealization of two possible (but uncommonly symmetrical) population distributions that might be found in application.The vertical lines are **one standard deviation apart**. Unless stated otherwise, standard deviations are in terms of raw data.

Bottom: Distribution of infinitely many randomly drawn means of size $n = 4$ from the same two populations. The vertical lines are **one standard error apart** (horizontal distance in standard errors of the mean). Standard errors are standard deviations of groups of parameters calculated as though the parameters were the data. For distributions of means, distances are usually measured in standard errors of the mean, Z, measured from $Z = 0$ at the hypothesized mean, μ_0, or at the true mean, μ. (They can also be simply called *standard errors* instead of standard errors of the mean.) The two sets of curves are scaled to show that the means are the same distance apart in the original units of measure. Even with this small sampling of means of random samples of size $n = 4$, there are $\sqrt{n} = 2$ times as many standard error units (standard deviations of the means) between the two means as compared with the number of standard deviations of the raw data units between the same two means in the top graph.

The distributions of sample means (bottom graph) are more easily separated (distinguished) from each other and their distributions are also closer to normal than are the distributions of the raw data populations (top graph) from which they came. Since the bottom graph distributions are of means, it is appropriate to measure the distance between the means in standard deviations of the means (standard error units, Z), rather than in standard deviation units of the raw sample data.

For most distributions found in common applications, means with sample sizes at or above $n = 30$ will have about 68% of the means within $Z = \pm 1$ standard error of the mean, and about 95% will be within $Z = \pm 2$ standard errors of the mean.

If you take samples from two different populations with overlapping distributions, the populations of their means will also overlap and have the same means as the parent populations. **Increasing** the sample sizes, n, will **increase** the amount of separation of the distributions of the means and **decrease** the amount of overlap, Figure 7.1. This standardization becomes disconnected from the original units of measure of the raw data and makes a new unit of measure that is defined by the variability of the data. The increased separation is a major reason why you use distributions of the sample means, rather than of the individual scores, to challenge **H0**: THERE IS NO DIFFERENCE. Such distributions with their enhanced separation are the basis for parametric statistical tests that use tabulated standardized tables to evaluate values such as $Z = D/S_{\bar{x}}$, where $D = \hat{\delta}$ which estimates $\delta = \mu_2 - \mu_1$ or $\delta = \mu - \mu_0$, the true difference (if any) between the population means. It also relates to how the confidence intervals tend to become narrower as the sample sizes become larger. Notation varies from book to book and from publication to publication. Some authors use δ in standard deviation units, but in this book it is in original units of measure unless stated otherwise.

So, this section is to help you visualize what 'happens to' the curves representing the distributions of the means when viewed as though they were dynamic with respect to n. It also touches on the use of a standardized scale, such as Z (or t), and what happens as n increases. So the phenomena illustrated by Figure 7.1 demonstrates one effect of the Central Limit Theorem. In this idealized illustration of two populations of sample means, $\bar{x}_i$, with the same underlying standard deviation, σ, but different population means, μ_1 and μ_2, the right and left member of each of the first two pairs of curves differ in terms of the mean. The mean of a distribution and the distribution of all possible sample means are the same; **the mean of all the means** equals the mean of the parent population. However, for interpreting the results of testing, we may standardize the distribution of the sample means in Z units in such a way that the center of the hypothesized parent mean is at $Z_0 = 0$, as in Figure 7.1.

Notice the **reduced overlap** and improved separation of the distributions created by considering means rather than individual scores, even at this **tiny sample size**, $n = 4$. If I had started with asymmetrical (maybe bumpy or jagged) curves, the distribution of the means would have become more symmetrical and a little more like a normal curve with each increase in the sample size, n.

A uniform distribution is one in which all the values are equally probable (no bumps). Figure 7.2 is a set of graphs of sample distributions resulting from generating large numbers of means from a uniform distribution of different sizes, n, using the R language `runif()` function to get random numbers.[7] Each little graph represents the distribution of an independent sample of $100,000$ means of sample size n, starting with 'raw', $n = 1$, plain random numbers. This distribution of single random values ($n = 1$ raw) is roughly a flat distribution and not normal at all. However, as the sample size increases, the distribution of the resulting means quickly st..arts to become symmetrical and less platykurtic (less flattened). The results support the CLT in that the results **approach** normality. They do so rapidly at first, but the trend seems to slow, never stopping or quite reaching a perfect asymptotic normal curve as the sample grows larger. For instance, the curve with $n = 27$ is just a little less close to normal than the curve with $n = 30$.

Figure 7.2: Effect of Sample Size on Random Distribution

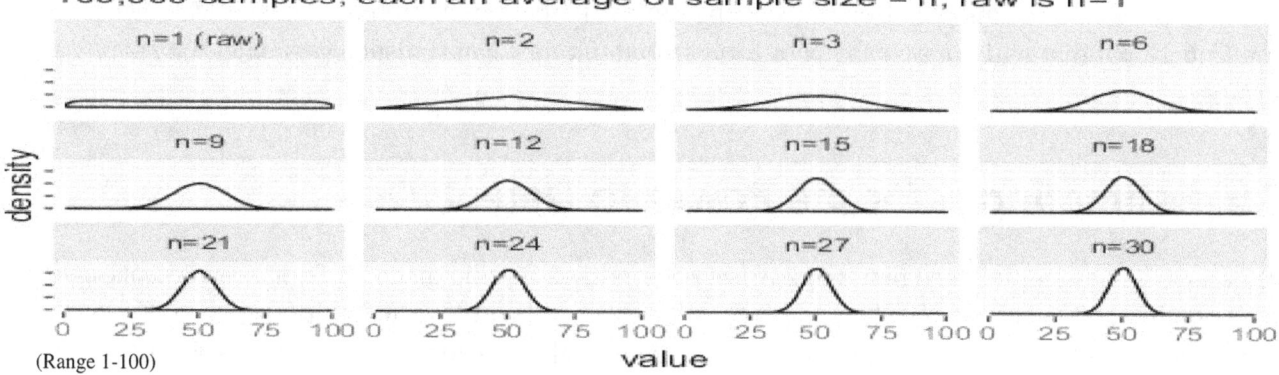

These results are from machine produced pseudo-random values, and even though $100,000$ seems big, it is still a rather restricted population. Better truly random samples can be had from Geiger counter clicks or from static received by radio telescopes pointed in appropriate directions. Even so, these results are reassuring in that samples as large or larger than $n = 30$ are sufficiently close to normal for most applied uses.

The curves in Figure 7.3 represent the observed distributions of $100,000$ different means of various sample sizes, n, from the same machine-generated source of random numbers, but before the

---

[7]Violin plots equivalent to Figure 7.2, using the same data, appear in Figure 15.6

Figure 7.3: Effect of Sample Size on Distribution Log$_{10}$ Random Means

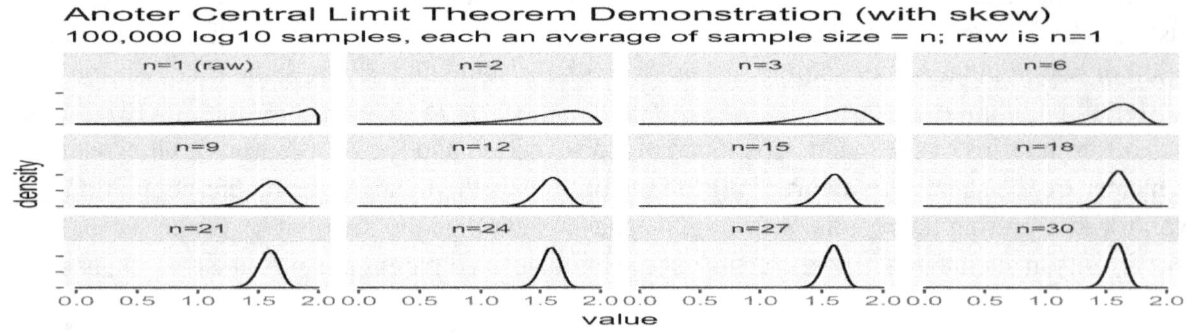

(Log10 of 50 = 1.698970)

means are found, the log to the base ten of each number is used in place of the number itself. The original log_{10}-transformed pseudo-random numbers with $n = 1$ produce a somewhat skewed distribution. However, the distribution of the means becomes less skewed as they approach a 'large' sample size, $n = 30$, at which the symmetry is rather good and the distribution is, in application, close to a sufficiently normal curve[8] (also might see Figure 15.6 and Figure 15.2). Again, if you look carefully, there is still a discernible difference between $n = 27$ and $n = 30$, even at this size.

Very extreme outliers that result in one or a very few of the data points accounting for most of the distribution in terms of their contribution to the grand mean, or to their contribution to the total variance, can invalidate the CLT (Lumley, Diehr, Emerson, and Chen 2002). Their effect is as though they violate the assumption of a finite mean and finite standard deviation, but the reality might be that they do not really belong in the same population. For instance, there is a debate about whether the stock market and common economic indicators have a finite mean and standard deviation. They probably do, under conditions where millions of individual small players are taking part in a stable environment. However, a single large insurance company can suddenly buy or sell on such a scale as to swamp the market; likewise, the environment, such as a media craze, or a sudden war, can cause most players to act together. In such cases, the resulting data is alienated from the usual population. The message here is that a population must have some cohesion in order to be a population. Such outliers can often be eliminated from the data by removing them with such methods as MAD (Section 15.6.2), but that will not save the poor fools who think they can statistically predict the stock market.

7.5 Estimating n for the Power to Detect D_{min}

Let's review one sample, one tail t (and Z) tests comparing a sample mean, $\bar{x}$, from a random sample to a hypothesized mean, μ_0. The hypothesized mean, μ_0, can be **any value including zero**. Keep in mind that, in application, and with sufficient sample sizes, the distribution of the means as under the CLT, the probability distribution, can be informatively represented as involving areas under a normal

curve. The shaded area of the tail of the small figure shows a part of such a curve from a particular value of t (Z for very large samples, and for normal approximations) all the

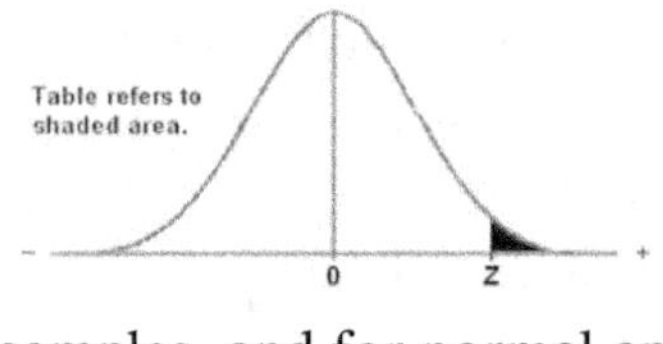

---

[8]The third moment skew for the 'raw' ($n = 1$) data in Figure 11.10 is $\widehat{skew} = -0.9891363$. The sign is not important regarding sample size, but to compensate for this skew the sample should have a size of at least $25 \times 0.9891363 = 24.73$ which rounds up to a minimum of about $n = 25$ (Subsection 15.2.1).

way to the right 'end' ($+\infty$) of the distribution. This one tail representation characterizes the most commonly available printed tables such as Appendix 4. The direction is unimportant because the symmetry of a normal distribution and the fact that $Z = 0$ at the center means that the area cut off by $-Z$ all the way to $-\infty$ is the same as from $+Z$ all the way to $+\infty$. Also note that if the real mean, μ, **equals** the hypothesized mean, the real difference is $\delta = \mu - \mu_0 = 0$. So, chance aside, if the equal means were really equal, a 'Z test' (or a t test) should result in $Z = 0$, but of course you cannot put chance aside.

Using a one tail test, you might ask, "Is the average life span of Pacific Islanders 68 years?" Then $\mu_0 = 68$ years is the value against which you test by taking a very large random sample from, say, death certificates. The closer the truth is to 68 years, the closer $\bar{x} - \mu_0$ will be to zero, whatever the divisor. In planning and visualizing Z or t tests, you can consider areas under appropriate standard curves as **summed or integrated frequencies** or, equivalently, as **summed probabilities**. So, especially in light of the fact that the researcher has no better starting place, the calculations of statistical significance tests are built on the assumption that **H0** is true. This assumption that **H0** is 'true' is **only** for the purpose of **calculating** that the observed difference between the means has a probability too tiny to be true, in which case **H0** is rejected. If **H0** is **not rejected**, it is not considered to be 'true', it is considered to be **unknown**.

If the true mean, μ, is not equal to μ_0, the further apart they are, the further the estimated Z or t is likely to be from $Z = 0$ or $t = 0$. So, the greater the Z or t, the less the probability of **H0** being true. In application, normal curve based statistical tests examine the estimated distance to challenge the provisional assumption that μ is equal to μ_0. This hypothesized equality is rejected if the estimated distance measured by Z or t is sufficient, meaning that the estimated difference is as great as or greater than the reasonable minimum difference (Subsection 3.1.2), which if not chosen by you will be chosen by your experimental design anyway.

In application, three considerations remain. One is, what is a distance between μ and μ_0 that is sufficient to be considered different enough for practical purposes? Another is, what should the acceptable sensitivity, the power, be? The final one, partly answered by Section 7.4 above, is how can you design your test to have an acceptable probability, the power, $1 - \beta$, of detecting this distance?

You cannot put a number on β, the probability of getting a Type II error, failing to reject **H0** when in fact it is false, without somehow implying the existence of an alternative mean, μ_A. One of the first things you do for a power analysis is **choose** a reasonable minimum difference, D_{min}, your targeted value of $\delta = \mu_A - \mu_0$ (or $\delta = \mu_2 - \mu_1$), on the original scale of measurement. This can provide you with a reference point, $\mu_A = \mu_0 + D_{min}$, where μ_0 can even equal zero. That is, for a power analysis, you have to find β with respect to some particular value, D_{min}, that you directly or indirectly supply. Indeed, each value of β is only defined in respect to (and cannot exist without reference to) a reasonable minimum difference. Although this discussion refers to the difference between means, reasonable minimum differences are also needed for power analyses comparing probabilities, variances, etc.

For a **single sample one tail test**, your hypothesis pair could be **H0**: THERE IS NO TRUE DIFFERENCE versus a one tail alternative. You can choose either the + tail, **HA**: THE MEAN OF THE SAMPLED POPULATION IS GREATER (by at least D_{min}) or you can choose the - tail, **HA**: THE MEAN OF THE SAMPLED POPULATION IS SMALLER (by at least D_{min}). Because β, and hence the power, $1-\beta$, can only be defined in terms of a particular reference point, $\mu_A = \mu_0 + D_{min}$, the challenge is to get the power, $1 - \beta$, you will need to detect a particular reasonable minimum difference, $D_{min} \neq 0$.

The normal curves in Figure 7.4 all have the same underlying population variance, σ^2. The leftmost curve in each pair have the null hypothesis mean, μ_0, and the rightmost curves in the pair have an alternative mean, μ_A, which **might** be the unknown truth ($\mu = \mu_A$). Under the model for the Z or t tests, if the populations happen to not be the same, there are two distributions, and the members of each pair of curves are assumed to differ by $\delta = \mu - \mu_0$ original units of measure. If they could be perfectly alike, $\mu - \mu_0 = 0$, then the hypothesized curves would overlap completely and there would be only one curve (one population) instead of two. But experiments and observations are never perfectly alike, nor do they perfectly estimate the true mean.

You can estimate $\mu - \mu_0$ on a standardized scale, $Z = (\bar{x} - \mu_0) / \sqrt{S^2/n}$, as though **one standard error of the mean** was a 'new unit of measure'. You might choose the usual criterion, $\alpha = 5\%$, to mark off the rightmost gray shaded area in each pair. On the standardized Z scale, the mean, μ_0, of the hypothesized distribution is set at $Z = 0$ (the center of the leftmost member of each pair in Figure 7.4). Then, if your sample resulted in a $|Z| \geq Z_\alpha$ in the chosen direction due to **chance alone** when **H0** was true, you would be experiencing a Type I error, declaring **H0** to be untrue when it is, in fact, true (false statistical significance). In application, a Type I error would be declaring a statistical difference when the true difference was either too small to be meaningful, or (rarely) nonexistent. All Z values as extreme or more extreme than Z_α (in the chosen + or - direction) are to be declared statistically significant; that is, they imply that **H0** is too unlikely to be true. (Notice that the larger the Z value, the smaller the area left in the tail; larger Z values mean smaller probabilities that **H0** is true).

Recall that, **on the Z scale**, the means and their distributions tend to be moved further apart as the sample size, n, gets larger (Section 7.4). For an a priori power analysis, after choosing α, you must choose a reasonable minimum difference, D_{min}, that you accept as being a meaningful true difference, on the original scale. You must also choose a probability, β, which is the probability of failing to reject **H0** when the true mean, μ_A, differs from the hypothesized mean, μ_0, by a particular (exact) amount, $|D_{min}| > 0$. Then, for **that** α, you can seek the sample size, n, at which there is a probability, $1 - \beta$, the power (that you also choose) of getting statistical significance if $\delta = \mu_A - \mu_0$ **exactly equals** D_{min}.

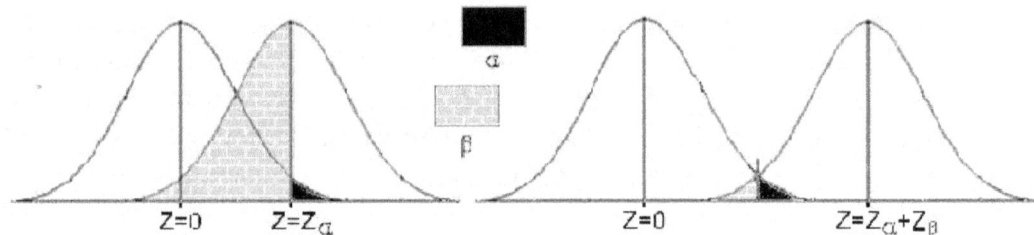

The left curves in each pair represent the distribution if the null hypothesized mean were true. But what if it were not true? Then β would be the probability, by chance alone, of **not** detecting a true mean, with a distribution represented by the right curves, which have mean $\mu_A = \mu_0 + D_{min}$. The leftmost whole pair is what you would have if your separation was only adequate to put the center of the alternative distribution on Z_α. Half of it would be below Z_α, so you would only have a 50% probability of detecting a difference equal to D_{min} in original units of measure. But if your sample size was large enough that only Z_β of the alternative distribution was below Z_α you would have a $(1 - \beta)\%$ probability (the power) of detecting it. You want **the** sample size that will create a separation such that $Z_{D_{min}} = Z_\alpha + Z_\beta$, **not merely** Z_α.

Figure 7.4: Achieving Power Requires Both α and β.

Even if $\mu_A - \mu_0 = D_{min}$, the first or **leftmost pair** in Figure 7.4 represents the situation in which a researcher has chosen a sample size, n_{left}, such that the center of the right overlapping curve, $\mu_A = \mu_0 + D_{min}$, is in a position equivalent to Z_α on the Z scale. This sample size, n_{left}, would result in the probability of getting a Type II error (failure to detect a reasonable minimum difference) of $\beta = 50\%$ (failure to detect half the time). So, the power would only be $1 - \beta = 50\%$. With respect to inference, if your sample size is such that $Z_{D_{min}} = Z_\alpha$, you might as well **flip a coin** to estimate if your null hypothesis is false, and if your null hypothesis is true, you still take an α probability of false rejection. This sample size, or even less, is common! Some sample size calculators on the internet have actually calculated sample size this way. Changing n **changes** the population of differences between means which is being tested. That is, you can choose a population of means, using a new sample size per mean, for which the test is **more powerful**.

Increasing from the sample size, n_{left}, that was used to get the leftmost pair of curves, you could continue on to the **rightmost pair of curves** in Figure 7.4, reaching a larger sample size, n_{right}, where the distance between the means on the Z scale will be large enough that β is squeezed down, as in the rightmost pair of curves, to the probability, β, needed to get your chosen power, $(1 - \beta)$. You can view this as squeezing the probability, β, of getting a Type II error down to the little brick patterned area in the tail by moving Z_β units further in the positive direction as shown on the right side. In Figure 7.4, where $\beta = 0.05$, you would obtain a power of $1 - \beta \geq 95\%$, assuming $\delta \geq D_{min}$. Conceptually combining these two steps moves the original center of the alternative curve further away to the correct distance, $Z_\alpha + Z_\beta$ on the Z scale. Choosing the correct sample size, n, targets an ideal,

$$(Z_\alpha + Z_\beta)\sqrt{\sigma^2/n} = D_{min}, \tag{7.2}$$

in the original units of measure. Hypothetically, if we really knew the true variance, σ^2, and if it so happened that $D_{min} = \delta$, the exact unknown and unknowable difference, the cumulative probability of getting a Type II error really would equal β, and the cumulative probability of detecting statistical significance at criterion level, α, really would equal $1 - \beta$.

> **IMPORTANT FACT:** Although α is never usefully equal to $1-\alpha$, the unsigned values of Z_α and $Z_{1-\alpha}$ are equal, due to the symmetry of the normal curve. They are points located the same distance from the mean, $Z = 0$. Likewise, β is never usefully equal to $1 - \beta$, but the unsigned values of Z_β and $Z_{1-\beta}$ are equal. Because of this, there are many ways to write the above Equation 7.2; that is, the equation also appears with the term $(Z_\alpha + Z_\beta)$ replaced by $(Z_\alpha + Z_{1-\beta})$ or $(Z_{1-\alpha} + Z_{1-\beta})$ or even $(Z_{1-\alpha} + Z_\beta)$. Assuming that you are considering the absolute value (the unsigned value) of $Z_{whatever}$, these all work out to the same value. There are various ways of explaining this (informal derivations). The first, the one I use, is based on a concept of $Z_{...}$ representing distances on the horizontal axis of the normal distribution graph. By this approach of literally looking at it, the middle of the alternative distribution based on D_{min} is viewed as reaching a power of 50% if it is centered on a criterion based cutoff which is Z_α (or $Z_{\alpha/2}$). From there, I picture this alternative center as being moved Z_β units **further** away from the hypothesized mean (zero on the Z scale). So the distance $(Z_\alpha + Z_\beta)$ is what I want. However, some authors see using the equivalent $(Z_\alpha + Z_{1-\beta})$ as superior for explaining power from a verbal logic based view.

Looking at the testing procedure that uses the Neyman-Pearson approach from an **applied** view, in practice the 0 point is merely a reference point and the power is associated with a very attainable point that is at D_{min}, which is located with respect to this unattainable 0 reference point. Above D_{min} the power achieved will be progressively greater, and below D_{min} it will progressively drop off. The estimate of power is only precise at D_{min}. The literature (See the Bibliography) cited in this book includes publications of authors who supported (often with some reservation), and authors who did not support (sometimes with little in the way of reservations), the Neyman-Pearson frequentist approach to inference. However, whether you go so far as to use it for inference or not as when only presenting CIs (or p_values), you still have no other way to either control or estimate the sensitivity of parametric analytical methods. That means that you must understand the concept of the Neyman-Pearson inferential approach, **particularly with respect to power**, whether you use it for actual yes-no inference or not.

7.5.1 Actually estimating n for one sample

The rightmost pairs of curves in Figure 7.4 illustrate the advantage of using a sufficiently large sample. To achieve sufficient power, you have to increase the sample size, n, until your results on the scale, Z, of what might be called units of $\sigma/\sqrt{n}$ are out where β is squeezed down to as small as you want it to be. As linguists can tell you, the more important something is, the more names it has. You might call the processes featured in this chapter, **a priori power analysis**, **sample size estimation**, or **power analysis for** n, and all of these are correct.

So, how are you supposed to actually do an a priori power analysis to estimate the **sample size**, n, that is sufficient to squeeze the β down to where the power $1 - \beta$ is what you are after? First you have to find a starting place, often by using Equation 7.3 (below) with S an estimate of the standard deviation, σ, and your chosen α, β, and D_{min}.

Starting as though using an estimate of σ from a huge data source, let's consider an approach derived from $Z = D/\sqrt{S^2/n}$. Assuming it is true that the true difference really equals the reasonable minimum difference ($\delta = D_{min}$), and substituting $Z_\alpha + Z_\beta$ as the observed Z, as well as substituting S as the estimated σ, gives

$$(Z_\alpha + Z_\beta) = \frac{D_{min}}{\sqrt{S^2/n}},$$

which is also a rearrangement of, and in part justified by, Equation 7.2 (above). Then squaring both sides, you get

$$(Z_\alpha + Z_\beta)^2 = \frac{D^2_{min}}{S^2/n}.$$

Writing $\hat{n}$ (n-hat) to indicate that it is an estimate of the sought after sample size n, and rearranging, gives the equation you will apply,

$$\hat{n} = \frac{(Z_\alpha + Z_\beta)^2 S^2}{D^2_{min}}, \tag{7.3}$$

where you choose your particular value of β to attain the desired power (with respect to the reasonable minimum difference, D_{min}). This Equation 7.3 can also be obtained directly by rearranging Equation 7.2. Note that β is the complement of $1 - \beta$, the power; $\beta = 1 - (1 - \beta)$. Also notice that the mean does not directly enter into the estimation of the desired sample size.

One of several possible rearrangements of the Equation 7.3 for finding an estimate, $\hat{n}$, of a one tail sample size that squares only once, is

$$\hat{n} = \left(\frac{(Z_\alpha + Z_\beta) S}{D_{min}} \right)^2.$$

Of course you get the values of Z_α and Z_β (or, $t_{\alpha, df}$ and $t_{\beta, df}$) from software or tables such as Appendix 4 (or, Appendix 3).

Because the variance, S^2, is an estimate, you usually have eventually to use t in place of Z. However, the value of t depends on the df, which in turn depends on n. You may well wonder how you can adjust the estimation process using a yet to be known df. So you usually use Z first to estimate $\hat{n}$, your starting estimate of n. Then you will have to use $t_{\alpha, df}$ and $t_{\beta, df}$ in place of Z_α and Z_β to get

$$\hat{n} = \frac{(t_{a, df} + t_{\beta, df})^2 S^2}{D^2_{min}}. \tag{7.4}$$

Then you may have to use this equation to *iterate*, to repeat the estimation again using $t_{\alpha, df}$ and $t_{\beta, df}$ repeatedly with the df adjusted by your new estimate, $\hat{n}$, of n, for each iteration, until your estimate, $\hat{n}$, **stops changing by** $\geq |0.5|$. If you know the df (the denominator of S^2) you can start with Equation 7.4, using that df, but you will still have to iterate at least once. In any case Z is an OK starting place and if your first estimate is huge, roughly meaning $n > 4500$, you can stop there.

Always round the value $\hat{n}$ **up** at the **last iteration**, but if it is not the last iteration, then round in whichever direction is called for, according to ordinary arithmetic or algebra rules, and continue. For a one sample Student's t test you will adjust your t values by using a new $df = \hat{n} - 1$ for each successive value of $\hat{n}$ until you reach a stopping point as mentioned above.

As is illustrated in the examples which follow, you should, in general, stop when the $t_{\alpha, df}$ values appear to no longer be growing larger in a table of t probabilities, which is accurate at least to the number of figures that you want. You might not have to iterate more than once or twice. Note that for normal distribution based tests, this requires you to have an estimate, S (or S^2), for the standard deviation **and** to a priori choose what you and your peers will see as a plausible reasonable minimum difference, D_{min}. Obtaining such an estimate of σ, as mentioned in Section 7.2, may require sampling the target population, maybe from relevant records or some sort of pilot study. Your estimate of σ or σ^2 must be based on a source of data that duplicates your planned experimental or observational technique. The larger the df of S (or S^2) the more accurate will be your estimate of the sample size needed to obtain the power you desire.

Equation 7.3 and Equation 7.4 also apply for paired t tests, where you are estimating n_{d}, the number of pairs. Sometimes you will find it easier to use other approaches to choosing your initial sample size, such as Subsection 7.9.1. If you intend to use small sample sizes, $n < 30$, then be ready to defend your assumption of underlying normality. Do **not** use t for a normal approximation; the t distribution does **not** pertain to the normal approximation. Z will be used for the normal approximation; it will only require Equation 7.3 and no iteration.

> **CAUTION:** The values of Z_α and Z_β (or $t_{\alpha, df}$ and $t_{\beta, df}$) are absolute values unless stated otherwise. That is, $(Z_\alpha + Z_\beta)$ is the sum of two positive values, the unsigned (thus treated as positive) **distances** from the mean on the Z scale, regardless of direction. Do not be confused by online calculators, etc.

Variances, or standard deviations, with higher degrees of freedom (denominators), are better and may produce more accurate, and sometimes even slightly smaller, sample size estimates. Also, larger pilot samples will often lead to smaller final samples.

7.5.1.1 Example: Preliminary One Sample A Priori Sample Size Estimate, $\hat{n}$

One Tail A Priori $\hat{n}$, One Sample, $\bar{x}$ vs μ_0 (Also see Subsection 5.2.0.1) The values in these fructose examples are fictitious. Such research would be a good idea, but I do not know if it has really been done, or what the standard deviation or average weight of male college students might be. Let's return to our public health student who suspected that the average weight of male students at Little U College who drank high fructose sweetened soft drinks is more than the average for those male students who did not. In this scenario, she was interested in a one tail test to compare against a mean, μ_0, corresponding to the average weight of male students in the university age range from a national survey. However, the hypothesized mean, μ_0, does not enter into the actual estimation of the desired sample size. Before even collecting the local students data she needed to make a sample size estimate, $\hat{n}$. She had what she considered a good estimate, S, of the standard deviation, σ, that was based on a national sample, which she understood was from a vast random sample of tens of thousands. So, since the degrees of freedom for this S were virtually unlimited, she was justified in using Z for her initial estimate of the necessary number of male students to be weighed at random. She decided she needed at least an 80% chance of detecting a $D_{min} = 5$ lb difference between the two groups. She chose a criterion, $\alpha = 5\%$, and to get a power of 80%, she was willing to accept a $\beta = 20\%$ probability of not detecting that 5 lb difference if it existed $((1 - \beta)\% = 100\% - 20\%)$.

Methods: Using this estimate based on the government estimate (remember this is fictitious) of σ of $S = 26$ lb for males of college student age, she formed a hypothesis pair,

H0: THERE IS NO TRUE DIFFERENCE BETWEEN THE AVERAGE WEIGHT OF MALE STUDENTS DRINKING HIGH FRUCTOSE SWEETENED SOFT DRINKS AT LITTLE U AND THE CURRENT NATIONAL AVERAGE FOR COLLEGE AGE MALES.

HA: THE WEIGHT OF MALES DRINKING HIGH FRUCTOSE SWEETENED SOFT DRINKS AT LITTLE U IS MORE THAN THE NATIONAL AVERAGE FOR THEIR AGE GROUP

Analysis: She got the desired values of $Z_{\alpha=0.05} = 1.645$ and $Z_{\beta=0.20} = 0.8416$ from the table in Appendix 3. Using the above variant of Equation 7.3, $\hat{n} = \left(\frac{(Z_\alpha + Z_\beta)S}{D_{min}} \right)^2$, she calculated

$$\hat{n} = \left(\frac{(1.645 + 0.8416)\,26\,\text{lb}}{5.00\,\text{lb}} \right)^2 = 167.193\,.$$

The pounds, above and below, cancel out even before squaring. She thought she was finished, so she correctly rounded up to the nearest integer, resulting in a sample size estimate, $\hat{n} = 168$. The squaring strongly magnifies small computational differences. For instance, I wrote a computer routine to do this, but input $Z_{0.20}$ and $Z_{0.05}$ more accurately, and got a rather noticeable effect in the second decimal place. However, the additional decimal places are only of interest in that they **tell you to round up**.

Results and Discussion: At first she thought, assuming her sources of information about the variance were sufficiently accurate, that she needed to weigh $n = 168$ male college students to achieve the power, $1 - \beta = 0.80$, that she desired. Notice that this was a large enough sample that even if the population under consideration was not almost normal, she could feel reasonably certain that the CLT would provide a sufficiently close to normal distribution of the means. However, the **sample size** in the actual experiment will **ultimately determine** its degrees of freedom. So, because she needed to adjust for her estimated moderate sample size, she still needed to use t. Otherwise the estimate of n might be too small. So she **still needed** to refer this original $\hat{n} - 1 = df = 166$, where $\hat{n}$ was rounded the common arithmetic way to 167 according to the rules of ordinary arithmetic, to a table of t with $\alpha = 0.05$, and to **repeat** the process, substituting $t_{0.05, 166}$ for $Z_{0.05}$, in order to approach a satisfactory estimate.

Of course it would be nice if she could have had the **true** standard deviation parameter, σ, instead of an estimate, the statistic S, but the true standard deviation, σ, is almost never known. The question of obtaining a sufficient estimate of S needs careful consideration, as does the estimation of the sample size, especially because once you have analyzed the data you cannot readily go back and change n. Unless she was reasonably certain that the estimate, S, was rounded up, she might have considered weighing $n = 170$ male students, because one of her inputs (S) was accurate to only two figures. At the very least, if your estimate of the variance is to, for instance, two figures, you have to round **up** to the nearest three figures for reporting. So, even though it is bad news, something like $\hat{n} = 10,561$ would have to become $\hat{n} = 10,600$, or if your two figure estimate, S, is weak, $\hat{n} = 11,000$. If a large or trustworthy sample variance is itself rounded **up**, the rounding will justify proceeding as though it had an additional figure. (But did she know whether this estimate of $S = 26$ was conservatively rounded up rather than carelessly down? Maybe the student should have researched the original source, or contacted the reporting entity.)

However, even assuming 26.0 (with a significant trailing zero) is an adequate estimate of σ, she still had some more work to do before she could estimate an acceptable sample size for a Student's t test based on her previously estimated sample size close to 167.[9] For now, assume that she assumed correctly that her S was rounded up and that it was thus safe to assume that this estimate was unlikely to be more than 26.00 pounds, even in the third or fourth decimal place. (A proper literature survey, possibly a telephone call, could have eliminated this difficulty, but it is left here to warn the unwary.)

Question 3 gives you an opportunity to try this and to examine the approach further by using the odd numbered answers at the back of this book.

> **CAUTION:** For surveys, **certain agencies** will **not** accept for publication the reporting of the means or their CIs with sample sizes resulting from the commonly accepted minima, $\alpha = 5\%$ and $\beta = 20\%$ (see Section 7.14). Such surveys are required to have precision equivalent to a very high power, and a very low probability of getting a Type I error. For this and for many other reasons, some of which underlay these agencies' requirement for publication, the research community should consider new minima of maybe $\alpha = 1\%$ and $\beta = 5\%$ (Section 4.10 and Benjamin et. al.2020).

7.5.1.2 Example: Adjusting Less than Huge $\hat{n}$, 1 Sample A Priori 1 Tail

One Tail a priori $\bar{x}$ vs μ_0 estimation of n (Subsection 7.5.1.1 continued) Our public health student **had already** used Z_α and Z_β to estimate that she should weigh $\hat{n} = 168$ male students to achieve her chosen power, $1 - \beta = 0.80$, at the $\alpha = 0.05$ level that she desires. This would be a large enough sample size to enable her to meet the normality requirement of the t test, **but**, it is necessary to repeat the calculation, using **t**, to **adjust** for having only an unadjusted estimate of σ.

Analysis: She had already calculated $\hat{n} = \left(\frac{(Z_\alpha + Z_\beta)S}{D_{min}}\right)^2 = \left(\frac{(1.645 + 0.8416)26.00\,\text{lb}}{5\,\text{lb}}\right)^2 = 167.193$,

but because it was based on Z, this might be too small to provide an appropriate df for testing. Rounding (not always up, because this is not the last step) according to the usual rule for a difference less than 0.5, she gets $df = 167 - 1$ for the next step. She looks up $t_{0.05,\,\hat{n}-1=166} = 1.654$ and $t_{0.20,\,166} = 0.844$. So she runs this calculation again using these values using $\hat{n} = \left(\frac{(t_{\alpha,\hat{n}-1} + t_{\beta,\hat{n}-1})S}{D_{min}}\right)^2$,

$$\hat{n} = \left(\frac{(1.654 + 0.8435)\,26.00\,\text{lb}}{5\,\text{lb}}\right)^2 = 168.662,$$

which rounds up to $\hat{n} = 169$.

Results and Discussion: Because her original sample size estimate was a little short and would have produced less than 80% power, she used this new **larger estimate** which rounded up to

---

[9]You might think that because her initial estimate was based on a very large sample, the public health student could continue with Z values. However, she is going to be looking for statistical significance, or a CI based, in part, on the estimated variance based on her estimated sample size that rounds up to $\hat{n} = 168$. In addition to the fact that she is sampling a local population rather than the nation-wide population, such a sample size could be underestimated by Z. So she would be well advised to be as conservative as possible, and to use a t distribution which adjusts for the estimation of variance by the testing procedure.

$n = 169$ replications. However, because the t tables (such as Appendix 2) make little distinction about one more or less at this sample size, $\hat{n}$ would not have changed by as much as 0.5, so she stopped there. She did not bother to calculate the sample size again with this new estimate, $\hat{n} = 169$. To avoid the possibility of a dropout or some other difficulty such as a new estimate of $S \gg 26$ lb, it would be a good idea (reasonably conservative) to weigh $n = 170$ male college students. If she weighs 170 instead of 169, the slight power increase would not lead to excessive sensitivity or a ridiculously tiny CI, but more importantly it would provide the power to estimate its centering.

She calculated this for one tail at $\alpha = 5\%$, but she may also wanted a confidence interval, and that requires two tails. So, to be consistent, she could report a 90% CI (5% in each tail), but **only** following a statistically significant one tail test with $\alpha = 5\%$. That is, she may be interested in only one tail, but a confidence interval has a beginning and an end in hopes of trapping the true mean between them by chance. Because this is not a particularly common confidence interval, she might wish to consider reporting the margin of error instead (Subsection 7.13).

> **CAUTION: It is always better to overestimate sample size, $\hat{n}$, than to underestimate it. Overestimates can be costly, but they will only be a problem if they are vastly too large.**

How powerful is a test going to be, if, as is usually the case, the true (unknown and unknowable) difference, δ, is not equal to the reasonable minimum difference, D_{min}? For instance, suppose the real difference was $\delta = 4.00$ pounds rather than 5.00 pounds, would that mean that your test would have no chance of detecting that there was a difference? No, that would mean the same sample size would result in **less probability** of detecting that there was a difference. Well, would you then always detect a difference of 6.00 pounds? No, not always; the fact that it was larger would only **increase** the probability that you would detect that there was a difference.

Let's contemplate the 's-shape' graph, Figure 7.5, assuming a one sample one tail test with sample size, $n = 170$, criterion $\alpha = 5\%$, and $\beta = 20\%$. This curve was calculated using t with $df = 169$, because using Z would inaccurately shift the whole curve down a tiny amount. The shape would be the same.

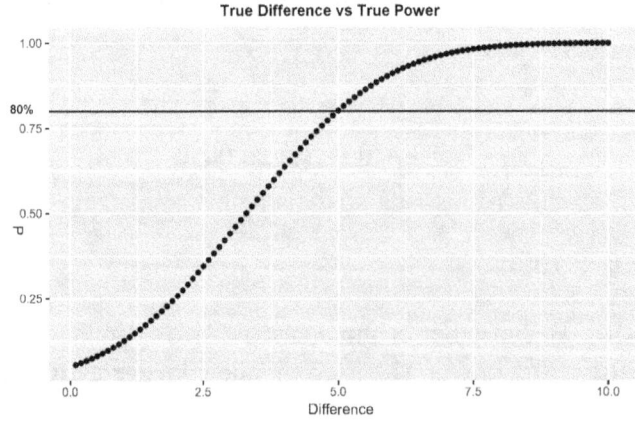

Figure 7.5: Power tends to drop rapidly when $D_{min} < \delta$ and to have diminishing increases when $D_{min} > \delta$. (Horizontal line is 80% power.)

The actual values do not matter; the point is the shape of the curve. As the true difference moves to the right above, the effective power of the test to detect that there is really a difference tends to increase rapidly, approaching, but never quite reaching, certainty. But, for differences below the reasonable minimum difference, you see a steady somewhat steeper drop, eventually leveling off to still claim statistical significance 5% of the time even if the true difference were zero, $\delta = 0$. Yes, chance is still going to trick you some small percentage of the time.

7.5.1.3 Example: One Sample, One Tail A Priori, Larger D_{min}

What if our public health student is interested in a much larger true difference, δ, than in the earlier example and has not taken any samples? Recall that she had an estimate, $S = 26$ lb, of the standard deviation, σ, from a huge national survey and wished to obtain a sample size estimate, $\hat{n}$, that would allow her to attain an $\alpha = 5\%$ and a power equal to $1 - \beta = 80\%$ of detecting a larger difference, D_{min}. Assume she decides to treat S as accurate to any number of places.

One Tail A Priori $\bar{x} - \mu_0$ estimate with larger D_{min} What if she uses a much larger $D_{min} = 15$ lb, but with the same 80% power, etc.?

Methods: She based her calculations on $S = 26.00$ lb for males of college student age from a nationwide survey. She wished to test:

H0: THERE IS NO TRUE DIFFERENCE BETWEEN THE AVERAGE WEIGHT OF MALE STUDENTS DRINKING HIGH FRUCTOSE SWEETENED SOFT DRINKS AT LITTLE U AND THE CURRENT NATIONAL AVERAGE WEIGHT FOR COLLEGE AGE MALES. VS

HA: THE AVERAGE WEIGHT OF MALES DRINKING HIGH FRUCTOSE SWEETENED SOFT DRINKS AT LITTLE U IS MORE THAN THE NATIONAL AVERAGE FOR THEIR AGE GROUP, at the $\alpha = 5\%$ level, with $\beta = 20\%$. $D_{min} = 15$ lb was the targeted reasonable minimum difference.

Analysis: She used Z for her initial analysis, she got the desired values of $Z_{\alpha=0.05} = 1.645$ and $Z_{\beta=0.20} = 0.8416$ from the table in Appendix 4. Using one of the arrangements of Equation 7.3, $\hat{n} = \left(\dfrac{(Z_\alpha + Z_\beta) S}{D_{min}} \right)^2$, she calculated

$$\hat{n} = \left(\frac{(1.645 + 0.8416)\, 26.00\,\text{lb}}{15\,\text{lb}} \right)^2 = 18.58 \,.$$

This is a small sample estimate using a larger than large sample approach, so, because she must iterate, she rounded according to arithmetic custom. In this case, the decimal part is > 0.5, so she rounded up the preliminary initial (under) estimate to $\hat{n} = 19$. She knew that for such a small estimate she should iterate using t for a more accurate estimate, $\hat{n}$. She referred to the table in Appendix 3 and repeated the process, using $t_{a,\hat{n}-1} = t_{0.05, 19-1} = 1.7291$ and $t_{\beta, \hat{n}-1} = t_{0.20, 19-1} = 0.8601$ with Equation 7.4 to calculate

$$\hat{n}_t = \left(\frac{(1.7291 + 0.8601)\, 26.00\,\text{lb}}{15\,\text{lb}} \right)^2 = 20.1416.$$

this would round up to 21 but she rounded the usual way and used $19\,df$ of freedom, on her next iteration with Equation 7.4 to get $\hat{n} = 20.15541$, so because this still conservatively and correctly **rounded up** to the same value $\hat{n} = 21$, she stopped. She then decided, given that the normal distribution assumptions of the t test were true, she could use a sample size of $n = 21$. Because she had to round up so much, she did not worry about the rather unlikely problem, in this design, of having a dropout.

Results and Discussion: This D_{min} is a rather large true difference, and it is doubtful that this would really be the reasonable minimum difference in which she would have been interested. I contrived this example to show you how to handle small sample size estimates, such as when you really are interested only in large true differences. Assuming her source of information about the standard deviation is sufficiently accurate and that it is no greater than 26.00 lb, she felt she should weigh $n = 21$ male college students to attain the power, $1 - \beta = 0.80$, that her professor is likely to require. This assumes normality, and such an assumption might be accepted because she claimed that she had heard that the human weight distribution, although truncated, is often accepted as otherwise being almost normally distributed and therefore suitable for comparisons using the small sample t test. (I do not know that human weight is normally distributed, or if the present trend toward more obesity has distorted whatever evidence might have previously been considered. Be very careful whenever making such an assumption based on hearsay!) Assume she appropriately discussed this assumption of underlying normality in the Results and Discussion she provided in her report. If she were worried that her professor would not accept this assumption, she should consider increasing her sample size to at least $n = 30$ so that the mean could be assumed to be part of a distribution of large sample means and thus almost certain to be normally distributed.

Even though the very last calculation step in the above example did not change the rounded estimate, that does not mean that it might not have been necessary. You often will not know if it is needed until you have done it.

7.5.1.4 Example: One Tail, One Sample A Priori, Moderate D_{min}

This example will give you a little more feeling for the cost of power by using a power of $1 - \beta = 95\%$ instead of $1 - \beta = 80\%$.

More expensive one tail a priori $\bar{x} - \mu_0$ estimate with moderate D_{min} She decided to see how big a sample size she would need if she had done the example in Subsection 7.5.1.3 with $\beta = 1 - (1 - \beta) = 5\%$, and $\alpha = 5\%$. She used the standard deviation estimate of $\sigma = 26.00$ lb, from a huge sample size $\hat{n}$, that targeted a $\delta = 5$ lb. Of course, because she set it, $D_{min} = 5$ lb can be considered to be exact.

Methods: As before, she wishes to test

H0: THERE IS NO TRUE DIFFERENCE BETWEEN THE AVERAGE WEIGHT OF MALE STUDENTS DRINKING HIGH FRUCTOSE SWEETENED SOFT DRINKS AT LITTLE U AND THE CURRENT NATIONAL AVERAGE FOR COLLEGE AGE MALES.

HA: THE WEIGHT OF MALES DRINKING HIGH FRUCTOSE SWEETENED SOFT DRINKS AT LITTLE U IS MORE THAN THE NATIONAL AVERAGE FOR THEIR AGE GROUP,
however, she was now estimating testing at the $\alpha = 5\%$ level with power $(1 - \beta) = 95\%$, and the targeted $D_{min} = 5$ lb.

Analysis: She started with $Z_{0.05} = 1.645$, for both Z_α and Z_β in Equation 7.3, $\hat{n} = \left(\frac{(Z_\alpha + Z_\beta)S}{D_{min}} \right)^2$, to calculate

$$\hat{n} = \left(\frac{(1.645 + 1.645)\, 26.00\,\text{lb}}{5\,\text{lb}} \right)^2 = 292.63\,.$$

Rounded to $n = 293$, this was not an impossibly large sample size for a graduate study or a professional project. However, she was not finished because it is necessary to try at least one iteration using t. So, with $\hat{n} - 1 = 292\, df$, she had to repeat the process, using $t_{0.05,\,292} = 1.650$ from a table of t such as Appendix 3, and this works out to

$$\hat{n} = \left(\frac{(1.650 + 1.650)\, 26.00\,\text{lb}}{5\,\text{lb}} \right)^2 = 294.50.$$

She consulted a table of t for the next iteration, but the change was way out in the fifth decimal place and was not enough to change the outcome so she rounded **up** to get a final estimate of $\hat{n} = 295$. Always round up at the last because you cannot have a fraction of a trial and it is always better to overestimate sample size, $\hat{n}$, than to underestimate it.

7.5.1.5 What about normality when estimating n?

Total sample sizes of **n** ≥ 30 generally are sufficient for attaining normality and standard deviations that are themselves based on total sample sizes of **n** ≥ 30 are generally sufficient for attaining normality and thus sufficient for the estimation of sample sizes, $\hat{n}$. However there are distributions where this is not sufficient, such as when there is extreme skew. Nonetheless, the larger a sample size for estimating S, the less affected by non normality the calculation of the sample size estimate, $\hat{n}$, will be. Sawilowsky and Hillman (1993) examined sample sizes up to $n = 80$ and reported that at $n \geq 80$, power calculations based on the t-test were appropriate, even when the data were decidedly non-normal..

7.6 Estimating n for 1 Sample 2 Tail Power to Detect D_{min}

To plan a **single sample two tail test**, your hypothesis pair could be **H0**: THERE IS NO DIFFERENCE, meaning $\mu - \mu_0 = 0$ (a point of reference for the purpose of calculation), versus **HA**: THERE IS A DIFFERENCE, implying $|\mu - \mu_0| \geq D_{min}$ original units of measure, where $\mu_A = \mu_0 + D_{min}$ (another point) is the alternative mean, given $D_{min} = \delta$. Note that the absolute operator, $|\,|$, always means that either + or - is taken as +, or rather it means either sign does not apply, so **HA** refers to a difference in either direction. Now, referring to Figure 7.6, if the true difference had been in the less than hypothesized tail (- tail), the alternative curve on the far right would still apply, but would be flipped over on the left because a two tail difference can be statistically significant in **either direction**. On the standard normal, or Z, scale, the center of the hypothesized curve under **H0** of each pair has the mean, μ_0, at $Z = 0$. All Z values greater than or equal to $+Z_{\alpha/2}$ or less than or equal to $-Z_{\alpha/2}$ are statistically significant. Assuming **H0** is true, if you found a Z value tail-ward of the Z cutoff, $Z = \pm Z_{\alpha/2}$, you would be experiencing a Type I error, false statistical significance due to chance alone. Recall, on the Z scale, that the means and their distributions tend to be moved further apart as the sample size, n, gets larger (Figure 7.1). For a power analysis for planning a test, you must a priori choose a reasonable minimum difference, D_{min}, something that you accept as justifiably meaningful, on the original scale. Then you can estimate a sample size, n, such that there is a probability, $1 - \beta$, of getting statistical significance if $|\delta| = |\mu_A - \mu_0|$ were to **exactly equal** D_{min}.

    Different possible outcomes of two tail one sample power are illustrated in Figure 7.6 (above). You only have to split the criterion into a suitable cutoff by changing from α to $\alpha/2$, thus dividing the

Figure 7.6: $\bar{x}$ versus μ_0 : Two Tail Power with Increased Sample Size,

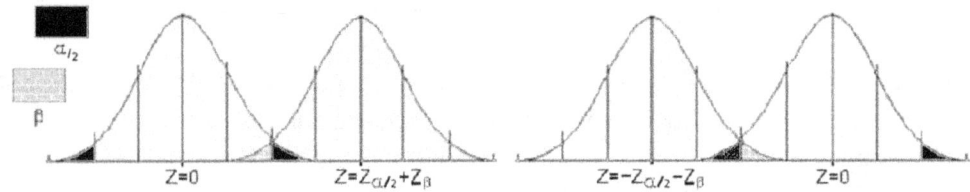

The pair of overlapping curves on the left represents a two tail test and is the result of taking β into consideration as well as $\alpha/2$ when estimating a sample size sufficient to squeeze β, the probability of a Type I error, down to achieve the power $|1 - \beta|$ as large as you choose in the upper tail as would be the case if the actual difference turned out to be positive. But if the difference turns out to be negative, then the result is represented by the pair of overlapping curves on the right, as though the left pair were flipped over. **For a two tail test**, either a positive Z or a negative Z could represent a meaningful difference.

criterion equally between the two tails. The power you want is associated with the single alternative power and is thus still $1 - \beta$. So the correct equation to estimate the sample size for a two tail test is

$$\hat{n} = \left(\frac{(Z_{\alpha/2} + Z_\beta) S}{D_{min}} \right)^2 .$$ (7.5)

It is used in much the same way as for a one tail estimate, except it does not heed direction at all. Sample size estimation based on power is also the preferable accepted way to estimate the size that will justify your confidence interval, no matter how short or long it might be.

A majority of estimates involve df sizes that, unlike airline arrival records or public health records, are not sufficiently huge to justify the use of Z. However, you may start with Z, but for samples that are not ultimately going to be huge (say $n > 4500$), you must use a modification of the above Equation 7.5 after your first estimate of $\hat{n}$. This is done by substituting t values that use the df from the estimation of S or S^2 ($df = n - 1$ for a one sample Student's t test) for the Z values. This merely requires a substitution of $t_{\alpha/2, df}$ for Z in Equation 7.5 to get

$$\hat{n} = \frac{\left(t_{a/2, df} + t_{\beta, df} \right)^2 S^2}{D_{min}^2} .$$ (7.6)

You can always start with Z (Equation 7.5) but you can alternatively start with using the df from your **source of** S^2 if it is not conceptually almost infinite, **using t** (Equation 7.6). But if the df of your source of S^2 (or S) is $df < 60$, definitely start with Equation 7.6. **After your first estimate**, $\hat{n}$, iterate and continue iterating as in the preceding Section 7.5.1, using the new values of $\hat{n}$ in $t_{\alpha/2, n-1}$ and $t_{\beta, n-1}$ for a one sample, two tail t test, until the estimate, $\hat{n}$, stops changing.

> **Note:** Whatever the hypothesis pair you specify, β **never needs** to be divided between the tails, not even for two tail tests; it is not affected by the tails.

Question 9 provides both an exercise and an example using this approach.

7.6.1 Sample size for paired t tests

Suppose you doubted that there was a difference in the strength of the average person's right hand versus their left hand. Since each person has two hands and people differ, you can use the sampled difference d_i for each person i, and find the mean $\bar{d}$. Such a paired t test, Section 5.4, might be thought of as a two sample test, but it is calculated **as though each pair is just one individual sample**. So, it is included here, before the less structured tests of two independent samples. If you measure the difference between the hands you could test such hypotheses as **H0**: THERE IS NO DIFFERENCE , $\mu_d = 0$, versus **HA**: THERE IS A DIFFERENCE. But, as always, it is impractical to actually target a true difference exactly equal to zero. So, zero is merely a reference point. If you have a sufficiently justified estimate, S_d, of σ_d for the **one tail** paired tests, then substituting $\hat{n}_d$, denoting the fact that it is about pairs, into Equation 7.5 gives

$$\hat{n}_{\mathbf{d}} = \left(t_{cutoff, df=n_{\mathbf{d}}-1} + t_{\beta, df=n_{\mathbf{d}}-1}\right)^2 S_{\mathbf{d}}^2 / D_{\bar{\mathbf{d}}\,min}^2, \tag{7.7}$$

where $D_{\bar{\mathbf{d}}\,min}$ represents your chosen reasonable minimum difference (from $\mu_{d_0} = 0$) for pairs, and $S_{\mathbf{d}}^2$ is the standard deviation for the pairs,[10] and the degrees of freedom are $df = n_{\mathbf{d}} - 1$ from the variance $S_{\mathbf{d}}^2$. You select α as the $cutoff$ for one tail, or $\alpha/2$ for two tails.

> **NOTE:** This standard deviation, $S_{\mathbf{d}}$, is the standard error of the differences within the pairs. If the pairing is appropriate, a great deal of the unpaired apparent variance can be eliminated. That is the purpose of the pairing. Using S from unpaired tests when you are planning a paired test might cause you to overestimate the sample size, but that would at least be conservative.

In the unlikely situation in which you have an estimate of $S_{\mathbf{d}}^2$, you can use

$$\hat{n}_{\mathbf{d}} = \left(Z_{cutoff} + Z_{\beta}\right)^2 S_{\mathbf{d}}^2 / D_{\bar{\mathbf{d}}\,min}^2,$$

and then iterate Equation 7.7 (above) and then iterate using the $df = \hat{n}_{\mathbf{d}} - 1$ from your first, and successive, estimates of $\hat{n}_d$. Or you can start using Equation 7.7 with the df of your estimate of $S_{\mathbf{d}}$. If you do not have an estimate of $S_{\mathbf{d}}$, as from a pilot study with, say, $n_{\mathbf{d}} = 10$ to 20 paired measurements, you might get a **crude** estimate using an estimated S from, say, $n = 25$ to 30 measurements of the unpaired control data. (If such data is available, it might overestimate the sample size a bit.)

This question of paired variances and standard deviations can be a challenge for paired t tests and for the RCBD ANOVA tests (Subsection 9.4), which can be seen as an expansion of the paired t test (Subsection 9.4). A few such $S_{\mathbf{d}}$ standard deviations might be known, such as maybe the standard deviation for paired blood pressure readings as taken from the two arms of each patient in a particular class of patients using a particular protocol and cuff.

Aside from the difficulty of estimating an initial $S_{\mathbf{d}}$, the paired t test, which tends to be robust with sufficient sample size, might be the ideal for many on-farm studies, particularly in light of the sensitivity of the ANOVA to non normality and variance differences between the treatments (Chapter 9). Instead of investing in additional treatments in some vague hope of additional discovery, you can invest in more replication and also avoid the pitfalls of multiple comparisons (Subsection 10.2.0.1).

7.7 Estimating n for Two Independent Sample, $\bar{x}_2 - \bar{x}_1$, Power to Detect D_{min}

When you can be sure that the two sample sizes you will be using will be very close to equal, $n_1 \approx n_2$, the following equations are suitable. It is widely believed that a difference in sample sizes of one is still sufficiently close to equal when $(n_1 + n_2)/2 \geq 30$. But, in Figure 7.2 and Figure 7.3 there is

---

[10]It might be acceptable to substitute $S_{\mathbf{d}}^2 = MSE$ from a randomized complete block ANOVA (Subsection 9.4), with $df_{\mathbf{d}} = df_W$ into Equation 7.7. This may work if you can duplicate the conditions under which you obtained the data for the randomized complete block ANOVA. Using t tests with an ANOVA will be covered later when we come to multiple comparisons.

a slight but detectable difference in the effect of the CLT even between $n = 27$ and $n = 30$. This is one of the several reasons why you should always strive for equal sample sizes. Eventually, in Subsection 7.11.2, we will see how to correct for unequal sample sizes.

Sometimes you will need to think about costs and returns when you choose what power, $1 - \beta$, you want (see Subsection 7.9.1). If you have a measure of third moment *skew*, as you ideally (if possible) should, then the minimum total sample size should be whichever is **greater**, your sample size estimate (below) or $25 \times skew$. If, as is usual, you cannot get an estimate of skew, your sample size estimate is almost always adequate anyway, it's just that having an estimate of skew further insures that the sample size estimate is adequate (Subsubsection 5.2.1.1). If you decide to use small overall sample sizes ($n_1 + n_2 < 30$), then be ready to defend your assumption of underlying normality because the CLT might not provide sufficient help with such small sample sizes.

If your test's actual two sample sizes cannot be equal, $n_1 \neq n_2$, your estimate of the total sample size, $\widehat{n}$, will only approximately equal the necessary $\hat{n}_1 + \hat{n}_2$ sample size, and you will have to **adjust for unequal sample sizes** via the method in Section 7.11.2.

7.7.1 Two equal samples, power analysis to estimate n

To derive a normality based power analysis equation for two equal random samples, recall that the process of calculating the pooled standard error of the mean uses S times $\sqrt{1/n_1 + 1/n_2}$, and if the sample sizes are equal, this equals $\sqrt{2/n}$. Starting with Student's t test for equal sample sizes, Equation 5.3, $t_{cutoff, df} = (\bar{x}_2 - \bar{x}_1)/S_{1,2}\sqrt{2/n}$ (with $df = 2n - 2$ and $_{cutoff}$ equals **either** $_\alpha$ or $_{\alpha/2}$), rearrange and square to get $\dfrac{2S_{1,2}^2 t^2}{(\bar{x}_2 - \bar{x}_1)^2} = n$. Then substitute $(t_{cutoff,n-1} + t_{\beta,2n-2})$ for t, D_{min} for $(\bar{x}_2 - \bar{x}_1)$, and $\hat{n}$ for n to get

$$\hat{n} = \frac{2 \left(t_{cutoff, 2n-2} + t_{\beta, 2n-2}\right)^2 S^2}{D_{min}^2}, \tag{7.8}$$

the estimated sample size for **each** of the two equal samples being compared. When the sample size of your source of S (or S^2) is **not** from a huge sample, and you know the df of S^2 (and $df < 4500$, meaning not huge), you can start with this Equation 7.8.

But for initial estimates when you do not know the df of your estimate of S^2 **or** for an initial $df > 4500$, you should start with Z values instead of t values. That is, substituting Z for t in Equation 7.6, you obtain the big sample size estimation for **each** sample in a two sample test,

$$\hat{n} = \frac{2 \left(Z_{cutoff} + Z_\beta\right)^2 S^2}{D_{min}^2}, \tag{7.9}$$

and the total sample size estimate is $\widehat{n} = 2\hat{n}$.

Regardless of which of these two equations you got $\hat{n}$ from, if the sample size estimate, $\hat{n}$, is not huge (say $\hat{n} > 4500$), you need to return to Equation 7.8 and **iterate** (do the estimation again, and maybe yet again, with new values of $t_{cutoff,2\hat{n}-2}$) using each new estimate, $\hat{n}$, to get the new df, until the change in the sample size estimate, $\hat{n}$, is less than 0.5. Notice that β is the **same** for both one and two tail testing.[11]

If for some reason you have to go on with $n_1 \neq n_2$, that is, if one of the samples cannot be large enough but the other one is still adjustable, you will have to correct your estimated total sample size, $\widehat{n}$, for unequal sample sizes by using the method in Section 7.11.2.

---

[11] Even though you use tables of Z for huge sample sizes and approximations such as the binomial approximation, **two** sample tests of the general form, $Z = \frac{\bar{x}_2 - \bar{x}_1}{\sqrt{S^2/n}}$, are historically called *Student's t tests*.

Then for the actual test **when** $n_1 \neq n_2$, modern statisticians usually use Welch's t test, which in turn includes the Satterthwaite approximation of S to better handle sample size differences (Section 5.3), but at a cost. However, the correction for unequal sample sizes in Section 7.11.2 should be sufficiently conservative to compensate for that cost and provide an adequate sample size for the **Welch's t test**.

7.7.1.1 Example: Two Tail, Two Sample A Priori Sample Size

Two tail a priori $\bar{x}_1$ vs $\bar{x}_2$ estimate of n Assume the student (Subsection 7.5.1.3) had only suspected that the average weight of male students at Little U College using soft drinks with high fructose sweetening is **different** than the average for those male students at Little U who were not using soft drinks with high fructose sweeteners. 'Different' could be either higher, +, or lower, -. She chose a criterion of $\alpha = 5\%$, so she only risks $\alpha/2 = 2.5\%$ probability of getting a Type I error in either direction (the + and the - tails). She used an updated standard deviation estimate of σ, $S = 26.00$ lb, from a national survey. She wished to obtain a sample size estimate, $\hat{n}$, that would allow her to attain an acceptable power equal to $1 - \beta = 80\%$ probability of detecting a $D_{min} = 20$ lb.

Methods: She formulated an appropriate hypothesis pair with $\alpha = 5\%$, and $\beta = 20\%$:

H0: THERE IS NO DIFFERENCE WITH RESPECT TO WEIGHT ASSOCIATED WITH WHETHER MALE STUDENTS AT LITTLE U DRINK OR DO NOT DRINK HIGH FRUCTOSE SWEETENED SOFT DRINKS.

HA: THERE IS SUCH A DIFFERENCE. 'Difference' is defined as a reasonable minimum difference of at least $D_{min} = 20$ lb.

Analysis: The national survey estimate of σ is based on tens of thousands of weighings. So for her a priori analysis, she got values of $Z_{\alpha=0.025} = 1.9600$ and $Z_{\beta=0.20} = 0.8416$ from a table like Appendix 4. Starting with these popular values of Z, she used a variant of Equation 7.9,

$\hat{n} = 2 \left(\dfrac{(Z_{\alpha/2} + Z_\beta)S}{D_{min}} \right)^2$ for $\bar{x}_1$ vs $\bar{x}_2$, which gave her

$$\hat{n} = 2 \left(\frac{(1.9600 + 0.8416)\, 26.00 \,\text{lb}}{20\,\text{lb}} \right)^2 = 26.53 \,.$$

Note that the multiplier, 2, in this equation was **not** because of the two tails; it was necessary because she was estimating from **two samples**. This estimate was rounded to the nearest integer to get $\hat{n} = 27$. Even though this was a two sample estimate, it only gives $\hat{n}$, the sample size for **only one** of the two samples.

The next step was to iterate until the estimate proved to be stable using values of t. She had to get the degrees of freedom using the total sample size for two samples together, $\hat{\mathbf{n}} = 2\hat{n} = 54$, so she went to a table of t such as Appendix 3, with $df = \hat{\mathbf{n}} - 2 = 52$, and used Equation 7.8,

$$\hat{n} = \frac{2 \left(t_{\alpha/2=0.025,\, df=52} + t_{\beta=0.20\, df=52} \right)^2 S^2}{D_{min}^2} \,,$$

with $t_{\alpha/2=0.025,\, df=52} = 2.00665$ and $t_{\beta=0.20,\, df=52} = 0.848588$ from a suitable table or online calculator. This worked out to be

$$\hat{n} = 2\left(\frac{(2.00665 + 0.848588)\,26.00\,\text{lb}}{20\,\text{lb}}\right)^2 = 27.55506\,,$$

which rounds to $\hat{n} = 28$ so $\widehat{\mathbf{n}} = 56$. This was a change in sample size, so she iterated until the integer (counting) value had stopped changing. Using $\widehat{\mathbf{n}} - 2 = 54\,df$, she looked up $t_{\alpha/2=0.025,\,df=54} = 2.004879$ and $t_{\beta=0.20,\,df=54} = 0.848328$, to get

$$\hat{n} = 2\left(\frac{(2.004879 + 0.848328)\,26.00\,\text{lb}}{20\,\text{lb}}\right)^2 = 27.51587\,,$$

which rounds up to $\hat{n} = 28$. That did not change by a whole integer, so $\widehat{\mathbf{n}} = \hat{n}_1 + \hat{n}_2 = 56$ total samples, with equal sample sizes $\hat{n}_1 = 28$ and $\hat{n}_2 = 28$.

Discussion: In application, for a public health study of weight, $D_{min} = 20\,\text{lb}$ is a rather large reasonable minimum difference, and it is doubtful that this would really be the minimum in which she would be interested. It appears that she should weigh a total of $\widehat{\mathbf{n}} = 2\hat{n} = 56$ male college students to test with $\alpha = 0.05$ and a power of $1 - \beta = 0.80$, which is the smallest power customarily sought, and which her professor is likely to at least require. She will particularly need to make it clear that $D_{min} = 20\text{lb}$ was what she a priori had targeted as the minimum value of $\delta = \mu_2 - \mu_1$ that would be sufficiently meaningful.

7.7.2 Where exactly did you get this ES?

The effect size, symbolized ES (Subsection 4.5), is the estimated difference between two things, based on your experimentation or observation. However, in the confused world of research jargon, "Where did you get this effect size?" often (although incorrectly) is referring to how you chose your reasonable minimum difference, D_{min}, instead of referring to the actual observed ES (Section 3.3.1). Effect size depends on the results of analyzing the data, whereas the reasonable minimum difference is a reference point that you select (a priori) before the data is obtained. (This relationship, reasonable minimum difference before the analysis, and effect size after the analysis, might be why some authors call both the "effect size".)

Note that you should choose D_{min} beforehand, and you might have to defend why you see it as reasonable. If D_{min} was somehow preexisting but you feel justified in using it after the fact, you should be able to explain how it qualifies as preexisting. Then be afraid, be very afraid. Often, in fields such as agricultural and industrial research, you should choose your D_{min} in consideration of **cost** (and if cost goes up, that might be an acceptable reason to change it, provided you make it clear what you have done). If a new agricultural treatment or procedure costs $5 per hectare, then your reasonable minimum difference, D_{min}, might be enough crop improvement in harvest per hectare value to at least pay for the added cost. But how would you explain to another political scientist how you chose a D_{min} with respect to even an established index of 'conservativeness'?

7.7.2.1 Avoid accusations

Leland Wilkinson and the Task Force on Statistical Inference APA Board of Scientific Affairs, August 1999, wrote, "Because power computations are most **meaningful** when done **before** data are collected and examined, it is important to show how effect-size estimates [sic: reasonable minimum differences] have been derived from previous research and theory in order to **dispel suspicions** that they might have been taken from data used in the study or, even worse, constructed to justify a particular sample size.“ (bold type face added).[12]

CAUTION: Your choice of D_{min} had best be justifiable, and also best be recorded with a verifiable date, if it is larger than the D_{min} chosen by other researchers, or if it is larger than the ES actually observed by previous researchers. Why shouldn't it just be larger? Well, a lot of people cheat, and it might look like you had post hoc increased your D_{min} to create the illusion (the deception) that you had designed for a reasonable minimum difference that was an after the fact D_{min}.

Of course cheaters could also manipulate α and β after the fact. So how is it possible to avoid the impression of cheating, or being accused of cheating? There is already a system in place in many commercial chemical research institutions. At the end of each day, the researchers take their bound record books to a notary to be notarized, verifying the date by which these pages were done, under penalty of perjury. Such records have been crucial in many patent suits, but the dates that are most admissible in court are the dates of notarization, not merely some past date in a notarized statement. In particular, the plan for each experiment with your selected α, β, which tail or tails you selected and your reasonable minimum difference, should be notarized before you start to collect data. The resulting planned sample size and the outcome, including deviations from the plan such as falling short of the planned sample size, should be in the same (preferably bound) record, and it should be notarized as soon as possible. In doing this, remember that it is the same as testifying in court. If your technician handled crucial details, then your technician and you must both sign the notarization. In your absence, your stand-in should sign the notarization under the supervision of an administratively responsible person. Frequent notarization will also reduce the tendency of some groups to use 'data cleaning' to manipulate the results.

7.7.2.2 Seeking what is reasonable

As mentioned several times in this book, doing an adequate literature review is essential before deciding to fund or initiate tests or research observations. While focusing on medical trials, Chalmers (2012) estimated that over 50% of studies were designed without reference to systematic reviews of existing evidence. Chalmers goes on to further estimate that 85% of research is waste, and that costs roughly \$85 billion per year. I wish Chalmers was wrong, but I fear otherwise. Cohen (1962), and Freedman and Bernstein (2001) have likewise reported finding large numbers of papers which had little chance of detecting a meaningful difference.

7.7.3 Can you support accepting H0?

No, you cannot ever accept H0. Recall that the intent of inferential statistics is to uncover evidence leading to the rejection of something. Some older texts say that for an inferential process, such as a t test, you should accept **H0** as true until proved otherwise. **But**, in application, such acceptance is only **provisional**, in order to provide a starting reference point, generally zero difference. Note that the zero is a starting point; the reasonable minimum difference is with respect to it and is implied by the alternate hypothesis, **HA**. Then, after the estimation process (after all calculation is finished)

---

[12]The next sentence in the above quote from Leland Wilkinson and the Task Force on Statistical Inference (1999), "Once the study is analyzed, confidence intervals replace estimated power in describing results", is perhaps an acceptable idea but a matter of style. It is **not accepted** as replacing "estimated power in describing results" by all journals. Be aware of the requirements of your profession's customs and its journal's requirements for statistical reporting.

such provisional acceptance becomes meaningless. You either reject **H0**, or you know nothing new regarding the acceptance or rejection of the hypothesis.

However, it is harmful when no one knows that many other researchers have failed to get statistical significance using likewise well powered designs. We need to know how many times an experiment has been done and what the results were, whether statistically significant or not. Recall that most research projects use $\alpha = 5\%$, meaning that they take a one in 20 chance of getting false statistical significance. Researchers who are desperate for statistical significance to prove what they sincerely believe, or what is necessary for their continued funding, sometimes run essentially the same test over and over, ultimately reporting the one in the series that finally comes out statistically significant. This is cheating regardless of whether the researcher is deceiving him or herself, or deceiving the funding agency. However, when a number of independent researchers are running experiments asking essentially the same question, the effect will be the same as one researcher running the same experiment again and again. In other words, if they don't know what the others are doing, it will be as if each is contributing a replication of the same experiment. In the present scientific environment, that falsely statistical significant outcome would be the only one published.

All should not be lost

Your editor might feel justified in refusing the publication of negative results. That is up to journal policy. However, there are journals of statistically nonsignificant results in a few fields; for instance, the *Journal of Negative Results* ISSN: 1459-4625, for ecology, evolution and biology, the *Journal of Negative Results in BioMedicine* ISSN: 1477-5751, and the unfortunately named psychology oriented *Journal of Articles in Support of the Null Hypothesis* ISSN: 1539-8714. Hopefully there will be more (Keats and Jonathon 2013).[13] A sufficient number of such journals would assist in surveys of the literature of very similar research to integrate the results by various processes called meta-analysis. Presently, meta-analysis suffers from a bias resulting from the fact that almost nothing but statistically significant results are being published. Anderson, Sprott, and Olsen (2013) point out that journals of negative results also provide valuable information for future research.[14] For instance they provide a valuable service by including the results of tests, **but not with respect to inference**. They can be a great help to researchers looking for such information as an estimate of variance for power analysis. However, authors publishing negative results must **not** be allowed, in any way, to imply that these negative outcomes support acceptance of **H0**, let alone support its rejection.[15] Reporting distribution information (including a measure of variability, such as the estimated variance, S^2, or the sample standard deviation, S, with well defined methods and materials) would be a contribution to your field. (It might also be useful to include a CI for σ, see Section 8.2.5, and a p_value if you make

---

[13] A.T. Trout, T.J. Kaufmann and D.F. Kallmes (2007) have an excellent editorial in the AJNR that is a clear explanation of some of the difficulties involved with attaching too much meaning to a failure to find statistical significance. However, I am not certain I agree with some of the authors they cite, who suggest you can present a CI for δ after a test which is not statistically significant. At the very least, it would be very unacceptable if you did not have an a priori power analysis or had not attained a power of 80% for an a priori D_{min}. The best policy is to be aware that the presence of a CI for the difference might suggest that a meaningless result is somehow meaningful after all (that there is evidence that a difference exists). Remember that, like significance tests, CIs are subject to chance and that a CI can **fail to include the true value** (in this case, δ) by chance alone.

[14] The targeted value of the true difference, δ, meaning the reasonable minimum difference, D_{min}, should always be stated whether you found statistical significance or not. If your journal's style demands that you minimize statistical details, you may simply say that you designed the experiment to at least detect a true difference equal to D_{min}.

[15] Note that a very large percentage of research outcomes are the result of designs that have little more than a 50% power to detect an appropriate reasonable minimum difference; less power is not uncommon. No one could possibly believe that negative outcomes could, under such circumstances, support any conclusion. But, even with high power tests, lack of statistical significance still fails to support any yes-no conclusion.

it clear that it in no way relates to statistical significance.) It would help researchers plan future research. Other potential information of value would be a discussion of pitfalls to be avoided, a single boxplot (Subsection 15.1.1), or a stem and leaf plot (Subsection15.1.2), or maybe estimates of skew and kurtosis (Section 15.2) for either your control or your combined data. When a test is not statistically significant, your data might suggest a new hypothesis, but you definitely must not imply that your sample data was in any way evidence of a lack of difference, because no evidence of any difference does not mean that there was in fact no difference.

7.7.4 A priori musts for power seekers

If you want to succeed, being a power seeker is unavoidable. For an *a priori power analysis* to estimate the appropriate sample size, n, you will need a **preliminary estimate**, S, **of the standard deviation**, σ (Section 7.2 or, if desperate, Subsection 7.2.3). If possible, before considering power, you might wish to consider what is the minimum required sample size to overcome skew and kurtosis (Subsubsection 5.2.1.1), to justify the assumption of normality. Then it can be daunting to consider the choices you personally must make when doing an experiment or survey.

- The better the accuracy of S, the estimate of σ, the more accurate your sample size estimate, $\hat{n}$ will be.

- You must choose a reasonable minimum difference that you wish to detect, assuming the alternative hypothesis, **HA**, is true. The **smaller** the reasonable minimum difference you choose, the **more expensive** the experiment will be in terms of sample size, n.

- You must also choose the power, $1 - \beta$, that you want, but you will have to reach it via its complement, β, the probability of getting a Type II error, because $\beta = 1 - power$. The smaller the β (the probability of getting a Type II error), the more the cost in terms of sample size, n. It is common to choose to target the minimally acceptable, $1 - \beta = 80\%$ power, meaning $\beta = 20\%$. However, this leaves you a probability of one out of five times of not detecting a real true difference as small as the reasonable minimum difference, when it is exactly the true difference. Rein et al say that some researchers are justified in calling for $1 - \beta = 90\%$ (2005), but maybe you really could **consider** a power of 95% when your budget allows it. Even at $1 - \beta = 95\%$ you still have a one in twenty chance of not detecting a difference as small as your reasonable minimum difference (where $\delta = D_{min}$). If your budget restricts you to 80% power, do not completely despair; recall that a true difference, δ, greater than D_{min} is more likely to be detected than $1 - \beta$.

- You must also choose an acceptable maximum probability, α, of getting a Type I error (rejecting **H0** when it is in fact true, by chance alone). The smaller your chosen α, the greater the cost in terms of sample size, n. However, α is a **necessary reference point** even if you do not intend an ultimate yes-no inference. For instance, there is a possibility that future journals in some fields will allow you, or require you, to report a confidence interval (Chapter 6) rather than declaring statistical significance.

- Only reporting CIs without testing would be valid, when allowed by editorial policy, provided you did not use an inferential test, the result of which was left out for whatever reason. It would be prudent to use an a priori, specified critical α and β (implying likewise an a priori D_{min}) in order to choose a sample size, $\hat{n}$, which would provide a confidence interval that will be **properly centered** with respect to the probability of enclosing the true parameter of interest (e.g. the mean, μ, or a measure of variability such as σ^2) and have a reasonable width. But on the other hand, reporting a CI for an effect size, δ, **without** a statistical test could be **misleading** because it would imply that you were justified in concluding that an effect (a difference) existed via statistical significance.

- Providing an alternative hypothesis is an **essential** part of the Neyman-Pearson approach; it allows you to squeeze the most, or rather, the most useful, information out of your data. However, the more qualifiers you place in your hypothesis pair, the more likely you are to paint yourself into an unintended corner. It is best to keep both hypotheses (**H0** and **HA**) simple and succinct. For instance, the hypothesis pair,
  H0: THERE IS NO DIFFERENCE
  HA: THERE IS A MEANINGFUL DIFFERENCE.
  Although the latter will require you to define, also a priori, what numerical difference (D_{min}) will be meaningful to this particular study, there are a number of reasons why it is both useful and essential to do so. You will also need to be ready to defend your choice of D_{min}.

- If your sample sizes must be unequal, or you cannot control one or more sample sizes, you should adjust your power analysis accordingly (Section 7.11.2).

- A minimum sample size for normal statistics, such as Student's t test, is $25 \times skew$, referring to third moment skew; the classical $n = 30$ will almost always cover kurtosis, but it is not necessarily enough to cover extreme skew.

- If you do not have a lot of available extra data to estimate $skew$, a total sample size of $n = 60$ to 80 will almost always be adequate to compensate for skew. But it might not always be large enough to obtain the power you desire. So your sample size depends on both overcoming any skew that might be present as well as being large enough to obtain the power you want.

Incidentally, you will sometimes be asked for an 'effect size' when you should be asked for a reasonable minimum difference; this is an awkward nomenclature error that you may occasionally need to overlook. Simply answer the question in the context that was intended.

Your estimate, $\hat{n}$, of each sample size, n, is likely to seem somewhat large even when you are using the excessively liberal but most commonly used $\alpha = 5\%$ and the perhaps weaker standard, $\beta = 20\%$ (for a given reasonable minimum difference). That's the way chance is.

Do not rush to accept a very small final sample size estimate, $\hat{n} < 30$, unless you can guarantee normality. You should always try to minimally use $n = 30$ samples. If you must use very small samples, $n < 25$, you should consider some sort of nonparametric approach (Chapter 14), maybe accompanied by a stem and leaf plot (Subsection 15.1.2).

Publications about the controversy and debate associated with **sample size and power** include Cohen (1977), Hulley, Cummings, et al. (2001), Kraemer and Thiemann (1987), Paterson, et al. (2016), and Lipsey (1990). Although placing excessive faith in standardizing, Lipsey, et al. (2012) discusses the pitfalls of using 'small', 'medium', and 'large' in educational interventions. With respect to these three supposed indexes, Glass, McGaw, and Smith (1981) were among the first to argue that the effectiveness of a particular intervention can only be interpreted in relation to other interventions that seek to produce the same effect. The point is that you **cannot use** 'small', 'medium', or 'large' for sample size estimation!

> **CAUTION: Beware** of a priori power analysis programs, or online sites, which **fail** to ask you **all** of the following: the power $(1 - \beta)$, **or** the probability of getting a Type II error (β); the reasonable minimum difference (in some form such as D_{min}); and, for comparing means, the variance or standard deviation, S or S^2 (p may provide equivalent variability information for binomial based tests). If they fail to require these inputs they are likely to give you sample size estimates which **only** provide for a $(1 - B) \approx 50\%$ probability of finding some not-clearly-defined difference. Said difference will theoretically be zero, but in application something meaningfully larger.

Some sort of pilot sample or pilot study may often be essential to find an acceptable estimate of σ (or σ^2) for a priori sample size estimation. If you have enough data, a very good approach is to take a random sample of it to estimate the variance (and maybe *skew*), and then actually do the test to compare the means, using a sample of size $\hat{n}$ from the remaining unexamined data. Other potential sources of S or S^2 are covered in Section 7.2.

7.8 Subsequent Power Analysis

A subsequent, in the sense of after the rest of the test, estimation of the power, $1 - \beta$, that you have achieved is not post hoc if you have a priori laid an adequate foundation to **enable** it to be a **continuation** of the analysis. If possible, you should always run an a priori power analysis to design for the power you want. An acceptable reason for a subsequent power analysis is that when you actually do the experiment you might find that some of your measurements or observations are missing, leaving you with a total sample size, **n**, that is less than your intended sample size, $\hat{n}$. Similarly, some research provides no way to control your sample size. As with an a priori power analysis, you ideally should make arrangements for the subsequent analysis in all cases in which you cannot control sample sizes. That is, you should choose an a priori reasonable minimum difference (D_{min} or $\mathcal{D}_{min}$), and cutoff (α, or half of it, $\alpha/2$). A post hoc analysis might also be useful for roughly evaluating the meaning of the outcomes of experiments that were reported by other researchers without a power analysis.

> **CAUTION:** In order to do a valid subsequent power analysis, you must have specified a reasonable minimum difference as well as a criterion, α, before the examination (or analysis) of the data. Your choice of α and D_{min} had best be justifiable, and also best be recorded with a verifiable date; perhaps it should be notarized as is the practice in the chemical industry.

If you fail to find statistical significance and cannot reject **H0**, your failure to reject **H0** does **not** in any sense support **H0** (Subsection 7.7.3). However, especially if your sample size is large enough for the Central Limit Theorem to kick in, usually $n \geq 30$ (maybe even $n \geq 25$), your estimate of variability, S or S^2, will be worth reporting, along with the methods and materials which define what you are really measuring. Whatever your sample size, a boxplot of the control (or combined) data would also be appropriate to provide future researchers with an idea (for instance the skew) of the underlying distribution (Subsection 15.1.1). With only a few exceptions (possibly involving adequate confidence interval estimates; see Chapter 6), you should always report the power associated with an inferential test. More and more editors are requiring any report of statistical significance to also report a confidence interval, and an estimate of the effect size. With respect to the means, for instance, in the case of a 'Z test' or a Student's t test, the effect size is $D = \bar{x} - \mu_0$ or $D = \bar{x}_2 - \bar{x}_1$ in the original units of measure (Equation 7.10).

7.8.1 Direction of power, determined not decreed

Power is best considered in terms of a distance (an a priori minimum reasonable difference) and it is also best reported using the scale of the original units of measure as D_{min} (or $\mathcal{D}_{min}$ for the normal approximation). The word **distance** does not imply direction unless specifically designated to have direction, for instance, the $+$ direction or the $-$ direction. There are other scales than the units of measure, but, if possible, they should be only reported for reasonable minimum distances or effect sizes if unavoidable.

   If you consider the direction of Z, as in Figure 7.6, you may find yourself contemplating the fact that a direction exists and that this direction has not been chosen by you, but by the sample, that is, by the population or populations sampled, in combination with chance. A negatively directed Z value can occur by chance alone in a one tail test even though your alternate hypothesis was **HA**: $\mu > \mu_0$. In some contexts negative Z value cuts off more than half the hypothesized population centered at Z_0. Whenever any Z value is in the direction opposite the expected direction, the result includes a $p\_value > 50\%$, part of one half plus all of the other half of the normal curve. That is, $-Z$ says there is more than a 50% chance of this happening by chance alone.

7.8.2 Subsequent one sample power analysis

This division into topics is somewhat artificial, to simplify the explanations. All one sample 'Z tests' (and t tests) consider a standard scale with the location $Z_0 = 0$, which represents the mean, μ_0, corresponding to your null hypothesis, **H0**, and all other Z (or t) values are measured from it, whether **H0** is true or not. This includes the cutoff, Z_α or $Z_{\alpha/2}$ (or $t_{...}$). If **HA** were true in such a way that the distribution were represented by the targeted distribution that is really centered at $\mu_A = \mu_0 \pm D_{min}$ on the original scale, this alternative center would be at a new $Z_A = Z_{cutoff} + Z_\beta$ in standard error units. On the original unit scale its distance from μ_0 in original units is **also** D_{min} original units from μ_0. But, you must **not even check** the real direction, whether the observed difference, D, represented by $\pm Z_\beta$, is $+Z_\beta$ or $-Z_\beta$ **until** the data is being analyzed.

7.8.2.1 One tail one sample subsequent power analysis

The basic test equation for t or Z actually estimates the overall probability that a random sample would produce a given outcome, iff the real difference, δ, really equals your chosen minimum reasonable difference, D_{min} (or $\mathcal{D}_{min}$ for the normal approximation) whether we are aware of this targeting of D_{min} or not. We can derive a post hoc equation for the estimation of β (hence $1 - \beta$) by rearranging Equation 7.3 for a Student's t test with huge samples, or for a normal approximation. Let's assume that $\widehat{Z}$ estimates $(Z_\alpha + Z_\beta)$ (and using our observed difference D to mean either $\mathcal{D}$ or D) so that we can write

$$\widehat{Z} = (Z_\alpha + Z_\beta) = \frac{D}{S/\sqrt{n}},$$

where n is the sample size you actually attained. Subtracting Z_α from the outcome, $\widehat{Z}$, of a test gives a measure of how much the probability estimated by $\widehat{Z}$ exceeds, or falls short, of the Z_α value for your chosen cutoff, the criterion, α. We will use an absolute value because we are interested in an absolute distance, $|D|$ (not a directed distance), and substitute D_{min} for D. The resulting estimate of Z_β is just that, an estimate, so let's mark it as such with a hat, $\widehat{}$. Finally we rearrange the above equation to obtain an equation for a **one sample**, **one tail** *subsequent power analysis*, to get an estimate of

$$\widehat{Z}_\beta = \frac{|D_{min}|}{S/\sqrt{n}} - Z_\alpha iff \frac{|D_{min}|}{S/\sqrt{n}} - Z_\alpha > 0.$$

The subtraction of Z_α removes the probability of false statistical significance. Note that this **only** calculates $\widehat{Z}_\beta$ when this outcome is **positive**, in which case we could find the power, $1-\beta$, by reading β from common tables of Z and then subtracting from 1.

But this outcome is not always positive, so we have not yet arrived at the complete algorithm for the estimation of power. When obtained this way, $|Z| - Z_\alpha$ has a potential for a certain amount of **ambiguity**; it can occasionally really represent the power, $Z_{1-\beta}$. If you consider the direction of Z, as in Figure 7.6 you may find yourself contemplating the fact that a direction exists and that this direction has not been chosen by you, but by the sample, that is, by the population or populations sampled, in combination with chance. It may surprise you that a negatively directed Z value can occur by chance in a one tail test even though your alternate hypothesis is **HA:**$\mu > \mu_0$. A negative Z value cuts off more than half the hypothesized population centered at Z_0. Whenever any Z value is in the direction opposite the expected direction, the result includes a $p\_value > 50\%$, part of one half plus all of the other half of the normal curve. That is, $-Z$ says there is more than a 50% chance of this happening by chance alone. Therefore we must substitute a **preliminary value**, (lowercase) $z = \frac{|D_{min}|}{\sigma/\sqrt{n}} - Z_\alpha$, and acknowledge that it is in terms of a $\pm$ directed distance resulting from the subtraction.[16]

This lowercase z is not always a direct estimate of what you might find in a table of Z because the most common tables delimit an area (probability) no greater than the right hand half of the normal curve (the distance from where $Z = 0$). So, for application to power analysis, this book uses lowercase $\hat{z}$ to represent a $\pm$**directed** estimate of a distance from the alternative mean, $\mu_A = \mu + D_{min}$, in **standard errors of the mean** (Section 4.9).

Beware: in many **other** contexts, lowercase z is an estimate in reference to σ, not to $\sigma/\sqrt{n}$.

As will soon be discussed, this lowercase z can refer to either β or $1 - \beta$, depending on whether it is positive or negative, respectively. For small and smallish samples, in the sense of not huge, this lowercase z must be interpreted as a t value (as $z = |t| - t_\alpha$).

If your sample is huge, $df > 1000$, you start the subsequent power analysis by **obtaining a directed distance estimate**,

$$\hat{z} = \frac{|D_{min}|}{S/\sqrt{n}} - Z_\alpha, \tag{7.10}$$

on the way to an eventual estimate of the power attained by your test. This $\hat{z} = |Z| - Z_\alpha$ may result in either a positive or negative outcome and the interpretation of the sign requires a little caution. Note that the resulting estimate might only be accurate to **three figures**; use Equation 7.11 (below) for $df \leq 4500$ if you want four figure accuracy.[17]

In application to power, $\hat{z}$ estimates the directed standardized distance from μ_A, the **center of the alternative distribution** to the criterion (or cutoff for two tails), and cuts off an area of the alternative distribution which might include **up to one half of it**, or when z is negative, one half of it **plus part** of one half of it. It can then be interpreted by applying certain rules to obtain either Z_β, or $Z_{1-\beta}$; delimiting the area of interest under the alternative distribution is equivalent to either β or $1 - \beta$, depending on whether $\hat{z}$ is positive or negative respectively.

If your degrees of freedom, df, is not huge, calculate this standardized distance estimate, $\hat{z}$, on a slightly different scale, the t scale for t at the degrees of freedom you obtained:

---

[16]For inferential tests, D_{min} is a distance in original units of measure. Although referring to standard errors, Z values are treated as unitless and they are looked up from the tables as positive, $|Z|$. Any Z value **times** S is a distance in whatever units of measure you are using. Recall that $|D_{min}|$ must be in the same units of measure and refers to the un-directed, therefore absolute value of D_{min}. Also, in this application to power estimation, Z_α is assumed to be directionless, and thus treated as positive unless stated otherwise.

[17]The degrees of freedom for most one sample Student's t tests are $df = n-1$, but there can be situations in which the df are other than $df = n - 1$.

$$\hat{z} = \frac{|D_{min}|}{S/\sqrt{n}} - t_{\alpha,\,df=n-1}. \tag{7.11}$$

This could also handily be written

$$\hat{z} = |t| - t_{\alpha,\,n-1}, \tag{7.12}$$

where $|t|$ is **from your test**. To get the value, $t_{\alpha,\,df=n-1}$, using either of these, you would refer to t tables such as Appendix 3 (more precise values can be found from software). When you use t, the power estimate will usually be noticeably different from the results of using Z. You still **need to interpret** $\hat{z}$, but first some explanation relevant to the rules for its interpretation follows:

Common tables such as Appendix 3 and Appendix 4 (but not all tables) that were designed for t tests (or 'Z tests'), give probabilities that can be represented as shaded tail areas in either direction, as per the two following graphs:

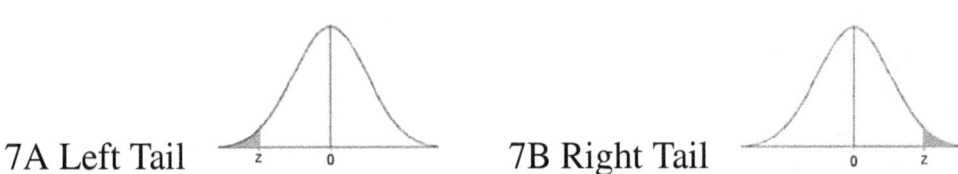

7A Left Tail   7B Right Tail

The horizontal scale is in terms of Z or t as shown. Distance has no sign unless it is stated to be *directed* (either explicitly, or by showing a + or - sign). In calculating tests of statistical significance, the Z scale is the distance from the hypothesized mean under **H0**, μ_0, and it is measured in standard errors of the mean. Because this standardized mean, μ_0, is 0 standard errors of the mean from itself, it is at $Z = 0$ (Section 4.9).

The right tail, the shaded area in 7B, is for a one tail test and it is about **HA**: IT IS MORE THAN. The left tail, the shaded area in 7A, is about **HA**: IT IS LESS THAN. The shaded area of either represents the probability of randomly getting a sample mean that far, or farther, away from $Z = 0$ by chance alone *iff* the **H0**: THERE IS NO TRUE DIFFERENCE describes the true distribution. Tests such as Student's t test or the normal approximation estimate the probability of the observed difference between the sample mean and the hypothesized mean being due to chance alone (the probability of getting a Type I error). The probability of getting Z (or t), as shown in 7A or 7B, is generally obtained from the tables as though positive because the Z (normal) and t distributions are symmetrical.

However, when using (lowercase) $\hat{z}$ to estimate β, you are generally interested in a shaded area of a particular **overlapping** alternative, **HA**, distribution that is centered at $\mu_A = \mu_0 + D_{min}$ (see illustrations below in the next two subsections). The + (or -) directed distance, $\hat{z}$, is calculated from μ_A (which is at $Z_{D_{min}}$) to the common boundary, α (which is at Z_α). Equation 7.10 and Equation 7.11 both use $\hat{z}$ to measure a distance from μ_A to that common boundary defined by Z_a (or $t_{\alpha,\,df}$) that cuts off a tail of the alternative distribution. Of course you a priori choose D_{min} with respect to any of the following: an a priori D_{min} justified by you, a D_{min} based on cost-return, or a D_{min} corresponding to literature you can cite. Recall that the probability α is with respect to the distribution implied or specified by **H0**, whereas the probability β is with respect to an alternative distribution with a hypothetical mean equal to D_{min}.

7.8.3 Interpreting positive $\hat{z}$ (lowercase)

Once you obtain the value of $\hat{z}$ from either Equation 7.10 or Equation 7.11, the **rule for a positive** (the most common sign) value of $\widehat{Z}$ or $\hat{t}$ is[18]

$$\textit{iff } \hat{z} \geq 0, \text{ then } \hat{z} = \widehat{Z}_{\beta} \text{ or } \hat{z} = \hat{t}_{\beta, df} . \tag{7.13}$$

Then you may find the probability, β, by using tables like Appendix 4 to interpret $\hat{z}$ as though it were a $\widehat{Z}$ or a $\hat{t}$ resulting from a test, but in terms of the alternative distribution. However, if any but huge sample sizes are involved, interpreting $\hat{z}$ values using t software such as found on the internet will

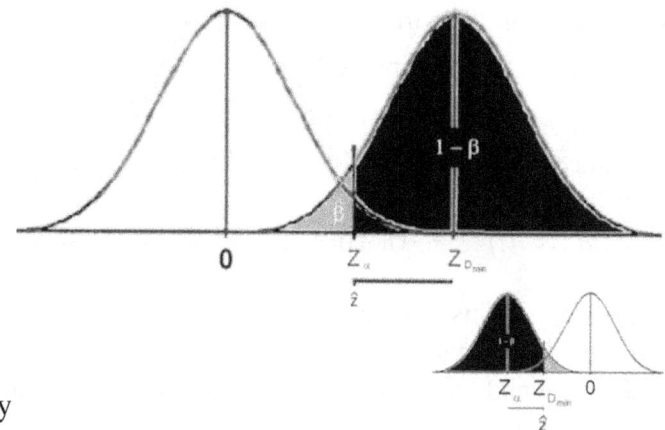

provide the best numerical accuracy

Also consider the **smaller** pair of curves, which represent a one tail test in the opposite direction, with **HA**: IT IS LESS THAN. It represents the situation in which the true mean, μ, is really equal to $\mu_A = \mu_0 - D_{min}$, in which case the true distribution, with that mean, would have its own set of probabilities due to chance alone. This small graphic is simply to indicate that when $\hat{z} > 0$ the rule works the same way, with **no regard to which tail**.

So far the discussion has only been about one tail, one sample tests. But later it will be shown that this rule applies for both **one and two tail** tests, and **also** to two sample tests.

7.8.4 Interpreting negative $\hat{z}$ (lowercase)

The $\hat{z}$ value is estimated as a distance from Z_{α} to the absolute reasonable minimum difference, $|D_{min}|$, in Equation 7.10 and Equation 7.11 via the subtraction of Z_{α}. Regardless of the sign of the true difference, positive $\hat{z}$ is in the direction of whichever tail to which it refers. So, if the sign of $\hat{z}$ is not about the direction of the tail, what is the **meaning of a negative value**, $\hat{z} < 0$? A negative $\hat{z}$ means that the area representing the probability, β, of getting a Type II error, extends clear back from the cutoff, Z_{α} or $Z_{\alpha/2}$, of the tail of interest **across** the vertical center line representing $Z_{D_{min}}$, to also include the other half of the alternative distribution centered at $Z_{D_{min}}$. So it defines an area, including part of one half of your chosen reasonably different alternative distribution, clear back to also include all the other half (large gray area). Thus when $\hat{z}$ is negative, β includes **more than 50%** of an idealized alternative distribution **centered at** D_{min}.

This situation where $\hat{z} < 0$ can be graphed as shown,

---

[18]Recall that *iff* means 'if and only if'.

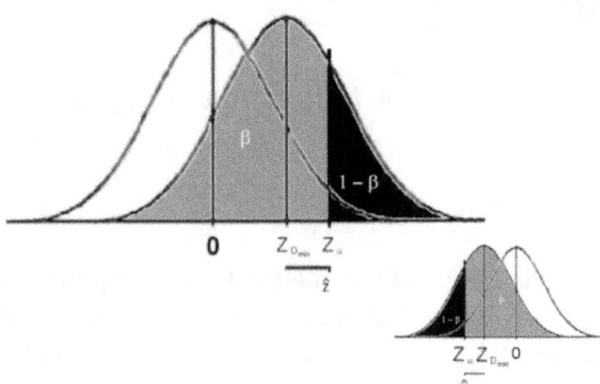

where $\hat{z} < 0$ cuts off β, the gray shaded area.

When $\hat{z} < 0$, it marks off the furthest (darkest) tail of the alternative distribution with its mean at $\mu_0 + D_{min}$, representing the probability $1 - \beta$. So you can look up the **absolute value** of negative values of $\hat{z}$, either using a table of Z or using software for t, as is appropriate, to directly find the power, $1 - \beta$. This provides us with the **rule for negative** $\hat{z}$:

$$\text{iff } \hat{z} < 0, \text{ then } |\hat{z}| = \hat{Z}_{1-\beta} \text{ or } |\hat{z}| = \hat{t}_{1-\beta, df} . \tag{7.14}$$

The smaller graphic is to illustrate that the rule works the same way **without regard to the direction of the tail** of interest.[19]

However, weeping and tearing of one's raiment can be in order when $\hat{z}$ is a negative number. This is because, at $\hat{z} < 0$, you have moved to the negative side where $z = Z_{1-\beta}$, and moving on in that direction results in a power, $1 - \beta$, that is increasingly less than 50%. When $\hat{z}$ is negative, you have less than a 50 : 50 probability of detecting the reasonable minimum difference. You should avoid this situation that requires the application of Equation 7.14. One way to avoid it is to use an a priori power analysis whenever possible. You usually can use an a priori power analysis to at least estimate the minimum sample size for which you should strive.

7.8.5 Two tail one sample subsequent power analysis

For the moment we will stay with one sample tests. Two tail subsequent power analysis is the same as one tail subsequent except that you divide the criterion, α, between two tails and use $\alpha/2$. So, start with

$$\hat{z} = \frac{|D_{min}|}{S/\sqrt{n}} - Z_{cutoff} \text{ or } \hat{z} = \frac{|D_{min}|}{S/\sqrt{n}} - t_{cutoff} ,$$

where $_{cutoff}$ **is** $\alpha/2$. Then, the sign of D_{min} can be either positive or negative, and although $\sqrt{n}$ and S are both positive, we need to use the absolute value, $|D_{min}|$, of the reasonable minimum difference. This leads to

$$\hat{z} = \frac{|D_{min}|}{S/\sqrt{n}} - Z_{\alpha/2} , \tag{7.15}$$

---

[19]The classical ($p >$) tables are constructed in such a way that all the values of the tabulated probabilities in the distribution are for cumulative probabilities corresponding to $Z \geq 0$ or $t \geq 0$. So, the rules presented here, Equation 7.13 and Equation 7.14 are, at least operationally, compatible with the common convention used in these tables, and with most software used to automate their function. That is, the center line $Z_{D_{min}}$ is the $\hat{z} = 0$ reference point for the distribution defined by the reasonable minimum difference, and $\hat{z}$ is a distance from it, the interpretation of which depends on applying these rules for the use of the commonly available tabulations. If you use the **R or S languages**, pnorm($\pm z$) or pt($\pm z, df$) gives you the answer you need without needing to consider the rules about z (when you put in $\hat{z}$ with its sign whether + or -).

for when the source of S is from a huge sample. Otherwise it will need the substitution of t, which is corrected for the independent estimation of S, instead of Z, giving

$$\hat{z} = \frac{|D_{min}|}{S/\sqrt{n}} - t_{\alpha/2, df}.$$

(7.16)

Because normal distributions are continuous, and theoretically go on forever in either direction, two of them on the same scale must overlap no matter how far apart their means are located. They graph as alternative probability bumps in the same unending continuum. An alternative distribution can only have a central measure (center) that is in **one direction** from $Z = 0$, the direction in which it most overlaps the distribution that would exist if it were true. Although for two tail tests you do not know the direction of $\delta = \mu - \mu_0$ before the fact, it conceptually exists with respect to the direction of the expected sample mean, μ, from the hypothesized mean, μ_0. That is, D is a distance, and distances can be given direction by the context in which they are observed. Finding if its estimation, $D = \hat{\delta}$, is statistically significant involves seeing if the *cutoff*, $\alpha/2$, equals or exceeds $Z_{\alpha/2}$ or $t_{\alpha/2}$. But the estimation of β is about a probability that is **never split** into two different tails. So, provided there really is a difference, it could result in either of the conditions shown,

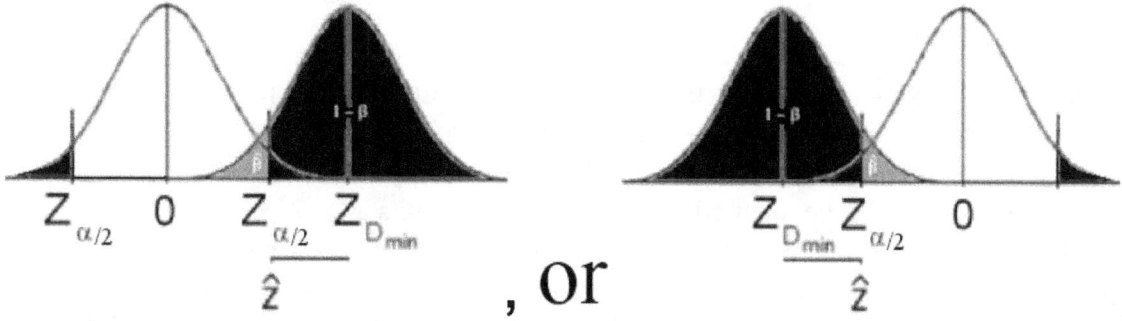

, or

Note that the power, $1 - \beta$ (or β), is about the probability of **detecting** (or **not** detecting) the reasonable minimum difference, and can **only** exist in one direction or the other.

Question 13 is a worked example of a two tail one sample subsequent power analysis and some associated challenges. Question 21 explores related issues. Question 15 can both be used as an exercise and as an additional example, but with $n_1 = n_2$.

7.8.6 Two sample subsequent power analysis

The principles are the same for two sample subsequent power analysis as for the one sample analysis, except that you have to account for two samples. If the sample sizes come out unequal, the calculation attempts to adapt for the difference. Except for analyzing published outcomes, the only appropriate reason to perform a subsequent power analysis is a change in sample size, due to inability to control it, or due to mishap. So, for **two samples** you can substitute $\sqrt{1/n_1 + 1/n_2} = \sqrt{\frac{n_2 + n_1}{n_1 n_2}}$, or, **if** $n_1 = n_2 = n$, you **can use** $\sqrt{2/n}$,

$$\hat{z} = \left|\frac{D_{min}}{S\sqrt{2/n}}\right| - Z_{cutoff} \text{ or } \hat{z} = \left|\frac{D}{S\sqrt{\frac{n_2 + n_1}{n_1 n_2}}}\right| - Z_{cutoff},$$

(7.17)

when the total sample size you obtained provides a huge (almost unlimited) number of degrees of freedom, df.

Otherwise, use the smaller sample version,

$$\hat{z} = \left| \frac{D_{min}}{S\sqrt{2/n}} \right| - t_{cutoff,\,2n-2} \text{ or } \hat{z} = \left| \frac{D_{min}}{S\sqrt{\frac{n_2+n_1}{n_1 n_2}}} \right| - t_{cutoff,\,n_1+n_2-2}, \tag{7.18}$$

where the degrees of freedom are $2n - 2$ **or** $n_1 + n_2 - 2$. Throughout this section, $\hat{z}$ is translated to Z or t by applying the appropriate rule, Equation 7.13 or Equation 7.14, and then refer to a table of Z such as Appendix 4 or to a software source of t (tables of t are not accurate to a sufficient number of decimal places).

7.8.6.1 Example: Two Tail Two Sample Subsequent Power

Two Tail two sample subsequent $\bar{x}_2$ *vs* $\bar{x}_1$ power analysis George Copycat, also a public health student, but two years after our heroine in Subsection 7.5.1.1, did a two tail experiment similar to hers, except he weighed $n_1 = 15$ students and $n_2 = 17$ students.

Methods: He had formulated an appropriate hypothesis pair,

H0: THERE IS NO DIFFERENCE.

HA: THERE IS A DIFFERENCE IN THE WEIGHTS OF MALE STUDENTS DRINKING OR NOT DRINKING HIGH FRUCTOSE SWEETENED SOFT DRINKS AT LITTLE U.

Considering the small, unequal samples, it should not be a surprise that he had not done an a priori power analysis. After he reported his nonsignificant results at $\alpha/2 = 2.5\%$, his professor sent him back to do a post hoc power analysis to explore what the power was, assuming that the true effect, δ, was $|\bar{x}_2 - \bar{x}_1| = 20$ lb, that is, using $D_{min} = 20$ lb. (Do **not** do this for publication; you **must** have an a priori basis for D_{min}. Because the choice of D_{min} was arbitrary, almost capricious, this could not have gone beyond an academic exercise.)

Somehow, Copycat's standard deviation estimate from his data just happened to come out $S = 26.0$ lb., which corresponded to the pre-existing national survey used by the previous student. So, he set out to find β, the probability of getting a Type II error, and from it to estimate the power of his experimental design. From a suitable source, he found that $t_{0.25,30} = 2.042272$.

He substituted these values into Equation 7.18, $\hat{z} = \left| \frac{D_{min}}{S\sqrt{\frac{n_1+n_2}{n_1 n_2}}} \right| - t_{\alpha/.2,\,n_1+n_2-2}$, to calculate

$$\hat{z} = \left| \frac{20}{26.0\sqrt{\frac{15+17}{15\times17}}} \right| - 2.042272 = 0.129188\,.$$

Because this resulted in $\hat{z} > 0$, he applied the rule, Equation 7.13, that *iff* $\hat{z} \geq 0$, $\hat{z} = t_{\beta,\,df}$. He then returned to the table of t, but the range of available values of t at $df = 30$ was inadequate, so the obvious thing to do was to **use software** to find β from his estimate of $t_{\beta,\,30\,df}$. Checking an online calculator, he found that $\beta = 0.551$. (When doing this, recall that whether the criterion is one tail or two tail, β and $1 - \beta$ are both one tail.) He was **horrified** to estimate that he had only a 0.449 probability of having detected a true difference of 20 lb if it really existed. Properly rounded down, that is only about a $1 - \beta = 44.9\%$ probability of having detected his chosen reasonable minimum difference.

Results and Discussion: As the professor informed him, the probability of detecting a difference would increase as the real difference, δ, increased and would drop off as the real difference decreased.

However, this $D_{min} = 20$ lb was a rather large reasonable minimum difference. Also $n > 30$ sounds great for approaching normality, but that alone does not guarantee power. George really stumbled into a pitfall when he failed to do an a priori power analysis to see whether he could find anything with what he mistook for an adequate sample size.

Incidentally, you can speculatively use Equation 7.10 or Equation 7.11 for an a priori estimate of power if you have a fixed (pre-chosen) sample size, n.

7.8.7 An unofficial post hoc $\hat{z}$?

For a crude idea of the post hoc power of research published without a power assessment but with a Z or t value, you might use the fact that the above equations for $\hat{z}$ include what is algebraically like an absolute Z or t test outcome with a subtraction. So a meta analysis could use α to find $\hat{z} = |Z| - Z_{cutoff}$, or $\hat{z} = |t| - t_{cutoff,\ df}$ where $_{cutoff}$ can be either α for one tail tests or $\alpha/2$ for two tail tests and use the above rules, Equation 7.13 or Equation 7.14, to crudely interpret $\hat{z}$.

7.9 Exploring and Planning for Power

Figure 7.7: n vs $1 - \beta$ power curve

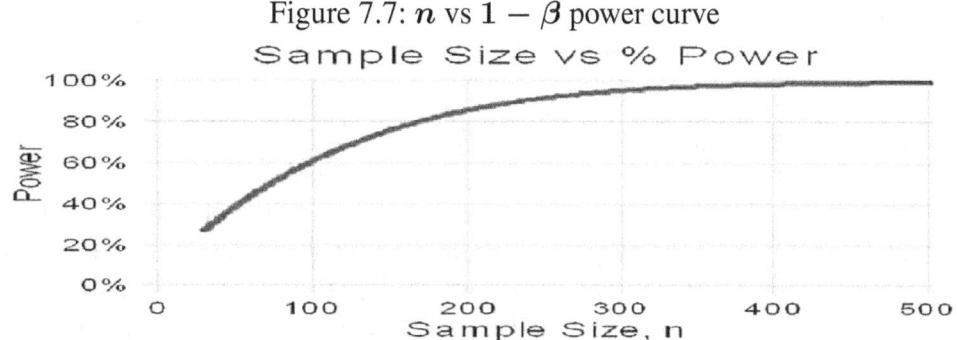

$\alpha = 0.05$, $\sigma = 26$ lb, $D_{min} = 5$ lb. Fig. derived from the Example 'One tail, one sample a priori, moderate D_{min}' (Subsection 7.5.1.4)

Given that you have an estimate, S^2, of the variance, σ^2, and you have chosen a reasonable minimum difference, D_{min}, and the probability of getting a Type I error, α, you may wish to explore the cost of minimizing β, the probability of getting a Type II error, in terms of the sample size, n. With increasing sample size, n, the power is theoretically asymptotic to 100% (Figure 7.7). However, unless you can sample the **entire** population, it could never reach 100%, and depending on the researcher's choices, it may tend to persistently stay **quite a distance** from 100% for all practical sample sizes. Ideal α and β probabilities are likely to be expensive when D_{min} is small enough to be useful. You might consider yourself fortunate to afford sufficient resources to test with $\alpha = 5\%$ and $\beta = 10\%$. Ideally when reporting negative results, the variance (or standard deviation) of the combined data, or the control, the targeted power and associated D_{min} should be reported in the abstract as well as in the body of the paper. If your results are not statistically significant and you publish in a journal that allows statistically nonsignificant results, your discussion section can optionally include a graph of power, $1 - \beta$ versus D_{min} (for example Figure 7.7 is for $S = 26.0$ lb, $\alpha = 5\%$, and $D_{min} = 5$ lb). It is especially important that your estimate of σ or σ^2 **should also** be published if your methods and materials are **completely described**; it might someday be used for planning future research.

7.9.1 Managing $1 - \beta$ and n graphically

Historically researchers have had too few samples and never suspected how little power they were achieving. Ironically, you can pay for too much as well as for too little and waste your investment either way. One solution is to use what-if graphs to choose the power, $1 - \beta$, you can afford in terms of sample size, n.

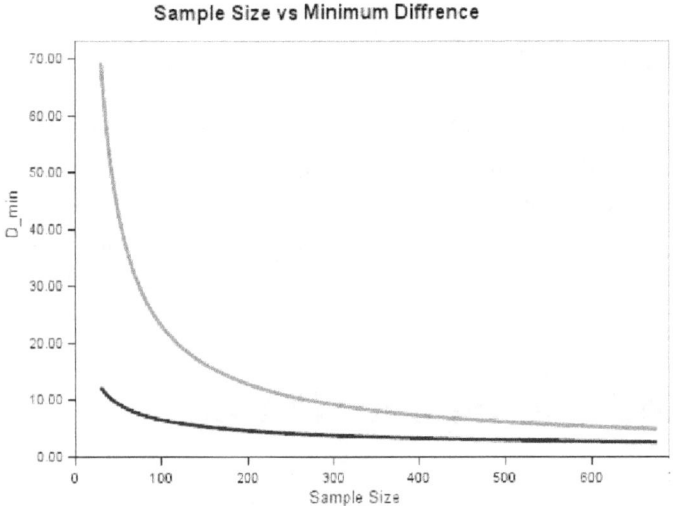

Figure 7.8: n vs Attainable D_{min}

The **upper curve** represents one tail power, an $\alpha = 1\%$ probability of getting a Type I error and a $1 - \beta = 90\%$ power **given the standard deviation,** $\sigma = 26\,\text{lb}$. The **lower curve** represents an $\alpha = 5\%$ probability of getting a Type I error and a $1 - \beta = 80\%$ power with the same standard deviation, $\sigma = 26\,\text{lb}$. Unfortunately it is much more expensive to attain a small D_{min} using the safer choices (upper graph) than using the weaker and riskier choices (lower graph). The reasonable minimum difference, D_{min}, is also in pounds.

The power, $1 - \beta$, that would be in the realm of economic practicality could be found by drawing a graph such as Figure 7.8 that (like Figure 7.7) refers all the way back to the example of the public health student weighing male students in a study of the effects of high fructose drinks (Section 7.5.1.1). This figure shows sample sizes for $D_{min} = 5$ with $\sigma = 26\,\text{lb}$ and a criterion, $\alpha = 5\%$. As the graph shows, the public health student in the example from which the values are drawn would need $n = 166$ to have $1 - \beta = 80\%$, and almost $n = 250$ for $1 - \beta = 90\%$. The latter power would be great, but would it be affordable in terms of available resources for your research (time and space for instance)?

Many researchers are so underfunded or shorthanded that there is a question of whether they can afford to attain a sample size sufficient to have a reasonable power, $1 - \beta$, to find what they are looking for. The return to your investment in terms of sample size actually tends to **level off** as sample size **increases**, but researchers rarely reach large enough samples for that to become an issue.

7.9.2 Sample size vs power

Ideally you should be able to report that you had 90% power to detect some particular D_{min} at $\alpha = 1\%$. However, many researchers are settling for the minimal commonly accepted 80% power and $\alpha = 5\%$. Given that you have an estimate, S, of σ and that you have chosen a targeted difference, D_{min}, you should at least always check to see whether you can, in terms of total sample size, **n**, afford $\alpha \leq 5\%$ and $\beta = 10\%$. Then if something does not go as planned, you still have a hope of at least being at or below a barely acceptable $\beta = 20\%$. You may wish to check one or two specific choices of α (or $\alpha/2$) and of β with individual calculations, but an examination of graphs of power,

$1 - \beta$, versus sample size is much more informative. So, how do you draw such graphs as Figure 7.7 and Figure 7.9.1?

You can either find β, the probability of getting a Type II error, or the power, $1 - \beta$, for various sample sizes, using the rules Equation 7.13, or possibly Equation 7.14, and one of Equation 7.10, Equation 7.11, Equation 7.17, or Equation 7.18 from Section 7.8.

You can then produce such a sample size vs power graph with adequate visual accuracy using a spreadsheet.[20] First put a column of integer values of sample sizes n; perhaps start with $n = 10$ and increase by increments of 5 or 10. Next, use a built in statistical function to put in a corresponding column of t values. Then you can put the right side of one of the above mentioned equations in a column called 'beta' or 'power' that links to n and t columns to get the values you want. Do not forget that you can estimate power by subtracting β, the probability of a Type II error from 1.

Given an estimate, S, of σ and having chosen a reasonable minimum difference, D_{min}, you could also approach this from the other direction, inputting β or $1 - \beta$ to find and graph corresponding values of n. This could be done using Equation 7.3 or its t version, or Equation 7.8 or its Z version Equation 7.9.

7.9.2.1 I know what α and β I would like, but ...

You may hope to achieve some ideal α and β and wonder what compromise between sample size, n, and the reasonable minimum difference, D_{min}, you might be able to accept. The steepness, that is, the slope of such graphs is often a clue as to what strategy you might use. If you draw curves like the one in Figure 7.8 you will see that your ideal can be costly to achieve unless you target an uncomfortably large reasonable minimum difference, D_{min}. In the steep areas of such curves you get a **lot of return** for a **little change** in sample size. However, as n gets larger, the achievable reasonable minimum difference decreases less with each increase.

Another challenge, in many applications, is that, especially if your sample size is huge, you might have to justify why you choose a particular reasonable minimum difference (Section 4.5). Huge sample sizes can cause hypersensitivity so that drug sellers can claim significance where whatever the test detected was infinitesimal (hence without useful effect). This awkwardness may often be avoided by dated notarized laboratory records, but explaining is always appropriate.

A handy application of a curve such as Figure 7.7 is to see what reasonable minimum difference might have been detected for different values of power, $1 - \beta$, given a particular α and n. Such a graph could be very useful for the post hoc evaluation of research which has had no a priori power analysis, as is unfortunately still common in the literature of many fields.

On the other hand, assuming you can justify selecting your own reasonable minimum difference, you might want to a priori draw such a graph for at least 80% power and consider how far you want to go beyond where the curve approaches horizontal. The closer to horizontal the curve gets, the less return in terms of power you will get by increasing the sample size. That is, the return from increasing sample size tends to diminish. This part of the curve that is leveling out can make even a small increase in power very expensive in terms of sample size.[21]

The curves for each given power, $1 - \beta$, in Figure 7.8 do not necessarily suggest a small enough sample size to make a researcher happy. Neither is cheap in terms of the cost of being able to detect a

---

[20]Beware of precise statistical estimation or estimates involving tiny probabilities on spreadsheets. But most spreadsheets will suffice for power graphs if you do not need extreme accuracy. You may eventually highlight and export the graph to be dressed up using Gimp, PaintShop Pro, Krita, or whatever. The various R language packages produce wonderful and diverse graphics, but even with the aid of RStudio, learning it is a challenge (likewise for Python).

[21]Power, $1 - \beta$, is the complement of β (and vice versa), so the graph of one is the graph of the other flipped over. Figure 7.7 refers to the power, $1 - \beta$, rather than β. Either tends to level off as the sample size increases, indicating that the effect of sample size decreases more rapidly as n grows larger.

comfortable small true difference, δ. The lower curve shows the relation between sample size, n, and the reasonable minimum difference, D_{min}, the value of the true difference, δ, that you could target for the minimally popular $\alpha = 5\%$ and power, $1 - \beta = 80\%$, with the standard deviation $S = 26$ lb (Subsection 7.5.1.1). It shows that increasing small sample sizes has a greater relation to the minimum distance D_{min} that you could target than does increasing larger sample sizes. The top curve is for the same S, with a more conservative alternative with $\alpha = 1\%$ and $\beta = 90\%$. Observe that the higher starting point and slower leveling off of this curve, this alternative, will take a very large D_{min}, or a very large sample to attain this power and criterion, and the return to increased sample size, n, levels off later and more slowly. The leveling off of such curves may also be seen as Mead's (1988) law of diminishing returns.

So, how do you get the $\widehat{D_{min}}$ which estimates the expected $\delta = \mu_1 - \mu_0$ for each of the points, $\left(n, \widehat{D_{min}}\right)$, as for such a graph? Start with $\hat{n} = \left(\frac{(Z_{cutoff} + Z_\beta)S}{D_{min}}\right)^2$ and take its square root to obtain $\sqrt{\hat{n}} = \frac{(Z_{cutoff} + Z_\beta)S}{D_{min}}$, where $_{cutoff}$ depends on the criterion and number of tails and $\hat{n}$ estimates the sample size, $\hat{n}$. Then, for a $_{cutoff} = \alpha$ for a one tail, or $\alpha/2$ for a two tail, huge **one sample** test estimate of the D_{min}, it is only necessary to rearrange this equation, and substitute $\widehat{D_{min}}$ and n for $\hat{n}$ to get

$$\widehat{D_{min}} = \frac{(Z_{cutoff} + Z_\beta)S}{\sqrt{n}}. \tag{7.19}$$

Of course, most readers will not commonly be working with sample sizes of many thousands. So, you would generally substitute $t_{cutoff,df}$ in the above equation instead of Z_{cutoff}, and $t_{\beta,df}$ instead of Z_β to estimate

$$\widehat{D_{min}} = \frac{\left(t_{cutoff,df} + \widehat{t}_{\beta,df}\right)S}{\sqrt{n}}. \tag{7.20}$$

Note that β **is the same** (undivided) regardless of the number of tails, so you always use Z_β or $t_{\beta,df}$.

For a **two sample test**, rearrange $Z = \frac{\bar{x}_2 - \bar{x}_1}{S_{1,2}\sqrt{1/n_1 + 1/n_2}}$ with unequal sample sizes, and substituting $(Z_{cutoff} + Z_\beta)$ for Z, the estimate is

$$\widehat{D_{min}} = \frac{(Z_{cutoff} + Z_\beta)S}{\sqrt{1/n_1 + 1/n_2}}, \tag{7.21}$$

where $cutoff$ is either α or $\alpha/2$. If you assume equal samples, $n = n_1 = n_2$, then use

$$\widehat{D_{min}} = \frac{(Z_{cutoff} + Z_\beta)S}{\sqrt{2/n}}. \tag{7.22}$$

For continuous data, as long as you are not working with many thousands in your sample, you would use

$$\widehat{D_{min}} = \frac{\left(t_{cutoff,n_1+n_2-2} + \widehat{t}_{\beta,df}\right)S}{\sqrt{1/n_1 + 1/n_2}} \quad \text{or} \quad \widehat{D_{min}} = \frac{\left(t_{cutoff,n_1+n_2-2} + \widehat{t}_{\beta,df}\right)S}{\sqrt{2/n}}. \tag{7.23}$$

For two tails you substitute an $\alpha/2$ for $_{cutoff}$ and for one tail use α for $_{cutoff}$. (Stick to Z **if** this involves a normal approximation.)

7.9.2.2 Example: Explore the Reasonable Minimum Difference

Let's consider a one tail, one sample, t test, $\bar{x}$ vs μ_0.

Methods: Assume yet another scenario, that our public health student started by forming a hypothesis pair,

H0: THERE IS NO TRUE DIFFERENCE BETWEEN FEMALE FACULTY DRINKING HIGH FRUCTOSE SWEETENED SOFT DRINKS AT LITTLE U AND THE CURRENT NATIONAL AVERAGE WEIGHT OF FEMALE WHITE COLLAR WORKERS.

HA: THE WEIGHT OF FEMALE FACULTY DRINKING HIGH FRUCTOSE SWEETENED SOFT DRINKS AT LITTLE U IS <u>GREATER</u> THAN THE NATIONAL AVERAGE FOR THEIR GROUP.

Criterion: She selected a popular criterion, $\alpha = 5\%$.
From a then recent national database she accepted that officially, $\mu_0 = 162.7$ pounds for the appropriate age group.

Data: Her raw sample data in pounds was:
$x_i =$189.4 191.6 126.4 197.2 153.8 224.6 106.8 241.8 156.7 151.7 104.1 197.4 183.8

Analysis: She then calculated that $\bar{x} = 171.18$ lb and $S = 42.480$ lb. (In this example the sample standard deviation was quite different from the preceding. Chance does things, and populations may vary in unexpected ways.)
So, $t = \frac{(171.2 - 162.7)\,\text{lb}}{42.48\,\text{lb}/\sqrt{13}} = 0.721$ with $df = n - 1 = 12$. (The pounds over pounds canceled out, so, as always, t is unitless. And she rounded to four figure accuracy in order to correspond to the mean, μ_0.) She referred $\alpha = 0.05$ and $df = 13 - 1 = 12$ to a table of t (Appendix 3) and found that the one tail value of $t_{0.05,\,12}$ required to reject **H0** is 1.782. She did not reject **H0** because she only found $t = 0.7195$.

Data Exploration Data exploration is not for reporting or inference, it is merely for future reference such as sample size estimation, as when the results are from a pilot study. Upon getting such an obviously nonsignificant result and realizing that $df = 12$ is doomed to being ridiculously weak, she decided to treat her experiment as little more than a pilot study. She decided to investigate how big a D_{min} she would have needed for a $1 - \beta = 80\%$ power at this sample size, using Equation 7.20,

$$\widehat{D_{min}} = \frac{(t_{\alpha,\,n-1} + t_{\beta,\,n-1})\,S}{\sqrt{n}}\,.$$

She obtained the values of t with $12\,df$ with $\beta = 20\%$ and $\alpha = 5\%$ from a table such as Appendix 3, to calculate

$$\widehat{D_{min}} = \frac{(1.782 + 0.873)\,42.8\,\text{lb}}{\sqrt{13}} = 31.52\,\text{lb}\,.$$

So, she rounded up to estimate that $\delta = 32$ lb might have produced a statistically significant result. Obviously 32 lb would seem too large and not a very reasonable minimum difference.

Discussion: She could only validly report that she failed to reject **H0** at an $\alpha = 5\%$ level. Because she did not get statistical significance, she should not have shown a CI for the difference, but it would be embarrassingly large anyway. Because it was only a learning exercise, she should mention the fact that she did not do an a priori power analysis, but as it was, she claimed that it was unlikely that she could have attained a sample size that could lead to the detection of any truly reasonable $D_{min} = \delta$. She might point out, that if seen **as a pilot study**, it did provide a standard deviation estimate that future researchers could use for an a priori determination of sample sizes to target their desired minimum differences, if they were focusing on the weights of female faculty specifically at Little U. But it is unlikely that a sample size so much smaller than the $n = 30$ minimum necessary to have an almost reasonable expectation of normality would have provided a useful estimate of σ. Otherwise, if her sample had been large enough, she could then have used her estimated standard deviation to provide graphs of either n vs β or of n vs D_{min}, based on repeated calculations. (Beware, all the data is really **fictitious**.)

In this case, her estimate of σ only applies to Little U, and such a small sample size requires an **assumption** of normality for Student's t test to be valid, or for a valid estimate of D_{min}. A shortcoming of her project was that she was careless in not noticing that the very literature she consulted actually contains a reasonable estimate, S, of the standard deviation. She could have used it to make an a priori estimate of the necessary sample size (Equation 7.3) in the first place. Another shortcoming is that $df = 12$ is rather small for obtaining a useful estimate of the standard deviation or of a suitable D_{min}. An estimate of the standard deviation based on $df \geq 25$ would have been more certain to be of use for future research planning.

The fact that such reports, even with adequate sample sizes, would eventually become trash is a shame. Their potential as a basis for future experimental designs would seem to point to the need to publish in a journal of negative results, if such is available in her field, as opposed to the Journal of Irreproducible Results, http://www.jir.com/, which might still be available.

Question 17 is about a similar matter.

7.9.2.3 Minimum possible D_{min} given limited sample size

Sometimes you want to estimate the smallest D_{min} for a fixed maximum sample size, n, that is possible, given your resources. This can be estimated, using the appropriate choice of the above Equation 7.19 through Equation 7.23. To be credible, your α should be no more than 0.05 and your β should be no more than 0.20. However, this exploration of possible reasonable differences is only acceptable for a priori consideration.

7.10 Power for the Normal Approximation

This section refers clear back to Section 2.4 and it also relates to Section 3.5. You cannot use t for situations involving the binomial. The t distribution is intended for the additional error included in separately estimating the variance. Note that the normal approximation does not estimate the variance separately from the mean. For a given n, the probability, p, determines the variance. That's why **t does not apply** to the normal approximation. Therefore, Z is what you use with the normal approximation.

Sample size estimates for the normal distribution might work best when $\mathcal{D}_{min}$ is not very big; even $|\mathcal{D}_{min}| \leq 0.25$ is rather extreme, and the hypothesized probability should be $0.25 \leq p_0 \leq 0.75$. These limits are not fixed in stone, but outside of them, the results might not be very accurate and these should be considered rough minima. However, in terms of sample size, a $\delta = 2\%$ is extremely easier to discover (using $\mathcal{D}_{min} = 0.02$) than a $\delta = 1\%$ (using $\mathcal{D}_{min} = 0.01$) and a $\delta = 10\%$ is a rather extreme difference.

7.10.1 $\hat{n}$ for one sample a priori normal approximation

Suppose you were sampling a population to see whether the probability of a certain binomial outcome (such as the probability of having a girl baby) was equal to a certain hypothesized probability,

p_0. Using the normal approximation (Section 5.6), you could calculate a one sample normal approximation test, using

$$Z = (\hat{p} - p_0) / \sqrt{p_0 (1 - p_0) / n},$$

where the true value, p, is usually **estimated as a relative frequency** (a proportion), $\hat{p} = r/n$, where r is the number of successes, say, butter up and n is the sample size, say, the number of times you have to observe the toast seeming to have leaped off the counter. Before getting the data, you should decide how big a difference the reasonable minimum difference, $\mathcal{D}_{min}$, should be in order to be meaningful, because, if you do not choose a $\mathcal{D}_{min}$, the test will choose one for you. Then, using the normal approximation, if $\delta_p = p - p_0$ is really equal to the $\mathcal{D}_{min}$ that you choose, how large should your sample be to have a power, $1 - \beta$, to attain statistical significance with your chosen cutoff (α or $\alpha/2$)?

Recall that the power, $(1 - \beta)$ refers to the ability to detect the reasonable minimum difference if such a difference exists or is exceeded. Then starting with

$$\hat{n} = \frac{(Z_{cutoff}S + Z_\beta S)^2}{D_{min}^2} = \frac{(Z_{cutoff} + Z_\beta)^2 S^2}{D_{min}^2},$$

remember that the variance is

$$S^2 = pq = p(1 - p),$$

according to the normal approximation. Then, starting with p_0, your hypothesized probability, you can calculate $p_1 = p_0 + \mathcal{D}_{min}$, the probability plus the minimum reasonable difference. These two probabilities can be used to find the two different variances, p_0 to find the variance if **H0** is true, and p_1 to find the variance if the difference is exactly $\mathcal{D}_{min}$. Then Z_{cutoff} relates to the approximated normal curve if p_0 is true, and Z_β relates to the approximated normal curve if p_1 is true. Therefore we can associate the two variance approximations with the appropriate Z_{cutoff} and Z_β values to estimate the required sample size[7.24],

$$\hat{n} = \frac{\left(Z_{cutoff}\sqrt{p_0(1 - p_0)} + Z_\beta\sqrt{p_1(1 - p_1)}\right)^2}{\mathcal{D}_{min}^2}, \quad (7.24)$$

where $_{cutoff}$ may equal either α or $\alpha/2$ for a one tail or a two tail test, respectively (adapted from NIST 2013). This sample size estimate is one potential approximation among several and none is perfect, but this one is good because it acknowledges that there are potentially two underlying distributions and one normal approximation will be taller and more compact than the other (there will be different variances).

Any calculation involving the binomial approximation works best when

$$p_1 = p_0 + |\mathcal{D}_{min}| \quad (0.25 \leq p_1 \leq 0.75),$$

however, it might be adequate when the sample size is enormous and p_1 is outside the limits of this range.[22]

---

[22]Unfortunately the NIST 2013 publication had an example with extreme $p_{...}$ and $q_{...}$ values beyond the range considered accurate for the normal approximation. If any probability, $p_{...}$ (or $q_{...}$), is outside the range $0.25 \leq p \leq 0.75$, you may wish to consider some approach other than a normal approximation.

7.10.2 $\hat{n}$ for two independent sample normal approximation

If you hypothesize **H0**: THE PROBABILITIES ARE EQUAL, meaning you are challenging $p = p_0$, you can estimate the size necessary for each sample, assuming equal sample sizes, $\hat{n} = \hat{n}_1 = \hat{n}_2$, to obtain the desired power, using,

$$\hat{n} = (Z_{cutoff} + Z_\beta)^2 (p_0 q_0 + p_1 q_1) / \mathcal{D}_{min}^2, \qquad (7.25)$$

for each sample (Snedecor and Cochran 1967, p221-222) where $1 > |D_{min} = p_1 - p_0| > 0$ using

$$p_1 = p_0 + \mathcal{D}_{min} \qquad (|p_1| < 1)$$

where p_0 is what you hypothesize and $\mathcal{D}_{min}$ is what you choose for a reasonable minimum differ-ence (in this book's notation).[23] Note the notation uses $q = 1 - p$, so that $p(1 - p) = pq$ and that $(p_0 q_0 + p_1 q_1)$ involves pooling the variance. Unlike Student's t test where we assume that the two variances can remain equal when the means are different, the means and variances of the binomial distribution are both linked to the value of p, so we need to consider the pooled variance because, if there is a difference $p_0 q_0 \neq p_1 q_1$. That is if you wish to detect a difference of $\mathcal{D}_{min} = 10$ and your null hypothesis is **H0** THERE IS NO DIFFERENCE meaning that $p_1 = 1/2$, then $p_2 = 0.50 + 0.10 = 0.60$. Of course you can hypothesize a different p_0 under the null hypothesis and it is up to you to supply the reasonable minimum difference, power (hence β), and the $_{cutoff}$ may equal either α or $\alpha/2$ for a one or a two tail test, respectively. This sample size equation provisionally assumes **H0** is true, as is customary for calculation in most situations. You always **can use** 0.50 for p_0 (and $q_0 = 0.50$) and that will always produce an adequate sample size estimate, however it might be some-what larger than you need. If $n_1 \neq n_2$, this estimate itself will be **provisional** and it would need further adjustment as in Subsection 7.11.2.

Pearson's chi square provides a different test, using what is called 2×2 tables in Chapter 12. In reality, it is a variation on the same theme, and you may eventually learn to mix or substitute these approaches, one for the other.

7.10.2.1 Continuity in sample size estimation

Some very pragmatic authors differ on whether or not to use the continuity correction when estimat-ing sample size. Van Belle and Millard (1998) say you should not use a continuity correction for es-timating sample size. Grunkemeier and Jin (2007) suggest that maybe there might be an advantage in using it for sample size estimation because it is conservative, in that it increases the estimate a little.

7.10.3 Subsequent binomial power for uncontrolled n

The following power analysis is for **when** you have **a priori specified** a reasonable minimum dif-ference, $|\mathcal{D}_{min}| \leq 0.25$, and β, or $1 - \beta$, as well as α, but the sample size was uncontrollable (as when collecting wildlife data), or was not subject to your control (dropouts, etc.). As in the original test, you get p_0 from the null hypothesis, **H0**. Because your test was a normal approximation, Z is the measure of interest. So, this uses a standardized estimate, $\hat{z}$, which, in accordance with the proper rule, may be translated into $\widehat{Z}_\beta$ or $\widehat{Z}_{1-\beta}$.

---

[23]Note that Equation 7.25 estimates the variance two different ways, one for **H0** and one for **HA** and then combines them. Less accurate approaches might just use only the variance for **H0** and then put a 2 in the denominator.

7.10.3.1 One sample

Given a hypothesized p_0, a subsequent power equation for a one sample normal approximation based test (a sampled result against a hypothesized result) may be obtained by making simple substitutions, using the α, p_0 and the $\mathcal{D}_{min}$ used in the planning of tests involving continuous data in Equation 7.10 to get

$$\hat{z} = \frac{|\mathcal{D}_{min}|}{\sqrt{(p_0 q_0 + p_1 q_1)/2n}} - Z_{cutoff}, \tag{7.26}$$

where $_{cutoff}$ equals either your a priori α or $\alpha/2$, and p_0 is the probability under your **H0**, and $p_1 = p_0 + \mathcal{D}_{min}$ under **HA**. The 2 in the denominator of Equation 7.26 is part of the variance (pooling and averaging) process to compensate for calculating the variance twice once for **H0** and the other for **HA**.

7.10.3.2 Two equal samples

If the samples turned out equal, $n = n_1 = n_2$, then this formula is

$$\hat{z} = \frac{|\mathcal{D}_{min}|}{\sqrt{(p_0 q_0 + p_1 q_1)/n}} - Z_{cutoff}, \tag{7.27}$$

where p_0 is the probability under your **H0**, and $p_1 = p_0 + \mathcal{D}_{min}$ $\quad (0.25 \le p_0 \le 0.75)$ and the sum under the radical sign is the variance of the difference (adapted from Snedicor and Cocharan 1967 page 220).

7.10.3.3 Two unequal samples

For two samples when you cannot control the sample size, and the total sample sizes are large ($n \ge 30$), you can use your **a priori** values of $\mathcal{D}_{min}$, p_0 and $p_1 = p_0 + \mathcal{D}_{min}$ $\quad (0.25 \le p_0 \le 0.75)$ to estimate

$$\hat{z} = \frac{|\mathcal{D}_{min}| \sqrt{\frac{2}{1/n_1 + 1/n_2}}}{\sqrt{(p_0 q_0 + p_1 q_1)}} - Z_{cutoff}. \tag{7.28}$$

It correctly adapts for $n_1 \ne n_2$ by using the **attained sample sizes,**[24] but do **not use** the observed probabilities; they are a source of added chance variation. This is also a potential power estimator for, say, meta studies, whether statistical significance is achieved or not.

7.10.3.4 Interpreting z for either one or two samples

For any of these equations, the $_{cutoff}$ can be α or $\alpha/2$, depending on whether it is a one tail or a two tail test, respectively. Then it is merely a matter of applying the rules listed as Equation 7.13,

$$iff\ \hat{z} \ge 0,\ \text{then}\ \hat{\boldsymbol{z}} = \widehat{Z}_\beta,$$

and Equation 7.14,

$$iff\ \hat{z} < 0,\ \text{then}\ |\hat{\boldsymbol{z}}| = \widehat{Z}_{1-\beta}.$$

Depending on which rule is applicable, you can find β and calculate $1 - \beta$, or find the power, $1 - \beta$, directly, either from a table of Z or by using software. As Grunkemeier and Jin point out post hoc

---

[24]The subscript numbering of the probabilities (p_0 and p_1) and of the sample sizes (n_1 and n_2) are **not related**.

power analysis **after failure to reject** is a **questionable** practice (at least stick to your **a priori inputs**, the sample size in the only thing that is validly different; the rest is due to chance).

7.11 Can or Can't Control One Sample n (Approximation or Not)

7.11.1 Dropouts

Data loss during experiments is often unavoidable. Therefore you should always consider putting in some extra subjects or experimental units after you do an a priori power analysis to estimate a minimal sample size. All dropouts need explanation and a careful consideration of why they occurred.

In order to maintain sufficient power, especially in human trials, the researcher must consider including extra subjects to **provide for dropouts**. Human dropouts in a drug test may demonstrate how difficult it will be to persuade the patients to actually continue taking a medication. The rate of dropping out should be viewed as part of the efficacy of the protocol because it is going to reduce the effectiveness of the medication in the field. For instance, it is very difficult, almost impossible, to get schizophrenia outpatients to continue taking any presently available medication. This, in effect, renders all of the medications for schizophrenia almost useless for outpatient treatment.

Assume that in your field you and your colleagues tend to observe about a 7% loss of your experimental subjects (they could be plants, humans or whatever). Then you should use a sample size at least equal to $(1/(1 - 0.07)))\,\widehat{\mathbf{n}} = 1.07527\widehat{\mathbf{n}}$. So if your estimate $\widehat{\mathbf{n}} = 100$ subjects, you should use at least 108 subjects.

Some references in the extensive literature on this problem in medical trials include Kemmler, Hummer, Widschwendter and Fleischhacker (2005), Verbeks, Lesaffre and Spiessens (2001; the WHO (World Health Organization) dropout data might still be found by searching
http://www.who.int/entity/mediacentre/en. Additional material should be available via Google Scholar or Semantic Scholar.

7.11.2 Adjust for a priori efficiency, unequal two sample

For a two sample test, you sometimes already know that one sample has a maximum, n_f. The subscripted f as used here means fixed. Then for a particular reasonable minimum difference, D_{min}, you can use the methods drawn from Subsection 7.8 to estimate the power, and see if the project might be worth its cost.

Avoid deliberately using unequal sample sizes for inferential tests, such as a two sample Student's t test. But what if you know that the **smallest** of your two samples **cannot exceed** some size, n_f, and that is less than you would have used if you could achieve equal sample sizes, $\hat{n}$? When we initially estimate sample size we proceed as though the final sample sizes will be equal, and we estimate $\widehat{\mathbf{n}} = (\hat{n}_1 + \hat{n}_2)$, but when we turn out to have unequal sample sizes, the actual calculation of t or Z partially deals with unequal sample sizes via $\frac{1}{n_1} + \frac{1}{n_2} = \frac{n_2 + n_1}{n_1 n_2}$. However, when $n_1 \neq n_2$ we still need to adapt for a resulting loss of relative efficiency with respect to sample size. You can, in application, adjust for this by adding in an estimate of the lost relative efficiency. First find a provisional estimate of $2\hat{n} = \widehat{\mathbf{n}}$ as above with Equation 7.8, or Equation 7.9, keeping in mind that $n_1 \neq n_2$. A crude estimate of the adjustable sample size is not $\hat{n}$ but rather,

$$\hat{n}' = \widehat{\mathbf{n}} - n_f.$$

(Please do not be confused by the accent marks, $'$, they only exist to alert you that there has been a change from an earlier value. I will try to remember to add an accent mark each time I adjust or correct $\hat{n}$.)

Kifle and Desta (2012) stated, "The **loss of relative efficiency** due to unequal sample sizes in two groups can be calculated as follows: $1 - \bar{n}_h/\bar{n}$." The **arithmetic mean** in the denominator is $\bar{n} = \sum_{i=1}^{i=m} sample\_size_i/m$, but because there will be only $m = 2$ samples, $\bar{n} = (n_f + \hat{n}')/2 = \hat{\mathbf{n}}/2$. The ***harmonic mean*** is the mean of the inverses,[25] in this context, of the sample sizes,

$$\bar{n}_h = \frac{m}{\sum_{j=1}^{j=m} 1/sample\_size_j}. \tag{7.29}$$

. However, because we are only interested in these $m = 2$ samples, this simplifies to

$$\bar{n}_h = \frac{2}{1/n_f + 1/\hat{n}'}.$$

The numerator, $m = 2$, is the **number of samples**; so in this application only $m = 2$ sample sizes, the adjustable and the fixed, are being averaged. When sample sizes are **unequal**, the estimate of the harmonic mean, $\bar{n}_h$, is always smaller than the arithmetic mean, $\bar{n} = \hat{\mathbf{n}}/m$. So you can estimate the ***relative efficiency lost*** in our initial estimate of the sample size as

$$effficiency\_lost = 1 - \frac{\bar{n}_h}{\bar{n}}. \tag{7.30}$$

But Equation 7.30 is combined into the actual Equation 7.31, below, for correcting the total for the two samples. It, in effect, adds the estimated loss back, using $\hat{\mathbf{n}}'' = \hat{\mathbf{n}} + \hat{\mathbf{n}}(1 - \bar{n}_h/\bar{n})$, which can be rearranged to

$$\hat{\mathbf{n}}'' = \hat{\mathbf{n}}\left(2 - \frac{\bar{n}_h}{\bar{n}}\right). \tag{7.31}$$

This must be **rounded up to an integer**. So, for a two sample test with unequal sample sizes, the estimate of the **adjustable sample size**, $\hat{n}'$, is

$$\hat{n}' = \hat{\mathbf{n}}'' - n_f. \tag{7.32}$$

Example: You Correct for $\hat{n}_1 \neq \hat{n}_2$

Assume: You can only get a sample size of $n_f = 20$, but you have provisionally estimated that you need 2 equal sample sizes, each of $\hat{n} = 50$, to compare samples. That is, your provisional total sample size is $\hat{\mathbf{n}} = 2\hat{n} = 100$.

Analysis: You provisionally estimate that you need $\hat{\mathbf{n}} - \hat{n}_f = \hat{n}' = 80$. This would seem to be a rather large difference; hopefully your differences will often be considerably smaller. But, you are aware that unequal sample sizes weaken the test, so you want to adjust your total, $\hat{\mathbf{n}}$, to avoid the loss of power due to unequal sample sizes. For a comparison of 2 samples, you first estimate the arithmetic mean, $\bar{n} = \hat{\mathbf{n}}/2 = 50$, and the harmonic mean,

$$\bar{n}_h = \frac{2}{1/20 + 1/80} = 32,$$

where the 2 in the numerator is the number of values for which you are finding the harmonic

---

[25]The harmonic mean of 2, 4, 3, 1 is $\bar{x}_h = 4/(\frac{1}{2} + \frac{1}{4} + \frac{1}{3} + \frac{1}{1}) = 1.92$, but the arithmetic mean is $\bar{x} = \frac{2+4+3+1}{4} = 2.5$

mean. Then, if you wish to know $efficiency\_lost = 1 - \frac{\bar{n}_h}{\bar{n}}$ or,

$$efficiency\_loss = 1 - \frac{32}{50} = 1 - .64 = 0.36,$$

or 36%. But you want to correct your **total** size estimate $\widehat{n}'$ for this pairwise comparison,

$$\widehat{n}'' = \widehat{n}\left(2 - \frac{\bar{n}_h}{\bar{n}}\right) = 100\,(1.36) = 136,$$

which would have been rounded **up** to the nearest integer if it had not already been an integer. So, the **total sample size** is $\widehat{n}'' = 136$ for this sample pair, and the **adjustable sample size** must be adjusted to

$$\hat{n}'' = \widehat{n}'' - n_f = 136 - 20 = 116,$$

where $n_f = 20$ is the sample size you could not control.

Discussion: So, you will be using a total of $n''' = 136$ samples; that is quite a big difference from $n = 100$. **Un-equalness is expensive.**

Strive to avoid such extremes; efficiency may not be the only problem with such an extremely unbalanced design; the extreme unbalance of the correction may itself be a problem if such a disproportionate correction is needed. Can you change the question to address something more workable?

The above approach can also be used for the normal approximation.

7.12 Finite Sample Size Adjustment

Recall that the hat, $\widehat{}$, indicates that something is an estimate, and that **n** can stand for n, or $n_1 + n_2$, whichever is the total. If the total estimated sample size, $\widehat{n}$, is **greater than** 5% of the whole population size, N, then it should be corrected. Use the following to **correct a sample size estimate** $\widehat{n}$:

$$\widehat{n}' = \frac{\widehat{n}}{1 + \frac{\widehat{n}}{N}}. \tag{7.33}$$

The resulting estimate, $\widehat{n}'$, the **corrected sample size**, will be smaller than the original estimate, $\widehat{n}$. Even so, you should round up (possibly negating a small correction). Question 11 provides both an exercise and an example of this finite sample size correction (answer at back of book). For the **actual testing** you should refer to Section 5.7.

7.13 The Margin of Error

The *margin of error*, $\widehat{E}$, of a statistic, such as the sample mean, is the half-width of a (symmetrical) confidence interval (Lohr 1999). It is the $\pm$ part, $\left|Z_{\alpha/2}S_{\bar{x}}\right|$, of the CI, the *length of one whisker* of a confidence interval,

$$\widehat{E} = \sqrt{Z_{\alpha/2}^2 S^2/n}, \tag{7.34}$$

in the original units of measure when possible. Engineers for instance, might like to be $(1 - \alpha) \times 100\%$ confident that their sample estimate is included within, at most, a certain number of units of the central measure (such as the mean). That is, they may want to be 95% confident that it is within $E = 2.5\,\text{mm}$ of the mean. They are thus seeking a margin of error of $Z_{\alpha/2}S/\sqrt{n} = 2.5\,\text{mm}$.

7.13.0.1 Controlling the width of the CI

You can target a certain width of a CI by doubling the margin of error, either starting with an a priori known p, a hypothesized p_0, or an estimate of S or of S^2 from a pilot survey or some other source (Subsection 7.2). Unfortunately you can only attempt to estimate a particular sample size by using a priori knowledge of the variability.

The following sample size estimates do not apply to sample sizes for statistical inference. Testing requires larger sample sizes and therefore gives narrower (more accurate) CIs.

A prospective parametric CI width

The width of a CI is twice the estimated margin of error ($width = 2\widehat{E}$). So to target a parametric CI width, you rearrange Equation 7.34 to estimate the necessary sample size, **being aware that** $E = \frac{1}{2}width = onewhisker$ and calculate

$$\widehat{\widehat{n}} = t^2_{\alpha/2,df}S^2/E^2 \text{ or } \widehat{\widehat{n}} = Z^2_{\alpha/2}S^2/E^2 . \tag{7.35}$$

The latter may lead to a need for iteration, using t.

7.13.0.2 A prospective binomial margin of error

To control the width of a binomial (fraction or percent of success) CI using normal approximation (Section 5.6), you will need to provide an a priori probability, p_0 or p, of success to use the normal approximation of $S = \sqrt{np(1-p)}$ (Section 7.10). Then you can use that in Equation 7.35.

7.13.0.3 CI width, $2E$, or width via a priori power

You can a priori choose the width of a confidence interval by actually calculating a chosen margin of error and doubling it, or you can indirectly choose the margin of error of a CI indirectly by an a priori power estimate. If we were to (somehow reconcile the two universes and) compare the sample size necessary for the power to detect a reasonable minimum distance, D_{min}, with the sample size necessary to target an equal margin of error, $E = D_{min}$, we would find that there were intervening factors. The power analysis **for a test would require** from about twice as big a sample size (at $\alpha/2 = 0.025$ and $\beta - 0.20$) up to four or five times as big (at say, $\alpha/2 = 0.005$ and $\beta = 0.01$). So what is the advantage of a power analysis if you are only interested in the CI? It estimates the sample size you need to estimate the **centering** of the confidence interval. Whether we are considering CI width or centering (location), chance is going to have an effect.

7.14 RSE, Especially Health Surveys

Reports of highly variable results can be a challenge to assess, but the Central Limit Theorem leads to a reduction of the standard error of the mean through sufficient sample sizes. The *relative standard error,* RSE, is the standard error of the mean divided by the mean and reported as

$$RSE = 100\left(S_{\bar{x}}/|\bar{x}|\right)\%,$$

where the $100\ (\ )\%$ notation is taken to mean that the fraction is always reported as a percent. RSE is used mainly for survey data, as opposed to testing. Some data organizations set reliability standards that their data must reach before publication. The U.S National Center for Health Statistics, NCHS, typically requires at least $n = 30$ observations, if not more, for publication of simple reports of statistical surveys, and does not report an estimate if the relative standard error exceeds 30% (Klein et al. 2002). The lower the relative standard error, the more precise the estimate, such as the mean, tends to be.

Adhering to such a requirement is thought to usually prevent you from ending up with inconveniently wide, or even unrealistic, confidence intervals. Unfortunately there is a tendency in some fields to use RSE in place of more easily understood CIs. For publication under such strict requirements, if a confidence interval based on counts has a width more than, say, 8%, you may want to check to assure that its $RSE \leq 30\%$.

Let's see how you could find a **minimum sample size** for the RSE for counting data with the normal approximation, using

$$RSE = 100 S_{\bar{x}} / |\bar{x}| \leq 30.$$

Starting with a normal approximation approach, $\bar{x} = np$, and $S^2 = npq$ where $q = 1 - p$, and keeping p and q separate,

$$0.30 = \sqrt{\frac{npq}{n}} / np,$$

$$0.30 = \sqrt{pq} / np,$$

then,

$$n_{min} = \frac{\sqrt{pq}}{0.30p}.$$

Round up. This might be seen as especially of interest for smallish values of p such as the tuberculosis morbidity in N. Korea. If you guess too high a value of p when designing your survey, your estimate might be too small for acceptance under the NCHS rule. Especially for larger values of p this might not provide enough power in itself to use the data for inference. Beware, if your initial estimate of p is too large, your sample will be too small.

Question 19 involves estimating an RSE and can be used as both an exercise and a worked example.

7.15 Fisher and Neyman-Pearson Separately in the Same Experiment?

Fisher saw the p_value as a way for researchers to check for possible evidence against their preferred hypothesis and then make up their own minds as to what was true or false. However, Neyman and Pearson set up an a priori criterion, an a priori barrier against emotional involvement or other bias such as pressure from sponsors and prevailing opinion. The Neyman-Pearson approach not only demands that you use double blind random sampling, but also that you do not change your criteria after the fact. Even though this greatly lessens the potential for bias, it is a somewhat rigid quality control approach. Although you absolutely cannot change your approach while or after analyzing the data, there is no theoretical barrier to keep you from a priori opting to use Fisher's approach to challenge hypothesis development after you used the Neyman-Pearson approach to design the experiment, particularly with respect to sample size. Then either could validly be used in the actual test, so long as you only use one, and **keep them separate**. That is, if you absolutely cannot accept null hypothesis testing, you could still use the Neyman-Pearson approach to only estimate a reasonable sample size (Section 7.3). However, you must decide **a priori** to do the analysis using **Fisher's approach**, rather than the Neyman-Pearson, before starting the analysis (before seeing the data).

Whether you decide to use the Neyman-Pearson approach for actual testing or not you would still have to provide, at least provisionally, an α and a β with reasonable minimum difference, $\mathcal{D}_{min}$ or D_{min} (Subsection 4.7.4). The reasonable minimum difference, D_{min} or $\mathcal{D}_{min}$, could be in terms of differences between means or between proportions (fractions, maybe as %; or binomial probabilities), even between other statistics. That usually requires a sufficient a priori estimate of variability, but properly done, it would give you a suitable sample size for using Fisher's approach to obtain a meaningful p_value. You could follow that with an interval estimate, such as a boxplot (Subsection 15.1.1).

This independently stepped approach could assure that a confidence interval would, through the resulting sample size, managed power, and targeted reasonable minimum difference, be expected to be correctly centered and roughly conveniently dimensioned.

Although using the Neyman-Pearson approach to choose a sample size and then simply reporting a p_value and an effect size, such as $\bar{x}_2 - \bar{x}_1$, using Fisher's approach may seem attractive, and maybe justifiable, you want to be certain, **in advance**, that your editor and reviewers will accept it.[26] But, unfortunately, there is potential for considerably more confounding in reporting a p_value than in reporting statistical significance.

7.16 Justification of the Technique

You are going to be randomly sampling means from distributions which hopefully have a finite mean and a finite standard deviation. If the sample size is large enough, the Central Limit Theorem, Subsection 1.1.4, will reassure you that the distribution of the means is likely to be normal.[27] Then, assuming that the two populations have the same variance, only their true means, μ_1 and μ_2, are likely to differ. This assumption of equal variances is not too outrageous in application for many sampling situations, such as when comparing two groups of fast food places in the same chain, which only differ by the factor in which you are interested, such as what brand of coffee they offer. Other examples where this assumption is particularly good include comparing yields of melon fields irrigated at dawn versus fields randomly chosen to be irrigated in the late afternoon,[28] or comparing two different algebra texts randomly assigned to classes in the same large school, etc.

As it does today, the power of statistical tests has always depended on both the sample size and the difference between the things being sampled. However, for more than half a century, most researchers either ignored that fact, or were ignorant of it. The resulting tests were generally only likely to detect very large differences. The probability of finding statistical significance really relates to the true distribution of the means, but when inference is via the Central Limit Theorem it is accessed on a standardized scale that is measured from the middle of the hypothesized distribution under **H0**, the null hypothesis. In application, **HA**: THERE IS A DIFFERENCE does not look for ridiculously small differences, and the developers of the various tests do not appear to have intended it to do so. So, a major objective of this book is to enable you to know how to take control of what you target and of the probability of your detecting it.

---

[26]If you report the variance or standard deviation estimate someplace in any paper it may help your colleagues to design their work, or to confirm your conclusions. Journals should encourage this, even if only in footnotes.

[27]The average grade of an individual student is like a personal mean, but not like a mean drawn at random from the whole population, as is usually assumed in 'grading on the curve'. Such individual scores do not meet the specifications of the CLT sufficiently to justify such practices as grading 'on the curve'. A major fault is treating the score or scores of an individual as random samples from the whole population, which they are not; they refer to particular individuals rather than to means of scores of groups of individuals randomly chosen from the whole population.

In addition, the scores of tests like the various IQ tests are confounded by factors affecting the class; they are not a random sample of the class itself. 'Intelligence' tests are attempts at measuring a hodge-podge of different things as though they were one, and they are widely criticized in popular literature (eg. Herrnstein and Murray, 1994).

[28]Although, in most cases, irrigation at dawn minimizes plant diseases, this may not be true in all climates and seasons.

Although **H0**: THERE IS NO DIFFERENCE seems to suggest the impossible, it is semantically correct in that, as the difference moves away from the impossible zero difference, the probability of detecting that difference increases. Controlling power, $1 - \beta$, to detect a particular minimum difference, D_{min}, is about setting up a situation in which you can at least be assured of a given probability of detecting a reasonable minimum difference from zero difference which you have accepted as being large enough. If the true difference is larger, your ability to detect it will be greater. Recall that 'Z tests' (and t tests) involve a transformation to a standardized Z (or t) scale with the location, $Z_0 = 0$, determined by μ_0 from your null hypothesis, **H0**, and all other Z (or t) values are distances measured from 0, whether **H0** is true or not. This being measured from $Z = 0$ includes test scores such as the cutoffs, Z_α or $Z_{\alpha/2}$. Assume **HA** is true in such a way that the distribution is represented by the targeted curve that is really centered at $\mu_A = \mu_0 \pm D_{min}$ on the original scale. This center will be at a new $Z_A = Z_{cutoff} + Z_\beta$, and its distance from $Z_0 = 0$ (or $t_0 = 0$) will represent the untransformed distance from μ_0 that equals D_{min} in the original units of measure. Distances do not have direction unless so stated, so because this is a distance from 0, its length is $|Z_A|$ and thus usually treated as positive, as when consulting the tables. Whether this distance of interest is directed in the $+$ or in the $-$ direction depends on your choice of direction for D_{min} for one tail tests, but it depends on the unsigned (absolute) direction of the real unknown true difference for two tail tests.

Why not include D_{min} in the working hypothesis, **H0**? The answer is that **H0**: THERE IS NO MORE DIFFERENCE THAN D_{min} would imply mathematical challenges (noncentral distributions etc.) which add to the complexity of the tests and which are not fully worked out for the range of widely accepted applications that form the relatively coherent set of applied procedures presently available. So, like the criterion, α, and the power, $1 - \beta$, we treat the reasonable minimum difference, D_{min}, as part of the specifications for accepting the alternative **HA**, not part of the null hypothesis, **H0**, upon which we base our calculation of the Z or t test itself.

7.16.0.1 Revisiting and revising the way to do research

Cohen, a pioneer in power analysis, did a review of the of the Journal of Abnormal Psychology Volume 1960 and found that the mean power to detect 'medium' effect sizes was only about 48% (1962). Eventually his concept of "small", "medium" and "large" proved to be impractical. However, decades later, Sedlmier and Gigerenzer et al. (1989) reviewed papers in similar areas and found even more extremely weak outcomes when looking at articles that implied that the null hypothesis was true. Even when the null hypothesis is false, researchers using experiments with such limited power were lucky to detect anything, and such approaches tend only to detect very large differences. It is reasonable to at least insist on taking no more than the recently popular one in five, 20% probability that **H0** is really false (80% power).

Now you may exclaim, 'Only 80% power! Does that mean I would only have an 80% probability of detecting a given reasonable minimum difference?' Yes, if you choose to use the presently popular levels, you have $\beta = 0.20$, a **one in five probability of not detecting** your chosen reasonable minimum difference, D_{min}, when the real difference is $\delta = D_{min}$. However, the larger the real difference, δ, is than D_{min}, the more chance you have of detecting it. There is even a small chance of detecting differences a little smaller than D_{min}, but it drops off rapidly below $\mu_A = \mu_0 + D_{min}$.

What chances you take regarding Type I and Type II errors are determined by you and your budget. In terms of effort and resources, more power costs more, and it is up to you to decide how much you want to buy. It would be a very good idea to calculate a variety of tentative power values and then sketch a power curve like Figure 7.7 (Subsection 7.9.1), which might help you decide what power you can afford.

Questions for Ch 7

Write the equations where appropriate and fill in the values. **Odd numbered questions** have worked answers at the back of the book and can be studied as additional examples. The data tends to be **fictitious**.

1. Assuming that you intend to sample what is possibly two distributions, how can you, in effect, increase the apparent separation between means on the Z scale?

2. In a t test you perform a standardization of the difference between two means (one of which might be hypothetical), using what?

3. Assume that a national survey using the latest kind of cuff[29] shows that the systolic blood pressure of a particular category of retired male patients had a standard deviation of $S = 14.530$ mmhg and an average of $\mu_0 = 130.50$ mmhg. You are interested in whether the side effects of a new medication will reduce the mean by at least $D_{min} = 2.000$ mmhg. **What sample size** do you need for a trial, using a criterion of $\alpha = 5\%$ and a power of $1 - \beta = 80\%$?

4. You want to do the same systolic blood pressure test as in Question 3, but against $\mu_0 = 128.100$ for females in the same retired group of diabetes patients. However, your records estimate that the standard deviation for their systolic blood pressure is $S = 10.346$ mmhg. You are interested in whether regular meditation can reduce the mean by at least $D_{min} = 2.000$ mmhg. What sample size do you estimate is needed, using a criterion of $\alpha = 5\%$ for a one tail test and a power of $1 - \beta = 80\%$? (In reality you would have to provide a few additional measurements to allow for attrition when using human subjects, but ignore that in answering this question.)

5. What is the most popular power used by researchers who consider power?

6. You divide α in half as $\alpha/2$ so it can be shared by two tails. What about β ?

7. Often data from two populations that have different means overlap inconveniently. How does Student's t test, **in effect**, reduce this overlap?

8. What theorem regarding the normal distribution that was applied repeatedly in this chapter is important regarding sampling?

9. A diabetes clinic whose past records show that the systolic blood pressure of tens of thousands of retired male patients with a similar condition taking the presently most successful blood pressure medication has a standard deviation $S = 14.530$ mmhg. You wish to see if a particular diet will alter the average blood pressure by $D_{min} = 2.000$ mmhg, using a criterion $\alpha = 5\%$ and $\beta = 20\%$. What sample size do you estimate that you need for a trial?

10. A national survey shows that the systolic blood pressure of thousands of retired female patients with a similar condition taking the presently most successful blood pressure medication has a standard deviation $S = 10.346$ mmhg. Suppose you wish to see whether an apple a day will **change** the average by as much as 2.000 mmhg, using $\alpha = 5\%$ and a power, $1 - \beta = 80\%$. Assume you can get equal sample sizes for the treatment and the control. (This question is not the same as Question 4.)

---

[29]Be careful of blood pressure records. The kind of cuffs has changed and the new ones might not produce data in the same population as those of a decade or two ago.

11. You are comparing the adult triglyceride (TG) levels between the tribes in a Native American nation with only two tribes of about the same size, and your budget is very limited. You have estimated that you need to sample a total of $\widehat{n} = 178$ out of a total population with only $N = 870$ adult subjects. This sample size estimate $\widehat{n}$ is a large part of the total population N. How do you **correct** this estimated sample size?

12. Claude Baud is a senior cat interviewer supervising voluntary cat interviewers who go door to door and ask to visit people's mother cats if they have a litter in order to study the genetics of domestic cats. Yes, although all this data is fictitious, this job title is authentic and has existed at least once. Such things as the frequencies of the color pattern were actually related to the human dispersal of these inquilines from populations elsewhere. Imagine that, after weighing a few dozen mama cats, Claude suspected that the cats in his team's region were heavier. Since then, Claude's team has weighed $n = 1080$ mother cats in Claude's region and found an average weight, $\bar{x} = 3.61$ kg. Claude had a priori hypothesized
H0: THERE IS NO DIFFERENCE vs **HA**: MY DISTRICT'S MOTHER CATS ARE HEAVIER. His control was the mean of the EU combined sample of millions of mother cats, $\mu_0 = 3.49$ kilograms with an estimated variance, $S^2 = 0.842$ kg^2, targeting an **a priori** reasonable minimum difference of 0.100 kg. What was the **subsequent power** of the test? Assume Claude was happy with three figure accuracy and used Z rather than t tables to evaluate the result (permissible for three figures because $df > 1000$).

13. Suppose a regulatory agency required that the concentration of benzene in the air of the central city of some metropolitan area be no more than $\mu_0 = 1.0$ ppb, and they were willing to accept 0.50 ppb as a reasonable minimum difference. The historical standard deviation estimate was $S = 1.3$ ppb. They form a hypothesis pair, **H0**: THERE IS $\mu_0 \leq 1.0$ ppb AIRBORNE BENZENE vs **HA**: THE CONCENTRATION IS LARGER (than $\mu_0 = 1.0$ ppb), $\alpha = 0.05$ and take a random sample of size $n = 25$. They find an average level of $\bar{x} = 2.6$ ppb airborne benzene. Should they have used Z, or t, to evaluate the result? **What is the estimated subsequent power** of their test? Is there a difficulty regarding assumptions here?

14. A laboratory has compared the nicotine content of two independently owned brands of cigarettes. They hypothesized **H0**: THERE IS NO DIFFERENCE vs **HA**: THERE IS A DIFFERENCE and they a priori chose $\alpha = 5\%$. They were interested in a difference of $D_{min} = 0.05$ mg or more per cigarette. However, due to a bureaucratic decree, they were only allowed to sample equal numbers of each brand at random for a total sample size of $n = 200$ cigarettes. The nicotine content had a pooled standard deviation of 0.200 mg. One brand produced a mean of 0.92 mg. of nicotine and the other produced a mean of 0.94 mg. What was the **power** of their test?

15. Flooded rice is perhaps the most important crop in the world. Its yields reach well over 5000 kg/ha (avg$\approx$ 3629 kg/ha, about 4 tons/acre); a failure of the world's rice production would bring apocalyptic famine and war. Suppose an agronomy team examined the effect of two trace element additives on a total of 2000 rice paddies in a small region of Southeast Asia before their funding ran out. Equal numbers of paddies with each treatment were sampled. Treatment A produces a mean of 3796 kg/ha and treatment B produces a mean of 3851 kg/ha. From earlier even larger data they got a standard deviation of $S = 1187$ kg/ha and had used that for planning to get a much larger sample size which they could not reach because of a funding change. They are especially interested in whether there is a real difference of 50 kg/ha or more in favor of either treatment (their a priori MRD). What would be the power of such a study if they chose a criterion of $\alpha = 5\%$, and the requirements of randomness, etc. were met? (**It is OK to assume**

they accepted a final three figure accuracy. Be sure to study the explanation of the answer after you make your estimate.)

16. The standard deviation for the diameter of a certain imported plastic part is estimated to be $S = 5.12$ thousandths of an inch. The company that handles its distribution wants to compare parts from two manufacturers with a criterion of $\alpha = 1\%$ and a power of $1 - \beta = 95\%$, using random samples of $n = 128$ for each manufacturer. They are interested in the parts that are too small as well as the ones that are too large, so a **two tail test** is indicated. An interview of the board of directors reveals that they are inflexible about these requirements, including the sample size. They are only interested in whether the parts are different, and if so, they fondly imagine they will be able to sort it out from there. Assuming randomness, calculate the $\widehat{D_{min}}$ which they are implying. Assume a normal distribution of the means here.

17. A few decades ago a French jawbreaker company thought that their 100 gram jawbreakers, which they sold for 1.02 francs wholesale, were a better deal than the imported 1.00 franc Algerian imitation. They decided to use a criterion of $\alpha = 1\%$ and formed a hypothesis pair, **H0**: THERE IS NO DIFFERENCE VS **HA**: THE FRENCH PRODUCT IS HEAVIER. They randomly selected a sample, $n_1 = 250$, of their own product. However, the candy boat from Algeria was overdue and they could only find and weigh $n_2 = 172$ of the competitor's product. They a priori knew the variance for the jawbreakers was $S_{1.2}^2 = 1.6210$ grams. Even though they did not formally choose an a priori D_{min}, what would be considered an acceptable reasonable minimum difference? (Hint: Subsection 7.7.2 and Subsection 7.11.2) Then, assuming they could call the sample of the Algerian product random, what would be the power?

18. You select a 5% criterion and form a hypothesis pair, **H0**: THERE IS NO DIFFERENCE VS **HA**: THERE IS A DIFFERENCE, and then run an $\alpha = 5\%$ criterion t test at great cost and find a $p\_value = 5.01\%$. Is it OK to go ahead and publish an article saying, in your opinion, that this is sufficient evidence that **H0** is false?

19. The U.S National Center for Health Statistics, NCHS, typically does not allow the reporting of an estimate if the relative standard error, RSE, exceeds 30%. Suppose you suspect that the Sweaty Archipelago has a tuberculosis morbidity of 1% ($p = 0.01$), but you want to check this by sampling. What would be the very smallest sample size you could reasonably take? (No inference needed since the fictional funding source is only interested in morbidity data.)

20. Suppose you assess the cost of an experiment targeting a particular reasonable minimum difference, either D_{min} (or $\mathcal{D}_{min}$ for proportion or probability) by graphing the sample size against the power; what happens when the graph starts to level out?

21. Many (but not all) species of fish produce equal numbers of offspring of each gender, as does the one under consideration. However, some workers think this fish species, that produces equal numbers of male and female fry, might have a gender dependent rate of survival to adulthood. So your computer guru, Mega Binary, suggests testing this using a normal approximation with criterion $\alpha = 1\%$ and $n = 1024$, in hopes of detecting a reasonable minimum difference of $\mathcal{D}_{min} = 0.16$ when p_0 is the probability of a randomly sampled fish being a male under **H0**: THERE IS NO DIFFERENCE. The alternative hypothesis is **HA**: SURVIVAL TO ADULTHOOD IS ASSOCIATED WITH GENDER. If Mega's plan were used, what would be the power (to three figure accuracy)? You can make an a priori estimate of the power using Equation 7.26. But, would there be something that is less than ideal, in application, about Mega's choices?

22. Americans tend to vote close to 50% for each side of a political issue, no matter how strongly debated. When voting for persons, they tend to vote a little more for the incumbent. However, an election for sheriff includes both an incumbent and association of the candidates with some very hot issues. A group of volunteers is interested in getting people registered to vote, and they also intend to ask their clients for which of the two sheriff candidates they intend to vote. They hypothesize that **H0**: THERE WILL BE ALMOST NO DIFFERENCE IN THE VOTE (it will be decided by chance) vs **HA**: THE INCUMBENT WILL WIN. Estimate the sample size, n, needed for a one sample normal approximation with a criterion of $\alpha = 1\%$ to have a $1 - \beta = 0.90\%$ probability of detecting a difference of $\mathcal{D}_{min} = 1\%$. Should you use t or Z ?

23. You have estimated that each of the two samples for your test needs to have a size of $\hat{n} = 86$, but the maximum **available** sample size for **one** of the samples is $n_f = 58$. What is the adjusted total sample size for the pair being compared, and what is the final adjusted size estimate for the **adjustable** sample?

24. There is about to be a plebiscite in which ethnicity is expected to decide how people vote. It is known that 40.2% of the people in this region are of ethnicity A and 59.8% are of ethnicity B, and everyone is registered for the first time. You are doing a survey before the election to ask the residents how they will vote.The hypothesis pair is **H0**: HOW THEY SAY THEY WILL VOTE WILL NOT DIFFER FROM THE PROPORTIONS OF THE ETHNIC GROUPS vs **HA**: HOW THEY SAY THEY WILL VOTE WILL DIFFER FROM THE PROPORTIONS OF THE ETHNIC GROUPS. Estimate the sample size, n, you will need under these difficult conditions, for a one sample two tail normal approximation test to have 80% power to detect a reasonable minimum difference, $\mathcal{D}_{min} = 0.050$, and using a criterion of $\alpha = 0.050$. What are you measuring here?

25. What is the advantage of a power analysis if you are only interested in the CI?

8 χ^2 and F

Chi square and F tests are analogous to one and two sample Z tests comparing means, except chi square and F respectively refer to one variance and two variances.

8.1 Chi Square

Chi square, χ^2, is a distribution that is related to, but different from the normal distribution. The means tend to be normally distributed, and variances of random samples tend to be chi square distributed. The ***chi square test*** compares the variance of a sample to the true variance, σ^2, or in application, a hypothetical variance, σ_0^2. This test is useful for physicists, engineers, plant breeders, etc. who have a theoretical σ^2 they need to test. It is also a basis for other inferential tests, including the F test (Equation 8.15), many nonparametric tests (Chapter 14), and the Pearson's chi square test (Equation 12.3). It also allows us to construct confidence intervals for the variance (Equation 8.7). The chi square distribution is **continuous**, and is useful for a variety of tests, many of which are not restricted to questions of variability. Chi is the Greek letter χ and is pronounced 'keye' (rhymes with 'eye'); ***Chi square*** is usually defined by

$$\chi^2 = \frac{(n-1)\,S^2}{\sigma^2} \qquad df = n-1\,. \tag{8.1}$$

Although the divisor can be, and usually is, a **hypothesized** variance, σ_0^2, it is **never** a sample estimate, such as S^2. For another group of applications, Equation 8.1 can be simplified to

$$\chi^2 = \frac{\sum_{i=k}^{i=k}\left(x_i - \bar{x}\right)^2}{\sigma^2},$$

where the numerator, $\sum_{i=k}^{i=k}\left(x_i - \bar{x}\right)^2$, is called the **sum of squares of error**, SSE. The word 'error' in such names as 'sum of squares of error' only refers to chance variations from the expected. It is not about errors in the sense of mistakes. The denominator (divisor, or bottom) can, in application, be a hypothesized variance, σ_0^2. But the denominator cannot simply be another sample variance. The chi square distribution **approaches** normality when the number of sample size approaches[1] large ($n \geq 30$, often only $n > 25$).

The sum of squares of error, $SSE = \sum_{i=1}^{i=n}\left(x_i - \bar{x}\right)^2$, is useful for many applications, with or without being part of χ^2. If it is the sum of more than a few squared errors, it can be broken into smaller sums which still add up to the original sum. Or conversely, a sum of squares can be added to a sum of squares to get a bigger sum of squares. That is, sums of squares are ***additive***.

Suppose you had the resources to take **n** $= 300$ random measurements, a sample from a single source, such as the weights of pedigreed miniature schnauzer dogs. Then assuming that they are still in random order, you could calculate $k = 10$ independent sums of squares of size $n = 30$. Because

---

[1]There is a less widely used continuous distribution, the ***chi distribution***, where $\chi = \sqrt{\chi^2}$. It is the distribution of the positive square roots of random chi square values. Like the χ^2 distribution, the χ distribution approaches normal as the sample size grows larger (but it does so more rapidly).

they are from the same distribution, you would add them together to get a total sum of squares,

$$SST = SS_1 + SS_2 + \cdots + SS_{10} = \sum_1^{30} (x - \bar{x}_1)^2 + \sum_{31}^{60} (x - \bar{x}_2)^2 + \cdots + \sum_{271}^{300} (x - \bar{x}_{10})^2.$$

Individually, these sums of squares will each have a chi square distribution, and $df_1 + df_2 + \cdots + df_{10}$ for a total $df_t = 290$ (ten df having been lost in estimating the separate means).

At an even more general level, Theorem 8.8 in Freund and Walpole states, "If $x_1, x_2, \ldots, x_n$ are independent variates having standard normal distributions, then

$$\boldsymbol{y} = \sum_{i=1}^{i=n} \boldsymbol{x}_i^2 \tag{8.2}$$

has a chi square distribution, χ^2, with n degrees of freedom." (1980, p 267). Note that this theorem assumes a definite μ and σ, so no df are used up). So, the result of substituting sums of squares for $\sum_{i=1}^{i=n} x_i^2$ results in a chi square distribution, and because division by a constant such as σ_0^2 does not alter a variate distribution, then, as when $\boldsymbol{x}_i^2 = (x - \bar{x})^2$,

$$\chi^2 = \frac{\sum_{i=1}^{i=n} (x_i - \mu)^2}{\sigma_0^2} = \frac{SSE}{\sigma_0^2}$$

is chi square distributed.

As you should recall from the binomial approximation, the normal, binomial processes with p not too far from $1/2$ may, with some dependence on having a sufficient sample size, n, also approximate a normal distribution. All this leads to various characteristics that make both the sums of squares and the chi square so useful.

Does some of this look like the square of something familiar? First recall that $S_2 = \sum_{i=1}^{i=n} (x_i - \mu)^2 / (n - 1)$, and consider that $(n - 1) S_2 = \sum_{i=1}^{i=n} (x_i - \mu)^2$, so,

$$\chi_{n-1}^2 = (n - 1) S^2 / \sigma^2 = \sum (x - \mu)^2 / \sigma^2.$$

Therefore, when we apply this to the deviations and variance of means, $\chi^2 = \sum_{i=1}^{i=n} (x_i - \mu)^2 / \sigma_{\bar{x}}^2$. So, with a sufficient sample size (conceptually approaching infinity, but far less in application) $Z = (\bar{x} - \mu) / \sigma_{\bar{x}}$ is the square root of χ^2 with 1 df, and $Z^2 = \chi_{1df}^2$.

There are many ramifications, more to follow, relating to the fact that the 'Z test' so nicely arises from the definition of the CLT (Subsection 4.3). The CLT can be demonstrated to apply to χ^2 with the result that the chi square distribution will also approach being normally distributed as n grows larger. It is usually fairly close to normal at about $n = 25$. Also be aware that dividing by a constant, such as σ^2, does not change the distribution of random samples, whether for a chi square distribution, a normal distribution, or whatever. Snedecor and Cochran (1967) agree with the above derivations, that for a normally distributed population y of means with a population mean of μ_y,

$$\chi^2 = \frac{\sum_{i=1}^{i=n} (y_i - \mu_y)^2}{\sigma_y^2} = \frac{(\bar{x}_1 - \mu)^2}{\sigma_{\bar{x}}^2} + \frac{(\bar{x}_2 - \mu)^2}{\sigma_{\bar{x}}^2} + \ldots + \frac{(\bar{x}_k - \mu)^2}{\sigma_{\bar{x}}^2} \quad \text{iff } y \sim \mathcal{N}(\mu, \sigma^2), \tag{8.3}$$

where $\sim \mathcal{N}(\mu, \sigma^2)$ means approximately normally distributed (They seemingly expect us to infer that $y = \mu_x$, or simply μ.) This can be seen as an expansion of the summation in Equation 8.1. It will be **important** to try to keep Equation 8.4 and Equation 8.3 in mind while studying the uses of the chi square. Equation 8.3 leads to another often **useful** definition of chi square,

$$\chi^2 = \sum_{i=1}^{i=k} Z_i^2 \qquad (k-1)\ df, \tag{8.4}$$

which says that χ_{k-1}^2 is the **sum of** k **independent standardized** Z^2 **values**. In the initial definition of $\chi^2 = S^2\,(n-1)\,/\sigma^2$, Equation 8.1, the numerator is S^2, the sample variance, with $n-1$ degrees of freedom taken out by multiplication. But when you consider χ^2 as the sum of k summed Z^2 values, you have $k-1$ degrees of freedom (Equation 8.4), the k being used to emphasize that it refers specifically to the distribution of k summed Z^2 values. So be aware that in order to avoid a notational quagmire in this chapter, n is **used for** sample size within Z^2 values, and k is **used for** the number of summed values of Z^2. As you shall eventually see, some applications of the chi square distribution will lose additional degrees of freedom because of the way the data is collected. Given perfectly random sampling, and **assuming** $S^2 = \sigma^2$ or σ_0^2, then the sample distribution of a given χ^2 has an expected value (a mean) **equal to the degrees of freedom**.

The chi square distribution has many uses. Even though nonparametric tests, such as the ones described in Chapter 14, require unbiased random data, they do not themselves require normally distributed data, and many of them include calculations that produce values whose distributions approximate the assumptions of the chi square distribution. The extremely useful Pearson's chi square test is commonly and **incorrectly referred to** as 'the chi square test' (Chapter 12). It uses the model of the chi square distribution but is technically **not** the same as the chi square test. It uses χ^2 to evaluate the sum of k normal approximations, after the fashion of Equation 8.4.

8.2 Will the Real Chi Square Test Step Forward

I am tempted to designate the applications in this section the *true chi square* test, or real chi square test, because the χ^2 distribution is continuous, and χ^2 applies specifically to continuous data, such as variances and standard deviations. The approximations that use χ^2 to approximate probabilities associated with discrete data such as binomial data are also useful (Chapter 12). Some authors may vary their notation to emphasize the distinction between chi square and its approximations (Section 8.2.8), but this book generally uses χ^2 to denote the outcome, whether it refers to a true chi square test or to an approximation which happens to approximate a chi square distribution. Hence there is some quibble about the naming of such tests and of a few applications based on the summation of Z^2 values.

Keep in mind that the variance, $\sigma^2 = \frac{\sum_1^N (x-\mu)^2}{N}$, is the true, but unknown, mean of the **squared deviations from the population mean**, and that this is estimated by the variance estimate, $S^2 = \frac{\sum_1^n (x-\bar{x})^2}{n-1}$ (the sample variance), when using a sample of size n.

The chi square distribution is **not** a 'squared t distribution' as is often mistakenly stated. The t distribution is a result of using the chi square to adjust Z for the effect of having to estimate the variance. That is, for a sample of size n,

$$t = Z\sqrt{(n-1)\,/\chi_{n-1}^2} \qquad (n-1)\,df$$

(Student 1908, Fisher, R. A. 1925). The term, $(n-1)\,/\chi_{df}^2$, under the radical sign, $\sqrt{\ }$, approaches 1 as n grows larger. It is an adjustment for the fact that not only is the mean $\bar{x}$ an **estimate**, but S is also an **estimate**. The resulting t is generally valid, assuming that the randomly sampled standard deviation and mean are independent. Given sufficient sample size ($n \geq 30$, maybe even 20), t tends to be applicable to continuous data, but it is not applicable to the binomial distribution and the Poisson distribution (Section 15.4), which are common distributions involving discrete data.

The whole chi square distribution is positive, and its graphed curve tends to be asymmetrical, **skewed** to the right (drawn out to a more pronounced right tail). Due to squaring in the calculation of χ^2, it is always positive regardless of the direction of the true difference. That is, χ^2 is the sum of both $(+Z)^2$ values and $(-Z)^2$ values as positive chi square components. Aside from the fact that squares of real numbers are always positive, **distances have no direction** (unless specified otherwise). So, this summation of absolute Z^2 values makes sense if you think of variation as a measure of a distance, the extremeness of the jiggling back and forth to either side of an expected value such as the sample mean. So, before they are squared, the Z values are two tailed, plus and minus, but chi square has two tails of its own: **more** (positive) **variation**, and **less** (positive) **variation**.

Recall that a statistical test generally gives you an estimate of how likely your results are if **H0**: THERE IS NO DIFFERENCE is true. Given that your alternative hypothesis is something like **HA**: THE VARIANCE OF THE SAMPLED POPULATION IS GREATER, then if the p_value found by the χ^2 estimate is more **unlikely** than your preselected cutoff (α or $\alpha/2$), you can declare statistical significance and conclude **H0** is too unlikely to be true. Given that the null hypothesis is true, the upper tail (right part) of a graphed χ^2 distribution can be seen as the probability of observing a larger variance than the hypothesized variance, σ_0^2, by chance alone. However, the lower tail can be seen as the probability of observing less variance than the hypothesized variance by chance alone.

The sum of the probability of occurrence A and probability of !A (not A) is 100%. The probability of all the possibilities in a population happening is 1 for 100%, and if you have a probability, p, of something happening, its complement is $1 - p$. The $p >$ values, **also** called p_more values are complementary to $p <$ values which would be the cumulative probability of the value of, say, chi square that you are looking up. Unless stated otherwise, this book and its tables **only refer to** complementary cumulative $(p >) = 1 - (p <)$ **tables such as Appendix 5** for the chi square distribution. When using unfamiliar software or tables you need to experiment a little to see if you are getting $p >$ values that have decreasing outcomes for larger inputs, or $p <$ values that have increasing outcomes for larger inputs.

Tables of chi square have α values shown along the top margin. The degrees of freedom are shown for each row. The body of each table contains the chi square values you are looking up for a given alpha and degrees of freedom. (In some applications, the α values can be seen as a small set of *complementary cumulative probabilities*, the p_values, which are the probabilities of randomly obtaining a value as **large as** or **larger than** an observed value of χ^2 by chance alone.)

Figure 8.1: Half Full or Half Empty?

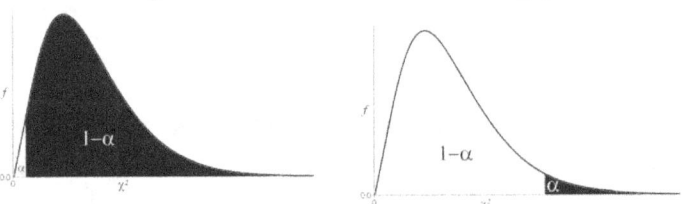

Here are two versions of the same chi square distribution (this one for 6 *df*); the dark areas represent the probabilities shown in the common χ^2 tables. Chi square values are on the horizontal axis, and relative frequencies are represented by the heights. χ^2 distributions are **asymmetrical** starting at zero, but the tables are only tabulated in **one direction**, usually $p >$, what this book calls p_more. This is handy when you are testing for **more-than** using the right tail. Notice that the width of α depends on the tail of interest. Unfortunately the dark area in the version of the figure on the left is the cumulative probability the tables give you when you are testing for **less-than** in the left tail and that is big and gets bigger as χ^2 decreases! So, because the total frequency equals 1 (100%), to get a probability for a **left tail** (light area), you must **subtract the** p_more **value** (dark area labeled $1 - \alpha$) from 1 after using a common $p >$ table such as Appendix 5. That is, for the lower tail we use the complement of the complementary cumulative frequency.

Squared values and therefore their sums are always positive and since χ^2 values are the sums of one or more Z^2 values, they are always positive. All this leads to a set of asymmetrical distributions (one for each *df*) that are truncated at zero on the left. So the small values of χ^2 are on the left and increase from zero toward the right as shown in Figure 8.1 (and everything is positive).

8.2.1 Tables and one tail chi square tests

Chi square in the form $\chi^2 = (n-1)S^2/\sigma_0$ was originally intended to be used for **one sample tests** to compare a sample variance to a hypothesized variance. As for other $p >$ tables, the **greater** the estimator, the **smaller** the p_value, as shown in the rightmost six columns of Appendix 5 for χ^2. Looking up an $\alpha \le 5\%$ for the **right tail** that is greater than hypothesized by **H0** is like using most other $p >$ tables. So, in application, for a right tail test you can use $\chi^2 \ge \chi^2_{\alpha\,df}$ for statistical significance in the upper tail using these six columns.

However, for the **left** (lower or less than) **tail**, these $p >$ tables only give the cumulative values on the right. So for small values of χ^2 you must, in effect, look up the equivalent $1 - p\_value$ in the table for the whole right tail, and then subtract it from 1 to find your answer, the $p\_value' = 1 - (1 - p\_value)$ that is to be compared to $_{cutoff}$ to see if the resulting p_value' is as **small** or smaller than $_{cutoff}$ when testing for a variance less than hypothesized by **H0**. So in application, for a left tail test, you can use $\chi^2 \le \chi^2_{(1-\alpha),\,df}$ for statistical significance in the lower tail. Section 8.2, and Figure 8.1 may help you with these concepts.[2] As you go through this chapter and some of the later chapters, you will see that there are many other uses for both the χ^2 test and the χ^2 distribution of variances.

8.2.2 Two tail chi square tests

For either a confidence interval or a two tail test with **HA**: THERE IS A DIFFERENCE, you would be using $\alpha/2$ as a cutoff. Although two tail tests of the variance are less common, they can be very useful in some situations.

For a two tail test with a 5% criterion, to see if the variance of the population being sampled is either greater than σ_0^2 or less than σ_0^2, split the difference between these tails by using the 2.5% cutoff value and its complement, the 97.5% cutoff. Consider that the subscript $\alpha/2 = 0.025$ is on the right for greater variation than expected if **H0** were true, and $1 - \alpha/2 = 0.975$ is still positive, but what you are really looking for, the other 2.5%, is way over there on the left for less variation. Like other inferential tests, chi square needs an adequate, but not extremely excessive, sample size.

If the hypothesized variance, σ_0^2, really equals the true variance, σ^2, then the sample distribution of χ^2 has a **mean** (an expected value) equal to the degrees of freedom. A **smaller variance is implied by** $\chi^2 < (n-1)$ and would be statistically significant if χ^2 is smaller than the value shown for the $1 - \alpha/2$ values in the body of a table such as in Appendix 5. A **larger variance is implied by** $\chi^2 > (n-1)$ and would be statistically significant if χ^2 is larger than the value shown for the $\alpha/2$ (Figure 8.1).[3]

---

[2] Referring to using tables, Subsection 8.2.1, the two sets of six columns of values in the tables of Appendix 5 represent p_value and $1 - p\_value$, or α and $1 - \alpha$, depending on what you are using them for. As you can see, the α values of the left hand tail are more horizontally compressed than the alpha areas of the right hand tail; thus they (or their complements) have to be individually treated when considering both tails, as when finding a CI.

[3] This book chooses to use greater than or equal to in places where there might be a question of significance, because the border between greater than or equal to and simply greater than is infinitesimal. However, this assumes **fair practice**, meaning **rounding <u>counter to significance</u>** at every step.

8.2.3 More about the χ^2 distribution

Figure 8.1 is the common way to sketch the chi square distribution. However, as the degrees of freedom, df, grow small, the chi square grows more right skewed until down at $df = 1$, it looks like this

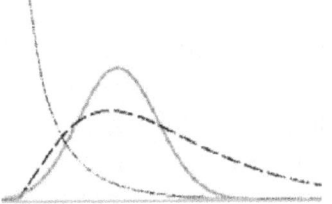 with about $1 - p\_more = 68\%$ of the total distribution (between $\chi^2_{1\,df} = 0$ and $\chi^2_{1\,df} = 1$). That leaves about $p\_more = 32\%$ of the distribution between 1 and ∞. As the degrees of freedom become large, the chi square distribution approaches normal[4] (Figure 8.2).

Figure 8.2: χ^2 approaching normal as k increases

So, the **upper tail**, which runs to infinity, pertains to **how much greater**, regardless of direction, the observed variability is than the hypothesized variability (a signless characteristic). However, the **lower tail** of χ^2, which is truncated at zero, handles differences in terms of variability as though positive and pertains to **how much less** the observed variability is than the hypothesized (or true) variability. It is also **signless**, tends to be **more compact** (to bulge upward more) **than the upper tail**. So, tables of χ^2 are used **differently** with regard to the **lower tail**. Do not be intimidated because the equations and explanations jump between the sample standard deviation, S, and its square, the sample variance, S^2; **if you have one**, the other is implied and can be simply evaluated, when and if needed.

Most of the material so far is to give you background for a multitude of procedures and explanations regarding tests based on χ^2 or returning a χ^2 value. In application the true χ^2 test is only occasionally used as a one sample test to compare a simple variance, S^2, to a hypothesized variance, σ_0^2. But understanding this little used test, when applied to a single sample variance, will **enable you** to understand and find confidence intervals for variances and standard deviations (Section 8.2.5).

8.2.4 Effect size for variance using chi square

Effect size, ES, values are drawn from samples and are therefore **estimates** attempting to quantify differences. The fictional example in Subsection 8.2.5.1 will involve a plausible situation with sufficient reason to use a hypothetical σ_0, because you usually cannot know the true σ. At the time of this writing, editors in many fields demand an effect size, ES, and prefer it to be in the original units of measure for every report of statistical significance (or of a p_value). However, we and they often have to settle for a unitless ratio with respect to an ideal or hypothesized ideal such as σ_0^2.

---

[4]The χ^2 distribution can approach normal almost forever, but it is never going to be **perfectly** normal no matter how big the sample. For one thing, it is a square and always positive and positive numbers can get larger forever but can never be smaller than zero. So the χ^2 distribution, no matter the number of degrees of freedom, is always truncated at zero ($0 \leq \chi^2 < \infty$). If you think about this, the peak of the chi square moves away from zero as the df increase. In application, $\chi^2 = 0$ can end up so far away from the expected as to approach being 'infinitely' improbable. That is, in application and given sufficient sample size (even $df \geq 25$) this truncation tends to become unimportant; it includes too little of the total probability to be of consequence. (Note that the notation, $X \sim \mathcal{N}\left(\mu, \sigma^2\right)$, means that population X is roughly normally distributed with mean μ and variance σ^2.)

The estimated variance of the sampled population is represented by (not equaled by) the whole numerator of $\chi^2 = \frac{(n-1)S^2}{\sigma_0^2}$, that is, by the sum of squares of error, $SSE = (n-1)S^2 = \sum(x - \bar{x})^2$ (Equation 8.1). But, by division of χ_0^2 by n you get a ratio, called **phi square**, which quickly approaches the estimated variance divided by the hypothesized variance

$$\phi^2 = \chi^2/n \approx S^2/\sigma_0^2 . \tag{8.5}$$

This ratio is a **handy effect size** estimating σ^2/σ_0^2, a comparison between the variance of the sampled population and the hypothesized population. That it is an approximation is indicated by $\approx$. But this approximation is close to equality and it improves quickly with increased n. So, phi square, ϕ^2, can be used as an ES, reporting the fraction of the original variance that is present after treatment. It can also be expressed as a percent. Note that this ϕ^2 is with regard to a known, or at least a hypothesized population variance, σ_0^2. This is also an unusual way to use ϕ^2 because, unlike most of its other applications, in this application it can validly exceed $\phi^2 = 1$ (100%). This is a comparison between a known (or at least hypothesized) variance and a sample variance, and because variance is what the true χ^2 is about, it is a very **appropriate** ES for single sample testing.

If you were comparing **two sample variances** (two estimates), you would use a test value called F (Equation 8.15) rather than chi square. F is a (sample size weighted) ratio between sample variances, **not** a ratio between the hypothesized variance and a single sample variance.

Another way to report the ES is as the **difference** between the true standard deviation estimate and the hypothesized standard deviation,

$$\widehat{D}_\sigma = (S - \sigma_0)\text{ original units.} \tag{8.6}$$

The notation is mine, but it corresponds to common practice.[5] Because $\widehat{D}_\sigma$ is in **original units**, researchers might find it useful when studying differences in variation from an additive ($\pm$) approach rather than a ratio approach (Subsection 8.2.6). Either this or the ϕ^2 (above) approach is appropriate for reporting an ES. Both might be included in the discussion part of a publication, however, wherever you put it, one or the other is likely to be required as part of the researcher's formal results.

A researcher might, if permitted by editorial policy, simply report a confidence interval (Section 8.2.5) for σ, which is in the original units of measure. However, it is difficult to target a correctly centered CI which is not ridiculously large (or small) without an ES-like reasonable minimum difference as part of an a priori power analysis.

> **CAUTION**: Do not use either ϕ or ϕ^2 with a Yates corrected (Subsection 12.7.1) χ^2 (Grissom and Kim 2005).

The chi square test compares variances, but it can also be used to indirectly compare their square roots, the standard deviations. That is, if one is significant, the other is likewise significant; they can even be considered as transformations, one of the other. However, a confidence interval for either S or S^2 might conveniently, and validly (as part of the description of your results rather than a test), avoid the need for a χ^2 test as such.

8.2.5 Confidence intervals for variability

The chi square distribution provides an excellent approach to produce confidence intervals. Keeping in mind that the chi square distribution is **asymmetric,**

---

[5]The hat, $\frown$, in Equation 8.6 indicates that something is an estimate; the letter D indicates a difference, and its subscript σ indicates that it is a difference involving standard deviations.

you can estimate a confidence interval for the variance using the usual tables, such as Appendix 5 and equation

$$\frac{(n-1)\,S^2}{\chi^2_{\alpha/2,\,n-1}} \le \sigma^2 \le \frac{(n-1)\,S^2}{\chi^2_{1-\alpha/2,\,n-1}}, \tag{8.7}$$

where $1 - \alpha/2$ and $\alpha/2$ are looked up independently because of the asymmetry. The two whiskers of a CI for variance are **different lengths**. The smaller the sample size, the more extreme this effect. Unfortunately, they are in squared units of measure and thus their interpretation might be confusing.

But, the confidence interval for the standard deviation,

$$\sqrt{\frac{(n-1)\,S^2}{\chi^2_{\alpha/2,\,n-1}}} \le \sigma \le \sqrt{\frac{(n-1)\,S^2}{\chi^2_{1-\alpha/2,\,n-1}}}, \tag{8.8}$$

is in the **original units of measure**. Always be aware that square roots are assumed to be positive (as they are in this equation) unless stated otherwise. A sample size of at least $n = 25$ would be necessary for an estimate of variability. With increasing sample size, the whiskers will approach equal lengths, but they will never actually get there. The necessary chi square values for the corresponding probabilities may be found in a $p >$ table such as Appendix 5.

Question 1 is another example of calculating a CI for variance.

BEWARE: Once again, some tables and software will read from the opposite end of the distribution and give the $1 - \alpha/2$ a higher value than the $\alpha/2$ value of chi square (the dark areas in the illustration above reversed). Best experiment a little to see which end is which. In that case, the divisors above must be reversed.[a] Remember to put the largest **divisor** (denominator) on the left.

---

[a] The qchisq(x, df) function in the R language, for instance, must be set to qchisq(x, df, lower.tail=F) to correspond to the CI for variance equations in this book.

8.2.5.1 Example: CI for Variability

Close to identical sizes of agricultural products, castings, fibers, and even grains of sand are highly desirable, because they are handy for relatively simple machines to handle. That is, the less variability, the better for mechanization. This example shows both a not quite ideal, but good for illustration of the methods, application of the chi square and a confidence interval for the variability.

Tomatoes in Uniform Hardball Tomato Company (motto: educate the public about what they want to eat and then stuff it down 'em) prides itself on very uniformly sized tomatoes. However, Hardball is still getting some large or small extremes which mess up their highly mechanized handling of their tomatoes for the produce market. Over the years they have striven to get below their persistent standard deviation of $\sigma_0 = 12\,\text{mm}$.

Management was persuaded to allow a soil sterilization company to treat a large demonstration field using a brief exposure to radiation. Any sort of soil sterilization tends to increase yield. However, Dr. I.M. Veggie, their chief of horticultural science, wondered what effect this inexpensive form of soil sterilization was having on variability. Dr. Veggie checked with the quality control labs and was assured that the test field and the conventionally treated fields around it had been getting the same variance the last many years. Although primarily interested in a standard deviation estimate, Dr. Veggie formed a hypothesis pair, testing the variance as a **stand-in** for the standard deviation in the actual calculation,

H0: THERE IS NO DIFFERENCE BETWEEN THE STANDARD DEVIATION FROM THE RADIATED SOIL AND THE EXPECTED STANDARD DEVIATION ($\sigma_0 = 12\,\text{mm}$) FROM THE UNTREATED SOIL.

HA: THERE IS A DIFFERENCE

Criterion: Dr. Veggie selected a 5% criterion, meaning that there is a 2.5% cutoff for each tail, **less** variance and **more** variance.

Methods: Dr. Veggie collected $n = 100$ almost tasteless hard green tomatoes in as random a fashion as possible from the treated soil plot, that were waiting to be shipped and chemically reddened (his company calls it ripened) in the shipping containers. Unfortunately this sample is from only one demonstration plot, and thus not really a formal experiment. But it was an opportunity to explore before seeking funding to proceed to further investigation. A laboratory technician happened to be available to measure all of them and reported $S = 10.0\,\text{mm}$ for the tomatoes from the treated plot.

Analysis: Thinking in terms of statistical significance testing, using Neyman-Pearson inference in which an a priori cutoff is set, Dr. Veggie had a priori decided to test using Equation 8.1

$$\chi^2 = \frac{(n-1)\,S^2}{\sigma^2} \qquad (n-1)\,df\,,$$

with an $\alpha = 5\%$ criterion (giving $\alpha/2 = 0.025$ in each tail). The tail rules for using the chi square (and its tables) seemed different from those to which he was accustomed. However, when he sketched a crude two tail version of Figure 8.1, it reminded him that the χ^2 distribution is a skewed distribution. He calculated that

$$\chi^2 = \frac{(99)\,10.0^2}{12^2} = 68.750\,.$$

He looked in a table and found $\chi^2_{1-\alpha/2=0.975,\,df=99} = 73.3611$ was the lower tail cutoff on the χ^2 scale, and $\chi^2_{\alpha/2=0.025,\,df=99} = 128.422$ was the upper tail cutoff. But, since the observed χ^2 was less than the degrees of freedom (the expected, or middle value) this answer was in the lower tail. So, because $(\chi^2 = 68.750) < \left(\chi^2_{1-\alpha/2=0.975,\,df=99} = 73.361\right)$, he declared statistical significance to reject **H0**, and accept **HA** instead.

However, Dr. Veggie saw confidence intervals as both more informative than other approaches and as representing an effect size in themselves. He had also heard that some theorists are skeptical about using chi square to compare a sample variance to a hypothetical variance. The CI requires two separate chi square values, $\chi^2_{\alpha/2,\,df}$ for the lower tail and $\chi^2_{1-\alpha/2,\,df}$ for the upper tail. He wished to present a $1 - \alpha = 0.95$ confidence interval to Hardball's board of directors, and referred to Equation 8.7

$$\frac{(100-1)\,S^2}{\chi^2_{.0.025,\,n-1}} \leq \sigma^2 \leq \frac{(100-1)\,S^2}{\chi^2_{0.975,\,n-1}}\,.$$

This requires the appropriate chi square values from a table such as Appendix 5. For $p\_more = 0.025$ and $1 - n = 99\,df$, his rounded values of the above give $\chi^2_{0.025,\,99} = 128.422$. Still at

99 df, but in the lower tail, $1 - p\_more = 0.975$, the table gives $\chi^2_{0.975,\,99} = 73.361$. Notice that p_more in this application was $\alpha/2$ because it is a two tail interval.

However, the CI for variance would be in mm^2 and the board would be more comfortable with the original units of measure which is mm. So Dr. Veggie applied Equation 8.8 and he calculated

$$\sqrt{\frac{(99)\,10^2}{128.42}} \leq \sigma \leq \sqrt{\frac{(99)\,10^2}{73.361}}$$

in mm. Conservatively rounding down on the left and up on the right, his desired CI for the standard deviation was

$$8.780\,\text{mm} \leq \sigma \leq 11.62\,\text{mm}.$$

Other than insisting that a CI represented an effect size, what could Dr. Veggie have done about the requirement to report an ES? He could have simply subtracted the standard deviations, thereby preserving the original units of measure. So, he could report the estimated difference between the standard deviations,

$$\widehat{D}_\sigma = S - \sigma_0 = 10.0\,mm - 12.0\,mm = -2.0\,mm.$$

This could also be an acceptable ES since this is also in the original units of measure.

Discussion: First let's consider this exploratory process as though it was a valid experiment. This researcher obtained statistical significance and both a CI and an effect size. Notice that the CI was more informative than the ES, especially regarding the lopsided world of the standard deviation. The CI for the ES came awkwardly close to (or depending on how you got there, reached) $\sigma_0 = 12$.

However, the sample was **not** a masterpiece of randomization; sampling a single field would **not** be accepted by peer reviewers. An appropriate approach would have included a randomized way of collecting a number of variance measurements from a sufficient number of treated fields. The data from this crude experiment and already existing Hardball data could be used a priori to consider transformations, and to estimate sample sizes.

Postscript The ES could have been reported as a percent in terms of the ratio-like measure ϕ^2, called phi square (Equation 8.5). It is the square of the approximate measure of correlation, phi ($\phi = \sqrt{\chi^2/n}$). Then $\phi^2 = \chi^2/n$ estimates the effect size as the **estimated fraction** of the **hypothesized variance**, σ_0^2, which is observed after the treatment, with data from Dr Veggie's sample,

$$\phi^2 = \chi^2/n = 68.75/100,$$

or about 69%. To roughly check this, compare the ratio square of observed standard deviation to $(\sigma_0)^2$:

$$\phi^2 = S^2\text{mm}^2/\sigma_0^2\text{mm}^2 = 100/144,$$

or about 69% **as much variance** as without the treatment.

That would have sounded good, but maybe too good; it would be better to report results in terms of standard deviation. That is, maybe ϕ itself should be used to report

$$\phi = \sqrt{\chi^2/n} = \sqrt{68.75/100} = 0.8292,$$

or, conservatively rounding up, about 83% **as much standard deviation** as the treatment. After all, standard deviation was what was started with and ϕ is less likely to exaggerate the effect in the minds of those stockholders who are less familiar with statistics.

However, these last two measures are not in the original units, and they might not have been as well received outside of theory oriented audiences as would have been the estimated difference between standard deviations, which is symbolized above as $\widehat{D}_\sigma$. Some folks might also have been discomforted by the fact that if this treatment had **increased the variability**, it would have produced a ϕ^2 **over one hundred percent** (greater than 1.0). That is reasonable in terms of variance, but any correlation implied by $\phi > 1$ would be meaningless. Variances are convenient in lots of situations, but standard deviations are often easier to explain to your client.

CIs can be subject to problems of either lack of power (to correctly center them) or of extremes of sensitivity (too wide or too narrow). One of the major weaknesses of this example was that the researcher did not plan for power. Depending on the targeted α, β, and D_σ, Dr. Veggie should have, a priori, determined an appropriate sample size, but we will return to that in the next section. However, for this demonstration of calculation approaches, we will agree to disagree with the trustworthiness of his investigation of round red objects and move on.

Question 5 is another example and exercise regarding the use of phi as an effect size.

For a formal inferential experiment, Dr. Veggie could do a **power analysis** and consider using a nonparametric test with Siegel-Tukey ranking (Sub subsection 14.2.2).

8.2.6 A priori power analysis for true chi square

CIs and tests are in different conceptual universes, the first attempting to trap the parameter of interest and the second estimating the probability that the observed statistic represents a hypothesized value. However, the underlying probabilities which are of relevant interest to us tend to be the same in application, and are similarly related with respect to sample size. So, an a priori power analysis is much to be preferred for a chi square test, but even if you are not going to look for statistical significance, an a priori sample size estimate will **enable you** to obtain a more accurately located and conveniently sized CI.

The **standard deviation of the sample standard deviation** may be estimated to a quite satisfactory accuracy, using

$$\sigma_S \approx \sigma/2\sqrt{n} \qquad (n \geq 25),$$ (8.9)

where the $\approx$ means approximately. David Lane (1999) mentioned a form of this approximation and stated, "The standard deviation is positively skewed for small n but it is approximately normally distributed if $n \geq 25$."

To estimate the sample size for comparing a sample variance with a hypothesized variance, and by implication, their square roots, the corresponding standard deviations, you can use your selected α and β, and you can then choose a reasonable minimum difference between standard deviations,

$$D_{\sigma\_min} = \sigma - \sigma_0.$$

The estimation process resembles the estimation of the sample size for a large single sample Student's t test, but note that the actual t distribution does not apply here, so use Z for the normal distribution. This approximation and a priori size estimation also works for CIs.

So, for **one tail** with appropriate arrangement and substitution of $\widehat{\sigma_S} \approx \sigma_0/2\sqrt{n}$ into Equation 7.3, $\hat{n} = \frac{(Z_a + Z_\beta)^2 S^2}{D_{min}^2}$, we have

$$\hat{n} = \frac{(Z_a + Z_\beta)^2 \sigma_0^2}{4D_{\sigma\_min}^2},$$ (8.10)

the sample size estimate for challenging a hypothesized variance with a sample variance, using the chi square test.[6] Then, when you use this sample size to test, you will only declare statistical significance if the difference is in the preselected direction (tail). The tails represent either more variability or less variability, but not both. Note that standard deviations underlay these equations even though the results will be used for comparing their squares, the variances.[7]

For a **two tail** chi square a priori estimate of n, use

$$\hat{n} = \frac{\left(Z_{a/2} + Z_\beta\right)^2 \sigma_0^2}{4D_{\sigma\_min}^2}. \tag{8.11}$$

Like other two tail estimates, this divides α between the two tails, so this uses $a/2$ with two tails, but β does not change, because it is in one direction. This equation is derived from Equation 7.5 by substituting standard deviations and their standard deviation (for means and their standard deviation) to provide a suitable sample size for a confidence interval at, say, $\alpha = 0.05$ and $1 - \beta = 0.80$.

Even if you do not intend to actually run a χ^2 test, a sample size selected this way would mean that your 95% CI for the standard deviation would tend to exclude any statistically significantly different (outside) value 80% of the time. Notice that a priori sample size determination and chance will affect the width and especially the width of any CI.

Note that if $\hat{n} < 25$, you must use at least $n = 25$ anyway.

8.2.6.1 Example: An A Priori Power Analysis for Variability

Tomatoes in uniform (continued) Suppose that Dr. Veggie really had decided to do an a priori power analysis to estimate an appropriate sample size for the hypothesis pair,

H0: THERE IS NO DIFFERENCE BETWEEN THE STANDARD DEVIATION OF THE RADIATED AND THE CONVENTIONALLY TREATED SOIL.

HA: THERE IS A DIFFERENCE BETWEEN THE STANDARD DEVIATION OF THE RADIATED SOIL AND THE CONVENTIONALLY TREATED SOIL, WHICH HAS AN EXPECTED VARIANCE OF $\sigma_0 = 12\,\text{mm}$.

Criteria: This is a two tail test, so Dr. Veggie selected an $\alpha = 5\%$ criterion, meaning that there was a 2.5% cutoff for each tail, and he sought a power, $1 - \beta = 80\%$. He also needed to define what is meant by an 'effect' (or 'DIFFERENCE') by specifying what would be a reasonable minimum difference. That was easy because upper management had been demanding an improvement of $D_{\sigma\_min} = 1.5\text{mm}$.

Analysis: For the **upper tail** of the chi square, he would have used $Z_{\alpha/2=0.025} = 1.9600$, $Z_{\beta=0.20} = 0.8416$ and $D_{\sigma\_min} = 1.5\text{mm}$ is the reasonable minimum difference. He used Equation 8.11, $\hat{n} = \frac{\left(Z_{a/2}+Z_\beta\right)^2 \sigma_0^2}{4D_{\sigma\_min}^2}$, to calculate

$$\hat{n} = \frac{(1.9600 + 0.8416)^2\,144}{4(1.5)^2} = 125.583.$$

---

[6] The chi square test for variance can only be a one **sample** test.

[7] The above sample size estimator for χ^2 is a good approximation even when underlying normality is not involved, as I found when I ran a Monte Carlo simulation using flat random values (Equation 8.12). The exact expected σ_S involves the fourth moment, etc., all outside the scope of this book and awkward to work with algebraically. So, I do not present an approach to estimating a suitable small sample size, $n \leq 25$, for the chi square test of variance. But such a sample size is unlikely to have enough power to be of any value in application.

He would have **rounded up** the estimate to $\hat{n} = 126$ to achieve an 80% power of detecting D_{σ_min} with his a priori chosen $\alpha = 5\%$.

Discussion: He would have achieved more power if he had proceeded as in this example, rather than as in Subsection 8.2.5.1, because the required sample $\hat{n}$ would have been much larger. Either sample size would have produced an acceptable approximation of normality, but the larger sample would be a bit better at that too. Most experiments need greater numbers to attain a reasonable power than they do to merely meet the assumptions. However, the true difference may really have been greater, providing an increase in effective power, and even if the effective power were lower than 80%, that does not preclude achieving statistical significance with some samples. Recall that $Z_{\alpha/2=0.025} = 1.959964$, which rounds to $Z_{\alpha/2=0.025} = 1.9600$ for five digit accuracy, and is handy for sample size estimation to 5 figures (decimal places). This is especially useful when summing $Z_\alpha + Z_\beta$, especially before squaring, which magnifies small differences.

Question 3 is both an exercise and example of a similar calculation.

8.2.7 Subsequent power analysis, chi square for variance

Because we need to estimate an underlying variance, σ^2, by partitioning the variances, finding a sample size for an ANOVA will often require a pilot study that is almost a complete full dress rehearsal except for being smaller and not requiring the usual F test. Similarly, a subsequent power analysis is sometimes necessary, when evaluating an experiment in which the sample size deviated from your plan, or when (post hoc) roughly evaluating someone else's report. A subsequent power analysis looks at how much power your test would have at the α level of statistical significance with an a priori justifiable reasonable minimum difference, $D_{\sigma\_min} = \sigma - \sigma_0$. As for the a priori power analysis, the subsequent power analysis is based on the fact that the standard deviation is approximately normally distributed for $\hat{n} \geq 25$ and **the standard deviation of the standard deviation**, $\sigma_S \approx \sigma/2\sqrt{n}$. For $\hat{n} \geq 25$ it is possible to apply a method similar to Section 7.8 along with material from the preceding Subsection 8.2.6. It assumes that the outcome of your test, when expressed in terms of Z, equals $Z_{\alpha/2} + Z_\beta$, where α is your test criterion, but β is unknown. So, you can estimate the **standard error of the sample standard deviation** as $S_S \approx S/2\sqrt{n}$. The CLT will improve this approximation for larger sample sizes. Substituting this standard error of the sample standard deviation, S_S, into a rearrangement of Equation 7.11, $\hat{z} = \left|\frac{D_{min}}{S/\sqrt{n}}\right| - Z_\alpha$, gives $\hat{z} = \left|\frac{D_{min}}{S/2\sqrt{n}}\right| - Z_\alpha$ (where the denominator represents S_S). So we can write

$$\hat{z} = \frac{|D_{\sigma\_min}|}{S/2\sqrt{n}} - Z_\alpha \tag{8.12}$$

for **one tail** (greater variation than or less variation than, but not both), where n is the sample size you achieved (or to which you are limited).

For **two tail** testing,

$$\hat{z} = \frac{|D_{\sigma\_min}|}{S/2\sqrt{n}} - Z_{\alpha/2}, \tag{8.13}$$

where $\alpha/2$ splits α between the tails. As for the a priori sample size estimation, this will **not be likely to have sufficient power** if $n < 25$; much larger sample sizes, $n \gg 30$, would be better.

To interpret the distance, $\hat{z}$, for either Equation 8.12 or Equation 8.13 use the appropriate rule, either

$$iff \ \hat{z} \geq 0, \ then \ \hat{\mathbf{z}} = \hat{Z}_\beta$$

or

$$iff \ \hat{z} < 0, \ then \ |\hat{\mathbf{z}}| = \hat{Z}_{1-\beta}.$$

Refer this to a standard table of Z (such as Appendix 4). But recall that you ultimately want to estimate the power, $1 - \beta$. When reporting the final results it would be prudent as well as conservative to **round** z **down** to two figure accuracy (three figure accuracy for very large sample sizes).

8.2.7.1 Example: Subsequent Power Analysis of Variability

Tomatoes in uniform (continued) As you recall by now, Hardball Tomato Co. is striving to get their tomatoes below their persistent standard deviation of $\sigma_0 = 12$ mm. Dr. I.M. Veggie, their chief of horticultural science, is interested in whether the sterilization of the soil with radiation will have any effect either way. Assume he has done no a priori power analysis. So, now he decides to attempt a subsequent power analysis to determine the power $(1 - \beta)$ of the test that he has already run with a sample of $n = 100$, using the original (and thus only applicable) $\alpha = 5\%$, and the hypothesis pair,

H0: THERE IS NO DIFFERENCE BETWEEN THE STANDARD DEVIATION FROM THE TREATED AND THE UNTREATED SOIL.

HA: THERE IS A DIFFERENCE BETWEEN THE STANDARD DEVIATION AND THE EXPECTED STANDARD DEVIATION ($\sigma_0 = 12$ mm),

Criterion: Dr. Veggie had selected $\alpha = 5\%$ for his two tail test, meaning 2.5% cutoff for each tail, less variance and more variance. He now chose a reasonable minimum difference of $D_{\sigma\_min} = 1.5$ mm in accord with what the board of directors originally wanted.

Methods: In conformity with the original test, he performed a two tail retrospective power analysis.

Analysis: For this two tail analysis, he used $Z_{\alpha/2=0.025} = 1.9600$ and $D_{\sigma\_min} = 1.5$ mm, which is in the original units of measure. So, he applied Equation 8.13, $\hat{z} = \frac{|D_{min}|2\sqrt{n}}{S} - Z_{\alpha/2}$, to estimate

$$\hat{z} = \frac{|1.5| \ 2\sqrt{100}}{12} - 1.9600 = 0.5400 \,,$$

and then using the rule, Equation 7.13, that when $\hat{z} > 0$, then $\hat{z} = Z_\beta$, and referencing a suitable table or program, gives $\beta = 0.294599$. That means that the power is estimated to be $(1 - \beta) = 0.705$ at this sample size, $n = 100$. So, assuming proper randomization, Dr. Veggie did have a moderate probability (a power of 70% before the test) of detecting an effect of $D_\sigma = 1.5$ mm. His overall chances would have been better if his true difference happened to have been larger. But, as always, due to chance, the exact true difference is unknowable.

Discussion: Dr. Veggie definitely came close to stepping over the line here! This would clearly have been a post hoc analysis and would only have been applicable in the privacy of his closet, as an exploratory procedure, had he not found an excuse to use a preexisting a priori D_{min}. However, suppose Hardball had instead been a priori interested in a reasonable minimum difference of 2.0 mm. Then

$$\hat{z} = \frac{|2.0| \ 2\sqrt{100}}{12} - 1.9600 = 1.3733 \,,$$

and in accord with the positive $\hat{z}$ rule, $\beta = 0.08483$. Because $(1 - \beta) = 0.914$, Dr. Veggie's 'experiment' would have had a lot of power if the data could have been collected from random plots or by some other appropriate method.

Unfortunately, as you may recall, the design lacked sufficient randomness for a conclusion about anything other than, maybe, that particular field the samples were from. Such exploratory activities provide potential hypotheses for future research and little more.

Even if there were a real difference, equaling or exceeding the reasonable minimum difference, any test would only have a particular **probability** of detecting a difference equaling or correspondingly exceeding the power to declare statistical significance for the chosen reasonable minimum difference. Certainty is unattainable.

Question 7 is an exercise similar to the above example.

> **CAUTION:** when interpreting Z values, remember that the classical style table in Appendix 4 reads in terms of $p > P(Z)$, whereas some other tables (and some calculators) read in terms of $p < P(Z)$.

8.2.7.2 If you cannot increase n

If you have an inflexible ceiling on n, your sample size, and suspect it could be too small to be useful, you could substitute your hypothesized σ_0 for S in either Equation 8.12 or Equation 8.13 (Subsection 8.2.7) to see if the experiment would be powerful enough to bother doing.

8.2.8 The symbol, χ^2

What I call the *true chi square test* is true in that it uses χ^2, chi square, as a continuous distribution, that is, as defined, rather than an approximation for discrete data. It is an application of the characteristics of a continuous distribution to continuous data. However, the most widely used simple applications use approximations of chi square to handle categorical discrete data. The approximation of chi square using the properties of the binomial distribution and its relatives for use with discrete data in Chapter 12 is especially popular and useful. It is incorrectly but commonly called and even indexed as the 'chi square test' (it's really Pearson's chi square test, $P\chi^2$) and is confusingly but generally also written as χ^2.

Goodness of fit tests are another example of these so-called chi square tests. They test whether the distribution of your data is the same as you hypothesize it to be. Such goodness of fit tests use the chi square to compare data to some distribution by breaking the data into n pieces (**binning**) to be used as categories. Such an application of Pearson's chi square test (Chapter 12) can compare the observed counts in each of the categories to their corresponding theoretical, or 'expected', values. In application, true goodness of fit tests, including some genetics examples that follow, are really just a single test. That is, the researcher does not use the counts to test between the separate categories, but rather uses them to test for one hypothesized pattern within a category.

Other tests, which are not always called chi square tests, but which are chi square based, are also very useful. In these applications, randomly collected data is analyzed in such a way as to be broken into subsample means, that, given sufficient subsample size, individually approach having normal distributions, even though the underlying distribution does not have to be normal, or chi square distributed. The resulting contrived squared Z values are then combined so as to have a chi square distribution. Many of these chi square based tests are incorporated into the evaluation of nonparametric tests that, as per the definition of nonparametric, do not require the underlying distribution to actually be normal (Chapter 14).

Capitalized variable names such as X or Y occasionally stand for whole populations of x_i and y_i values, but what does the bold-sans-serf symbol, X, represent? When referring to test result values rather than theoretical values, some authors use a sans serf X^2 to symbolize some estimate that approximates, or is contrived to approximate, a χ^2 distribution. These authors reserve the Greek letter chi (χ) for referring to the true (continuous), or theoretical, value of chi square. Maybe they consider X^2 as a statistic versus considering χ^2 as a parameter. Unfortunately what X^2 is used to symbolize sometimes varies from author to author (myself included). So consider context whenever you see this X^2 notation and do not use it without clarification.

It is useful to know that the mean of chi square distributed values approaches the degrees of freedom, and that the median approaches this mean as the df increases (the skew decreases, meaning the distribution becomes more symmetrical). The additivity of such distributions underlies many applications, including the chi square distributed approximations that are crucial to Chapter 12.

8.2.9 Truths about the χ^2 distribution

The chi square distribution is skewed, and assuming randomness, it has the following characteristics:

Its **mean**, $E\left(\chi^2_{df}\right)$, is equal to the degrees of freedom, df.

Its **median** is approximately $((3 \times df) - 2)/3$, $df > 2$.

Its **mode** is $df - 2$, $df > 2$.

Its **variance** is $2\,df$.

Because chi square values are squared, they are **positive** (in general their square roots are assumed to be positive), thus the distribution is **truncated at** 0.

Given two or more random variables with chi square distributions, the **chi square values are additive**, meaning that

$$\chi^2 = \chi^2_{1,\,df} + \chi^2_{2,\,df} + \ldots \text{ and } df = df_1 + df_2 + \ldots .$$

By considering the first three of these truths you will see that the mean, the median, and the mode tend to approach each other as the df grows large.

8.3 The F Test

Whereas the true chi square test (Chapter 8) is a one sample comparison of the sample variance, S^2, with a fixed σ^2 or σ^2_0, the F test is a **two** sample test that compares two sample variances, S^2_1 with S^2_2. The F distribution (named after R.A. Fisher) and this leads to his mathematical definition,

$$F_{1,2} = \left(\chi^2_{df_1}/df_1\right) / \left(\chi^2_{df_2}/df_2\right), \tag{8.14}$$

where $\chi^2_{df_1}$ and $\chi^2_{df_2}$ are **independent** and

$$\chi^2 = df \times S^2/\sigma^2 = \frac{df \times S^2_1}{\sigma^2} = \sum (x - \bar{x})^2/\sigma^2.$$

Recall that χ^2 is used to compare only one sample variance to an ideal or hypothesized population variance. But, the **F test compares two sample variances**, S^2_1 to S^2_2, or compares equivalent variance estimates, mean squares, MS_1 to MS_2. To informally derive the test, where, as hypothesized by **H0:** THERE IS NO DIFFERENCE, the underlying population variances σ^2_1 and σ^2_2 are **assumed to**

be equal to a shared underlying variance, σ^2. This assumption is very important for tests such as the ANOVA and it can restrict the potential for their use.

Assuming this assumption is true, that $\sigma_1^2 = \sigma_2^2$, then σ_1^2/σ_2^2 cancels out as do the identical $df_{\ldots}$ values. This can be seen in the expansion of Equation 8.14,

$$F_{1,2} = \left(\chi_1^2/df_1\right)/\left(\chi_2^2/df_2\right) = \left(\frac{df_1 \times S_1^2}{\sigma_1^2 \times df_1}\right) \Big/ \left(\frac{df_2 \times S_2^2}{\sigma_2^2 \times df_2}\right) = S_1^2 \big/ S_2^2. \qquad (8.15)$$

Because F refers to two sample estimates of variance, you must now consider two degree of freedom values, df_1 and df_2 for S_1 and S_2 respectively. So, $F_{1,2}$ is the value that you submit to a table of F such as Appendix 6 **with respect to** df_1 and df_2 in order to evaluate whether this particular F value, the ratio of one variance to another, indicates a statistically significant difference too unlikely to be due to chance alone.[8]

In applications such as the ANOVA (Chapter 9), deviations from the mean in either direction are treated as absolute distances, and they are evaluated by the entire value of the criterion, α. That is, the common use of the F test, and the usual tables, is to see if the **variance** in the denominator is **greater** than in the numerator in $F = S_1^2/S_2^2$.

However, F distributions also have two tails, more variation and less variation, and because they are very asymmetrical and tend to have a higher cumulative frequency on the left (lower F values) than on the right, their critical values are very different for each tail.

Where practical (when the samples are not too large) in application, you can have a **better** test than the F test for comparing the variability of two samples **by using Siegel-Tukey ranking** (Sub subsection 14.2.2) in a Wilcoxon test (Subsection 14.1.1). This approach avoids the assumptions of the F test, especially the assumption of normality.[9]

In what follows be aware that the **sum of squares** (SS) of something is the numerator (top of the fraction) involved in calculating a variance,

$$S_x^2 = \frac{\sum_{i=n}^{i=1}(x - \bar{x})^2}{df} = \frac{SS_x}{df_x}.$$

The variance, S^2, is often called the **mean square**, MS, because it is a sort of mean of the sum of squares. In some parts of the ANOVA, you must multiply the denominator by something to scale a SS and the resulting variance estimates, but hopefully that will become clear as we proceed to Chapter 9.

8.3.1 One and two tail F

Assuming the underlying distribution sufficiently approaches normality, the F test in the form

$$F = S_1^2/S_2^2$$

can be used to compare variability based on two estimated variances, each with its own number of degrees of freedom. When the assumption of normality does not hold, F suffers from a loss in power as compared to non-parametric counterparts (e.g. Subsubsection 14.2.2). Depending on your design,

---

[8]Given no effect (pure random outcomes), the mean of the distribution of F depends on the denominator and is expected to equal $(df_2 - 2)/df_2$ when $df_2 > 2$ (NIST 2006, 2013).

[9]Unfortunately the F test and Levene's test are themselves rather dependent on the assumption of normality, and they often lack power to determine if the difference between the variances is sufficient to invalidate the assumptions of such tests as Student's t test (Hayes and Cai, 2007).

you must be careful when looking up the value F in a table because the order of the degrees of freedom is important.

> **CAUTION:** Some tables of F read in the opposite direction from the common tables like the ones in this book.

One tail:

One could set up a one tail test with an a priori hypothesis pair,

H0: THE FEMALE AGE OF MARRIAGE HAS THE SAME VARIANCE IN ORTHODOX CATHOLIC FAMILIES AS IN ISLAMIC FAMILIES IN TURKEY

HA: THE FEMALE AGE OF MARRIAGE IS MORE VARIABLE IN ORTHODOX CATHOLIC FAMILIES THAN IN ISLAMIC FAMILIES IN TURKEY.

Before estimating the variances, it is necessary to choose whichever variance is a priori **hypothesized** to be **larger**. That is the one to choose to be in the numerator (on the top) for a one tail test. Call this hypothesized larger one S_1^2, in this instance Orthodox Catholics, and then the variance for the hypothesized lower alternative, Islamics, is S_2^2. Being a one tail test, referring only to the tail representing more variance, it will use the whole criterion, α, as the cutoff. When looking up F_{α, df_1, df_2}, always be careful to note that df_1 refers to what is put in the numerator (top), and df_2 refers to what is put under it in the denominator (the divisor if you wish to call it that). If $F = S_1^2/S_2^2$ is **greater than** $F_{\alpha, df_1 df_2}$, then you can, assuming randomness and a normal distribution, declare statistical significance.

Two tail:

The F distribution is asymmetrical. But, most tables of F are presented in terms of **only the upper tail** because the upper (one) tail of the F test is most often used in the ANOVA, its popular use (Section 9.1). So how can you proceed if you a priori decide to test

H0: THE FEMALE AGE OF MARRIAGE HAS THE SAME VARIANCE IN ORTHODOX CATHOLIC FAMILIES AS IN ISLAMIC FAMILIES IN TURKEY

HA: THE FEMALE AGE OF MARRIAGE IS DIFFERENT WITH RESPECT TO VARIANCE IN ORTHODOX CATHOLIC FAMILIES VS ISLAMIC FAMILIES IN TURKEY.

That would be a two tail test, either more variance, or less variance. You can use the same tables as used for an ANOVA for two tail testing if you take into account the relationship where $F_{1-\alpha/2}$ is the **lower tail** cutoff (and $F_{\alpha/2}$ is for the **upper tail** cutoff) for a **two tail** F test. Then if $F = S_1^2/S_2^2$ is either **greater than** $F_{\alpha/2, df_1 df_2}$, or **less than** $F_{1-\alpha/2, df_1 df_2}$, you can, assuming randomness etc., declare statistical significance. Notice that this decision depends on two different values in a table of F like the one in Appendix 6. Some tables might not have F values for large values of alpha such as $\alpha = 0.975$, in which case you may have to look online for tables or online calculators of F.

Question 9 illustrates a simple F test comparing variances.

Questions for Ch 8

Write the equations where appropriate, and then fill in the values. **Odd numbered questions** have worked answers at the back of the book and can be studied as additional examples. The data tends to be fictitious.

1. Toll roads can make a minimum speed estimate by timing when a car entered a gate, and when it took an exit. Assume that the standard deviation of the calculated speed of cars entering a certain expressway at a toll gate in a tourist area, and exiting at a toll gate in a city suburb, has for decades been calculated from random samples to be $\sigma_0 = 6.552$ mph during the month of July. However, income level has changed drastically since last July, and such behaviors as drinking, and who visits tourist areas, varies with the health of the economy. Your hypothesis pair is **H0**: THERE IS NO DIFFERENCE vs **HA**: THE VARIANCE FOR JULY OF THIS YEAR IS DIFFERENT with a criterion of 5%. You randomly sample the entry and exit times of $n = 150$ cars that pass that way in July; their estimated standard deviation is $S = 6.945$ mph. Show the calculations for the **appropriate test** and present a **CI** for the **variance**. (Pretend you do not know better and neglect any consideration of power.)

2. Estimate the CI for the observed standard **deviation** (rather than for the variance) for Question 1.

3. Assume you have not yet made the observation in Question 1, but you have a hypothesized $\sigma_0 = 6.552$ mph and your hypothesis indicates a two tail test and you make **an a priori sample size estimate** with a reasonable minimum difference of $D_{\sigma\_min} = 0.3$ mph. Show how you would do this with $\alpha = 0.05$ and $1 - \beta = 0.90$, and give your answer.

4. A common measuring device in machine shops, labs, and for inspection, is an outside screw micrometer, little more than a c-clamp with a carefully calibrated screw mechanism based on the concepts of so many units of measure, say fractions of an inch or of a millimeter, per turn. Assume you have purchased a metric micrometer which the manufacturer claims has a **variance of** $\sigma^2 = 2.5\,\mu m^2$ where a micrometer, μm, equals 0.001 millimeter. A μm is a millionth of a meter. However, you start to think about this thing warming unevenly, part of it in your hand and part in the cooler air, so that expansion and contraction might affect the accuracy. Even though you do not doubt that it might have as low a variance as stated when in a controlled environment, you wonder what its operational variance might be when it is in your hand. You want a confidence interval, and choose $\alpha = 5\%$ and $\beta = 5\%$ with a reasonable minimum difference of $0.5\,\mu m$ in terms of **standard deviations**, and you need to know what sample size is necessary to test this measuring tool. Assume you are going to measure a typical workpiece or specimen repeatedly under typical working conditions. (Hint: work in μm units. When the context would be confusing or when Greek letters are not available, μm is sometimes written um. You might wish more information on what this device looks like or about the units of measure. A web browser is a good source of such information.) **What should be your sample size?**

5. Calculate an ES for Question 1 as the percent change in the variance that is presumed to be due to the effect of different income conditions this year (what might be called the 'treatment').

6. Calculate an ES for the standard deviation in Question 1 in **original units of measure**.

7. Estimate a subsequent power for Question 1 with an $\alpha = 0.05$. Assume that you had a priori decided on a reasonable minimum difference of $D_{\sigma\_min} = 0.3$.

8. Your client only wishes to pay for a sample of $n = 25$ to test if the variability of his area's water was higher than the rest of the watershed for a certain trace mineral. The long running records show a standard deviation, $\sigma_0 = 1.32\,\mu g/L$. The client insists on a minimum difference, $D_{\sigma\_min} \geq \sigma - \sigma_0$, of $0.50\,\mu g/L$. What was the power of the resulting test with respect to this reasonable minimum difference, D_{σ_min}? Assume $\alpha = 0.05$, and use the slightly rounded down $Z_{\alpha=0.05} = 1.6449$. (Note that μg means micrograms, and that $n = 25$ is just adequate for meeting the distribution assumptions required by this application. This is an a priori question, but it can be answered by using an equation generally used for after the fact power analysis.)

9. For illustration only, assume you know your data distribution is symmetrical and mound shaped, approaching normality, and you wish to compare the variance of the individual fruit mass, in grams, of a particular variety of cantaloupe grown under organic conditions, $S_A^2 = 926.56\,g^2$ with a sample size of $n = 50$, versus when grown with commercial pesticides, $S_B^2 = 815.18\,g^2$ with a sample size of $n = 100$ fruits. You had a priori hypothesized **H0**: THE VARIANCES ARE THE SAME vs **HA**: THE VARIANCES ARE DIFFERENT. Assuming randomness, are they statistically significantly different at the $\alpha = 0.05$ level? Hint: $F_{0.025,49,99} = 1.5974$.

10. Prior to designing a thesis project, a nursing student hypothesized that **H0**: THERE IS NO DIF-FERENCE BETWEEN THE WEIGHT VARIABILITY OF TWO MALE COLLEGEVILLE POPULA-TIONS vs **HA**: THE WEIGHTS OF THE MALE POPULATION VISITING THE LOCAL SUPER-MARKET IS MORE VARIABLE THAN THE POPULATION OF MEN AT THE COLLEGE ITSELF. The student weighed a random sample of $n_2 = 30$ men, including students, faculty, and staff, at the college and, also at random, $n_1 = 75$ men visiting a supermarket. The supermarket sample had a standard deviation of $S_1 = 28.1\,lb$ and the college sample had a standard deviation of $S_2 = 25.9\,lb$. Is the difference statistically significant at the $\alpha = 5\%$ level? (Recall that when $n > 25$ the distribution of standard deviations is close to normally distributed.)

9 ANOVA

Random samples only produce statistics (estimates of parameters), and statistics are variable due to chance. In this chapter we view the standard deviation as an average variability of all the data. But we also take the variability among[1] the means of a number of random samples into consideration. Even the difference between a mean, $\bar{x}$, of a random sample and an actual parameter, μ or hypothesized μ_0, under random sampling, can be viewed as likewise a variability. From such a viewpoint, even Student's t test and the 'Z test' could be seen as attempts to decide whether there is one population or two by comparing variabilities. In statistics, we often measure variability as variance, the sums of squared differences divided by their degrees of freedom.

The *analysis of variance*, that has the acronym *ANOVA*, is actually a family of inferential tests that compare the variability among treatments, that feature a **separation of variances** due to different factors (Section 9.1). As a test, the ANOVA examines the differences between the resulting partitioned variance estimates from samples. It was, like most parametric tests, conceived with normally distributed raw data in mind . It is customarily evaluated by using a distribution family called the 'F distribution. Soon after its appearance, workers assumed that because the ANOVA involves a variety of means, that the CLT would lend it sufficient robustness to be freely used as an overall test for three or more treatments at a time. However, there is controversy concerning how well this assumption of robustness is met by the ANOVA, and it may get worse. Likewise, the global forms of the ANOVA have additional assumptions not included in simpler tests. Although false significance may occur more often when the assumptions of a test are not met, it is commonly assumed that violations of the assumptions of a global ANOVA usually manifest as varying degrees of reduction in power. However, violations of the assumptions of a global ANOVA can also result in false significance. In any case, the requirement that there be a statistically significant ANOVA before proceeding to **post hoc** (in the sense of not preplanned) multiple comparisons may need to be reconsidered, but remains (in 2020) a widely accepted requirement for now (Chapter 10). Effect sizes for the ANOVA are discussed in Section 9.5.

Because the ANOVA is sensitive to non normality, it might be best to consider using only the simpler ANOVA designs with a lower significance level of $\alpha < 5\%$, or the nonparametric alternatives to the ANOVA, especially the Kruskal-Wallis test (Section 14.3). If you must use blocks, consider Friedman's test (Section 14.5), instead of the RCBD ANOVA. Although these non-parametric tests are sensitive to unequal variances between the treatments, they are much less sensitive than the ANOVA. Understanding this chapter will help you to also understand and apply these nonparametric tests.

Even if you never actually use an ANOVA in your own research, this chapter will provide you with many useful concepts. Understanding the processes involved in the ANOVA will also get you on the right track if you later consider many of the wonderful up and coming procedures in the field (or subfield) called 'data science'. At the moment, data science is a powerful tool for anyone, but especially for those who seek to rule and influence the world, and for those who seek to curb the greed of those who seek to rule the world.

---

[1]The explanations for the use of 'between' and 'among' are well described online.

9.1 Partitioning Sums of Squares for Testing

The ANOVA is a simultaneous inference test, also called a global test or, because variances are in question, an 'omnibus' test. It is used for inference to test more than two treatments at a time to see if any differences exist; it does not tell which treatment is statistically significant from which. A global test, or omnibus test, might used to try to discover if at least one of the five most common medications for controlling the effects of schizophrenia in outpatients is more cost effective than the others, but if one or more are more cost effective, it won't show which. However, technically, the phrase **omnibus test** has a more restricted definition. It refers to a global test that specifically involves the comparison of variances; however, global and omnibus are commonly used interchangeably, although sometimes incorrectly.

But then it is only logical that you would still want to know which treatments, if any, are statistically significantly better than which other treatments. Pairwise comparisons can follow, but they are in addition to the global test itself.[2] Such multiple comparisons, that compare effects of one treatment to another, tend to be expensive in terms of the number of replications required. They become more expensive as the number of comparisons increases (Chapter 10).

The final testing stage of the ANOVA assumes that all the variances are the same and that any difference between the overall underlying variance of all the data and the apparent underlying variance among the treatments is only due to differences in their means. If the treatments (or other factors of interest) have different variances from each other, it messes up the test, so meeting the assumption of the homogeneity of the variances is essential. However, let's look at how the various variances are partitioned out for testing.

The **sum of squares**, $SS = \sum (x - \bar{x})^2$, also called the **sum of squared deviations**, is the numerator of the equation defining any estimated variance, $S^2 = \sum (x - \bar{x})^2 / df$ (the whole also called a **mean square**, MS). So, you can calculate the mean squares, the variance estimate, for each component. The process called **partitioning** separates the **total sum of squares**, SS_T, into two or more sums of squares, each with its share of the available degrees of freedom. This involves subtraction and often a multiplication to scale the various sums of squares, one to another. This is followed by division by the appropriate degrees of freedom to complete a variance estimate, the $MS_{whatever}$. The total of all the partitioned sums of squares together is $\sum SS_{whatever} = SS_T$. Likewise, the total of all the degrees of freedom of the partitions equals the total sum of squares, df_T.

For instance, $MS_T = SS_T / df_T$ is the total variance, the unaltered S^2, and is calculated much the same as any variance for all the data. Because the numerator of MS_T is a sum, it can theoretically be taken apart into component parts using subtraction. The calculation of one component, the **sum of squares among** the treatment means, SS_A, is an important step in most Model I ANOVAs. It is weighted by being multiplied by n to scale it to $\sigma_A^2 = n \times \sigma_{\bar{x}}^2$ to then be compared to the estimated σ^2. So, the estimator of the sum of squares among is

$$SS_A = n \sum_{i=1}^{j=m} \left(\bar{x}_j - \bar{\bar{x}} \right),$$

where m is the number of treatments, $\bar{\bar{x}}$ is the mean of all the data, and n is the (equal) sample size per treatment. Component sums of squares such as SS_A have their own degrees of freedom such as $df_A = m - 1$. That is, $MS_A = SS_A / df_A$ is a way of estimating σ^2 based on the sample variance among the treatments, assuming **H0** is true, meaning that they are all alike (no effect).

---

[2]Depending on approach, Students t and an F test with one degree of freedom can be equivalent. Pairwise comparisons subsequent to the global ANOVA can use F tests, each with one degree of freedom as though continuing the ANOVA, but modern practice tends to favor using subsequent t tests based on a variance estimate from the ANOVA.

Partitioning variance appears to be where the ANOVA excels. Although, in common application, the comparing of one treatment to another is not usually part of the ANOVA itself, the ANOVA family of tests feature this rather elegant process of partitioning. Partitioning allows you to get at the factors in which you are interested, and especially to eliminate the effects (the variance) of some factors in which you are not interested. It starts with the total sum of squares, SS_T, which is simply the denominator of the variance estimate for the whole sample before it is subdivided by partitioning. The partitioning is most commonly used to provide a better estimate, MS_W, of the theoretically underlying overall variance, σ^2. The mean square within would also seem to be a better estimate of the variance for comparing treatments, one to another at the df_w in subsequent 'multiple comparison tests' (Chapter 10).

Factors of interest can be chosen by you, but, the *sum of squares within*, SS_W, does not refer to a factor, but like factors, it refers to a component of the total sum of squares and is obtained by one or more subtractions from the total sum of squares. It is used to find the *mean square within*, $MS_W = SS_W/df_W$, which is an estimate of the underlying variance, σ^2, after the complication of a chosen factor, or factors, has been partitioned out. So, in application, MS_w is an estimate of the **effect of chance alone**. In any common kind of ANOVA, you estimate it by subtracting the factor sums of squares from the total sums of squares. So, what is left over after partitioning out various accessible factors (and their components, such as treatments), is the *sum of squares within*,

$$SS_W = SS_T - SS_{every\,thing\,of\,interest},$$

also called the *within sum of squares*. In a simple one way ANOVA, you would simply subtract the sum of squares among, SS_A, which represent the one factor of interest, the treatments, to find $SS_W = SS_T - SS_A$.

The degrees of freedom are also additive, and they correspond to their sums of squares. So, *degrees of freedom within* are left after other degrees of freedom are subtracted from the total degrees of freedom,

$$df_W = df_T - df_{every\,thing\,of\,interest}.$$

They are also called the *within degrees of freedom*.

Now you can complete the estimate of the variance of any particular factor or of the remaining variability to get the *mean square* for whatever factor,

$$MS_{whatever} = \frac{SS_{whatever}}{df_{whatever}}.$$

The sum of squares among, SS_A, refers to the treatment factor and is calculated from the means. The sample size n is used to finish the estimation of the assumed underlying variance, σ^2, by

$$MS_A = \frac{SS_A}{df_A} = \frac{n \sum_{j=1}^{j=m} \left(\bar{x}_j - \bar{\bar{x}} \right)}{m-1}.$$

In a simple ANOVA, the mean square among, MS_A, could be seen as an attempt to estimate the underlying $\sigma^2 = n\sigma_{\bar{x}}^2/df_A$ if **H0** is true, otherwise it estimates the effect of chance plus the effect of the treatments. The $_A$ is for among and the MS_A is the overall variance estimate from the difference *among* the treatment means. After partitioning, the sum of squares $SS_W = SS_T - SS_A$, and $df_W = df_T - df_A$, and the mean square within is found by

$$MS_W = \frac{SS_W}{df_W}.$$

So, for the simple one way ANOVA, $df_W = df_T - df_A$ could also be written $df_W = (\mathbf{n} - 1) - (m - 1) = \mathbf{n} - m$, where $\mathbf{n}$ is the total number of things in the whole experiment and m is the number of treatments. Because partitioning provides for so many different approaches and designs, the ANOVA is really a whole family of tests.

The final testing stage uses an F test to see if the variance among the means, MS_A, is statistically significantly greater than MS_W.

$$F = MS_A / MS_W$$

tests whether the there is a difference due to the treatments. This F is usually evaluated for statistical significance based on a criterion, α, by using $p >$ tables of F_{α, df_1, df_2}, such as Appendix 6, in which the subscript $_1$ is for the numerator (top) and $_2$ is for the denominator (bottom). In some of the applications they are $_A$ and $_W$ respectively. In application, when sample sizes are large enough, $F \approx 1$ would be taken as meaning no difference.

The ANOVA, when viewed as an omnibus test, proceeds as though the **variance** of all the data, **before** treatments are applied, starts off the **same**, $\sigma_1^2 = \sigma_2^2 = \cdots = \sigma$, regardless of the partition. This equality of all the variances is called **homoscedasticity**. Dividing the sums of squares by their degrees of freedom gives their mean squares, which are variance estimates. In the one way ANOVA, any effect of the treatments increases the among treatment variance estimate, $MS_1 = \widehat{n\sigma_{\bar{x}}^2}$, by changing the means in either the + or the - direction. That is, changes (deviations from the mean) increase the variance regardless of their direction. The object is to see if the means are different via their effect, if any, on the mean square, as compared to an assumed overall distribution of mean squares (that are due to chance). It is intended to reject **H0** if MS_1 is statistically significantly greater than MS_2, which can be tested by $F_{1,2} = MS_1 / MS_2$. That is,

$$F_{1,2} = \left(\frac{SS_1}{\sigma^2 \times df_1} \right) \Big/ \left(\frac{SS_2}{\sigma^2 \times df_2} \right) = \left(\frac{SS_1}{df_1} \right) \Big/ \left(\frac{SS_2}{df_2} \right),$$

by canceling out σ^2. Although the F test when used in an ANOVA is in one direction, that of greater variability, the underlying directions in which the means can vary are larger or smaller; either will increase MS_1. So, we use α as the criterion (**not** $\alpha/2$).

The assumption of equal underlying variances for all the factors (homoscedasticity), such as treatments in an ANOVA, is sometimes too large a leap of faith. The **global ANOVA** has also proved to be **more sensitive to deviations** from the **assumption of normality** than have other popular tests such as the two sample t test and its equivalent, the two sample ANOVA. For relatively small samples the global ANOVA is sensitive to any deviation from normality, but even for larger samples, a particular problem seems to be a lack of robustness with respect to skew. This might be due to whatever is causing the skew being distributed unequally among the partitions, or maybe it involves additional sources of confounding. More complex ANOVAs such as the two way ANOVA (Section 9.4) include more partitioning and seem to experience more confounding due to non normality such as skew.

In several places in their classic and still accepted text, *Statistical Methods*, Snedecor and Cochran (1967) indicated that they did not trust the Central Limit Theorem to provide the assumed normality for any form of the ANOVA if the $df_W < 20$. Even this may be too liberal. Unfortunately versions of the ANOVA using alternatives to the F test need further consideration.[3] In many, perhaps most instances, it might be better to use a nonparametric substitute for the ANOVA (Section 14.3 or Section 14.5). However, the very widely used classic one way ANOVA with a final F test will be presented in Section 9.2 with more detail and with a worked example.

---

[3] An extensive collection of papers on this matter can be found by entering the words *normality* and *ANOVA* into Google Scholar. Some of these papers suggest that even when the sample size of the treatment samples is reasonably big, the harm is most extreme with respect to lack of symmetry of the distribution.

9.1.1 Model I, fixed effects ANOVAs

A vocabulary is necessary for what follows and to read related material about the ANOVA. As you may recall, a parameter in the context of inferential statistics (testing by sampling) describes a population. Examples of parameters are the mean μ, a central measure, defining where the center (in the sense of the middle) is, and the standard deviation σ (or its square the variance, σ^2) which describes its variation (or squared variation), both of which refer to a sort of average absolute distance that chance, ϵ, makes the sample values dance around a center such as μ. ***Effects*** are the outcomes of different conditions and treatments. Although our attempts to measure them are distorted by chance because we cannot sample everything, we generally accept estimates of parameters such as μ and σ as numerical measures (or descriptions) of effects. A model I ANOVA is a ***fixed effects model***, meaning that these parameters are unchanging throughout because the treatments are assumed to be applied in a consistent manner, with no additional variation. That is, we assume that chance has an effect that is independent of the treatments and hence their effects. (There are many ways to design and partition an ANOVA, giving rise to references to the model II, or random effects model ANOVA, in which we assume that the various means and variances each contain their own element of chance. There is a debate about the definition of a model II ANOVA, but the 2012 edition of the Sokal and Rohlf textbook show how a version of what might fairly be called a model II random effects model can be used to sort out the contribution of different aspects of a combination of treatments in a relatively large pilot study, and hence to decide where to place the emphasis in designing the final experiment.)

A ***factor*** is a class of things that potentially affects an outcome, In fixed effect models (model I), factors can include two or, usually more treatments within them. For instance a factor could be different family structures potentially affecting the percent of the male children that eventually serve terms in prison. Another factor could be the particular awards an artist has received as potentially affecting the price of that artist's art. In fixed effect models we proceed as though there is no variation in the treatments within such factors. The treatments in a factor should not be such that they can be ordered such as income levels. The simplest fixed effects model is the ***one way*** ANOVA, a one factor example of a ***Model I ANOVA,*** which is generally used to compare samples of $m \geq 3$ treatments (where m is the number of treatments) to see whether they all represent the same population or separate populations (Eisenhart 1947, Snedecor and Cochran 1967).

If we number the treatments $j = 1, \ldots, m$ so that the sample mean $\bar{x}_j$ represents the effect of a particular single treatment $_j$ in a factor, let's call it factor A (its effect in x_j is a_j), then the fixed effects model says that the sample mean for treatment $_j$ is

$$\bar{x}_j = \mu + a_j + \epsilon, \tag{9.1}$$

where μ is the overall population mean (including all treatments and factors) estimated by the grand mean (the overall mean), the **random effects** are represented by ϵ (epsilon), and a_j is the effect of treatment $_j$. The random effects, ϵ, are assumed to be present with or without the treatments and they theoretically average out to 0. Any treatment $_j$ is assumed to always have the same effect, a_j, on everything to which it is applied.[4] This model is used for a **one way ANOVA** because there is only **one** factor, that includes two, or (usually) more individual fixed (non varying) effects. In this simple model, factor A is the treatment factor, which can include actual treatments, or may include observational classes, but they must be repeatable by you or others. This model requires that they be sufficiently under the control of the researcher. That is, you could be measuring the growth rate of plants

---

[4]A one way model II ANOVA would assume that the treatments contain variation within them, so its model is $\bar{x}_j = \mu + (a_j + \varepsilon_j) + \epsilon$.

that you expose to different colors of light, a wavelength factor, or you could be timing how long it takes self presenting pedestrians with differing mobility impairments to cross Broadway at the corner in front of Macy's, a traffic factor. Either one can be duplicated by you or by another experimenter or observer.

If **H0** is true with respect to treatment $_j$ of factor A, then $a_j = 0$; the treatment has no effect and has an expected mean equal to the hypothetical population mean $\mu_{a_j} = \mu$. However, the effects of chance, ϵ, are assumed to be **normally distributed**. So, even though the fixed effect $a_j = 0$, the effects of chance, ϵ, remain and that means $\bar{x}_j$ is still unlikely to come out exactly μ_j. Therefore, the ANOVA generally finishes with a stochastic (meaning somewhat chance dependent) estimate from which you might infer statistical significance if a real difference exists.

If there is another factor, B, the effect of its treatment $_k$ can be called b_k, and, although still Model I, it is a **two way ANOVA**. Its fixed effects model is

$$\bar{x}_{j,k} = \mu + a_j + b_k + \epsilon.$$

One factor could be hospital wards and the other could be sanitary regimes. The number of factors in an ANOVA can, **in theory**, be expanded to include as many factors as are economically and logistically practical to examine, so it could be a three way or four way etc. ANOVA.

The Model I ANOVA tends to ignore the possibility of **heteroscedasticity**, the possibility that the variability is unequal across the range of values of the included variables (such as treatments). There exist tests for this condition, given sufficient data and an amenable experimental design, but that requires additional data that cannot be included in the analysis. However, if the treatments can be defined and repeated with sufficient precision (minimizing outside sources of variability), this problem might be reduced in application.

The ability of the sums of squares to be partitioned, although limited by the availability of degrees of freedom and by the limitations of practical experimental design, seems to offer a veritable Lego or Meccano set for analytical ANOVA models. However, if the design is too elaborate, mishaps (plant or animal deaths, mistakes, equipment breakdowns, etc.) may cause both analytical and validity issues.[5]

9.2 One Way ANOVA

In application, the **fixed effect *one way***, or ***single factor***, **ANOVA** tests for differences among a group of classes or treatments, that are variations on a single factor. The one way ANOVA **usually** uses a ***completely randomized design***, a **CRD**, meaning that the samples are taken completely at random and that the treatments are applied completely at random. For instance, if you were testing a new painkiller against a control, such as a placebo, you could select your subjects at random and then apply the treatments at random in a double blind manner. It is a Model I analysis of variance, meaning that you assume that there is an overall variance, σ^2, that is the same for all the treatments, but that the total variance can include variance due to the effects of differences in the centering (the means) among treatments.

The original idea by R.A. Fisher[6] (1925) to create the ANOVA was quite clever. Recall that the Central Limit Theorem says that given that a distribution has a finite (as in fixed) mean, μ, and finite standard deviation, σ, then as the sample size, n, becomes larger, the distribution of the sam-

---

[5]Cochran, G.G. and G.M. Cox's *Experimental Designs* (1950) is a classical masterpiece on the subject of experimental design, in particular as it relates to the ANOVA. Although, in common with many older works, it tends to neglect the issue of Type II error, and power, it is still informative.

[6]Fisher also made many outstanding contributions to the study of evolution.

ple means will approach a normal distribution with mean $\mu_{\bar{x}} = \mu$ and standard error of the mean, $\sigma_{\bar{x}} = \sigma/\sqrt{n}$, even when the underlying frequency distribution is not normal (Subsection 4.3). This implies that the resulting distribution of means will have a variance of $\sigma_{\bar{x}}^2 = \sigma^2/n$. **If the means are really from the same population**, multiplying the sample variance of the means by the sample size would be yet another estimate of the underlying variance, σ^2. That is, $\sigma^2 = n(\sigma^2/n)$ **when** there is no difference between the treatment means, $\mu_1 = \mu_2 = \cdots = \mu_m$. Having designed his experiments or observations to obtain a suitable sample, Fisher then tested the resulting variance estimates to see if there was statistical significance between his estimate, MS_W, of the σ^2 of the untreated data, and his estimate, MS_A, of n times σ^2/n (where MS_A is a variance that can be directly estimated from the sample means of the treatments).

It is rather surprising that the ANOVA, with more than two treatments, is somewhat sensitive to non normality, because the CLT upon which it is founded should help; the sums of squares divided by the df seem to qualify as means. But since it is sensitive to skew, it is alleged to be necessary to observe the rule that requires $df_W \geq 20$. Actually the rule should probably require that $df_W \gg 20$, and **even that might not prevent problems**.

In application it is best to get the same sample size for each treatment, so let's initially use the assumption that all treatments have the same sample size, n. Later we will adapt for unequal sample sizes. The principles are the same. Note that MS_W is an estimate of σ^2 that is found by eliminating other factors, while MS_A is also an estimate of σ^2 based on the means, but MS_A is subject to the effects of differences among them. If there exists variation that is due to a difference of one or more means from the others, that difference is additional variation in the mean square deviation among the means. **It will increase** the estimate, $MS_A = n\widehat{(S^2/n)}$, and in that case it will **over estimate** σ^2.

So, after partitioning, the idea that there is only one population, hence only one source of variance, can be tested, using $F = MS_A/MS_W$. In this simplest ANOVA, the one way ANOVA, the idea is to test whether the MS_A is statistically significantly larger than MS_W:

$$F = \frac{Mean\ Squares\ Among\ treatments}{Mean\ Squares\ Within\ treatments} = \frac{SS_A/df_A}{(TSS - SS_A)/df_W} \tag{9.2}$$

Put more succinctly, $F = \frac{MS_A}{MS_w}$, $\quad df_A = m - 1$, $df_w = \mathbf{n} - m$. This is used to test

H0: THERE IS NO EFFECT (no differences other than variations due to chance).

That is, one challenges the hypothesis that there is no difference, other than that due to chance, ϵ, between the variance within, σ_W^2, and the variance among, σ_A^2. If the F test is statistically significantly different, one rejects **H0** and accepts

HA: AT LEAST ONE OF THE TREATMENTS HAS AN EFFECT.

That is, this alternative is that there is no difference among the means. But if F is not statistically significant, then, as always, the absence of statistical significance proves nothing, it does not even lend support to **H0**. When entering the most common tables of F, $df_1 = df_A$, and $df_2 = df_W$, with cutoff equal to the criterion, α.

For an explanation of how to apply a one way ANOVA, we will initially accept that the power is largely controlled by the choice of sample sizes. Later we will consider β and $1 - \beta$ as they relate to the ANOVA, with respect to the actual choice of sample size (Section 10.4). Do not forget that where possible, ANOVA experiments, including observations, should be designed to have **equal sample sizes**. However, that occasionally cannot be done, as when it is only possible to get a limited sample size for one treatment, or when lost data must be adjusted for. Then a different n_j must be used for each treatment:

$$SS_A = \sum_{j=1}^{j=m} n_j \left(\overline{x}_{A_j} - \overline{\overline{x}} \right)^2. \tag{9.3}$$

In more ambitious Model I ANOVA designs, it is sometimes possible to do more subtractions of sums of squares and of degrees of freedom on the way to a theoretically even purer estimate of σ^2 as MS_W.

The partitioning of effects to get a more specific estimate of σ^2, the MS_W, seems to be the great advantage of the ANOVA. Note that in some publications, the **mean square within**, MS_W, can also be called the ***mean square error***, MS_E, a potentially misleading designation.

The ANOVA is generally presented as a table; Table 9.1 presents two notations for a formal ***one way ANOVA table*** for **equal sample sizes**:

Sources of Variation	df	SS	MS	F
Among treatments	$df_T - df_w$	$n \sum_1^m \left(\bar{x}_i - \overline{\overline{x}} \right)^2$	nSS_A/df_A	MS_A/MS_W
Within	$n - m$	$TSS - SS_A$	SS_W/df_W	
Total	$n - 1$	$\sum_1^m \sum_1^n \left(x - \overline{\overline{x}} \right)^2$	$MS_T = SS_T/df_T$	

Sources	df	SS	MS	F
Among	df_A	SS_A	MS_A	$F = F_A$
Within	d_W	SS_W	MS_W	
Total	df_T	TSS	MS_T	

Table 9.1: Equivalent One Way ANOVA Tables

Note that the mean square total, MS_T, is not directly used in the F test.[7]

Instead of the word 'among', many authors use the **less** confusing but **less** correct word 'between'. However, there is a potential notation, or mnemonic, problem involving the first letter of the word between and it later confuses rather than helps students.[8] Notation may vary from table to table, but the evaluation of F is based on both the MS_A and the MS_W, so it can be designated F_{df_A, df_W}, and it is also commonly represented by F_{df_1, df_2}, as it is in tables such as Appendix 6.

Question 3 provides you with a chance to try this yourself and with another example of a one way ANOVA.

9.2.0.1 Example:Worked Model I One Way ANOVA

The search for and identification of human pheromones is an ongoing area of research. (But this data is fictitious.)

Objective and Hypothesis: Amadora's perfume research laboratory was trying to discover a new pheromone to attract males to the users of their perfume. They had developed an index of attention, and while developing their instrumentation, they found that attention, their measure of attraction, was relatively close to normally distributed. (Well, it was mound shaped and almost symmetrical.) They used the same simple cologne base as the carrier for each pheromone treatment and by itself for the control. They had two candidate pheromones with the romantic names D and E. The control was called C. They formulated the following hypothesis pair:

---

[7]The MS_T is larger than the MS_W because it includes the variance that was partitioned out. That is, MS_T is assumed to be a poorer estimate of the underlying σ^2.

[8]In discussing a one way ANOVA, avoid using the uppercase letter B carelessly; for instance, it could, depending on the author, mean between treatments or between blocks or simply block. Treatments definitely are not blocks.

Table 9.2: Perfume Test Data Means and Sums of Squares

C	SS	D	SS	E	SS	
28.74	219.4581	17.69	1.8162	11.92	0.7877	
18.32	19.3375	16.91	0.3188	16.00	24.6800	
11.30	6.8690	20.39	16.3231	12.13	1.1933	
12.93	0.9861	6.49	97.1856	19.39	69.7838	
15.42	2.2210	12.30	16.3479	6.02	25.1146	
12.26	2.7818	15.22	1.2639	9.62	2.0002	
11.51	5.8516	18.85	6.2634	10.77	0.0711	
9.80	17.0495	29.07	161.9760	17.60	43.1136	
20.32	40.9310	15.54	0.6414	6.00	25.4089	
25.49	133.8105	25.26	79.4650	13.00	3.8435	
14.38	0.2109	9.36	48.8171	17.91	47.3110	
17.61	13.5949	19.39	9.2818	5.71	28.3315	
12.10	3.3402	8.03	69.1838	7.35	13.5788	
10.75	10.1089	12.20	17.1742	9.67	1.8787	
8.56	28.8248	10.73	31.5161	6.86	17.4517	
12.96	0.9388	17.37	1.0531	3.77	52.7797	
7.50	41.3008	19.45	9.6391	3.15	62.1968	
21.71	60.5774	8.41	63.0287	9.13	3.6467	
11.11	7.9326	15.58	0.5868	9.86	1.3876	
16.00	4.2862	21.84	30.1760	17.41	40.5912	
12.16	3.1007	15.95	0.1549	16.15	26.1993	
13.95	0.0009	16.06	0.0819	3.80	52.3406	
13.68	0.0612	20.24	15.1375	12.10	1.1401	
18.47	20.6490	15.47	0.7652	13.76	7.4126	
8.77	26.5624	11.84	20.2866	14.02	8.8856	
6.09	61.3510	19.60	10.6115	10.27	0.5797	
8.21	32.6992	19.66	11.0207	8.99	4.1844	
15.43	2.2736	20.95	21.2302	13.12	4.3452	
11.00	8.5459	14.80	2.3747	16.05	25.1829	
11.23	7.2584	15.70	0.4210	9.54	2.2510	
417.75	782.9139	490.35	744.1421	331.07	597.6719	Sums
13.9249		16.3450		11.0358		Means

H0: THERE IS NO DIFFERENCE AMONG THE TREATMENTS.

HA: THERE IS A DIFFERENCE AMONG THE TREATMENTS.

Criterion: $\alpha = 5\%$. (The F test has a potential for 2 tails for more and less variance, but, as in the ANOVA, it can be seen as referring to **increased deviations** from the means **regardless of direction**. That is, in the ANOVA, you are generally only interested in greater + or - deviations from the mean, so you **just use** α.)

Methods: They carefully scrubbed $n = 90$ women of the targeted age range and appearance and randomly assigned them to $m = 3$ equal size treatment groups of $n = 30$ each. The women had been screened to be certain they and their family members had never worked for Amadora or its competitors. This eliminated certain types of bias. Depending on which treatment group they fell into, they were anointed with treatments C, D, or E in a double blind random manner. They were then dressed identically and tested by being randomly exposed to the same randomly selected panel of similarly screened men who were all attached to the same attention-o-meter that showed a collective level of attention. The data appears in Figure 9.2.

Analysis: The results are presented in Figure 9.2. The individual treatment means were $\bar{x}_C =$

13.9249, $\bar{x}_D = 16.3450$, and $\bar{x}_E = 11.0358$. So the grand mean, their average, $\bar{\bar{x}}=13.7686$, was found by adding them together and dividing by $m = 3$. That is, the mean of all the means is the mean of the whole sample. The estimate of $SS_A = n \sum_{j-1}^{j=m} \left(\bar{x}_j - \bar{\bar{x}} \right)^2$ (where $n = 30$ here) may seem a little strange, but remember that MS_A estimates $n \times \sigma_{\bar{x}}^2$, which is **only** expected to equal σ^2 **if there is no treatment effect**. The rest of the details were recorded in the following ANOVA table:

Sources of Variation	df	SS	MS	F
Among treatments	2	423.91074	211.95537	$F = 10.8419$
Within	87	1700.81722	19.54962	
Total	89	2124.72796	23.87335	

Results and Discussion: Referring to a table of F such as Appendix 6 with $\alpha = 5\%$, $df_A = 2$ and $df_W = 87$, the researchers noted that any $F_{\alpha=0.05, df_1=2, df_2=87} \geq 3.1013$ was statistically significant **with these** numbers of degrees of freedom. So, there is a statistically significant difference at the $\alpha = 5\%$ level. In this kind of table it was very useful to have entered all sums of squares and mean squares to the same number of decimal places. Rounding was avoided, at least until all the sums of squares were calculated. Otherwise one of the estimates could have been seriously damaged in a subtraction procedure. **NOTE:** Statistical significance in a global test such as an ANOVA simply indicates that there is difference **among** treatments, **not between which** treatments. The difference could be between an apparent attractant and an apparent repellent, or between one of the treatments and the average of the other two, or between one and the grand mean. In order to look for more specific statistically significant differences, it is necessary to consider multiple comparisons, which will be covered in Chapter 10. The hypothesis was kept very simple so that the researchers could say they at least observed a difference, even if they were mistaken about the cause (Subsubsection 3.4). Not all statisticians agree with this view, but it is not always known what is causing the difference, even if it can be proved that a difference exists.

The researchers chose a panel of male applicants whose families had never worked for Amadora or for their competitors and who could demonstrate a sense of smell to be used with the attention-o-meter. Contrast this with what might have happened if they had used men from their own laboratory, or from their suppliers. These men might have had biases such as foreknowledge, a wish to please, or a specialized taste in smells. If the experiment is done correctly, a technician without a sense of smell would be chosen to apply the pheromone without knowing what treatment was what.

CAUTION: This is a common way to analyze data where researchers have created their own scale **without knowing** if it is an **equal interval scale**. But this is **usually wrong**, as is likely in the above example, because we have no way of being certain it is on an equal interval scale. Parametric tests assume an equal interval scale; that is, they need a measurement scale upon which, say, 10 units is really twice 5 units, and 3 units plus 8 units actually is a meaningful 11 units. (Maybe someday technical advances will make it possible to determine such a measurement scale for attention.) Nonparametric tests such as the Kruskal-Wallis test (Section 14.3), followed by subsequent comparisons between treatments using the Wilcoxon test (Subsection 14.1.1) would be **more likely to be valid**. All they require is a reasonable tendency towards symmetry and that the values be ordered; 10 has to be less than 11, and 5 has to be less than 10. But let's let it go on through all the processes that are otherwise correct and then we can see how well it works out using **preferable** nonparametric tests (Chapter 14).

Under **H0** the value of F would ideally approach $F \approx 1$ because if the means are equal, then the estimates of the variances, MS_A and MS_W, should be estimates of the same[9] variance. If $F < 1$, MS_A is less than MS_W, and the researchers might suspect some sort of non randomness in part of the experiment. It often means there is a computational error (such as premature rounding), but it is sometimes associated with an error in the planning of the experiment. To check this the researchers might even see if MS_W/MS_A is so large as to be equivalent to significance at the $\alpha = 5\%$ level of F_{df_W, df_A}, in which case there is reasonable justification in suspecting that something other than chance has gone wrong.

A frustrating fact about ANOVAs with more than two treatments is that they do not really compare individual treatments with other individual treatments. After running a global ANOVA, you are likely to wish to break the results down into pair by pair multiple comparisons (Chapter 10). You can do this either as Student's t tests or as equivalent two treatment F tests. Either will use df_W, and the MS_w as the denominator. However, chance is again going to rear its ugly head and require you to make expensive (in the sense of sample size) cutoff adjustments (Section 10.1.2 and Section 10.1.3).

> **CAUTION:** The rules regarding rounding during statistical calculations do not necessarily follow the rules of thumb used in the calculations for some engineering applications. In calculating any ANOVA, minimize or better yet, avoid rounding until **after the subtraction** to find SS_W or SS_A. Although it did not happen in the above example, leading figures can be used up in the subtraction so that only the trailing figures remain. **If you have rounded too soon**, statistically significant results could mistakenly be called statistically nonsignificant.

9.2.0.2 One way ANOVA table for unequal sample sizes

The MS_A is calculated a little differently when the sample sizes are unequal. In that case the ANOVA table changes to:

Source	df	SS	MS	F
Among	$df_A = m - 1$	$\sum_{j=1}^{j=m} n_j \left(\bar{x}_j - \bar{\bar{x}} \right)^2$	$\dfrac{\sum_1^m n_j \left(\bar{x}_j - \bar{\bar{x}} \right)^2}{df_A}$	MS_A/MS_W
Within	$df_W = \mathbf{n} - m$	$SS_T - SS_A$	$MS_W = SS_W/df_W$	
Total	$df_T = \mathbf{n} - 1$	$\sum_{j=1}^{j=m} \left(\sum_{i=1}^{i=n_j} \left(x_i - \bar{\bar{x}} \right)^2 \right)$	$MS_T = SS_T/df_T$	

Table 9.3: Unequal Sample Size ANOVA Table

Especially note the subscripted n_j and its location in the calculation of MS_A, nothing else is changed. Incidentally, this formula for MS_A also works for equal, meaning unchanging, n_j. In the table for unequal sample sizes, the mean square within is found by subtraction, after using $\sum_{j=1}^{j=m} n_j \left(\bar{x}_j - \bar{\bar{x}} \right)^2$ to calculate the SS_A, which is then subtracted from the TSS to get the sum of squares within, SS_W.

---

[9]Because of the effects of estimation, under **H0:** THERE IS NO TRUE DIFFERENCE, the value of F for the ANOVA actually is approximately $\mathbf{n}/(\mathbf{n}-2)$, but as $\mathbf{n}$ becomes large, then this ratio approaches 1.

9.2.0.3 Alternative formula for one way sum of squares within

Before going on, let's examine an alternative way to calculate the sum of squares within directly,

$$SS_W = \sum_{j=1}^{j=m} \sum_{i=1}^{i=n_j} (x_i - \bar{x}_j)^2, \tag{9.4}$$

where m is the number of treatments and n_j is the number of individuals in treatment $_j$. As before, the mean square within, or mean square error, $MS_w = SS_w / (\sum_1^m (n_i - 1))$. Of course, when sample sizes are equal, $MS_w = SS_w / (\mathbf{n} - m)$. Notice that by using the **actual sample mean**, $\bar{x}_i$, of each class (treatment) instead of the grand mean, this estimate of the variance actually sums the errors from within the individual treatments. It also produces the same result as is obtained by subtracting the sum of squares among, in the single factor ANOVA. That is, it estimates the underlying variance, σ^2, without including the effect that can be introduced from among treatment means. You can even use this estimate of SS_W in this simple one factor situation to find the sum of squares among treatments, using $SS_A = TSS - SS_W$ in a single factor (one way) ANOVA. However, it is still widely believed that when you get to the two way ANOVAs, subtracting the sums of squares for more factors, when appropriate, and with the appropriate formula modifications, may aid in perfecting your estimate, SS_W, as an estimate of the true σ^2. The ANOVA is mathematically elegant, but in application, data rarely if ever meets the assumptions of this test.

9.3 Heteroscedasticity

The word *heteroscedasticity* means that there exist sub-populations, such as some of the variables, that have different variances than others.[10] This violates the assumption of the ANOVA that all the variables have the same variances. Unequal sample sizes magnify any effects of heteroscedasticity and of non normality.[11] Minimize heteroscedasticity by minimizing outside sources of unrelated variation in the design of your experiment or observation plan.[12] The **sensitivity of the ANOVA** to heteroscedasticity and to skew are reasons why you might want to consider nonparametric equivalents of the ANOVA (Section 14.3 and Section 14.5).

9.3.0.1 Two Papers Seeming to Justify the ANOVA

A commonly cited paper in defense of the ANOVA is Blanca et al. (2017) who ran massive Monte Carlo simulations. "In order to examine the effect of non normality on F-test robustness, a one-way design with 3 groups and homogeneity of variance was considered." Like most other studies they focused on Type I error; they did not consider Type II error (the power aspect). However, they used Bradley's (1978) liberal criterion that a statistical test is considered robust if the empirical Type I error rate is between 0.025 and 0.075 for a nominal alpha level of 0.05. Thus they went from accepting one false positive in twenty tests to accepting **one false positive in thirteen tests**. By accepting this, they concluded, "The results showed that in terms of Type I error the F-test was robust in 100% of the cases studied, independently of the manipulated conditions."

---

[10] You also want to avoid heteroscedasticity in linear regression.

[11] Papers on the subject of heteroscedasticity relative to the one way ANOVA include Rice and Gaines (1989).

[12] To check that two variances are equal to each other, some researchers use Levene's test which is available in popular software suites, but it is **not recommended** because it is only valid under the assumption of normality, and it collapses as soon as one variable deviates even slightly from the normal distribution (Hayes and Cai 2007, Rakotomalala 2008).

A less cited study, also by Blanca et al. (2018), focuses on heterogeneity of variance with respect to Type I error, using massive and comprehensive Monte Carlo simulated data in the ANOVA. They found a necessity to study, and usually manipulate, the variance structure of the data before performing the actual ANOVA. Then in the majority of cases (in some instances the data proved totally unsuitable) the ANOVA, using this pre-manipulated data, was robust (using the Bradley's 1978 definition of 0.05) with respect to Type I error. They acknowledged that neither their study nor the literature in general considered such matters as power, and they were also aware that some researchers would see this process as unacceptable pre-test data exploration. They concluded:

> "Whatever the case, we encourage researchers to analyze the specific characteristics of their design and the data obtained and, if their data do not meet the assumption of variance homogeneity, to choose the best alternative in order to obtain valid results. To this end, the best approach is to perform a simulation study involving the specific conditions of the real data so as to determine whether or not F-test is robust in the situation being considered. We are aware, however, that applied researchers are not usually familiarized with this procedure."

Upon careful reading and even a rereading, **neither** of these well researched and recounted papers should, as they seem to intend, encourage or justify using the ANOVA. Both papers refer to the simplest possible omnibus ANOVA (only 3 treatments) and both allow too extreme a deviation from even the extravagant $\alpha = 5\%$ (1/20) maximum; the 2018 paper also proposes a questionable a priori processing procedure that is beyond any but the most math oriented researcher. Indeed I am referring to these two papers as **evidence against using the ANOVA** for inference. It is likely that, as Blanca et al. extend their Monte Carlo simulations to include power, they will produce additional evidence that the ANOVA should not be used in this way[13].

9.4 Two Way Randomized Complete Block ANOVA Design

Early agricultural scientists compared treatments like different crop varieties in little squares like gardens that were called blocks. If they were comparing a group of chosen potato varieties for yield, each block got one row each of all the varieties chosen for comparison. They noted that due to soil and micro climate differences, different blocks got different results even if they were close together. So, they came up with the idea of replicating the blocks to have groups of blocks that were as identical as possible, except that the treatments were independently randomized within each block. Each block got all the same treatments, in this case, the chosen potato varieties, but they were each arranged in a different random order. The order in which the blocks were planted, hoed, etc. was also random.

This is now called a randomized complete block ANOVA design (a RCBD) and is analyzed by what is called a two way ANOVA, which reduces the variability of the results of the treatment comparisons by partitioning out (subtracting out) the effect of the variation due to the differences between these little garden-like plots. This makes the tests more powerful and is intended to compensate for the fact that the chosen varieties would be grown on different farms. So, this complete block test was an attempt to choose the best variety for growing under many different conditions by detecting a statistically significant difference. Statisticians would say that there were two factors here, the variety factor and the location (plot) factor.

---

[13]Researchers such as Blanca et al. have a right to discuss their conclusions and opinions so long as their procedures and research methods are adequately described, as they have been. However, I am of the opinion that their conclusions are grasping at straws on behalf of the ANOVA, a long established procedure, which when confirmed to be inadequate will leave a multitude of those of us who have used it and published its results for decades, with very red faces.

Now some agricultural researchers do experiments like comparing many potato varieties, with each of many farms getting all the treatments (all the chosen potato varieties) but they still refer to the farms as 'the blocks'. Later, medical researchers comparing different medicines would use the same approach, calling a reasonably large enough number of clinics 'blocks' and having the doctors apply a variety of different medicines (maybe with a placebo included) as treatments. Some research uses classrooms, factories, years, etc. as blocks. (Some publications refer to the treatments as the primary factor, and to the blocks as the secondary factor.)

9.4.0.1 Analysis:

Given that **H0** is true, the Model I two way ANOVA for the results of a given treatment in a given block of the RCBD is

$$x_{j,k} = \mu + a_j + b_k + \epsilon, \tag{9.5}$$

where μ is the true (untreated) population mean, a_j is the effect of the treatment factor ($_j$) of the main effect; b_k is the effect of block $_k$; and ϵ is the effect of chance. The effect of chance ϵ (epsilon) is the source of the underlying variance, σ^2, that is assumed to be the same throughout all the experiment.

The analysis is described by the RCBD ANOVA table, where the $_B$ subscript refers to blocks, and $_A$ refers to treatments, so m_A is the number of treatments, and m_B is the number of blocks. So, an additional source of variation, the blocks, is considered in the ANOVA table:

Sources of Variation	df	SS	MS	F
Among treatments	df_A	SS_A	MS_A	$F = F_A$
Blocks	df_B	SS_B	MS_B	(F_B)
Within	df_W	SS_W	MS_W	
Total	df_T	TSS		

This may be expanded to show how each cell arises:

Sources Variation	df	SS	MS	F
Among Treatments	$m_A - 1$	$m_B \sum_{j=1}^{j=m_A} \left(\bar{x}_{A_j} - \bar{\bar{x}} \right)^2$	$\frac{SS_A}{m_A - 1}$	$F_A = MS_A/MS_W$
Blocks	$m_B - 1$	$m_A \sum_{k=1}^{k=m_B} \left(\bar{x}_{B_k} - \bar{\bar{x}} \right)^2$	$\frac{SS_B}{m_B - 1}$	$F_B = MS_B/MS_W$
Within	$(m_A - 1)(m_B - 1)$	$TSS - SS_A - SS_B$	$\frac{SS_W}{(m_A - 1)(m_B - 1)}$	
Total	$n - 1$	$TSS = \sum \sum \left(x - \bar{\bar{x}} \right)^2$		

The additional contributor to variance is the **sum of squares for blocks** ,

$$SS_B = m_A \sum_{k=1}^{k=m_B} \left(\bar{x}_{B_k} - \bar{\bar{x}} \right)^2.$$

Notice that jargon also calls the different contributors to the variance, the treatments and the blocks, 'factors'.

The total sample size is $n = m_A m_B$, and the total degrees of freedom are $df_T = n - 1 = m_A m_B - 1$, where $m_{...}$ is the number of things (blocks, or treatments) in factor $_{...}$. So, with a little algebra, $df_w = (m_A m_B - 1) - (m_A - 1) - (m_B - 1) = (m_A - 1)(m_B - 1)$ is the within degrees of freedom. The gain from using the RCBD instead of a one way, *completely randomized design*, ANOVA is a reduction in error variance, while the cost is a decrease in error (within) degrees of freedom.[14] Or you

---

[14] Because $df_T = n - 1$ is the total of all the other degrees of freedom, $df_W = df_T - df_A - df_B$, you can demonstrate that $df_w = (n - 1) - (m_A - 1) - (m_B - 1) = (m_A - 1)(m_B - 1)$. Start at the end rather than at the beginning by multiplying out $(m_A - 1)(m_B - 1) = m_A m_B - m_A - m_B + 1$. Then take advantage of $-1 - (-1) = 0$ by adding, and subtract extra ones to get $(m_A m_B - 1) - (m_A - 1) - (m_B - 1)$, which is $(n - 1) - (m_A - 1) - (m_B - 1)$. So,

could say that some of the degrees of freedom are moved to the blocking factor. The gain in power associated with the decreased error variance (MS_W) outweighs the loss in power due to its reduced degrees of freedom. The resulting MS_W is almost always an **improved estimate** of the true variance, σ^2. A statistically significant F_B would be further evidence of this improvement, but its lack is not evidence against it. It should not affect your conclusions regarding the main effect, $_A$.

The above RCBD ANOVA table also serves as the format for the formal description in your research notes, and possibly, depending on your editor and policy, in your publication. Ideally, such material should at least be made available to your colleagues after your research is published. Even if the editor does not wish a complete ANOVA table, it could still be made available either online, as on a web page, or upon inquiry.

9.4.0.2 Example: Randomized Complete Block Design ANOVA

Let's look at an example from a not so well known journal to see how to do such an analysis. The student completed his master's degree, but eventually left his doctoral program to go into the antihero industry.

RCBD worked Gollum's thesis research involved using four kinds of vegetable waste as fertilizers and no waste as a control. All four were potentially good long term fertilizers, but unless vegetable waste is adequately composted (rotted), its decay when applied to the soil may take up all the nitrogen and some other nutrients to feed the decay organisms for the first season. Often, even though yields are depressed the first season, the second season for such 'green manures' can be excellent because this decay process locks up nutrients and thus reduces their loss through leaching[15]. However, the local agricultural stakeholders, although allegedly organic farmers, were not patient beings. They were more interested in applying fertilizer material for immediate cropping than for future gain. Gollum set up a complete block experiment with Abomocrop, a new patented crop that was all the rage for feeding prisoners. He formed a hypothesis pair,

H0: THERE IS NO TRUE DIFFERENCE AMONG THE TREATMENTS.

HA: THERE IS A TRUE DIFFERENCE AMONG THE TREATMENTS.
(Notice that there is really no difference between this and a hypothesis pair for a CRD ANOVA.)

Criterion: $\alpha = 5\%$

I	II	III	IV	V	VI
A5	A3	A4	A1	A1	A3
A1	A4	A1	A3	A3	A4
A3	A5	A2	A2	A2	A5
A2	A1	A3	A5	A4	A1
A4	A2	A5	A4	A5	A2

Methods: Gollum used a table of random numbers, and for each randomization he marked off the numbers he used, so he never used that table sequence again. He drew up a randomization plan (a map, above) of each of $m_B = 6$ experimental plots, randomizing the positions of a completed set of $m_A = 5$ treatments separately for each of these blocks. One of the treatments,

---

it comes back to where it started, and it demonstrates that $(m_A - 1)(m_B - 1) = df_T - df_A - df_B = df_W$.

[15]If the farmer has enough land to leave enough of it fallow on a rotation plan, applying a suitable amount of low grade vegetable waste just before fallowing reduces or eliminates the need for some kinds of chemical fertilizer. But you can overdo the amount of straw, etc. unless you intend a very long fallow.

which was designated 'A2', was the one Gollum chose to be the control; it would get just as much rototilling at planting time, but no vegetable waste. The order within the plots in which the seed bed rows were rototilled, with or without vegetable waste, was randomized. The order in which the $m_B = 6$ plots were planted and serviced was also completely randomized, as were any other operations. As per standard operating procedure (SOP), he went to the university's statistical consulting service to check that his plan was correct and if any improvements were possible. Even though the plan showed no evidence that Gollum had performed an a priori power analysis of any sort, the affable statistician was behind on his publication rate and just glanced at Gollum's work and said, "It seems OK."

It takes time to harvest each plot, so Gollum harvested one complete plot at a time, in random order, in case a few hours made a difference in crop weight, or in case a longer interruption occurred. Gollum's graduate adviser, Professor Snaggletooth, finished up for Gollum after he was taken away between plot harvests with sunstroke, but the harvest was otherwise uneventful.

Analysis: The results in kilograms per plot were as shown in Table 9.4. Filling in the ANOVA table

Table 9.4: Yield in Kilograms

Treatments		Block 1	Block 2	Block 3	Block 4	Block 5	Block 6	$\bar{x}_A$
A1		52.16	57.89	58.78	79.57	80.06	46.93	**62.56**
A2		37.50	35.44	51.36	51.07	37.21	39.94	**42.08**
A3		35.09	66.57	80.53	64.66	48.41	37.83	**55.52**
A4		49.12	56.36	49.91	42.18	56.58	48.82	**50.49**
A5		50.00	49.79	49.43	53.38	37.51	32.33	**45.41**
$\bar{x}_B$		**44.77**	**53.21**	**58.00**	**58.17**	**51.95**	**41.17**	$\bar{\bar{x}} = 51.21$

using data from Table 9.4 and a few calculations that should be familiar by now (preceding subsection), produced the following ANOVA table:

Sources Variation	df	SS	MS	F
Among treatments	4	1589.2551	397.3138	3.6984
Blocks	5	1206.4268	241.2854	2.2461
Within	20	2148.5698	107.4285	
Total	29	4944.2518		

So $F = 3.6984$ for the main effect, the treatments. Referring to a table of F such as Appendix 6, Gollum found the critical value $F_{\alpha=5\%, df_1=4, df_2=20} = 2.8661$, so the difference among treatments was statistically significant.

However, it is unclear if he was even referring to the blocks when he set the criterion, and at $\alpha = 5\%$ the difference among the blocks would not be statistically significant because $F_{\alpha=5\%, 5, 20} = 2.7109$, whereas he only obtained $F = 2.2461$. Some might take that as meaning that blocking is not terribly important. However, there is a small chance that there had not been enough power associated with testing the block differences to formally verify an existing useful advantage sufficient to warrant using up 5 degrees of freedom. Blocking is theoretically the best policy where there is an obvious possibility of a meaningful additional source of variation, as in horticultural global tests.

Results and Discussion The important fact is that the difference among treatments is **statistically significant** at the $\alpha = 5\%$ level, meaning that Gollum may reasonably accept that at least one of the 'fertilizers' produced a stronger effect than at least one of the others. Other than missing the Maying festival there in Mordor, Gollum had reason to be pleased because his results were statistically significant. However, the ANOVA itself did not tell him which treatment or treatments were statistically significantly different. Other than the fact that the weights were reported to a rather unlikely level of precision for field plot yield measurements, this design, including its replication rate, m_b (often just called b for blocks), is a rather commonplace agricultural research experimental design. The F value for the main (treatment) effect indicated a $p\_value = 2.07\%$ that could be intellectually interesting to the researchers, however, it is not something they should put in the publication because they **targeted** $\alpha = 5\%$, and it could have been chance that carried them that much lower than 5%. "However, one wonders, one does, my precious.[16] What was its power, $1 - \beta$, and for what increase or decrease of yield, D_{min}? Well we are not out of Mordor yet and we will return to this matter.

Most researchers ignore testing block effects, and that can be OK, unless they are wondering about modifying designs. The point is that the sum of squares within equals the total sum of squares minus both the sum of squares among treatments, SS_A, and the sum of squares among blocks, SS_B:

$$SS_W = TSS - SS_A - SS_B.$$

Statistical tests based on the normal distribution, at least internally, are dependent on the Central Limit Theorem. Thus Snedecor and Cochran (1967) repeatedly mention that the effect of the Central Limit Theorem on assuring normality for any ANOVA might be insufficient if the degrees of freedom $df_W < 20$. By that standard, they would consider this example to be just adequate with respect to assuming normality without further supporting evidence.

Question 6 provides an exercise with an answer at the end of the book for the RCBD ANOVA.

Multiway processes allowing more than two factors are possible, but in most applied situations, their experimental designs are difficult to manage because it becomes increasingly difficult to compensate for data loss. As the number of factors increases, there may also be a yet greater difficulty when skew is present, especially when the cause of the skew is distributed unequally among the factors (hence the numerical partitions). Heteroscedasticity and various sources of confounding may also be exacerbated by increased numbers of factors.

The RCBD can be seen as an extension of the paired t test (Section 5.4), which can be viewed as an RCBD with just two treatments. However, the paired t test is more robust and resistant to confounding. Unfortunately, too many treatments can sometimes also sabotage the assumption of homogeneity. That is, maybe you should not include additional treatments just to see what the outcome will be, or to try to make the experiment seem more impressive.

Ironically, experiment station blocks were contrived to be as like as possible with respect to variance, but actual farms are **not** likely to be as similar to each other as to have equal variances. Note that the Friedman test (Section 14.5) provides a nonparametric two way analysis with very little potential for confounding caused by interactions between the blocks. The Friedman test is somewhat weakened by heteroscedasticity, especially with respect to the treatments, but less so than the RCBD.

9.4.0.3 Missing data in the RCBD

One of the seemingly weird aspects of this design is that the number of blocks act as the sample size for testing the treatments, and the number of treatments act as the sample size for testing the blocks. Missing data is serious because it loses degrees of freedom as well as some of the information content, and it reduces the power. Thus a careless farm worker, a mechanical malfunction, or a human

---

[16] Apparently this researcher was suffering from PTSD, and lived in an antediluvian time variously chronicled by J.R.R. Tolkien, Wagner's Ring Cycle and various folk tales.

subject dropping out is always a serious threat to the validity of a complete block experiment. But, if there are dropouts or some plants or animals die during the experiment, it could be that the treatment contributed to that.

There exist various controversial analytical approaches to coping with data losses. Statistical computer packages often contain routines for computing ANOVAs with missing entries. However, be aware that results from corrected data are often considered less authoritative than an analysis from a complete data set.

A common approach to chance data loss is to estimate a surrogate for a missing value, $x_{j,k}$, where j is the row (the treatment as this book lays it out) number, and k is the column (the block) number. To estimate such a value (that will play a relatively neutral role in your test), multiply the sum of the row by the number of cells in that row and multiply the sum of the column by the number of cells in the column, while **temporarily setting the missing cell equal to** 0 for each summation. Then subtract the grand total (of all the cells) and divide the result by one less than the number of cells in the row times one less than the number of cells in the column, to get

$$\hat{x}_{j,k} = \frac{\left(m_A \sum_{i=1}^{i=m_B} x_{i,k} + m_B \sum_{i=1}^{i=m_A} x_{j,i}\right) - \sum_{h=1}^{h=m_B} \sum_{l=1}^{l=m_A} x_{h.l}}{(m_A - 1)(m_B - 1)}. \tag{9.6}$$

Now you can put your estimate, $\hat{x}_{j,k}$, in the empty cell before calculating the sums of squares. However, in order to estimate the mean square, you must **reduce the degrees of freedom** for the TSS **and**, as a result, for SS_W. That is,

$$df'_W = (df_T - 1) - df_A - df_B$$

(Snedecor and Cochran 1967). The $'$ mark is used here to indicate that it is an adjusted value.

Modern computer software can handle more than one missing value. However, if these multiple missing values total more than 10% of your data, there can be some question about the value of your results. Try not to have missing values, especially not in any approach more complicated than a one way ANOVA. The power loss will be greater than you might expect. One possible solution, if you can afford it, is to estimate how many blocks you need and add one if you have reason to fear the effect of data loss. That way you could even drop a whole block and simplify the analysis and how it is viewed. Even so, the dropping of a block would have to be explained in your publication or report.

Question 7 can be used both as an exercise and as an example of missing data in the RCBD ANOVA.

9.5 ANOVA, Fraction of Variance, ES Measures

Editors are sometimes quite rigid about requiring effect sizes. However, they might not be aware that a simple two sample F test can be seen as an effect size in itself because it is the fraction (which might be presented as a %) that one variance is estimated to be with respect to another variance. When you consider, in an ANOVA, that $F = S^2_{effect}/S^2_{within}$ is a ratio of the mean squares that estimates the variances, F could **itself** be viewed as an index, an ES, for whatever you are targeting. That is, you could say that $F = 1.9862$ estimates that the variance, $n\sigma^2_{\bar{x}}$, which includes the effect of the factor of interest in an ANOVA, is about 198%, or 1.98 times σ^2, the variance without the treatment effects, and with other effects partitioned out. Well, that is not very informative, but it might be useful for evaluating the difference between other variances using the F test (but maybe not for a whole ANOVA).

Effect size:

In light of the many contexts in which an F value may occur, that may not seem very satisfying. A variety of other more commonly used unitless effect size measures for use with the ANOVA are also presented below. Unfortunately, all of them might be more difficult for the layperson to appreciate than even F itself, and they are not necessarily interconvertible with each other.

The following ES measures are not much more informative, but they are each popular for use with the ANOVA in their own circle of admirers, so they are generally accepted when journal policy requires reporting an effect size for the ANOVA. The $_{effect}$ subscript usually indicates the factor of interest:

Eta squared,

$$\eta^2 = \frac{SS_{effect}}{TSS} \tag{9.7}$$

is generally as good a descriptor as any of the first three measures included here. It is the fraction of the total observed variability, with one or more factors, other than the factor of interest, removed. In the one way ANOVA the numerator is generally the SS_A. Eta squared is somewhat sample size dependent, but as the sample sizes get larger, this bias gets smaller. (When based on massive data sets, this sort of descriptor is a good lead-in to modeling, as long as you are not too quick to equate models with reality before demonstrating repeatability.) The example in Subsection 9.5.0.1 includes working out an η^2.

Partial eta squared might be more suitable than eta squared for an ANOVA with two (or more) factors,

$$\eta^2_{partial} = \frac{SS_{effect}}{SS_{effect} + SS_W}, \tag{9.8}$$

the fraction of the variance **attributable** to the effect of interest (usually the effect of the treatments), **divided by** the total variance, **less** the variability of any other factors (such as the additional factor in a two way ANOVA). When the number of effects is > 2, this ES is somewhat less dependent on the actual sample sizes than is eta squared. Also see the coefficient of determination below.

Omega squared is another ES estimator for the ANOVA,

$$\omega^2 = \frac{SS_{effect} - (m-1)\, MS_W}{TSS + MS_W}, \tag{9.9}$$

where m is the number of treatments in the effect (factor). It is **less biased** than either the above or the below, but (as defined here) it is intended for **equal sample sizes**.[17] However, it is not provided in many of the popular statistical packages. The possibility for omega squared to come out negative due to the effects of chance is seen as a defect; **if you get a minus value,** call it 0. The example in Subsection 9.5.0.1 includes working out an ω^2.

---

[17]Cohen (1988), who also created Cohen's d, proposed the following categories to interpret the strength of omega squared, ω^2:

- ~~small = 0.099~~

- ~~medium = 0.0588~~

- ~~large = 0.1379~~

However, these are struck out here because they have not, in general, proved to be equivalent to the small, medium and large widely published for Cohen's d (Section 5.9) as was formerly commonly believed. Likewise, they and the ones for d were very field and design specific, and their use can be very **misleading** (e.g. Barnette and McLeann 1999, 2002, 2006, Paterson, et al. 2016). **Do not use** any such 'small', 'medium', or 'large' designations published as standardized effect sizes to estimate sample size. This **ongoing error** has tended to give power analysis an undeserved bad name.

RMSSE is the ***root-mean-square standardized effect***; it was suggested by Fleischer (1980), who discusses using it as an effect size across several classes (along with a technique for setting a confidence interval) for the analysis of variance. To calculate this ES, apply it across all m of the treatments,

$$RMSSE = \sqrt{\frac{1}{m-1} \sum_{j=1}^{j=m} \frac{\left(\bar{x}_j - \bar{\bar{x}}\right)^2}{MS_W}},$$

where $m - 1$ is the df for treatments (Steiger 2004). This was also proposed as an estimate of the average value of the deprecated Cohen's d for the classes. For example, $RMSSE$ avoids an experimental effect in pounds from being different from the same effect in kilograms. But $RMSSE$, unlike d, is **not distorted** by sample size effect, and it is not readily converted to d. The F value in the F test itself could be considered as a ratio sort of ES, because $\boldsymbol{RMSSE = \sqrt{F}}$. However, effect sizes in **original units of measure** are to be **preferred** whenever possible. Unlike Cohen's d, $RMSSE$ is a numerically valid ES among treatment means, but its standardization and lack of units implies a return to the **disastrous invalid concept** that there is some sort of 'small', 'medium', and 'large' that is standard across all experiments, even when they are only moderately similar. **Do not** use this "small", "medium" and "large" concept for estimating a sample size.

The coefficient of determination, R squared,

$$R^2 = 1 - (SS_W/TSS) = SS_{all\,effects\,and\,interactions}/TSS,$$

is an ES drawn from linear correlations (Section 11.3), it is a measurement used to explain how much variability of one factor can be caused by its relationship to another related factor.. It estimates the fraction of the variance that is due to factors, **without** SS_W. That is, it can be seen as the part of the variance that arises from causes (hopefully the treatment) **other than the natural variance**, σ^2. Although tending to favor straight line relationships, the coefficient of determination, R^2, is widely accepted as a measure of the fraction of the variation in **one variable**, say y, that is explained by the variation in **another variable**, x. It is useful and intuitively satisfying in this role. With respect to the ANOVA, it can also be written

$$R^2 = (TSS - SS_W)/TSS.$$

There have been futile and misleading attempts to convert ω^2 and $\eta^2_{partial}$ to R squared.[18] The coefficient of determination may look somewhat like η^2, but there is a difference between the singular $_{effect}$ in η^2 and the more inclusive $_{all\,effects\,and\,interactions}$ in $R^2 = SS_{all\,effects\,and\,interactions}/TSS$. Its square root, R, the *correlation coefficient*, is subject to the difficulties illustrated in Figure 11.3 as is clarified and covered in more detail in Section 12.5.1. That is, R refers to the tendency of two vari-

---

[18]*Adjusted R^2 is useful in the causal variable selection stage of model building. It adjusts R^2 for the number of causal variables (factors). It is*

$$R^2_{adj} = 1 - \frac{(1 - R^2)(n-1)}{(n-k-1)},$$

*where k is the number of causal variables (Theil and Henri 1961). Its use is advised in multiple regression (not treated here) because R^2 can, and usually does, become quite **inflated with more than one variable**. Its interpretation is a little different from the coefficient of determination, R^2. In this adjusted form, R^2_{adj} is not the square of the correlation coefficient, R. Unlike R^2, R^2_{adj} tends to increase **only if** the new term improves the model more than would be expected by chance. R^2_{adj} can be negative, and will always be less than or equal to R^2. Even for simple regressions, some researchers, especially some geneticists and economists, use it rather than R^2 for reporting the strength of simple correlations if the line clearly, or theoretically, does not pass through the intercept (where the line crosses the y axis.). That is, they see the slope and the intercept as representing 2 separate causal variables ($k = 2$ rather than $k = 1$). Maybe R^2_{adj} (adjusted R^2) could even be another potential ES value for global ANOVAs.*

ables to vary together in a straight line relationship.[19] R^2 reports the percent that one variance estimate is of another, and as such can be seen as a linear correlation-like estimator (for more explanation see Equation11.15). But, depending on the context, it might be seen as a fixed estimate rather than a dynamic relationship such as is always reported by R. The ANOVA is about 'yes, there is a difference' versus 'no, it is not known if there is a difference or not'. If a difference is detected, don't assume that it is necessarily due to a straight line relationship (you can use R^2, but not R).

All the above ES measures are included here only because they are accepted, sometimes demanded, as effect sizes by some social science and psychology journals, and they are effect sizes in the sense of measures providing you with a rough estimate of the contribution of a particular effect on the data. But these effect size measures should be used with care. Olejnik and Algina (2003) provide considerable information on these measures, and suggestions regarding what these authors see as the proper application of caution when using them in reporting.

They are also seen as standardized measures, which they are in a mathematical sense, but, as for Cohen's d (1988, Section 5.9), that **does not imply** that they can be assumed to provide a standard for the comparison of measures on different scales or among different experimental designs. However, viewing η^2, $\eta^2_{partial}$, or ω^2 as measures of something that could be described in terms of non-linear regression, with due thoughtful caution, has a place in more complex ANOVA designs.

With the **exception of ω^2 and R^2**, the above attempts to evaluate the effect size are best seen as **tentative** rather than definitive in the sense of inference. They could be regarded to be in the context of the combined **contribution of treatments** to observed **variances**. As is obvious from their equations, they **do not agree** with each other. However, especially for a one way ANOVA, ω^2 gives the smallest (hence most conservative) values, and it might be the least misleading if sample sizes are not large. At sufficiently large sample sizes, η^2, $\eta^2_{partial}$, and ω^2 approach similar values.

9.5.0.1 Example: Effect Sizes for Global ANOVA

Although the author has serious doubts about the application of such unitless ES values and especially their interpretation, we will explore how to calculate two of them using a one way ANOVA as an example.

Pheromone effect (continued, from Subsection 9.2.0.1) Amadora's perfume research laboratory was working on a new pheromone to attract males to their customers. They used the same simple cologne base as the carrier for each pheromone treatment and by itself for the control. They had two candidate pheromones with the romantic names D and E. The control was called C.

Analysis: The grand mean is $\bar{\bar{x}}$=13.7686 and the individual treatment means are $\bar{x}_C - 13.9249$, $\bar{x}_D = 16.3450$ and $\bar{x}_E = 11.0358$. Some of the analysis is summarized in the following ANOVA table:

Sources of Variation	df	SS	MS	F
Among treatments	2	423.91074	211.95537	$F = 10.8419$
Within	87	1700.81722	19.54962	
Total	89	2124.72796	23.87335	

Referring to a table of F, with $\alpha = 5\%$, $df_W = 87$, and $df_A = 2$, the researchers noted that any $F > 3.1012$ was **statistically significant** at the 5% level, with this number of degrees of freedom.

---

[19]When the number of categories (say, treatments) $m \neq 2$, or when sample sizes are not large, interconversion of any of the above, whether to Cohen's d, or to R^2, is often ineffective as well as misleading (Barnette and McLean 2006).

But their editor insisted on an 'effect size' for any declaration of statistical significance. So they decided to use Equation 9.7 to estimate **eta squared**,

$$\eta^2 = SS_{effect}/TSS.$$

The $_{effect}$ refers to a single factor in a one way ANOVA. So they used the treatment factor, which includes all the treatments, including the control. In this case they used $SS_A = SS_{effect}$, which gives

$$\eta^2 = 423.9107/2124.7280 = 0.19951295.$$

This estimate allows them to report their **treatments effect size** to be about 20.0% of the total observed variability. This eta squared measure might have given the researchers a feel for the contribution of their treatments to the overall data.

However, suppose they had decided to use Equation 9.9 to find **omega squared**,

$$\omega^2 = [SS_{effect} - (m-1)\, MS_W]\, /\, (TSS + MS_W).$$

Note that $SS_A = SS_{effect}$, and m is the number of treatments. So, they could have calculated omega squared,

$$\omega^2 = (423.91074 - 2 \times 19.54962)\, /\, (2124.72796 + 19.54962) = 0.1794597,$$

to get an adjusted estimate, that only about 17.9% of the total estimated variability is due to the treatments.

That's somewhat different, but what if they had decided to use the coefficient of determination,

$$R^2 = (TSS - SS_W)/TSS$$

$$(2124.72796 - 1700.81722)\, /2124.72796 = 0.1995129.$$

Again, the estimate is that about 20.0% of the variability is due to chance. Sometimes there is a little more difference between η^2 and R^2.

Results and Discussion: Note that you must avoid rounding any more than you have to before any subtraction, or you can often get an extreme error (even a zero in the denominator). But the ω^2 and η^2 give somewhat different effect sizes! How are their investors to interpret that?

All I can advise is to stick with whichever seems to be popular in your field, or the measure of effect size in someone else's peer reviewed paper which relates to your research. Some authors would agree that ω^2 is the one you should take most seriously as an estimate of the percent of the variance that was due to treatments for a simple one way ANOVA. However, R^2 might be more to the point as an estimate of the percent of the variance due to all effects and interactions.

9.6 It Costs to Break Global Tests Into Comparisons

Global tests, including what are called omnibus tests, include many independent treatments and as many probabilities, or even more, of getting a Type I error or a Type II error into one test. That is, if you are going to want to separate them into differences between pairs, you are going to be saddled with the overall difference. That is, you will not just be able to consider the probability of a false positive, or of missing a true meaningful difference for treatment A versus B, then for C versus D, then E versus F... . The probabilities are all part of the same analysis and the multiplication law (Section 1.4.3) applies. For instance, if you are going to do three comparisons, A to B, C to D, and E to F, with a $1 - \alpha = 95\%$ chance of not getting a false positive for each one and they are all part of the same analysis (like all being part of the same gambling bet), you are going to have only a

$$P_{A,B} \times P_{C,D} \times P_{E,F} = 0.95^3 = 0.857375$$

chance of avoiding a false positive. That would be an overall $\alpha = 0.142625$ or about a 14% chance of getting a false positive, a Type I error. This gets much worse as you go on with more treatments or more comparisons.

Even if you only have 4 treatments and want to compare every one of them to every other one of them, that would involve $n = 4$ things taken $r = 2$ at a time to get $^nC_r = \frac{n!}{r!(n-r)!} = 6$, and

$$0.95^6 = 0.7350919,$$

giving you an overall $\alpha = 0.264908$ or about a 26% chance of at least one false positive. This is not the whole story, in situations where sample sizes relate to preplanned pair by pair comparisons, you are also going to have an equivalent problem with β, the chance of getting a Type II error, not detecting a true meaningful difference.

> **If you need Multiple Comparisons for an ANOVA,** see Section 10.4 for sample size estimation. But be aware that the only method of estimating sample size for the ANOVA presented in this book assumes that the ANOVA is intended to lead to multiple comparisons.

Unfortunately, the popular global ANOVA power calculations incorporated in most software packages are questionable, and **without preplanned multiple comparisons** I cannot find an approach suitable to present here.[20]

9.7 CIs From the ANOVA

Confidence intervals for the means might be the most informative approach to reporting your results, instead of all the possible effect sizes in the sense of differences between pairs of means. Simply reporting the means as point estimates would be the least informative. After all, sampling does not really obtain the underlying means, μ_i, from a sample. Without CIs, all you know after an inferential test is a probabilistic estimate of a true mean, μ_i, or maybe also a probabilistic estimate of an effect size such as $\mu_C - \mu_A$. However, confidence intervals for the means or the effect sizes show a more realistic range in which you have a probability, the confidence, of having captured the true mean. To estimate this CI requires a variance, or rather a variance estimate. The ANOVA gives you the mean square within, MS_W, an attempt at a more accurate and specific estimate of the underlying overall variance, σ^2, than you get from simply using $S^2 = \frac{\sum(x-\bar{x})^2}{n-1}$ $(= MST)$, as in, say, a t test.

The MS_W appears to be a handy variance estimate for further comparisons between pairs of means and for **obtaining informative confidence intervals**. The estimation of the underlying σ^2 with the effects of treatments removed, and, when possible, such other effects as block effects removed, is an attractive claim of the ANOVA. This theoretical advantage of the ANOVA is that its estimate of σ^2 as MS_w facilitates comparisons and CIs, with df_W for each comparison or CI.

---

[20]Most of the methods of a priori power analysis available in major statistical packages are based on the reckless approaches published by Cohen. They even tend to include the 'small', 'medium', and 'large' folly, which even Cohen would not have applied to all sorts of research (Cohen 1988, p25) such as implied by a general computerized algorithm.

9.7.0.1 Example: CIs After ANOVA

So, let's **see how this works** with a continuation of the previous perfume example. This example is crucial to your understanding of some of the nuances of CIs.

Discerning scents (continued from Subsection 9.2.0.1) Amadora's perfume research laboratory was looking for a new pheromone to attract males to their customers. They used the same simple cologne base as the carrier for each pheromone treatment and by itself for the control. They had two candidate pheromones with the romantic names D and E. The control was called C. Let's look at how they would calculate CIs for the treatments.

Analysis: A one way ANOVA has been run and a mean square within (mean square error to some folks) estimate has been obtained. The individual treatment means are $\bar{x}_C = 13.9249$, $\bar{x}_D = 16.3450$ and $\bar{x}_E = 11.0358$, all with equal sample sizes, $n = 30$. The rightmost value, the width of a whisker, for each mean would be different if the sample sizes were different (wider for smaller samples). The mean square within, $MS_W = 24.4222$ with 87 df, is the estimate of the underlying variance, σ^2, to be used for individual CIs. The research team of Amadora's could present their financiers or stockholders with a simple set of $1 - \alpha = 95\%$ CIs. They would neither be setting some rigid standard, the use of a rigid $\alpha = 5\%$, nor explicitly avoiding it. They would be using the CI to provide an estimate (limited by chance) of the location of each treatment mean of a population of treatment means for which they could never know the exact value.

As per Chapter 6, because $df_W \geq 60$ and $\alpha = 5\%$, it is reasonable and conservative to use a modification of Equation 6.1,

$$\bar{x} - 2\sqrt{MS_W/n_j} \leq \mu \leq \bar{x} + 2\sqrt{MS_W/n_j}, \tag{9.10}$$

(which can also be written $\bar{x} - 2S_{\bar{x}} \leq \mu \leq \bar{x} + 2S_{\bar{x}}$) for each treatment with sample size n_j. That is, because $df_W \geq 60$, they were able to use the conservative approximation, $Z_{0.025} \approx 2.00$. (For $df_W < 60$ at a criterion of $\alpha = 5\%$ or for more accuracy or smaller df_W, they would have to revert to t.) The substitution of $\sqrt{MS_W}$ for the estimator, S, of σ is to eliminate the effect of the estimated differences among the means on the total sums of squares. The Central Limit Theorem hopefully makes it possible to use this best available estimate of σ^2 and the sample size, n (which is not an estimate), to get the best available estimate of $\sigma_{\bar{x}_j} = \sqrt{\sigma^2/n_j}$, the standard error of the mean $_j$. So, because $MS_W = 24.4222$, and given that **each sample** has $n = 30$ measurements, $S_{\bar{x}} = \sqrt{19.54962/30} = 0.8072509$.

Although centered differently, by having different estimates of their mean, these CIs are all the same width because the sample sizes are equal and the same estimate of σ^2 is being used for each of them. So, $2S_{\bar{x}} = 1.614502$ is the whisker length for any of these means. **Rounding is not done yet**; when adding or subtracting, the number of decimal points is what is limiting, whereas when multiplying, the number of 'significant figures' is limiting. Then the number of decimal places in the sum will be determined by the least number of decimal places in the numbers being added. So, the unrounded CIs will be: $\bar{x}_C \pm 2S_{\bar{x}} = 13.924912 \pm 1.614502$, $\bar{x}_D \pm 2S_{\bar{x}} = 16.344970 \pm 1.614502$, and $\bar{x}_E \pm 2S_{\bar{x}} = 11.035792 \pm 1.614502$. Some authors carelessly publish these as: $\bar{x}_C = 13.92 \pm 1.61$, $\bar{x}_D = 16.34 \pm 1.61$, and $\bar{x}_E = 11.04 \pm 1.61$, although this is not the preferred representation (see below). But it is necessary to **clearly define** what the rightmost value refers to if this $\pm$ notation is used for publication, and it must correspond to the particular journal's style. Especially over time, different journals, and even different authors in the same journal, use $\pm$ with different meanings.

These CIs could be written in a number of other ways. However, because they used the conservative $Z \approx 2$, the final sums have to be rounded to three figures. This 95% CI could best be presented as an inequality with specific confidence limits, $L_1 \leq \mu \leq L_2$, representing $\bar{x} - Z_{\alpha/2}S_{\bar{x}} \leq \mu \leq \bar{x} + Z_{\alpha/2}S_{\bar{x}}$. Although not the case in this instance, too many places could have been lost in subtraction had they rounded too early. Confidence intervals are rounded down on the left and up on the right. So, they round to:

$$12.3 \leq \mu_C \leq 15.6 \,,$$

$$14.7 \leq \mu_D \leq 18.0 \,,$$

$$9.42 \leq \mu_E \leq 12.7 \,.$$

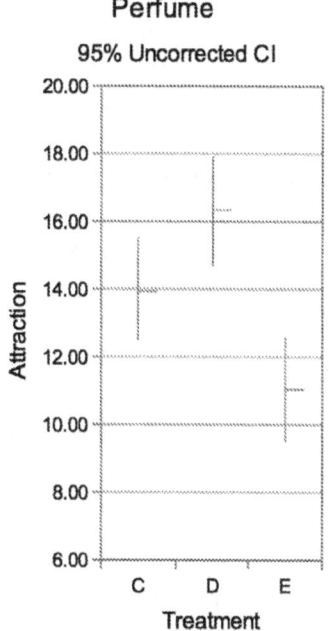

Discussion: Graphic presentations that have the mean marking the division between the whiskers are popular (especially if the sample sizes are different). However, although many researchers mark an initial estimate of the mean, they can only be 95% certain that the mean is anywhere within the whole CI, and they certainly would not be certain where.

You might also want to try, and then to study, Question 9.

CIs for randomized complete block ANOVAs can be done the same way as for a one way ANOVA using MS_W and df_w.

Even if you are not testing in the sense of yes-no inference, you still have to choose the criterion (for instance, $\alpha = 5\%$) in advance. However, in order to attain a reasonably centered a CI you will also need to specify a power and a reasonable minimum difference to choose a sample size. One difficulty with drawing a conclusion from a CI is that CIs are **usually** not corrected for multiple comparisons (Chapter 10). Testing would be needed in order to render a verdict as well as a set of descriptions. In general, CIs are reported at the 95% confidence level and thus use $Z_{\alpha/2} \approx 2.00$. But, if a big gamble is involved, a much wider 99% CI might be worthy of consideration instead. For very large samples ($df > 4500$), researchers may wish to a priori choose $Z_{0.0250} = 1.9600$ for greater accuracy at $\alpha = 5\%$. If $df_W \leq 60$, the confidence interval might be awkwardly wide, and one must use $t_{\alpha/2, df_W}$ instead of 2.

9.7.0.2 Example (continued from CIs after ANOVA, Subsection 9.7.0.1): Effect Size CIs

CIs for effect size after the ANOVA are a little trickier because they have to account for two means and two sample sizes.

Perfume effect sizes Assume Amadora's perfume researchers a priori choose to have a control, C, and a 95% confidence for estimating the limits, $L_1 \leq \mu_x - \mu_C \leq L_2$, for the differences (individual effect sizes). This effect size CI can be calculated by

$$(\bar{x}_2 - \bar{x}_1) - 2\sqrt{MS_W\left(\frac{1}{n_1} + \frac{1}{n_2}\right)} \leq \mu_2 - \mu_1 \leq (\bar{x}_2 - \bar{x}_1) + 2\sqrt{MS_W\left(\frac{1}{n_1} + \frac{1}{n_2}\right)}, \qquad (9.11)$$

where $Z_{\alpha/2=0.025} = 1.960$ is rounded up slightly to $2.00 \approx t_{0.025, 4500 \geq n \geq 60}$. When, as in this example, $n_1 = n_2 = n$, this can also be written

$$(\bar{x}_i - \bar{x}_C) - 2\sqrt{MS_W\,(2/n)} \leq \mu_i - \mu_C \leq (\bar{x}_i - \bar{x}_C) + 2\sqrt{MS_W\,(2/n)}.$$

So, $2\sqrt{MS_W\,(2/n)} \approx ZS_{\bar{x}}$, and given $MS_W = 19.54962$, then $S_{\bar{x}} = \sqrt{19.54962\,(2/30)} = 1.141625$, with the result that $2S_{\bar{x}} = 2.28325$. The differences between means, the 'plain language' ESs, in **original units of measure**, are $\bar{x}_D - \bar{x}_C = 2.42005$ and $\bar{x}_E - \bar{x}_C = -2.88912$. The use of $Z \approx 2$ means that only three digits can be reported, given the following CIs for the effect sizes of the two comparisons in the original units of measure (proprietary 'attraction units' in this example):

$$2.42005 - 2.28325 \leq \mu_D - \mu_C \leq 2.42005 + 2.28325\,,$$

$$0.137 \leq \mu_i - \mu_C \leq 4.70\,, \text{ and}$$

$$-2.88912 - 2.28325 \leq \mu_E - \mu_C \leq -2.88912 + 2.28325\,,$$

$$-5.18 \leq \mu_E - \mu_C \leq -0.605\,.$$

Be careful, rounding down on the left and up on the right, the correct way, can be tricky when minus signs are involved (-5.173 on the left rounded down to -5.18 **because -5.18 is less than -5.173**). (Because chance is always against us, we must conservatively round the CI towards wider rather than narrower.) These CIs that refer to the ESs are larger than the CIs that refer to the individual means, because there was not one estimated mean, but two estimated means. In general, confidence intervals are made for the ES of each pair at the 95% confidence level. The researchers considered two different kinds of graphic presentations for publishing the CIs (below); either is OK. Regardless of how presented, if either of the CIs for effect size had included zero, some readers might **wrongly have used that** to declare statistical significance at the $\alpha = 5\%$ level. If one wishes to declare statistical significance, one must **use a suitable test** such as a Student's t test **with correction** for multiple comparisons where appropriate. A **failure to entrap zero** does **not** necessarily imply statistical significance. It is interesting that the CI for the effect of pheromone E is entirely negative. The researchers might suspect a repellent effect, but that would require a new experiment and it is hard to imagine a market for that.

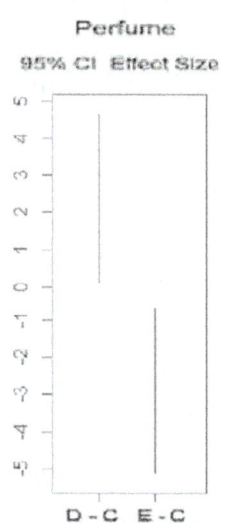 or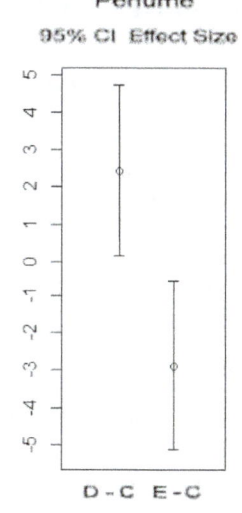

Too much rounding? If you really want more significant places you will have to use a t value, instead of $Z \approx 2.00$, which will allow your numerical reporting accuracy to only be limited by the numerical accuracy of the table of t. In this example, $t_{\alpha=0.025, 87df} = 1.98761$, because MS_W provides $df_W = 87$ degrees of freedom. Then you can report each CI in this example to 4 figures (one more than the data).

If the stakes are high, or the data is plentiful, you may wish to find a 99% CI using

$$(\bar{x}_2 - \bar{x}_1) - t_{\alpha/2, df_w} \sqrt{MS_W \left(\frac{1}{n_1} + \frac{1}{n_2} \right)} \leq \mu_2 - \mu_1 \leq (\bar{x}_2 - \bar{x}_1) + t_{\alpha/2, df_w} \sqrt{MS_W \left(\frac{1}{n_1} + \frac{1}{n_2} \right)}. \tag{9.12}$$

9.7.0.3 Arbitrary scales

As already mentioned, parametric tests assume an equal interval scale; that is, they need a measurement scale where twice some reading or grade is really twice as much effect. This is a common challenge in the social sciences, where data that is not on an equal interval scale may arise from a machine or a test or from an interview script. For instance, perfume researchers in Example 9.2.0.1 would have been more certain of a valid result if a nonparametric test were used. Likewise, difficulties involving other arbitrary scales can occur in other fields of science and engineering. The visual presentation provided by graphics such as the ones in the example will help clarify the outcomes.[21] Therefore, do not use ranks (ordered positions) as though they were the raw data in an ANOVA; for rank tests in lieu of the ANOVA, see either Section 14.3 or Section 14.5.

> **CAUTION:** It is necessary to reiterate that **CIs from global tests** are typically not adjusted for multiple comparisons and can neither be assumed to be indicative of statistical significance, nor can they refute it.

---

[21] Unfortunately, attempts at standardizing scales (other than those of means) have not succeeded for inference (e.g. Section 5.9), and standardized scales from the scores themselves are especially not suitable for estimating sample sizes. They also subtly lead to attempting to compare apples to oranges. It is presently unclear how to find a desirable transformation, or if any is needed, for ratios such as percentages taken from continuous data that includes a potentially wide range.

9.8 Familiarity With the Correction Term Can be Obligatory

Many authors and computer program manuals use a seemingly arcane equation,

$$\sum_{i=1}^{i=n}(x_i - \bar{x})^2 = \sum_{i=1}^{i=n}x_i^2 - \frac{\left(\sum_{i=1}^{i=n}x_i\right)^2}{n}, \tag{9.13}$$

for any sums of squares ($SS_{...}$). It was an easier way to calculate when the best available computer was a wide carriage mechanical comptometer. Although not used for that in this book, **it is very useful** for explaining the derivation of some processes that involve the sum of squares.

To derive this alternative approach itself, you need to remember that $\sum cx = c\sum x$, where c stands for any constant, and that $\bar{x}$ can be treated as a constant (like c), and that $\sum_{i=1}^{i=n}c = nc$ because you are just summing c with itself n times. Another basic idea is that $\sum(a+b) = \sum a + \sum b$. Then, with all this you can, using $(x - \bar{x}) \times (x - \bar{x})$, expand the sum of squares equation, $SS = \sum_{i=1}^{i=n}(x_i - \bar{x})^2$, into a quadratic (just carrying the $\sum$ along), and then simplify,

$$SS_{...} = \sum_{i=1}^{i=n}(x_i - \bar{x})^2 = \sum_{i=1}^{i=n}x_i^2 - \sum_{i=1}^{i=n}2x_i\bar{x} + \sum_{i=1}^{i=n}\bar{x}^2$$

$$= \sum_{i=1}^{i=n}x_i^2 - 2\bar{x}\sum_{i=1}^{i=n}x_i + n\bar{x}^2 = \sum_{i=1}^{i=n}x_i^2 - 2\left(\frac{\sum_{i=1}^{i=n}x_i}{n}\right)\sum_{i=1}^{i=n}x_i + n\left(\frac{\sum_{i=1}^{i=n}x_i}{n}\right)^2$$

$$= \sum_{i=1}^{i=n}x_i^2 - 2\frac{\left(\sum_{i=1}^{i=n}x_i\right)^2}{n} + n\frac{\left(\sum_{i=1}^{i=n}x_i\right)^2}{n^2} = \sum_{i=1}^{i=n}x_i^2 + \left[-2\frac{\left(\sum_{i=1}^{i=n}x_i\right)^2}{n} + \frac{\left(\sum_{i=1}^{i=n}x_i\right)^2}{n}\right]$$

$$= \sum_{i=1}^{i=n}x_i^2 - \left\{\frac{\left(\sum_{i=1}^{i=n}x_i\right)^2}{n}\right\}. \tag{9.14}$$

$$\therefore \sum_{i=1}^{i=n}(x_i - \bar{x})^2 = \sum_{i=1}^{i=n}x_i^2 - \left\{\frac{\left(\sum_{i=1}^{i=n}x_i\right)^2}{n}\right\},$$

where $\therefore$ is read *therefore*. The fraction on the right in {*curly brackets*} is the **correction term**, CT (means correction term, not C times T.).

So you may estimate the variance using the machine formula,

$$S^2 = \frac{\sum_{i=1}^{i=n}x_i^2 - \frac{\left(\sum_{i=1}^{i=n}x_i\right)^2}{n}}{n-1}. \tag{9.15}$$

This is called the **machine formula** for estimating the variance because it was popular for use with old mechanical calculating machines.

9.9 Must You Go ANOVA?

Because of the many uses and mathematical magnificence of the ANOVA, it is seen as one of the great achievements in the evolution of statistics. It can be used to partition out an underlying variance estimate, MS_W, with which you can, at least in theory, run pair by pair Student's t tests using an appropriately, but expensively corrected cutoff for multiple comparisons (Subsection 10.2.0.1). Like most parametric tests, the ANOVA was created assuming normality, but in application it is **more sensitive to non normality** than are most other parametric tests. The CLT is alleged to possibly compensate for this requirement when the sample size is big enough that $df_W \geq 20$ (Snedecor and Cochran 1967). However, after researching and writing this chapter, I am skeptical of its compensating for a lack of normality at any df_W or sample size. In any case, it is always a good idea to exceed your estimated sample size for an ANOVA to provide both for the likely loss of efficiency due to non normality, and to provide for the possibility of data loss. If you expect your data to have a problem with outliers you may wish to clean it using MAD (Section 15.6.2), but this will entail information loss.

However, there is mounting evidence that global nonparametric alternatives should be used instead of ANOVAs in most or maybe all cases. Then you can use nonparametric tests such as the Wilcoxon test (Subsection 14.1.1) for the multiple comparisons, or a t test for each comparison, but using the usual number of degrees of freedom for each treatment rather than df_W (Chapter 10).

If your data is particularly **non-normal**, say a kind known to graph as a U-shape, you should definitely use **nonparametric** alternatives. If you do not trust the ANOVA to be sufficiently robust, or if you see that you cannot achieve a sufficient number of degrees of freedom within, then you should consider the nonparametric alternatives such as the Kruskal-Wallis test (Section 14.3), or maybe for complete blocks, Friedman's test (Section 14.5).

When considering a global test, you should ask yourself if you can gain by testing merely to see if there is **a difference** somewhere among $m > 2$ treatments without comparing one or more pairs of said treatments. If answering **only** that minimum question is **not sufficient**, the cost of separately comparing the pairs of means (multiple comparison techniques) tends to be very expensive (Section 9.6). Assuming that you use maximum publishable $\alpha = 5\%$, you will be limited to $k = 5$ comparisons unless you really are willing to compare each individual pair at $\alpha < 1\%$, a process that demands very large to huge sample sizes and investment.[22] Thus if you a priori plan and document your global test to compare 5 treatments to a control ($m = 6$) it might be practical and affordable to run your comparisons at $\alpha = 1\%$. Making all two sample tests with $\alpha = 1\%$ is a great if somewhat costly way to avoid having your publications and decisions someday disproved. However, even with such a conservative approach, the question of the limitations of various global tests, especially the limitations of the ANOVA, need further literature review and study via computer simulations (also see Subsubsection 9.3.0.1).

9.9.0.1 Dependent factors mean you cannot go ANOVA

If your **factors**, such as treatments or blocks, are dependent, meaning they somehow interact or are even particularly relevant to each other, as when they can be ordered (as can be the elements of a Likert scale such as 'liked', 'neutral', 'disliked', 'hated') or were created by binning, they cannot be used for either the ANOVA or even its alternatives such as the Wilcoxon test (Subsection 14.1.1 or the Friedman's test (Section 14.5). You might still study such factors from different viewpoints using some sort of linear trend analysis (for instance estimating R^2 and also a p_value; some texts cover trend analysis in detail), using simple correlation, or using binning with Pearson's chi square (Chapter 13).

---

[22]Despite my personal skepticism about most global tests in application, I have extensively detailed how to use accepted approaches for using global tests and particularly for subsequent multiple comparisons. Because of the present popularity of such approaches I have gone into careful detail as to how to apply them in accordance with accepted theory.

Questions for Ch 9

Write the equations where appropriate, and then fill in the values. **Odd numbered questions** have worked answers at the back of the book and can be studied as additional examples. The data tends to be fictitious.

1. What might be a disadvantage of a randomized complete block experiment when analyzed as an ANOVA?

2. What is the advantage of a randomized complete block approach?

3. Mega Juiced had a problem with too much sodium in their soil and water. It affected the quality and yield of the grapes in their vineyards. They decided upon a rather ambitious set of trials involving $m = 4$ different gypsum and irrigation treatments. They decided to use a one way ANOVA with an $\alpha = 5\%$ criterion and hypothesis pair, **H0**: THERE IS NO DIFFERENCE IN SUGAR CONTENT, vs **HA**: THERE IS A DIFFERENCE IN SUGAR CONTENT. They had a lot of individually irrigated vineyards with similar soils and the same water source to choose from, and they were able to use 40 eighth hectare plots, each of which was chosen **at random** to receive only one of the $n = 4$ treatments, such that there were $n = 10$ plots per treatment. This means that there was a total sample size of $n = 4 \times 10 = 40$. After a few years of these regimes they obtained the following sugar content data from the grape harvest, in grams per liter (about ten times what the wine industry would call Brix degrees):

	Treatment				
	A	B	C	D	
	254	222	225	228	
	192	205	261	212	
	166	197	214	183	
	197	259	205	231	
	166	200	201	187	
	198	223	188	212	Grand
	224	177	219	237	Total
	206	185	231	220	&
	199	219	190	208	**Grand**
	246	206	224	212	**Mean**
Total	2048	2093	2158	2130	8429
Mean	204.8	209.3	215.8	213.0	**210.725**

Present an **ANOVA** table for this completely randomized design, and report whether there is a statistically significant difference.

4. Mega Juiced made an a priori decision to view the same plots and their treatments as in Question 3, as a separate experiment to compare yields. That is, there were $m = 4$ different gypsum and irrigation regimes, each with $n = 10$ plots, giving a total sample size of $n = 40$. They chose an $\alpha = 5\%$ criterion and made a hypothesis pair, **H0**: THERE IS NO DIFFERENCE IN YIELD vs **HA**: THERE IS A DIFFERENCE IN YIELD. After a few years of these regimes they obtained the following yield data, in kilograms of grapes per eighth hectare single treatment plot:

Treatment				
A	B	C	D	
651.36	914.71	1597.37	1239.68	
556.86	631.50	2086.32	1348.93	
883.08	282.69	1033.92	743.38	
1044.10	834.14	2296.33	648.65	
1018.78	519.96	2083.65	470.94	
1192.80	1281.30	1471.64	1626.77	
1326.06	638.97	2335.26	507.44	
640.18	1004.70	2176.65	1156.94	
373.54	770.48	1275.36	1264.44	**Grand**
558.17	529.08	1171.00	1217.50	**Mean**
Total	8244.93	7407.53	17527.5	10224.67	10851.158

The weighing method was reported to the nearest 10 grams (0.01 kg). List the means and an **ANOVA** table for this completely randomized design, and report if it resulted in a statistically significant difference. (Attempting such extremes of accuracy might have been particularly useful in avoiding ties, had they chosen to use a nonparametric analysis.)

5. A food scientist is concerned about the parts per million of a certain substance found only after processing chocolate candy bars. Each start up of a new candy bar batch is costly and produces a lot of waste, so the researcher chooses to use an RCB design. This consists of $m_A = 4$ variants of the candy bar recipe run at random order in $m_B = 8$ almost identical factories (blocks). The hypothesis pair is **H0**: THERE IS NO DIFFERENCE vs **HA**: THERE IS A DIFFERENCE, with a criterion of $\alpha = 0.01$. The resulting sums of squares are shown in the following table. **Replace the $* * *$ marks** with the proper values from calculation, and the preceding information as needed.

Sources Variation	df	SS	MS	F
Among Treatments	$* * *$	29304.17	$* * *$	$* * *$
Blocks	$* * *$	36216, 73	$* * *$	$* * *$
Within	$* * *$	$* * *$	$* * *$	
Total	$* * *$	107519.40		

Was the **Among Treatments** result statistically significant at the $\alpha = 1\%$ level? Did the **blocking** prove to be useful at the $\alpha = 5\%$ level? (The 5% level is fine for verifying the effect of the blocks; such a comparison is generally optional anyway.)

6. A start up micro-satellite company was attempting to develop rail guns to launch their units into orbit. The shape of the projectile was expected to affect both the initial acceleration along the rails and the air resistance. They tested $m_A = 6$ shapes of projectile, all of equal mass. They had $m_B = 5$ launch units; the capacitor banks of the launch units were supposed to be identical, but each tended to have its own slight quirks in its surge characteristics. The projectiles were expensive, so the engineers decided to use a complete block design, assigning all the shapes in random order to each launch unit, expending a total of **n** $= 30$ units. Their hypothesis pair was **H0**: THERE WAS NO DIFFERENCE vs **HA**: THERE WAS A DIFFERENCE and they chose a criterion of $\alpha = 1\%$.
When they measured the altitude attained in meters they obtained the following sums of squares shown. **Replace the $* * *$ marks** with the proper values from the reported *SS* in the table and the preceding information.

Sources Variation	df	SS	MS	F
Among Treatments	∗ ∗ ∗	6232967.4	∗ ∗ ∗	∗ ∗ ∗
Among Blocks	∗ ∗ ∗	6560923.5	∗ ∗ ∗	(*optional*)
Within	∗ ∗ ∗	∗ ∗ ∗	∗ ∗ ∗	
Total	∗ ∗ ∗	15044791.0		

Was the result statistically significant?

7. Given the following randomized complete block data in kilometers,

Treatments	Block 1	Block 2	Block 3
A1	52.16	57.89	58.78
A2	37.50	∗ ∗ ∗	51.36
A2	35.09	66.57	80.53
A4	49.12	56.36	49.91

fill in the **missing** ∗ ∗ ∗ data value with a suitable substitute. Also, what are the adjusted degrees of freedom for the MS_w of this RCBD?

8. Calculate the variance, S^2, for

49.4	53.4	37.5	32.3

using the **machine formula**. Show your work, including the equation with the correction term, CT, in curly brackets { }. Note that these brackets are used in this question only for instructional purposes. (If this calculation ever comes out negative, there is a mistake. In many instances, rounding before subtraction will lose almost all the information contained.)

9. You are assisting with a robotics class. This includes measuring the computer memory space, in kilobytes, used by C language programs from a student programming assignment. From a sample of $m = 3$ tutorials, each with $n = 21$ students, your ANOVA has a $MS_W = 600\,\text{KB}^2$, with $df_w = \mathbf{n} - \mathbf{m} = 60$. The sample means are $\bar{x}_A = 76.60\,\text{KB}$, $\bar{x}_B = 104.3\,\text{KB}$ and $\bar{x}_C = 96.10\,\text{KB}$. What are the 95% CIs for the three treatment means? Because $t_{0.025,\,60} = 2.00029782$, you cannot honestly use $Z = 2$ (see answer at end of book).

10. You somehow **could not get equal sample sizes**. Your ANOVA has a $MS_W = 600\,\text{KB}^2$, with $df_W = 60$. Your sample sizes are $n_A = 11$, $n_B = 19$, and $n_C = 33$. The sample means are $\bar{x}_A = 76.60\text{KB}$, $\bar{x}_B = 104.3\text{KB}$ and $\bar{x}_C = 96.10\text{KB}$. What are the CIs for the treatment means? (For $df = 60$, $t_{0.025,\,60} = 2.00029782$, so you cannot honestly use $Z = 2$. You may wish to do Question 11 first).

11. From a sample of $m = 3$ tutorials with equal sample sizes, $n = 21$ students each, your ANOVA has a $MS_W = 600\,\text{KB}^2$, with $df_w = 60$. You are interested in the differences between the control mean, $\bar{x}_C = 96.10\text{KB}$, and treatment means, $\bar{x}_A = 76.60\text{KB}$ and $\bar{x}_B = 104.3\text{KB}$. What are the ES values for the two treatments? What are the 95% CIs for the ES values? $t_{0.025,\,60} = 2.00029782$ so you cannot honestly use $Z = 2$ (see answer at end of book).

10 MULTIPLE COMPARISONS (and CIs / ANOVA)

> Vanity of vanities, said Ecclesiastes: vanity of vanities, and all is vanity.
>
> Challoner's revised Douay-Rheims Version (Old Testament 1609 & 1610). The Whole Revised and
> Diligently Compared with the Latin Vulgate by Bishop Richard Challoner A.D. 1749-1752.

All but the smallest ANOVA, where $m = 2$ treatments, are omnibus tests (a class of global tests that specifically require normality). At best, they can only detect if at least one out of the whole group of included comparisons is statistically significant. But it is not usually enough to know that the lifespans of five different models of cars include at least one which is different from another; it is desirable to know which are different from what. Global tests tend to leave you with a need to evaluate what is going on by comparing single pairs. So, it is time to consider the subject of *multiple comparisons* (MCPs), the sorting out of what is different from what. In some publications, multiple comparisons are also called *family wide comparisons* or *pairwise comparisons*. Many of the things covered here also apply to Pearson's chi square tests, and to the nonparametric alternatives (Chapter 14) to the ANOVA which are becoming increasingly popular because of the sensitivity of the ANOVA to non normality and heteroscedasticity.

You **can only make unplanned** multiple comparisons if the global test is **statistically significant**. Then each comparison must be adjusted as though you were going to make every possible pair by pair comparison.

10.1 The Cost of Multiple Comparisons

If you do the same experiment over and over again you will eventually get a false statistical significance. In considering multiple comparisons, you immediately come up against the problem of *alpha inflation*; the more comparisons you make, the greater the probability that you will experience false statistical significance. If **H0** is true and you do only one comparison at $\alpha = 0.05$ and there is really no true difference, you will even then have no more than a $1 - \alpha = 0.95$ probability of avoiding a Type I error, false rejection. The *family wide error rate*, also called the *strict family wide error rate*, is the probability of suffering one or more false rejections, Type I errors, out of k multiple comparisons,

$$\alpha_{strict} = 1 - \left(1 - \alpha'\right)^{K},$$
(10.1)

where k is the number of comparisons to be made and α' is your intended cutoff for each comparison (Equation 10.1).[1] So, when $k = 2$, the family wide error rate, with $\alpha' = 5\%$, would be $1 - (0.95)^2 = 1 - 0.9025$, which equals 0.0975. That is, the family wide error rate would have an overall probability, the α_{strict}, of getting a Type I error equal to almost 10%, and this alpha inflation gets increasingly worse with more comparisons. Compensating for this can be very expensive in terms of sample size if you plan to make a lot of comparisons.

Historically, multiple comparisons for the ANOVA simply used the LSD (least significant difference), a simple twist on a Student's t test with MS_W as a stand in for S^2, and the **uncorrected** α as

---

[1]Equation 10.1 is the reverse of the Sidak correction, Subsection 10.1.3.

the cutoff for each comparison. This was **not satisfactory** because of alpha inflation due to repeated testing.

10.1.1 Oh, what can save me from my wretchedness?

Alas, nothing can save you from chance! You are doomed to have some false positives and also some false negatives. Your best defense is to constrain your approach as close to theory as is practical. A major underlying philosophy behind using inferential statistics for answering research questions is that you have to accept a reasonable probability of getting either a Type I or a Type II **error each time** you do an experiment. The common way of minimizing this is to increase the sample sizes.

Recall that the ANOVA is a global test, an all-inclusive test to see whether the differences between **any** (but **not between which**) of $m > 1$ pairs of treatments really refute the null hypothesis. A technically accurate, and informative, pair of hypotheses for an ANOVA or any global test ($n > 2$) is:

H0: $\mu_1 = \mu_2 = \cdots = \mu_m$

HA: SOME TWO, OR MORE, μ_j ARE NOT EQUAL.

If the difference between **just one** pair of the treatments is statistically significant, a global test should turn out to be statistically significant. Used with caution, the following corrections for multiple comparisons are sufficiently conservative to be trustworthy for breaking a global test into any number, k, of pair by pair comparisons, if you are willing to pay the price. These corrections are examples of **_closed test procedures_**, corrections that control the family wide error rate. You must consider both the Type I and Type II error rates. Global ANOVAs are quite commonly used in research, and another global test, the Pearson's chi-square test (Chapter 12), is also widely used. Because of the sensitivity of the ANOVA family to non normality, two nonparametric global tests, the Kruskal-Wallis (Section 14.3), and Friedman's (Section 14.5) test are likely to become more common, and multiple comparisons following them will be similarly handled.

CAUTION: You cannot do multiple comparisons that are **not preplanned unless** the global test was statistically significant (Sokal and Rohlf 1969) and then you must find a' as if you were going to test **all** the $k = \frac{m!}{2!(m-2)!} = \frac{m(m-1)}{2}$ possible pair by pair comparisons. You can then only declare statistical significance using α' for each comparison. You do not have to run all the comparisons, but the ones you do run must use the **same** α' as you would use to run them all. **Preplanned** comparisons are almost always a great deal more economical, and they do not require that the global test be statistically significant.

10.1.2 Bonferroni's correction

Bonferroni's correction, the most popular way to adjust sample sizes for doing family-wide comparisons, is based on **_Boole's inequality_**, also known as the **_Bonferroni inequality_**. Note that Boole's inequality is just that, an inequality, not an equality. Thus when working with it, do not assume it is reversible. It involves k hypotheses about the true differences among observations or treatment categories, $i = 1, ..., k$. Let $P(A)$ be the unknown overall probability that an event in category A occurs, and let $P\left(\cup_{i=1}^{i=k} A_i\right)$ mean the probability that at **least one** of k included events, $A_1, A_2, ..., A_k$, occurs under random selection when there is no true difference among their probabilities, then

$$P\left(\cup_{i=1}^{i=k} A_i\right) \leq \sum_{i=1}^{i=k} P(A_i). \tag{10.2}$$

This justifies the **Bonferroni correction** that is generally applied to the probability of getting Type I errors when k multiple comparisons are being made for an overall (family wide) probability of getting a Type I error, α. In this application it produces a corrected (per-comparison) cutoff,

$$\alpha' = \frac{\alpha}{k}, \tag{10.3}$$

for each comparison (where α refers to the **family wide criterion**, α_{strict}). Once you find a corrected cutoff, α', you can look up, or use suitable software to find, $Z_{\alpha'}$ or $t_{\alpha', df}$ as a cutoff for each comparison.

Bonferroni's correction is very useful wherever you need to distribute the probability among categories or factors, regardless of whether you want to make multiple inferences, produce a set of comparison corrected CIs, or whatever. It is becoming a common part of statistical analysis in numerous applications, and it is named after its creator, C.E. Bonferroni (1935, 1936)[2] It does **not** require the treatment classes to be **independent**. It can be applied to comparisons following most global statistical tests, including ANOVA, as well as Pearson's χ^2 (Section 12.1), the Kruskal-Wallis (Section 14.3), and Friedman's (Section 14.5).

The Bonferroni correction is conservative and is becoming the most popular correction for multiple comparisons. But as R.S. Rodger (1973) points out, you can only use it with less than all possible multiple comparisons *iff* **you a priori specify which comparisons**. Unplanned multiple comparison approaches really cut into power because there are

$$k_{max} = \frac{m!}{2!\,(m-2)!} = \frac{m\,(m-1)}{2}$$

possible comparisons. This is also true for the Sidak correction (below) and any valid approach to multiple comparisons. The Bonferroni correction is demonstrated in the second (alternative) part of Section 10.1.5.1.

10.1.3 Sidak's correction

For multiple comparisons, and **assuming no dependence**, with an overall criterion, α_{strict}, you can use the **Sidak correction** (Dunn 1961, Sidak, 1967), which[3] uses a per comparison adjusted cutoff,

$$\alpha' = 1 - (1 - \alpha)^{1/k}, \tag{10.4}$$

and can also be written $\alpha' = 1 - \sqrt[k]{1 - \alpha}$, if that is a more familiar notation. In either case, α, as used in these equations, refers to α_{strict}, which could (in a non standard usage) be other than the global criterion, α. Once you find a corrected cutoff, α', you can look up, or use suitable software to find $Z_{\alpha'}$ or $t_{\alpha', df}$ as a cutoff for each comparison. A formal derivation of the Sidak correction might start with the family wide maximum permissible probability of not getting a single Type I error, $(1 - \alpha)$, among any of the k comparisons. The use of the kth root can be seen as viewing, say, $(1 - \alpha)$ as the result of multiplying the independent probability $(1 - \alpha')$ times itself k times, as per the multiplication law of probability that refers to independent probabilities, and then working back.

---

[2]In some contexts the Bonferroni correction is called the Holm–Bonferroni method due to a later contribution by Holm (1979). But, the Holm approach to using the Bonferroni correction may need further theoretical consideration before deciding to apply it in research settings where a criterion, α, and the power, $1 - \beta$, are both of interest. Its proponents believe that it is better for large numbers of comparisons, but its detractors suspect it might be too liberal.

[3]Do not confuse Dunn's contribution to the Sidak correction with Dunnett's test, which may not be sufficiently conservative.

Like Bonferroni's correction, **Sidak's** correction can be used to adjust for testing k hypotheses such as for multiple comparisons, **but** only for **independent** comparisons. After rounding, for small numbers of comparisons, it gives the same result as does the Bonferroni correction, but, for **not-so-small to large** numbers of comparisons with no dependence, it is **slightly less costly** in terms of sample sizes.

Because the Sidak correction depends on the multiplication law (Subsection 1.4.3), it requires the treatments to be **independent** of each other. However, this is often a reasonable assumption in most well designed statistical analysis of the kind discussed here.

As is the case for the Bonferroni correction, if you want to use the Sidak correction with less than all possible multiple comparisons, you have to **a priori specify** which comparisons (Rodger 1973). Otherwise you will have to use such a correction with all $k_{mac} = \frac{m(m-1)}{2}$ possible comparisons, and unplanned multiple comparisons are **only permissible** following a statistically significant global test (Sokal and Rohlf 1969).

10.1.3.1 Decreeing the per comparison α'

Assuming no dependence, you can use a reversed Sidak correction, Equation 10.1,

$$\alpha_{strict} = 1 - (1 - \alpha')^k \ ,$$

to make k preplanned comparisons. Note that $\alpha = 5\%$ is the largest value that will ever be acceptable to your peers and sponsors. That is, you can a priori decree the probability, α' that you are willing to take for each comparison, provided that the resulting family wide $\alpha_{strict} \leq 5\%$, given k =the maximum number of comparisons you are committing to make.

10.1.3.2 Example Careless vs Sidak correction:

A cautionary tale Suppose you were to compare $k = 15$ related, but **independent**, treatments against the same control, and you made **each comparison** a statistical test, each with its own criterion, $\alpha' = 5\%$. This means a probability, $p = 0.05$, of getting a Type I error for each of the tests. That is, in this **atypical instance** the corrected cutoff is $\alpha' = 0.05$ of getting a false statistical significance and $q = 1 - p = 0.95$ for each comparison. Now use a capitol $P_{(...)}$ for the overall probability, including the summed probabilities of all the comparisons for ... comparisons. Assuming **H0:** THERE IS NO DIFFERENCE, the probability of getting $Q_{(0)}$ (the probability of getting no false statistically significant outcomes) in $k = 15$ trials is the complement of $P_{(r>0)}$ (the probability of getting any statistical outcome). Assuming randomness, you can find the probability of $r = 0$ by using the binomial theorem (Equation 3.1) to find Q_0, the probability of getting no statistically significant outcomes:

$$Q_{(0)} = \binom{15}{0} p^0 q^{(15-0)} \ .$$

Since the probability of all the outcomes has to equal 1 (100 percent), we can then use $P_{(r>0)} = 1 - Q_{(0)}$. The relatively easily calculated Q_0, the probability of finding just 0 outcomes, is the complement of $P_{(r>0)}$, the probability of all the other possible outcomes together.

Result Working this out requires you to recall that $\frac{r!}{0!(r!-0)!} = \binom{r}{0} = 1$, because any value to the ^{0}

power equals 1, and 0! is also defined to be equal to 1. So, with $\alpha' = 0.05$,

$$Q_{(0)} = \frac{15!}{0!(15! - 0)!} (0.05)^0 (0.95)^{15} = 1 \times 1 \times 0.95^{15} = 0.4633 \,,$$

or simply $(1 - \alpha')^k$. Since $Q_{(0)} = 0.4633$ is the probability of getting no false positives by chance alone, its complement, $1 - P_{(r>0)} = 0.5367$, estimates α_{strict}, the **comparison-wide** probability of getting at least one Type I error. This

$$P_{(r>0)} = 1 - (1 - Q_{(0)})^k$$

is equivalent to

$$\alpha_{strict} = 1 - (1 - \alpha')^k \,.$$

This α_{strict}, about 54%, is a rather disturbingly high probability of observing at least one false statistical significance.

This example also demonstrates a derivation of the Sidak correction, Equation 10.1.

Both Bonferroni's and Sidak's corrections are demonstrated, using the same inputs, in the second part of the example in Subsection 10.1.5.1

10.1.4 You have a problem if you do not preplan for multiple comparisons

Requirements for multiple comparisons that are **not preplanned** are that the global test must have been statistically significant (Sokal and Rohlf 1969) and you must find a' as if you were going to test all the possible comparisons. You can then only declare statistical significance using α' for each comparison. You do not have to run all the comparisons, but the ones you do run **must use the same** α' as you **would use to run them all**.

Actually, as in the following illustrations, you can do **unplanned** multiple comparisons if you take the **loss** in efficiency associated with doing all the possible comparisons. But, it would be folly to undertake a global test without proper preplanning of any multiple comparisons. This is especially true if you were intending the global test to be an ANOVA, where the chance of confounding through heteroscedasticity is introduced by one or more of the treatments. The ANOVA is also suspect in the presence of non normality (which is almost always present).

Applying statistical tests for which you have **not already planned** before collecting or examining the data is very poor science. The inferential method depends on the estimation of the probability that the rejection of a hypothesis via statistical significance when it is true is a coincidence that is too unlikely to be true. Examining the data and then looking for a test afterwards, without preplanning it, tends to defeat this goal.

10.1.5 Comparing the Bonferroni's with Sidak's

The Bonferroni correction amounts to a rather straightforward division of α into equal parts shared among the comparisons, but the Sidak correction applies the multiplication law (Equation 1.4) to the confidence, a derivation that assumes all the comparisons are independent. Sample size estimates using approaches that involve the Sidak correction sometimes require smaller (never larger) sample sizes than those that involve the Bonferroni correction. However, because it requires independence, the Sidak correction should be used only if there is reason to safely assume complete independence.

Because the Sidak correction allows you to assume a tiny bit more power for the same sample size, you might prefer the Sidak correction if you are **certain of independence**. Beware, even the Bonferroni correction might be messed up by excessive dependence. Otherwise, it is safest to consider the Bonferroni correction as a little more resilient to some types of dependence than the Sidak correction. Reality is complex, and the Bonferroni correction is more conservative, meaning that it is less likely to produce misleading results.

Spreadsheets and some Java applets tend to be **inaccurate** for very small probabilities. For instance, you may have some difficulty in finding $Z_{\alpha'}$ values if α' is equal to some very tiny probability. Maybe tables generated using the R language, or the very similar S language, would be better. Classical sets of tables such as Fisher and Yates' *Statistical Tables for Biological, Agricultural, and Medical Research* (1963) and Pearson's *Tables for Statisticians and Biometricians* (1914-1931) might be useful in some instances. Question 3 is an exercise that involves the Sidak correction. Here is an example involving counts:

10.1.5.1 Example: Multiple Comparisons, Preplanned vs Not Preplanned

Facing reality (subsequently) Assume a researcher counted the number of fish out of $n = 30$ salmon that had sea lice, Lepeophtheirus salmonis, from each of $m = 6$ locales, and a **global test** found the difference among sites statistically significant at the $\alpha = 0.01$ level. It was convenient that the fish populations from which the sample was taken were known to tend to average about 50% with sea lice. The researcher decided to make multiple comparisons, but had **not preplanned which comparisons**. Because this decision is unplanned, the effect in terms of power is as though the researcher had chosen to do all the k possible comparisons regardless of which the researcher then chose to actually calculate. That is, **after having seen** the outcome of the global test, the researcher must adjust for **all** the possible comparisons, regardless of the choice of which to compare. This required a much lower cutoff, α', for each comparison, and that means that a much greater difference between a pair of treatments was needed to declare statistical significance. Independence appeared to be obvious, suggesting that the Sidak correction would have worked. But, to be conservative, the researcher used the Bonferroni correction. The smallest reasonable probability of getting at least one Type I error is $\alpha = 0.05$, and this is a two tail test because it is about 'difference'. This unplanned choice to do multiple comparisons involves taking combinations of $m = 6$ things $r = 2$ at a time. So, $k = C_2^6 = m!/r!(m-r)! = 6!/(2!4!)$ or $720/48 = 15$ possible comparisons, and the researcher had to adjust α' as though for all $k = 15$ comparisons (Equation 2.3), whether they were to be made or not. The application of Equation 10.3 shows that to declare statistical significance, the researcher would have had to observe a $p\_value \le \alpha'/2 = (0.05/15)/2$ or $\alpha'/2 = 0.001667$ (which happens to be $\frac{1}{6}$ of 1%) for any individual comparison. That would require an awkwardly high $Z \ge 2.935$.

(An Alternative to having to face reality Assume that **before taking the data** (or at least before seeing it or discussing its contents), the sea louse researcher decided multiple comparisons would be needed and the researcher **recorded** what multiple comparisons were to be made **in advance**. In that case the number of comparisons would have been up to the researcher, and the global test would not have to be significant. For instance, this approach could include only a control (say, wild stock), and this would be compared to samples from each of 5 fish farms. The hypothesis pair for each comparison would be
H0: THERE ARE NO DIFFERENCES
HA: THE FISH FARMS ARE DIFFERENT FROM THE CONTROL (WILD).
This would have been an **a priori commitment** to do these particular two tail comparisons and

no others. Because sea lice studies often lead to civil suits and even prosecutions, as well as regulatory action, it would have been essential to have this plan notarized with a date before starting. This would have justified using only $k = 5$ comparisons. If the researcher had a priori chosen to use the **Bonferroni** correction, $(0.05/5)/2 = \alpha' = 0.005000$, this requires a $Z_{0.005000} = 2.5758$ cutoff for each comparison. However, if the researcher had chosen to use the **Sidak** correction, referring to Equation 10.4, the researcher would have found that given $k = 5$ comparisons, then $(\alpha'/2) = \left(1 - (1 - 0.05)^{1/5}\right)/2$ or 0.005103 (a little more than half a percent). This would require a cutoff of $Z_{0.005103} = 2.5688$, a small advantage seen in the second decimal place (third significant figure). When compared with the Bonferroni correction, the Sidak correction would, in effect, provide a tiny advantage in favor of declaring statistical significance (apparent power). However, the Sidak correction is only acceptable if the sea lice counts were really independent of one another.

Note that, regardless of which correction was used, these comparisons were for counts and therefore used the normal approximations ('Z test') and did not refer to a t distribution. In principle, the same techniques for calculating α' would be used whether Students t test, the binomial approximation, or some nonparametric test was used.

10.1.6 Best multiple comparison design

When I was a student, the first edition (1969) of Sokal and Rohlf's Biometry textbook was rightfully an ongoing sensation. One of the professors stood up in seminar and said, "I am not going to use a book that makes it more difficult to detect statistical significance." Well, reality is reality, and the previously accepted method of doing multiple comparisons was wrong.

Planning ahead is a big step toward success and that means choosing an appropriate design. So what is the best design for multiple comparisons? Simplicity is always best and that fits in with the increasingly popular Bonferroni correction. You may find it most convenient to designate an a priori control and then use either the Bonferroni correction or the Sidak correction. If you only compare the control with each non-control treatment then you will only have $k = m - 1$ comparisons, where m is the number of treatments, including the control.

Designating an **a priori control**, implying a commitment to $k = m - 1$ comparisons, and using the **Bonferroni correction** is going to work out **best** for most tests. However, the Sidak correction is useful for understanding some approaches and is sometimes a justifiable cost saving approach to analysis *iff* independence is indisputable.

There are a variety of other multiple comparison tests, presented in various other texts and offered by various computer packages. For instance, one of the more acceptable alternatives is the SNK used with the ANOVA (but not included here); even it has a number of pitfalls and may mislead you. Some, but not all, of the alternative multiple comparison tests require that the sample sizes are all equal. Regardless of the approach you use, equal sample sizes are to be preferred wherever they are possible, but that is not always possible. In short, Bonferroni's approach is best, but Sidak's is a little less costly and often quite acceptable, whereas the others are often a bad choice, or involve massive computer processing.

Question 1 is both an example of, and an exercise in, unplanned multiple comparisons (not the preferred design).

10.2 Multiple Comparisons from ANOVA

In application, global tests such as an ANOVA do not usually make much sense without pair by pair comparisons between at least some of the pairs of treatments. Most statistical packages have provision for this, but it is not clear how they relate to the power of the multiple comparisons.[4] The standard error of the mean for making pair by pair comparisons using Student's t tests for multiple comparisons following an ANOVA is

$$S_{\bar{x}} = \sqrt{MS_W \left(1/n_1 + 1/n_2\right)}, \tag{10.5}$$

with df_W as the degrees of freedom for each comparison. If the $df_W < 80$, the sample size should ideally be at least twenty five times the skew, $(n_1 + n_2 = \mathbf{n}) \geq 25 \times skew$ for each comparison. In any case, the accepted minimum requirement for assuming normality for the ANOVA also requires $df_W \geq 20$, which is barely adequate for making subsequent multiple comparisons using Student's t test. To do multiple comparisons of pairs of treatments following an ANOVA, it is necessary to use corrections such as the Sidak's correction or the Bonferroni correction. The $S_{\bar{x}} = MS_W$ estimate based on the ANOVA with df_W can also be used to derive a set of CIs from an ANOVA, even CIs with corrected cutoffs, if the correction is **made clear** to the reader.

This approach is in line with the classically accepted use and application of the ANOVA. Even so, there is evidence that the requirements of the ANOVA might not often be met even when these standards are met, and its use is likely to eventually be discontinued in most fields. However, the principles in this chapter that are not specific to the ANOVA itself still apply to any alternative approach that you may choose, such as nonparametric global tests.

10.2.0.1 Example: Multiple Comparisons ANOVA Results

Sorting out the effects (continued from Subsection 9.2.0.1) Assume that the Amadora team had, a priori, chosen to only compare their treatments to a control. That is, because C was a priori designated a control, they had committed to make only $k = 2$ comparisons. So they need a corrected cutoff, α', such that the overall comparison-wide criterion is α_{strict}. They can use either the Bonferroni correction, Equation 10.3, $\alpha' = \alpha_{strict}/k$, or the Sidak correction, Equation 10.4, $\alpha' = 1 - \sqrt[k]{1 - \alpha_{strict}}$, where k is the overall number of comparisons.

Analysis: Assume they a priori decided to use a family wide criterion of $\alpha = 5\%$ with a two tail test for each comparison and to use the Bonferroni correction, meaning $\alpha'/2 = (\alpha_{strict}/2)/2 = 0.0125$, for the $k = 2$ comparisons. Once again, the individual treatment means were $\bar{x}_C = 13.9249$, $\bar{x}_D = 16.3450$, and $\bar{x}_E = 11.0358$, all with equal sample sizes, $n = 30$. The mean square within, $MS_W = 24.4222$ with 87 df_W, was the best available estimate of the underlying variance, σ^2, to be used for such comparisons. They looked in a table of t such as Appendix 3 and found that $t_{0.0125, 87} = 2.2809$. They calculated $S_{\bar{x}} = \sqrt{MS_W \left(\frac{1}{n_1} + \frac{1}{n_2}\right)}$ to estimate

$$t_{\alpha/2, df_W} = \frac{\bar{x}_i - \bar{x}_C}{S_{\bar{x}}},$$

and because all the sample sizes were equal, $n = 30$, they only had to calculate

$$S_{\bar{x}} = \sqrt{MS_W \left(2/30\right)} = 1.275989$$

---

[4]BEWARE; many commercial packages include the deprecated 'small', 'medium', and 'large' approach to making multiple comparisons.

to estimate the treatment standard error of the means. Then for treatment D versus the control C,

$$t = \frac{16.3450 - 13.9249}{1.141625} = 2.1198,$$

and for treatment E vs the control C,

$$t = \frac{11.0358 - 13.9249}{1.141625} = -2.5306.$$

Because $t_{\alpha'/2, df} = 2.2809$, only E vs control C is statistically significant at an **overall** $\alpha_{strict} = 5\%$ level when proper adjustment is made for multiple comparisons. So, Amadora can only conclude that treatment E repels males. Perhaps because of the assumptions associated with an ANOVA, these comparisons come out a little differently when using a nonparametric test (Subsubsection 14.3.0.1).

Discussion: Let's note that they neither considered sample size nor had a pilot study to estimate skew. If they had not a priori designated $k = 2$ comparisons, and then went ahead anyway and made two comparisons, or even **any** number of comparisons, they would have to adjust for all the possible comparisons. That is, even if they had just made two comparisons, but not preplanned them, they would have to adjust as though they were making all $k = \frac{3!}{2!(3-2)!} = \frac{3 \times 2}{2} = 3$ possible comparisons. Then $\alpha'/2 = \alpha_{strict}/6 = 0.008333$ would have to have been used, which would mean that $t_{0.008333, 87} = 2.4412$ would have required an even greater effect size to discover if any of the pairs being compared were statistically significantly different.

The Sidak correction would have been a little less conservative and it tends to be less costly in terms of sample size when very large numbers of comparisons are planned. However, the slightly more conservative Bonferroni correction is more intuitive and involves less assumptions. It is also more likely to be available to you as a built in procedure in statistical computing packages.

Using a t test with MS_w^2 as the estimate of S^2 is really a continuation of the ANOVA with the equivalent of a subsequent F test, and is not as robust as a regular Student's t test. ANOVAs lose power when their assumptions are not met. Maybe that is why when this data was used in a nonparametric equivalent of this example the results came out somewhat differently (Subsubsection 14.3.0.1).

Not all real effects are big enough to be detected at $(\alpha/2)_{strict}/k$. Maybe they should use the MS_W from this experiment as a pilot study estimate of σ^2 and proceed to a new test. That way they would be able to a priori estimate a large enough sample size so that whatever reasonable minimum difference, D_{min}, they choose will be likely to be statistically significant at α' for each preplanned multiple comparison, assuming it is equaled or is exceeded. That is, if you only come close to statistical significance and cannot bear to let go, try a larger scale experiment using your latest estimate of σ^2 to estimate a more appropriate sample size. But note the following caution, especially concerning α and D_{min}.

> **CAUTION:** Running the same test over and over too many times is like flipping a coin until you finally get a tail. Repeating over and over until you get statistical significance **changes the probability**, so all the test results must be somehow combined and reported (Section 10.8). To do otherwise is cheating. Changes, as in experimental procedure, may change the variance and defeat anything that might have been gained by using your first test as a 'pilot experiment' and it will greatly reduce the accuracy if you combine the likelihoods (probabilities of the outcomes).

So, as the number of comparisons, k, increases, it becomes more difficult to detect statistical significance for multiple comparisons. For instance, it is not unusual for agronomists to use $m = 8$ treatments, meaning that they could do $k = m - 1 = 7$ comparisons if they used a preplanned control, but they would have to use $k = \frac{m!}{2!(m-2)!} = \frac{m(m-1)}{2} = 28$ if they had no verifiable a priori comparison plan ready before data collection. For simplicity, let's use Z to illustrate a comparison which would be similar if we more properly used t. For $\alpha' = 5\%$ and $k = 7$, an overall Bonferroni corrected α is $(\alpha'/2)/7 = 0.025/7 = 0.00357$, they would need $Z \geq (Z_{0.00357} = 2.690)$ for a single corrected comparison to declare it statistically significant. But, for $\alpha' = 5\%$ and $k = 28$, an overall Bonferroni corrected $\alpha'/2 = 0.025/28 = 0.000893$, and requires $Z \geq (Z_{0.000893} = 3.124)$ for any single corrected comparison to be statistically significant. Unfortunately, it would take a very large sample size per treatment to achieve the necessary power to detect any reasonable minimum difference, except a very large one, even with just $k = 7$ comparisons. That is, $k = 28$ would usually be impossibly expensive in terms of sample size even if the RMD were very large. (Properly using t rather than Z would demonstrate that this problem is even bigger than indicated by tentatively using Z.)

If the agronomists used the Sidak correction instead of the Bonferroni correction for the unplanned $k = 28$ comparisons, then $\alpha' = 1 - \sqrt[k]{1-\alpha} = 1 - \sqrt[28]{0.95} = 0.001830$, and they would have to use $Z_{\alpha'/2} = Z_{0.000915}$. Thus, assuming complete independence among the treatments (which seems obvious), they could declare statistical significance if $Z \geq 3.117$. The difference from the Bonferroni correction only appears in the third significant figure. So, even with $k = 28$ comparisons, assuming independence, there is only a little difference between Bonferroni's correction and the Sidak correction. Note that values of Z that are greater than 3 usually require impractical huge samples.

10.3 Why Global Tests?

As contrasted with tests which simply compare one treatment with one other treatment, ***global tests*** try to detect if there is at least one significant difference among any of more than two treatments. But, humans almost always wish to know how each thing relates to each other thing, or at least to a control (a standard).

Rosenthal, Rosnow, and Rubin (2000) state that "Global questions seldom address questions of real interest to researchers, and are typically less powerful than focused procedures." Their criticism is definitely true if the global test is used without multiple comparisons. But the ANOVA is supposed to be more than simply a global test in that it tries to find a more accurate estimator of the variance (hence the standard deviation) of the factor of interest by removing the effect of some factors that are not part of the question at hand. This improved estimator, the MS_W (also called the MSE or MS_E), assuming it works, can greatly improve the power and specificity of subsequent pair by pair tests. That is, by giving you a better estimate of σ^2, the ANOVA should be handy for both confidence intervals and for further contrasts. For pilot studies leading to sample size estimates, the MS_W potentially provides information for subsequent ANOVA testing, under the same conditions.

> **CAUTION:** the $MST = TSS/df_T$ from a pilot study would generally be a preferable estimate of the variance, σ^2, needed to estimate the sample size for a future single sample t test for sufficiently similar data. However, a MS_W would be appropriate for estimating the sample size of a subsequent ANOVA to facilitate a set of multiple comparisons.

It is hard to believe that the number of global tests being used in such areas as field crops is necessary. Are some of them mere copycats? Maybe, in some cases, the large expenditure used for testing $m = 4$ or $m = 5$ things at a time would be better spent just comparing the two most consequential things. The same investment would provide more power, with maybe a more reasonable cutoff than 5%, which equals one chance in twenty of a false positive. It would be logistically simpler too. The larger two treatment tests might also include regressions as well as inferential testing. For pairing or blocking, whether on farm or ward by ward, using paired t tests might be excellent, and also avoid the issue of the sensitivity of the ANOVA to deviations from normality.

An ANOVA or other global test can be an expensive luxury and possibly a trap, because if its assumptions are not met, the effective α might be unacceptably larger than the accepted maximum of 5%, or its power might be reduced, causing it to fail to detect an acceptable minimum difference (see Subsubsection 9.3.0.1).

10.4 Sample Size for ANOVA

Researchers are not likely to merely want to know that there is a difference in the average mileage of similar models of Fords, Chevrolets, Hondas, Toyotas, and Subarus when they must be scrapped. The researchers will usually ask such questions as which is going to go further, Hondas or Toyotas? So, the overall sample size, **n**, you need for your global test will be the sum of the individual sample sizes you need for the multiple comparisons. To find it, you need to specify an a priori targeted minimum difference, D_{min} , and then find a cutoff, α' or $\alpha'/2$, and β' for each individual comparison, out of the k comparisons you intend to make with criterion α and power $1 - \beta$.

Beta inflation must be taken into account when considering multiple comparisons, and you correct for it the same way as for alpha inflation, using either the Bonferroni correction, Equation 10.3, $\beta' = \beta_{strict}/k$, or the Sidak correction, Equation 10.4, $\beta' = 1 - \sqrt[k]{1 - \beta_{strict}}$, where k is the overall number of comparisons. This will enable you to control the power of each comparison, $(1 - \beta')$.

Some aspects of the following may also apply to the experimental design for other global tests such as Pearson's chi square test (Chapter 13), the Kruskal-Wallis test (Section 14.3), or Friedman's test (Section 14.5). However, as for Student's t test, you also need an appropriate estimate of σ^2 to find the sample size for the ANOVA.

Incidentally, there is also a sample size estimation procedure for the nonparametric Kruskal-Wallis test (Subsection 14.4) that can be used to **avoid having to find a way to estimate σ, or σ^2**
.

Ideally, for the estimate of σ^2, you need a MS_W **with** $df_W \geq 20$ **from a pilot study that used exactly the same** experimental conditions and design (except for the sample size, and without needing to find F), and the **same kind of ANOVA**. That is, to get a relevant estimate of the variance, σ^2, the pilot study should ideally be an ANOVA which is a miniature of the full scale test you ultimately intend to perform (an RCBD for an RCBD, a one way ANOVA for a one way ANOVA, etc.). A three treatment pilot study with the designated control (if any) and the other treatments chosen, if possible, to represent what you suspect will have representative variances, might be adequate. A t test with $n \geq 30$ and two treatments (or a paired t test with >15 pairs) might provide a usable but not ideal estimate of $S \approx \widehat{MS_W}$. One treatment should definitely be the control (if you plan to use a control). A variance estimate of the control alone could provide a rough estimate of the MS_T and give you a somewhat inflated sample size. Finally, and worst of all for overestimating the sample size, you could simply find the variance, $S^2 = \sum (x - \bar{x})^2 / df$, of one treatment, preferably the control and minimally with $n \geq 30$, under exactly the same conditions as those from which you intend to get your final data. Unfortunately the last two alternatives do not partition out the effect of differences among treatment means. That is, they might seem to work theoretically, but in application they might overshoot the mark. However, within reason, accepting a possible small overestimation of sample size is not a bad thing when you can afford it. This is especially true because any deviation (which is always likely) from normal will reduce the power of an ANOVA. Whichever alternative you use, round your variance estimates up.

Typically you would a priori, that is, before you obtain the data, choose the values of $\alpha = \alpha_{strict}$ and $\beta = \beta_{strict}$ and work from there. Be careful; it would not be considered prudent to accept a power less than 80% (the more savvy granting agencies would not be happy with less). It is good practice to choose a minimum reasonable difference, D_{min}, in original units of measure for making multiple comparisons. You could use the Sidak correction (Section 10.1.3), $\alpha' = 1 - (1 - \alpha_{strict})^{1/k}$, to estimate the **per comparison** cutoff to attain your chosen overall α_{strict}, if you can safely assume independence. At a possibly slightly greater expense, you can more conservatively use the Bonferroni correction (Section 10.1.2), $\alpha' = \alpha/k$, which is not as sensitive to a lack of independence. Likewise for $\beta' = 1 - (1 - \beta)^{1/k}$ and $\beta' = \beta/k$, to correct with respect to power, $1 - \beta'$.

If possible, use the df_W and MS_W from your pilot study as the variance estimate, and apply a modification of Equation 7.8,

$$\hat{n}' = \frac{2\left(t_{\alpha'/2, df_W} + t_{\beta', df_w}\right)^2 MS_W}{D_{min}^2}, \qquad (10.6)$$

to take account of t's adjustment for variance estimation. (Otherwise, it might be necessary to start with Z and then iterate using $t_{..., 2\hat{n}'-2}$ and S until $\hat{n}'$ stops changing.) The resulting value, $\hat{n}'$, is the sample size estimate **for each treatment** involved in the comparison.

The total sample size, $\widehat{\mathbf{n}}'$, is the sum of the individual estimates ($\hat{n}'$) as though they are all able to be equal. However, even if they cannot be equal, the total $\widehat{\mathbf{n}}'$ is still at least

$$\widehat{\mathbf{n}}' \geq m\hat{n}'. \qquad (10.7)$$

That is, in situations (later in this chapter) where you are **forced to use unequal sample sizes**, the estimates of the sample sizes of samples $\hat{n}'_j$ over which you have control might have to be **increased** (adjusted upward).

CAUTION: Do **not** use 'small', 'medium', or 'large' in any sample size estimation. Their appropriate values would, if known, change substantially for seemingly minor changes in experimental design, sampling protocol, subject area, or observational environment. The published values are from a very restricted area of educational testing which is almost never duplicated.

Cohen devised power analyses that involve various 'effect sizes' that were popular until recently. These effect sizes are an ongoing disaster. They are **not** appropriate for defining your reasonable minimum difference or for estimating a sample size or a power. Their 'small', 'medium', and 'large' values were **arbitrary** (Cohen 1988, p. 25). They should never be used for sample size estimation (e.g. Paterson, et al. 2016). Yet they appear in the most expensive and popular software suites.

10.4.1 ANOVA, $\widehat{\mathbf{n}}'$ with equal treatment sample sizes, n'

Note that the final result, $\widehat{\mathbf{n}}'$, is dependent on the number of treatments, m, and on your a priori choice of how many comparisons, k, you intend to perform. you might get by without considering skew if, say, the $df_W > 80$. Ironically, your estimate of skew might not be especially good if $df_W \ll 80$.

10.4.1.1 Example: A Priori $\widehat{\mathbf{n}}'$ for ANOVA With Multiple Comparisons

Example 10.4.1.1 (below) illustrates how to do this in practice.

Global ANOVA a priori equal n' Pretend that Amadora's team (from Section 9.2) had chosen to compare their treatments to the control, and that they first decided to make an a priori power analysis. Because treatment C was, a priori, designated as the control, and there were only $m = 3$ treatments, including the control, they were committed to make only $k = 2$ comparisons. The object was to find a corrected cutoff, α', such that the overall family wide criterion would be $\alpha = 5\%$ and the overall power would be $1 - \beta = 90\%$. Either the Bonferroni correction (Equation 10.3), $\alpha' = \alpha/k$, or the Sidak correction (Equation 10.4), $\alpha' = 1 - \sqrt[k]{1 - \alpha}$, where k is the overall number of comparisons, would be acceptable. The Sidak correction is a little less costly in terms of sample size, when very large numbers of comparisons are planned, but they were only going to make two comparisons. The Bonferroni correction involves less assumptions and is more likely to be available as a built in procedure in statistical computing packages. But even though the Bonferroni correction might be a slightly better choice, we are using this example to demonstrate how to use the Sidak correction as part of the sample size estimation.

Analysis: They had, a priori, decided on a criterion of $\alpha = 5\%$ for the overall probability of getting a Type I error, rejecting **H0** when it was, in fact, true. They had performed lots of calibrating experiments using Playboy magazine center pages and blank tax forms, both treated with the control mixture. So they were able to, a priori, decide on $D_{min} = 3.00$ attraction units as the reasonable minimum difference. They had previously obtained an estimate of $S^2 = MS_w = 24.5$ with 87 df from what they had come to consider to be just a pilot test, and they intended for all the sample sizes to be equal. Applying the Sidak correction[5], $\alpha' = 1 - \sqrt[k=2]{0.95} = 0.02532$. They then used $\alpha'/2 = 0.01266$ because **each comparison is a two tail test**. They found that $t_{\alpha'/2=0.01266,\, df=87} = 2.2757$. They wished to have what they saw as sufficient power, $1 - \beta = 90\%$, so $\beta' = 1 - \sqrt[k=2]{0.90} = 0.051317$, and the corresponding $t_{\beta'=\alpha/2=0.05132,\, df=87} = 1.6496$. These values were from an online calculator. However, because online programs are not always trustworthy, they checked a couple of them against each other, and checked the results against table entries near their input probabilities.[6]

Because the comparisons are based on MS_W, they used t with $df_W = 87$ to get a sample size estimate, $\hat{n}'$, **for each treatment**, using Equation 10.6,

$$\hat{n}' = \frac{2\left(t_{\alpha'/2,\, df_W} + t_{\beta'\, df_W}\right)^2 MS_W}{D_{min}^2} \text{ . So,}$$

$$\hat{n}' = \frac{2\left(2.2757 + 1.6496\right)^2 24.5}{9.000} = 83.888 \text{ .}$$

This estimates the sample size you should use for one column; the leading 2 is only there because this calculation is done as though finding the sample sizes for **each member** of a comparison. The result should be rounded up to $\hat{n}' = 84$. So, recalling that $m = 3$ treatments (including the control) in this example, they used Equation 10.7,

$$\widehat{\mathbf{n}}' = m\hat{n}',$$

so $\widehat{\mathbf{n}}' = 3 \times 84 = 252$ total samples for the ANOVA.

---

[5] $\sqrt[2]{1 - \alpha} = \sqrt{1 - \alpha}$

[6] The best and most convenient online calculator for t values, at the time of this writing, is http://www.danielsoper.com/statcalc3/calc.aspx?id=10 (site courtesy Dr. Daniel Soper). However, proper application of the R language would be as good.

Discussion: This is a reasonable total sample size estimate for this global test, the one way ANOVA, to attain an a priori family wide criterion of $\alpha = 5\%$, and a 90% power for an a priori selected $k = 2$ comparisons, given the stated estimate of σ^2 (as MS_W) and their chosen reasonable minimum difference of $D_{min} = 3$ attraction units.

Although it is not part of estimating a sample size, you might want to look ahead and see that they have already chosen the cutoff, $\alpha' = 0.02532$, and, with a little thought, they could estimate the df_W,

Source		df
Among	$3 - 1 =$	2
Within	$251 - 2 =$	249
Total	$252 - 1 =$	251

to use with it. However, they will still need to run the experiment to get SS_W, and use this new MS_W estimate of σ^2 with its df_W, for each comparison against $t_{\alpha'/2=0.01266, \, df=249} = 2.2495$ (after conservatively rounding t **up** in the fourth decimal place).

These researchers had initially missed a practical sample size for providing their hoped for power by 165 women (55 too few per treatment). Although a family wide $\beta = 10\%$ is widely accepted, they could have chosen $\beta_{strict} = 5\%$, which would be safer and require an even larger sample size. Or they could have been daring and used the minimal $\beta_{strict} = 20\%$ for an even lower sample size. As a pilot test, their preceding test was rather large as pilot tests go, but the larger the pilot test, the better the estimation of the sample size. Research often needs to be done in two steps, and it can be quite expensive in terms of time, sample size, and money.

A discussion, similar to the above **overall sample size estimation**, for another group of **global tests**, Pearson's chi square tests, is in Section **13.2**.

Question 7 provides another example of this estimation process as well as providing you an opportunity to practice.

10.4.2 A priori unequal sample sizes, one way ANOVA multiple

It is best to have equal sample sizes for an ANOVA. However, there might be some situation in which the available sample size of one or more treatments has a 'fixed' maximum that is less than the estimated sample size, $\hat{n}$. That would mean that the other treatments that are going to be compared with fixed sized treatments must have larger sample sizes in order to attain the targeted power. (Watch out; the italicized n and the bold **n** are very different symbols. The latter is a total, often the sum of two or more of the former.)

It is necessary to a priori specify D_{min} (and specifically describe the k comparisons that are intended). Either the Bonferroni correction or the Sidak correction can be used to estimate α' and β'. Then, $\hat{n}'$ can be **estimated** by using these **corrected inputs** in Equation 10.6. Provisionally assign this sample size to every non-fixed sample size.

You must choose the **smallest fixed sample size** and adjust the sample size of **every** adjustable (unadjusted) sample size treatment with which that treatment **is to be compared**. The **fixed** sample size in a comparison is designated n_f. The sample size of the **adjustable** provisional treatment sample size, $\hat{n}'$, must be increased to

$$\hat{n}'_A = 2\hat{n}' - n_f.$$

However, finding this provisional sample size, $\hat{n}'_A$, is only needed because it is necessary to correct for the loss of relative efficiency resulting from the fact that Equation 10.6 estimates sample sizes **as though** the sizes are going to be equal.

The next step requires two kinds of mean for each comparison, the arithmetic mean,

$$\bar{n} = \left(\hat{n}'_A + n_f\right)/2 = \hat{n}',$$

and the harmonic mean,

$$\bar{n}_h = \frac{2}{1/n_f + 1/\hat{n}'_A}.$$

If you are curious, you can estimate the *loss of relative efficiency* (Kifle and Desta, 2012; see Subsection 7.11.2) of this comparison with unequal sample sizes by

$$efficiency\_lost = 1 - \frac{\bar{n}_h}{\bar{n}}.$$

However, the loss of relative efficiency is already included, with rearrangement, in Equation 10.8 (below). That is, substituting in the equation for single comparison adjustment, $\widehat{\mathbf{n}}' = n_f + \hat{n}'_A = 2\hat{n}'$, gives a per comparison (2 treatment) sample size,

$$\widehat{\mathbf{n}}'' = \widehat{\mathbf{n}}' \left(2 - \frac{\bar{n}_h}{\bar{n}}\right), \tag{10.8}$$

which compensates for the relative efficiency that was lost by having unequal sample sizes. This must be rounded **up** to an integer. The adjusted sample size treatment of **each adjusted member** of the comparison will finally be

$$\hat{n}'' = \widehat{\mathbf{n}}'' - n_f.$$

Assign $\hat{n}''$ **to all adjustable** sample sizes of treatments to be compared with fixed samples of size n_f.

Now, *iff* there are **no more fixed** sample sizes, or *iff* **all** the corrected treatments sample sizes **have been estimated**, you are finished. You can **sum** all these to estimate the total sample size for the ANOVA,

$$\text{total } \widehat{\mathbf{n}'''} = \sum_{j=1}^{j=m} \mathbf{n}_j,$$

where this $\mathbf{n}_j$ can be any included sample size now assigned to one of the $j = 1, 2, \ldots m$ treatments, whether an **efficiency corrected**, a **fixed**, or an **unchanged** sample size.

But, *iff* there are **both** smaller samples and treatment sample sizes that have **not been included** in an adjustment process, you should **repeat** this process with the **next smallest** sample before summing the final sample size estimates. But **only** adjust sample sizes which were **not increased** by a preceding adjustment involving a smaller fixed sample. That way you **only make the estimates larger**, never smaller. Continue the process until either there are no unadjusted samples, or no more fixed treatment sizes. This will, of course, also increase the total, $\mathbf{n}'''$, for the whole ANOVA. **Unequal sample sizes** could mean that when you actually calculate the ANOVA, you will need to use Equation **9.4**.

If you have to correct for more than one size of fixed sample size, or if the sample size of the control is fixed, or if the difference between the fixed sample sizes and the estimated sample sizes seems large, then three place accuracy (attained by rounding up) in the final result might be the best you can hope for.

If the results seem ridiculous, put brain in reverse gear and start backing away from the situation. For instance, if fixed $n_f = 5$ and $\hat{n}'_i = 48$, you might not get a meaningful result comparing their treatments using a t test or any other test. Will 5 measurements contain enough information to make a valid decision in a test which requires such large total sample sizes? Also be aware that the targeted power can't be achieved by comparing two undersized fixed samples with each other. Unfortunately unequal sample sizes contribute to the harm caused by failures to meet the assumptions of the ANOVA, especially the assumption of homoscedasticity.

10.5 Unplanned ANOVA Comparisons

If you have analyzed unexamined preexisting data with built in sample size limitations, or if you have had data loss such as dropouts, you can only run an unplanned comparison by comparison power analysis of an ANOVA if you have also a priori set criteria[7] α, and have provided an a priori (or appropriately pre-existing) per comparison reasonable minimum difference, D_{min} (such as an increase in crop value equal to the treatment cost). If you have not, a priori, specified the comparisons you intend to make, you **must calculate** the corrected value of α' **as though** you were going to distribute the power among all possible comparisons. That might not be practical, because you would need to use $k = m!/ \left(2! \left(m - 2 \right)! \right) = m \left(m - 1 \right)/2$ samples **per comparison**. You also require a statistically significant ANOVA when you have not preplanned. **Do not use a criterion**, α, **larger than** 5%.

The principles are the same as for two sample subsequent power analysis, and you start with one of the following two variations, which are derived from Equation 7.18, by substituting the use of a Sidak or Bonferroni corrected α, and of MS_W as a best estimate of σ^2. For equal sample sizes this gives

$$\hat{z} = \frac{|D_{min}|}{\sqrt{MS_W \left(2/n \right)}} - t_{\alpha'/2, df_W} \quad n_1 = n_2. \tag{10.9}$$

But if you have **unequal** sample sizes, due to logistics, dropouts or otherwise lost data, then use

$$\hat{z} = \frac{|D_{min}|}{\sqrt{MS_W \left(\frac{n_1 + n_2}{n_1 n_2} \right)}} - t_{\alpha'/2, df_W}. \tag{10.10}$$

A positive lowercase $\hat{z}$ is **interpreted** according to the rule that

$$iff \ \hat{z} \geq 0, \ then \ \hat{z} = \hat{t}_{\beta', df_W} \ .$$

However, if your power is **disastrously** extremely low (it even can be that $\hat{z} < 0$), then you will have to use the rule that

$$iff \ \hat{z} < 0, \ then \ |\hat{z}| = \hat{t}_{1-\beta', df_W}.$$

This is the power of a single unplanned comparison between two treatments following an ANOVA.

Some authors will tell you that you can analyze the data and then decide to do an unplanned power analysis if you do this or its equivalent for every comparison, even if you have no a priori specifications. That would be mere **data exploration** unless you used some well established reasonable minimum difference very logically linked to the question at hand, such as a price-return relation. Likewise, you **could not use** any $\alpha > 5\%$. However, 5% is possibly acceptable because this maximum is established, and commonly expected anyway, whereas a greater value would be unacceptable. Even with 5%, the results would be open to question, as to whether they were fit for publication, or for remuneration as a conclusion forming the basis for a consulting report. Definitely **do not** report the results of such a completely post hoc procedure unless you clearly explain that it is post hoc.

---

[7]Generally α can reasonably be assumed to be the maximum $\alpha = 5\%$ if not specified.

10.5.0.1 Combining β for unplanned comparisons

For the **power of a whole ANOVA** that had comparisons, you could produce a family-wide un-planned power estimate; you must proceed as though you had planned to test every possible comparison. However, if you had a plan and some data was lost, you can, within reason (with respect to your plan), adjust your plan. Then to estimate the power of the 'family wide' group of comparisons for which you have estimated per comparison values of β', you could combine their probabilities using the Stouffer's z-score method in Section 10.8. For an unplanned subsequent estimate, it is handy to substitute the signed lowercase z estimate into Stouffer's equation, Section 10.8, to estimate

$$\widehat{z'} = \frac{\sum_{j=1}^{j=k} \hat{z}_i}{\sqrt{k}},$$

and finally apply the appropriate rule for interpreting $\widehat{z'}_\beta$ according to the sign of the result, as shown above (Section 10.5).

10.6 RCBD Multiple Comparisons

If a full dress pilot study is not practical for a random complete block ANOVA, then an S_d^2 from a paired t test, where the pairs use two of the intended treatments with, say, $n > 14$ pairs, might approach a satisfactory approximation of a MS_W if the pairing consists of an intended control with a key treatment or else with the most variable treatment in the group of treatments to be tested. However the sample size estimate, $\hat{n}$, will not be perfect, and will usually be a little large.

To avoid the sensitivity to the non normality problem of a RCBD ANOVA, you can use a non-parametric Friedman test instead(Section 14.5). You could then proceed to use more robust paired t tests of each comparison with its own variance. To estimate the sample size for Friedman's test, you could use the sample size estimated for the RCBD ANOVA (Subsection 10.6.1) times 1.0472 (see Subsection 10.6.1) to compensate for the slight difference in efficiency. It is possible that the effect of the non normality will be as large as it apparently was in the one way ANOVA example, in which case the use of a nonparametric test is likely to result in greater true power.

As with a one way ANOVA, subsequent multiple comparisons will **use** the MS_W and df_W of the randomized complete block ANOVA (Subsection 9.4) for **paired t tests** for multiple comparisons, **and for CIs** of the differences which also serve as the ESs. The **effect of pairing** (or rather, grouping) has **already been accounted for** by subtracting SS_B to obtain SS_W on the way to MS_W, the estimate of σ^2 with $df_W = df_T - df_A - df_B$ degrees of freedom. Then, for multiple comparisons after the simple RCB design covered in the preceding chapter, test between pairs, using

$$t = \frac{\overline{d}}{\sqrt{MS_W/m_B}} = \frac{\overline{d}}{S_{\overline{d}}},$$

where m_B is the number of blocks and $\overline{d}$ **is the average difference between the means**,

$$\overline{d} = \frac{\sum_{j=1}^{j=m_b} (x_A - x_B)_j}{m_b},$$

for that particular comparison (out of the k comparisons). If, as is usual, you are testing for no difference between means, then $\mu_0 = 0$. So, the procedure is nearly the same as using any other t for a multiple comparison in a simpler ANOVA. That is, allowance must be made for α and β inflation via the Bonferroni correction or the Sidak correction, and it is necessary to have a priori decided on

α, D_{min}, and, when possible, β. It is essential to a priori decide which comparisons to make, preferably using a control, to avoid having to test every possible comparison at an inflated cost. When you have a lot of these to do with the same calculation each time, you can make a decision rule as in the following example.

10.6.0.1 Example: Multiple Comparisons for RCBD ANOVA

Gollum was able to run multiple comparisons on the RCBD ANOVA in Example 9.4.0.2 because he has preplanned his experiment to have multiple comparisons against a control. This approach is, in part, adapted from Section 5.4 and Subsection .

Methods and Materials: The experiment had $m = 5$ treatments and $m_b = 6$ blocks and $MS_W = 107.4285$. Gollum had chosen A2 as the control, so he only has to do $k = m - 1 = 4$ comparisons (However, those specific $k = 4$ a priori planned comparisons are the only ones he can make without defaulting to, and expensively adapting to all the $k = m(m-1)/2 = 10$ possible comparisons.) He had chosen an overall criterion, $\alpha = 5\%$ and a power, $1 - \beta = 20\%$. Then he used Bonferroni's correction to find a corrected cutoff for **each two tailed comparison**, $\alpha'/2 = (0.05/4)/2$ which gives a cutoff of 0.00625.

Analysis: He found the mean difference between the pairs of treatments being compared and rearranged a common formula for the paired t test,

$$t_{\alpha/2,\, dfw} = \frac{\bar{d}}{S_{\bar{d}}},$$

to create a *decision rule* in the form of an inequality, so that when

$$\bar{d} \geq t_{\alpha/2,\, dfw} S_{\bar{d}},$$

there is statistical significance. Since he was comparing two at a time, m_b is the number of pairs, and $S_{\bar{d}} = \sqrt{MS_W/m_b}$, so

$$S_{\bar{d}} = \sqrt{107.4285/6} = 4.2314.$$

Then, Gollum used a suitable online calculator labled "Student's t" and found that

$$t_{\alpha'/2=0.00625,\, df=5} = 3.809666.$$

Then he was able to calculate

$$t_{\alpha/2,\, dfw} S_{\bar{d}} = 3.809666 \times 4.2314 = 16.1202,$$

the critical t value times the standard error of the mean differences from Example 9.4.0.2. So his decision rule worked out to say that there was statistical significance at an experiment wide 5% level if

$$\bar{d} \geq 16.1202.$$

From Example 9.4.0.2 he created the following table and finally marked the results of the comparisons in the $\bar{d}$ column:

Treat	Blk 1	Blk 2	Blk 3	Blk 4	Blk 5	Blk 6	$\sum_{all} d$	$\overline{d}$
A2= C	**37.50**	**35.44**	**51.36**	**51.07**	**37.21**	**39.94**	-	-
A1	52.16	57.89	58.78	79.57	80.06	46.93		
A1-C =d	14.66	22.45	7.42	7.42	42.85	6.99	122.87	**20.478**
A3	35.09	66.57	80.53	64.66	48.41	37.83		
A3-C =d	-2.41	31.13	29.17	13.59	11.20	-2.11	80.57	**13.428**
A4	49.12	56.36	49.91	42.18	56.58	48.82		
A4-C =d	11.62	20.92	-1.45	-8.89	19.37	8.88	50.45	**8.40**
A5	50.00	49.79	49.43	53.38	37.51	32.33		
A5-C =d	12.50	14.35	-1.93	2.31	0.30	-7.61	19.92	**3.32**

Only one comparison, A1 vs A2, was statistically significant.

10.6.1 Sample size for RCBD ANOVA

As already mentioned, given the appropriate information, the researcher can usually control power by managing sample size. The question of pragmatic reality and theoretical reality are somewhat at odds for power analysis for the randomized complete block ANOVA design. Shieh and Jan noted that no specific guidance for constructing a practical efficiency measure for the RCBD could be found in the literature (2004). The most popular approaches do not clearly and directly relate to power. Power can be handled to some extent by considering the comparisons between treatment pairs as though you were working with a one way ANOVA (Section 10.4), but using m_B for the number of blocks instead of n. The following recommendations might get you into the right ballpark. However, the subject needs further exploration by both theoreticians and applied statisticians.

10.6.1.1 RCBD a priori power

Whether you simply intend to find confidence intervals or to make multiple comparisons, you will need some way of calculating a reasonable overall sample size estimate, $\hat{n}$. That requires an estimate of the underlying variance, σ^2, for instance by using a pilot study,

$$\widehat{MS_W} = \frac{SST - SSB - SS_A}{df_T - df_B - df_a},$$ (10.11)

or a similar partitioning based on records. To attempt to justify the assumption of normality, ideally design the pilot or partitioning of records with a big enough sample size per farm or ward so that $df_W \geq 20$.

Before the actual test, it is necessary to define an a priori D_{min} for the k planned comparisons, and provide α' and β' (as by applying the Bonferroni correction). Strive to duplicate the conditions from which the variance estimate was derived. Then, having adjusted for the number of comparisons, proceed to estimate the total sample size per comparison, substituting this $\widehat{MS_W}$ into Equation 7.7 to estimate the **number of blocks**,

$$\widehat{m_B} = \left(t_{\alpha'/2', df_W} + t_{\beta', df_W}\right)^2 \widehat{MS_W}/D_{\mathbf{d}\,min}^2,$$ (10.12)

for the RCBD as though estimating replications. Then each treatment comparison can be **a paired Student's t test** (Section 5.4).

Although not an ideal pilot study, the standard deviation of a paired t test with df_d can provide an estimate of the variance to be used in place of $\widehat{MS_W}$ and df_W in Equation 10.12. The members of such a paired t test should only contain treatments you intend to use. If you intend to use a control, it should be one of the paired treatments. When using a paired t test for the variance estimate, iterate at least once, substituting the $\widehat{m_B} - 1$ for the df_W of both $t_{\alpha'/2', df_W}$ and t_{β', df_W}.

Whichever source of the variance you use, you can then estimate the total sample size n for the RCBD by multiplying the estimated number of blocks, $\widehat{m_B}$ times m, the planned number of treatments, to estimate the total sample size for the whole RCBD,

$$\widehat{n} = m\,\widehat{m_B}.$$

Because of the hazard of data loss and to compensate for sensitivity to non normality, maybe put in at least an extra complete block. Unfortunately, skew could come into this too; when possible, which it rarely is unless a **lot** of data on the treatment distribution is available, it is advisable that $\widehat{m_B} \geq 25 \times skew$. Perhaps big sample sizes resulting in $\widehat{df_W} > 80$ would avoid the problem of skew, but unlike the t test, this may not work for an ANOVA.

10.6.1.2 RCBD Subsequent power

As already mentioned, estimating power is not well understood for the randomized complete block design ANOVA. Maybe the whole subject of power for the RCBD should be investigated with massive and varied Monte Carlo studies (contrived data sets starting with random numbers and appropriately adding effects in a suitably controlled manner). However, a reasonable approximation of either overall power or of the power of individual comparisons should be obtainable by using $S^2_{main\,factor} = MS_W$ from the RCBD, and then using it to proceed as if for the one way ANOVA, Section 10.5.

Remember that (unless you actually have an a pririo multiple comparison plan with k) either the Sidak or Bonferroni corrected cutoff α', can only be estimated as though intending to make all the $k = \frac{m(m-1)}{2}$ possible comparisons where m is the number of treatments. Start with the following, which is derived from Equation 7.11, and includes substituting

$$\hat{z} = \frac{|D_{min}|}{\sqrt{MS_W/n_d}} - t_{\alpha'/2, df_W}. \tag{10.13}$$

A per comparison positive lowercase $\hat{z}$ is interpreted according to the rule that

$$iff\ \hat{z} \geq 0,\ \text{then}\ \hat{z} = \hat{t}_{\beta', df_W}.$$

However, if the power is disastrously low (it can even be that $\hat{z} < 0$ due to the subtraction of $t_{\alpha'/2, df_W}$), then use the rule that

$$iff\ \hat{z} < 0,\ \text{then}\ |\hat{z}| = \hat{t}_{1-\beta', df_W}.$$

This is the power of a single pairwise comparison following an RCBD ANOVA.

This should be an acceptable approximation if you round β' up, **or** $(1 - \beta')$ down, for $z < 0$, to no more than three figures (not including leading zeros) in the last step. Note that the prime superscripts,$'$, in this context merely indicate a single paired comparison.

10.6.1.3 Relative efficiency of estimation due to blocking

To estimate the improvement in the relative efficiency of the RCBD versus the one way ANOVA, consider the reverse of partitioning to get an estimate of what the sum of squares within of the one way ANOVA would be, divided by the sum of squares within of the RCBD:

$$REEB = \frac{(m_B - 1)\, MS_B + m_B\, (m_A - 1)\, MS_W}{(m_A m_B - 1)\, MS_W}. \tag{10.14}$$

If the blocking was effective, this will come out to be $REEB > 1$ (greater than 100%), indicating a gain. Question 5 illustrates and serves as an example of this (answered in back of book).

CAUTION: It is necessary to reiterate that CIs (even those resulting from global tests) are typically **not adjusted for multiple comparisons**, and even if they were, they should neither be assumed to be indicative of statistical significance, nor should they be assumed to refute it.

10.7 Arbitrary Scales

Although common in the social sciences and elsewhere, such scales are **not** suited for parametric statistics, such as Student's t test and the ANOVA. So, it's best to use a nonparametric procedure (Chapter 14). Examples of scales that absolutely should be analyzed with nonparametric methods are Likert scales, such as *don't like, almost like, like, like a lot;* or rating a service on a 1 to 10 scale; or contrived scales such as IQ, where a value twice as much is not really twice as smart, and adding IQs makes no sense. Their means and variances have somewhat limited meaning numerically, because adding and multiplying scores from such scales has little meaning. This is a common challenge in the social sciences where such data may arise from either an instrument that is a machine, or an instrument that is a test, or an interview script[8]. Likewise, difficulties involving other arbitrary scales can occur in other fields of science and in engineering.

10.8 Combining Test Results

Stouffer's Z-score method of combining outcomes (Stouffer, et al. 1949)[9] is generally the most foolproof approach for combining probabilities (Whitlock 2005):

$$\widehat{Z} = \frac{\sum_{j=1}^{j=k} (\pm Z)_j}{\sqrt{k}}, \tag{10.15}$$

---

[8]Unfortunately, attempts at standardizing scales using the standard deviation of the scores, whether arbitrary or not, have not succeeded for inference (e.g. Section 5.9), and were especially not suitable for estimating sample sizes. Maybe the ratio of each choice of the discrete choices such as percents would be preferable, especially as a way to present the effect size for some arbitrary scales.

[9]There are many ways to combine probabilities (Fisher 1925, 1948; Stouffer, et al. 1949; Chen 2011; Lancaster 1949). Section 10.8 shows how to combine **k** probabilities, in this case to find the overall β for the ANOVA using Stouffer's Z-score method.

R.A. Fisher demonstrated that, given **k** independent tests, and the null hypothesis is really true, then probabilities can be combined, using

$$\chi^2 = -2 \sum_{i=1}^{i=k} ln\,(p_i) \qquad df = 2\mathbf{k}.$$

This calculation produces a value that follows a chi square distribution with 2k degrees of freedom. **Note:** $ln\,(\ )$ refers to the natural logarithm ($ln = log$ to the base e) and ln is best obtained from a computer or a scientific calculator, as tables of such are cumbersome. **Fisher's method is only** presented here so you will know what it is. You may need to know because it is often offered by computer packages and is common in literature. In some situations, and with care, it can produce satisfactory results. Stouffer's method is to be preferred over Fisher's because Stouffer's method is not biased, as is Fisher's, with respect to score sizes. So Stouffer's method can be used in more varied situations.

where **the** $\pm$ gives the Z scores **direction corresponding to the tail** to which they refer. Stouffer's Z compares **k** independent one tail cumulative normal areas, Z_j, where $j = 1, 2, \ldots,$ **k**. If you have test scores other than Z, then Z values can be found from t tests, F tests, or any test that potentially produces a p_value, making it possible to substitute an equivalent Z_{p_value} as a Z score. Even the results of non-parametric tests (Chapter 14) can be equated to Z scores for this procedure. The division by $\sqrt{\mathbf{k}}$ results in the expected variance of $\widehat{Z}$ equaling one, as it should (because any Z is, by definition, standardized). This will require a table such as Appendix 4 or online calculators to obtain a Z value corresponding to each p_value and vice versa to get a final p_value from $\widehat{Z}$.

The objective of combining probabilities is usually not inference, but rather data exploration to be combined with the analyst's judgment, as in ***meta-analysis*** (evaluating results from many independent researchers). That is, except for possible use in power analysis, it is best used in discussing 'expert' opinions, and for making research budgeting decisions. Meta-analysts might be justified in requiring a minimal power, reasonably 80% in line with custom, for including results in their analysis, but they should make any such restriction clear in their methods description.[10] However, when combining your own results, you must include every available experiment, no matter how weak or lacking in potential for significance, or you will bias the results of this technique. Question 9 is a worked example of this technique (also see Question 8).

Questions for Ch 10

Write the equations where appropriate, and then fill in the values. **Odd numbered questions** have worked answers at the back of the book and can be studied as additional examples. The data tends to be fictitious.

1. I have run a global experiment with $m = 3$ **independent** treatments. It turns out to be statistically significant, and I decide to make multiple comparisons at a per comparison $\alpha' = 1\%$ level with what would be, in effect, a strict criterion of **no more than** $\alpha_{strict} = 5\%$. Is this even possible?

2. Find α_{strict} if someone post hoc chose to do multiple comparisons after a statistically significant global experiment with $m = 7$ independent treatments and used $\alpha' = 0.01$ for each. Also find α_{strict} if they used $\alpha' = 0.05$ for each comparison.

3. Calculate a **Sidak corrected** α' needed to achieve an overall (family wide) $\alpha = 5\%$, where you a priori plan to make $k = 8$ two tail comparisons. As usual, show and substitute into (fill in) equations. What would be the $t_{\alpha'/2, df}$ cutoff for each comparison, given $df_W = 16$? Are there any **special** assumptions?

4. Calculate a **Bonferroni corrected** α' needed to achieve an overall $\alpha = 5\%$, where you a priori plan to make $k = 8$ two tail comparisons. As usual, show and substitute into (fill in) equations. What would be the $Z_{\alpha'/2}$ cutoff for each comparison, given that the df was sufficiently large? What is the assumption associated with using Z?

5. Find the efficiency, REEB, for a randomized complete block that has $MS_W = 107.4285$ and $MS_B = 241.2854$, with $m_B = 6$ blocks, and $m_A = 5$ treatments (Subsubsection 10.6.1.3).

---

[10] Stouffer's method is particularly popular for meta-analysis that combines information across a body of literature to produce a descriptive overall p_value. However, that is a complete subject in itself and should not be undertaken without a good deal of consideration and review of associated literature. One of the greatest difficulties is that tests that are not statistically significant are rarely published, and this **creates a strong, frequently misleading, positive bias**.

6. Suppose in Example 10.4.1.1, Amadora's team had chosen a power of $1 - \beta = 0.80$. The number of comparisons is still $k = 2$, and they had a priori decided on an overall criterion of $\alpha = 5\%$, two tail comparisons, and a reasonable minimum difference of $D_{min} = 3.00$ attraction units. They had an initial estimate of $S^2 = MS_W = 24.5$, with 87 df from what they had come to consider to be a pilot study, and they intended for **all the sample sizes to be equal**. What would the total sample size estimate $\hat{n}$, for planning the next experiment, have been? (You might wish to do Question 9 first.) Do this two ways, once with the **Sidak correction**, and once with the **Bonferroni correction**; be sure to round up in both. (In addition to the Sidak correction and the Bonferroni correction you might need Equation 10.6, and Equation 10.7.

7. In Example 9.2.0.1, Amadora's team had a priori decided to make $k = 2$ comparisons with a reasonable minimum difference of $D_{min} = 3.00$ attraction units. They had an estimate of $S^2 = MS_W = 24.5$ with 87 df from what they had come to consider to be a pilot study, and they intended for all the samples sizes to be equal. Estimate the overall sample size if Amadora's had decided on a criterion of $\alpha = 0.05$ with a power of $1 - \beta = 0.95$ using the Bonferroni correction. (The important difference from Question 6 is the power.)

8. A friend of the author got a result that was statistically significant at the 5% level. The question was about an uncommon NZ moth. However, he was not able to get enough moths to feel comfortable about the experiment, and he suspected that its outcome may have been suspiciously close to $< 5\%$. The journal editor, a very polite product of Cambridge, agreed that the data seemed too thin. So my friend struggled to get more data, but the next time the results were not significant. Assume that referring back to his computations, and calculating to a greater accuracy, this researcher found that for the first experiment the $p\_value_1 = 0.04971$ and for the second experiment the $p\_value_2 = -0.1101$ (the minus sign indicates that the result was in the opposite tail). Show how my friend might have gotten a combined p_value, and state your result. Note that this is basically an exploratory procedure and would not be suitable for inference. (Such outcomes are quite possible with some non-normal distributions and smallish samples, $n < 60$.)

9. Suppose an industrial marine biology team was searching research results for data comparing the effectiveness of marigold (Tagetes) powder to hot pepper oil as an anti-fouling boat bottom paint additive. Combining the results involves scoring one treatment (marigold) as plus and the other (pepper oil) as minus because a score for one counts against the other. They found two peer reviewed journal articles and a laboratory record from their own company. The first journal article reported a test that had a $p\_value_1 = -0.42200$ (the minus indicating that the pepper oil got a better mean on the rating system than did the marigold powder). The second journal article had a $p\_value_2 = 0.05300$. The team's own lab record from their company showed a $p\_value_3 = 0.15771$. So, two sources favored marigold and one did not. What was the overall combined p_value for this mini meta-analysis? (This data is fictitious.)

10. What is the assumption of the Sidak correction that is not an assumption of the Bonferroni correction?

11. What might be a problem if you intended to have, say, $k > 10$ comparisons and hope to have a reasonable chance of avoiding false significance for every comparison?

11 CORRELATION and REGRESSION

Correlation is about the degree to which things vary together and regression actually attempts to model how they vary together. Correlation and regression are huge subject areas. Even the basic simple approaches presented in this chapter have many details and ramifications which are too extensive to be included here. In addition to the basics commonly taught, this chapter includes material to enable you to conveniently estimate the necessary a priori sample sizes to effectively use these simple procedures (Section 11.8).

Correlation and regression attempt to investigate if, and how, things vary together, but it can't determine if one causes the other, or if both are indirectly caused by the same thing. We often cannot adequately assess whether even a strong correlation might be a coincidence. Do patients become a little better each day because we treat them, or do they get a little better each day because they would anyway? Do the beggars in a tourist area increase as the weather becomes warmer, or do they increase because the tourists increase when it gets warmer? Regression is intended to predict the value of a factor when its value is connected to a potentially predictive value. The predictive value is sometimes called the *causal value*, but it does **not** have to be the **actual cause**[1]. However, neither regression nor correlation are a crystal ball. They only report an apparent trend in what has been observed; it remains to determine why there is a trend. There is also the real problem of Simpson's Paradox in which the trend is not consistent for smaller sample ranges (Subsection 11.14.1). Examining a scatter plot may alert you to this possible pitfall.

11.1 Descartes Revisited

In 1637, Descartes published two articles describing how to use a 2D representation to graph (plot) pairs of numbers, the **Cartesian[2] coordinate system**. He could be said to have invented graphs. It was a great breakthrough in mathematics because it represented the first merging of algebra and geometry. This coordinate system originally referred to two straight lines that intersect (cross) at right angles. The first is usually horizontal, the second is usually vertical, and they are often labeled **x** and **y** respectively. These have since been extended to 3D, and with proper slight of hand, even higher.

When there is a straight line relationship, the formula for the *dependent* y coordinate (the vertical distance from the x axis) to a straight line on a Cartesian coordinate system may be described as though from an independent x value using the *straight line formula*,

$$y = \mathbf{a} + \mathbf{b}x.$$

Some authors write $y = mx + c$, but the use of m might be confusing (even though quite appropriate) in some contexts; sometimes they use other letters. Variable x is called the *independent variable*, but is also called the *explanatory variable* (or *exogenous variable*), which is the variable upon which you or some outside, possibly random, effect may act, and y is seen as the result of that action. The variable y is called the *dependent variable* because it is seen as depending on x, but y is also called

---

[1]causal value, not always the cause

[2]Cartesian refers to Des**cart**es name.

the *response variable* or (*endogenous variable*). However, it is statistically very **difficult to prove dependence**.

The constant **b** is the *slope,* the average change in the y variable (upward when positive, downward when negative) with respect to changes in the x variable. It is equal to 0 when the line is horizontal. For a straight line, the slope is the change in the response variable (y axis) for a single unit of change in the explanatory variable (x axis).

The slope may also be seen as the tangent angle of a straight line with the horizontal. Or more intuitively, the *rise* (vertical signed distance) over the *run* (horizontal signed distance). Some would say **b** $= rise/run$.

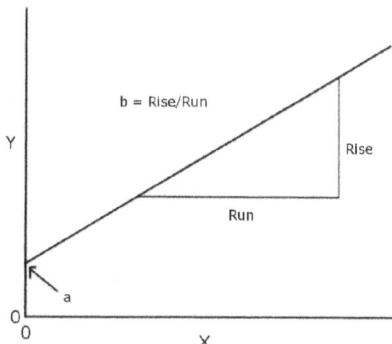

The *y intercept*, **a**, is the place where $x = 0$, and is where the line (sloping in the diagram) crosses the y axis. If there is no slope (a horizontal line), $b = 0$, then $y =$ **a** no matter what value of x you choose. The values, **a** and **b**, are called the *coefficients of the regression*. However, the phrase 'regression coefficient', without explanation, usually means **b**.

The pairs of values of x and y are written in parenthesis, (x, y), each pair to designate a *point*[3]. Points are infinitesimally small. The *origin*, $(x, y) = (0, 0)$, is the point where the axes cross each other. The line from the above formula would cross the y axis at the point $(0, \mathbf{a})$, the intercept, also called the *y intercept*. If a relation fits a **straight line** as a whole, or in part, that part of the relation which fits the straight line is called *linear*.

The *Pythagorean theorem* can be stated that given two points, (x_1, y_1) and (x_2, y_2), the distance between them is

$$d = \sqrt{(x_1 - x_2)^2 + (y_1 - y_2)^2}.$$

As mentioned, d is a *distance*, and distances do not specify a sign (or any other indication of direction). However, it can be given a sign (or some other specification of direction), in which case it becomes a *directed distance* (also called a *vector* and sometimes represented as $\overrightarrow{d}$, and drawn as $\rightarrow$, $\leftarrow$, $\nearrow$, $\searrow$, etc.).

Linear relationships[4] do not seem to be as common as nonlinear (curved line) relationships, also called *curvilinear* relationships. However, linear relationships are the easiest to work with. Fortunately, linear approximations work well for a lot of things even though they may **turn out to be curved** if you look at them on a sufficiently large scale. That is, we perceive a farm in Iowa or a real estate opportunity in Siberia as flat because we are looking at such a small piece of the round earth. This approximate flatness over short distances on long curves, sometimes referred to as *tangents*, is why, over reasonable distances, we lay out irrigation canals and building foundations using straight line measurements without practical difficulties. So, over relatively short distances, whether it is a three micron transect on the surface of an ommatidium, or a half kilometer of potential rail bed on

---

[3]Graphs of discrete data may be made of points and connecting lines, which can be called *nodes* and *edges* respectively.

[4]A relationship or *transformation*, $f(\ )$, is linear if it is additive, $f(x) + f(y) = f(x + y)$, and if it has scalability, $cf(x) = f(cx)$, where c is any real number.

Figure 11.1: No Straight Lines in Nature?

The common belief that there are no straight lines in nature is not always true, the sides of crystals are often straight, even in so called devil's post pile basalt formations; sea urchin spines have linear edges; force vectors such as gravity appear to be straight, etc. However, light traveling in interstellar space is assumed to be traveling in a straight line but is bent a little by the gravity of stars. This almost rectangular chunk of Antarctic Polar ice is about the size of Delaware. US public domain, if copied please give credit to source: NASA (Earth, Goddard Space Flight Center, Ice, IceBridge)/Jeremy Harbeck (2019).

the pampas in Argentina (part of the curvature of the earth), straight lines often are good approximations of reality.

11.2 How to Regress, at Least Linearly

Regression is useful to discovering linear relations, which can be represented as straight lines. Regression is a wonderful, vast subject. simple regression produces a descriptive estimate of a straight line, often with confidence intervals. Linear models based on the concepts involved in regression are the foundation of advanced mathematical statistics courses. However, this book only focuses on the simple two variable regression model.

As with all statistical procedures, chance prevents getting any more than a probable outcome. So let's consider a common regression model in a probability dominated world,

$$y_i = \mathbf{a} + \mathbf{b}x_i + \epsilon_i.$$

The many uses of this are sometimes intimidatingly referred to as ***regression analysis***. Well, it's not much of a change from Descartes' original formula for a point on a line. It just inserts epsilon, ϵ, to represent the chance, the noise, what statisticians call error.[5] The noise, ϵ, has an expected value, a theoretical mean, equal to zero, $E(\epsilon) = 0$ (***additive identity***). Why call 0 'additive identity'? Well, it

---

[5]The squared errors about the line can be estimated,

$$\sum_{i=1}^{i=n} \epsilon_i^2 = \sum_{i=1}^{i=n} (y_i - \mathbf{a} - \mathbf{b}x_i)^2 = \sum_{i=1}^{i=n} (y_i - \hat{y}_i)^2,$$

is simply a handy way to look at zero in mathematics; for example, $0 + anything = anything$. In the case of the above, the model assumes that the variation averages out to nothing. The addition of zero leaves what you added it to, identical to when you did not add zero. However, because you are sampling only part of the whole, your outcome will usually contain only part of the noise, and that part, by itself, is unlikely to average out to exactly zero.

In application, simple regression is commonly used to approximate straight lines and straight line models (equations for straight line relations). It may be used in situations in which the researcher controls (selects) the individual x_i values and observes the corresponding y_i results which are, in part, subject to random chance. It may also be used when the researcher collects points by observation, in which case both the x_i and y_i in each naturally occurring pair are subject to chance. However, the basic analysis is still viewed as though all the chance was associated with the y variable. Either is analyzed in much the same way in elementary situations involving simple linear regression.

Before starting your regression analysis, but **after forming any hypothesis** about the existence of a linear effect, plot all the n individual points, (x_i, y_i), as a scatter plot and look to see if there is a possible line, or maybe an oval representing a line through its main axis. Holding a hard copy (a paper scatter plot) at a slant with one corner pointed toward your eye may help to see a linear tendency if any is present.

If so, minimize the square of the vertical (assuming the y axis to be vertical) distance between n scattered data points and a predictive line through them using *simple linear regression*. The remaining distances are called the *residuals*. The formula for estimating the slope, **b**, of this line is

$$\hat{b} = \frac{\sum_{i=1}^{i=n} (x_i - \bar{x})(y_i - \bar{y})}{\sum_{i=1}^{i=n} (x_i - \bar{x})^2} = \frac{SXY}{SSX}, \tag{11.1}$$

with $df = n - 2$. The fact that the term involving x is squared in the denominator (the bottom), whereas the numerator (the top) involves the product of similar unsquared terms for each of x and y, may intuitively be seen as an estimate of y/x (but this intuitive simplification is excessively crude in most contexts). There are other possible ways to make this estimate, but Equation 11.1, also called the *least squares* method, is mathematically, and in application, the proven best for use in the presence of random variation. Its numerator, the *sum of the cross products*, can be divided by $n - 1$ to estimate the *covariance*, $\text{Cov}(x, y)$, of x and y. It is a result of variability, the contribution to the total variance (Equation 11.8) produced by the interaction of the two variables. So, an alternative way of writing Equation 11.1 could be

$$\hat{b} = \frac{SXY/(n-1)}{SSX/(n-1)} = \frac{\text{Cov}(x, y)}{S_x^2},$$

that is, the covariance divided by the sample variance of x estimates the slope.

Given a linear relationship, meaning $\mathbf{b} \neq 0$, then the middle, the sample mean, $\bar{x}$, of the explanatory variable, determines the middle, the sample mean, $\bar{y}$, of the response variable. So, after obtaining an estimate, $\hat{b}$, of the slope, **b**, you can use $\hat{b}$ to estimate the y intercept, **a**, where the line crosses the y axis (the place where $x = 0$), using

$$\hat{a} = \bar{y} - \hat{b}\bar{x}. \tag{11.2}$$

The equations for estimating $\hat{a}$ and $\hat{b}$ define the process of simple *least squares regression*, which

---

where ϵ_i is the effect of chance for point i.

provides the best estimate of the ***best fit line***, also called the ***least squares regression line***,[6]

$$\hat{y}_i = \hat{\mathbf{a}} + \hat{\mathbf{b}} x_i. \tag{11.3}$$

In literature this result is sometimes referred to as the ***LSL*** which is an acronym for *least squares line* (technically the least squares regression line, but often shortened for convenience). Also note that the explanatory variable can either be random, or it can be determined without changing either equation.

The least squares line is almost universally considered the **best** unbiased linear estimator **with or without** the involvement of normality. ***Best***, in this context of best fit, means to have the lowest possible mean squared error of the estimate.[7] The underlying distribution usually does not prevent you from estimating an LSL.

There is a vast literature, going back more than a hundred years, on the subject of finding the best estimates of lines. The approach shown here is best in most applications and it fits in nicely with the other subjects covered in this book. My experience and that of my students has confirmed that trying to draw a satisfactory line through a cloud of points by visual estimation, using a ruler, will tend to have a less accurate result than an LSL.[8]

Curvilinear relationships can be investigated with such transformations of the data as $ln(x)$. These are applied in such a way as to treat curves such as $ln(\hat{y}) = ln(a) + b \times ln(x)$ as though straight. (Also see Section 15.3, mostly for transformation to normal.) But it is going to take some thought to see what the results mean when transformed back to something that looks almost humanly familiar. In the case of the line representing transformed data, $ln(\hat{y}) = ln(a) + b \times ln(x)$, it needs to be transformed back to estimate the curve $\hat{y} = ax^b$. In addition to examples below in the text itself, Question 1 is both an exercise and an example of finding the slope and the intercept.

11.2.1 Considering epsilon, ϵ (residuals)

For each explanatory value, x_i, there is an observed response, y_i. The difference between the observed response, y_i, and the estimated $\hat{y}_i$ is the ***residual***, $\epsilon_i = y_i - \hat{y}_i$. The process of finding the **LSL**, the *least square line*, is the process known to **best minimize** the total squared error, $SS\epsilon = \sum(y - \hat{y})^2$. Let's examine these errors (not mistakes, but results of chance variation) using Figure 11.2, which is based on the data to be found in the example Subsection 11.4.0.1,

The slanted line is from the linear estimate (via Equation 11.1 and Equation 11.2) $\hat{y}_i = \hat{\mathbf{a}} + \hat{\mathbf{b}} x_i$ for every point, $(x_i, \hat{y}_i)$, on the line, whereas the scattered dots represent the observed points, (x_i, y_i). The arrows represent directed distances, the **errors**, or rather their estimates, the ***residuals***, the differences with respect to the Y variable. Up is for a positive distance, and down is for a negative dis-

---

[6]The process of finding the least squares line is sometimes called the **OLS** for **ordinary least squares**.

[7]The Gauss-Markov Theorem, named after Carl Friedrich Gauss and Andrey Markov, states that in a linear regression model in which the errors have expectation zero, are uncorrelated and have equal variances, the best linear unbiased estimator of the coefficients is given by the ordinary least squares estimator. The *errors* are the differences between the data points and the estimated points. To simply estimate and draw them, not test the relationship, they need not be normal. They do need to be homoscedastic, meaning that the variances of the values of y are the same for different values of the x variable (A. C. Aitken 1935, pp. 42–48). If the errors (differences between the line and the points observed) are correlated, there might be another linear effect tending to curve, or reshape, the linear effect you are looking for. (This footnote is largely from a 2013 entry in Wikipedia, which is an often good source of history and starting places for mathematics, and sometimes for statistics: http://en.wikipedia.org/wiki/Gauss%E2%80%93Markov_theorem .)

[8]A linear regression with either more than one dependent or one causal variable assumes a lack of heteroscedasticity in either the dependent or independent variables. There must not be any sub-populations that have different variances from others. Although such regressions are not difficult if you understand simple regression, this book does not describe such *multiple regressions*.

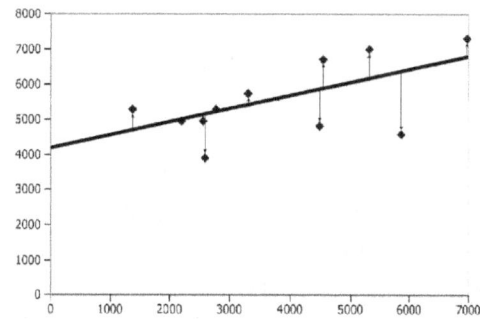

Figure 11.2: A Least Squares Regression Line

The errors (residuals) are shown as arrows (vectors). They are the amounts (in terms of y) by which the data points, (x, y), do not agree with the estimated points, $(x, \hat{y})$, on the line at the corresponding value of x. The sum of the squares of these residuals equals $\sum_{i=1}^{i=n} (y_i - \hat{y}_i)^2 = SS\epsilon$.

(Although just a demonstration of technique, Subsection 11.4.0.1 happened to turn out such that it was convenient to use graphics, including this one, and even data from it throughout the chapter, and not just in the example itself.)

tance. That is, they are the observed differences between the estimated line and the sampled data points.

The variance of the residuals is the **residual mean square**, the **RMS**, σ_ϵ^2; it is estimated by

$$S_\epsilon^2 = \frac{\sum_{i=1}^{i=n} (y_i - \hat{y}_i)^2}{n - 2} \approx \sigma_\epsilon^2, \tag{11.4}$$

and the numerator of this, $\sum_{i=1}^{i=n} (y_i - \hat{y}_i)^2 = SS_\epsilon$, is the **residual sum of squares**.[9] The square root of Equation 11.4, and its square root

$$S_\epsilon = \sqrt{\frac{\sum_{i=1}^{i=n} (y_i - \hat{y}_i)^2}{n - 2}}, \tag{11.5}$$

is called the **standard error of the estimate**. Having the same distribution at every point in the range is the assumption of **homoscedasticity**. To use a test that assumes a normal distribution of the errors **to see if a linear relation exists or to estimate a sample size**, you must assume homoscedasticity, and that the residuals are normally distributed at every point in the line over the range under consideration (Figure 11.4). However, such assumptions do not have to be met in order **to simply estimate** an untested slope and intercept (to estimate a least squares line).

11.2.2 Forced lines

The least squares line tends to pivot around the point $(\bar{x}, \bar{y})$. Especially when you believe the line to be straight, you may have sufficient reason to assume that the intercept is at the origin, $(x, y) = (0, 0)$. If, $\mathbf{a} = 0$ when $x_i = 0$ the line crosses the y axis at $(0, \mathbf{a}) = (0, 0)$. So, you can **force the line**

---

[9] A single error can be represented by $\epsilon_i = (y_i - \mathbf{a} - \mathbf{b}x_i)$. The sum of the sample errors,

$$\sum_{i=1}^{i=n} \epsilon_i = \sum_{i=1}^{i=n} \left(y_i - \hat{\mathbf{a}} - \hat{\mathbf{b}}x_i\right) = \sum_{i=1}^{i=n} (y_i - \hat{y}_i),$$

is assumed to equal zero if the entire population could be sampled, but because the residual sum of squares is only a sample, that is unlikely.

through the origin by substituting $0 = \bar{x}$ and $0 = \bar{y}$ into Equation 11.1,

$$\hat{\mathbf{b}} = \frac{\sum_{i=1}^{i=n} x_i y_i}{\sum_{i=1}^{i=n} x_i^2},\tag{11.6}$$

with $df = n - 1$(Snedecor and Cochran, 1967, p116). If you intend to force the line through the origin and your application will use discretionary as opposed to random values of x_i, you might find it helpful, when possible, to choose equal numbers of positive and negative values of x_i equidistant from $x = 0$. Of course, forcing may strongly affect both the slope and the residuals.

The least squares line can be similarly forced through **any point** other than the origin, (x_0, y_0), by substituting x_0 for $\bar{x}$, and y_0 for $\bar{y}$, in Equation 11.1, where you a priori choose the point, $(\bar{x}_0, \bar{y}_0)$. And, when possible, this can be further facilitated by choosing values of x_i equidistant in either direction from your forced mean, $\bar{x} = x_0$. **Do not** include $(\bar{x}_0, \bar{y}_0)$ or $(x, y) = (0, 0)$ as a **data point**, just because you expect it, even if you are forcing the line. Your slope estimate will be biased, and your colleagues will not be amused.

As with forcing through the origin, any forced LSL has

$$df = n - 1\,.$$

Question 1 is an illustration that shows the calculation while also showing what happens when forcing is unwarranted.

The question of degrees of freedom for simple regression strikes many users as a mystery. It is a result of the less-than-obvious hidden restriction (information consumption) associated with the estimation of the slope parameter, **b**, and intercept parameter, **a**. However, only one df is used when the LSL is forced, in which case **a** is decreed, not estimated (resulting in $df = n - 1$).

> **CAUTION:** Be warned that any conclusion resulting from regression becomes questionable beyond the range of the original data. This even applies to projecting, say, annual or monthly data into the future. That is, **beware of *extrapolation***.

11.3 Pearson's Product Moment Correlation

Pearson's R, the ***product moment correlation coefficient***, is a widely used index of correlation, the tendency of two measures, x and y, to vary together in a linear manner (Stigler, 1989). Karl Pearson was influenced by the work of Francis Galton when he developed it (Pearson 1930). Linearity can be seen as a tendency of two variables to form a straight line if graphed against each other. Pearson's R is popularly called the correlation coefficient or simply the correlation. It, and variations on the correlation theme, are common ES values for various tests, especially those involving discrete variables such as counts (Chapter 12).

The theoretical ideal of the expected value for the ***product moment correlation coefficient*** is

$$\rho = \frac{\sum_{i=1}^{i=N} (x_i - \mu_x)(y_i - \mu_y)}{\sqrt{\sum_{i=1}^{i=N} (x_i - \mu_x)^2}\sqrt{\sum_{i=1}^{i=N} (y_i - \mu_y)^2}},$$

where ρ is the Greek letter ***rho***. N is the size of the entire population of paired values. The radical sign (the $\sqrt{\text{square root}}$ thingy) always indicates a **positive** (+) square root, unless stated otherwise. So the numerator determines the sign.

Substituting the sample size, n, for N, and the estimated mean, $\bar{x}$ for μ_x, and $\bar{y}$ for μ_y, gives the equation for **estimating** the product moment correlation coefficient for x and y,[10]

$$R = \frac{\sum_{i=1}^{i=n} (x_i - \bar{x})(y_i - \bar{y})}{\sqrt{\sum_{i=1}^{i=n} (x_i - \bar{x})^2 \sum_{j=1}^{j=n} (y_j - \bar{y})^2}} = \frac{SXY}{\sqrt{SSX}\sqrt{SSY}}. \tag{11.7}$$

A regression tries to detect and quantify a directed slope (either increasing or decreasing) of y with respect to x. Equation 11.7, $\hat{b} = \frac{\sum_{i=1}^{i=n} (x_i - \bar{x})(y_i - \bar{y})}{\sum_{i=1}^{i=n} (x_i - \bar{x})^2}$, for the slope of a regression could be written as $\hat{b} = \frac{\sum_{i=1}^{i=n} (x_i - \bar{x})(y_i - \bar{y})}{\sum_{i=1}^{i=n} \left[\left(\sqrt{(x_i - \bar{x})^2} \right) \left(\sqrt{(x_i - \bar{x})^2} \right) \right]}$. However, correlation tries to detect, and to measure the degree of, a possible linear (straight line) relation, whether it is of y with respect to x, **or** of x with respect to y. So, instead we use $\hat{b} = \frac{\sum_{i=1}^{i=n} (x_i - \bar{x})(y_i - \bar{y})}{\sum_{i=1}^{i=n} \left[\left(\sqrt{(x_i - \bar{x})^2} \right) \left(\sqrt{(y_i - \bar{y})^2} \right) \right]}$, which somewhat explains Equation 11.7.

It works out that $-1 \leq R \leq 1$ because the numerator, the ***cross product***, can be either positive or negative. That is, R can have a direction, either plus or minus. Given pairs of measurements, (x_i, y_i), its calculation starts with estimating the deviation of the means, $(x_i - \bar{x})$ and $(y_i - \bar{y})$, for each pair, $_i$.

Another way to define R is as a ratio of the ***covariance***, itself a kind of variance,

$$cov(x, y) = \frac{SXY}{n-1} = \frac{\sum_{i=1}^{i=n} (x_i - \bar{x})(y_i - \bar{y})}{n-1}, \tag{11.8}$$

divided by the individual variances of X and Y to give

$$R = \frac{cov(x, y)}{S_x S_y} = \frac{\left[\sum_{i=1}^{i=n} (x_i - \bar{x})(y_i - \bar{y}) \right] / (n-1)}{\left(\sqrt{\sum_{i=1}^{i=n} (x_i - \mu_x)^2 / (n-1)} \right) \left(\sqrt{\sum_{i=1}^{i=n} (y_i - \mu_y)^2 / (n-1)} \right)}. \tag{11.9}$$

The $n-1$ divided by $\left(\sqrt{n-1} \times \sqrt{n-1} \right)$ can be seen as canceling each other out to get back to Equation 11.7. The **numerator** of the equation in either form **determines the sign** because the denominator is always positive.

If either $R = 1$ or $R = -1$, they vary completely together, and if $R = 0$, there is no observable linear relation. Furthermore, if $-1 \leq R < 0$, then x and y vary together in **opposite** directions (a **negative relation**), and if $0 < R \leq 1$, then they vary together in the **same** direction (a positive relation).

However, curves can be approximated by straight lines over short distances and what is a 'short distance' is relative. A few miles of rail track on the pampas of Argentina might be flat for all practical purposes, even though on a global scale it might be curved. Likewise, at micron scale a small cross section at 14,000 magnification on an electron microscope can be treated as though flat even though the cell is an elliptical shape. As can be seen in Figure 11.3, some curve-linear relationships will, to some degree, also result in correlation, $R \neq 0$. At least check a scatter plot when considering correlation results.

Spearman's rank correlation coefficient, Section 14.6, is better for curvilinear data and should be considered when effects that do not lead only to straight lines are important to you. It is dependent on the degree to which the relation tends to be generally + or -. There is some information loss in the ranking process for this nonparametric correlation, but it is quite small. If the relation is other than very close to straight, you will gain more than you lose.

---

[10]Some books use a lowercase r for the basic two variable correlation, but in this book the capitalized R is used for notational convenience.

Figure 11.3: Correlation vs Distribution

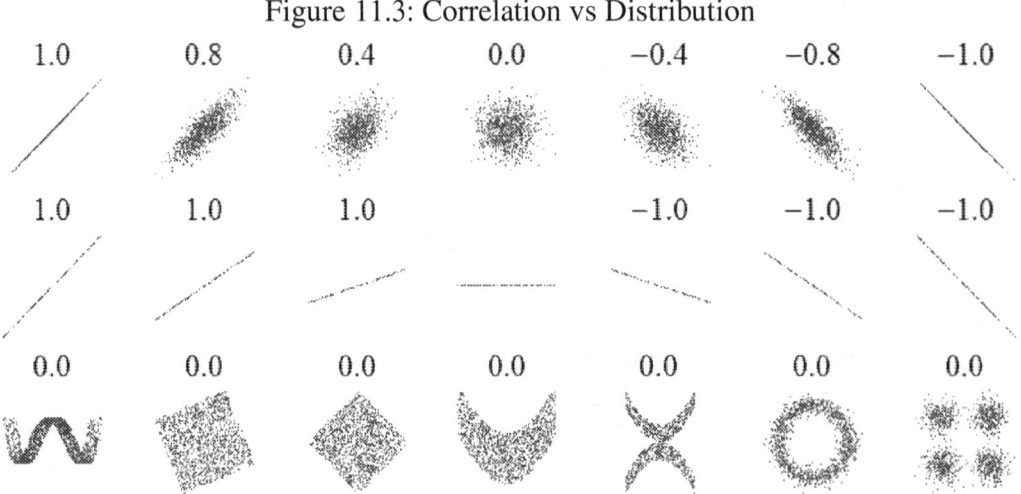

A collection of scatter plots with their Pearson's product moment correlations, R. The center figure is undefined because it would result in division by zero. This image came from Wikipedia and was contributed by Denis Boigelot. It is freely available for your use if accompanied by **a citation referring to the original source**, Denis Boigelot via Wikipedia.

Correlation does not mean causation. That is, things varying together do not necessarily demonstrate that one causes the other.[11] Correlation could be a result of an outside factor that affects both variables even though they do not interact, or interact relatively little. It is also possible that the observed correlation might be caused by two other things which vary together to some extent due to yet another outside factor, etc. Popularly reported correlations are often post hoc, so their varying together is often due to the fact that unusual things must happen sometimes and that when they are unusual, we are more likely to notice them.

The square of R is ***Pearson's coefficient of determination***,

$$R^2 = \left(\frac{cov(x,\,y)}{S_x S_y} \right)^2,$$

which, in many applications, is seen as the **contribution of the linear relation to the variance**. Since it is squared, R^2 is positive, only going from 0 to 1. It is handy to think of it as the percent of variance due to correlation (also see Equation 11.15). It estimates the average variability of the points from a horizontal no effect line (a line representing a constant, $\mathbf{a} = \mu$). It is also a popular measure of the **strength of the linear relation**, the part of the variance caused by the linear effect of the independent variables. It is useful for measuring a variety of relationships provided you realize that this measure implies linear relations (straight lines).

11.4 Clarifying and Expanding on Regression.

Maybe an example will assist your understanding.

11.4.0.1 Simple regression

Example Effect of Size on Cost:

---

[11] Notice that even the English word coincidence is comprised of two parts, meaning together-happening. Based on experience and reason, we regularly use this word in a way which would be pointless without the premise that correlation (things happening together) does not mean causation.

This data is abstracted from the Fiscal Times, "2013 Public Universities with the worst graduation rates".

Size X	Cost Y	Error Sq X	Error Sq Y	Cross Prod
2590	3906	1497953.4628	2555928.8926	1956696.8430
5311	7000	2241281.1901	2235840.5289	2238559.2066
1371	5288	5967804.8264	46970.7107	529445.0248
4486	4820	451706.1901	468851.4380	−460198.9752
6964	7312	9923072.7355	3266234.7107	5693073.3884
2199	4956	2607931.3719	301101.6198	886144.6612
2774	5288	1081410.9174	46970.7107	225376.6612
4542	6704	530116.3719	1438255.0744	873179.5702
5860	4590	4186488.0083	836725.9835	−1871615.1570
2558	4956	1577307.6446	301101.6198	689151.5702
3298	5732	266162.1901	51652.8926	−117252.0661
Mean 3813.9090909	5504.7272727			
SS		30331234.91	11549634.18	10642560.73 SXY
Var.		3033123.491	1154963.418	1064256.073 Cov.

An 'error' in this case is either $(x_i - \bar{x})$ or $(y_i - \bar{y})$. Although this table is visually represented as being rounded to ten figure accuracy, the numbers in my calculations were recorded to even greater accuracy. The cost column (Y) is in dollars, the size column (X) is in students. I was interested in demonstrating more than one approach to linear effects, starting with a regression. But first having a priori hypothesized the existence of a relation (statistically significant slope) I opted to look at a scatter plot:

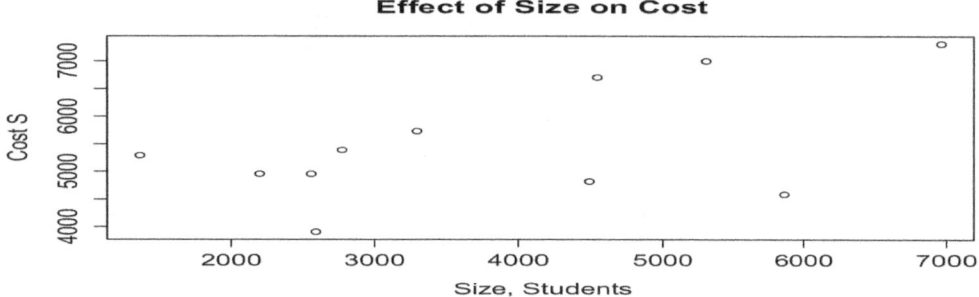

Effect of Size on Cost

Well, it did not look like much, but seeing (eye balling) can be deceptive and eleven points are not much to look at anyway. I might as well proceed, particularly because this is for purposes of demonstration and not research.

Finding the slope The existence of slope, meaning $\mathbf{b} \neq 0$, would indicate a linear relationship. I put size, x, on the horizontal axis and cost, y, on the vertical axis. So, I applied the regression Equation 11.1 to estimate the slope,

$$\hat{\mathbf{b}} = \frac{\sum_{i=1}^{i=n}(x_i - \bar{x})(y_i - \bar{y})}{\sum_{i=1}^{i=n}(x_i - \bar{x})^2} = \frac{SXY}{SSX} = \frac{10642560.73}{30331234.91} = \$0.3508779237/\text{student}.$$

Which, for calculation purposes, could be rounded to $\hat{\mathbf{b}} = \$0.350878\,per\,student$, about the added amount per student, before the final rounding of my eventual estimate $\hat{y}_i$. I had students times cost in dollars divided by students square, so by cancellation I got my estimate in dollars per student as a measure of slope. That was not too much of a surprise. However, I still needed a starting price estimate, $\hat{a}$ dollars, where the line crosses the y axis ($x = 0$ students is a theoretical reference point).

The intercept If there was no slope there could still be a constant, $a = \mu_y$, for any value of x. Then y_i would be expected to equal μ_y for every x_i and that could be represented by a horizontal line.[12] But there apparently is slope, $\mathbf{b} \neq 0$. So, I would have applied Equation 11.2 to find the

---

[12]In such mathematical contexts as Cartesian coordinate systems, lines at right angles could be said to be invisible to

y intercept,

$$\hat{a} = \bar{y} - \hat{b}\bar{x} = \$5504.7273 - (\$0.350878 \,/\, \text{student}) \times 3813.9091 \,\text{students} = \$4166.51 \,\text{tuition}.$$

Notice that we are doing algebra with the units; **b** is dollars per student unit, and x is student units. The student units (over student units) cancel out, so the results of the regression are in dollar units. This *fitted line* could be drawn as

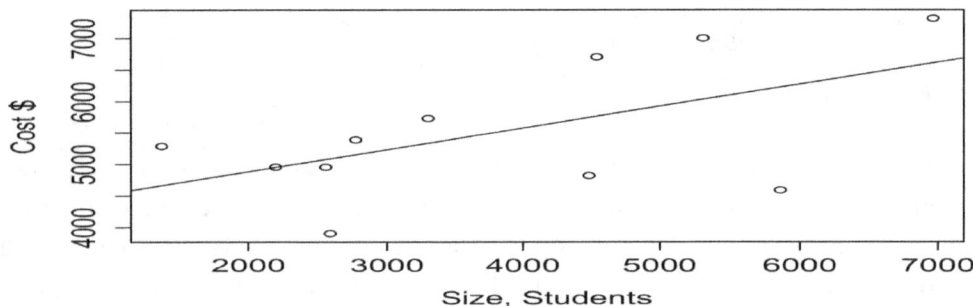

Effect of Size on Cost

Suppose I was rather naive and I had not actually tested to see if there was a statistically significant relationship as represented by the slope, but I now decided to do so. I could test the strength of the linear relationship by first calculating R, using Equation 11.7,

$$R = \frac{\sum_{i=1}^{i=n} (x_i - \bar{x})(y_i - \bar{y})}{\sqrt{\sum_{i=1}^{i=n} (x_i - \bar{x})^2 \sum_{j=1}^{j=n} (y_j - \bar{y})^2}} = \frac{SXY}{\sqrt{SSX \times SSY}}$$

$$= \frac{10642560.73}{\sqrt{30331234.91} \times \sqrt{11549634.18}} = 0.5686132513.$$

So, $R = 0.56861$ when rounded to a reasonable accuracy. This correlation coefficient could even be seen as a sort of effect size for linear effect, but how would my readers be able to interpret it? However, I could have squared R to calculate the coefficient of determination, R^2, to estimate the percent of the variation that is associated with a linear relationship. Squaring R (before rounding) gives the coefficient of determination, $R^2 = 0.32332$ (after rounding). This is a handy and meaningful **effect size**; $R^2 \approx 32\%$ of the observed variability is estimated to be due to linear effect.

Results and Discussion The resulting linear model corresponding to a straight line equation is estimated to be

$$\hat{y}_i = \$4166.51 + \$0.350878 \, x_i.$$

Accuracy to the nearest cent is too optimistic, in that the data is only to the nearest dollar. The final estimate $\hat{y}_i$ for a given x_i should at least be rounded to the nearest ten cents. That is, the slope and intercept should ultimately be reported to no more than five figure accuracy. Notice that the coefficient of determination suggests that the linear relation accounts for only about 32% of the variance. Aside from questions related to statistical significance, **think** before fully accepting the accuracy of predictions based **solely** on a linear relation, such as this, which may not account for enough of what might **also** be going on. Traditionally, many people have questioned any linear relation for which $R^2 < 50\%$ (more exactly, $|R| < 70\%$). However, that can be way too conservative for some mostly very large sample applications in which this standard

each other. That is, there is no relation between them.

is often not met.

Despite a number of shortcomings related to creating an easily understood question, this example is valuable as a demonstration of calculation and theory. Therefore it will be continued a number of times below.

This subject of college size vs expense might be an interesting area of study. However, there are discrepancies in the design and sampling I used in this example. It would be more reasonable to randomly sample from a more homogeneous group of colleges, such as all state university main campuses. However, the colleges in this journalist's study appear to mostly, but not all, be feeder colleges intended to provide service to students near their homes and then funnel many, if not most of them on to finish at their larger parent colleges.

Correlation is a good indicator of the existence of a straight line relation (such as the principal axis of an oval), and it is an estimator of the effect of that linear relation, but it does not describe said line. Linear variation, in the context of correlation, can be thought of as a measure of the average distance of the points above and below a horizontal line at $\bar{y}$. Some of the variation is due to an apparent line that could be described by a regression (x and y varying together in a linear fashion), and some of it is just due to chance. The coefficient of determination, R^2, is an estimate of how much of the variation (measured as variance) is estimated to be due to the actual linear part of the variance. Your understanding of this may improve as you read on through this chapter.

11.5 Tests of Linearity

Can the relation be approximated by a straight line that is sufficiently different from horizontal to be meaningful? In general, tests of the slope, **b**, are like tests of any other variable compared to another variable, in this instance a reasonable minimum difference from zero. That is, **b** is a continuous variable, not a category or a success. We consider such things as how much slope there is, or if there is enough slope to be of practical interest, with the intent of possibly rejecting the null hypothesis that there is no slope, or correlation, at all. Keep in mind that even though you are testing to see if slope exists, your results show whether the **amount of the slope** is statistically significant. In application that means that there is enough evidence to support the belief that there is more of a linear slope than some reasonable minimum difference, whether formally sought, or a result of your testing procedure.

Testing $\mathbf{b} \neq 0$ vs $\mathbf{b} = 0$ has many applications which often dovetail with other applications. Failure to reject $\mathbf{b} = 0$ might be because although a relationship exists, it is too curved (e.g. quadratic, maybe u-shaped) to be called linear, or that it is actually too random to have any meaning at all. Statistical significance supporting the alternative that $\mathbf{b} \neq 0$ implies a straight line relation. Although not obvious at first, each of the following three equivalent tests has its own implications for particular, although related, applications.

Regression, whether you are using it to estimate a slope, or an intercept, or both, is not in itself dependent on normality. However, for estimating CIs, and for estimating sample size, as well as for null hypothesis testing, it is commonly assumed that the residuals, $y_i - \hat{y}$, are normally distributed, with equal variance, for every x_i (Figure 11.4).

Otherwise, you can test the residuals for normality with a Q-Q plot (Section 11.10), but that usually requires a lot of points. However, if you do not believe the residuals are normally distributed around the line, Conover (1980) presents a nonparametric approach to testing linear regression and for confidence intervals. Even these techniques seem to at least require that the residuals are from the same distribution[13] that tests for statistical significance of the same least squares line, without the requirement of normality. However, you can easily verify that a monotonic (y always increasing with

---

[13]The technique called 'nonparametric regression' requires very large amounts of data, and is not covered in this book.

Figure 11.4: Normally Distributed Residuals About Regression Line

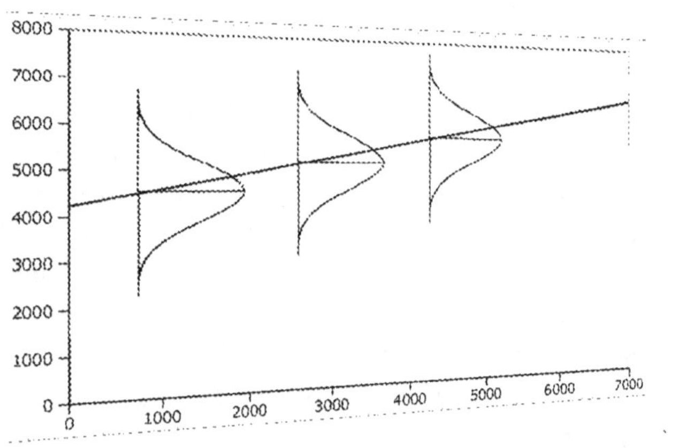

In the presently popular tests, the residuals are assumed to be normal, with equal variance, for every value of x.

x, or y always decreasing with x) relationship exists by putting the same data into a Spearman's correlation and then testing its significance (Section 14.6).

In this chapter, the example in Subsection 11.4.0.1 is used as the basis for many of the figures in this book.

11.6 F Testing (Also Leads to Adjustment)

Using F to test whether a straight line effect (a slope) exists touches on issues important to other processes associated with correlation and regression. It also relates to adjusting data for further analysis. This approach is incorporated into a variety of more complex designs, as well as into the briefly discussed *analysis of covariance*, ANCOVA (Section 11.12), which is like an ANOVA, but which allows linear adjustment for factors such as age.

After you have calculated $\hat{y}_i = \hat{a} + \hat{b}x_i$ for every point, (x_i, y_i), you can calculate the appropriate sums of squares corresponding to Figure 11.2, Figure 11.5, and Figure 11.6, and proceed to an F test for slope after the manner of the analysis of variance:

Sources		SS	df	MS	F
Regression	**(Fig 1.3)**	$\sum_{i=1}^{i=n} (\hat{y}_i - \bar{y})^2$	1	$SSR/1$	$MSR/MS\epsilon$
Residuals, ϵ	**(Fig 1.1)**	$\sum_{i=1}^{i=n} (y_i - \hat{y}_i)^2$	$n-2$	$SS\epsilon/(n-2)$	
Total	**(Fig 1.2)**	$\sum_{i=1}^{i=n} (y_i - \bar{y}_i)^2$	$n-1$	$SSY/(n-1)$	

As in the ANOVA, subtraction is a common way to find a remaining df. The F test combines the results of a two tail test, meaning it tests for greater slope, up (+) or down (-), into one value, F, and uses a cutoff equals α instead of $\alpha/2$. This works because the F test is a test of greater variance, and it is one tailed in that sense. Using standard tables such as Appendix 6, the resulting F value is compared with values based on the criterion, α. That is, you declare statistical significance if $F \geq F_{\alpha, df\_reg, df\_\epsilon}$.

Think before fully accepting the accuracy of estimations or predictions based solely on a straight line relation, which may not account for enough of what might also be going on. Assuming the apparent relation is true, you still should be aware that the lower the value of R (or R^2), the less the effect of a linear relation on either estimation or prediction (Section 11.3).

When drawn on Cartesian coordinates, $b = 0$ always produces a horizontal line (Figure 11.5).

Figure 11.6 includes such a line on the same coordinates as the least squares regression line. So, the figure shows deviations due to regression (not the residuals) between these two lines with respect to the sampled values on the X axis. When squared, the sum of squared deviations, $SSR = \sum_{i=1}^{i=n} (\hat{y}_i - \bar{y})^2$, is the **sums of squares due to regression**.

The residuals are the differences **between** what was **observed** and what was **estimated** (Figure 11.2). They represent the effect of chance, or at least the effect of non-linear effects. The **residual sum of squares** is $\sum_{i=1}^{i=n} (y_i - \hat{y})^2$, also referred to as the **total squared error** in Subsection 11.2.1 (above). For testing, for CIs, and for estimating sample size, the distribution of the residuals is assumed to be normal, with the same standard deviation for every value of x_i (Figure 11.4).

11.6.0.1 $\sqrt{\text{F}}$ can also be evaluated as a t test of b:

In order to understand another way to challenge **H0: b** $= 0$, including the approach we will use for the a priori estimation sample size, note that, with respect to regression, $F_{\alpha, df\_reg, df\_\epsilon} = F_{\alpha, 1, df\_\epsilon}$ (note that $df_{regression} = 1\,df$). Because in such contexts where $df_1 = 1$, $\sqrt{F_{\alpha, 1, df_2}} = t_{\alpha/2, df_2}$, we can write the square root of this F test as a two tail t test, using the ratio of two standard errors, the standard error of the regression, and the standard error of the residuals,

$$ t = \sqrt{\frac{MSR}{MS\epsilon}} = \sqrt{\frac{\sum_{i=1}^{i=n} (\hat{y}_i - \bar{y})^2 / 1}{\sum_{i=1}^{i=n} (y_i - \hat{y}_i)^2 / (n-2)}} = \frac{S_{reg}}{S_\epsilon} \qquad df = n - 2, \qquad (11.10) $$

where S_{reg} has $1\,df$ and S_ϵ has $n-2\,df$. The F test is a two tail test of slope, but the t test can be used for **either one or two tail tests**.

11.6.0.2 Example: F Test of Slope

F for slope of some college data (refers to Subsection 11.4.0.1) As you recall, I obtained a slope, an intercept, and even a correlation. However, assume I had not tested to see if there was really a relationship other than an apparent line solely due to the effects of chance. I could have used an F test with an $\alpha = 5\%$ and $n = 11$. The hypothesis pair would have been

H0: THERE IS NO LINEAR RELATION (**b** $= 0$),

HA: THERE IS A LINEAR RELATION (**b** $\neq 0$).

The sums of squared residuals (errors if you prefer) is estimated in the following table. Note that the check comes out very close to zero considering the estimate involved thousands. The tiny error could either have been hardware rounding errors in the computer, or rounding on my part. In any case the estimate is acceptable.

	Size X	Cost Y	a+bx	y − (a + bx)	Squared
	2590	3906	5075.284	1169.284	1367224.6573
	5311	7000	6030.023	−969.977	940856.0541
	1371	5288	4647.564	−640.436	410158.7397
	4486	4820	5740.548	920.548	847409.2936
	6964	7312	6610.024	−701.976	492770.5002
	2199	4956	4938.091	−17.909	320.7482
	2774	5288	5139.845	−148.155	21949.7973
	4542	6704	5760.198	−943.802	890763.1034
	5860	4590	6222.655	1632.655	2665561.1503
	2558	4956	5064.056	108.056	11676.0405
	3298	5732	5323.705	−408.295	166704.4866
Sum	41953	60552	60551.992	−0.008	7815394.571
Mean	3813.909	5504.727	5504.727	−0.0008	

The sum of squares of regression, SSR, can be found by subtraction from the total sum of squares, SST, as in Example 11.4.0.1.

Sources		SS	df	MS	F
Regression	(Fig 1.3)	3734239.611	1	3734239.61	4.3003
Residuals, ϵ	(Fig 1.1)	7815394.571	9	868377.175	
Total	(Fig 1.2)	11549634.18	10	1154963.418⋆	

⋆Not needed for calculating F, but sometimes of interest.

Figure 11.5: Deviations from $\overline{y}$

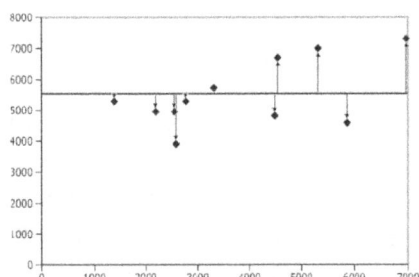

The arrows represent deviations, $y_i - \bar{y}$, about a constant line, $\bar{y} = a$, that is perfectly horizontal, no slope, hence no linear relation, **b** = 0. The summed squares of these deviations is $\sum_{i=1}^{i=n}(y_i - \bar{y})^2 = SSY$.

Results and Discussion: A table of the cumulative $p \geq F$ values, such as Appendix 6 gives $F_{0.05, 1, 9} = 5.1174$. The result, generally rounded to four decimal points, $F = 4.3003$, was found **not statistically significant**; **HA** was not rejected. So, I **cannot conclude anything about H0 or HA**. As will soon be explained, this F could be evaluated against $t_{\alpha/2, df} = \sqrt{F_{\alpha, df1, df_2}}$, using a table of t because it has only one degree of freedom in the denominator. That is, $t = \sqrt{4.3003} = 2.0737$ with $9\, df$. From a table such as Appendix 3, $t_{0.025, 9} = 2.2622$. So because the estimated t value was less than this, it was likewise not statistically significant from this alternative point of view (recall that a smaller t means a larger p_value).

Question 3 is also an example of using an F test to see if the slope is significant.

11.6.1 Just t test b and a (and standard errors of)

In most applications, **testing the slope**, **b**, is of greater interest than testing the intercept, **a**. We have already tested **H0: b** = 0 vs **HA: b** ≠ 0 using an ANOVA-like approach (Section 11.5) which could also be easily modified into a t test. However, it did not clearly include a standard error estimate that you will need to produce a CI for the slope (Section 11.7). The standard error of the slope, **b**, is

$$S_{\mathbf{b}} = \sqrt{\sum (y - \hat{y})^2 / (n-2)} / \sqrt{\sum (x - \bar{x})^2} = S_\epsilon / \sqrt{\sum (x - \bar{x})^2}. \qquad (11.11)$$

This may also be called the ***standard error of the regression***[14] (SER), see Figure 11.6.

So here is another t test of slope, one that includes such an estimate of the standard error of **b**,

$$t = \frac{\mathbf{b} - \mathbf{b_0}}{S_{\mathbf{b}}} \qquad df = n - 2, \qquad (11.12)$$

---

[14]Equation 11.10,

$$t = \cdots = \cdots = \frac{\sqrt{\sum (\hat{y}_i - \bar{y})^2}}{\sqrt{\sum (y_i - \hat{y}_i)^2 / (n-2)}},$$

Figure 11.6: deviations due to regression

In mathematical contexts such as Cartesian coordinate systems, lines at right angles, such as the y axis and the horizontal line at $\bar{y}$ could be said to be invisible to each other. That is, there is no relation between them (hence a lack of linear relation).

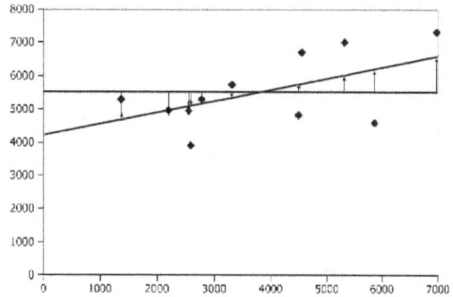

The deviations of the regression line values $\hat{y}_i$ from a no-slope line at $\bar{y}$ are shown as arrows for the values of x_i that were sampled. When the squares of these are summed, the result is the sum of squares **due to regression**, $SSR = \sum_{i=1}^{i=n} (\hat{y}_i - \bar{y})^2$.

where $\mathbf{b}_0$ **is usually** $\mathbf{b}_0 = 0$, **but** it is easy to test against some other hypothesized slope, $\mathbf{b}_0$. This is a two tailed test, so it is evaluated against $t_{a/2,\,n-2}$. **One tail tests** are also possible, using $\mathbf{HA}{:}\mathbf{b} > \mathbf{b}_0$ or $\mathbf{HA}{:}\mathbf{b} < \mathbf{b}_0$, and in either of these, the result would be evaluated against $t_{\alpha,\,n-2}$. Question 5 and its answer in the back of this book might be enlightening regarding this approach.

11.6.1.1 ES in original units if slope statistically significant?

Editors of many social science journals are supposed to require an effect size if you declare statistical significance. This trend could be spreading outside the social sciences.

Remember that the coefficient of determination, R^2, can be used as an effect size. However, it is unitless, and ES values in original units of measure are to be preferred. So, what would you use if an editor were to ask you for an effect size of the slope of a simple regression because you had declared it statistically significant?

A possibility is the root-mean-square deviation (RMSD) or root-mean-square error (RMSE),[15] a measure of the differences between values predicted by a model or an estimator and the values actually observed. This was introduced in Subsection 11.2.1 as the rather familiar, by now, standard error

shows that this is equivalent to $\mathbf{b}/S_\mathbf{b}$, use the substitution of $\hat{y}_i = \hat{a} + \hat{b}x_i$, and $a = \left(\bar{y} - \hat{b}\bar{x}\right)$, to expand the term under the radical sign in the numerator, $\hat{y} - \bar{y} = \left[\left(\bar{y} - \hat{b}\bar{x}\right) + \hat{b}x\right] - \bar{y}$. This simplifies to $\hat{y} - \bar{y} = -\hat{b}\bar{x} + \hat{b}x$. So its summed squares is $\mathbf{b}^2 \sum (x - \bar{x})^2$. Putting this altogether,

$$t = \cdots = \cdots = \frac{\sqrt{\mathbf{b}^2 \sum (x - \bar{x})^2}}{\sqrt{\sum (y - \hat{y})^2 / (n-2)}} = \frac{\mathbf{b}}{\sqrt{\sum (y - \hat{y})^2 / (n-2)}/\sqrt{\sum (x - \bar{x})^2}}$$

$$\therefore \sqrt{\sum (y - \hat{y})^2 / (n-2)}/\sqrt{\sum (x - \bar{x})^2} = S_\mathbf{b}$$

Another way to look at this is that the $\sqrt{\sum (x - \bar{x})}$ term adjusts for the effect of the x variable, which is not even required to be randomly selected.

[15]RMSE is available using regr.eval in the R language package called DMwR.

of the estimate, S_ϵ,

$$RMSD = S_\epsilon = \sqrt{\frac{\sum (y - \hat{y})^2}{n - 2}},$$

in original units of measure. Many texts misleadingly use $n - 1$ in the denominator of $RMSD$ and $n - 2$ in estimating S_ϵ; the difference is minimal when n is large. The $n - 2$ divisor also fits the F test, so let's accept that two df are used up by estimating $\hat{y}$, giving $n - 2$ for estimating $RMSD$ or $RMSE$. All these various names reflect the variety of fields and applications in which it is used. Whichever you call it, it combines all the residuals into one index. This applies to a variety of situations, including all the tests for slope in this chapter.

11.6.1.2 Example: Student's t Test for b

In addition to comparing slopes using $(\mathbf{b} - \mathbf{b_0})/S_\mathbf{b}$, the simple existence of slope can be tested by either an F test (Example 11.6.0.2) or by a Student's t test, resulting in the same outcome.

Some College data, tested for b using t: If, in Subsubsection 11.6.0.2 I had not used an F test to see if there was really a linear relationship other than an apparent line solely due to the effects of chance, I could have used a t test instead. I could have chosen an $\alpha = 5\%$. The hypothesis pair would be

H0: $\mathbf{b} = 0$ (THERE IS NO LINEAR RELATION),

HA: $\mathbf{b} \neq 0$ (THERE IS A LINEAR RELATION).
This would have used Equation 11.12, $t = \mathbf{b}/S_\mathbf{b}$ with $n - 2\, df$, where the standard error of the slope is

$$S_\mathbf{b} = \sqrt{\sum (y - \hat{y})^2 / (n - 2)} / \sqrt{\sum (x - \bar{x})^2}, \text{ which gives}$$

$$S_\mathbf{b} = \sqrt{SS\epsilon / (n - 2)} / \sqrt{SSX} = \sqrt{(7815395/9) / 30331234.91} = 0.16920342647526.$$

$$\text{So, } t = \mathbf{b}/S_\mathbf{b} = 0.350877924/0.16920342647526 = 2.0737,$$

rounded to four decimal places for the more common tables. This is the same result as in Subsection 11.4.0.1. In Appendix 3 I found that $t_{0.025, 9} = 2.2622$. Because the estimated t value is less than this, it is **not statistically significant**. That is, we cannot be certain there is any slope at all. If it had been statistically significant, the effect size could have been reported in original units of measure as the $RMSD = S_\epsilon = \sqrt{868377.175} = 931.868$, or about 931.87/student. However, quite a few editors might merely be interested in the slope, $\mathbf{b}$, itself.

11.6.2 t test for the intercept

Although it is not very common to test the value of the intercept, $\mathbf{a}$, it is sometimes useful to test **H0:** $\mathbf{a} = 0$ vs **HA:** $\mathbf{a} \neq 0$. Recall that the line has a slope but it rotates around a point representing the mean, $(\bar{x}, \bar{y})$, of both x and y. The ***standard error of the intercept***,[16] $\mathbf{a}$, is

$$S_\mathbf{a} = \sqrt{\frac{SS\epsilon}{(n - 2)} \left(\frac{1}{n} + \frac{\bar{x}^2}{SSX} \right)} = S_\epsilon \sqrt{\frac{1}{n} + \frac{\bar{x}^2}{SSX}}. \qquad (11.13)$$

---

[16]The $+ \frac{\bar{x}^2}{SSX}$ term under the radical sign compensates for the effect of an explanatory mean, $\bar{x} \neq 0$. In situations in which you control the value of the mean as μ_x, and where negative values are possible, you can contrive that $\mu_x = 0$, in which case Equation 11.13 becomes $S_\mathbf{a} = S_\epsilon \sqrt{\frac{1}{n} + \frac{0}{SSx}} = S_\epsilon / \sqrt{n} = S_{\bar{\epsilon}}$.

So, the t test for this hypothesis pair is

$$t = \frac{\mathbf{a}}{S_{\mathbf{a}}} \qquad df = n - 2\,, \tag{11.14}$$

and this is a two tailed test, so it is evaluated against $t_{a/2,\,n-2}$. However, one tail tests are **also possible**, using **HA**:a > 0 or **HA**:a < 0, and in either case the result would be evaluated against $t_{\alpha,\,n-2}$.

11.6.2.1 Example: Two Tail t Test of the Intercept

Now to actually test if the intercept fails to pass through the origin.

H0: THE INTERCEPT **a** = 0

HA: THE INTERCEPT **a** $\neq$ 0

Methods and Materials: Using the college data and $\alpha = 5\%$, I found the standard error for the intercept by substituting into Equation 11.13,

$$S_{\mathbf{a}} = \sqrt{\frac{7815394.571}{9}\left(\frac{1}{11} + \frac{(3813.909091)^2}{30331234.91}\right)} = 703.8392\,.$$

Then substituting into Equation 11.14, $t = \frac{\mathbf{a}}{S_{\mathbf{a}}}$ with $df = n - 2$ and $a = \$4166.51$,

$$t = \frac{4166.51}{703.8392} = 5.91969\,,$$

which is more than $t_{\alpha/2=0.025,df=9.} = 2.26216$.

Results and Conclusions: So, the intercept is statistically significantly different at the $\alpha = 5\%$ (two tail) level because $t > t_{\alpha/2=0.025,df=9.}$.

11.6.3 t test for correlation R (background for a priori n)

Tests comparing slopes, including this comparison against a hypothesized slope, are valid. However, some authors feel that tests of standardized variables, such as correlations, are open to question and should be avoided (Wright 1976). This is because standardization creates an almost infinite opportunity for ambiguity. Supposedly there is a correlation between the US stock market and weather conditions for people living on the sidewalks of Calcutta. The causal relationship, if a relationship exists, would be the season of the year. However all the above tests of slope give the same results as Equation 11.16 (below), which tests the existence of slope by testing the existence of a linear correlation (you cannot have one without the other). A horizontal line has 0 slope and is not varying **with** anything.

So, tests to see if a straight line exists and tests to see if there is a meaningful amount of slope, or a meaningful correlation, are really the same test in disguise. As already covered, Pearson's coefficient of determination (Subsection 12.5.1),

$$R^2 = \frac{\left[\sum_{i=1}^{i=n}(x_i - \bar{x})(y_i - \bar{y})\right]^2}{\sum_{i=1}^{i=n}(x_i - \mu_x)^2 \sum_{i=1}^{i=n}(y_i - \mu_y)^2}\,, \tag{11.15}$$

is, in many applications, seen as the contribution of the linear relation to the overall variance of y. Thus, its complement, $1 - R^2$, represents the part of the variance that is not due to a linear relation. It can be shown that you can run an F test for linear correlation, using

$$F = \frac{R^2}{(1-R^2)/(n-2)},$$

and compare it against $F_{\alpha, 1, n-2}$ (the 1 is included to indicate the associated df, resulting in a mean square). However, it is particularly useful to take the square root of this because it leads to a t test for linear correlation,

$$t = \frac{R\sqrt{n-2}}{\sqrt{1-R^2}}, \tag{11.16}$$

and this can be rearranged to be the basis for an a priori estimation of the sample size, n (Subsection 11.8.0.1). Note that this test, if statistically significant, can be seen as both validating the existence of a slope and that there is a correlation.

If a scatter plot of the residuals shows **no particular pattern**, you are relatively safe in assuming that there are no leftover non-random effects. Equation 11.16 can be a **two tailed test** with **H0**: THERE IS NO SLOPE (b = 0) **evaluated against HA**: THERE IS A SLOPE (b $\neq$ 0), which is evaluated by using $t_{a/2, n-2}$.

11.6.3.1 Example: Testing for Straight Line Relation

Is the college data straight? Let's test the correlation to see if there really is a statistically significant linear relation. Suppose I had actually used a two tail test of correlation with $\alpha = 5\%$. The hypothesis pair would be

H0: THERE IS NOT A LINEAR RELATION ($R = 0$)

HA: THERE IS A RELATION ($R \neq 0$)
From the original college example in Subsection 11.4.0.1, $n = 11$,

$$R = \frac{SXY}{\sqrt{SSX}\sqrt{SSY}} = \frac{10642560.73}{\sqrt{30331234.91} \times \sqrt{11549634.18}} = 0.5686132510,$$

and squaring this gives the coefficient of determination, $R^2 = 0.3233210292$. So using Equation 11.16,

$$t = \frac{R\sqrt{n-2}}{\sqrt{1-R^2}} = \frac{0.5686132510 \times \sqrt{11-2}}{\sqrt{1-0.3233210292}} = 2.0737,$$

which is the same value of t as for the tests of slope (even before rounding from nine figures). The critical value of $t_{0.025, 9} = 2.2622$. So, because my result is less than that, it is **not** statistically significant. I do not know if there is any slope at all, other than the appearance of slope due to chance. If this had been statistically significant, an ES for this could either be reported as the RMSD reported in the example in Subsection 11.6.0.2, or as $R^2 \approx 32\%$, meaning that this percent of the variation would be affirmed to be due to linear effects. Either or both might be interesting to the reader.

Discussion: The test did not provide evidence of a linear relation, so it does not say anything new about whether there is, or is not, a linear relation. Some workers might wonder if the sample is just too small. They might say that maybe about 32% of what is being observed is linear. However, if it is not significant, they would be on very thin ice. Also, any but a very strong correlation might be simply some other relation which only tends to increase in the general direction of a straight line over the range of the data. Some folks would refuse to accept $R < 0.7$, and say that unless $R^2 \geq 0.49$, meaning that at least almost half, or more, of the observed variance is accounted for by what appears to be a linear relation, it would be unreasonable to say the relation is linear. However, this rule breaks down with very **large** samples, so that even $R^2 < 0.49$ might still be acceptable; modelers (big data scientists), for instance, are more interested in how much the relation might be approximated by a linear model. So, in general, with more moderate or small sample sizes, it might be best to accept that $R > 0.70$ is necessary in order to be meaningful. This means $R^2 > 0.49$, which approaches or exceeds 50%, estimating that roughly half or more of the variance in the response variable is due to the causal variable. Question 13 at the end of this chapter is a worked sample size problem seeking an $R^2_{min} = 50\%$.

Tests for linear effect assume that the residuals, the contribution of random effects, will be normal. To assess whether this might be true, see Section 11.10 for a suitable technique (at least for big samples). Another issue is that such tests do not really detect just any linear relation, but only linear relations which have **large enough R or R^2_{min} to be detected**.

11.7 Confidence Interval for Slope and Intercept

You can usually estimate a confidence interval for slope if it is possible to do a corresponding t test. For either a t test or an F test, the distribution of the residuals must be normally distributed, and have the same variance for every value of x in the range under consideration. If so,

$$\hat{b} - t_{\alpha/2, n-2}S_b \leq b \leq \hat{b} + t_{\alpha/2, n-2}S_b, \tag{11.17}$$

using Equation 11.11, $S_b = \sqrt{\sum (y - \hat{y})^2 / (n-2)} / \sqrt{\sum (x - \bar{x})^2}$, and $t_{\alpha/2, n-2}$ is from a table such as Appendix 3.

Although not as often of interest, the CI for the intercept may similarly be estimated, using

$$\hat{a} - t_{\alpha/2, n-2}S_a \leq a \leq \hat{a} + t_{\alpha/2, n-2}S_a, \tag{11.18}$$

where the standard error of the intercept is from Equation 11.13,

$$S_a = \sqrt{\frac{SS_\epsilon}{(n-2)}\left(\frac{1}{n} + \frac{\bar{x}^2}{SSX}\right)} = S_\epsilon \sqrt{\frac{1}{n} + \frac{\bar{x}^2}{SSX}}.$$

The result might help you decide whether you really want to force the line through zero, as for calibrating some kinds of instrument, or even force it through some other point (Section 11.2.2).

11.7.0.1 Example: CI for Slope and for Intercept

College data Sub-subsection 11.6.0.2 and several continuations have not found statistical significance. That means I could neither accept **H0**, nor did I have sufficient evidence to accept **HA**. However, assume I was stubborn, or just curious, and wanted to see what a 95% CI for its slope and intercept would be if the test had indicated that a linear relation existed. So, I applied Equation 11.17 where $\hat{b} = \$0.3508779237$/student, $S_b = \$0.16920342647526$/student, and $t_{0.025, 9} = 2.2622$, to obtain

$$\$ (0.35088 - 2.2622 \times 0.16920) / \text{student} \leq b \leq \$ (0.35088 + 2.2622 \times 0.16920) / \text{student}$$

$$-\$0,31884/\text{student} \le \mathbf{b} \le \$0.73364/\text{student}.$$

To get a CI for the intercept, $\hat{\text{a}} = \$4166.51077$, I used Equation 11.18, but first I had to estimate $S_{\mathbf{a}} = S_{\epsilon}\sqrt{\frac{1}{n} + \frac{\bar{x}^2}{SSX}}$, by getting $S_{\epsilon} = \sqrt{868377.175} = 931.8676$, $\bar{x}^2$, and SSX from Subsection 11.4.0.1,

$$S_{\mathbf{a}} = 931.8676\sqrt{\frac{1}{11} + \frac{3813.909091^2}{30331234.91}} = 703.8392.$$

Now, the $df = n - 2 = 9$; the confidence interval uses the same $t_{0.025,9} = 2.2622$ to get

$$\$(4166.51 - 2.26215 \times 703.8392) \le \mathbf{a} \le \$(4166.51 + 2.26215 \times 703.8392)$$

$$\$2574.32 \le \mathbf{a} \le \$5758.7$$

Discussion The outcome for the slope looks rather uninformative. The lack of significance has already implied that the CI for **b**, if it exists, remains unknown. But this example demonstrates **how to do the calculations**. Assuming a line even exists, are the errors normally distributed? The issue of normality is discussed in Section 11.10.

Question 7 is another instance of finding CIs for slope and intercept and the answer is in the back of the book.

11.8 Sample Size for Either Correlation or Regression

If your sample size is too small, you have little chance of getting a significant correlation, but if it is sufficiently too large, you can obtain a **meaningless** 'statistically significant' correlation for anything (for instance, due to an infinitesimally small side effect). Therefore, you must a priori determine a suitable sample size which eliminates these difficulties. **You do not need a pilot study**.

In order to find an a priori sample size for either correlation or regression, you will only have to choose a criterion, α, a power, $1 - \beta$, and select a plausible or useful **reasonable minimum difference** (RMD), which is the R^2_{min} that you consider to be a minimal coefficient of determination. That is, you choose what proportion of the variance you want to be due to the interaction. In this application, the RMD can be a minimum fraction (for instance, a percent) of the variation due to linear interaction between x and y, or it can be the square of the minimum correlation coefficient R that you consider meaningful. Then assuming the residuals (independent random effects) will be normally distributed, the dependence of R^2 on sample size will allow you to estimate the sample size that will be required. As a simple derivation of an a priori estimator of sample size, n, start with Equation 11.16, $t = \frac{R\sqrt{n-2}}{\sqrt{1-R^2}}$ (which can be shown to equal $t = \hat{\text{b}}/S_{\mathbf{b}}$). Then:

- start rearranging to get $1/t = \frac{\sqrt{1-R^2}}{R\sqrt{n-2}}$, $\sqrt{n-2} = \frac{t\sqrt{1-R^2}}{R}$,

- then squaring and rearranging again, $\hat{n} = \frac{t^2\left(1-R^2_{min}\right)}{R^2_{min}} + 2.$
  However, as in estimating sample sizes for preceding t tests, and for the same reason, the t value you substitute in to find the sample size is $t = \left(t_{\alpha/2,n-2} + t_{\beta,n-2}\right)$.

- So, your sample size estimate is

$$\hat{n} = \frac{\left(t_{\alpha/2,\, n-2} + t_{\beta,\, n-2}\right)^2 \left(1 - R^2_{min}\right)}{R^2_{min}} + 2 \,. \qquad (11.19)$$

But, you cannot know $t^2_{\alpha/2,\, n-2}$ or $t^2_{\beta,\, n-2}$ without some idea of n.

- So, you must **initially substitute** $Z_{\alpha/2} + Z_\beta$ for $t^2_{\alpha/2,\, n-2} + t^2_{\beta,\, n-2}$.

- Then iterate (repeat), often twice or more, using the provisional $\hat{n}$ values to look up $t^2_{\alpha/2,\, \hat{n}-2}$ and $t^2_{\beta,\, \hat{n}-2}$ each time until you are satisfied that there is no longer a meaningful change in your sample size estimate.[17]

11.8.0.1 Example: Sample Size for Correlation or Regression

Assuming that the residuals are normally distributed, and the sample will be obtained in some reasonable and unbiased manner, you can estimate a correlation sample size or a regression sample size using only three values, R^2_{min}, $\alpha \geq 5\%$ and $\beta \leq 20\%$. You do **not** need preexisting information such as from a pilot study.

Sample size for correlation or regression at the UN11.8.0.1 Suppose you were a consultant to a UN advisory agency interested in the sample size required for $\alpha = 5\%$ criterion, and a power of 80%, to detect if there exists a linear relation that is strong enough to account for $R^2_{min} = 10\%$ of the variation of y in a sample of $(x_i,\, y_i)$ point estimates based on the ages and the wages of undocumented male workers in Israel. Let x be the age of each such worker in Israel versus y, their income.

Analysis Applying Equation 11.19, $\hat{n} = \frac{\left(t_{\alpha/2,\, n-2} + t_{\beta,\, n-2}\right)^2 \left(1 - R^2_{min}\right)}{R^2_{min}} + 2$, where $R^2 = 0.10$, the power equals $1 - \beta = 0.80$, and $\alpha = 0.05$, it is necessary to look up $Z_{\beta=0.20} = 0.841621$ and $Z_{\alpha/2=0.025} = 1.959964$, and sum them to find $Z_{\alpha/2} + Z_\beta = 2.801585$. Then applying Equation 11.19, with the substitutions of Z values,

$$\hat{n} = \frac{\left(Z_{\alpha/2,} + Z_\beta\right)^2 \left(1 - R^2_{min}\right)}{R^2_{min}} + 2 = \frac{(2.801585)^2 \times (1 - 0.10000)}{0.10000} + 2$$

$$= \frac{7.848879 \times 0.90000}{0.10000} + 2 = 72.63991 \text{ subjects.}$$

Now you round the customary arithmetic way to get the initial estimate, $\hat{n} = 73$, for the first iteration, using $t_{\alpha=0.025,\, \hat{n}-2=71} = 1.99394$ and $t_{\beta=0.20,\, \hat{n}-2=71} = 0.84671$ to find

$$\hat{n} = \frac{(1.99394 + 0.84671)^2 \times 0.90000}{0.10000} + 2 = 74.62363 \,,$$

which rounds to $\hat{n} = 75$ male immigrants. This is enough change that you should use another

---

[17]At the time of this writing there is a paper on finding sample sizes for multiple regressions (meaning regressions where there are several different factors suspected of contributing to $\hat{y}$) posted as http://psych.unl.edu/psycrs/942/q1/mrpower.pdf . It uses F instead of t, but the principle is similar. It was posted in support of a course, Psyc942, Research Methods and Data Analysis, offered each semester by Cal Garbin at the University of Nebraska-Lincoln.

iteration with $t_{\alpha=0.025,\,\hat{n}-2=73} = 1.99300$ and $t_{\beta=0.20,\,\hat{n}-2=73} = 0.84657$ to find

$$\hat{n} = \frac{(1.99300 + 0.84657)^2 \times 0.90000}{0.10000} + 2 = 74.56842\,,$$

which also rounds to $\hat{n} = 75$ male immigrants. The results of iteration have stopped changing so this is where you can stop. This is the estimate you are after.

Discussion Be careful to iterate enough to be certain that the sample size estimate is **not** still changing. When you consider this estimation process, you see that smaller values of R^2 would give increasingly greater sample size estimates. Smaller values than $R^2 = 10\%$ may not be readily accepted by your colleagues. However, at least in this case, it is possible to show (if you wish to do the calculations) that if the UN agency would agree to let you change to $\alpha = 1\%$ and $\beta = 10\%$, the results would be a lot more dependable and would only cost you a little less than twice as many samples.

11.8.1 Simple linear estimation and beyond

In many publications, a *simple correlation* or a *simple regression* is defined as 'simple' when there is only one explanatory variable. More complex multiple regressions, with more than one explanatory variable, are not covered here. Each such multiple regression tends to require a very big sample size. However, multiple regressions are a very powerful tool.

11.9 Checking Out the Errors

Especially in this day of computerization that makes it easy, you should **plot the residuals against the explanatory** variable to look for hidden trends (patterns). The residuals should not show a distinct pattern of changing (increasing scatter or decreasing scatter) with x because the variance of y is assumed to be the same at all levels of x. They are also assumed to be normally distributed about y at every level of x. The amount of data needed to clearly show such behavior to be true over the whole range would have to be very, perhaps impractically large, but we can sometimes detect extreme trends that would invalidate our tests. Let's look at a plot of the residuals against the explanatory variable from the simple regression in Subsection 11.4.0.1, **Regression assumes** there is **no au-**

Figure 11.7: Scatter Plot of Residuals from simple regression example

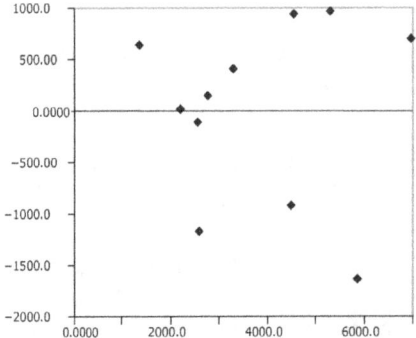

It is difficult to assess a scatter plot for such a small sample, but the distribution of the residuals should not show any obvious trend as the explanatory variable increases or decreases. A horizontal least squares line, such as shown, suggests that the original regression accounted for all straight line effects.

tocorrelation, meaning that the **explanatory variable**, x, **is uncorrelated to the residuals**, $\hat{y}_i - y_i$. So, you might check for correlation of the residuals with the explanatory variable. Straight line effects are indeed about slope, which is x and y changing together. The horizontal least squares line in Figure 11.7 shows that there is zero remaining regression slope, meaning that the original regression accounted for all straight line effects.

For t (or F) testing the slope, or for estimating CIs for **a** and **b**, the remaining question about the residuals is whether they are really normally distributed, as is commonly assumed. One approach to this is covered in Section 11.10. Also for t (or F) testing, there should be no straight lines or other obvious pattern in the scatter plot of the residuals, but you might need sufficient data to be likely to see such effects.

11.9.0.1 Example: Are the Residuals Correlated With x?

Cross correlation messes up linearity (college data continued) As shown in the simple regression using the crude media generated college data in Subsection 11.4.0.1 and its other continuations, I got values supposedly representing a linear effect, but could not present sufficient statistical evidence that it was due to anything other than chance alone. Suppose that I had gotten more definite results or suspected that straight lines were not involved. Then, after examining a scatter plot of the residuals (Figure 11.7), I could have gone on to examine the residuals about the regression line, using the correlation coefficient for the residuals, $res_i = \hat{y}_i - y_i$, vs the explanatory variables, x_i. Here is a table showing the calculation of the covariance for use in further calculations:

X	Errors	Cross Prod.
2590	1169.284	3028445.56
5311	-969.977	-5151547.85
1371	-640.436	-878037.76
4486	920.548	4129578.33
6964	-701.976	4888560.86
2199	-17.909	-39381.89
2774	-148.155	-410981.97
4542	-943.802	-4286748.68
5860	1632.655	9567358.30
2558	108.056	276407.25
3298	-408.295	-1346556.91
Σ 41953	-0.008	-26.486
Cov.		-2.9429

I could have done this using Pearson's R (Equation 11.7), but I used a modification of it, Equation 11.9, to get

$$R_{res} = \frac{Cov\,(x,\,res)}{S_x S_{res}} = \frac{-26.486}{\sqrt{3033123.491} \times \sqrt{1154963.418}} = -0.000014151,$$

almost zero cross correlation. So there is no evidence that this requirement for either linear correlation or linear regression was not met. Note that this R_{res} refers to a possible relationship between the residuals and the causal variable. It is different from the R due to a relationship between x and y that I calculated previously.

11.9.1 What if the residuals do vary with x?

If you suspect that there is correlation between the explanatory variable, x_i, and the residuals, $\hat{y}_i - y_i$, then a line and scatter plot might be sufficient to reveal this violation of the assumptions associated with a simple regression. If there is no relation violating the assumptions, the points should appear to be relatively uniformly distributed around the estimated line.

Otherwise counts, especially when small sample sizes are involved, tend to behave as though Poisson distributed, which means as though they have a frequency distribution that follows the equation $f(r, \lambda) = \frac{\lambda^r e^{-\lambda}}{r!}$ (Section 15.4). Regardless of sample size, if a correlation between the explanatory variable, x_i, and the residuals, $\hat{y}_i - y_i$, is observed, you might consider a model that assumes that the distribution resembles a Poisson distribution and use a log transformation of y. That is, you use the raw x_i and $ln(y_i)$ in a simple regression (Section 11.2) to obtain a model of the form, $log_e y = a + bx$. **Back transformation produces a predictive model,**

$$y = e^{a+bx} + \epsilon. \tag{11.20}$$

However, it can, at first, appear to be harder to evaluate than a simple linear regression[18]. This approach is sometimes called a 'Poisson regression', but that name might be seen as a misnomer because it can also work with continuous data, not merely counts. The name arises out of the fact that the Poisson distribution is the most common useful distribution in which the means and the variances (or x and the residuals) change together (are correlated).

Testing to see if this log converted regression is valid

Just as for the simple regression, you can test the slope of $ln(y) = a + bx$ to see if a relation exists using Equation 11.12.

11.10 Hoping for Normality:

If you could take every value, x_i, in a population and graph its value on the x axis against its frequency on the y axis, you would get a curve representing the distribution's true frequencies, which when you are drawing samples at random are equal to the probabilities:

$$f_{x_i}/N = (number\ of\ times\ it\ occurs) / (total\ number\ of\ things\ in\ the\ population) = P_{x_i}.$$

Each value, x_i, of a truly random distribution (a distribution that is **unaffected** by nonrandom effects), has an equal probability ($P_{x_1} = P_{x_2} = \ldots = P_{x_N}$) of being drawn by chance. Such a distribution is sometimes called *flat*, meaning that when $P_{x_i} (= f_x)$ is graphed against x_i, the result is a straight line. But if the distribution is a normal (*Gaussian*) distribution, the resulting bell shaped curve is a mound peaking at the mean (the values in such a bell shape curve tend to cluster around the central measure, a strong effect other than chance).

Random samples from either of these populations tend to produce *distribution graphs* that approximate the whole population's graph. However, randomly sampled data will have some raggedness and will tend to thin out to become even more ragged at the extremes.

11.10.1 Cumulative P_{x_i} for each x_i

Plotting probabilities as a **cumulative** frequency distribution curve puts the x_i values on the x axis and their cumulative probabilities, $y_i = \sum_{j=1}^{j=i} P_{x_j}$, on the y axis. That is, the first entry has the probability P_{x_1}, the second has the probability $P_{x_1} + P_{x_2}$, so on to $P_{x_1} + P_{x_2} + \ldots + P_{x_i}$.

---

[18] The value $e = 2.718281828459\ldots = \sum_{x=0}^{x=\infty} 1/x!$, so, as you might suspect, it could relate to both the slopes of graphs and to our counts, or in some applications, even to continuous variables. It is very useful in higher mathematics and statistics.

If you plot a flat (a truly random) distribution in this cumulative way, the result will be a slanted straight line. But if you plot a normal distribution in this cumulative way, the result will form a sort of *S* shape, but stretched out and steeper around the middle (Figure 11.8).

Figure 11.8: Cumulative Random vs Normal Distributions

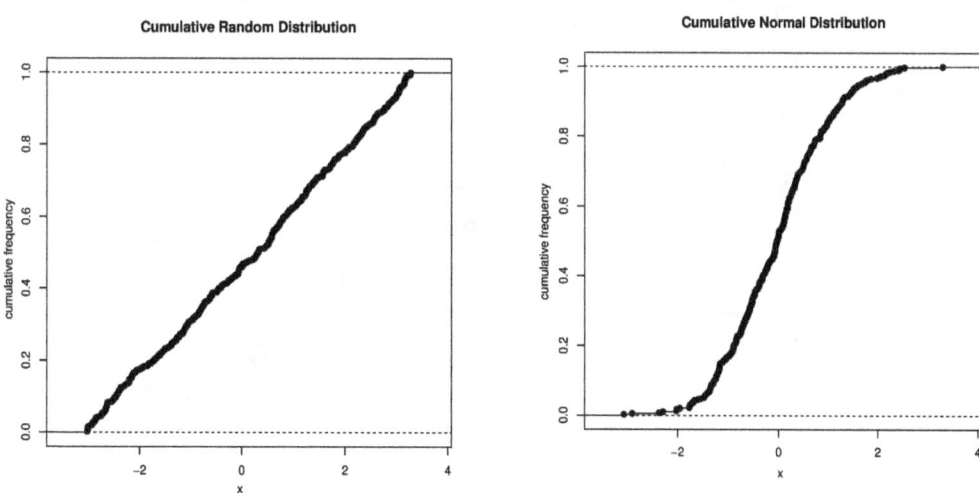

This plot against the raw data, x, gives a pattern seemingly the **opposite** of a Q-Q plot such as Figure .
From an R script using `plot()` and `ecdf()` functions.

11.10.2 Q-Q Plots (cumulative against cumulative)

Now we are going to make a completely different kind of plot. Instead of plotting x as the causal factor against the cumulative probability (which arises from the frequency when there is random sampling), we are going to plot the cumulative frequency of a hypothesized distribution of any kind, as though it were the causal factor, against the observed cumulative frequency, $\hat{f}_{x_i} \left(= \widehat{P}_{x_i} \right)$ of the random sample data. We will be interested **in the shape** of the result rather than the relationship between the scales of the axes.

There are a variety of ways to check for normality, all of which have limitations due to the effects of chance and therefore require suitably large sample sizes in order to be likely to see anything.[19] The word 'see' suggests what is perhaps the most useful of all applied statistical processes, visualization.

If you plot the cumulative frequencies from a random sample from any kind of distribution against cumulative frequency values from a distribution of the same kind, the result will tend to be a straight line, except for a little deviation at the extremes, where the sample data becomes sparse.

For instance, if you plot cumulative frequencies from a random sample from a normal distribution against cumulative frequency values from any normal distribution (such as from a table), the result will tend to be a straight line except for a little deviation at the extremes. Similarly, the result will be **curved** if you plot frequency values from two **different kinds of cumulative distributions** against each other.

One of the more satisfactory ways to check to see if your data meets the assumption of normality is to plot the cumulative distribution of the frequency estimates from said data against *acumulative Gaussian distribution* (a cumulative normal distribution) as in Figure 11.9 or Figure 11.10. If the data is randomly drawn from a normal distribution, this should produce a straight line, but if it is not

---

[19]You might also want to see Figure 11.10 for more evidence of the effect of sample size.

Figure 11.9: Random vs Approaching Normal

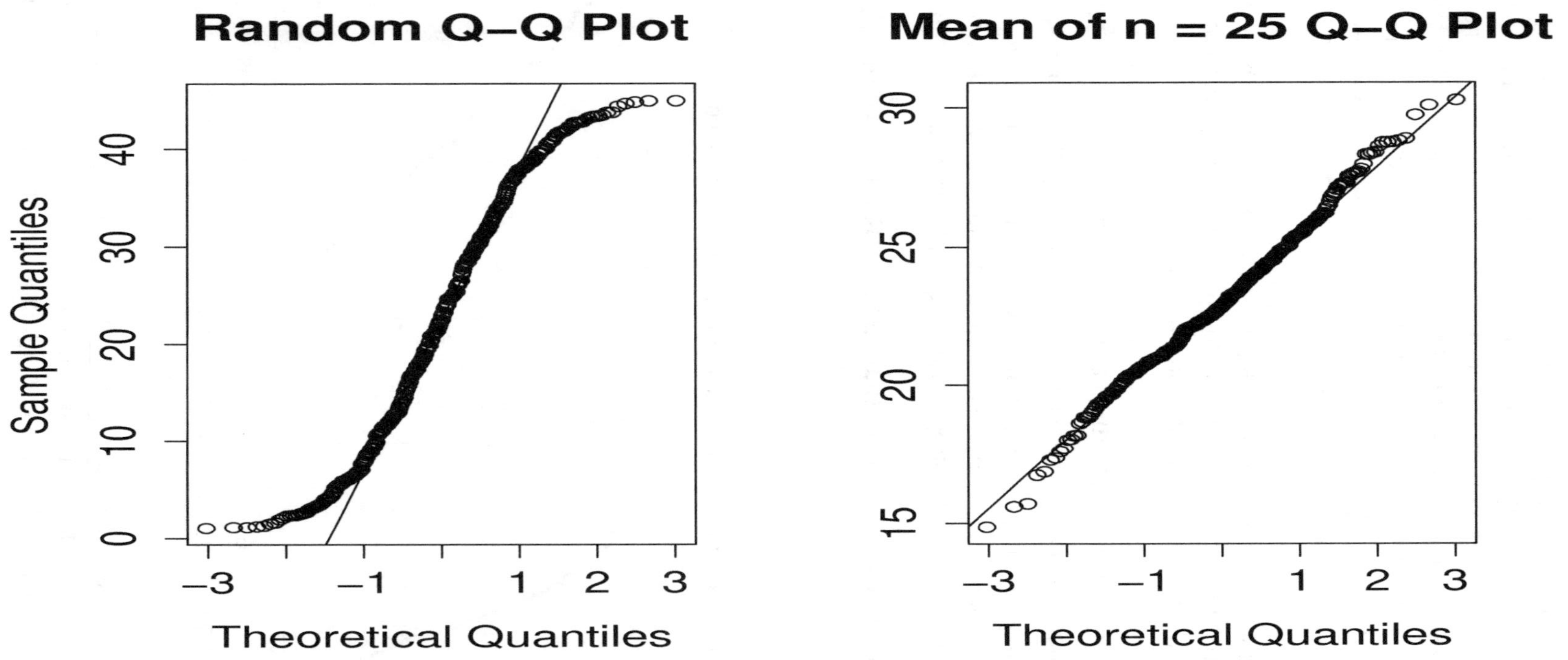

Random Q–Q Plot

Mean of n = 25 Q–Q Plot

A Q-Q plot is a way to examine whether your data approaches being normally distributed by plotting sample cumulative frequencies against cumulative normal distribution frequencies. This figure also demonstrates the effect of the Central Limit Theorem. On the left is a plot of the cumulative probability estimates of an ordered sample of 400 random numbers. It is *S* shaped because it is plotted against a cumulative Gaussian scale (horizontal). The plot on the right is of cumulative probabilities from 400 **means** of data sampled from the same source. Although the sample size per point was only $n = 25$, the plot is less steep and almost linear, showing that the distribution of the means is approaching a normal distribution. The actual values on the axes are of little interest in themselves. This illustrates only one of the effects of the CLT, the effect on the shape of the distribution of means as sample size increases. RStudio was used to create these plots. The **left one** is definitely **not normal**, but the **right one is close** enough for most applications.

Figure 11.10: Q-Q Plots of Log Random, $n = 1$, and Means of Log Random, $n = 30$

The shape of the left curve shows a great difference from normality, but the right curve shows a pretty good correspondence. Matters such as variance and centering are somewhat changed, hence the differing scales on the left, but the shape is the matter of importance here. Since the ragged ends of plots like these from random samples are not considered to really be part of the population (they are considered to be outliers), the ticks on the lower axis could be interpreted as quartiles.

from a normal distribution, it will usually produce an *S* shape curve. Serious deviations from normality tend to result in relatively smooth curves that are most obviously deviant near the ends (Geary 1947, Hart and Hart 2002, Shapiro 1990). Notice that the interpretation of such plots is a little subjective; the effect of non normality is most noticeable near the ends of the plot, but the very ends tend to be a bit ragged. You can plot these actual values against the cumulative normal distribution, or if $n < 30$, even against a cumulative t distribution. Various software packages and the R or *S* languages also provide for such plots, which are often called *Q-Q plots*.[20] Most versions of a *Q-Q plot* show one or both scales in *quartiles*, fourths of the distribution, but the shape of the Q-Q plot is what is of interest.[21] Figure 11.10 shows a couple of comparative Q-Q plots of the effect of sample size: the log_{10} of a random distribution of sample size $n = 1$ versus a corresponding log_{10} of means from the same random distribution with $n = 30$.

Such plots might be useful for examining the residuals resulting from regression. These cumulative plots are generally considered the **best available tests for normality** despite the fact that there is a **subjective** element in their interpretation. You should be aware that this graphical approach is best with continuous data, and is not generally applicable for the normal approximation (which is truncated at 0 and 1).

---

[20] A program by Tuttle, Oliver, and McGinnis (2007) for comparing small samples to a t distribution, <u>Cumulative Probability Plot</u>, plots against a presumed t distribution for sample sizes when $n < 30$. It was originally intended for quality control associated with monitoring potential radioactive contamination and is freely downloadable from the internet, courtesy of Boeing and the DOE. Apparently Cumulative Probability Plot was used to meet the State of California, Department of Health Services, Radiological Health Branch requirements for facility release in the 1980s. Using this program t might be more useful for $25 \leq n \leq 80$.

[21] *Quartiles* are three values cutting a probability distribution into four parts, each part including $1/4$ of the data. That is, quartiles can be used to cut the whole distribution into four equally probable sections (Figure 11.10). Quartiles are popular with statisticians, but unfamiliar to most folks who have no trouble with percents which cut a population into one hundred parts rather than into just four parts. Just like you can have a fourth or a tenth of a percent, you can have a fourth or a tenth of a quartile. Visually there is little difference unless you look at the numbers on the scale, which would range from one to four rather than one to one hundred. So, the quartiles of a normal distribution may be graphed as ticks cutting the bell shaped curve into four parts; the area under the curve between any two quartiles is the same as the area between any other two quartiles.

Quartiles are of particular interest with respect to box plots, Subsection 15.1.1. But, Cumulative Probability Plot is scaled as percentiles of the t distribution, or the normal distribution (conceptually cutting the distribution into 100 parts) on the lower axis.

For binomial data, or less than very large samples of discrete data (ie. counts), you can, rather subjectively, use other graphics, such as boxplots (Section 15.1.1). Such other graphics will show you if the data is suitably balanced between one end and the other, and hence might be adequate to stand in for a Gaussian distribution.

The particular plotting technique discussed here, Q-Q plotting, is one of the best ways to verify that your data approaches meeting the assumption of normality, but no technique is sufficiently powerful to justify using student's t test for small samples. For small samples you would have to use nonparametric approaches such as the Wilcoxon test to analyze your data (Section 14). If you have a very big data set (such as the whole US census) and can afford to sacrifice a random subsample of it, you can use that random subsample, sometimes called a training sample by modelers, by examining it before your analysis and using all sorts of graphical or analytical techniques, including a Q-Q plot, in order to form hypotheses and to select the ideal approaches to analyzing them. However, you **must not recombine it with the original data** that you intend to analyze.

11.10.2.1 Example: Are the Residuals Normally Distributed?

Distribution of errors for the college data I entered the residuals in a spreadsheet and exported them as a comma separated file (.csv) to <u>Cumulative Probability Plot</u> to see if I could determine if this very small sample of $n = 11$ residuals had an underlying normal (or rather a t) distribution. The resulting graphical output is:

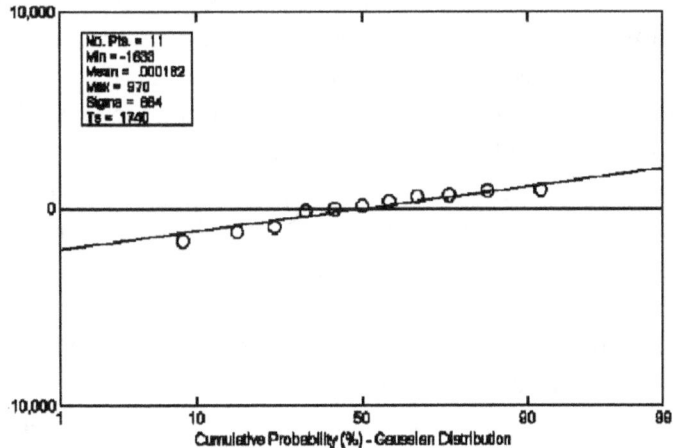

Does it really tell me 'Yes, the residuals are normally distributed' or does it exclaim, 'Are you mad? There is so little data'. Well ...

Discussion: The particular plotting technique discussed here, Q-Q plotting, is one of the best ways to verify that your data approaches meeting the assumption of normality, but it is not sufficiently sensitive to be used for small samples. For small samples you might get a <u>rough</u> idea of whether the data was sufficiently symmetrical or not using a boxplot (Subsection 15.1.1). Likewise, if you have reason to believe the residuals are not going to be normally distributed, or that some other assumption is going to be meaningfully unmet, you might consider examining their residuals versus the causal variable, x, using a nonparametric (not dependent on normality) Spearman's correlation (Section 14.6). That might at least tell you if there was some sort of auto-correlation-like effect, whether straight line or not.

11.10.2.2 Paper and pencil note:

Classical probability graph paper with a Gaussian probability axis versus a linear axis might still be available from some stationary suppliers for hand plotting cumulative results to obtain the same graphs. This will, with more work, give you the same results as a computer program.

11.11 Predictions are Not Observations

The model upon which a regression line is calculated assumes that the residuals are normally distributed around every observed point with the same variance, as shown in Figure 11.4. Thus there is only one CI that applies to every point on the line being observed. But the CIs for **predicting** response points are not the same for every point, but rather their **width becomes greater** as the explanatory variable is **further** from the mean, $\bar{x}$, of the explanatory variables (from the pivot point, $(\bar{x}, \bar{y})$). The difficulty involves the fact that both the sample mean of the response variable, $\bar{y}$, and the slope, **b**, are subject to chance. Recall that the slope of the line determines its rotation around a pivot point at the means, $(\bar{x}, \bar{y})$. If you were to hold a long straight stick in your hand and turn your wrist (the pivot point) a little, a point on the stick near your hand would only move a little, but the end of the long stick would move a goodly distance. This affects both the prediction of the line, and the absolute size of the variation about it. So, the *standard error of prediction* is

$$Se_{pred} = S_\epsilon \sqrt{1 + \frac{1}{n} + \frac{(x - \bar{x})^2}{SSX}}, \tag{11.21}$$

with $df = n - 2$. As the role of $(x - \bar{x})^2$ in this equation suggests, the standard error of the prediction will be larger as your values of x are further from the mean. So the predicted CI for y_x for one **particular** value of x is

$$\hat{y}_x - t_{\alpha/2,\,n-2} Se_{pred} \leq y_x \leq \hat{y}_x + t_{\alpha/2,\,n-2} Se_{pred}, \tag{11.22}$$

where the -2 is the *df* lost in estimating $\bar{x}$ and $\bar{y}$. This CI would have to be calculated for each point of interest. If you predicted the CI for every possible value of x in the sample range, the resulting plot would be shaped roughly like the curved (outermost) lines in Figure 11.11.

Figure 11.11: Prediction Confidence (gray curve)

The innermost line is the regression line. It is crossed by lines representing the effect of chance on the slope. The outermost flanking curves (gray) represent the combined effect of chance on slope and the mean, $\bar{y}$, associated with the intercept.

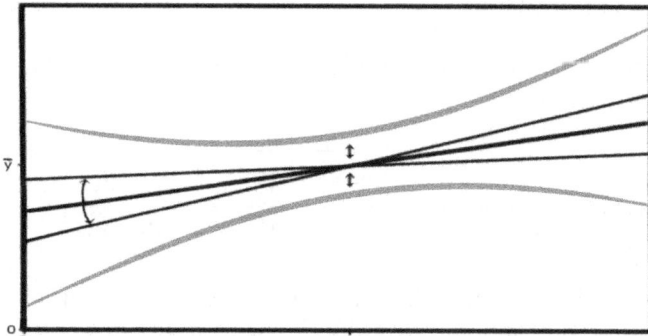

So, the vertical arrows indicate variation associated with the sample mean and the curved pointers indicate variation associated with the sample slope. Together the effects represented by the arrows and pointers result in the outermost confidence of prediction curves for a given α. (Not to scale.)

Modeling is about prediction, and modelers often take what is called a *training set*, a **random** sample of the original data set. They experiment with this training set to get what they feel is a reasonable model. They then compare this result with the result they get from the test set (remaining data) to see if the correspondence indicates that their model seems to be adequately predictive. The details are beyond the scope of this book, but some readers might wish to look into this very productive activity. Data that you have already been exploring should never be subsequently recombined with data you intend to use inferentially, as to confirm a model.

11.11.1 Prediction by regression

11.11.1.1 Example:Two Predictions

If you have a significant slope and intercept, you will be able to predict various values of y_i for each value of x_i. In the calculation, a subscript of $_x$ refers to a variable with respect to any particular x and a subscript of $_{pred}$ designates a prediction.

From the college data In the simple regression example (Subsection 11.4.0.1), the estimates were $\mathbf{b} = \$0.35087792$ per student and $\mathbf{a} = \$4166.5108$. The mean point where the slope pivots is at $(\bar{x}, \bar{y})=$ (3813.9091 students, \$5504.7273). Assume I was silly enough to use this result to make a prediction, even though the **test of the regression** was not statistically significant. At least it was not clearly non-normal, but the data was too limited for me to really be comfortable with the assumption of normality. But, I plowed on; the observed explanatory variable, x, ranged from 1371 students to 6964 students. I could choose anything in the range of the sampled explanatory variable. So, I chose to predict the cost of two universities, one with 3800 students, and another with 5500 students.

Analysis I found the predicted value for the first explanatory variable, 3800 students, using the simple linear model

$$\hat{y}_{3800} = \mathbf{a} + \mathbf{b}x = \$4166.5108 + (\$0.35087792/\text{student}) \times 3800\,\text{students} = \$5499.847\,.$$

The student units canceled out, leaving \$. I calculated the standard error of the residuals,[22] $S_\epsilon = \sqrt{MS\epsilon} = \$961.87537$, from the $MS\epsilon = 868377.175$ in Sub-subsection 11.6.0.2. It could also have been found using Equation 11.4, $S_\epsilon - \frac{\sum_{i=1}^{i=n}(y_i-\hat{y}_i)^2}{n-2}$. I still needed to find the standard error of prediction, $Se_{pred} = S_\epsilon\sqrt{1 + \frac{1}{n} + \frac{(x-\bar{x})^2}{SSX}}$. So, for the **first prediction**, $x = 3800$ students,

$$Se_{pred} = \$961.87537\sqrt{1 + \frac{1}{11} + \frac{(3800 - 3813.9091)^2\,\text{students}^2}{30331234.91\,\text{students}^2}} = \$1004.649,$$

where the $SSX = 30331234.91$ students^2 came from the original simple regression (Subsection 11.4.0.1). Notice that the students^2/students^2 cancel out, leaving just the dollars, \$. So, using $t_{(\alpha,,n-2)} = t_{0.025,9} = 2.26216$, the CI for this prediction of $y_x = y|(x = 3800)$ students

---

[22]You can find the standard error of the residuals with either $MS\epsilon \quad = \quad \sum_{i=1}^{i=n}(y_i - \hat{y}_i)^2/(n-2)$ or $MS\epsilon \quad =$ $SSR/(n-2)$, which are, of course, equal to each other.

is[23] $\hat{y}_{3800} \pm t_{0.025} S_\epsilon$. That is,

$$\hat{y}_x - t_{\alpha/2,\,n-2} Se_{pred} \leq y_{3800} \leq \hat{y}_x + t_{\alpha/2,\,n-2} Se_{pred},$$

$$\$5499.847 - 2.26216 \times \$1004.641 \leq y_{3800} \leq \$5499.847 + 2.26216 \times \$1004.641,$$

$$\$3327.20 \leq y_{3800} \leq \$7772.50 \,,$$

rounded to the nearest ten cents.

I found the **second prediction** and its CI the same way, but for $x = 5500$ students. So, the second predicted value of the response was found using the simple linear model,

$$\hat{y}_{5500} = \mathbf{a} + \mathbf{b}x = \$4166.5108 + (\$0.35087792/\text{student}) \times 5500 \,\text{students} = \$6096.339 \,.$$

The standard error of prediction for the resulting greater value of $(x - \bar{x})^2$ is

$$Se_{pred} = \$961.87537 \sqrt{1 + \frac{1}{11} + \frac{(5500 - 3813.9091)^2 \,\text{students}^2}{30331234.91 \,\text{students}^2}} = \$1046.9068 \,.$$

Because the Se_{pred} for $x = 5500$ students was **further from the mean**, $\bar{x} = 3813.9$ more students, than was the first, $x = 3800$, the CI for y_{5500} was wider. So, assuming normality etc., $y_x = y | (x = 5500 \,\text{students})$ is estimated to have a probability of $1 - \alpha = 0.95$ of being captured within the CI, $\hat{y}_{5500} \pm t_{0.025} S_\epsilon$. That is,

$$\$6096.339 - 2.26216 \times \$1046.915 \leq y_{5500} \leq \$6096.339 + 2.26216 \times \$1046.915 \,,$$

$$\$3728.10 \leq y_{5500} \leq \$8464.60 \,.$$

Rounding must be minimized during calculation, and maybe for further estimation. However, **when reporting** these results, round to no more than **one more figure than the original data**. In this case, that is to the nearest ten cents (one decimal place). Even then, you may wish to round or truncate in the direction which is least optimistic. Consider the loss that an overly optimistic estimate in the final decimal place might cause for a large cartel deciding whether to buy a private university.

Discussion I correctly limited myself to selecting explanatory values that were within the range of the original data. Some people extrapolate beyond the sampled range, but statistics is not a very dependable crystal ball. The curly brackets, { }, are only there to call your attention to the fact that Se_{pred} is larger for the second estimate even though both estimates are from the same sample. The second estimate for the larger explanatory variable, $x = 5500$ students, was much further away from the mean, $\bar{x} = 3813.9$ students, than was the first that was for $x = 3800$ students. The larger Se_{pred} resulted in a wider CI. However, as is to be expected when the slope is not significant, the width of either CI was so wide as to approach uselessness. This is yet another disadvantage of too small a sample size, n.

The whole of the simple regression example (Subsection 11.4.0.1) and its continuations is purely an

---

[23] As you may recall from the first chapters of this book, $y|x$ reads 'y given x'. It refers to your expected y given that x = whatever number of students, assuming that your model, the regression, is correct. You might exclaim, "But it's just an estimate!" Yes, think about that when you read articles focusing on estimates, especially estimates of what is supposedly going to happen in the future.

illustration of how such calculations work and how they are performed. Unfortunately, some people do perform such studies without any a priori consideration of sample size.

A serious **properly designed** study of this matter might be interesting. Are larger universities more expensive? There are a lot of universities to sample from. If size increases the expense per student, then why? What about the idea that larger is more efficient? At the moment, almost all the fees do seem to be so high as to be difficult to explain. How much do the journals that are now published by a small group of publishers rather than the societies themselves, affect the cost of maintaining adequate libraries, hence a major part of the cost of a university? Or, could American universities be approaching *entropic death* (failure due to the dedication of too much of the available resources to feedback)? Or, could the greater factor be too much government interference, a difficulty that, at least in private business, harms the small operations far more than the big ones? Uncovering linear (or transformed, thus curvilinear) relations might be informative, but correlation (hence regression) does **not prove** causation. Inferential testing might be a definitive approach to answering some, but not all, of these questions; even it only tells you that two things are different, which cannot directly prove your explanation of why they are different (causation a mystery yet again).

11.12 Adjusted Data

It is often a good idea to adjust the data for such continuous factors as age as part of the inferential ANOVA testing process. Unfortunately, the calculations for the resulting test, the ***ANCOVA***, to reduce the error variance, involve adjustment for the error in the estimate of the linear model, adjustment for the F test part, adjustment separately for any multiple comparisons, adjustment separately for reporting the means, and adjustment separately for the CI of the treatments. Nonetheless, the ANCOVA, which is available in many computer packages and facilitates all these adjustments, often increases the power of the test. And, the adjustment used in the ANCOVA only costs $1\,df$. It is especially likely to be useful if the coefficient of determination for the variable you wish to adjust for, and the variable in which you are interested, is $R^2 > 0.49$. Because of the complexity of presenting the ANCOVA adequately, and the intended scope of this book, further discussion is not included. Note that Snedecor and Cochran (1967) repeatedly mention that the effect of the Central Limit Theorem on assuring normality for any ANOVA might be insufficient if $df_W < 20$.

11.13 Caution When Evaluating Tests of Slope

Tests for slope, either t or F tests with **HA**: … IS NOT INCREASING LINEARLY, or **HA**: … IS NOT DECREASING LINEARLY, or **HA**: THE SLOPE $\neq 0$, rather commonly, perhaps uselessly, result in a declaration that the slope is statistically significant. Such hypersensitivity is because larger samples tend to increase the tendency to seem to show statistical significance based on minimal effects. In application, tests to determine if a meaningful linear relation exists are most useful if the sample is at least the size indicated by an a priori power analysis, but not a great deal larger. As a check on any simple linear regression, and especially if the sample size is somewhat larger than required by a power analysis, a coefficient of determination of $R^2 \geq 0.49$, equivalent to $(R \geq 0.70)$, strongly implies that at least a meaningful simple linear relation exists.. It is generally a good idea to include an R^2 value describing the fraction of the observed variance that was due to linearity in support of any simple linear regression.[24] Question 9 provides an exercise for finding both the product moment

---

[24]Sometimes in situations with categorical data, even those in which the categories can be ordered, you might ask whether you need a correlation (or a regression) or a Pearson's chi square. That is, some analysts might reconsider how they look at their data, or perhaps at what the question should be.

correlation and the coefficient of determination.

11.13.1 Adjusted coefficient of determination

Regression may have more than one independent causal variable, a subject only peripherally touched upon in this book, but added variables tend to inflate the coefficient of determination, R^2, excessively and misleadingly. Unless some action is taken to correct R^2 inflation, **embarrassingly** optimistic untrue outcomes can be, and are, **frequently reported**. *The Adjusted coefficient of determination* attempts to adjust for the number of causal variables.

The Adjusted coefficient of determination is calculated as

$$R^2_{adj} = R^2 - \left(1 - R^2\right) \frac{k}{(n - k - 1)} = 1 - \frac{(1 - R^2)(n - 1)}{(n - k - 1)},$$

where k is the number of independent causal variables (Theil, 1961).

Its interpretation is a little different from the coefficient of determination, R^2, and it is not really the square of the correlation coefficient, R. Unlike R^2, R^2_{adj} increases only if the new term improves the model more than would be expected by chance. Adjusted R^2 can be negative, and its size will always be less than or equal to R^2.

> **WARNING:** The R^2_{adj} adjustment is a less than perfect attempt to correct for the fact, as in multiple regression, that **added causal variables** sometimes make R^2 inflate meaninglessly. However, so far, R^2_{adj} seems to be the only widely accepted approach we have.

Likewise, the possible unintended inclusion of useless variables can be greatly reduced by techniques such as **factor analysis**, or **principal component analysis**. All this and multiple regression are outside this book's scope.

11.13.1.1 R^2_{adj} might be used with simple regression too

Some researchers, especially some geneticists and economists, conservatively use R^2_{adj} rather than R^2, even outside modeling, for reporting the strength of simple correlations, but **only** if the line clearly, or theoretically, **does not pass through the intercept**. They see the intercept, as well as the slope, as representing a causal variable. So they use $k = 2$ rather than $k = 1$.

11.14 Statistical Pitfalls

If you are not paranoid you do not understand the situation. I strongly suggest you read *How to Lie with Statistics* (Huff 1954), both to confirm why we need to thoughtfully apply statistics, and because it describes many pitfalls involved in drawing conclusions from it. The equations and operations shown in this textbook, or any statistics orientated book, are only that, equations and operations. Without painstaking mental precautions, you can come to all sorts of unfortunate conclusions, especially with respect to correlation. Correlations might be caused by chance or by one or more factors mutually affecting the variables under consideration and not by the effect of one of the variables on the other, or you might have mistaken the effect for the cause. Whether with correlation or regression, statistically significant tests, especially but not solely those tests of data that are not from some sort of highly controlled experiment, could simply be due to chance. Many, but not all tests are from fresh data collected after the hypothesis is made. However, correlations are often sought after the data is accumulated and the researcher has perhaps had most of their life to observe what appears to be a parallelism, in effect a post hoc correlation. The best policy is to challenge what seems to be the case and see if it can be demonstrated to be wrong beyond a reasonable doubt. The finger of fate is cruel indeed, if due to chance alone, it falsely disproves what you had initially chosen as true. But, correlations are usually run to confirm what seems to be true, and unless sample sizes are chosen with respect to power, they may confirm relationships which might be so weak as to lack meaning in application.

NOTE: It is sometimes said that the residuals (the deviations from the model) should be normally distributed for Pearson's correlation and for regression. However, this is **only required if** you intend to test for **significance with a parametric test** such as Student's t test or an F test (as in the most common and widely applicable applications).

Normality is only required for parametric testing to see if there is a meaningful amount of slope, not for the estimation of straight line correlation, R. Correlation tells you how well the data fits the model with straight lines regardless of deviations from a normal distribution of the residuals. Likewise the regression fits a valid straight line approximation regardless of deviations from a normal distribution of the residuals even though it might only be an approximation parallel to the tangent of a curve. An alternative, the Spearman's correlation coefficient Equation 14.11, is nonparametric and is less specific and only tests the tendency of two things to increase or decrease together regardless of whether they tend to follow a straight line together or if they follow parallel curves together.

11.14.1 Simpson's paradox

Figure 11.12: Simpson's paradox

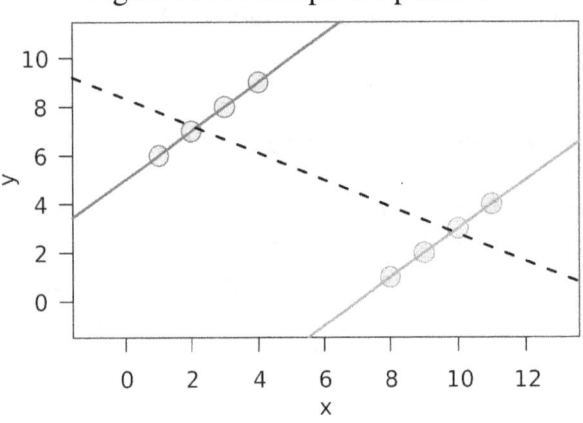

An example of Simpson's paradox for quantitative data: An artifact in the form of a negative sloping relation represented by the **downward sloping dashed line** is observed due to the inclusion of two separate groups that have positive trends (either solid line). These solid lines appear when the groups are sampled separately.

An example of misleading results is Simpson's paradox, or the Yule-Simpson reversal paradox, in which a trend appears in several different sets of data, but disappears, or even reverses, when the sets are combined. It is often encountered in the social sciences and medical science.

Edward H. Simpson first described this phenomenon in a technical paper in 1951, but the statisticians Karl Pearson et al. (1899) and Udny Yule (1903) mentioned similar effects earlier (Wikipedia, Simpson's paradox, 2017).

The pitfall is often a matter of the context in which the data was collected. For instance, the conclusion that being bilingual caused lower school performance could have been avoided if the data had been collected separately, one set for medium to high income families and the other for low income families. Separating the two groups or adjusting for income would actually slightly reverse these particular results. Modern educationalists now correctly encourage families to speak their original language when with their children. Simpson's paradox is often inadvertently encountered where one group (such as low income) is a good deal larger than the other. Sometimes a simple scatter plot of a large data set will reveal a tendency to form two or more clusters. They might be a clue to the causes of, or to the existence of, Simpson's paradox.

Questions for Ch 11

Write the equations where appropriate, and then fill in the values. You should do odd **Questions 3, 5, 7 and 9 in order**, so you can get information for the succeeding questions in that group. Similarly for even questions; you should also do **4, 6, 8, and 10 in order**. However, most of what you need might be right there in the answered Question 3, and in the unanswered Question 4. **Odd numbered questions** have worked answers at the back of the book and can be studied as additional examples. The data tends to be fictitious.

1. Is it necessary for the residuals to be normally distributed in order for you to make valid correlation or regression estimates?

2. What is the meaning of $R^2 = 1$?

3. This question and some of the other odd questions refer to the following table in Figure 11.13 in which each row is a state (1990 US Census)[25] and to the scatter plot of the same data immediately below. (**Before** any test to see if a linear relationship exists, you can only make a scatter plot from a subsample of the data not to be included in the test. **After** the test, you can make a scatter plot from any of the data.)

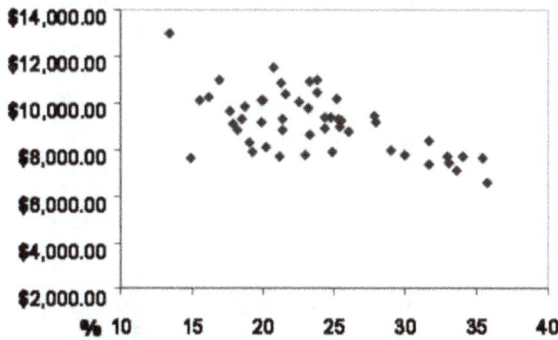

   Find the slope and intercept for the least squares line for the regression with $x = \%$ with '**no high school**' as the explanatory axis. Also write the resulting equation for the estimated line, using $y = \$$ **income** as the response variable. You can round for the final report, but retain an almost unreasonable number of figures until you finish all the calculations. (Keep the unrounded calculations **until** you do other questions which require additional calculation.)

4. This question and many of the other even questions refer to the table, Figure 11.14, and to the following scatter plot, in which each point is a state (US Bureau of Census 2007, 2008).[26] **Find**

---

[25] Per capita income derived from Bureau of the Census. Source: US Department of Commerce, Bureau of Economic Analysis, Survey of Current Business.

    Education data from US Census Bureau, 1990 Census of Population, CPH-L-96; 2000, American FactFinder; 2007 American Community Survey, R1501. "Percent of Persons 25 Years and Over Who Have Completed High School," R1502.

[26] Poverty data from: US Census Bureau, 2007 American Community Survey; "B17001. Poverty Status in the Past 12 Months by Sex and Age;" and "B17010. Poverty Status in the Past 12 Months of Families by Family Type;" by Presence of Related Children under 18 Years by Age of Related Children; using American FactFinder; <http://factfinder.census.gov>; (accessed 1 October 2008).

Figure 11.13: No High School vs Later Per Capita Income, $n = 50$ states

%No HS vs			Average Income for that State 28yr Later			
Year 1990				Year 2008		Cross
X, No HS %	x-mean_x	SX	Y, Income	y-mean_y	SY	Product
33.1362	9.420459	88.745052	7465.00	−1679.180	2819645.472	−15818.647
13.3705	−10.345288	107.024987	13007.00	3862.820	14921378.352	−39961.986
21.3404	−2.375382	5.642437	8854.00	−290.180	84204.432	689.288
33.6518	9.936008	98.724250	7113.00	−2031.180	4125692.192	−20181.820
23.8053	0.089580	0.008025	11021.00	1876.820	3522453.312	168.125
15.5693	−8.146482	66.365173	10143.00	998.820	997641.392	−8136.869
20.7920	−2.923783	8.548506	11532.00	2387.820	5701684.352	−6981.467
22.5139	−1.201827	1.444388	10059.00	914.820	836895.632	−1099.455
25.5545	1.838765	3.381056	9246.00	101.820	10367.312	187.223
29.0751	5.359375	28.722899	8021.00	−1123.180	1261533.312	−6019.543
19.9355	−3.780289	14.290581	10129.00	984.820	969870.432	−3722.904
20.2542	−3.461546	11.982300	8105.00	−1039.180	1079895.072	3597.169
23.7977	0.081954	0.006717	10454.00	1309.820	1715628.432	107.345
24.3594	0.643636	0.414267	8914.00	−230.180	52982.832	−148.152
19.9122	−3.803541	14.466924	9226.00	81.820	6694.512	−311.206
18.7282	−4.987541	24.875568	9880.00	735.820	541431.072	−3669.933
35.3863	11.670529	136.201237	7679.00	−1465.180	2146752.432	−17099.425
31.6860	7.970238	63.524695	8412.00	−732.180	536087.552	−5835.649
21.1736	−2.542154	6.462549	7760.00	−1384.180	1915954.272	3518.799
21.5820	−2.133815	4.553167	10394.00	1249.820	1562050.032	−2666.885
20.0054	−3.710405	13.767102	10103.00	958.820	919335.792	−3557.610
23.2217	−0.494096	0.244131	9801.00	656.820	431412.512	−324.532
17.6414	−6.074399	36.898317	9673.00	528.820	279650.592	−3212.263
35.7171	12.001329	144.031886	6573.00	−2571.180	6610966.592	−30857.576
26.0777	2.361932	5.578722	8812.00	−332.180	110343.552	−784.587
18.9955	−4.720233	22.280599	8342.00	−802.180	643492.752	3786.476
18.1790	−5.536740	30.655488	8895.00	−249.180	62090.672	1379.645
21.2285	−2.487303	6.186679	10848.00	1703.820	2903002.592	−4237.917
17.8490	−5.866757	34.418835	9150.00	5.820	33.872	−34.145
23.3285	−0.387239	0.149954	10966.00	1821.820	3319028.112	−705.480
24.9270	1.211234	1.467087	7940.00	−1204.180	1450049.472	−1458.544
25.1943	1.478520	2.186022	10179.00	1034.820	1070852.432	1530.002
30.0399	6.324174	39.995174	7780.00	−1364.180	1860987.072	−8627.311
23.3078	−0.407986	0.166452	8642.00	−502.180	252184.752	204.882
24.3313	0.615577	0.378935	9399.00	254.820	64933.232	156.861
25.4062	1.690414	2.857500	9018.00	−126.180	15921.392	−213.296
18.5197	−5.196053	26.998966	9309.00	164.820	27165.632	−856.413
25.3308	1.615052	2.608393	9353.00	208.820	43605.792	337.255
27.9752	4.259396	18.142451	9227.00	82.820	6859.152	352.763
31.7062	7.990415	63.846739	7392.00	−1752.180	3070134.752	−14000.646
22.9315	−0.784290	0.615111	7800.00	−1344.180	1806819.872	1054.227
32.9370	9.221229	85.031073	7711.00	−1433.180	2054004.912	−13215.682
27.8602	4.144472	17.176648	9439.00	294.820	86918.832	1221.873
14.8570	−8.858763	78.477682	7671.00	−1473.180	2170259.312	13050.553
19.2129	−4.502900	20.276109	7957.00	−1187.180	1409396.352	5345.753
24.8365	1.120693	1.255952	9413.00	268.820	72264.192	301.265
16.1779	−7.537905	56.820013	10256.00	1111.820	1236143.712	−8380.794
34.0108	10.295036	105.987759	7764.00	−1380.180	1904896.832	−14209.002
21.3970	−2.318750	5.376602	9364.00	219.820	48320.832	−509.708
16.9612	−6.754549	45.623934	11018.00	1873.820	3511201.392	−12656.809
Sum 1185.7883	0.0000000	1554.885093	457209.00	0.0000	82251119.380	−212506.751
Mean 23.71577	(Check)	31.73234884	9144.180	(Check)	1678594.2731	

In the final report assume that x_1 and y_i were only to four place accuracy.

the slope and intercept for the least squares line using regression with causal variable $x = \%$ of the residents in **poverty** and response variable $y_i = \%$ of the residents **vote Republican**. Subsequent questions about testing will follow from this as though assuming normality of the residuals at each point, (x_i, y_i). However, that assumption is likely to be untrue due to the fact that we are using percentages. Possibly, in a real research situation you could use a transformation, but that would make interpretation difficult. But, in application you could at least verify that a (monotonic) relation exists by following the regression with a Spearman's rank correlation and the test of its validity (Section 14.6).

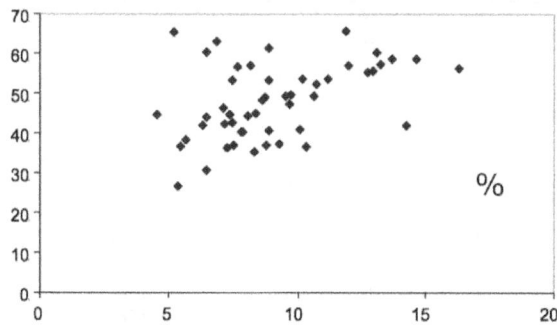

Write the resulting equation for the estimated line or major axis. You can round for the final report, but retain an almost unreasonable number of figures until you finish all the calculations. You will need to carry unrounded **values to subsequent questions** which require more calculation.

5. Use an **F test** with $\alpha = 5\%$ to decide whether there is a linear relation (a meaningful slope) between x and y, using the data from Question 3. **Then make a two tail t test** from your sums of squares in the ANOVA-like table. I have started this for you by using the linear model (the equation) found in Question 3 to estimate the sum of squares of error for the $n = 51$ states. I did this by creating three more columns in Table 11.13 (shown without them in Figure 11.13). In the first new column I used Equation 11.3, $\hat{y} = a + bx_i$, to get a column of estimates, $\hat{y}$, one for each x_i value. In the second, I subtracted this from y to get a whole column of **residuals**, $y - \hat{y}$. In the third I squared to get $(y - \hat{y})^2$. Here is a very small piece cut off of the **bottom** of these three columns and the y column, **with pertinent sums**:

y	..	$\hat{y}$	$y - \hat{y}$	$(y - \hat{y})^2$
...	...	...	...	...
7957.00		9763.57665134556	-1806.58	3263719.20
9413.00	...	8990.02306829325	422.98	178909.48
10256.00	...	10181.0569294797	74.94	5616.46
7764.00	...	7728.04582587062	35.95	1292.70
9361.00	...	9463.1338049402	-99.14	9827.91
11018.00	...	10073.3023689817	944.70	892453.61
Σ	...		0.00	53208956.02

The 0.00 sum serves as a check; if you round a little too much in some questions, you will miss this check on your calculations. You will not have to create your own columns; the **last** value in this table fragment and **some** values from the table in Figure 11.14 are all you need. Does either the F test, or a t test **derived from** F, distinguish between tails?

6. Use an F test with $\alpha = 5\%$ to decide whether there is a linear relation between x and y, using the data from Question 4. Then make a two tail t test from your sums of squares in the ANOVA-like table. I have started this for you by using the linear model (the equation) found in Question 4 to estimate the sum of squares of error for the $n = 51$ states. I did this by creating

Figure 11.14: % Below Poverty Level vs % for Republican Presidential Candidate 2007

State	x=% poverty	x-mean_x	SX	y=% vote R	y-mean_y	SY	Cross Prod
			Below Poverty Level vs % Popular Vote Republican, by State				
Alabama	13.0865	3.9834	15.8677	60.4	12.3760	153.1654	49.2989
Alaska	6.4548	−2.6483	7.0133	60.2	12.1760	148.2550	−32.2453
Arizona	10.1874	1.0844	1.1759	53.8	5.7760	33.3622	6.2633
Arkansas	13.6923	4.5892	21.0610	58.8	10.7760	116.1222	49.4535
California	9.2932	0.1901	0.0361	37.3	−10.7240	115.0042	−2.0387
Colorado	8.3795	−0.7236	0.5235	44.9	−3.1240	9.7594	2.2604
Connecticut	5.6866	−3.4165	11.6722	38.3	−9.7240	94.5562	33.2217
Delaware	7.5366	−1.5664	2.4537	37.0	−11.0240	121.5286	17.2683
Florida	8.5982	−0.5048	0.2549	48.4	0.3760	0.1414	−0.1898
Georgia	10.7058	1.6028	2.5690	52.2	4.1760	17.4390	6.6933
Hawaii	5.3636	−3.7395	13.9835	26.6	−21.4240	458.9878	80.1140
Idaho	8.8637	−0.2393	0.0573	61.5	13.4760	181.6026	−3.2252
Illinois	8.7898	−0.3133	0.0981	36.9	−11.1240	123.7434	3.4846
Indiana	8.7292	−0.3739	0.1398	49.0	0.9760	0.9526	−0.3649
Iowa	7.3564	−1.7467	3.0509	44.7	−3.3240	11.0490	5.8060
Kansas	7.6784	−1.4247	2.0297	56.8	8.7760	77.0182	−12.5029
Kentucky	13.2264	4.1233	17.0018	57.5	9.4760	89.7946	39.0726
Louisiana	14.6276	5.5246	30.5210	58.6	10.5760	111.8518	58.4280
Maine	7.8533	−1.2498	1.5620	40.5	−7.5240	56.6106	9.4034
Maryland	5.4407	−3.6623	13.4127	36.8	−11.2240	125.9782	41.1061
Massachusetts	7.2572	−1.8458	3.4070	36.2	−11.8240	139.8070	21.8247
Michigan	10.0555	0.9524	0.9071	40.9	−7.1240	50.7514	−6.7850
Minnesota	6.4620	−2.6410	6.9751	44.0	−4.0240	16.1926	10.6276
Mississippi	16.2759	7.1728	51.4494	56.4	8.3760	70.1574	60.0796
Missouri	9.5014	0.3983	0.1587	49.4	1.3760	1.8934	0.5481
Montana	9.7347	0.6316	0.3990	49.7	1.6760	2.8090	1.0586
Nebraska	8.1780	−0.9251	0.8558	57.0	8.9760	80.5686	−8.3035
Nevada	7.4886	−1.6144	2.6064	42.7	−5.3240	28.3450	8.5953
New Hampshire	4.5513	−4.5517	20.7181	44.8	−3.2240	10.3942	14.6747
New Jersey	6.3312	−2.7719	7.6832	42.1	−5.9240	35.0938	16.4204
New Mexico	14.2323	5.1293	26.3094	42.0	−6.0240	36.2886	−30.8987
New York	10.3140	1.2110	1.4664	36.7	−11.3240	128.2330	−13.7129
North Carolina	10.6331	1.5301	2.3411	49.5	1.4760	2.1786	2.2584
North Dakota	7.4741	−1.6290	2.6536	53.3	5.2760	27.8362	−8.5945
Ohio	9.6923	0.5892	0.3472	47.2	−0.8240	0.6790	−0.4855
Oklahoma	11.8634	2.7603	7.6194	65.6	17.5760	308.9158	48.5156
Oregon	8.8691	−0.2339	0.0547	40.8	−7.2240	52.1862	1.6899
Pennsylvania	8.0922	−1.0108	1.0217	44.3	−3.7240	13.8682	3.7642
Rhode Island	8.3126	−0.7905	0.6249	35.3	−12.7240	161.9002	10.0581
South Carolina	11.1593	2.0563	4.2282	53.8	5.7760	33.3622	11.8769
South Dakota	8.8535	−0.2495	0.0622	53.2	5.1760	26.7910	−1.2914
Tennessee	11.9745	2.8715	8.2453	56.9	8.8760	78.7834	25.4872
Texas	12.7502	3.6471	13.3016	55.5	7.4760	55.8906	27.2660
Utah	6.8514	−2.2516	5.0699	62.9	14.8760	221.2954	−33.4955
Vermont	6.4453	−2.6577	7.0634	30.6	−17.4240	303.5958	46.3078
Virginia	7.1157	−1.9873	3.9495	46.4	−1.6240	2.6374	3.2274
Washington	7.8405	−1.2625	1.5940	40.5	−7.5240	56.6106	9.4993
West Virginia	12.9522	3.8492	14.8162	55.7	7.6760	58.9210	29.5463
Wisconsin	7.1472	−1.9559	3.8255	42.4	−5.6240	31.6294	10.9999
Wyoming	5.1938	−3.9093	15.2825	65.2	17.1760	295.0150	−67.1459
Sums	455.1520	0.0000	**359.5204**	2401.2000	0.0000	**4379.5512**	**544.9204**
Mean (/n, /df)	**9.10304**	(Check)	**7.33715**	**48.0240**	(Check)	**89.3786**	

some more columns in the table shown without them in Figure 11.14. In the first new column, I used $\hat{y}_i = \mathbf{a} + \mathbf{b}x_i$ to get a column of estimates, one $\hat{y}_i$ for each x_i value (note that you found $\mathbf{a}$ and $\mathbf{b}$ in Question 4). Next I subtracted $\hat{y}$ from y to get a whole column of errors due to regression, $y - \hat{y}$. Then I squared to get $(\hat{y} - \bar{y})^2$. Here is a very small piece cut off of the bottom of the resulting expanded table including the y column, **with pertinent sums**.

y	...	$\hat{y}$	$y - \hat{y}$	$(y - \hat{y})^2$
...	..	...	...	...
55.5	...	53.551952	-1.948048	3.794890
62.9	...	44.611218	-18.288782	334.479550
30.6	...	43.995697	13.395697	179.444709
46.4	...	45.011814	-1.388186	1.927060
40.5	...	46.110384	5.610384	31.476408
55.7	...	53.858121	-1.841879	3.392518
42.4	...	45.059558	2.659558	7.073249
65.2	...	42.098815	-23.101185	533.664742
Σ	...		**0.000727**	**3553.617524**

The 0.000727 sum serves as a check (compared to the size of the values, it is **near** 0, as it should be. If it is not near 0 you have done something incorrectly. It only takes a little rounding to seriously fail this check on your calculations. You will not have to create your own columns; the **last value** in this table fragment **and** some values from the table in Figure 11.14 are **all you need**. The F test cannot distinguish between tails. (I did present the columns differently in Question 5, Figure 11.13; you should be able to cope with small changes in notation.) And, this is purely an illustrative problem; it is **not** the best practice to use normal distribution based tests on percentage data, even though it is common practice.

7. Test the data from Question 3 as a simple t test of slope, as opposed to a t test derived from an ANOVA-like approach. The hypothesis pair is **H0**: $\mathbf{b} = \$100/\%$ vs **HA**: $\mathbf{b} \neq \$100/\%$, with $\alpha = 0.05$. Details from the above given table in Question 5 might help. How many tails are involved? When must such a hypothesis pair be made? Report the RMSE.

8. Test the data from Question 4 as a t test of slope for the hypothesis pair, **H0**: $\mathbf{b} = 0$ vs **HA**: $\mathbf{b} > 0$, with $\alpha = 5\%$. (You need to refer to the calculations in Question 4 to do this and calculations from Question 6 would help too. If we really look at this as an inference, that there is a significant linear effect [as slope], there should be an a priori hypothesis to that effect before you examine the data you are actually analyzing.) Do the above, as an exercise, even though, in application, percentages tend to be skewed.

9. Provide a 95% confidence interval, to the nearest dollar, for both b and a in Question 3.

10. Provide a 95% confidence interval for b in Question 4 (assume a normal distribution).

11. Find Pearson's product moment correlation and the coefficient of determination for the data in Question 3. What percentage of the variation in the response variable is estimated to be due to a straight line effect?

12. Find the **product moment correlation coefficient** and the **coefficient of determination** for the data in Question 4 (also Figure 11.14). **What part** of the equation determines the sign, whether x and y vary together or in opposite directions? **What percentage** of the variation in the response variable is due to a straight line effect? Are these estimates **dependent on** a normal distribution of the residuals?

13. Do a regression that forces the line from the data in the example in Subsection 11.4.0.1 through the origin. How does the resulting $\hat{b}$ compare with $\hat{b}$ in the example that was not forced?

14. What does RMSE mean?

15. An environmental scientist wants to see if there is a linear trend in the thickness, y mm, of California Condor egg shells over the x decades since DDT was prohibited. The scientist will have a hard time finding enough museum egg shell specimens, but hopes to get enough to test for a correlation coefficient of $R > 0.7$. Because of potentially limited sample sizes, the researcher feels stuck with a minimally acceptable criterion, $\alpha = 5\%$, and a power, $1 - \beta = 80\%$, to even detect a linear relation that is strong enough to account for $R_{min}^2 = 50\%$. Estimate the required sample size, $\hat{n}$, as though you were that scientist.

12 CATEGORICAL DATA

Categorical data is discrete data that is in categories rather than continuous. Categorical data is countable, for instance, the number of yes answers or the number of no answers. However, it can also be in the form of percents. We will start off with the procedure, which was until relatively recently, the most commonly used to test for differences between sets of categorical data and is still the preferred way to test the goodness of fit of data to particular data distributions, and to data distribution-like models.

12.1 Introducing Pearson's χ^2

Pearson (1900) adapted the continuous chi square distribution to create tests of *nominal* data, meaning discrete data that consists of **counts** of things that fall into particular **distinct categories**.[1] Together this group of tests is called *Pearson's chi square test*, sometimes written $P\chi^2$. To remember the word nominal, note that each category (also referred to as each *factor*) can have a name. In context, nominal may also be called *categorical*. The categories are compared with each other or to a hypothetical pattern of counts. The procedure always consists of observed counts versus expected counts in each category.

Binning means pre-processing data to group continuous data together into equal size intervals called *bins*. That is, you could call $0.0 \leq A < 1.5$ 'category A' or 'bin A' and $1.5 \leq B < 3.0$ 'category B' or 'bin B'. Notice that A never includes exactly 1.5, but B includes and starts with 1.5. The number of things that fall into each bin is then countable and thus discrete. Binning can alternately use $\leq$ with $<$, or use $\geq$ with $>$ to clearly avoid overlap. An example of four bins is:

- T = the number of adults six feet tall or taller ($6 \leq T$).

- M = the number of adults five feet tall or taller, but less than six feet tall ($5 \leq M < 6$).

- S = the number of adults four feet tall or taller, but less than five feet tall ($4 \leq S < 5$).

- J = the number of adults less than four feet tall ($J < 4$).

So, binning can be used to group continuous data into ranges to create categories, which can then be counted, for Pearson's chi square, and that is what Pearson did for his published example of his $P\chi^2$ technique. That is, Pearson initially intended $P\chi^2$ to be used with **binning** as a test of the **goodness of fit** of data to particular statistical distributions. It can be used for binned continuous data such as the normal distribution, or for discrete data (with or without binning). Aside from making continuous data into categorical data, it is thought that binning discrete data can be useful for reducing the effects of small observational errors.

After many attempts to test whether a sample of data fit a normal curve, Pearson concluded, "It requires at least a sample of more than 100 individuals to determine whether it would be best to use a rectangle or some other curve!"(1935). Even back then, there was evidence of the need for much

---

[1] Pearson was most productive in an era without handy computers, or even very advanced computometers (massive adding machines). Therefore, he derived his approach using logarithms.

larger sample sizes than have mistakenly come to be thought of as adequate. Pearson's chi square can be used for either testing goodness of fit or for comparing two or more data samples.

Pearson's χ^2, $(P\chi^2)$, can be seen as an application of the definition of chi square as the summing of Z^2 values as per Equation 8.4,

$$\chi^2 = \sum_{i=1}^{i=k} Z_i^2 \, ,$$

from Chapter 8 and its results are evaluated using a table of χ^2. Although, in Pearson's chi square, the Z^2 values may arise from continuous data broken into categories, Pearson's approach is also popular for categorical analysis of data that is already discrete count data. When the Pearson's chi square is used for counts of categorical data it can be seen as the sum of a set of normal approximations.

Unfortunately you may come up against some difficulty with complex designs when you wish to report a truly meaningful effect size, or to make a power analysis, if you cannot define an actually **meaningful** reasonable minimum difference. Recall that a reasonable minimum difference is not the effect size, but it is the difference you target in order to define β for calculating power, $1 - \beta$. These challenges[2] are touched on in Chapter 13.

Be aware that, like most of this book, what follows tends to use a frequentist viewpoint, that if you sample n things at **random** and you find r things in a certain category, then the sample relative frequency, $\hat{f}$ (the proportion of successes), is an estimate of the probability,

$$\hat{p} = r/n,$$

or, to be semantically more precise, the probability of this happening. So, **Pearson's chi square** tests whether the observed counts, O, in each category equal the expected counts, E, for that category. It is calculated by (substituting c columns for k categories),

$$P\chi^2 = \sum_{l=1}^{l=c} \frac{(O_l - E_l)^2}{E_l} \, . \tag{12.1}$$

The Os are what you observe (from your sample) and the Es are what you expect (values implied by your null hypothesis). Since the expected values are hypothetical, they are seen as not representing a variable. Often it is convenient to just represent the result as χ^2 rather than $P\chi^2$. There are many testing applications that use Pearson's chi square. Some applications are often seen as different tests in themselves.

12.2 Goodness of Fit Testing and $P\chi^2$

Goodness of fit (GOF) was the original intent of Pearson's chi square and is where it excels. Goodness of fit tests attempt to reject the idea that your data came from a certain hypothesized distribution. A goodness of fit test challenges your hypothesis, **H0**, that your sample fits the distribution (in the sense of the pattern). The irony is that a goodness of fit test really only tests for badness of fit. That is, the null hypothesis is that there is **no meaningful difference** between the distribution from which your sample arises and the distribution implied by **H0**.

---

[2]Some unfortunate researchers use Cramer's V, or a standardized index by Cohen with the discredited 'small', 'medium', and 'large' to define either effect sizes or reasonable minimum differences. Cramer's V is also called Cramer's phi (Equation 12.8). For any but the very simplest designs, it is only vaguely defined in terms of variance and does not even approach a valid correlation as claimed.

The goodness of fit test is really a $1 \times c$ test, meaning one row by c columns (bins if the data is continuous), that can be used to challenge the goodness of fit of data to stochastic models such as genetic ratios, as well as to hypothesized distributions. It is commonly mistakenly called a $2 \times c$ table, but for consistency, the line of E **values should not be called a row**. In this role it has no more than $c - 1$ degrees of freedom because (as you will see) you use up a degree of freedom calculating the total, **n**, for finding the E_l values.[3]

Let's first consider the simplest Pearson's chi square test to see how it works. We will consider a generalized **goodness of fit** test, as is appropriate for Pearson's original intent. We will proceed as though challenging the hypothesis that the data, which is discrete, comes from a particular distribution, in this case a particular binomial distribution.

Recall that each binomial distribution of counts, r, is fully defined by a probability, p, and the sum, n, which is the same for either the Os or the Es. Also recall from the normal approximation, Section 5.6, that np is the mean of a particular binomial distribution, with variance is $np(1-p)$. The observed value, O, in a given category cell can be seen as the number of successes or counts, $O = r$, in that category. For a given n, each $r = n\hat{p} = O$, what we count in the sample. However, there is also an expected value, $E = np_0$, for each cell where p_0 is determined by the null hypothesis.

12.2.0.1 Example: Goodness of Fit

Mendel meets Pearson: Let's take an unhistorical look at Gregor Mendel's results using garden peas.[4] Assume that Mendel had already partly developed his famous model. When he crossed pea plants that had what he called 'simply inflated pods' with pea plants that had 'constricted pods', he only got offspring plants with simply inflated seed pods in what is called the F1 generation.. That fitted his expectation, but did the second generation, the F2? Then, based on the *phenotypes* (how the offspring appeared) he had reason to create a null hypothesis, **H0**, that there was only one gene with two forms (*alleles*) and that the inflated phenotype was *dominant* (took over when both genes were present). If so, aside from the effects of chance, under his theory, by crossing sibling plants from that first generation (or *selfing*, breeding them to themselves), he should expect 3 plants with the dominant simply inflated pods for each 1 with constricted pods in this inbred generation, F2.

Methods and Materials All his plant lines had been previously observed to breed true, and he had avoided accidental unintended breeding partly through his choice of a plant with a closed pollination system. Without human intervention, pea plants almost always pollinate themselves. From earlier trials, he has reason to run a rather large experiment with a pair of hypotheses based on his model,

H0: 75% WILL BE OF THE DOMINANT PHENOTYPE (implying that 25% will be of the recessive phenotype).

HA: THE $3 : 1$ MODEL DESCRIBED IN **H0** DOES NOT FIT.

Criterion: $\alpha = 5\%$

Analysis: Assume he designated the simply inflated phenotype, uppercase **I**, and alternative wild phenotype, lowercase **i**. The observed data was from Mendel's experiments (1866). The first expected entry, $E_{\mathbf{I}}$, was obtained by multiplying Mendel's total for that experiment, $\mathbf{n} = 1121$ by $3/4$ and the second was found by subtraction, $E_{\mathbf{i}} = \mathbf{n} - E_{\mathbf{I}}$, to get the contingency table,

---

[3]The goodness of fit test is generally thought of as limited to a set of fixed model predictions and a set of outcomes from one random sample. Some authors such as Liu (2012) might differ from this view.

[4]Mendel died before Karl Pearson published his work on modern statistical inference.

	I	**i**	**Total**
O	822	299	1121
E	840.75	288.25	
$\hat{p} = O_j/\mathbf{n}$	0.733274	0.266726	1

. Then the calculation was

$$\chi^2 = \sum_{j=1}^{j=2} \frac{\left(O - E_{\mathbf{I}_j}\right)^2}{E_{\mathbf{I}_j}} = \frac{(822 - 840.75)^2}{840.75} + \frac{(299 - 288.25)^2}{288.25}$$

$$\chi^2 = 0.8191, \qquad (\mathbf{c} - 1) = 1\,df\,.$$

This is the way such a test is usually written, but the χ^2 is obviously Pearson's chi square, $P\chi^2$. The inclusion of the probabilities, $\hat{p}s$, is not shown in all texts; they are simply the Os divided by the total sample size, $\mathbf{n}$. Note that in goodness of fit tables like the above, the E values **never** count as a row (or column) in the calculation; that is why the above calculation only estimates $df = \mathbf{c} - 1$ (at the most).

Results: This χ^2 value estimated from the sample was **not statistically significant** (**H0** was not rejected) because it was much smaller than the chi square value required for statistical significance with $\mathbf{c} - 1 = 1\,df$ from a common table based on $\chi^2_{0.05,\,1}$ like in Appendix 5. In this application this means that Mendel could find no reason to reject his hypothesis.

The effect sizes could be reported as an absolute difference in probability for this table with only $df = 1$. Because $p_I - p_{0_I} = 0.733 - 0.750 = -0.017$ and $p_i - p_{0_I} = 0.267 - 0.250 = 0.017$, then if the test **had been** statistically significant, the '**effect size**' could have been reported as $ES = 1.7\%$. A table with more columns might need multiple effect sizes in terms of likelihood of the results. However, we do not report effect sizes where there is no statistical significance. This unhistorical version of Mendel could have also presented a $(1 - \alpha)\,100\% = 95\%$ confidence interval for the probability (equal to the proportion) of the F2 plants in the theoretical population of all possible outcomes that would display the simply inflated phenotype using Equation 6.10,

$$\widehat{p_{\mathbf{I}}} - Z_{\alpha/2}\sqrt{\widehat{p_{\mathbf{I}}}\left(1 - \widehat{p_{\mathbf{I}}}\right)/\mathbf{n}} \le p_{\mathbf{I}} \le \widehat{p_{\mathbf{I}}} + Z_{\alpha/2}\sqrt{\widehat{p_{\mathbf{I}}}\left(1 - \widehat{p_{\mathbf{I}}}\right)/\mathbf{n}}.$$

Then, using $Z_{\alpha/2=0.025} = 1.9600$ for a 95% CI and referring to his results ($\hat{p}$ values below his above contingency table) to fill in the remaining values, the CI for the proportion of simply inflated pods is

$$0.73327 - 1.9600\sqrt{(0.73327)(0.26673)/1121} \le p \le 0.733274 + 1.9600\sqrt{(0.73327)(0.26673)/1121},$$

or,

$$0.70738 \le p \le 0.75916.$$

By reversing the order, and then subtracting each of the above limits of the CI from 1.0000, he could get a CI for the proportion, $q = 1 - p$, of constricted pod phenotypes,

$$0.24084 \le p \le 0.29262.$$

Discussion: The researcher failed to get statistical significance, but at least he can say that he had found **no evidence** against his model. Then he could calculate confidence intervals that might at least include the probabilities predicted by his model. Indeed, in the common classroom fruit fly experiment, the CIs might be seen as a useful way to demonstrate the model even if they are merely exploratory.

Note that this is a very simple Pearson's chi square, and that not all the calculations are as handy as when there are only two possibilities. A variety of effect size strategies might be needed for reporting different tests.

The results mean that there was almost no evidence contradicting his null hypothesis. However, remember that there is really nothing new to support it either. But, aside from the statistically nonsignificant inferential test, he could have been 95% confident that the interval between 70% and 76% enclosed the real percentage of plants in an infinite number of crosses that would display the simply inflated phenotype. Many authors round up and use $Z \approx 2$ in place of the value $Z_{\alpha/2} = 1.9600$ found in a $p >$ table of the normal distribution, Z, such as in Appendix 4. This is more conservative, in that it gives a slightly wider CI and is thus acceptable, even though it is accurate to no more than three figures. However, you would still have to avoid rounding the other numbers until you are ready to report the range (appropriately to no more than three figures, using $Z \approx 2$). Also be aware that in calculating a CI for the results of a Pearson's chi square, you cannot use t, because it does **not** apply to the normal approximation. Student's t test requires two estimates, $\bar{x}$ and S, but the normal approximation **only** requires one estimate, $\hat{p}$.

Mendel's choice of an organism with which to work, and inherited characteristics on which to focus, was remarkable. The inexperienced genetics researcher should be warned that due to modifying genes, outcomes rarely appear to fit Mendelian models sufficiently to avoid rejection by statistical tests with adequate sample sizes. Popularly cited fruit fly and pea allele models are the **exception, not the rule**. This is particularly evident if you are breeding for plant resistance to disease or insect attack, or if you are attempting to follow inheritance of human characteristics such as cleft chin, let alone looking for possibly heritable tendencies for schizophrenia or autism in humans. In reality, only **chromosomes**, strings of genes, not single genes, are distributed in a simple random fashion. Most applied crosses are subject to numerous modifiers and apparently to environmental switching of genetic expression. If not **linked**, meaning residing on the same chromosome, genes are inherited as per Mendel's models, but their expressions are often masked to varying degrees by other genes, and sometimes by the environment. **Linkage** (probability effects due to closeness on the same chromosome) comes into play when you are considering two or more genes. Although early to mid twentieth century fruit fly researchers did find a large number of genes that behaved nearly as conveniently as Mendel's pea plant genes when different chromosomes happened to be involved, they also got a lot of messy results that then led to more great discoveries, such as probability based chromosome maps. Rarely will you get results that are good statistical fits to the elementary Mendelian model when initiating genetic studies in areas such as human health or agricultural genetics.

This example only had $k = 2$ possible outcomes, but Pearson's chi square can apply to models that can have any **number**, $k > 1$, of categories (classes, outcomes).

Question 3 is an example of this test applied to a slightly different challenging of fit.

12.2.0.2 A little theory using a 1×2 table

Well, it takes $k = 2$ to tango, and because $df = k - 1$, it takes **at least** $k = 2$ categories (or bins) for a Pearson's chi square to be valid, so the Mendel model Example 12.2.0.1 is a simple place to start an explanation of $P\chi^2$.

Starting with just $k = 2$, we could write

$$P\chi^2 = \sum_{j=1}^{j=2} \frac{(O_j - E_j)^2}{E_j} = \frac{(\mathbf{n}\hat{p} - np_0)^2}{\mathbf{n}p_0} + \frac{(\mathbf{n}\hat{q} - nq_0)^2}{\mathbf{n}q_0}, \tag{12.2}$$

where $q = 1 - p$, and p_0 and q_0 are the hypothesized probabilities. So, $\mathbf{n}\hat{p}$ is the observed number of **successes** (r for one cell), and $\mathbf{n}\hat{q} = \mathbf{n}(1 - \hat{p})$ is the number of **failures** ($\mathbf{n} - r$ for the other cell) in each of the k columns (pairs of cells). With a 1×2 table we are only considering $k = 2$ classes because each cell has both an O and an E (even though often incorrectly drawn as though there are two rows). Because we know the total number, k, of categories (shown as columns) that we are sampling, one O cell is predetermined by the other (or others) in the row, hence the $(k - 1)\, df$.

But, is something missing in the denominator, the divisor of $\sum_{l=1}^{l=k} \frac{(O_l - E_l)^2}{E_l}$? Starting with a simple binomial approximation model, we have a row with two values, O and E in each column, the success and the failure column. We are using estimators from the normal approximation of the binomial to find a pair of Z^2 values ($\chi^2 = Z_1^2 + Z_2^2$), but the variance, npq, **seems** to be hiding in Equation 12.2. Recall the notation that $(1 - p) = q$. So, let's put both fractions over a **common denominator** ($np_0 q_0$) by multiplying top and bottom of Equation 12.2 accordingly by p_0 or q_0 so they can be added together[5],

$$P\chi^2 = \frac{(\mathbf{n}\hat{p} - np_0)^2 q_O}{\mathbf{n}p_0 q_0} + \frac{(\mathbf{n}\hat{q} - nq_0)^2 p_0}{\mathbf{n}p_0 q_0} = \frac{(O_{succcess} - E_{success})^2 q_0}{E_{success}} + \frac{(O_{failure} - E_{failure})^2 p_0}{E^{\iota}_{failure}}.$$

Well, even if you write it as $P\chi^2 = qZ_1^2 + pZ_2^2$, that still looks awkward until you realize that

$$(O_{succcess} - E_{success}) = (\mathbf{n} - O_{failure}{}^{\iota}) - (\mathbf{n} - E^{\iota}_{failure}) = (-O_{sucess}) - (-E_{success}),$$

meaning that the denominators only differ in sign, $(O_{succcess} - E_{success})$ equals $-(O_{failure} - E_{failure})$. Then, when both sides of this are squared as in $\sum_{j=1}^{j=2} \frac{(O_j - E_j)^2}{E_j}$, the signs are lost, giving $(O_{succcess} - E_{success})^2$ equals $(O_{failure} - E_{failure})^2$. So, we can simplify and generalize

$$(p_0 + q_0)(O_{succcess} - E_{success})^2 + (p_0 + q_0)(O_{failure} - E^{\iota}_{failure})^2$$

$$\text{to get } (p_0 + q_0)\sum_{j=1}^{j=2}(O - E)_j^2 \text{ so } P\chi_{df}^2 = \frac{(p_0 + q_0)\sum_{j=1}^{j=2}(O - E)_j^2}{\mathbf{n}p_0 q_0}.$$

Now, recall that $\mathbf{n}p_0 q_0 = \sigma^2$, and recalling that $p_0 + q_0 = 1$,

$$\therefore P\chi_{df}^2 = \frac{\sum_{j=1}^{j=k}(O - E_j)^2}{\sigma^2} = \frac{\sum_{j=1}^{j=k}(r - r_0)_j^2}{\sigma_r^2} = \sum_{j=1}^{j=k}\frac{(O - E)^2}{E}\left(\approx \frac{\mathbf{n}S_r^2}{\sigma_r^2}\right), \tag{12.3}$$

at least for $k = 2$. Note that $\therefore$ means **_therefore_**.[6] The nS_r^2 is of interest because $(O - E)^2/\mathbf{n} \approx (r - \bar{r})^2/(\mathbf{n} - 1)$, so $\sum_{j=1}^{j=k}(r - \bar{r})^2/(\mathbf{n} - 1) = S_r^2$, which justifies accepting the result of this derivation, which approximates nS_r^2/σ_r^2, which quickly approaches one way of defining a chi square distribution, Equation 8.1, $\chi^2 = (n - 1)S^2/\sigma^2$, even though it did not, at first, look like that was true. All this is in accordance with the definition of chi square itself, in Equation 8.3, as well as with Equa-

---

[5]Recall that you do not change the value of a fraction or ratio by multiplying both the numerator and the denominator by the same value.

[6]You can put $\therefore$ where you have demonstrated your claim to be true.

tion 12.1. Pearson's chi square, $P\chi^2$, is a χ^2 value, and because χ^2 values are sums, they are additive. The number of successes is predetermined by the number of failures in such a table, and similarly, any one category of $k = 3$ or $k = 4$ categories would be predetermined, as would one of however many more categories. So, although we could laboriously go into more algebra, suffice it to say that, even for larger tables, $P\chi^2 = \sum_{j=1}^{j=k} \frac{(O_j - Ej)^2}{E_j}$ can be generalized to any number, $k \geq 2$, of categories, not just successes and failures as for $k = 2$. So, for a given row (or column) of cells, there is always one that is predetermined by the fact that it is easy to estimate the necessary row sum (and column sum in larger tables). This summing can be seen as using a degree of freedom, as would estimating a mean, which is just a weighted sum. $P\chi^2$ can then be evaluated by referring to a common χ^2 table with no more than $(k - 1)\,df$ for a single row or column of k cells. The degree of freedom that is used up can either be viewed as used by the summing or simply by being lost to the predetermined cell.

12.2.0.3 More rows and applications

Recalling that this test can be used to compare continuous distributions, for instance the normal curve to a sample, you will need to cut continuous data into categories, *bins*. A **rule of thumb** is that every bin should have at **least five** data points. The degrees of freedom are often reduced when there are increased numbers of calculated estimates,

$$df = k - number\ of\ totals - number\ of\ other\ estimates.$$

Thus, if you want to do a goodness of fit test to see if your sample of size $n = k$ fits a normal curve, you have to estimate the mean and that costs a degree of freedom, and you have to estimate the variance and that costs you another degree of freedom. Then you could find a set of k numbers representing a normal curve with that mean and standard deviation as did Pearson (1935). But you would only have $(k - 1 - 2)\,df$. Similarly, if your sample was suitable, you could try a goodness of fit test for a binomial distribution. But then you would have to estimate **one value**, the probability, p, and that would cost you one degree of freedom. So, your test of a sample of size k counts would only have $(k - 1 - 1)\,df$. Be aware that zero degrees of freedom would be meaningless.

$P\chi^2$ has been extended to applications other than just goodness of fit. It has so many applications that some folks think it is the 'real' and even the only χ^2. So, many authors carelessly refer to Pearson's chi square test as 'the chi square test'. However, others pedantically symbolize its test value as X^2 instead of χ^2, emphasizing that it is not just the basic chi square, and that it only combines k approximations of the normal curve that together tend to follow the χ^2 distribution.

12.3 $P\chi^2$ with Contingency Tables

You can also use Pearson's chi square test, Equation 12.1, $P\chi^2 = \sum_{l=1}^{l=k} \frac{(O_l - E_l)^2}{E_l}$, where $k = \mathbf{r} \times \mathbf{c}$, to see if two or more sets of data come from the same population, or are from populations that started out as the same population but have somehow been changed by a treatment or treatments. That is, $P\chi^2$ can be applied to nominal or ordinal data, meaning categorical or categorized counts enumerated (recorded or visualized) into tables called *contingency tables* (often simply called 'tables'; e.g. Table 12.1), with the counts of each category in its own *cell*, usually the intersection of a row and a column. The number of *observed* things, denoted O, that fall into a particular cell, is really a member of a category pair that includes E, the *expected* number of things, which is not always shown. The expected values, E, arise from the null hypothesis, **H0**, and total sample size, **n**. The null hypothesis is the equivalent of **H0**: THERE IS NO DIFFERENCE BETWEEN THE OBSERVED POPULATION

AND THE EXPECTED POPULATION other than the effects of chance. The Os can be seen as counts of successes in a binomial situation in which $O = r$ for that cell. The sum of all the Os equals the sum of all the Es and the sum of either equals **n**, the sum of all the counts of either kind in a table.[7] Contingency tables are also called ***cross tabs***.

Let's consider the common (not about fit) **r** × **c** tables where the **bold r** and **c** refer to **r** rows and **c** columns of cells. Note that this book uses a bold **r** for the number of rows (**r** is **not** the italicized r, the number of successes).

In addition to the grand total, **n**, the sum of the sample sizes, there are also marginal row and column totals, $n_{i\ or\ j}$. So, if you are given all the counts except one in a row or column, you can find the remaining one. One cell in each row or in each column is ***predetermined*** because, if you have $n_{i\ or\ j}$, you do not need any more input. Predetermined **does not mean** dependent, it merely means that through knowing the sample size, you already know an amount of information equivalent to one outcome. You can view the degrees of freedom as being the total number of cells minus the predetermined cell or cells. For a simple contingency table it works out that $df = (\mathbf{r} - 1)(\mathbf{c} - 1)$. If there is **only one row or column, r = 1 or c = 1**, then $df = k - 1$.

12.4 $P\chi^2$ Challenging Independence or Homogeneity

In statistics, ***association*** means that the outcomes relate to the categories; it does not say in what manner they are related. Unfortunately, the word association is ambiguously defined throughout the social sciences, and the definitions sometimes contradict each other. To avoid ambiguity, let's define association as a **failure** of the hypothesis of either **homogeneity** or of **independence** to be true.[8] In application, which is which, homogeneity, or independence, may depend on the sampling method.

Homogeneity means that the statistical properties of any one part of an overall data set are the **same** as any other part. When Pearson's chi square is used as a test of homogeneity, the data is collected by randomly sampling from **each group separately**. This sampling approach is an example of ***restricted randomization***. If you take a sample of female fish at random to count the hard fin rays and then you sample an equal number of male fish at random, you are restricting the randomization by gender. Thus a Pearson's chi square to see if the number of hard dorsal fin rays are the same for both genders, using this restricted random design, would challenge the homogeneity of hard fin ray numbers among genders. With restricted randomization for testing homogeneity, one category, say gender, is not at random. I prefer this restricted randomness to be from column to column (alternatively some texts might randomize from row to row).[9]

Independence means that events in one category **do not affect** the outcome in another category. When Pearson's chi square is used as a test of independence, a sample of **events** is collected at random from the **whole** population and **then** it is classified into the distinct categorical variables observed for each unit. This **randomization is unrestricted**; both categories are from the same random population. That is, if you collect a random sample of fish and then separate the completely random sample into genders before seeing if gender and the number of hard dorsal fin rays are related, you

---

[7]Snedecor and Cochran (1967) call attention to the fact that in a single row or column of a multicolumn χ^2 table, the sum of all the deviations equal zero, $\sum_{l=1}^{l=k}(O_l - E_l) = 0$, and that this is equivalent to having one value predetermined (resulting in one less df).

[8]Some people see 'nonparametric' as meaning not directly using the normal distribution. So, they ambiguously say that 'the chi square is nonparametric'. Apparently they are referring to Pearson's chi square, but they should consider that Pearson's chi square is no more, or less, nonparametric than is the normal approximation. Indeed, some people dispute this and seem to think in terms of the definition of χ^2 in relation to Z (Equation 8.4). In any case it is a half-full half-empty issue. The distinction is barely relevant; do not let it distract you.

[9]In other forms of testing, such as ANOVA, the word 'homogeneity' refers specifically to the homogeneity of **variances** regardless of class (factor, or category), in contrast to heterogeneity.

have obtained both the genders and the hard fin ray counts at random and your test challenges independence. A simple two category question might ask if the same proportion of alcohol intoxicated pedestrians tend to end up with broken bones when hit by a car as do sober pedestrians when hit by a car.[10] Pearson's χ^2 can (as can a two sample binomial) be used to compare the proportions in each category pair, such as intoxicated and broken bones, to test if they would be statistically significantly different from each other if only random chance were involved. Unfortunately testing independence does not allow you to control the number of items in each category.

If two categories are free from association (independent or free from association because they are homogeneous, meaning the same), you should be able to find the probabilities of both of them happening by multiplying the probability of one happening times the probability of the other happening (probability law for independent variables Equation 1.4). But, if they are associated (meaning different categories are statistically different), your results will be different from each other to such a degree that a Pearson's chi square test with a sufficient sample size should be statistically significant.

So, these terms relate to your **experimental technique**; often nature chooses which, homogeneity or independence, for you. The calculation of Pearson's chi square is not concerned with the nature of the association, whether dependence or lack of homogeneity; the kind of randomization of the sampling determines that difference.

Often you are interested in whether the proportions of two or more categories are somehow associated. In challenging independence, you will find an expected count, E, for every observed count, O. If you multiply the hypothesized independent probabilities times the total sample size of their source, you will get an expected count, E. Pearson's χ^2 uses the chi square distribution to estimate the probability, by chance alone, of getting differences between the Os and the Es as large or larger than in your sample. If that probability is less than your a priori criterion, α, then it is statistically significant evidence of non-independence or non-homogeneity. That is, it is evidence of association, suggesting that there is some non random relationship (association) between the categories. In application, the distinction between independence and homogeneity becomes murky when the researcher has no opportunity to control sample size, but for instance, suspects dependence. In reality the experimenter may have good reason to report an effect size for one kind of association when the experimental conditions suggest the other (Subsection 12.5.1).

Pearson's chi square allows us to use tables of counts in rows and columns to look at whole experiments even if they involve **counts** of values that are themselves on a Likert scale. Let's imagine a two row table with five columns. Say the two rows contain randomly obtained counts of folks who say they are happy and of those who say they are sad, and the columns are from widely separated geographical regions.[11] If there was no effect of location on reported happiness versus sadness, we would expect the percent (the proportion relative to the grand total, **n**) of all the people in a particular cell would be proportional to the number of people in the row in which it occurs. Likewise the percent in that cell should also be proportional to the number of people in its column. After all, the law of large numbers tells us that as the sample size increases, the sample mean tends to represent the true mean (the expectation). Aye, there's the rub; even though the law says that there is less interference from chance as the sample size increases, chance does still interfere. Well, Pearson's chi square estimates the probability of getting differences between the observed and expected values as large or larger than in your sample, by chance alone.

The effect of happiness or sadness on where people locate can also affect the difference between

---

[10]Folklore has it that 'drunks are less likely to be as seriously injured when in an accident because they are relaxed'.

[11]Surveys of happiness or sadness across cultures are common and yet they illustrate the great difficulty sociological studies face due to unexpected issues. In N. Korea your life might be in danger if you told some foreigner that you were 'sad'. Likewise in England or Scandinavia it might make people uncomfortable to tell a stranger that they were anything but 'happy'. However, in much of the US, telling your troubles to all comers willing to listen seems to be a widely enjoyed pastime.

the observed versus the expected, which tends to be larger when there is either dependence or lack of homogeneity. It uses common χ^2 tables to discover either dependence (such as correlation resulting from interactions between rows and columns) or non-homogeneity (such as differences in the variance among the rows, or among the columns). So, to challenge either independence or homogeneity, $P\chi^2$ uses **expected** values calculated from the corresponding **row and column** totals relative to the **grand total**, n. For inference, $\mathbf{r} \times \mathbf{c}$ tables have $(\mathbf{r} - 1)(\mathbf{c} - 1)\ df$. Recall that this book uses a bold $\mathbf{r}$ for the number of rows (it is **not** the italicized r, the number of successes).

	C_1	C_2	C_3	C_4	C_5	Row Totals
Row 1	$O_{1,1}$	$O_{1,2}$	$O_{1,3}$	$O_{1,4}$	$O_{1,5}$	$n_{\mathbf{r}_1}$
Row 2	$O_{2,1}$	$O_{2,2}$	$O_{2,3}$	$O_{2,4}$	$O_{2,5}$	$n_{\mathbf{r}_2}$
Col. Totals	$n_{\mathbf{c}_1}$	$n_{\mathbf{c}_2}$	$n_{\mathbf{c}_3}$	$n_{\mathbf{c}_4}$	$n_{\mathbf{c}_5}$	$\mathbf{n}$

Table 12.1: 2 x 5 Contingency Table

Table 12.1 merely demonstrates a very common notation of a 2×5 table showing only observed values represented by Os. The $\mathbf{c} = 5$ columns represent different groups (like treatments). The $\mathbf{r} = 2$ rows each represent one of two mutually exclusive outcomes of interest in each category. A *cell* is the intersection of a row and a column. It contains your counts, the *observed* values, $O_{i,j}$, in a cell for a particular category i (as in $\mathbf{r}_i$) and a particular group j (as in $\mathbf{c}_j$). The range of these subscripts are $1 \leq i \leq \mathbf{r}$ and $1 \leq j \leq \mathbf{c}$. In practice the Os will be replaced by actual counts. All the Os in a row i add up to $n_{\mathbf{r}_i}$ and all the Os in a column add up to $n_{\mathbf{c}_j}$. These sums, $n_{...}$, along the right and bottom are called the *marginal* values. The $\mathbf{n}$ in the lower right corner is the *grand total* (some authors use N but that is confusing because N is the size of the whole population). Such a **contingency table** is often referred to as an $\mathbf{r} \times \mathbf{c}$ table, or a *two way table* (Pearson 1904). It is also referred to as *cross tabulation* or cross tab, and can be described as "a type of table in a matrix format that displays the (multivariate) frequency distribution of the variables".

A much more detailed version of such a contingency table with the observed and the expected values and some of the necessary operations is shown in Figure 12.2. It will enable you to easily complete the analysis. The marginal sum, n, of all the Os in a row or column will **equal** the sum of all the Es in that particular row or column despite the fact that the two values in each cell are often quite different. The marginal values on either the right, or the bottom, will sum to the same value, the grand total $\mathbf{n}$.

		c_1	c_2	c_3	c_4	c_5	$\sum_{all\,c} O_r = \sum E_r$
$\mathbf{r}_1$		$O_{1,1}$ $E_{1,1}$	$O_{1,2}$ $E_{1,2}$	$O_{1.,3}$ $E_{1.,3}$	$O_{1,4}$ $E_{1,4}$	$O_{1,5}$ $E_{1,5}$	$\sum_{j=1}^{j=5} O_{1,j}$ $= \sum_{j=1}^{j=5} E_{1,j}$
$\mathbf{r}_2$		$O_{2,1}$ $E_{2,1}$	$O_{2,2}$ $E_{2,2}$	$O_{2,3}$ $E_{2.,3}$	$O_{2,4}$ $E_{2,4}$	$O_{2,5}$ $E_{2,5}$	$\sum_{j=1}^{j=5} O_{2,.j}$ $= \sum_{j=1}^{j=5} E_{2,j}$
$\sum_{all\,r} O_c = \sum E_c$		$\sum_{i=1}^{i=2} O_{i,1}$ $\sum_{i=1}^{i=2} E_{i,1}$	$\sum_{i=1}^{i=2} O_{i,2}$ $\sum_{i=1}^{i=2} E_{i,2}$	$\sum_{i=1}^{i=2} O_{i,3}$ $\sum_{i=1}^{i=2} E_{i,3}$	$\sum_{i=1}^{i=2} O_{i,4}$ $\sum_{i=1}^{i=2} E_{i,4}$	$\sum_{i=1}^{i=2} O_{i,5}$ $\sum_{i=1}^{i=2} E_{i,5}$	$\mathbf{n}$ $= \mathbf{n}$

Table 12.2: **Observations** and Expectations

However, **when reporting**, there customarily is only one entry, an observed total, $\sum_{j=1}^{j=c} O_{i,j}$, in each of the cells of the marginal values on the far right, and an observed total, $\sum_{i=1}^{i=r} O_{i,j}$, in each of the cells of the marginal values along the bottom.

Assuming either independence or homogeneity, a common way to **find** $E_{i,,j}$, the *expected* value for the cell at the intersection of row i and column j, is by **multiplying** the corresponding total row and column marginal values **by each other** and then **dividing by** the total sample size, **n**. This division is because **n** occurs twice in the multiplication of the marginal values. So, you can **calculate** the **expected value** of each cell, using

$$E_{i,,j} = \frac{\left(\Sigma_{h=1}^{h=c}O_{i,h}\right)\left(\Sigma_{l=1}^{l=r}O_{l,j}\right)}{\Sigma_{l=1}^{l=r}\Sigma_{h=1}^{h=c}O_{l,h}} = \frac{n_{r_i}n_{c_j}}{n}, \tag{12.4}$$

given randomness and that the **H0** is correct. Technically, this is equivalent to

$$E_{i,j} = \left(np_{c_i} \times np_{r_j}\right)/n = np_{c_i r_j},$$

but this expectation is **only true** if there is **no association**, that is, assuming that **H0** is true. Otherwise, and aside from chance, the difference between the observed and the expected values in a cell will be due to association. Any association (a lack of independence or a lack of homogeneity) will violate the null hypothesis, **H0**, and inflate the Pearson's chi square value. If the association is sufficient, the test will strongly tend to be statistically significant.

$P\chi^2$ has $(r-1)(c-1)$ **degrees of freedom** (when $r > 1$ and $c > 1$). Each row has one predetermined value and each column has one predetermined value, because either the design or the sampling process provides the row and column totals.

χ^2 values, or any equivalent thereof (such as Z^2 or $P\chi^2$), are sufficiently additive that not only can **c** be greater than 2, but it is possible to have $r > 2$. Thus you can perform a great variety of tests based on Pearson's chi square.[12]

If just one comparison pair (for instance, any two treatments or two conditions) is statistically significant, a global test should, in theory, be statistically significant and this is a reasonable assumption when working with Pearson's chi square tests that are larger than 2×2.

12.4.0.1 Example: Pearson's Chi Square Challenging Independence

Homicide: Year vs victim's gender ('other races'): This example is an application of Pearson's chi square to a question of independence because we have no control over the randomization. That is, although we assume the crimes are randomly committed (and detected), we have no control over the number of crimes in either gender or in the various years. Let's examine 4 years data from the Uniform Crime Reporting Program regarding the gender of the victims in a particular subgroup to see whether there is an association between the number of victims in each gender and the number of victims in each year.

H0: THE NUMBER OF FEMALE MURDER VICTIMS IS INDEPENDENT OF BOTH GENDER AND YEAR.

HA: THERE IS AN ASSOCIATION BETWEEN NUMBER OF FEMALE MURDER VICTIMS AND YEAR.

Criterion: $\alpha = 5\%$ (one tail, more variation for statistical significance)

Methods: This data came from the US Department of Justice, The Uniform Crime Reporting (UCR) Program, as reported by the Federal Bureau of Investigation. It pertains to a grouping of a

---

[12]Pearson's chi square assumes underlying binomial distributions that approximate normality, but it can handle even more than just rows and columns, using the multinomial distribution, an extension of the binomial distribution. You can even have $r \times c \times d \times \ldots$ tables, but such multidimensional contingency tables are challenging to represent on paper and are generally handled by computers.

number of minorities listed as 'Other Races' as opposed to Black, White and Unknown. Most of these minorities report a low per capita murder rate, but the data is handy for demonstrating the calculations involved.

	2006	2007	2008	2009	Totals
Male	406	345	321	360	1432
Female	301	238	215	232	986
Totals	707	583	536	592	2418

Analysis: In the following analysis table, the estimated values were found from the marginal totals, as though assuming **H0** to be true. The expected values were calculated based on the assumption that the gender ratio was unrelated to the year using Equation 12.4,

$$E_{i,j} = \frac{n_{r_i} n_{c_j}}{n}.$$

	2006	2007	2008	2009	Totals
Male O	406	345	321	360	1432
E	418.703	345.267	317.433	350.597	1432
$O - E$	-12.703	-0.267	3.567	9.403	0
$(O - E)^2/E$	**0.38540**	**0.00021**	**0.04009**	**0.25218**	(cont.)
Female O	301	238	215	232	986
E	288.297	237.733	218.567	241.403	986
$O - E$	12.703	0.267	-3.567	-9.403	0
$(O - E)^2/E$	**0.55973**	**0.00030**	**0.05823**	**0.36625**	1.66238
O Totals	**707**	**583**	**536**	**592**	**2418**
E Totals	*707*	*783*	*536*	*592*	*2418*
$\widehat{p}_{male}$	0.574	0.592	0.599	0.608	0.5922
$\widehat{p}_{female}$	0.426	0.408	0.401	0.392	0.4078

There are $(\mathbf{r} - 1)(\mathbf{c} - 1) = 3$ degrees of freedom, because one cell per row and one cell per column is predetermined. The separated **bottom 2 rows are optional** and estimate probabilities, but they are not part of the test itself.

The differences between the expected values and their observed values were then used to test (challenge) independence using Pearson's chi square,

$$P\chi^2 = \sum_{All} \frac{(O - E)^2}{E} = 1.66238.$$

Results and Discussion: Referring to a $p >$ table of χ^2 at $\alpha = 5\%$ with $3\,df$, we find that it requires at least $\chi^2_{\alpha=0.05,\,df=3} = 7.8147$ to reject. Therefore, because the estimated $\chi^2 = 1.66238$, **H0** was **not rejected**. Especially if statistical significance were observed, it would be handy and informative to present such an analysis with a **stacked column graph**.

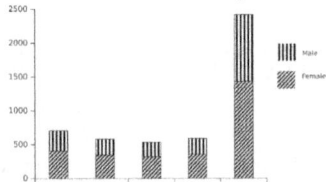

Was there really any randomization? Some folks think that only one in five homicides get classified as such, but that still doesn't change randomness of the fewer reported ones, does it?

This classification may be random, particularly with respect to gender of victims within each year, but who can be certain? The best practice is to take a random sample of the available data.

You and I might see 2418 as being a reasonable sample size for a Pearson's chi square test for this kind of question. However, some folks who work with this test all the time suspect it of becoming hypersensitive when $n > 400$. One way to avoid this would be to take a random sample of, say, 400 from the 2418 reports that are available. The effect of the hypersensitivity of actually making a decision might also be reduced by using a smaller value of α than you would otherwise choose. However, whether this concern is valid or not, the results of this example did not come anywhere near suggesting statistical significance even though the total sample size was a lot more than $n = 400$.

12.4.1 2×2 tables for association

The smallest Pearson's contingency tables are 2×2 and are both useful in themselves, and are handy for demonstrating some key ideas. They are also handy for multiple comparisons following global Pearson's chi square tests because larger contingency tables can easily be broken into simpler 2×2 contingency tables.

With a table of only the observed counts, we can calculate everything we need (such as the expected values) from the **marginal values** (the row and column totals). Each cell of a 2×2 contingency table contains a count of observations, which can fall into either of two categories, 'success' or 'failure' for each of two factors. Counts of randomly sampled fish from a chemical spill can include fish that are obviously cancerous, C, or **not** cancerous, $!C$, and dead, D, or **not** dead, $!D$. Each of the rows and columns sum to the total for one of these single categories. For instance, in the first column, adding such observations as the number of dead and cancerous fish, $O_{D \cap C}$, to the number of dead but not cancerous fish, $O_{D \cap !C}$, you get the observed total of the **dead fish**, $O_{D \cap C} + O_{D \cap !C} = n_D$, as a column total. This works out for the other column, and for the rows too:

Observed

	D	$!D$	$\sum C$
C	$O_{D \cap C}$	$O_{!D \cap C}$	n_C
$!C$	$O_{D \cap !C}$	$O_{!D \cap !C}$	$n_{!C}$
$\sum D$	n_D	$n_{!D}$	$\mathbf{n}$

As shown, the sum of the marginal row totals, which are on the right side of the table, always equal the grand total, $\mathbf{n}$, and the sum of the marginal column totals which are at the bottom of the table, also equal the grand total, $\mathbf{n}$.

The observed values (the Os) in the body of the table arise from the effect of chance plus the effect of any association between the two factors. However, the proportion of the total of all the counts, $\mathbf{n}$, that is represented by the row totals and column totals associated with each cell will provide us the needed information to calculate the expected values, $E_{whatever}$, that we would expect **if there were no association**:

How E is Calculated

	D	$!D$	
C	$n_C n_D / \mathbf{n}$	$n_C n_{!D} / \mathbf{n}$	n_C
$!C$	$n_{!C} n_D / \mathbf{n}$	$n_{!C} n_{!D} / \mathbf{n}$	$n_{!C}$
	n_D	$n_{!D}$	$\mathbf{n}$

Expected Values

	D	$!D$	
C	$E_{D \cap C}$	$E_{!D \cap C}$	n_C
$!C$	$E_{D \cap !C}$	$E_{!D \cap !C}$	$n_{!C}$
	n_D	$n_{!D}$	$\mathbf{n}$

Notice that the marginal values, the totals, do not change when we go from observed to expected. We usually combine the above two tables into one table, especially when we are reporting the results,

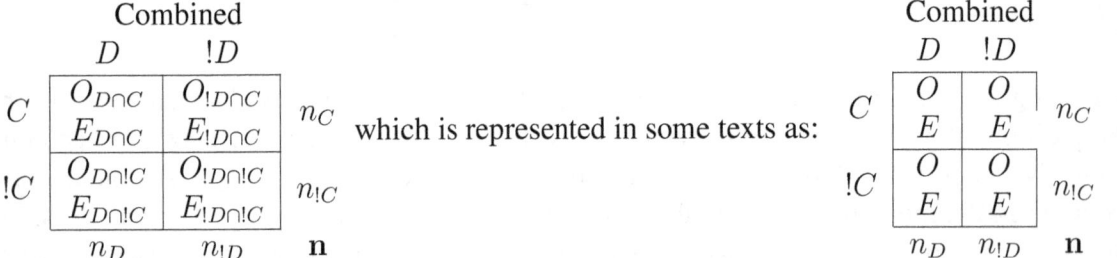

Sometimes the expected values are simply left out when reporting:

Terse Report

	D	$!D$	
C	O	O	n_C
$!C$	O	O	$n_{!C}$
	n_D	$n_{!D}$	$\mathbf{n}$

So, how does this calculation of an expected value, E, work? Well, we have to look at a hidden explanatory step involving probabilities and their associated laws. Using the frequentist model with random sampling, the probability of any individual drawn at random having characteristic C is the corresponding marginal value divided by the total number of individuals, $P_C = n_C/\mathbf{n}$. Under the multiplication law, when there is **no association**, the probability of the randomly drawn individual having **both** C and D is equal to their probabilities multiplied together, $p_C p_D$. So, multiplying the combined probability by the overall sample size, $\mathbf{n}$, would, under the assumption of no association, give the expected counts.

That is, since $p_C p_D = n_C/\mathbf{n} \times n_D/\mathbf{n} = n_C n_D/\mathbf{n}$, you can, under the assumption of no association, multiply by $\mathbf{n}$ to get $\mathbf{n} p_C p_D = E_{C \cap D}$ (or simply E_{CD}), the expected value that a randomly selected fish will have both C (be cancerous) and D (be dead). As we shall see, effects of association would change the observed values so that $P\chi^2$ would be larger than expected, and given an adequate sample size, that should result in statistical significance.

Expected Counts, E

	D	$!D$	**Totals**			D	$!D$	$\sum C$	
C	$\mathbf{n} p_C p_D$	$\mathbf{n} p_C q_D$	$\mathbf{n} p_C = n_C$	$=$	C	$E_{C,D}$	$E_{C,!D}$	n_C
$!C$	$\mathbf{n} q_C p_D$	$\mathbf{n} q_C q_D$	$\mathbf{n} p_{!C} = n_{!C}$		$!C$	$E_{!C,D}$	$E_{!C,!D}$	$n_{!C}$
$\sum D$	$\mathbf{n} p_D = n_D$	$\mathbf{n} p_{!D} = n_{!D}$	$\mathbf{n}$		$\sum D$	n_D	$n_{!D}$	$\mathbf{n}$

This brings us back to what we will report, either

Observed Counts

	A	$!A$	Total
C	O_{CA}	$O_{C!A}$	n_C
$!C$	$O_{!CA}$	$O_{!C!A}$	$n_{!C}$
Total	n_A	$n_{!A}$	$\mathbf{n}$

or

Observed Counts

	A	$!A$	Total
C	O_{CA} / E_{CA}	$O_{C!A}$ / $E_{C!A}$	n_C
$!C$	$O_{!CA}$ / $E_{!CA}$	$O_{!C!A}$ / $E_{!C!A}$	$n_{!C}$
Total	n_A	$n_{!A}$	$\mathbf{n}$

So, the marginal values, the row and column counts from the sample, can provide the expected values; this applies to any size tables. Therefore, using

$$E_{i,j} = \left(n_{\mathbf{r}_i} \times n_{\mathbf{c}_i}\right)/\mathbf{n} \tag{12.5}$$

is an obvious way **to calculate** the expected values for any size tables.[13].

To fill in the observed values, you only need the **marginal totals** and the **observed value of one cell** for a 2 × 2 table. The degrees of freedom are **equal** to the number of cells you **need** to know, along with the marginal values, in order to find the rest by subtraction. That is, you know the **row and column** totals are the **marginal** totals, so after enough cells in a row or column are filled in, you can easily find the others for that row or column. For a 2 × 2 table, you only need to know **one cell**, so $df = 1$. (For larger tables, the degrees of freedom are $df = (\mathbf{r} - 1)(\mathbf{c} - 1)$, the number of cells you must **fill before finding the rest** by subtraction.)

Let's consider 2 × 2 tables from the point of view of challenging the independence of a variable that has two alternatives with respect to the behavior of pet owners regarding the care of dogs vs cats. For instance, you could be comparing the proportion of animals of each species that are brought to emergency veterinary clinics in a disaster area on the day after a hurricane due to emergencies or causes definitely related to the storm, versus how many were brought for inappropriate reasons. People will bring animals because they had planned have the pet's teeth cleaned, or their nails clipped, and do not want to wait for the regular clinics to reopen, do not want to pay, etc. Such a study may in large part, but not totally, relate more to narcissism among owners who choose different kinds of small animals rather than animal health per se. Be careful to consider what you may really be studying before rushing to conclusions. Maybe you hypothesize

H0: THE SPECIES OF PET IS INDEPENDENT OF WHETHER THE OWNER INAPPROPRIATELY BROUGHT IT VS **HA**: IT IS NOT INDEPENDENT.

If you were testing more than two kinds of animals (dogs, cats, ferrets, parrots, etc.), you may eventually have to test each kind of animal against every other kind of animal unless you a priori choose which comparisons you intend to make before the data is obtained. Sometimes the question is no more than whether inconvenient pet owner behavior and choice of animal species are associated. How to obtain a random sample in such a situation is a logistic challenge. But assuming you call the sample random, you would be unlikely to have control over the number of animals in each of the two categories (storm or species). That defines this example to be a test of independence (neither category under your control).

Let's look at a probability structure of a possible two by two contingency table for testing independence or homogeneity from several viewpoints. For 2 × 2 tables, you can find one cell in each row and one cell in each column by subtraction. To calculate the expected values in a 2 × 2 table, you need to know only one cell, subtracting it (and one result of its subtraction) from the appropriate marginal values,

---

[13]In the days of hand calculation, yet another way of looking at 2 × 2 tables used a popular notation for considering them for a number of applications,

a	b	$a + b$
c	d	$c + d$.
$a + c$	$b + d$	**n**

Then Pearson's chi square was often found by

$$P\chi^2 = \frac{\mathbf{n}(ab - bc)^2}{(a+b)(c+d)(a+c)(b+d)}$$

(Pearson, 1900). This may also be of interest to some workers who have an affinity for linear algebra, though that is a notation system that is outside the scope of this book. I will not provide a further derivation here because this formula is non-informative, and not as handy, as the shortcut that uses Equation 12.6. However, you will see this left to right a to d **pattern** in related contexts.

	Cat, A	Dog, $!A$	Totals
Emergency or Due to Storm, B	$np_Ap_B = E_{A,B}$	$? = E_{!A,B}$	n_B
Not Emergency or Due Storm, !B	$? = E_{A,!B}$	$? = E_{!A,!B}$	$n_{!B}$
Totals	n_A	$n_{!A}$	$\mathbf{n}$

(where $p_A = n_A/\mathbf{n}$ and $p_B = n_B/\mathbf{n}$). So, the **expected number** in the upper left hand cell, the intersection of row B and column A, is $E_{A,B} = np_Ap_B$, which can be simplified to Equation 12.5,

$$E_{A,B} = n_An_B/\mathbf{n}.$$

Whichever of these approaches you use, you will get the results as the expected counts:

	Cat A	Dog $!A$	$\sum$
B	$E_{A,B}$	$E_{!A,B}$	n_B
$!B$	$E_{A,!B}$	$E_{!A,!B}$	$n_{!B}$
$\sum$	n_A	$n_{!A}$	$\mathbf{n}$

The example of cats and dogs vs owners' behavior after a storm happens to be a test of independence because the numbers falling into the rows and columns are out of our control.

However, as an example of a test of homogeneity, we could have used restricted randomization to obtain equal numbers of high school graduates versus non high school graduates with respect to filing bankruptcy. That would be a test challenging homogeneity.

Large tables (for omnibus or global tests) can be filled out from your marginal totals in much the same way as for the first cell in a 2×2 table. Multiple 2×2 tests may compare two groups at a time, and they can, with appropriate adjustment of the significance cutoffs, be used for multiple comparisons following global tests.

12.4.1.1 2×2 Shortcuts for Association

Before going on, let's look at a **shortcut** for 2×2 tables to test for association. We will see this particular 2×2 contingency table again as four cells split out of a global table for **association** of hate crime and location (office building vs jail-prison). The **easiest way** to calculate expected values is

$$E_{i,j} = (n_{\mathbf{r}_i} \times n_{\mathbf{c}_i})/\mathbf{n},$$

as seen in the cat and dog example above. This is Equation 12.5 and it even works for **large** contingency tables. The data is real, but the labels are irrelevant for the moment. To its right is another version with just the **observed** minus the **expected** values, $O - E$,

	J/P	Off. B	$\sum$
Race	$28\,O$	$55\,O$	83
	$22\,E$	$61\,E$	
Rel.	$5\,O$	$38\,O$	43
	$11\,O$	$32\,E$	
$\sum$	33	93	126

$O - E$

	J/P	Off. B	$\sum$
Race	6	−6	0
Rel.	−6	6	0
$\sum$	0	0	0

Isn't that remarkable? Well, no, it really isn't remarkable, you will get + and - the same number every time. Mathematics tends towards patterns. Music also has patterns, and musicians seem to pick up on math quite quickly once they get started. You only have to calculate one cell and take it from there, but if you do it by hand, you might want to calculate at least two cells to be certain you have not made a mistake. (I did round, but it still comes out equal even when I do it with six figures.)

12.4.1.2 Shortcut for 2×2

The calculation of Pearson's chi square tests for four cell 2×2 contingency tables, or treatments, could use $P\chi^2 = \sum_{all} \frac{(O-E)^2}{E}$. However, the above regular pattern and the fact that squares are always positive justify a shortcut for 2×2 tables, using

$$P\chi^2 = (O - E)^2 \sum_{all} \frac{1}{E}, \qquad (12.6)$$

where $(O - E)^2$ refers to **any one cell**, and $\sum_{all}$ means sum all of the reciprocals of the expected values (for **all** 4 cells) in the 2×2 table.

In the case of the above table, $P\chi^2 = 36 \left(\frac{1}{22} + \frac{1}{61} + \frac{1}{11} + \frac{1}{32} \right)$, so $\chi^2 = 6.6243$ with $1\ df$. From an online calculator this has a $p\_value = 0.0100$. Pearson would be happy with that, but you are better off first specifying a criterion, α.

12.5 Effect Size for $P\chi^2$

As you recall, an effect size, ES, is merely a single value, a point estimate of an effect due to something other than chance. An ES is appropriate whenever you declare statistical significance, and should ideally be reported in terms of original units of measure. It is useful to consider the effect size when trying to detect some form of association (relationship). Most psychology journals are now demanding an effect size, ES, whenever the researcher declares statistical significance, and it is reasonable that this trend is likely to spread to other research areas. All such journals prefer the effect size in original units of measure. If you report that a difference exists among central measures such as means, it makes sense to report the ES, the observed difference in the original units of measure whenever you have statistical significance. However, some procedures, such as is often the case with Pearson's chi square, can only produce effect sizes in terms of correlation or of probability.

Pearson's χ^2 is primarily sensitive to linear (straight line) relations (Section 11.1). So, correlation based indexes (Subsection 12.5.1.1), which are measures of linear relations, are often used to report the ES for Pearson's χ^2 results when they are statistically significant. Ideally, when challenging independence, *correlation* can be seen as a measure of dependence and is used to estimate how much categories affect each other. When the test is challenging homogeneity, the squares of the estimators otherwise used for correlation can be used for the ES values in terms of any difference in variance (even other statistical characteristics in general) among 'treatments'. That is, you might want to say that you estimate 40% of the variability in A is associated with B.

12.5.1 R-like values or probabilities for ES

Either Pearson's coefficient of determination, R^2, or for simple experimental designs, its equivalent, phi square, ϕ^2, are suitable as ES values, and as measures they **both** approximate

$$\frac{sampled\ population\ variance}{hypothesized\ population\ variance}.$$

These values can also be seen as estimating the proportion (eg. percent) of the **variance due to effects other than chance** (as when challenging goodness of fit). In application, these squares are used to detect heterogeneity (challenge homogeneity), simply implying that the distributions are different, not necessarily that they cause each other to be different. The square roots, phi ϕ and R, of these effect size estimators are popular effect size estimators for dependence, and imply the degree to which

the categories affect one another. But unlike the impression you may have had on your first exposure to Pearson's chi square, it really tends to detect linear (straight line) relations, implying that the
strength of the detected association (the effect size) can be measured as linear correlation. Thus there
may be some blurring of the precise relation between $P\chi^2$ and the interpretation of these preferred
ES values, making the choice of which to use, the squares or their roots, a little more flexible than
simply a matter of whether you could control the sampling procedure or not.

Pearson's coefficient of determination, R^2 (Subsection 12.5.1), is the square of the correlation
coefficient, R, and estimates the effect of one factor on the variance of the other. Phi, ϕ, **is an approximation** of the estimated correlation coefficient and is calculated directly from χ^2 itself (Equation 12.7). It is quite popular in many fields. Its square, ϕ^2, can be used as an estimation of the effect
of one variable on the variance of another variable. Except for values at the extremes of its range, ϕ
equals Pearson's correlation coefficient, R, for a 2 × c table with c < 3, and can be reported as an
ES in those cases (Section 12.5.1.1). R, and R-like measures, as well as their squares, treat the principal axis of oval-like relations as a straight line relation, the major axis. The other axis detracts from
the resulting estimate.

But, unfortunately, phi square, ϕ, increasingly deviates from R as the number of rows and columns
(above 2 × 2) gets larger, and it will exaggerate whatever effects are present with large numbers of
categories (similarly for the behavior of ϕ^2 with respect to R^2). So, it is best to restrict the use of ϕ or
ϕ^2 as effect sizes to 1 × 2 and 2 × 2 applications of Pearson's chi square.

Pearson's chi square and Pearson's correlation coefficient are similarly **weak for nonlinear effects** and **strongest for linear effects**. For instance, if two variables change together but the data
could be graphed to form a 'U' shape, the association can be very strong, but the linear correlation
will be reported by your calculations as almost 0. You can evaluate simpler ordinal effects via Spearman's rank correlation.

12.5.1.1 Which ES for Pearson's chi square?

Pearson's χ^2 is frequently used as a test for association. (In this book the word *relationship* is sometimes substituted for *association*.) Unfortunately, the word 'association' is poorly and inconsistently
defined in some fields, even in psychology when discussing experimental or observational outcomes.
Because some popular measures of effect size in this subsection and some of the ES values in the
next section are analogous to and include the commonly used **product moment correlation coefficient**, R, their squares have a meaning corresponding to (but not always equaling) and including the
coefficient of determination,

$$R^2 = \frac{\left[\sum_{i=1}^{i=n} (x_i - \bar{x})(y_i - \bar{y})\right]^2}{\sum_{i=1}^{i=n} (x_i - \bar{x})^2 \sum_{i=1}^{i=n} (y_i - \bar{y})^2}.$$

R^2 can be used to estimate how much of the variance is due to non-homogeneity (heterogeneity), or
lack of fit, so it can be used as an ES for 2 × c and 1 × c tables when c > 3 and especially if c is 6 or
more (Subsection 12.5.2.1). Likewise its square root, R, can be reported as a correlation (interaction)
effect size.

Phi (Greek letter ϕ) is an ES estimate that **equals** the correlation coefficient, R **for 2 × 2 tables**.
It is calculated by

$$\phi = \sqrt{\frac{\chi^2}{n}}. \tag{12.7}$$

Its range of estimated values is absolute, theoretically $0 < \phi < 1$, meaning almost no association
(no relationship) to almost complete association (completely related), but it can sometimes take on a

value greater than 1. There are also ways you can assign a sign ($\pm$) to it. For 1×2 or 2×2 tables, phi seems a very reasonable effect size because it is drawn directly from χ^2 (or $P\chi^2$) itself. Ideally ϕ can be seen as a measure of dependence, whereas its square, ϕ^2, is ideally a measure of non-homogeneity (differences in distribution).

Phi is reported more often than phi square, because phi square can seem to be rather tiny (the effect of squaring fractional values). Even so, when your randomization is restricted, it might be preferable to report and interpret ***phi square***,

$$\phi^2 = \frac{\chi^2}{\mathbf{n}},$$

as an *ES* representing the fraction of the variance that is due to difference. For tests of fit, ϕ^2 might more often be the better of the two.

Despite its tendency to relate to linearity, ϕ^2 can also be seen as $\phi^2 = \chi^2/\mathbf{n} = \frac{(\mathbf{n}-1)S^2/\sigma^2}{\mathbf{n}} \approx S^2_{treat}/\sigma^2$ as the total sample size, $\mathbf{n}$, grows larger. That is, it can be seen as addressing the effect of the treatments on variance. In application it can be reported as a ratio (or percent).

Large tables having **both** more than two rows and more than two columns can be used as global tests via $P\chi^2$, and (with appropriate corrected cutoffs) can be broken into smaller tables for multiple comparisons. However, with increasing numbers of variables, ϕ, and therefore ϕ^2, tend to become larger on their own, even when there is no association. So, ***Cramer's V***, more **commonly** referred to as ***Cramer's phi***, ϕ_C (Cramer 1946), was devised as an effect size for a global test with any number of cells. Cramer's phi, ϕ_C, can be seen as **an attempt** to correct for the effect of **more than** two rows or more than 2 columns.

The formula for Cramer's phi is

$$\phi_C = \sqrt{\frac{\chi^2/\mathbf{n}}{\min{(\mathbf{r}, \mathbf{c})} - 1}} \quad 0 \le \phi_C \le 1, \tag{12.8}$$

where min ( ) means **whichever** of the things in the parentheses ( ) **is less**, in this case, the number of **rows or columns**. Phi and Cramer's phi are the **same** when $\mathbf{r} \le 2$ **and** $\mathbf{c} \le 2$ (Liebetrau 1983). Note that many authors, and some software manuals, incorrectly call phi "Cramer's phi." The size of Cramer's phi is only an approximation of a correlation coefficient and does **not fully compensate** for the inflation problem.

An alternative to variance or correlation based effect sizes for a $2 \times c$ table might be to report the column probabilities as a percentage of one of the two categories of interest. However, when either category has only two alternatives and the other has six or more, maybe you could consider a correlation (Section 11.3), following a significant $P\chi^2$, or you might just consider a correlation and maybe report the coefficient of determination, R^2.

For Pearson's chi square, a tendency toward $\phi = 1$ is evidence, not necessarily proof, of a strong relationship, but a weak ϕ might still be meaningful in some contexts. The ***on diagonal*** in 2×2 tables include the cells where both categories are successes or both are failures $\begin{array}{|c|c|} \hline ++ & \\ \hline & -\,- \\ \hline \end{array}$. If the data falls **equally on** the diagonal and **off** the diagonal, $\begin{array}{|c|c|} \hline a & b \\ \hline c & d \\ \hline \end{array}$ $(a + d = b + c)$, **almost any** ϕ between $+1$ and -1 is **possible**, and then this coefficient is untrustworthy (Davenport and El-Sanhury 1991). See Sub-subsection 12.5.3.2 for more details specific to 2×2 tables. The correlation-like *ES* values, ϕ and ϕ_C, which are drawn from χ^2 are not more (or less) free of dependence on linearity than is Pearson's chi square itself. For $2 \times c$ tables where, say, $\mathbf{c} > 5$, and most cells have large counts, you might consider if you should do an ordinary correlation instead of a χ^2, or better yet, for $\mathbf{c} > 5$ you could consider a nonparametric Spearman's rank correlation which does not require a straight line effect (Section 14.6).

12.5.2 ES for goodness of fit

One way to look at a Pearson's chi square for a goodness of fit test is as though it is challenging homogeneity among the groups (the expected and the observed). So, restricted sampling is the preferred approach. Then the effect size could be reported as a coefficient of determination, R^2, or, for 1×2 tables, phi square, ϕ^2 (or maybe if you want a ballpark approximation, ϕ_C^2 when $c > 2$). Then the effect size could be considered in terms of differences in variability, rather than some sort of cause-effect interaction. That way you can report, as an ES, an estimated measure of variance about the model due to effects other than chance.

So, in the following example and its continuations, we will consider the effect size for a rather small apparent lack of fit of observed data to an expected model using a $1 \times c$ Pearson's chi square for fit. This example shows you the calculations involved, and the resulting effect sizes can be interesting data **exploration**, but you can only report an effect size for publication when you have a **statistically significant** lack of fit.

12.5.2.1 Example: Effect Size for Fit

Mendel meets Pearson (refers to Subsection 12.2.0.1) Let's continue the unhistorical tale about Gregor Mendel's results. When he crossed pea plants that had what he called 'simply inflated pods', with pea plants that had constricted pods, he only got offspring plants with simply inflated seed pods in what is called the F1 generation. Based on the *phenotypes* (how the offspring appeared), there was reason to create a null hypothesis, **H0**, that there was only one gene with two forms (*alleles*). The pollen and the ovum each contributed a copy of the gene, but, when present, the inflated phenotype was *dominant* (took over when both alleles were present). If so, aside from the effects of chance, under his theory, by crossing sibling plants from that first generation (*selfing*, breeding them to themselves), he should expect 3 plants with the dominant simply inflated pods for each 1 with constricted pods (both genes recessive) in this inbred generation. So, his hypothesis pair would have been

H0: 75% WILL BE THE DOMINANT PHENOTYPE (implying that 25% will be the recessive phenotype).

HA: THE 3:1 MODEL DESCRIBED IN **H0** DOES NOT FIT.

Criterion: $\alpha = 5\%$

Analysis: Assume he designated the simply inflated pod phenotype, **I**, and constricted pod phenotype, **i**. The first expected entry, E_I, was obtained by multiplying the total, $n = 1121$, by 3/4 and the second was found by subtraction, $E_i = n - E_I$, to get $1121 - 840.75 = 280.25$. So the augmented table was:

	l	**i**	**Total**
O	**822**	**299**	**1121**
E	840.75	288.25	
$\hat{p} = O_j/\mathrm{n}$	0.733274	0.266726	1

and the calculation was

$$P\chi^2 = \sum_{j=1}^{j=2} \frac{(O - E_j)^2}{E_j} = \frac{(822 - 840.75)^2}{840.75} + \frac{(299 - 288.25)^2}{288.25}$$

$$P\chi^2 = 0.8194, \qquad (\mathbf{c} - 1) = 1\, df.,$$

where $P\chi^2$ represents the Pearson's chi square value attained. Note the probabilities as possible ES values below the table.

Results: H0 was not rejected because the chi square value estimated from the sample was smaller than the chi square required for statistical significance with 1 degree of freedom, $\chi^2_{1,\,0.05} = 3.8414$ (from a suitable table). Without significance there is no effect size. If the chi square had been statistically significant, the effect size could have been reported as an absolute difference in probability,

$$ES = O_{\mathbf{I}}/\mathrm{n} - E_{\mathbf{I}}/\mathrm{n} = p_{O_{\mathbf{I}}} - p_{E_{\mathbf{I}}} \qquad (12.9)$$

$$ES = 0.75 - 0.733274 = 0.016726$$

or $p_{O_i} - p_{E_i} = 0.250 - 0.267 = -0.016726$. Either way, the effect size, if the test had been statistically significant, could be estimated to be about $\pm 1.7\%$ of either expected outcome. That means that it is positive for one category and negative for the other.

Or the effect size could also be reported as a correlation-like effect size with the correlation coefficient-like phi value (Equation 12.7),

$$\phi = \sqrt{\chi^2/\mathrm{n}}.$$

However, the concept of correlation-like effect size with correlation coefficients like R and ϕ is not intuitively informative here. Although $\phi = \sqrt{0.8194/1121} = \sqrt{0.00073099} = 0.027037$, what does this number say to you about the proposed model, other than that the degree of association, whatever it might be, appears to be very small? For an ES, when reporting statistical significance for this sort of simple model, it might be informative to estimate

$$\phi^2 = P\chi^2/\mathrm{n} = 0.8194/1121 = 0.00073096.$$

This estimates that about 0.07% of the observed variance appears to be accounted for by **nonrandom** (non-homogeneous) **effects**. In the complexity of a biological situation, this estimate that $\phi^2 = 7.31 \times 10^{-4}$ of the variance is due to non random effects (deviation from the model) is obviously too small to have relevant meaning. In reality there was almost no evidence contradicting his null hypothesis.[14]

However, **a 95% confidence interval** is definitely appropriate because he is interested in the proportion of one phenotype, and by implication (subtraction from 1), the proportion of the

---

[14]There appear to be upcoming new measures, such as Brownian correlation, which are less dependent on linearity than are Pearson's correlation and phi. However, introducing these newer tests is outside the scope of this book.

other. So, referring to Equation 6.4 and calling phenotype **I** a success,

$$\hat{p} - Z_{\alpha/2}\sqrt{\hat{p}\hat{q}/n} \le p \le \hat{p} + Z_{\alpha/2}\sqrt{\hat{p}\hat{q}/n}\,,$$

$$0.733274 - 1.9600\sqrt{0.733274 \times 0.26626/1121} \le p \le 0.733274 + 1.9600\sqrt{0.733274 \times 0.266726/1121}\,,$$

$$0.733274 - 0.025889 \le p \le 0.73327 + 0.733274 - 0.025889\,,$$

$$0.70738 \le p \le 0.75916\,.$$

So, he could be 95% confident that the ratio of the dominant phenotype, **I**, was between 70% and 76% and, by subtraction, that the recessive phenotype, **i**, was between 24% and 30%.

Discussion: Without statistical significance, reporting an ES is not trustworthy, because a lack of statistical significance concedes that differences could have been due to chance alone. The above example relates to categorized discrete data. Despite its being largely responsive to linear relations, ϕ^2, which estimates the part of the variance which is due to non-random deviation from the hypothesized model or distribution, might be a more easily understood measure of effect size than would ϕ when $P\chi^2$ is significant for a $1 \times c$ table challenging goodness of fit. However the difference between the expected probabilities and the observed probabilities provide a good effect size measure for these 1×2 tables challenging fit, because the absolute value of the effect size for one category will always equal the absolute value of the effect size for the other. **Maybe** the CI was what he really needed rather than focusing on the failure to detect a statistically significant difference and its associated ES. However, in application, finding the CI observed here could be seen as an estimation procedure rather than a testing procedure.

Question 3 is an example as well as an exercise for and about variance based effect sizes that at least meet the requirements of some journals for goodness of fit.

Do not fear, we will eventually provide for power analysis regarding Pearson's chi square in Chapter 13.

12.5.3 ES for Pearson's χ^2, 2×2 or 1×2

There are a number of effect size measurements that are especially suitable for, or which refer specifically to 2×2 contingency tables for detecting association, and often for 1×2 tables for fit. They are useful because larger global tables can often be broken into smaller tables.

12.5.3.1 2×2 or 1×2, ES from likelihood of the outcome

In either a 2×2 or a 1×2 the probability difference, $p - p_0 = O_i/n - E_i/n$, can be calculated from any one cell. Its **absolute value** is an appropriate ES for many applications of the χ^2,

$$ES = |O_i/\mathbf{n} - E_i/\mathbf{n}| = |p_{O,i} - p_{E,i}|\,.$$

A 2×2 table has only one degree of freedom. Recall that if you know the observed value of one cell, and the marginal totals, the row totals and the column totals, then it works out that you can fill in all the values in the other cells. If the contents of one cell deviates from the expected by a certain amount, the other cells will deviate alternatively, plus or minus, **by the same amount**. So, if you know only the contents of one cell, and if you have observed statistical significance, you can use

$ES = |O - E|/\mathbf{n}$ from that one cell for the effect size. That is, the categories varied from the expected by $\pm ES$. This is like the probability effect size you have already seen for 1×2 tables, which also have only one degree of freedom. This is probably the **simplest** and **most intuitive** approach.

When this $ES = |p_{O_i} - p_{E_i}|$ is treated as a probability and the original **n** is used to estimate the standard deviation, it can be presented as as CI (Section 6.4 or Section 12.6). It can be multiplied times a reference value, say 100,000 as for presenting an estimate of the effect per 100,000 as in public health or in demographics. Likewise the limits, $L1$ and $L2$, of the CI can then be multiplied by a reference value (such as 100,000) to present the CI in say, original units per 100,000 .

12.5.3.2 Pearson's 2×2 (or 1×2) χ^2, ϕ as ES for association

The phi coefficient, already touched upon in Subsection 8.2.4, is commonly used as an ES because it measures mostly linear association, as does Pearson's correlation coefficient, R. Given statistical significance of Pearson's chi square, then phi square, ϕ^2, is a good estimate of R^2 for 2×2 tables, and can **also** be considered as an estimate of the **ratio** between the **variance due to association** (the mostly linear effect of one on the other) and the **total possible variance**.

A **signed phi coefficient**, $\phi_\pm$, the *mean square contingency coefficient,* is calculated as the square root of the Pearson's chi square value divided by the sample size,

$$\phi_\pm = \pm\sqrt{\frac{\chi^2}{\mathbf{n}}} \tag{12.10}$$

where **n** is the total sample size and the result is between $-1 < \phi_\pm < 1$ (Guilford, J. 1936). This is the same as Equation 12.7 except that the sign ($\pm$) is included for emphasis of the fact that in this application it has a sign. If most of the data falls **on the diagonal** (a and d), the relation, $\phi_\pm$, is considered **positive**, **+**, meaning that an increase in one variable is associated with an increase in the other variable (a direct relationship). However, if most of the data falls **on the off-diagonal** (b and c), the relation, $\phi_\pm$, is considered **negative**, **-**, meaning that an increase in one is associated with a decrease in the other (an inverse relationship).

a	b
c	d

Although computationally the Pearson correlation coefficient reduces to the phi coefficient, the interpretation of versions of the phi coefficient must be evaluated cautiously. Likewise for its square, which might make more sense intuitively. A value close to $\phi^2 = 1$ is **evidence**, not necessarily proof, of a strong relationship. For instance, if the data **falls equally** on the diagonal (a and d), **and** on the off diagonal (b and c), **almost any** ϕ between $+1$ and-1 is **possible**, and then this coefficient and its square are **untrustworthy** (Davenport and El-Sanhury 1991). You must **always check** the total of the diagonal versus the total of the off diagonal of your 2×2 tables to avoid erroneous conclusions. Although different for different indexes, this difficulty is among the many difficulties associated with the use of indexes for reporting ES. Editors should consider favoring CIs and simple differences in original units of measure wherever possible.

12.6 Pearson's χ^2 for Association as a Gestalt

Gestalt is not the sound of sneezing; it is a word that originated from German for a whole concept or picture, and without multiple comparisons to complete them, most pictures involving global tests would be incomplete[15]. Example 12.6.0.1 demonstrates what has been covered so far regarding association and a bit more. It includes a larger $2 \times c$ global table with CIs, ESs, and the use of 2×2 tables for planned comparisons. Unlike the ANOVA which appears to provide an improved variance estimate for subsequent multiple comparisons, the global Pearson's chi square does not provide an improved basis for multiple comparisons. But because this global test appears to be more dependable than an ANOVA, it is reasonable and conservative to require significance of it before initiating multiple comparisons. However, the individual comparisons are merely a series of independent chi square tests with a correction of the criterion by the Bonferroni (or Sidak) correction.

If possible, estimate the a priori sample size needed to attain a desired power for a reasonable minimum difference before obtaining the data. Even if you have no control over the sample sizes, a minimum sample size estimate will enable you to determine if you will need to combine data classes, if you need to obtain more resources, or if the test is even going to be practical. The actual power that controls the sensitivity of a Pearson's chi square test for a given reasonable minimum difference will be covered in the following chapter.

12.6.0.1 Example: Pearson's Chi Square for Association

Association of hate and location Let's include multiple comparisons to consider data from the Uniform Crime Reporting Program regarding hate crimes motivated by race or religion and see whether motive is associated with where they happen. Notice that this could be classified as a test of independence because the agencies collecting the data had no control over the amount or randomization of data in either the rows or columns. Hate crime related to race and religion is one factor ($r = 2$ categories), and where they occur is the other factor ($c = 4$ locations),

H0: THE FREQUENCIES OF HATE CRIMES RELATED TO RACE OR RELIGION ARE NOT ASSOCIATED WITH THE LOCATION WHERE THEY OCCUR.

HA: THE FREQUENCIES ARE ASSOCIATED WITH LOCATION.

Planned Comparisons: Assume that the **Office Building location** has been **a priori chosen** as the **control**, meaning that only its possible association with each of the other categories (locations) will be tested following the global test. So there will be only $k = 3$ a priori (preplanned) comparisons.

Criteria and Cutoffs: There are $(c - 1)(r - 1) = 3 \times 1 = 3$ degrees of freedom, and we a priori choose an experiment-wide $\alpha = 5\%$ for the global test, using $\chi^2_{0.05,\,3df} = 7.815$. The Sidak correction Equation 10.4 for $k = 3$ gives $\alpha' = 1 - (1 - \alpha)^{1/k}$ for each comparison. This gives $a' = 1 - \sqrt[3]{0.95} = 0.01695$. So from an online calculator[16], $\chi^2_{0.0170,\,1df} = 5.7013$ is the value that must be equaled or exceeded for statistical significance for **each** pre-planned comparison. (The Bonferroni correction would have gotten $\alpha/3 = 0.016667$, which would have required a slightly larger chi square value, $\chi^2_{0.0169,\,1df} = 5.7311$.) Or, assuming no dependence, we could a priori have chosen a comparison-by-comparison fixed $\alpha' = 1\%$ for an overall $\alpha = 1 - (1 - \alpha')^k = 2.71\%$, (via a reversed Sidak correction, Equation 10.1) rounded up to 3%, but we did not do that here. Also remember that when making such a choice in advance, we must be certain not to choose an α' such that $1 - (1 - \alpha')^k > 5\%$ overall.

---

[15] Pearson's chi square for association differs from Pearson's chi square for goodness of fit, and the two have to be treated separately in some instances.

[16] The online χ^2 calculator is from the Fourmilab Switzerland,
http://www.fourmilab.ch/rpkp/experiments/analysis/chiCalc.html

Methods: This data came from the US Department of Justice, The Uniform Crime Reporting (UCR) Program, as reported by the Federal Bureau of Investigation. It pertains to a grouping of locations and just two of many motives for hate crimes.

Analysis: First the contingency table,

Hate Crime Incidents *2009*

Bias/Locale	Office Bldg.	Jail/Prison	Conv. Store	Dept./Disc. St.	Total
Religion O	38	5	9	10	62
E	26.9439	9.5608	11.8785	13.6168	
Race O	55	28	32	37	152
E	66.0561	23.4393	29.1215	33.3832	
Total	93	33	41	47	214

Bias/Locale	Office Bldg.	Jail/Prison	Conv. Store	Dept./Disc. St.	Mean $\hat{p}$
P(race) = $\hat{p}_j$	0.5914	0.8485	0.7805	0.7872	$\bar{p} = 0.7103$

with the expected values calculated from the marginal totals. The probabilities in the detached lower table row are **not** the probability of the crime; they **are** the probability that a hate crime, in a **particular place, is motivated by race**. It could be seen as a sort of effect size (but not part of the contingency table).

The next table is for the calculation of $P\chi^2$ (not the actual contingency table).

Calculations

Bias/Locale	Office Bldg,	Jail/Pri.	Conv. Store	Dept./Disc. St.	$\sum(O-E)^2/E$
Religion $O-E$	11.0561	−4.5608	−2.8785	−3.6168	
$(O-E)^2/E$	4.53672	2.17561	0.69754	0.96068	8.37054
Race $O-E$	−11.0561	4.6085	2.8785	3.6168	
$(O-E)^2/E$	1.85050	0.88742	0.28452	0.39186	3.41430
$\sum(O-E)^2/E$	6.38721	3.06302	0.98207	1.35264	**11.78484**

The $O-E$ values in the columns sum to zero, and likewise for the rows. This is a handy check for hand calculation. The resulting global total is

$$P\chi^2 = 11.785,$$

with $(\mathbf{r}-1)(\mathbf{c}-1) = 3$ degrees of freedom. This is much larger than $\chi^2_{0.05,\,3} = 7.815$, so the global test is **statistically significant** at the 5% level. So, this rejects **H0** and accepts **HA**.

A graph such as the following stacked column graph, summarizing the data, should be plotted and included as part of the gestalt, **after** the analysis. Its inclusion will give the reader a better intuitive understanding of the results:

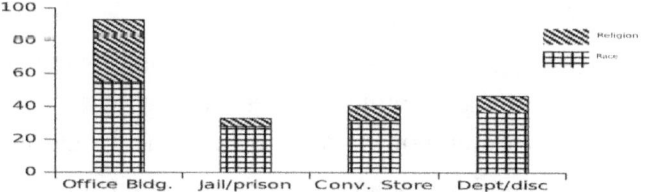

Because editors want graphs that stand alone, this might need an explanatory caption added to it. The lower part of each column is for race, the upper part is religion, and the locations are labeled at the bottom. However, their proportions are what is of interest. Graphics should be post hoc, meaning after the researcher has committed to a particular analytical strategy and completed all inferential tests. Otherwise the process, when applied to the data you intend to analyze, would be a priori data examination rather than true inferential statistical testing. Examining the data before analysis will often show apparent trends that are actually due to chance;

therefore in order to tell if these trends are real, it would be necessary to collect and analyze new data in a double blind fashion (the **original pre-examined data** should **never** be used for this).

Phi and Phi Square: The **effect size** reporting requirements of some journals might require a popular ES, and if **both** the number of rows and the number of columns are greater than 2, you could use Cramer's phi, $\phi_C = \sqrt{\frac{\chi^2/n}{\min(r-1,c-1)}}$. But the number of rows above is not greater than 2, so it reverts to (is equal to) the correlation-like phi, $\phi = \sqrt{\chi^2/n}$. So, using the $P\chi^2$ from above,

$$\phi = \sqrt{11.78/214} = 0.2347\,,$$

or an effect equal to the lack of independence that would be observed if there were roughly a 23% correlation. This is small for a correlation, but in light of the statistically significant χ^2, it might be seen as a meaningful ES. On the other hand, in light of the inadequacy of ϕ_C, is it a product of hypersensitivity due to an inadequately compensated large sample size?

Squaring this gives another ES, that is possibly more appropriate because there was no restriction on randomization. This one, **phi square**,

$$\phi^2 = 0.05505,$$

crudely estimates the proportion of variance which was not due to chance. It estimates that about 5.5% of the total variation was due to (roughly linear) association between the locales. Looking at this index, you might not think this a strong association. Unfortunately, because of their tendency to meaninglessly increase with the number of cells, the quantitative meaning of ϕ and ϕ^2 might seem obscure when applied to tables greater then 2×2. Even so, recall that the χ^2 value was statistically significant and that these ES values are drawn directly from it, so most journals might still see them as OK. **Do not use either** of them (ϕ or ϕ^2) for tables greater than 2×2 if you can find a **satisfactory** substitute (as when $c > 5$ in a $2 \times c$ table, in which case you might run a correlation and report R or R^2 as an ES).

CIs for p (uncorrected): If editorial policy will allow it, probability based CIs are better than ESs as points. For race, the estimated probabilities, $\hat{p}_j$, and the corresponding probabilities for religion, $\hat{q}_j = 1 - \hat{p}_j$, for each locale j, are:

Probability That Violence is Related to Race, $\hat{p}$, Versus Religion, $\hat{q}$		
Office Building	$\hat{p} = 0.5914,$	$\hat{q} = 0.4086$
Jail /Prison	$\hat{p} = 0.8485,$	$\hat{q} = 0.1515$
Conv Store	$\hat{p} = 0.7805,$	$\hat{q} = 0.2195$
Dep./Disc. Store	$\hat{p} = 0.7872,$	$\hat{q} = 0.2128$

You might report only a $\hat{p}$, or its CI, as the probability of the crime at each kind of site being motivated by race. These probability estimates are descriptive for possible effect sizes when $P\chi^2$ is significant, but they **do not** describe the strength of the reported statistical significance.

Although not part of the testing, a set of 95% confidence intervals may still be meaningful descriptions of effect size and thus of value to the readers. Because they are never adjusted for multiple comparisons, and because they are CIs, such things as whether they overlap, or include zero, **do not** have meaning with respect to statistical significance. Using

$$S = \sqrt{\hat{p}\,(1-\hat{p})\,/n} = \sqrt{\hat{p}\hat{q}/n}$$

as the estimate of each individual σ, the 95% probability CI that a hate crime at a given location is motivated by race is

$$\hat{p} - Z_{0.025}\sqrt{\hat{p}(1-\hat{p})/n} \leq p \leq \hat{p} + Z_{0.025}\sqrt{\hat{p}(1-\hat{p})/n}.$$

Because this is a normal approximation (Equation 6.10), t cannot be used. So, using $Z_{0.025} = 1.960$, the CI for the probability that a hate crime occurring in an office building (in effect, the control) is motivated by racial prejudice is

$$0.5914 - 1.960\sqrt{0.5914(1-0.5914)/93} \leq p \leq 0.5914 - 1.960\sqrt{0.5914(1-0.5914)/93}.$$

If you were using a calculator, you should minimally use at least four (better five) places to the right of the decimal point during the calculation. (Recall that counting numbers are considered accurate to an infinite number of figures.) However, you are justified in conveniently rounding up on the right and down on the left when you actually report the result.

The CI for the probability that this type of hate crime was motivated by racial prejudice in an **office building** is

$$0.4914 \leq p_{O.B.} \leq 0.6914.$$

The CI for the probability that this type of hate crime was motivated by racial prejudice in a **jail/prison** is

$$0.7261 \leq p_{J/P} \leq 0.9709.$$

The CI for the probability that this type of hate crime was motivated by racial prejudice in a **convenience store** is

$$0.6538 \leq p_{C.S.} \leq 0.9072.$$

The CI for the probability that this type of hate crime was motivated by racial prejudice in a **department/discount store** is

$$0.6702 \leq p_{D.S} \leq 0.9043.$$

Figure 12.1 is a graphical presentation of the 95% CIs for the proportion (the probability under our frequentist approach) of the hate crimes that were racially motivated.

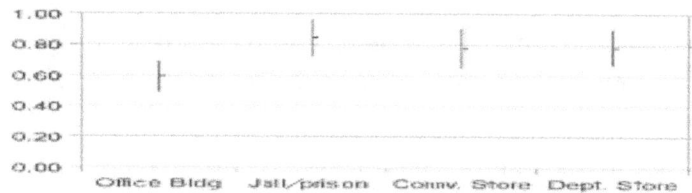

Figure 12.1: CIs for 2009 Racially Motivated Hate Crime Incidents

2×2 **Multiple Comparisons:** A global test can only detect that at least one of the pairs of factors (or treatments) is associated. Thus you will have to use a separate 2×2 table, a separate test, for each $P\chi^2$ comparison made. This, and especially the obvious lack of interaction between the locales, supports going forward with preplanned comparisons, using the slightly less conservative Sidak correction (rather than the slightly more expensive Bonferroni correction). Using 2×2 Pearson chi square contingency tables is the classical approach and is appropriate. Off. B (Office Building) can be compared with J/P (Jail/Prison) using a 2×2 table with

$1df$ freedom. Similar contingency tables will subsequently be constructed for each of the other comparisons to the Off. B control. They can each be tested using the 2×2 shortcut Equation 12.6,

$$P\chi^2 = (O - E)^2 \sum_{all\ i} \frac{1}{E_i},$$

and an **overall** (family wide) criterion, $\alpha = 5\%$ (the $(O - E)^2$ can refer to any one cell). **However**, this overall criterion must be split between $k = 3$ comparisons. So **applying the Sidak correction**, Equation 10.4, $\alpha' = 1 - (1 - \alpha)^{1/k}$, we share the probability among the individual 2×2 comparisons, using

$$\alpha' = 1 - (0.95)^{1/3} = 0.016952$$

for each. The critical $\chi^2_{\alpha'}$ value,

$$\chi^2_{\alpha'=0.016952,\ df=1} = 5.7013,$$

was found using an online calculator (http://www.fourmilab.ch). So, creating separate tables for O and E, the first χ^2 multiple comparison (separated into two parts) gives:

O	Off. B	J/P		E	Off. B	J/P	
Race	55	28	83	Race	61.26190	21.73810	83
Rel.	38	5	43	Rel.	31.73810	11.26190	43
	93	33	**126**		93	33	**126**

Each expected value was found from the first 2×2 contingency table (not from the global test), under the assumption of no association (there is both homogeneity and independence, one does not affect the other),

$$E_{i,j} = \frac{\left(\sum_{i=1}^{i=r} O_{i,j}\right)\left(\sum_{j=1}^{j=c} O_{i,j}\right)}{\sum_{i=1}^{i=r} \sum_{j=1}^{j=c} O_{i,j}},$$

for cell i, j. So

$$E_{1,1} = \frac{93 \times 83}{126} = 61.2619$$

for the first cell ($\mathbf{r}_{i=1}$, $\mathbf{c}_{j=1}$), upper left. The rest of the E values were found by subtracting from row or column totals. Then applying the shortcut, Equation 12.6,

$$P\chi^2 = (55 - 61.26190)^2 \left(\frac{1}{21.73810} + \frac{1}{61.26190} + \frac{1}{11.26190} + \frac{1}{31.73810}\right)$$

which gives $\chi^2 = 7.16111$,

which is statistically significant because it is **greater than** $\chi^2_{\alpha'=0.016952,\ df=1} = 5.7013$. Then, for the remaining two individual comparisons,

O	Off. B	C S		E	Off. B	C S	
Race	55	32	87	Race	55.75676	33.24324	87
Rel.	38	9	47	Rel.	39.24324	7.75676	47
	93	41	**134**		93	41	**134**

$$P\chi^2 = 4.46778 \quad NS$$

O	Off. B	D/DS		E	Off. B	D/DS	
Race	55	37	92	Race	53.81250	38.18750	92
Rel.	38	10	48	Rel.	39.18750	8.81250	48
	93	47	**140**		93	47	**140**

$$P\chi^2 = 5.31454 \quad NS$$

where NS signifies not statistically significant. As predicted by the statistical significance of the global test, at least one of the comparisons, in this case exactly one comparison, J/P, was statistically significantly different from the control, Off. B.

The corresponding phi effect sizes: Phi ES values in terms of correlation were found for the individual comparisons, using Equation 12.7, $\phi = \pm\sqrt{\chi^2/n_j}$, where n_j is the total for each 2×2 comparison, presented in the following table:

	ϕ	ϕ^2
Off.B J/P	0.2384	0.05683
Off.B CS	0.1626	0,03334
Off.B D/DS	0.1948	0.03795

The squares, ϕ^2, represent the proportion of variation due to association. This use of these effect size estimators is valid and accurate for 2×2 tables. You would only publish either ϕ or ϕ^2, **not both**.

Discussion: Even though the observed value in the cell with the smallest number in the global contingency table is just 5, the expected value in that cell is 11.2, which is greater than 3 (or 10 for more conservative analysts), as are all the expected values in all the smaller tables. So the χ^2 analysis is valid (Section 12.8).

The effect size in terms of phi, ϕ (or ϕ_C for three or more columns **and** three or more rows), which is correlation coefficient-like (R-like) is a commonly used ES for a Pearson's chi square test. Some researchers place little value on correlation estimates $< 70\%$, implying that they favor a coefficient of determination $\geq 49\%$ (or maybe 50%). This may relate to the alleged hypersensitivity of Pearson's chi square with very large samples. That is, these tests can be statistically significant **long before** the effect size is at all near such a value. Note that because the researcher has no control over the randomization in this example, it would be technically classified as a challenge of independence. Therefore, an estimate like the coefficient of determination estimate (R^2 or ϕ^2) of the effect on variance would be preferable to a correlation based ES. Phi square, ϕ^2, would be especially appropriate if you were providing an ES for every 2×2 multiple comparison, as was done in this example.

When using the normal approximation, the effect sizes of multiple comparisons are often more conveniently reported in terms of the probabilities (which, in a sense are the original units of measure). The normal approximation and Pearson's 2×2 tables both produce the **same result**. In either case, probabilities should be reported for the categories (represented as columns in this book) because of their value to other researchers, and **if there is statistical significance**, they (or their differences) might be seen as ES values, whether you choose to report ϕ or ϕ^2 for each comparison or not.

An ordinary table like Figure 12.3 of the probabilities that such race or religion related occurrences in each locale were motivated by race or religion might also be handy for your readers. Notice that every column sums to one (100%). Confidence intervals, **especially** for the probabilities, are very informative, but you had best include a caveat that they do not relate to the statistical significance (as though they were some form of multiple comparisons, which they

Bias/Locale	Office Bldg.	Jail/Prison	Conv. Store	Dept./Disc. St.
Race	0.5914	0.8485	0.7805	0.7872
Religion	0.4086	0.1515	0.2195	0.2128

Table 12.3: Probability that a hate crime is motivated by race or religion in each locale.

are **not**). However, depending on the context of the research, CIs for these probabilities might be much more informative than the best correlation-related ES indexes.

12.6.0.2 Even the above example is not a full gestalt.

This is still not a complete picture. This example is not fully under your control because the nature of the data leaves you with no control of the sample size, meaning that you have no control of the power. A problem that critics are likely to point out is that an **a priori, reasonable minimum difference** has not been stated. For publication of a power, you always need a reasonable minimum difference and a β or $1 - \beta$ as well as **proof that they were a priori chosen** (Chapter 13). Because $\beta = 20\%$ is as large as anyone would use, it might stand as an a priori β. Sometimes, in application, an a priori reasonable minimum difference might **preexist** in the form of an **economic value** (such as cost-return) **or similar logical value**. However, if not clearly preplanned, it must be the smallest possible such logical value.

But wait, we need to consider power (Chapter 13); is this data totally uncontrollable? Maybe, after running an a priori analysis, and finding that the total size of your otherwise unexamined sample for your uncontrolled yearly data is inadequate, you could combine several years, so as to reach a n estimated sample size above but as close to your estimated sample size, $\hat{n}$, as is possible. Better yet, combine a sufficient number of years so that you can sub-sample them at random. Otherwise, critics could challenge the lack of randomness.

12.6.1 An alternative approach:

In tests of association with random sampling and p values not very close to zero or one (ideally $0.25 < p < 0.75$, but closer is OK for very large sample sizes), the proportions follow a binomial model when used for pair by pair comparisons of counts. The expected values of these tests of association are calculated on the assumption under **H0** that the proportion of successes remain the same from group to group. So, it is possible to do these multiple comparisons using tests based on the normal approximation (Section 5.13). Instead of performing multiple Pearson's 2×2 chi squares, let's alternatively examine **multiple 'Z tests'** using the normal approximation.

Theory: Consider a group of tables, starting with a contingency table as for a Pearson's chi square and ending with a table relating to a normal approximation.

	c1	c2	$\sum_{\mathbf{r}}$
$\mathbf{r_1}$	$O_{1,1}$	$O_{1,2}$	$O_{1,1} + O_{1.2}$
$\mathbf{r_2}$	$O_{2,1} = n_1 - O_{1,1}$	$O_{2,2} = n_2 - O_{1,2}$	$\mathbf{n} - (O_{1,1} + O_{1.2})$
$\sum_{\mathbf{c}}$	n_1	n_2	$\mathbf{n}$

Binomial	$\mathbf{c_1}$	$\mathbf{c_2}$	$\sum_{\mathbf{r}}$
success	r_1	r_2	$r_1 + r_2$
failure	$n_1 - r_1$	$n_2 - r_2$	$\mathbf{n} - (r_1 + r_2)$
$\sum_{\mathbf{c}}$	n_1	n_2	$\mathbf{n}$

Group Probabilities of Success		
c_1	c_2	pooled
$\hat{p}_1 = r_1/n_1$	$\hat{p}_2 = r_2/n_2$	$\hat{p} = (r_1 + r_2)/n$
$(\hat{p}_1 + \hat{p}_2$ does **not** equal 'pooled')		

You can make each comparison in consideration of the last table of the three tables above (Group Probability). Barnes (1994) says the pooled probability, $\hat{p} = \frac{O_{1,1}+O_{1,2}}{n_1+n_2} = \frac{r_1+r_2}{n}$, is the best estimate of the probability of a success. This pooled probability, $\hat{p}$, including both locales, tentatively defines a binomial distribution upon which the variance estimate will be based. It is possible to test for association using a normal approximation,

$$Z = \frac{\hat{p}_2 - \hat{p}_1}{\sqrt{\hat{p}(1-\hat{p})\left(\frac{1}{n_1} + \frac{1}{n_2}\right)}}, \quad \hat{p} = \frac{r_1 + r_2}{n_1 + n_2}, \tag{12.11}$$

where we arbitrarily chose to put the successes in row 1. The number of successes, $r_j = O_{1,j}$, and each individual probability, $\hat{p}_1 = O_{1,1}/n_1$ and $\hat{p}_2 = O_{1,2}/n_2$, where n_2 is the total number of observations in category $_2$ (for instance, Off.B). The second display equation, $\hat{p}$, is the **pooled estimate** of the overall probability of a success including both columns (both categories). This approach uses a pooled standard deviation because the null hypothesis is that the distributions are the same (Snedecor and Cochran 1967). It is generally accepted that the sample size is sufficient for a normal approximation if $np \geq 5$ and if $n(1-p) > 5$ (some sources say ≥ 10 for both, and I agree with the more conservative 10 if $p < 0.25$ or $p > 0.75$) (Section 5.6). Well, this hate crime data at least meets the $np > 10$ requirement. By now, some readers may more clearly realize how Pearson's chi square can be seen as applying the normal approximation to obtain Z values and then summing the squares of these results. The following example will use the normal approximation directly (Subsubsection 12.6.1.1).

Equation 12.11 could also be written

$$Z = \frac{\frac{O_{1,2}}{n_2} - \frac{O_{1,1}}{n_1}}{\sqrt{\hat{p}(1-\hat{p})\left(\frac{n_2+n_1}{n_1 n_2}\right)}}, \quad \hat{p} = \frac{O_{1,1} + O_{1,2}}{n_1 + n_2},$$

but this expansion is not necessary for your calculations. A semi-verbal presentation of Equation 12.11 is

$$Z = \frac{difference\,between\,probabilities}{\sqrt{pooled\_p \times pooled\_q \left(\frac{1}{n_1} + \frac{1}{n_2}\right)}}.$$

So, for an $\alpha = 0.05$ level comparison of Office Building to Jail/Prison to determine if there is a difference in probability that a hate crime was associated with race rather than religion,

$$Z_{\text{Off.BJ/j/p}} = \frac{0.2571}{\sqrt{0.6587(1 - 0.6587)\left(\frac{1}{93} + \frac{1}{33}\right)}} = 2.6761,$$

which is obviously larger than $Z_{0.025} = 1.960$, and therefore statistically significant for association if you had a priori chosen $\alpha = 5\%$ and did **not** adjust for more comparisons.

12.6.1.1 Example: Further Examination of Pearson's Chi Square Alternative

Association of Hate and Location: Normal Approximation Here are some tables that use multiple comparison corrected values for $k = 3$ comparisons of office building (Off Bldg) with other places. These values for calculation are from the global table in the example in Subsection 12.6.0.1 to form a 2×2 subset where n_{r_1}, n_{r_2}, n_{c_1} and n_{c_2} are the marginal row and column totals, respectively. The first row was **arbitrarily chosen** to be treated as successes, but it could just as easily have been the second row, in which case the first row would have represented failures.

	Off. Bldg.	J/P	Σ_r
race	$O_{1,1} = r_1 = 55$	$O_{1,2} = r_2 = 28$	$n_{r_1} = 83$
religion	$O_{2,1} = !r_1 = 38$	$O_{2,2} = !r_2 = 5$	$n_{r_1} = 43$
Σ_c	$n_{c_1} = 93$	$n_{c_2} = 33$	$\mathbf{n} = 126$

Analysis: The chosen a priori family-wide criterion was $\alpha = 5\%$. However, a correction was necessary to evaluate such multiple comparisons. The Z values were calculated as

$$Z_{\text{Off.B J/p}} = \frac{0.2571}{\sqrt{0.6587\left(1 - 0.6587\right)\left(\frac{1}{93} + \frac{1}{33}\right)}} = 2.6761,$$

$$Z_{\text{Off.B CS}} = \frac{0.1891}{\sqrt{0.6493\left(1 - 0.6493\right)\left(\frac{1}{93} + \frac{1}{41}\right)}} = 2.1139,$$

$$Z_{\text{Off.B D/D}} = \frac{0.1958}{\sqrt{0.6571\left(1 - 0.6571\right)\left(\frac{1}{93} + \frac{1}{47}\right)}} = 2.3048.$$

The inputs and results of the calculations appear in this table,

RACE Other Place	$\hat{p}_1$ Off. B	$\hat{p}_2$ Other Place	$\mathcal{D} = \hat{p}_2 - \hat{p}_1$ Effect Size	$\hat{p}$ (Pooled for Locale)	**Z**
Off.B J/P	0.5914	0.8485	0.2571	$\frac{83}{93+33}$=0.6587	2.6761
Off.B CS	0.5914	0.7805	0.1891	$\frac{87}{93+41}$=0.6493	2.1139
Off.B D/DS	0.5914	0.7872	0.1958	$\frac{92}{93+47}$=0.6571	2.3048

where calligraphic $\mathcal{D}$ estimates the true difference, δ_p, between the probabilities, and $\mathcal{D}$ may be reported as the ES, the point estimate of the difference, when statistical significance is found. To evaluate such Z values when used for multiple comparisons, we should have, a priori, made one of two corrections:

Bonferroni's correction: The Bonferroni correction provides a tiny bit less power than the Sidak correction but is also less susceptible to dependence. It produces very similar results, is the most common in computer packages, and is slightly more intuitive than the Sidak correction. As mentioned before, it corrects an overall α divided between two tails. In this two tail, $\alpha = 5\%$ example, this gives $(\alpha/2)' = (\alpha/2)/k$. So, $(\alpha/2)' = 0.025/3 = 0.0.008333\ldots$ for **each** of the three comparisons. I used RStudio to find that $Z_{(\alpha/2)'/3} = Z_{0.008333,1} = 2.3940$ for 3 multiple comparisons with an overall 5% probability of getting a Type I error.

or Sidak's correction: You might decide that surely there is no dependence between the results for one location and for another. That is, they do not seem capable of affecting each other's outcomes. Then, if you are justified in assuming the absence of interaction, the Sidak correction could be used instead. So, the Sidak correction for each tail of the comparison is $(\alpha/2)' = 1 - (1 - \alpha/2)^{1/k}$, giving $(a/2)' = 1 - \sqrt[3]{0.975} = 0.008476$. From the same trusted online calculator, $Z_{0.008476} = 2.3877$ is the value that must be equaled or exceeded in order to declare statistical significance for a given comparison out of the $k = 3$ multiple comparisons. The Sidak correction gives a little bit more power with a greater number of comparisons. However, the finding of statistical significance would not be noticeably more likely for this small sample size.

Two tail multiple comparison results: Given that a Bonferroni corrected cutoff of $Z_{0.008333} = $ **2.3940** was a priori chosen for each comparison, you would get the following results,

compare	Z	Result
Off.B J/P	2.6761	**Sig**
Off.B CS	2.1139	NS
Off.B D/DS	2.3038	NS

where **Sig** is used to mean statistically significant and NS means not statistically significant, at the $0.008333\ldots\%$ per comparison level used in this example. In this **two tail comparison**, only one 'treatment', J/P, was statistically significantly different from the control, Off.B. You definitely can also report the direction (larger or smaller) of the statistically significant difference. Note that for two tail multiple comparisons, Z or χ^2 will give the same result. Although proportions are not exactly units of measure, the estimated effects are equivalent to proportions being estimated, $\mathcal{D} = \hat{p}_2 - \hat{p}_1$, seen in terms of probabilities from a frequentist viewpoint. **Two tail** multiple comparisons are the most **common** with Pearson's chi square, but you are not restricted to two tail comparisons.

One tail multiple comparison results: Of course the decision to use one tail comparisons must be made **before** the data is examined or analyzed and you have **a priori selected which tail**, via a hypothesis pair, for instance:

H0: THERE IS NO DIFFERENCE IN HATE CRIMES FROM THE CONTROL, OFFICE BUILDING

HA: A GREATER PROPORTION OF SUCCESSES (HATE CRIMES) OCCUR IN THE OTHER LOCALE
Then, for a **one tail** set of $k = 3$ comparisons, the Bonferroni corrected cutoff, $\alpha' = 0.05/3 = 0.01667$, means that $Z_{0.01667} = \mathbf{2.1281}$ (rounded up) is used to test for statistical significance at an overall 5% probability of getting a Type I error. Given statistical significance, a positive $Z \geq 2.1281$ indicates that the locale being compared has a statistically significantly **greater** proportion than does the control, Off.B. Summary results appear in the following table:

compare	Z	Result
Off.B J/P	2.6761	**Sig**
Off.B CS	2.1139	NS
Off.B D/DS	2.3038	**Sig**

All of the differences are in the positive, $+$, direction ($Z > 0$), but observed direction only has meaning if your hypothesis includes that direction (in this instance, $+$) and then only for those comparisons which are statistically significant. Notice that one tail statistical significance is always easier (in terms of sample size cost) to obtain than is two tail statistical significance. However, unless you can otherwise justify it, you might wish to have dated notarized records to verify that you a priori chose a particular tail.

ES: The **effect size**,

$$\mathcal{D} = \hat{p}_2 - \hat{p}_1 , \tag{12.12}$$

can now be reported for **each** statistically significant comparison as an ES in terms of probability, or proportion. Although relative frequencies are not exactly original units of measure, the probability difference being estimated, $\mathcal{D} = \hat{p}_2 - \hat{p}_1$, can be seen from a frequentist view as representing a difference in the relative frequency, which is the best available estimate of the true difference, $\delta_p = p_2 - p_1$. The probabilities being compared can be estimated and reported using $\hat{p}_1 = n_{r_1}/\mathbf{n}$ and $\hat{p}_2 = n_{r_2}/\mathbf{n}$ for each comparison, $\mathcal{D}$. If you multiply p_1 and p_2 times 1000, then you could report the difference between them as an ES in terms of hate crimes by motivation per thousand crimes at different kinds of locale.

CIs for differences Reporting confidence intervals for the true difference, δ_p, for each comparison might be a particularly informative way to both view and present results, combining concepts and approaches from Equation 6.7 and from Section 6.4. Start with a pooled (overall) probability estimate, $\hat{p} = \frac{r_1 + r_2}{n_1 + n_2} = \frac{O_{1,1} + O_{1,2}}{n_1 + n_2}$, for each comparison in order to estimate a normal approximation based CI. Then you need an equation which arises from having put the pooled standard error estimate (denominator) of Equation 5.6.0.2 into Equation 6.10 to get the CI,

$$\mathcal{D} - Z_{\alpha/2}\sqrt{\hat{p}\hat{q}\left(1/n_1 + 1/n_2\right)} < \delta_p < \mathcal{D} + Z_{\alpha/2}\sqrt{\hat{p}\hat{q}\left(1/n_1 + 1/n_2\right)}. \tag{12.13}$$

Given a suitable sample size, especially in a situation like this where you do not control the sample size, many readers might be happy just getting uncorrected multiple comparison confidence intervals for the differences, whether statistical significance testing occurs or not. Consider the following table with its equivalent graph:

ES	L1	L2	ES
Off.B J/P	0.06999	0.4242	0.2471
Off.B CS	0.04228	0.3359	0.1891
Off.B D/DS	0.05377	0.3378	0.1958

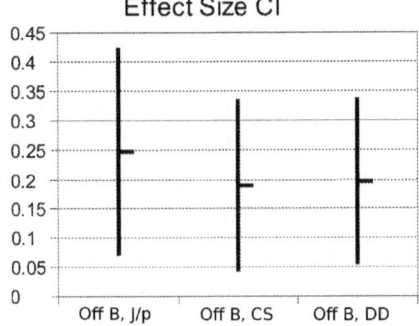

The point estimates, $\mathcal{D}_j$, of the effect sizes are marked as ticks on the lines representing their CIs. But the truth is that you can only be 95% confident that a particular CI even happened to **entrap** the actual effect size; you do **not really know** its true value. That is, it really only gives you a minimum boundary, $L1$, and a maximum boundary, $L2$, that you can be 95% certain include the difference. Of course you are taking a 5% risk, which equals a **one in twenty** risk, of being wrong **each time**. It might be prudent to use much larger sample sizes and maybe aim for 99% CIs based on a priori power analysis. That would seem particularly appropriate where there are $k \leq 5$ comparisons.

Discussion: Pearson's test for 2×2 tables are equivalent to simpler normal approximation tests. So, the squares of the normal approximation Z values **each equal** the corresponding χ^2_{1df} values. The one tail versus two tail comparison of office building (Off.B) vs department/discount store (D/DS) illustrates the fact that it is easier to get significance with a one tail test; it was statistically significant for one tail, but not for two tails. This might be useful just as long as the decision to test one tail or two is **before** you have any knowledge about the sample data (it must be an a priori decision).

The products of the total sample sizes and the estimated values are all $np > 10$, which would be conservative and would tend to compensate for some probabilities being awkwardly low ($p < 0.25$). So, most authors would agree that the normal approximation, or $P\chi^2$, should work if this is true.[17] However, for comparing the differences, pooling the probabilities might introduce some small inaccuracy. Although these results are only accurate to three figures, ES values based on either small sample sizes or extreme probabilities are made a little more acceptable by appropriately rounding up the $Z_{\alpha/2=0.025}$ to a conservative $Z \approx 2$ for each two tail 95% CI.

This example is still not a complete picture (not a gestalt), particularly because it **lacks** a consideration of power (coming soon).

> **CAUTION:** There would be **pitfalls** to interpreting **either** the **inclusion or non inclusion of zero** in a confidence interval as justifying a declaration of statistical significance. CIs and statistical significance are in separate universes; we could call them descriptive, and inferential, respectively.

Even though CIs are not practical to associate with statistical significance, CIs for the effect size in terms of proportions (probabilities from a frequentist view) might be more informative for Pearson's chi square than the best regression-related ES indexes, such as phi. Question 5 illustrates this approach for ES values estimated as $\mathcal{D}$ values for a goodness of fit test.

12.6.2 Measures that might relate to Pearson's 2×2 χ^2 and ES values

In specific fields, 2×2 tables can also provide a variety of measures which could be used as effect size measures. These descriptors are generally measures of association, and some can be tested or presented as CIs. However, you will have to refer to specialized books and other resources in fields such as epidemiology or machine learning to expand your understanding and application of some of them.

On the other hand, some authors question whether there is any generally applicable ES such as Cohen's Kappa for 2×2 categorical tables (e.g. Sim and Wright 2005). Some critics even make a more general statement that effect sizes regarding categorical data should **not**, in general, be emphasized by authors or taken overly seriously by readers (also see: Kilem 2002).

Measures which might be viewed as resembling phi:

The following might be useful as effect sizes or as reasonable minimum differences, or in some applications they might be the information you want. Although not covered in detail here, it is possible to estimate CIs for most such measures as well as for phi. Especially for some of them, such as the risk ratio and odds ratio (both below), this is often a very good idea.

---

[17]Box, Hunter and Hunter (1978) say that for Pearson's chi square to be valid it is necessary that

$$\left| \frac{1}{\sqrt{n}} \left(\sqrt{q/p} - \sqrt{p/q} \right) \right| < 0.3 .$$

So, for the most extreme pooled probability here, the one associated with the ES for Off. B, D/DS,

$$\left| \left(1/\sqrt{140} \right) \times \left(\sqrt{0.3429/0.6571} - \sqrt{0.6571/0.3429} \right) \right| = 0.005061,$$

and that is less than 0.3.

12.6.2.1 The odds ratio

Recall that the **odds** is a size measure that is historically a gambling term; it is the ratio of the probability that an event, A, will happen, to the probability that it will not happen, $\mathcal{O} = p_A / (1 - p_A) = p_A / q_A$. It originated as a popular way to discuss either chance or returns versus risk in an intuitive manner. Odds must not be mistaken for frequentist probabilities. If there is 1 chance in 5 of sampling a red candy from a bowl in a dark room, then the probability is $p = 1/5 = 0.20$, but if you make a bet at the horse track at $1 : 5$ odds (also ambiguously written $1/5$), then in terms of probability, the odds can be seen as $\mathcal{O} = p / (1 - p) = 0.20/0.80 = 0.25$. If you bet one dollar at the race track at $1 : 5$ odds and win, you get it back with 5 more.

The ***Odds ratio*** is the ratio between two odds. It is popular in medical fields, especially epidemiology and patient outcomes. If you are in such a field you would be well advised to become familiar with odds and the odds ratio. It is important because of its common use in medicine. It is a ratio of the odds of two related occurrences, and is often used in medicine to compare the relative odds of the occurrence of the **outcome** of interest (e.g. disease or disorder), given **exposure** to the treatment of interest (e.g. coming into contact with a toxin, using a medication, exposure to high temperatures, etc.). The odds of success in the **treatment** group, $\mathcal{O}_1$, divided by the odds of success in the **control** group, $\mathcal{O}_2$, is an **odds ratio**,

$$\mathcal{O}_R = \mathcal{O}_1/\mathcal{O}_2 = (p_1/(1 - p_1))/(p_2/(1 - p_2)) = (p_1 q_2)/(p_2 q_1).$$

This is more easily calculated as

$$\mathcal{O}_R = \frac{a\,d}{b\,c}, \tag{12.14}$$

where the a, b, c, and d lettering refers to the table:

		Outcome	
		+	-
Exposure	+	a	b
	-	c	d

Compared with $\mathcal{O}_R = 1$ when there is no (observed) effect:
If $\mathcal{O}_R > 1$, exposure is associated with higher odds of outcome.
If $\mathcal{O}_R < 1$, exposure is associated with lower odds of outcome.
Odds ratios are positive, and an odds ratio equal to $ad/bc = 1$ suggests that the condition or event under study is equally likely to occur in both groups. That is, the odds ratio equals one if the categories are independent. The odds ratio, $\mathcal{O}_R$, does not depend on which categorical input you choose for the columns and which for the rows because $ad/bc = ad/cb$. That is, rotating the table about its principal axis does not change the result of this particular calculation. Without further explanation here, the log to the base e (ln) of standard error for the odds ratio is found via

$$S_{ln(\mathcal{O}_R)} = \sqrt{\frac{1}{a} + \frac{1}{b} + \frac{1}{c} + \frac{1}{d}}.$$

So, the 95% CI for the odds ratio is[18]

$$\exp\left(ln\left(\widehat{\mathcal{O}_R}\right) - 1.96\, S_{ln(\mathcal{O}_R)}\right) < \mathcal{O}_R < \exp\left(ln\left(\widehat{\mathcal{O}_R}\right) + 1.96\, S_{ln(\mathcal{O}_R)}\right) \tag{12.15}$$

---

[18]Recall that $ln(x)$ is the natural log of x and $\exp(x) = e^x$. That is, $\exp(ln(x)) = x$.

(Daly 1998). Notice that the hat, $\widehat{}$, distinguishes the estimate from the true $\mathcal{O}_R$. The two extremes of Equation 12.15 are often called $L1$ and $L2$, as for any CI.

There is a problem even when publishing for an audience of medical practitioners, etc., because the odds ratio is often **confusingly large** (A'Court, Stevens, and Heneghan 2012). Viera (2008) indicates that this problem of exaggeration occurs when the outcome occurs in **more than** 10% of the unexposed population. It has been suggested that the odds ratio should **only** be presented as a measure of effect size when the risk ratio (below) **cannot** be estimated directly (Taeger, Sun and Straif 1998, Holcomb, Chaiworapongsa, Luke and Burgdorf 2001, Sinclair and Bracken, 1994).

> **CAUTION:** The odds ratio is just that, a ratio of the initial odds to the resulting odds. If this ratio is large and the initial likelihood is minute, the result can still be minute. Odds of two in ten million is double the odds of one in ten million, an odds ratio of 200%. However, both of these odds are very small. The public, especially at the hands of the media, may get the idea that a huge threat is looming, when the threat is still minute; this has actually happened in situations involving rare maladies and side effects of beneficial medications.

12.6.2.2 Risk ratio (relative risk)

The *risk ratio*, also called *relative risk*, is a less potentially misleading and more intuitive effect size commonly used in epidemiology. It is a single positive number referring to a particular outcome when exposed versus the same outcome when not exposed. It can be written

$$R_R = P(exposed \to outcome)/P\left(not\_exposed \to outcome\right).$$

It describes the strength of association or non-independence between two binary data values. It is the ratio of the probability of success or failure for each group. So, enter the results in a 2×2 table as

	Effect	No Effect
Exposed	a	b
Not Exposed	c	d

,

where the rows are the numbers exposed and the columns are the numbers of presumed outcomes. (The rows could also be treatment and the columns could be success and failure.) Then

$$R_R = \frac{a/\left(a+b\right)}{c/\left(c+d\right)} \tag{12.16}$$

(Cornfield, J. 1951). An $R_R < 1$ means the event is **less likely** to occur in the exposed (or experimental) group than in the not exposed (or control) group; and an $R_R > 1$ means the event is **more likely** to occur in the exposed group than in the not exposed group. Without further explanation here, the log to the base e of standard error for the risk ratio is found via

$$S_{ln(R_R)} = \sqrt{\frac{1}{a} + \frac{1}{b} + \frac{1}{a+b} + \frac{1}{c+d}},$$

then the 95% CI for the risk ratio is

$$\exp\left(ln\left(\widehat{R_R}\right) - 1.96\, S_{ln(R_R)}\right) < R_R < \exp\left(ln\left(\widehat{R_R}\right) + 1.96\, S_{ln(R_R)}\right)$$

(Daly 1998).

Although commonly applied in medicine, especially epidemiology, R_R can be used for binary outcomes anywhere when the probability of the outcome is **low**. When the outcome occurs in less than 10% of the unexposed population, the odds ratio, $\mathcal{O}_R$, provides a reasonable approximation of the risk ratio, R_R (Viera 2008). Question 7 is about the risk ratio.

> **CAUTION:** Relative risk can be a scientifically very illuminating value even with especially rare events. But when the denominator is very small (as for the chance of a rare cancer), a very high relative risk can be of little concern in application. An example is the fact that in a population of young women taking birth control pills, the risk of a thrombosis is very rare, but newspapers can get great sales and cause a panic by claiming that a new birth control will almost double the 'risk' of it happening.

12.6.2.3 Inter-rater reliability

A popular index of *inter-rater reliability*, *inter-rater agreement*, or *concordance*, is Cohen's kappa (Cohen 1960). It attempts to score how much homogeneity, or consensus, there is in the ratings given by judges. It was proposed to try to refine scales for measuring a particular variable, or to assess the need to retrain the raters (judges) so they choose in the same manner. Its equation is

$$\kappa = \frac{p_O - p_E}{1 - p_E},$$

where p_O (that's an O, not a nought) is the probability calculated from the observed number of yes choices, and p_E is the hypothesized, or expected, probability of a yes if **both raters choose independently** (κ is the Greek letter *kappa*). So, this estimator is yet another ratio; it can be verbally expressed as

$$kappa = \frac{P\,(observed\,agreement) - P\,(hypothesized\,agreement)}{100\% - P\,(hypothesized\,agreement)},$$

which gives correlation like values between -1 and 1. In application, negative values are counted as zero agreement and only values of $\kappa \geq 0.80$ are considered potentially useful as matched judges (McHugh 2012). To see how to fill in the above probabilities, consider the following table format for recording how judge X and judge Y each rate the same **n** things, or contestants:

		Judge X		
		yes	no	Σ
Judge Y	yes	a	b	a+b
	no	c	d	c+d
	Σ	...	...	n

Then $p_O = (a + d)\,/\mathbf{n}$, the estimated (from observation) probability of both raters agreeing (both yes or both no). And $p_E = [(a + b)\,/\mathbf{n}] \times [(a + c)\,/\mathbf{n}]$, the product of the probabilities of yes if the raters were really choosing independently. **Beware:** if used as an ES, kappa cannot be readily and validly converted to other ES measures (contrary to what Cohen claimed in 1968), and especially must not be used for estimating sample size.

12.6.2.4 Matthews correlation coefficient

In **machine learning** research, a homology coefficient (another estimator of agreement) is generally reported as M_{cc}, the **Matthews correlation coefficient** (Matthews 1975). It is intended for comparing the positive vs negative category to whether it is true (correctly classified) or false (incorrectly classified) in a table such as:

<table>
<tr><td></td><td></td><td colspan="2">Machine says</td><td></td></tr>
<tr><td></td><td></td><td>Positive</td><td>Negative</td><td></td></tr>
<tr><td rowspan="2">Really were</td><td>Positive</td><td>TP</td><td>FN</td><td>TP + FN</td></tr>
<tr><td>Negative</td><td>FP</td><td>TN</td><td>TN + FP</td></tr>
<tr><td></td><td></td><td>TP + FP</td><td>TN + FN</td><td></td></tr>
</table>

This is evaluated using the equation

$$M_{CC} = \frac{TP \times TN - FP \times FN}{\sqrt{(TP + FP)(TP + FN)(TN + FP)(TN + FN)}}, \tag{12.17}$$

where T =true, F =false, P =positive, and N =negative. So, you could say that it is the difference between the cross products divided by the product of the marginal totals. A value near $M_{cc} = +1$ tends to mean close to complete agreement, and an outcome near $M_{cc} = -1$ tends to mean close to total disagreement. Question 9 is about this, and the answer, after you try it, is an example of using it.

12.7 $1 \times c$ With $c > 2$ Pearson's χ^2, Genetic Ratios

Section 9.2 of Snedecor and Cochran (1967) mentions an instance in which Lindstrom (1918) suspected that the homogeneous recessive (which had the largest deviation) in a cross expected to fit a 9:3:3:1 model had a reduced survival rate leading to statistical significance for lack of fit. So, he reran the calculation without including it in either the O or the E, and the ratios of the other phenotypes failed to be significant. Unfortunately, as Snedecor and Cochran point out, this last step was clearly post hoc and thus invalid, except to suggest a hypothesis for future crossing experiments.

Low survival rate homogeneous recessives are not uncommon. So, since this might be a source of dependence, you might wish to use the Bonferroni correction for preplanned multiple comparisons for some genetic models.

12.7.1 Yates' correction

This is something that you need to know about because it is so common in literature. Yates (1934) proposed a continuity correction to adapt for the fact that when using χ^2 after the fashion of Pearson's chi square, you are using a continuous distribution to test discrete data. He offered the following continuity correction of Pearson's chi square:

$$P\chi^2_{Yates} = \sum \frac{\left(|O - E| - \frac{1}{2}\right)^2}{E}. \tag{12.18}$$

Snedecor and Cochran (1967) recommend it for tables no bigger than 2×2. Based on computer simulations, Veldman and McNemar (2007) claim that Yates' correction with one degree of freedom continually yields a better estimate of the exact probability of the discrete event concerned when used in conjunction with the usual tables of cumulative chi-square values. However, Camilli and Hopkins (1978) reported that "Yates correction decreases the accuracy of probability statements when either or both marginals are not fixed." In any case, this continuity correction is now generally considered to overcorrect and to result in an excessive loss of power. But some authors suggest that you might consider using it for small samples, $\sum O < 30$, with one degree of freedom. So, for a 1×2 or 2×2 contingency table with small sample size, it used to be common to use a Yates-corrected shortcut equation (Equation 12.6),

$$P\chi^2 = (|O - E| - 0.5)^2 \sum \frac{1}{E_{all\,j}},$$

where the first $O - E$ is from any one of the four cells. If at all possible, you should avoid using such small samples because they tend to waste resources, with little probability of a meaningful return. With larger samples, the effect of the Yates' correction tends to be minimal. In the classroom this is an additional detail for students to deal with, and may not be of much use in application anyway. **Do not use Yates correction if you intend to use ϕ or ϕ^2 as an ES.**

Incidentally, the typical simple phenotype table appearing in the Mendelian example (Subsubsection 12.5.2.1) for a simple Mendelian model is **not** a four cell table. It is a **two cell** (1×2) table.

12.8 Validity of $P\chi^2$

The validity of Pearson's chi square test when it involves counts, which are, of course, discrete data, is sometimes debated on the basis of the continuous nature of the χ^2. The truth is that the approximation of the chi square distribution for counts via $P\chi^2$ requires the expected values to be large enough. For this to satisfactorily meet the requirements of the approximation, many sources agree that you must design to meet the following minimum requirements:

- no cells can have an expected count of zero,

- as a general rule, the expected count should be greater than 5 in each cell of a two by two table (some authors say 10)

- there should be 5 or more in 80% of the cells in larger tables

Sometimes these requirements can be accommodated by combining some of the categories. If these requirements are not met for one tail 2×2 tests, or two tail 2×2 tests, **but** both of the row marginal totals or both of the column marginal totals are the same, you might be able to use Fisher's Exact test (Section 14.8). When you have no control over n, There is controversy, justified by widespread experience, over whether Pearson's chi square is attaining statistical significance too often when sample sizes are much above $n = 400$ or 500 subjects. Thus the above crime data might have been at or pushing the limits of acceptability. A discussion of this issue follows in the next section.

12.8.1 $P\chi^2$, Testing when $n \gg 400$?

The symbol $\gg$ means much greater than. Epidemiologists and some other researchers which traditionally used Pearson's chi square have become concerned that it is so often statistically significant when **n** $\gg$ 400. Indeed, SAS sometimes gives a warning regarding this phenomena when sample size calculations produce very large values of **n**. With a bit of scratch calculation you could avoid this apparent hypersensitivity to large sample size by an a priori adjustment of your reasonable minimum difference to keep your samples smaller, but then, especially for tables larger than 2×2, your minimum reasonable difference will usually be too big. So, this observation by the folks who most often use Pearson's chi square for inference is alarming.

Even more alarming, Kamran Siddiqui (2013) says, "The [Pearson's] chi-square is sensitive to sample size, its significance becoming less reliable with sample sizes above 200 or less than 100 respondents. In large samples, differences of small size may be found to be significant, whereas in small samples even sizable differences may test as nonsignificant." Hair, et al suggest that even if the $P\chi^2$ is a bit hypersensitive at the sample size you use, a high p_value is good because "it indicates that the observed model is not significantly different" from what was expected (1998). But be careful not to publish or report such a purely speculative acceptance of the null hypothesis.

A lot of folks actually use Pearson's chi square merely to get a p_value as easily obtained evidence against the outcome for which they hope (as a check). If you select the sample size for the Pearson's Chi square, and the overall sample is greater than $n = 400$ or less than $n = 100$ (or maybe outside $100 \leq n \leq 200$), you might be better off using a G test (Section 12.10) or maybe you can adapt some sort of big data technique (I just thought of this July 2, 2021).

When your sample size is larger than about $n = 400$, some statisticians might suggest that you consider whether your question can be better answered by an ANOVA, maybe an RCBD, or by a correlation. A huge variety of often complex-design-oriented ANOVAs are possible, using log transformed ratios of percents of counts. However, parametric tests, especially the ANOVA, are dependent on normality, which is supposedly attained via the Central Limit Theorem, and if that were true, then very big, not merely large, samples would be needed. Because of the sensitivity of ANOVAs to non normality (as is evidenced by the Monte Carlo tests reported by Blanca et al in Subsubsection 9.3.0.1, when the reader applies a strict maximum cutoff of 5.0% as opposed to the liberal cutoff definition, which can be as high as 7.5%, proposed by Bradley in 1978), statisticians are not agreed on the appropriateness of using an ANOVA as a substitute for Pearson's chi square. Some related difficulties can be reduced by keeping the number of treatments as small as practical.

The G test (Section 12.10) is an alternative to using Pearson's chi square for large sample inferential testing. It is very much like the chi square but is not presently widely suspected to be hypersensitive at large sample sizes. It is also less affected by low cell counts. But it is not readily converted to a coefficient of determination-like effect size.

So, we see that two of the most popular global tests, ANOVA and large $P\chi^2$, appear to be of limited accuracy. Perhaps we should try to keep it simple rather than trying to compare everything at once.

12.9 Which Table Based Test for Inference?

It is in conformity with common practice to use Fisher's exact test (Section 14.8) for certain unavoidable very small samples, Pearson's chi square for moderately big samples, and the G test (Section 12.10, below) for really big tests. However, I would estimate the sample sizes for any of these by **using the method** suggested for Fisher's chi square (Section 13).

12.10 The G Test

The **G test** is an alternative to Pearson's chi square that might have less of a tendency to hypersensitivity and that is becoming more popular than Pearson's chi square. It is calculated by

$$G = 2 \sum_{all\, i} \left(O_i - ln \left(\frac{O_i}{E_i} \right) \right),$$

where ln is the natural logarithm. Although G comes at the question from a different viewpoint than the chi square, the statistical significance of G can be found from an ordinary χ^2 table. Except for very small samples, it is at least as good as Pearson's chi square, but not as familiar. (For more on its application see Sokal and Rohlf 2012. Wikipedia, https://en.Wikipedia.org/wiki/G-test, gives a bit of theoretical information and some handy citations.) Power analysis, either before or after the test, should be satisfactory if you proceed as you would for a Pearson's chi square.

Pearson's chi square became popular back when the use of log tables was so tedious as to be onerous, and Pearson found a way around needing them. Although Pearson's approach remains superior for moderate sample sizes, G is superior to Pearson's chi square for larger samples and is a better approximation of what Fisher intended (Harremoës and Tusnády 2012). Deriving it here would go beyond the assumed knowledge of many readers of this book. A derivation of how the chi-square test is related to the G-test can be found in Hoey (2012). Pete Hurd and Ian Fellows have provided a true G test as the `likelihood.test()` function in the **Deducer package** of the R language. Other statistical software programs don't always deliver the actual G test even when they indicate they will.

Questions for Ch 12

Write the equations where appropriate, and then fill in the values. **Odd numbered questions** have worked answers at the back of the book and can be studied as additional examples. The data is mostly fictitious.

1. An online entity, APUS07, at the American Public University, had the idea of applying statistics to counts of candy colors versus customer color preferences. I took a slightly different approach to test to see if that distribution corresponded to an informal customer survey of colors of M and M candies, using a 'random sample' of $n = 361$ candies (OK, it was really a locally purchased bag). http://mms.com/us/about/products presented a customer preference survey on May 5, 2012 with the following percentages of customers preferring various colors: yellow 16%, red 25%, green 20%, blue 31%, and orange 8%. It seemed strange that they did not ask about customer preference for brown. It appears that the M and M site originally intended to include brown and that another part of the page covered it up so that it never had a chance of getting any votes. Thus the percentages reported added to 100%. Because of that I did not include 46 candies that were brown in the sample; they did not relate to the experiment as shown in the table. Assume for illustration that this really was a random sample taken after forming the hypothesis and designating a criterion, **H0:** THE POPULATION REPRESENTED BY THIS SAMPLE CONFORMS TO THE DISTRIBUTION OF AD HOC CUSTOMER PREFERENCE vs **HA:** THE POPULATION SAMPLED DOES NOT CONFORM TO THE AD HOC CUSTOMER PREFERENCE SURVEY, with $\alpha = 5\%$.

I counted the number of each surveyed color in the entire sample of size $n = 381$, and produced the following table of expected, E, and observed, O:

	Yellow	Red	Green	Blue	Orange	Total
E	0.16×381 = 60.96	0.25×381 = 95.25	0.20×381 = 76.20	0.31×381 = 118.11	0.08×381 = 30.48	381
O	41	51	67	116	106	381

I applied Pearson's chi square via Equation 12.1, $\chi^2 = \sum_{j=1}^{j=k} \frac{(O_j - E_j)^2}{E_j}$, to challenge whether the sample fits the customer survey. What do you conclude? You are not required to present a column graph or histogram in the answer to this question, but here is a column graph of the actual counts, with E on the left of each pair and O on the right.

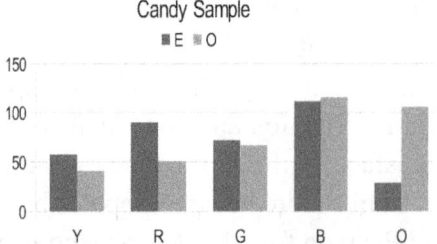

Candy Sample
■ E ■ O

You probably would put such a graph in a report.

Remember that **for testing**, graphs should be made **after** hypothesis, criteria and calculation; no peeking. (My wife and I ate the evidence.) The posting was much appreciated:
www.youtube.com/user/APUS07

2. A research group is studying courtship flights of two species of noctuid moths of the genus *Mocis*. They believe that one species flies high and the other flies low and that this is an adaptation which minimizes interbreeding. Their hypothesis pair is **H0**: THE TWO SPECIES ARE EQUAL WITH RESPECT TO HOW MANY FLY TO THE HIGHER LEVEL vs **HA**: THE TWO SPECIES ARE DIFFERENT WITH RESPECT TO HOW MANY FLY TO THE HIGHER LEVEL where high was over 6 meters and low was under 6 meters, with criterion $\alpha = 1\%$. They then released individual pairs at random and obtained **409** successful mating flights of species A, and **486** of species B. Of these successful flights, a total of **459** got up to the higher level; **350** of those were species B.

Make a Pearson's chi square table starting with the above given numbers. Then use the shortcut Equation 12.6 to test the hypothesis pair. Show the calculations in detail.

(The above is fictional. However, I have tried to follow the flights of two such *Mocis* species without recently developed micro electronics in an enormous cage and the pairs appeared to not have enough flying space. There is a relatively rare 'species' within the overlapping range of these two species which externally looks like a hybrid. If I had to do it now, I would start with a chromosome study, followed by a DNA examination, if necessary.)

3. Refer to your results for Question 1; estimate and report an effect size for each color as a **difference in probability**.

4. For Question 2, find a commonly used correlation-like ES. Be certain to include the sign. Then find the variance ratio-based ES.

5. For Question 1, produce a group of 95% CIs for the effect sizes, in terms of $\mathcal{D}$, the difference in (decimal) probabilities. (Hint: Equation 12.13 .)

6. Find a 95% CI for the Question 2 effect size in terms of **difference in probabilities** $(\mathcal{D}_H = P_B - P_A)$, as percents, of a moth flying high, given that it is in species B, as opposed to flying high, given that it is in species A.

7. Suppose, in a long term industrial hygiene study by regulators of worker exposure to a certain toxic chemical, there were 1190 workers exposed to an accidental spill and 70 of them eventually developed debilitating lung problems. However, out of 3150 on other shifts not exposed to the toxic spill, only 15 eventually developed such problems. This is not a random sample, but it is a typical case history and could lead to further studies. Estimate the **risk ratio**, based on the observation of this incident, of workers developing debilitating lung problems if exposed to such a spill, **and** briefly describe the results. (This might be seen as a descriptive statistic rather than an effect size.)

9. Suppose you are training a battle robot to fight in an arena with other robots, where the rule is, if your machine attacks stationary objects, it loses points. It is supposed to go around stationary objects. While programming (teaching) your machine, you need an appropriate index of machine learning, such as the Matthews correlation coefficient, to keep track of your progress. You randomly provide it with 1000 stationary objects and 100 moving objects. However, at this point in your program and test cycle it only attacks 78 moving objects, and it attacks 120 stationary objects. What is the **Matthews correlation coefficient** for this data (Equation 12.17)?

10. This contingency table (also called a cross tab) contains an approximation of what you might find in the records of a small city hospital after the first wave of COVID 19, for the deaths of people age 85 and older vs for people under 85, out of everyone who was hospitalized there for COVID 19. Use Pearson's chi square test to see if the difference is statistically significant.

COVID 19 Deaths Age $\geq$ 85 vs < 85			
Age	Lived	Died	Total
85 and older	5	38	**43**
Under 85	324	26	**350**
Total	**329**	**64**	**393**

Your hypothesis pair is **H0**: THERE IS NO DIFFERENCE IN THE NUMBER OF HOSPITALIZED COVID 19 PATIENTS OVER 85 YEARS OLD DYING THAN FOR YOUNGER PATIENTS vs **HA**: THERE IS A DIFFERENCE with $\alpha = 1\%$.

11. This table contains an approximation of what you might find in the records of a largish rural hospital after the first wave of COVID 19, for the deaths of people age 65 and older vs for people under 65, out of everyone who was hospitalized there for COVID 19. Use Pearson's chi square test to see if the difference is statistically significant.

COVID 19 Deaths Age $\geq$ 65 vs < 65			
Age	Lived	Died	Total
65 and older	124	50	**174**
Under 65	192	14	**206**
Total	**316**	**64**	**380**

The hypothesis pair is **H0**: THERE IS NO DIFFERENCE IN THE NUMBER OF HOSPITALIZED COVID 19 PATIENTS OVER 65 YEARS OLD DYING THAN FOR YOUNGER PATIENTS vs **HA**: THERE IS A DIFFERENCE with $\alpha = 5\%$.

12. Estimate a correlation like effect size as the fraction of the variance observed in Question 10 that is due to the effect of age equal to or over 85 vs younger.

13. Estimate a correlation like effect size as the fraction of the variance observed in Question 11 that is due to the effect of age equal to or over 65 vs younger.

13 POWER, For and Of $P\chi^2$

Controlling and assessing power for Pearson's chi square, $P\chi^2$, with 1×2 and 2×2 tables is pretty straightforward. We will also consider sample size for $2 \times c$ tables, testing for association even when $c > 2$. Power for more complex designs is a challenge, and will for the most part require you to go to specialists or to the literature, and possibly encounter a variety of pitfalls. In some situations, there is some debate about what is a meaningful effect size for tests involving categorical data (Kilem 2002). This is rather alarming; if we cannot define a suitable effect size for many of the popular designs used by Pearson's chi square, how can we define a suitable sample size for targeting them?

In any case this chapter will only cover sample sizes for relatively simple applications of Pearson's chi square. As in all research, complexity can lead to unexpected or unobserved confounding, so avoid unnecessary complexity. Note, you are working with counts, so, for inferential $P\chi^2$ testing, the per treatment (class or category) sample size is the treatment total (usually the column total in this book) estimated by $\hat{n}_i$.

> **CAUTION:** The internet is full of **misleading** answers to the question of how large should a sample be for Pearson's χ^2. Yes, it is essential that no cell has an expected value less than 5 (or more conservatively, 10). But, that will not promise you the power you may need to make your experiment worth doing. The statistics sites where people ask questions, and others, not a moderator, answer statistical questions are often worse than useless.

Hopefully you are not still confused over the α for Pearson's chi square, **unlike** $\alpha/2$ for the underlying normal approximation, Z. A difference in probability in either direction **increases the variance**, which is a directionless squared parameter, for either the + or - normal approximation as measured by Z. Pearson's chi square looks at both probability tails of the intrinsic normal approximations at once without any effect of sign. It sums the squared Z^2 values, however for 2×2 tables it is only concerned with the overall cutoff for the upper tail of the $\chi^2_{df=1}$ distribution. So, $Z^2_{\alpha/2} = \chi^2_{1df}$.

13.1 $\hat{n}$ for 1×2 Pearson's χ^2 Goodness of Fit

When you have a fixed model for estimating sample size for a 1×2 **goodness of fit** test you do **not** need prior knowledge as you do with many other tests, which often require you to first estimate the underlying variance, σ^2 (or maybe the probability, p_0). For testing fit, you get the only parameter you need, p_0, from the expected probability of either cell of the model being challenged. However, as for any sample size estimate, you need to a priori decide on a criterion, α, a power, $1 - \beta$, and a reasonable minimum difference, $\mathcal{D}_{min} = p_A - p_0$, a difference in probabilities[1]. Pearson's goodness of fit tests for 1×2 tables only have one degree of freedom. Calling $(Z_{cutoff} + Z_\beta)^2 = Z^2$ and remembering that $Z^2 = \chi^2_{df=1}$, you can use Equation 7.24 as though for a one sample normal approximation to get

$$\hat{n} = \frac{\left(Z_{cutoff}\sqrt{p_0(1-p_0)} + Z_\beta\sqrt{p_1(1-p_1)}\right)^2}{\mathcal{D}^2_{min}}.$$

---

[1] $p_A = p_0 + \mathcal{D}_{min}$ is the targeted minimum probability that you would see as meaningfully different. $\mathcal{D}_{min}$ **cannot** be zero, and as with any reasonable minimum difference, the smaller it is the more costly it is.

This *cutoff* usually equals $\alpha = 2(\alpha/2)$ because χ^2_α is for **either** tail when considering a normal approximation, and this equation targets a power, $1 - \beta$, for which β is, by its nature, only one direction.

13.1.0.1 Example: Pearson's 1×2 Goodness of Fit

Mendel Meets Pearson (alternative to Subsection 12.2.0.1) We return to Gregor Mendel's
results when he crossed pea plants that had what he called 'simply inflated pods' with pea
plants that had constricted pods and he only got offspring plants with simply inflated seed pods
in what is called the F1 generation. Assume that he already had partly developed his famous
model. Based on the *phenotypes* (how the offspring appeared), he thus had reason to create
a null hypothesis, **H0**, that there was only one gene with two forms (*alleles*) and that the inflated phenotype was *dominant* (took over when both alleles were present). If so, aside from
the effects of chance, under his theory, by crossing sibling plants from that first F1 generation from *selfing*, breeding them to themselves), he should expect 3 plants with the dominant
simply inflated pods for each 1 wild type recessive plant in the resulting inbred generation, F2
(refer to Subsection 12.2.0.1). How many samples would he need to test the following hypothesis pair using $\alpha = 5\%$ ($Z_\alpha = 1.6449$), $1 - \beta = 90\%$, and $\mathcal{D}_{min} = p_A - p_0 = 0.01$?
($Z_{1-\beta=0.10} = 1.28156$).

H0: 75% WILL BE OF THE DOMINANT PHENOTYPE (implying that 25% will be of the recessive
phenotype).

HA: THE 3 : 1 MODEL DESCRIBED IN **H0** DOES NOT FIT ($\alpha = 5\%$; $Z_{5\%}$ cutoff because chi square
two tail).

Calculation: Using Equation 7.24, $\hat{n} = \dfrac{\left(Z_{cutoff}\sqrt{p_0(1-p_0)} + Z_\beta\sqrt{p_1(1-p_1)}\right)^2}{\mathcal{D}^2_{min}}$,

$$\hat{n} = \frac{\left(1.9600\sqrt{0.75\,(0.25)} + 1.28156\sqrt{0.76\,(0.24)}\right)^2}{0.00010} = 19489.21$$

which should be rounded **up** to $\hat{n} = 19490$ plants. However, the accuracy of the source of
Z can make a difference of up to a little more or less than one in the last figure. This $\hat{n}$ **is the**
total because it is all one sample.

Discussion: Our imaginary Mendel might have thought it would take a lot fewer samples than
19490 to test something so simple with both a modest criterion and power of $(1 - \beta) = 0.90$.
Could this high cost be because $p_0 = 0.75$ is quite a ways from the binomial ideal of 0.5?
No, the big cost here is required by the very small $\mathcal{D}_{min} = 0.01$, or just one plant in 100 that
doesn't fit the model.

If he had chosen $\mathcal{D}_{min} = 0.05$, or just 5 plants per 100 plants (1/20) that don't fit the model,
the sample size estimate would have been only 825 plants.

If Mendel were planning to breed the **F1 hybrids** to the recessive wild type, he would expect to
get **equal numbers** of each phenotype. So, using the same $\alpha = 5\%$, $\beta = 10\%$, and $\mathcal{D}_{min} = p_A - p_0 = 0.01$, he would be testing

H0: 50% WILL BE OF THE DOMINANT PHENOTYPE (implying that 50% will be of the recessive
phenotype).

HA: THE 1 : 1 MODEL DESCRIBED IN **H0** DOES NOT FIT.

Calculation: He would need

$$\hat{n} = \frac{\left(1.9600\sqrt{0.50\,(0.50)} + 1.28156\sqrt{0.51\,(0.49)}\right)^2}{0.00010}$$

$$= 26265.12,$$

which rounds **up** to $\hat{n} = 26266$ total plants to do the whole experiment. This approximation is certain to be more accurate than the approximations for models with other p values because the binomial is truly symmetrical at $p = 1/2$.

(But **if he had chosen** $\mathcal{D}_{min} = 0.05$, he would only have needed a total of 1047 plants). So, the high cost here mostly relates to the very small minimum 'reasonable' difference, $\mathcal{D}_{min} = p_A - p_0 = 0.01$. For instance, if he had targeted a **much larger minimum difference**, $\mathcal{D}_{min} = p_A - p_0 = 0.10$, the estimate would be

$$\hat{n} = \frac{\left(1.9600\sqrt{0.50\,(0.50)} + 1.28156\sqrt{0.60\,(0.40)}\right)^2}{0.010},$$

which rounds **up** to $\hat{n} = 259$ pea plants. Since he is hoping for a $50:50$ relation ($1:1$), he also needs to round the odd estimates to even sample sizes. Notice that even though the approximation is better, the $1:1$ model, $p = 0.50 : q = 050$, needs more samples than the $3:1$ model, $p = 0.75 : q = 0.25$. It might be a very good idea to do some a priori **scratch calculations** before you decide on a reasonable minimum difference for your design.

Small criteria and high power are costly, but detecting **tiny differences** is exponentially **more expensive**. If you can justify it, or a priori document it, just doubling your reasonable minimum difference can greatly reduce your cost (in labor and space, as well as money).

13.2 $P\chi^2$ **A Priori** $2 \times c$ **Global Power**

Pearson's $2 \times c$ omnibus chi square tests are easily broken down into 2×2 tests which are equivalent to two sample normal approximations. Like normal approximations, they feature probability values found from the sample size, n, and number of successes, r, in the given sample size. All this makes power analysis for $2 \times c$ Pearson's tests rather straightforward, even if some of the steps may seem a little convoluted at first sight.

13.2.1 2×2 **a priori power for** $P\chi^2$

Pearson's chi square is mostly sensitive to linear effects, and as the degree of linearity (the goodness of the linear approximation) drops off, it becomes increasingly less powerful.

13.2.1.1 $\hat{n}$ **for** $2 \times c$ **frequencies and probabilities**

Most Pearson's chi square designs for association, especially a $2 \times c$ table, can be seen as $k - 1$ normal approximation tests, given that you are going to make k subsequent comparisons. So, a Pearson's 2×2 test is equivalent to a single normal approximation test, and much of the rest of this subsubsection parallels its equivalent Subsection 7.10.2.

You usually have no choice but to treat proportions as though they were the original units of measure. So, it is appropriate to estimate the sample size of 2×2 **tests** for association by using a reasonable minimum difference based on a frequentist concept that **probability equates to relative frequency**, $\mathcal{D}_{min} = p_2 - p_1$. You do not need to know the unknowable p_1 and p_2 to select a reasonable minimum difference, $\mathcal{D}_{min}$, but it is a big advantage to know a pooled probability, p, as from literature or past records. This is the same requirement as finding a standard deviation for other sample size estimates. **When in doubt**, you can always use the most conservative possible $p = 1/2$, meaning equality, or $1:1$. Because $p = 1/2$ is the most conservative possible p and produces the largest possible variance, you might overestimate the sample size by using it, but that is not a serious problem like underestimation. Nonetheless, larger sample sizes cost more time, resources, and money. For projects that have more than two variables, it might be best to run a pilot study, that is, a complete preliminary test, but without the actual testing.

Unfortunately the normal approximation does not work well outside the range $0.25 \le p \le 0.75$ with less than very big samples. If p is outside this range, no cell should be less than 10, and you might consider partly compensating by using $p = 1/2$ for sample size estimation. If it is true that $p = 1/2$, there is an added advantage that the normal approximation is going to be highly symmetrical and facilitate an acceptable approximation of normal even when n is a little less than 30 (say, as small as 25).

You also must a priori select the α and β in which you are interested, as well as the reasonable minimum difference. A common approach is derived from Equation 5.13, $Z = \frac{\mathcal{D}}{\sqrt{p(1-p)(2/n)}}$, for two sample tests roughly equivalent to a 2×2 Pearson's chi square test. The derivation of the estimator, $\hat{n}$, involves substituting $Z_{\alpha/2} + Z_\beta$ for Z, squaring and rearranging to get the equivalent chi square with one degree of freedom. That means you can estimate $\hat{n}$, the size for each sample, using Equation 7.25, as in assessing any similar binomial approximation. Having chosen the value p (designated p_0 here), you can estimate

$$\hat{n} = \left(Z_{\alpha/2} + Z_\beta\right)^2 (p_0 q_0 + p_1 q_1) / \mathcal{D}_{min}^2$$

for **each** sample where $1 > |\mathcal{D}_{min} = p_1 - p_0| > 0$. The **total** sample size is, of course, $\hat{n} = 2n$ for a given 2×2 table (with $\chi^2_{\alpha,1df}$). Binomial sample size estimation might work best when $\mathcal{D}_{min}$ is small, $|\mathcal{D}_{min}| \ll 0.5$, or when neither p_0 or p_1 are too far from $1/2$.

13.2.1.2 Example: Finding A Priori Sample Size for a 2×2 Table

Pub customers: Your company owns two tap-beer-only pubs, each in a different city, and it focuses on cultivating regulars. Its board of directors is divided over whether the consumption habits of their pub regulars vary from city to city. So you want to study the association between the two cities and the number of regulars who drink more than two glasses of beer in a single visit. Assume this question only regards there being **any** meaningful association between consumption habits of regulars and the cities. Given equal sample sizes per city, what is the total **number of regular** customers you need to observe? Your computerized checking system shows group size and credit and debit card information allowing you to identify persons coming alone to a given pub an average of four or more times a week, and you will take equal size random samples of these. But how large should the overall sample be to assure that you can detect a reasonable minimum difference, $\mathcal{D}_{min} = 10\%$, between the cities?

You have data from previous records to enable you to estimate an **expected** $p_0 = 0.6400$, **rather than** having to default to $p_0 = 0.5000$, which would have made the estimate a bit larger.

Methods: Your hypothesis pair is

H0: THERE IS NO ASSOCIATION BETWEEN CITY AND THE NUMBER OF SOLO REGULAR CUSTOMERS BUYING MORE THAN TWO GLASSES.

HA: THERE IS AN ASSOCIATION (a lack of homogeneity).

You choose an $\alpha = 5\%$ criterion (meaning an $\alpha/2 = 0.025$ cutoff in terms of Z) and you target $1 - \beta = 90\%$ power.

Calculation: Past records combining the two locales showed $p_0 = 0.6400$. So, to find p_1 you add your chosen minimum difference, $\mathcal{D}_{min} = 0.1000$, to this to calculate the alternative probability, $p_1 = p_0 + \mathcal{D}_{min} = 0.7400$. Then use Equation 7.10.2,

$$\hat{n} = (Z_{cutoff} + Z_\beta)^2 (p_0 q_0 + p_1 q_1) / \mathcal{D}_{min}^2 .$$

From a table, software, or an online source $Z_{\alpha/2=0.0250} = 1.9600$ and $Z_{\beta=0.1000} = 1.2816$. Substituting all this into the formula gives

$$\hat{n} = (1.9600 + 1.2816)^2 (0.6400(0.3600) + 0.7400(0.2600)) / 0.0100 ,$$

all calculated to four or more figures. The raw result is $\hat{n} = 444.277$, and that rounded up to the nearest even whole number estimates that you will have to randomly sample $n = 445$ customers for **each** city and count them as *successes* if they purchase more than two drinks.

Discussion: Note that it is essential that all this hypothesizing, criteria setting, and sample size estimation is done a priori. The experiment is yet to come. The approximation and hence the estimate would have been weakened an increasing amount as the maximum $p_{...}$ or $q_{...}$ grows larger than 0.05, and the resulting sample size estimate could be too small if one of the probabilities exceeded 0.75. However, Snedecor and Cochran were content to use an example in which the maximum was 0.80; in any case, definitely **round up** and if your maximum p value is too huge, then either use $p_0 = 0.5$, or **consider** a smaller $\mathcal{D}_{min}$.

The hypothesis pair is an example of **ambiguity**. Even if there is a difference, you do not know if it is only due to the locale or to the ambiance of each pub itself, or to other factors. In this experiment these possibilities are confounded in that they are inseparable. Such confounding is a potential pitfall in studies of schools, businesses, army units, impacted habitats, etc.

You must also realize that, due to the limitations of your automated sampling approach (computerized checking system), you can **only** assess individual regulars, **not** groups of regulars. Another difficulty associated with this example is that the answer, or at least p_0, regards the regulars who use **plastic** rather than **paper** money. The results of any research is defined by the details of the research techniques and these are often defined by available resources. Try to make your hypothesis as specific as possible, but you often cannot be certain that your design will test what you intended to test. Such challenges are not restricted to social science and marketing studies. They also plague many 'hard science' questions.

13.2.2 $\hat{n}$ for k multiple comparison $P\chi^2$

It makes sense that global tests usefully lead to multiple comparisons. If your global test is going to be followed by further testing, you need to choose your sample size accordingly. You need to estimate a sample size that will allow k comparisons. Correction for multiple comparisons can use **either** the Bonferroni or the Sidak correction to estimate α' and β', the per comparison cutoffs you will need in order to get a family wide α and β. Unfortunately, the global Pearson's chi square does not actually contribute to the power of the multiple comparisons, as apparently does the ANOVA, which potentially provides your comparisons with an improved estimate of the variance.

13.2.2.1 $P\chi^2$ a priori k equal multiple comparisons

You should always have an a priori plan for the specific pairwise comparisons you intend to make. If possible, you should also design for equal sample sizes, meaning that the total count in each treatment (column category) will be equal, $n' = n'_1 = n'_2 = \cdots = n'_c$. To **facilitate** k preplanned multiple comparisons for a $2 \times c$ test, you can use the techniques of Section 13.2.1, but use them with either a Sidak correction or a Bonferroni corrected α' and β', to estimate the corrected sample size, $\hat{n}'$, for each treatment (column). You need to multiply the number of groups (columns), c, times this estimate of the corrected sample size estimate, $\hat{n}'$, for each treatment, to estimate

$$\widehat{\mathbf{n}}' = \mathbf{c}\hat{n}',$$

the grand total. This assumes that each of the c columns will have the same total in it (meaning the sample sizes will be equal). Unequal sample sizes will require much larger total sample sizes, but the estimate of $\widehat{\mathbf{n}}'$ is still possible (Subsection 13.2.3).

13.2.2.2 Example: Global A Priori for $2 \times c$ Pearson's Chi Square (association)

Whether you are simply interested in probabilities or are interested in variances, if you can estimate sample sizes for a 2×2 Pearson's chi square test, it is relatively easy to estimate the sample sizes you need for whole $2 \times c$ tests, where $c > 2$ 'treatments'.

Prenatal Pregnancy One of my favorite research organisms is the melon aphid, *Aphis gossypii*, a tiny plant juice sucking insect. A winged female will mate with a winged male and then lay eggs on any of several seemingly unrelated plant families. These eggs hatch into wingless asexual females that do not mate; they give live birth to wingless females, clones of themselves, if their leaf remains healthy. Looking through a dissecting scope at the living green or yellow asexual females you can see the unborn aphids within, sort of like looking at insects trapped in yellow or greenish amber. One of the remarkable things about these organisms is that you can see into the unborn young and they often contain their own unborn young. Colonies develop very fast, 3 aphids can result in 100 or more in ten days.

Suppose I had a big colony of these asexual wingless aphids that arose from a single asexual female. Further suppose I wished to see if there was a difference between aphids on different plant species with respect to the prevalence of unborn within unborn. I intend to run the test for one week on each of $c = 7$ **species** of plants: red pepper, watermelon, cotton, asparagus, hibiscus, lemon, and okra. I will test using a Pearson's chi square with two categories, no unborn within unborn, and yes unborn within unborn, for each of the seven species of plants, one of which, watermelon, is to be the control. If I am planning to make $k = 6$ **comparisons**, how many aphids will I have to use to have a **family wide probability** of no more than $\alpha = 5\%$ of mistakenly concluding that there was a difference? I would want a **power** of $1 - \beta = 80\%$. I would only score the first unborn female that falls under the cross hairs as pregnant (gravid) or not pregnant while I am focusing on each randomly sampled already mature female. Assume I had enough accumulated data to know that when using this clone, $p_0 = 0.30$ of an unborn female having babies within it on the control, watermelon. Otherwise I would have to run a pilot study to get a probability for at least the control. My **reasonable minimum difference** would be $\mathcal{D}'_{min} = 10\%$. The hypothesis pair is

H0: THERE IS NO DIFFERENCE BETWEEN THE PROPORTION, ON DIFFERENT PLANT SPECIES, OF MOTHERS WITH GRAVID UNBORN FEMALES.

HA: THERE IS A DIFFERENCE.

Note that this example is only about the estimation of sample size.

Calculations: For this I chose the Bonferroni correction for comparing the control to each of the remaining 6 treatments, to calculate that $\alpha'/2 = 0.025/6 = 0.004167$ and $\beta' = 0.20/6 = 0.033333$. So, from a suitable source, $Z_{\alpha'/2} = 2.638257$ and $Z_{\beta'} = 1.833915$. To estimate the sample size of **one** column (plant species), substitute these adjusted values and the minimum reasonable minimum difference into Equation 7.10.2, as

$$\hat{n} = \left(Z_{\alpha'/2} + Z_{\beta'}\right)^2 \left(p_0 q_0 + p_1 q_1\right) / \mathcal{D}_{min}^2, \text{ to get,}$$

$$\hat{n}' = \frac{(2.638257 + 1.833915)^2 \left(0.30\,(0.70) + 0.40\,(0.60)\right)}{(0.10)^2}$$

$$= 900.0145.$$

(Depending on what you used to do the calculation, and your source of Z, there might be some disagreement after the decimal point, but rounding **up** tends to lose that effect.) Note that $Z_{\alpha'/2}$ was used because $P\chi^2$ is about both tails. Now, rounding up to an integer, there needs to be $\hat{n}' = 901$ in each column (aphids per each plant species). So, because **c** $= 7$ plant species, there will be a total of

$$\widehat{\mathbf{n}}' = \mathbf{c}\hat{n}' = 6307 \text{ aphids}$$

in the whole experiment. Notice that with all **c** $= 7$ treatments having a sample size of $\hat{n}' = 901$, all of the $k = 6$ multiple comparisons can then be performed with a **total** probability of getting a Type I error equal to 5% (experiment wide).

Discussion: What are young bugs coming to these days? Well, that is a lot of aphids for me to examine. Such a huge sample might not be ideal for Pearsons chi square. Choosing a larger $\mathcal{D}_{min}$ would ease my burden (for instance $\mathcal{D}_{min} = 0.20$ would bring the size per sample down to $n' = 231$ aphids per plant species). Applying the results of this approach would be satisfactory for any number of comparisons in terms of $2 \times c$ tables, if you can afford it.

Question 5 is another worked example of this procedure.

13.2.3 A priori $\hat{n}'$ for $P\chi^2$ 2×2 with unequal sample sizes

Pearson's chi square is the commonest approach to testing for association with discrete data. But, if the columns cannot have the same totals (meaning the columns have unequal sample sizes), things become a little more complicated.

In this book, the 2×2 tables for association usually use columns for treatments, and their **totals** are the **sample sizes**. The rows can be seen as replications, and can be yes or no, success or failure, or any other category (occasional publications use rows for treatments and columns for replication; sometimes which is which is ambiguous). Unfortunately, the way we estimate the sample sizes is designed for equal sample sizes, and we must therefore correct our estimates for inequality. If one sample has a **fixed** maximum sample size, n_f, that is less than your initial estimate of the required sample size $\hat{n}$, and if you cannot do anything to increase it, the unequal sample sizes will cost you power. Kifle and Desta (2012) state, "The loss of **relative efficiency** due to unequal sample sizes in two groups can be calculated as follows:

$$1 - \frac{\bar{n}_h}{\bar{n}},\text{"}$$

The denominator is the arithmetic mean sample size and the numerator, $\bar{n}_h$, is the harmonic mean sample size.[2] A harmonic mean for 2 values of n is

$$\bar{n}_h = \frac{2}{1/n_1 + 1/\hat{n}_2}.$$

So let's use a relative efficiency based correction to adjust back to the desired power with unequal sample sizes. First estimate the sample size as though for a regular test with equal sample sizes (Section 13.2.1). If the 2 × 2 table is to be used for multiple comparisons, this initial estimate should be adjusted for multiple comparisons to get $\hat{n}'$ using the Sidak or Bonferroni correction. The mean of the individual treatment is just the initial unadjusted for inequality mean sample size estimate, $\bar{n} = \hat{\mathbf{n}}/2 = \hat{n}$ (or $n = \hat{n}'$). The **unadjusted-for-inequality** estimate of the size of the adjustable member of the pair is $\hat{\mathbf{n}} - n_f$. So, the corresponding version of the harmonic mean after the appropriate substitution of n_f for n_1 and of $(\hat{\mathbf{n}} - n_f)$ for $\hat{n}_2$ gives

$$\bar{n}_h = \frac{2}{1/n_f + 1/(\hat{\mathbf{n}} - n_f)} \tag{13.1}$$

(do not round yet). (If you wish, you can use these two kinds of means to estimate the relative **efficiency lost** $= 1 - \frac{\bar{n}_h}{\bar{n}}$, **before** adjustment, due to the way we find these sizes when assuming they are equal.)

Now you can adjust the size of the total, $\hat{\mathbf{n}}$, up to $\hat{\mathbf{n}}''$ to **compensate** for the relative efficiency lost. By adding this estimated loss back using $\hat{\mathbf{n}}'' = \hat{\mathbf{n}} + \hat{\mathbf{n}}(1 - \bar{n}_h/\bar{n})$, and rearranging, you can obtain

$$\hat{\mathbf{n}}'' = \hat{\mathbf{n}}\left(2 - \frac{\bar{n}_h}{\bar{n}}\right). \tag{13.2}$$

This must be **rounded up to an integer**. For a test with only two samples, the final estimate of the final **adjusted sample size** approximation, $\hat{n}''$, for the sample with an adjustable size will be

$$\hat{n}'' = \hat{\mathbf{n}}'' - n_f.$$

If the difference between the fixed sample size, n_f, and $\hat{n}$ (or $\hat{n}'$) seems very large, then three place accuracy, rounded up, might be the best you can hope to attain.

Even then, the adjusted sample size, and the Pearson's chi square itself (and any normal approximation) with different sample sizes is **imperfect** because of the fact that different sample size binomial distributions have unequal variances. The word '*approximation*' implies inexact, but I am not aware of any better approach, and the above should get you very close to a correct sample size, if the difference in sample sizes is not too extreme and the smallest cell is ≥ 5 (some say ≥ 10). This suggests that treatment sample sizes should have a minimum total sample size of at least > 10 (or maybe better, ≥ 20) per column.

13.2.4 A priori unequal n, k multiple association

Sometimes you cannot control all of the treatment sample sizes. Consider a Pearson's 2 × c chi square test with c $>$ 2 columns, with multiple comparisons in which I equals one or more treatments, $(1 < I < c)$, that have a fixed maximum sample size, $n_{f_{1\ldots I}}$, that is less than your per treatment sample size estimate. In such a case you will have to correct your estimate for unequal sam-

---

[2]The harmonic mean of $x_1, x_2, \ldots, x_n$ is $\bar{x}_h = n/\sum(1/x_i)$. The harmonic mean is always the smaller mean if the sample sizes are not equal

ple sizes as for the 2×2 tests (Subsection 13.2.3). However, there is more than one comparison to be made, so you will have to estimate each of the adjustable treatment sizes in accordance with any comparison in which a fixed sized treatment takes part. .

But first, estimate the Bonferroni or the Sidak corrected α' and β'. Then using the 2×2 sample size estimation technique of Section 13.2.1 with the corrected α' and β', estimate $\hat{n}'$ for each column as though there were no fixed sample sizes. This $\hat{n}'$ is only temporary for any column that will take part in the comparisons. That is, **provisionally** assume every comparison will have a total sample size of $2\hat{n}'$. However, you cannot change the fixed sample sizes, so you must change the adjustable sample sizes which are not fixed.

Find the smallest fixed treatment size, say, n_{f_1}, and calculate the unadjusted size required for each changeable column that has to be compared with it, $2\hat{n}' - n_{f_1}$. But this still has to be adjusted on a comparison by comparison basis to correct the paired unequal sample size estimates. Use $\hat{n}'$ as $\bar{n}$, the arithmetic *mean* of the individual treatment (or column) sample size estimates:

$$\bar{n} = \frac{(2\hat{n}' - n_{f_1}) + n_{f_1}}{2} = 2\hat{n}'/2 = \hat{n}'.$$

Then use Equation 13.1 to find their *harmonic mean*,

$$\bar{n}_h = \frac{2}{1/(2\hat{n}' - n_{f_1}) + 1/n_{f_1}}.$$

Now you need to upwardly adjust the provisional size of the total per comparison (of two treatments), $\widehat{\mathbf{n}}' = 2\hat{n}'$, to compensate for the relative efficiency lost by having unequal sample sizes when you have estimated for equal sample sizes. To do this, add back the estimated *efficiency_lost*, $1 - \bar{n}_h/\bar{n}$, using $\widehat{\mathbf{n}}'' = \widehat{\mathbf{n}}' + \widehat{\mathbf{n}}'(1 - \bar{n}_h/\bar{n})$, which can be rearranged to get the per comparison sample size (Equation 13.2),

$$\widehat{\mathbf{n}}'' = \widehat{\mathbf{n}}'\left(2 - \frac{\bar{n}_h}{\bar{n}}\right),$$

and round it **up to an integer value**. For any two sample comparison where one member is of size n_{f1}, the **adjustable individual sample size** will finally be

$$\hat{n}'' = \widehat{\mathbf{n}}'' - n_{f_1}.$$

Assign this to all changeable sample sizes for treatments which take part in the **comparison with this fixed sample size** (or any equal fixed sample size). If you are making every possible comparison, then all the adjustable sample sizes must be this big. Adjustable sample sizes which are involved in no comparisons with fixed maximum size samples can stay the original estimated sample size, $\hat{n}'$.

Iff you have **one or more additional fixed sample sizes**, say, n_{f_2}, and you **also** have any **columns** to be compared to each other, for which you **have not already adjusted** the sample size for efficiency loss, repeat this process with the next smallest fixed sample size. Continue until you run out of fixed sample sizes, or out of columns which have not already been adjusted for loss of relative efficiency. Always calculate for the smallest fixed sample size before the next smallest.

However, **if you are only comparing to a single control** and the sample size of the control is adjustable, you will only have to **adjust the control to $\hat{n}''$** for the **smallest fixed sample size** to which it will be compared. If there are two or more controls which have no overlapping comparisons, you can adjust each according to the smallest fixed sample size of treatments to be compared with them.

You can sum all column sample sizes to estimate the **grand global total** sample size,

$$\widehat{\mathbf{n}'''} = \sum_{j=1}^{j=c} \left(\widehat{final\_sample\_size_j} \right) = \sum_{j=1}^{j=c} \left(\hat{n}''_j \right) ,$$

where **c** is the number of columns (treatments).

A similar approach for another group of global tests, the ANOVAs, can be found in Section 10.4.

13.3 Subsequent $2 \times$ c Association in Terms of Probability

Of course you are better off if you can preordain the design of your tests, including equal sample sizes. Unfortunately, this is not always possible and even when it is, subjects might be lost. Subsequent tests of power can be used to estimate how much power your test had with respect to the original hypothesis at the α level of statistical significance[3], and your (justifiable) a priori **reasonable minimum difference**, $\mathcal{D}_{min} = p - p_o$.

13.3.1 Subsequent 2×2 (or 1×2)

If your sample sizes change during the experiment or you could not control them, then if you have provided an α, and a $\mathcal{D}_{min}$ before obtaining the data, you can estimate the power attained using these a priori values and your achieved sample size. Of course β, and thus power, is only defined with regard to a reference point, the reasonable minimum difference, $\mathcal{D}_{min} = p_1 - p_0$, chosen by you, so β would only be the probability of not detecting the difference if the **true difference**, δ, actually happened to equal $\mathcal{D}_{min}$.

You can still use your initial estimate of the p_0 from your original hypothesis, **H0**, from either your model such as for testing the fit of Mendelian ratios, or from hypothesized probabilities such as $p_0 = 1/2$ as when you hypothesize equality.

13.3.1.1 For 2×2 $P\chi^2$

For 2×2 $P\chi^2$ as for tests of association with **unequal** treatment sample sizes, you can use Equation 7.28,

$$\hat{z} = \frac{|\mathcal{D}_{min}|}{\sqrt{(p_0 q_0 + p_1 q_1) \frac{1/n_1 + 1/n_2}{2}}} - Z_{cutoff} ,$$

where $p_1 = p_0 + \mathcal{D}_{min}$.

When the sample sizes are **equal**, you can use Equation 7.27,

$$\hat{z} = \frac{|\mathcal{D}_{min}|}{\sqrt{(p_0 q_0 + p_1 q_1)/n}} - Z_{cutoff} .$$

The value of $\hat{z}$ is interpreted in accordance with Subsection 7.8 by using either the rule that

$$iff \; \hat{z} < 0, \; then \; |\hat{z}| = \widehat{Z}_{1-\beta} ,$$

---

[3]Liu (2012) studied the power of r × c Pearson's chi square tests, seeming to see all of them as goodness of fit. However, Liu focused mainly on a theoretically more accurate non-central approach to Pearson's chi square for goodness of fit, and found only a slight difference between it and the normal approximation approaches.

or the rule that

$$\textit{iff } \hat{z} \geq 0, \text{ then } \hat{z} = \widehat{Z}_{\beta}.$$

Of course β, and thus power, is only defined with regard to a reference point, the reasonable minimum difference, $\mathcal{D}_{min} = p - p_0$, chosen by you, so β would only be the probability of not detecting the difference if the **true difference**, δ, actually happened to equal $\mathcal{D}_{min}$. The cumulative probabilities of $Z_{\beta'}$ or $Z_{1-\beta'}$ can be found from a $p >$ table as in Appendix 4. If much of this looks familiar, it is because 2×2 tables and simple normal approximations are equivalent.

13.3.1.2 For 1×2 **goodness of fit**

Assuming that you have **a priori** defined D_{min}, you can estimate the power for either a normal approximation or for a 1×2 $P\chi^2$ such as a **goodness of fit test**, using Equation 7.26,

$$\hat{z} = \frac{|\mathcal{D}_{min}|}{\sqrt{(p_0 q_0 + p_1 q_1)/2n}} - Z_{cutoff},$$

with its two rules as in the above subsection. The value of $\hat{z}$ is interpreted in accordance with Subsection 7.8 as in the preceding Subsection 13.3.1.1.

The following example illustrates this:

13.3.1.3 Example: Subsequent Pearson's for Fit Based on p_0 or q_0

Mendel Meets Pearson (another referral to Subsection 12.2.0.1): Again, we continue with the unhistorical tale about Gregor Mendel (Subsection). When he crossed pea plants that had what he called 'simply inflated pods', with pea plants that had 'constricted pods', he only got offspring plants with simply inflated seed pods in what is called the F1 generation. Based on the *phenotypes* (how the offspring appeared), there was reason to create a null hypothesis, **H0**, that there was only one gene with two forms (*alleles*) and that the inflated phenotype was *dominant* (took over when both alleles were present). If so, aside from the effects of chance, under his theory, by crossing sibling plants from that first generation (or from *selfing*, breeding them to themselves), he should expect 3 plants with the dominant simply inflated pods for each 1 plant with the recessive constricted pods in the resulting inbred generation. Pretend he then used a Pearson's chi square with the hypothesis pair,

H0: 75% WILL BE OF THE DOMINANT PHENOTYPE (implying that 25% will be of the recessive phenotype).

HA: THE $3:1$ MODEL DESCRIBED IN **H0** DOES NOT FIT.

Criterion: $\alpha = 5\%$

First Analysis: Now it is necessary to estimate the power. The hypothesized probability of a success was $p_0 = 0.75$ and he now, after the test, knows that the sample size obtained was $\mathbf{n} = 1121$ (where $\mathbf{n}$ is the grand total). He found that $\chi_{\alpha=0.05,\,1\,df} = 3.8415$ (using an anachronistic table).

	I	**i**	Total
O	822	299	1121
E	840.75	288.25	
$\hat{p} = O_j/\mathbf{n}$	0.733274	0.266726	1

But his results were $\chi^2 = 0.8191$, and that, being less, is not significant. So, the results were not statistically significant, but how could he know that he had made a reasonable attempt at challenging the fit? Well, knowing the power of his test would at least reassure him, if it was sufficiently high. (However, reassured or not, lack of significance really means nothing inferentially.)

Subsequent Power Analysis: Although not required due to the lack of significance, he sought to estimate the power, $1 - \beta$, of such a design. He knew $p_0 = 0.75$, so $q_0 = 1 - p_0 = 0.25$, and the sample size obtained was $n = 1121$. He needed to know $Z_{0.025} = 1.9600$ (now available in common tables), and we will suppose that it was his custom to **a priori** choose to use a (maybe unrealistically small) reasonable minimum difference of $\mathcal{D}_{min} = 0.01 = p - p_0$. So, after using $p_1 = p_0 + \mathcal{D}_{mom} = 0,76$, he applied the above Equation 7.26,

$$\hat{z} = \frac{|\mathcal{D}_{min}|}{\sqrt{(p_0 q_0 + p_1 q_1)/2n}} - Z_{cutoff} = \frac{|\mathcal{D}_{min}|\sqrt{2n}}{\sqrt{(p_0 q_0 + p_1 q_1)}} - Z_{cutoff},$$

to estimate

$$\hat{z} = \frac{0.01\sqrt{2 \times 1121}}{\sqrt{(0.75 \times 0.25) + (0.76 \times 0.24)}} - 1.95996 = -1.181429.$$

Because $\hat{z}$ was negative, then according to the appropriate rule, $|\hat{z}| = Z_{1-\beta}$. He checked a trustworthy source and found that $1 - \beta = 0.1187$, a power of about 11.9%.

Discussion: Well, it turns out that he had very little chance of detecting a difference of 1%. This is a very weak test if you are trying to detect tiny differences. Be cautious, there is some ambiguity with respect to how Z is entered into the tables. And if you use the R language, `pnorm(z)` gives you the answer you need without the rules about z (that is, just put $\pm z$ into `pnorm( )` with its sign). (Incidentally, if he had sought to detect a larger difference of $\mathcal{D}_{min} = 0.02$, he would have had a power of about 36%.)

Obviously the reasonable minimum difference is an important part of what defines the power of the test. Even if you cannot do an a priori power analysis, you should have a credible **a priori** reasonable minimum difference so you can estimate the power, $1 - \beta$. This is, in effect, a one sample test, so there is no issue of equality of sample sizes.

13.3.1.4 Fruit flies in the classroom

Genetics instructors sometimes have the students do a fruit fly cross and then have them calculate a 1×2 table with 2 observed and 2 expected values. They sometimes **mistake this** for a 2×2 table, which it is **not**. It is really only a 1×2 table, the second row only being the theoretically expected values put there for reference and calculation. So, this might be a good place to examine the common classroom fruit fly experiment, a cross that is expected to produce the same F2 generation ratio as the above pea plant experiment. Assume a student randomly counted out say, $n = 100$ flies and used an $\alpha = 0.05$, and of course, $p_0 = 0.25$ as the model's probability of a success, that a fly is homozygous recessive (maybe the albino-like form, called 'yellow'). Let's look at a graph of the power versus the reasonable minimum difference from the expected number of successes. So a calculation like the preceding would apply to this one sample binomial approximation, tested with χ^2 with one sample and two tails.

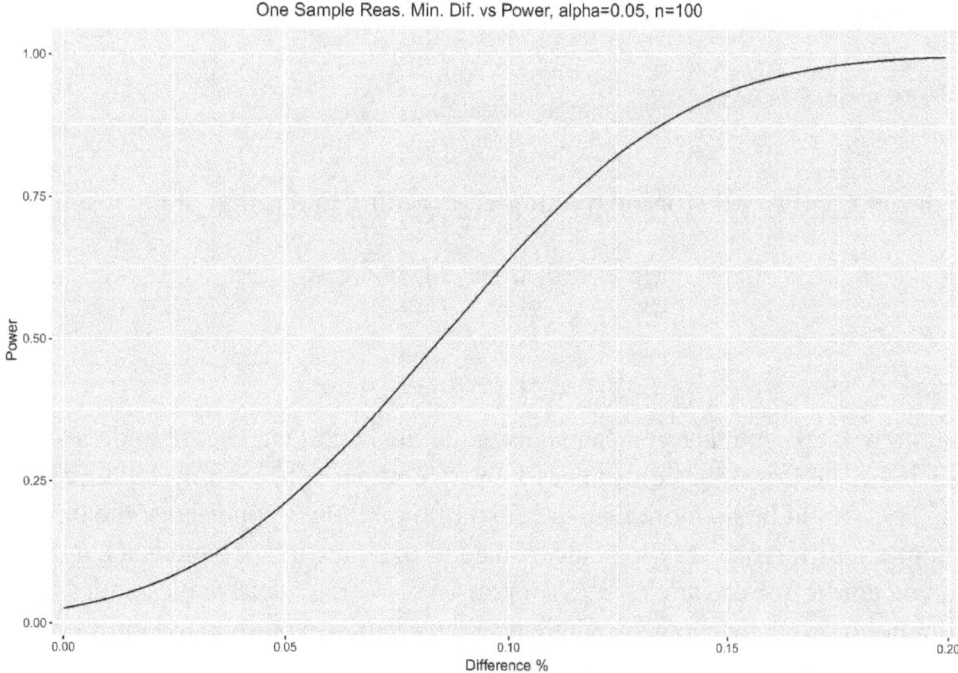

One Sample Reas. Min. Dif. vs Power, alpha=0.05, n=100

Figure 13.1: RMD, vs Power

Well, if you were looking for a **reasonable minimum difference** of 5% this test might appear to be very weak; there would be only a $1-\beta = 20\%$ chance of detecting that 5% difference. And even with considerably over a thousand plants, in the above sample the power would only be a little over twice this, at 5%. But, the student is not looking for tiny differences in machine parts or in blood pressure. The difference between the outcomes of simple crosses can easily be 15% or more if the model does not fit. Notice that the graph in Figure 13.1 tends to asymptotically approach a power of $1 - \beta = 100\%$ above a difference of about 20% with $\alpha = 5\%$ and $\mathbf{n} = 100$. So long as it is made **clear to the students** that they are checking for **big discrepancies**, the procedure is OK. Some instructors also often include a Yates correction in the Pearson's chi square with the effect of sacrificing some of the limited power that would be available to detect small differences; that is not necessary.

As in Example 12.2.0.1, a 95% confidence interval using Equation 6.10,

$$\left(\widehat{p}_{\mathbf{I}} - Z_{\alpha/2}\widehat{p}_{\mathbf{I}}\left(1 - \widehat{p}_{\mathbf{I}}\right)/\mathbf{n}\right) \leq p_{\mathbf{I}} \leq \left(\widehat{p}_{\mathbf{I}} + Z_{\alpha/2}\widehat{p}_{\mathbf{I}}\left(1 - \widehat{p}_{\mathbf{I}}\right)/\mathbf{n}\right),$$

might be more informative for such elementary genetic studies. But small sample sizes might tend to produce somewhat wide CIs; the instructor might do best to finish with a CI for the combined data of the whole class.

13.3.2 Subsequent power, r × c with k multiple

For a subsequent power estimate of a Pearson's r × c test with k multiple comparisons, you would need to, a priori, very clearly specify α, a targeted β, **and** the comparisons you intend to make. All this must be done **before** you obtain the data or information about its content. Then, if you have done all this and your sample size changes during the data collection, or if you preplan before encountering the data, as when you have no control over the treatment randomization (as when challenging independence) you can estimate the power of each comparison as in Section 8.2.7.

To estimate the power of the 'family-wide' group of comparisons you can combine the probabilities using the Stouffer's Z-score method, Section 10.8. For a subsequent estimate it is handy to substitute the signed lowercase z estimate into Stouffer's equation, Section 10.8 to estimate

$$\acute{z} = \frac{\sum_{j=1}^{j=k} \hat{z}_i}{\sqrt{k}},$$

and applying the appropriate rule for interpreting $\hat{z}'$ according to the sign of the result,

$$iff \; \hat{z}' < 0, \text{ then } |\hat{z}'| = \widehat{Z}_{1-\beta},$$

or,

$$iff \; \hat{z}' \geq 0, \text{ then } \hat{z}' = \widehat{Z}_{\beta}.$$

Some authors will tell you that you can analyze the data and then decide to do a subsequent power analysis if you do this or its equivalent for every comparison, even if you have no a priori specifications. This would be no more than data exploration unless you used some highly established reasonable minimum difference very logically linked to the question at hand, such as cost (= return). In that case you **could not use** any $\alpha > 5\%$ or any $\beta > 20\%$, because these maxima are established, and commonly expected anyway, whereas greater values would be too unreasonable. Even using these procedures, the results would be open to question, as to whether they were post hoc and therefore unfit for publication or for remuneration as a conclusion for a consulting report. Definitely **do not** report the results of such a completely post hoc procedure unless you **clearly explain the degree** to which it is post hoc.

Questions for Ch 13

Write the equations where appropriate, and then fill in the values. **Odd numbered questions** have worked answers at the back of the book and can be studied as additional examples. The data tends to be fictitious.

1. Let's use a new reasonable minimum difference to **recalculate** the example in Subsubsection 13.2.1.2 that is about testing whether the probability of a party of one ordering more than two beers is the same for the two cities. Find the estimated sample size $\hat{n}$ for each of the two samples to test if there is at least a reasonable minimum difference in probability equal to $\mathcal{D}_{min} = 0.05000$, still using $\alpha = 5\%$ and $\beta = 10\%$. Recall that your previous year combined receipts for the two locales suggest that $p_0 = 64\%$ of such customers in a party of one buying more than two drinks. Your hypothesis pair is **H0**: THERE IS NO ASSOCIATION BETWEEN CITY AND THE NUMBER OF REGULAR CUSTOMERS BUYING MORE THAN TWO GLASSES VS **HA**: THERE IS AN ASSOCIATION (a lack of homogeneity). This is clearly a two sample test, so use Equation 7.10.2,

$$\hat{n} = (Z_{cutoff} + Z_\beta)^2 \, (p_0 q_0 + p_1 q_1) \, / \mathcal{D}_{min}^2,$$

to make your estimate. (Remember to use $Z_{\alpha/2}$ even though considering χ_α^2.) Was a **larger or smaller** sample size needed than for the original example?

2. To what sort of effects is Pearson's chi square most sensitive? How does this relate to ϕ^2 as a measure of effect size? (Hint, you might want to include a simple equation for the latter part of the answer for this question.)

3. Gregor Mendel (1866) crossed F1 hybrid (Yy) pea plants which had yellow seeds with pure recessive (yy) plants which had green seeds. In accord with his theory, he expected

a $p_0 = 1/2$ of the offspring having each color of seeds. Assume that before he crossed these, he hypothesized **H0**: THE NUMBER OF PLANTS IN EACH CATEGORY WILL BE THE SAME vs **HA**: THE NUMBERS WILL NOT BE THE SAME. He chose $\alpha = 5\%$, $1-\beta = 80\%$, and a reasonable minimum difference of $\mathcal{D}_{min} = 2\%$. Estimate an appropriate total sample size. (Note that this is a one sample test that can also be seen as a goodness of fit test.)

4. Suppose for a 2 × 2 Pearson's chi square you can only get a sample of $n_1 = 30$ in the first category, but your initial a priori sample size estimate says you need a total of $\widehat{n} = 126$. Because our way of estimating $\hat{n}$ assumes equal sample sizes, you must adapt for the resulting loss of efficiency due to unequal sample sizes. Estimate the **efficiency lost** (Subsection 7.11.2). What is the corrected estimate of the **adjusted total sample size**, $\widehat{n}'$, and **also of the adjustable sample size** estimate $\hat{n}_2'$? (Once again, inequality is expensive.)

5. Assume I have lab records indicating that the probability of parthenogenetic female aphids on cotton leaves containing at least one unborn parthenogenetic female with unborn parthenogenetic females within it, after ten days, is $p_0 = 0.60$. For $k = 7$ comparisons with the cotton control, estimate n', the number of adult female aphids needed to be sampled **per plant species**, using an experiment wide $\alpha = 0.05$, and experiment wide power, $1 - \beta = 0.80$, with a reasonable minimum difference, $\mathcal{D}_{min} = p - p_0 = 15\%$. Including the control, there are $c = 8$ species; what is the **total** estimated sample size? Use the Sidak correction.

6. A political science student hopes to access good historical as well as recent records, and wants to test the null hypothesis, **H0:** THE FREQUENCY OF CONGRESS MEMBERS THAT HAVE NEVER CHANGED THEIR VOTE AFTER LARGE CORPORATE CAMPAIGN CONTRIBUTIONS IS 60% (alleged value by a news source), versus **HA:** IT IS EITHER GREATER OR LESS. Estimate the a priori total sample size for a simple 2 × 2 Pearson's chi square. This is obviously a one sample test so use $\hat{n} = \dfrac{\left(Z_{cutoff}\sqrt{p_0(1-p_0)} + Z_\beta\sqrt{p_1(1-p_1)} \right)^2}{\mathcal{D}_{min}^2}$, with $p_0 = 0.60$, $\alpha = 1\%$, $\beta = 10\%$, and $\mathcal{D}_{min} = 5\%$. Use Equation 7.24.

(If the result is very large, the student might need to sample all the way back to the increase in large corporations, and reduced regulation of their formation, just after the war between the states.)

7. Assume I am planning to do the same experiment as Question 5, still planning $k = 7$ comparisons, but with an expensive and hard to get species in the genus Sedum as the **control**. My initial estimate of the per treatment column (plant species) sample size is $\hat{n}' = 220$. But I have reason to doubt that I can depend on getting more than $n_f = 100$ parthenogenetic female aphids on this species. How might I adjust for this situation?

14 NONPARAMETRIC APPROACHES

The word **nonparametric** refers to procedures that do not directly use parameters such as the mean, μ, and the variance, σ^2, and which are not associated with the normal distribution. Nonparametric procedures do not require the raw data to approach a normal-like distribution. The sign test and other tests based on calculating directly from counts such as Pearson's χ^2, Chapter 12, can be seen as nonparametric by some people. On the other hand, the true χ^2 test (Section 8.2) is parametric in the sense that it is about variances.

Many nonparametric tests are order based and evaluate the order positions, $i = 1, 2, ..., n$, of the combined data points, x_i, when they are **arranged in order** from smallest to largest (or largest to smallest), instead of evaluating the x_i values themselves. It might seem strange to you that with random selection these tests depend on the fact that the sums and means of these order values quickly converge to normal. Such tests (Section 14.1) substantially reduce the exaggerated effect of outliers (suspiciously large or suspiciously small data values) with very little information loss with respect to correlation, or testing for statistical significance. So, nonparametric methods are worth considering when you are confronted with a lot of outliers in data that you would otherwise test with parametric tests. Lieges says that nonparametric tests give "less misleading results" when sample sizes are small than do parametric tests (1956).

Extreme cases of heterogeneity in grouped data, such as blocks, which could otherwise be considered for an ANOVA which is very sensitive to non normality, might be less of a problem using order based nonparametric tests, such as Friedman's test for a complete block design (Section 14.5).

Rank based nonparametric tests definitely do not compare means. Textbooks and some expensive software packages assume that they compare medians, but this is only true when the distributions are merely shifted versions of each other, and that is rare. Likewise textbooks and some software refer to the mean ranks but their separation increases as the sample size increases. The proper phrasing of a conclusion is usually, "the distributions were statistically significantly **different**", or "the categories were statistically significantly different." Instead of CIs, range based presentations and graphics such as boxplots and violin plots (Subsection 15.1.1) could be seen as, and used with, nonparametric procedures.

Unfortunately (as for parametric tests), many nonparametric tests are presented in the available texts as nil-testing vehicles, as though challenging absolutely no difference is a meaningful activity (Subsubsection 3.0.0.1). Statistically testing for absolutely no (nil) difference is impossible. Like all statistical tests, nonparametric tests only find a sample size dependent minimum difference, below which the effect is null (no effect with respect to application rather than absolutely no difference). For almost any rank based nonparametric test, strategies are (or eventually will be) available for choosing sample sizes for a desired power, provided that you choose an appropriate **reasonable** minimum difference.

The calculations in many nonparametric tests involve rearranging and scaling the data so that the likeness can be challenged by using what amounts to a Pearson's chi square in the final step. For these, when global, Cramer's phi (Equation 12.8) can be an acceptable alternative ES where the original units are unavailable or are in terms of an arbitrary index (Acock and Staving 1979). For sufficiently simple tests, you can confidently use phi (Equation 12.7). More effect size information follows in this chapter.

Monte Carlo procedures (enormous simulations involving repeated random sampling) indicate that most nonparametric tests have only slightly less power than corresponding parametric tests. Unfortunately the issue of estimating power and sample size in application is not well developed for all the seemingly useful nonparametric tests. So, not all nonparametric tests mentioned in this chapter include a nonparametric procedure for sample size estimation.

14.0.0.1 Nonparametric tests vs Student's t test

For situations where you cannot use a t test, Sawilowsky and Hillman (1993) showed that power calculations **based on the t-test** were appropriate even when the data was decidedly **non-normal**. They examined sample sizes up to 80. I would be a little cautious and increase these estimates by multiplying **any parametric sample size estimate by about** $1/0.95 \approx 1.05$, to be safe in using them for nonparametric procedures. This precaution is because if the data is normal, the equivalent nonparametric tests tend to be about 95% as powerful as is Student's t (e.g. Mood 1954). However, when the data fails to sufficiently approach normality, they tend to be more powerful than Student's t.

Leech and Onwuegbuzie make a very good argument for a greatly increased use of, and teaching of, nonparametric statistics (2002, with 52 references).

14.1 Rank Concepts

Statistical procedures based on ranking continuous data before analysis, rather than analyzing the raw data itself, are potentially very useful in applications where you have **lots of outliers** (data points that are too extreme to accept as quantitative values). However, sometimes the data are already in the form of ranks, or sometimes the only way to handle the data might be by arranging it in order, such as least desirable to most desirable. Another important use of ranking is to avoid the assumption of normality[1].

The first step in the most common approach to rank tests is to arrange all the values in order, such as from smallest to largest and then number them (give them ranks) starting with $r_1 = 1$ all the way up to finish with $r_n = \mathbf{n}$. For **ties**, say $x_{r_4} = x_{r_5}$, the 4 and 5 are averaged together so that you would rank each of them the same, in this case, $r_4 = r_5 = 4.5$. Subsubsection 2.6.3.1 might help with the ranking. But, you need to record the data to **as many** figures as possible to avoid having ties, because ties reduce the power of any ranking procedure. The **variance** of the first **n** non-zero integers (thus the first **n** ranks) is

$$\sigma^2 = \left(\mathbf{n}^2 - 1\right)/12,$$

and the **mean** is

$$\mu = (\mathbf{n} + 1)/2.$$

So, the underlying mean and standard deviation of the ranks is already determined by the sample size.[2]

Some tests, like the Wilcoxon rank test, rank all the data together and then put each rank value back into the treatment groups in place of the original raw data value, and then analyze that. Other tests, like the Spearman's rank correlation, cleverly use the data ranked individually for each treatment. Either way, these tests are usually designed so that we can either test for expected outcomes

---

[1]With a bit of arithmetic and a little experimenting, we can find all sorts of information about the distribution of ranks. For instance, the sum of n unsigned ranks is $\frac{n(n+1)}{2}$.

[2]The mean of the first n numbers can be worked out from the fact that the **sum** of the first n numbers is $\mathbf{n} = n(n+1)/2$, and the variance can be found by using the machine formula, Equation 9.15 and using the fact that the **sum of their squares** is $n(n+1)(2n+1)/6$.

using χ^2 somewhat like Pearson's test of the goodness of fit, or take advantage of the combined effects of summing ranks and taking advantage of the CLT to obtain a Z value (note that $Z^2 = \chi^2_{1df}$). Perhaps the most important pioneering nonparametric approaches involving ranks are Spearman's correlation (Subsection 14.6) and the rank tests by Wilcoxon.

Although the ranking process greatly improves the situation in which you have extreme outliers, by reducing the magnitude of their effects, it does not necessarily eliminate all their effects. Big outliers will be moved to the upper end of the rankings and small outliers will be moved to the lower end. But do these rank values always represent real data, let alone extremes? Bill Gates would produce an outlier in an income survey, and his income could be legitimately, and correctly, ranked at the upper end. But a burst of static in an analysis of a recorded sound print of a sax horn would not belong anywhere, because it simply covered up the real data value. Unless you knew what it was, you might be stuck with its reduced but still harmful effect at the upper extreme.

According to Fay and Michael (2010) a statistically significant difference in nonparametric tests refers to just that, a difference. They are obviously referring to the fact that we cannot specify a difference in original units of measure (e.g. see Section 14.1.0.1). In some applications, you are clearly targeting *ordinal differences* (e.g. differences in pecking order or order of arrival).

14.1.0.1 Preservation of differences between distributions under ranking

It is obvious that taking two samples (as from different treatments) and ranking all of them together, before separating the ranks back into their original samples, would tend to preserve some (but unfortunately not all) representations of differences in central measures such as the median. However, ranking can, **to some degree**, preserve differences in variance between two or more tests. Donald W. Zimmerman (2011) published a study highlighting the fact that when comparing two or more treatments, ranking preserves many characteristics of the underlying distributions. These characteristics, which are all preserved to **some** degree by ranking, include differences in skew, differences in variability, bimodality, and even tendency towards outliers. Each of these effects tends to be reduced **but** they may still have some effect on the probability of getting Type I and Type II errors. He also states, "Because the ranks are bounded by the total sample size, $n_1 + n_2 \, [+\dots]$ of the combined treatment groups, the frequencies of ranks in one group are constrained by the frequencies in the other. For this reason the shapes of the distributions of ranks are modified somewhat and do not precisely reproduce the shapes of the distributions of scores"[3]. So, ranking reduces the effects of a number of problems, but does not entirely eliminate them.

Some authors argue that when there is a difference due to treatments, the difference in medians 'only suggest a change in centering', and that the difference between means and medians are 'identically shifted' by the effects of the treatments. They are wrong; for instance, **outliers** will usually **shift the medians** much **less than** they will shift the **means**. However, despite the fact that rank tests are likely to be sensitive to populations that show differences between medians, rank tests (especially those which rank all the data **together** before breaking the ranks into treatments) do **not** really qualify as comparisons of the medians. Although there exists a very weak test specifically for comparing medians, the difference between medians, $\widetilde{x}_B - \widetilde{x}_A$, is **not a dependable effect size** for rank tests with small to medium samples; it is even possible to have a **statistically significant difference when the medians are equal**.

---

[3]square brackets, [ ], for correction of the Zimmerman quote.

14.1.0.2 Equal interval scales and not equal interval scales

Scales such that $2 + 3 = 5$ and $10 * 3 = 30$ are called *equal interval scales*. Measures of IQ, Likert scales of preference such as 1 to 10, or very disturbed, somewhat disturbed, a little disturbed, not disturbed, and many other psychological scales are purely ordinal. Larger IQ values are larger than lower IQ values, but you cannot say that an IQ of 75 is 3/4 of an IQ of 100. Such scales **fail to be equal interval scales**. However, because we can meaningfully arrange these non equal interval scores in order, they are **suitable for** the Wilcoxon Rank test and most other nonparametric **tests based on order**. They can even, with a little careful thought, be used in nonparametric correlation (Section 14.6). But measures like IQ are definitely not suitable for parametric procedures such as regression when tests of intercept or slope are involved (Section 11.2), or ANOVAs or for Student's t test.

14.1.1 Wilcoxon (-Mann-Whitney) summed rank test:

A nonparametric counterpart of the t-test, the Wilcoxon (1945) two sample summed rank test[4], extended by drawing from the *Mann–Whitney U test* (Mann and Whitney 1947), is a nonparametric test based on ordinal position (rank). Conceptually, the workings of the Wilcoxon test is as though you took a score from population A, and then compared it to every score in population B, one at a time. If, 'most of the time', the score of population A is higher, that would indicate a tendency for *superior* performance in population A (Grissom and Kim 2005, and Subsubsection 14.1.1.1).

The two sample Wilcoxon's test is often as good or better than Student's t test for small samples, even on scales of measurement that are suitable for parametric tests (Section 5.2.1). Mood appears to have been the first to prove that, in general, two sample rank tests are not excessively expensive in terms of power (hence sample size). The Wilcoxon test has $3/\pi = 95.493\%$ as much power with respect to sample size as the corresponding Student's t test (1954). Subsection 14.1.1.4 suggests that if you **multiply the sample size** you would need for a Student's t test by $\pi/3 = \mathbf{1.0472}$, you would get an improved estimate of the sample size needed for a Wilcoxon test. However, Equation 14.3 provides a more suitable a priori estimator for the necessary sample size for the Wilcoxon test.

At small sample sizes, the Wilcoxon test is more robust than Student's t test. Another advantage is that you do **not need a pilot study, or additional data**, to estimate the required sample size. Because Student's t becomes more robust as sample size increases and the Wilcoxon appears not to, it might be advisable to use the Wilcoxon test for combined sample sizes **n** < 60 when sample sizes are equal, but use Student's t test for **n** ≥ 60.

If skew is unusually extreme, you might consider the Wilcoxon test for sample sizes all the way up to **n** ≤ 80. Richard B. Darlington examined the Wilcoxon test with regard to asymmetry and concluded that ". . . Wilcoxon test has good validity as a test on W, even if the distribution of scores is quite skewed." (2018, http://node101.psych.cornell.edu/Darlington/wilcoxon/wilcox3.htm). Under skewed conditions, the Wilcoxon rank sum test is three to four times **more powerful** (Blair and Higgins, 1980; Bridge and Sawilowsky, 1999; Nanna and Sawilowsky , 1998).

Zimmerman and Zumbo (1992) concluded that rank methods are as influenced by unequal variances as are parametric tests, and recommended the t-test if differences in variance were a problem; Skovlund and Fenstad (2001) found that the t-test was superior to the Wilcoxon when variances were different. When $n_1 \neq n_2$, the Welch's t test, which has similar requirements to Student's t test, but is

---

[4]Wilcoxon's test may have become more commonly used after a publication by van der Vaart (1950). A proof by Noether (1987) alluded to testing each such randomly drawn pair of values via a sign test. This proof showed how the Wilcoxon two sample test circumvented a lot of this potentially huge task. You can also refer to Mood (1954) for a derivation which involves the expectations of n ranks.

modified for unequal variances, could cope with unequal variances even better when $\mathbf{n} \geq 30$ (Section 5.6).

These reports call for careful consideration before choosing between the Wilcoxon test and Student's t test (or Welch's t test). If you decide in favor of Student's t test, and you have sufficient information about the underlying distribution of your data, the sample size should ideally at least be $25 \times skew$ (Subsection 15.2.1). However, you cannot pre-examine the data you are actually going to test and estimating the skew (Subsection 15.2.1) would require a sufficient amount of data that would not be included in the testing process.

Calculating the Wilcoxon Rank Sums, two sample test

The Wilcoxon rank sum based **two sample test** combines the data from two samples ranked together, $(1, 2, \ldots, \mathbf{n})$, where $\mathbf{n} = n_1 + n_2$, the sum of the two treatment sample sizes (Langley 1971, Conover 1980). If possible, an a priori sample size should be estimated and used (Subsection 14.1.1.4). The null hypothesis is best stated as **H0**: THE DISTRIBUTIONS ARE THE SAME. As with other tests, you must have chosen α and whether one tail or two tails in advance. However, Mann and Whitney (1947) extended it for unequal sample sizes. If $n_1 \neq n_2$, it **requires more ranking** because you have to rank **both ways**, from smallest to largest and from largest to smallest (Mann and Whitney 1947). The **smallest rank sum** is symbolized W_R. The ties are averaged as described for ranking in general, but it is always better to collect your data a **little beyond** the limits of plausible accuracy in order to **break ties**. Ties between ranks tends to weaken the results.

First choose whether you are planning a one tail test (criterion α) or a two tail test (cutoff $\alpha/2$). To provide for power analysis you must also **a priori** choose a β or $1-\beta$, and a reasonable minimum difference, let's call it $\mathcal{P}_{S_{min}}$. Subsubsection 14.1.1.1 will explain $\mathcal{P}_S$ as an effect size that is in terms of the probability of drawing a success (a superior outcome) at random. It avoids the very tedious but equivalent use of sign tests to compare every score with every other score, which would take $n!$ comparisons, a very large number of comparisons in most applied situations. Unfortunately reasonable differences available for reporting are not in original units of measure. For publication you might even be able to use $\mathcal{P}_S$ as an effect size, or more conventionally you might be interested in choosing ϕ^2_{min} as the effect size in terms of the proportion of variance that is due to the treatment (also Subsubsection 14.1.1.1). After ranking the entire combined data set, assign each rank value back to the treatment received by the corresponding data value (Section 14.1). The sums of such ranks behave somewhat like means and converge rather rapidly to a normal distribution. Thus a Z estimate for the two tail Wilcoxon summed rank test can be calculated using

$$Z_w = \frac{n_R(n_1 + n_2 + 1) - 2W_R}{\sqrt{\frac{n_1 n_2 (n_1 + n_2 + 1)}{3}}}, \tag{14.1}$$

where the denominator is the standard error for this two sample test and W_R (sometimes just W in literature) is the **smallest rank total** of the samples, and n_R is the size of the sample that has the **smallest rank total**. The result, Z_W, can be evaluated using ordinary tables of Z when both $n_1 > 10$ and $n_2 > 10$.

With $4 \leq n_1 \leq 10$ and $4 \leq n_2 \leq 10$, W_R can be evaluated using a table such as Appendix 8, and statistical significance can be declared if W_R is less than the shown value that is often called 'R'. However, it is doubtful that such small sample sizes would have sufficient power to be useful unless you were only looking for a very big reasonable minimum difference. If the sample size is $\mathbf{n} \geq 60$, you might want to consider Welch's t test instead (Section 5.3). A few tied ranks might not have much effect on this test, but a lot can inflate the intrinsic variance estimate of the rank sum and reduce the power of the test. Unfortunately I was not able to find a simple, or an easily explained way to correct for tied ranks in this two sample test. If you observe a lot of tied ranks in this test, try to find a software package that automatically corrects for them (of course using the same cutoff, α or $\alpha/2$).

If you are only interested in **comparing variabilities** instead of differences that roughly equate to differences in central measures, you could use Siegel-Tukey ranking before running Wilcoxon's test (Subsubsection 14.2.2); it is available in R (eg. PMCMRplus package) and in Stata.

14.1.1.1 Effect size for the Wilcoxon two sample test

The Wilcoxon test tries to detect a reasonable minimum difference, $P_{S,min}$, from a $50 : 50$ chance under **H0**, of a single randomly selected value being larger when selected from one population than a single randomly selected value from the other. That means that the null hypothesis is that each treatment has a 50% probability because the populations are alike. This observed ***probability of superiority*** (Grissom 1994), $\mathcal{P}_S$, is the most appropriate effect size for a nonparametric Wilcoxon rank sum test. Estimating $\mathcal{P}_S$ requires you to first estimate a value,

$$U = W_R - \frac{n_R\,(n_R + 1)}{2},$$

which originates as part of an alternative, but equivalent, calculation of the Wilcoxon test. Then ,

$$\mathcal{P}_S = \frac{U}{n_A n_B},$$

the **additional probability** exceeding 50% (no difference) that a score randomly drawn from population A will be **greater** than a score randomly drawn from population B **or vice versa** (Grissom and Kim 2005). So the Wilcoxon's test is used to discover a statistically significant likelihood ($p\_value \geq \alpha$) of having a probability, $\mathcal{P}_S$, above equality of finding a higher score on a single draw.

Would an NBA scout have a greater probability of finding a superior high school athlete in Los Angeles or in Chicago? If the observed $\mathcal{P}_S = 3\%$ is statistically significant (at an a priori alpha of say 5%) then we will estimate that if the scout chose the correct city there would be a $50\% + \mathcal{P}_S = 53\%$ chance of getting the more qualified athlete, the other city providing a chance of only $50\% - \mathcal{P}_S = 47\%$. Although the relation of the median with $\mathcal{P}_S$ varies with the particular distribution of the raw data, the direction of the difference can **usually** be determined by which, A or B, had a **larger** median. (A post hoc look at a scatter plot may help with interpreting statistical significance with this test.) As is the case with most other tests, the larger the effect size, in this case $\mathcal{P}_S$, the smaller the p_value that will be compared to the criterion α.

If you have statistical significance, you can report an alternative effect size as phi square, an estimate of the fraction of the observed ordinal variance that is **due to the difference** between the treatments,

$$\phi^2 = Z_W^2/\mathbf{n},$$

which is equivalent to $\phi^2 = \chi^2/\mathbf{n}$ (Equation 12.7). However, be aware that the test is technically about the probability of randomly selecting a larger value from one of the sampled distributions than from the other.

Correlation based effect sizes are sometimes reported, and ϕ can mathematically be interpreted as a correlation. But, for many experimental designs, it would be presumptuous to assume that the values of the two populations vary together in some linear or curvilinear manner merely because you have statistically significant evidence that there are two populations instead of one. They could simply be different!

14.1.1.2 Example Wilcoxon's Two Sample Summed Rank Test

This example refers back to the example in Subsection 5.2.0.2

Methods and Materials The public health student suspected that the average of the weights of male students at Little U College that use soft drinks with high fructose sweetening is more than for those who do not.. She hypothesized:

H0: THERE IS NO TRUE DIFFERENCE BETWEEN THE POPULATION OF WEIGHTS OF MALE STUDENTS AT LITTLE U DRINKING HIGH FRUCTOSE SWEETENED SOFT DRINKS AND THE POPULATION OF WEIGHTS OF MALE STUDENTS AT LITTLE U WHO DO NOT.

HA: THERE IS A DIFFERENCE BETWEEN THE POPULATIONS.
$\alpha = 5\%$

She randomly sampled the weights of $n_1 = 13$ high fructose soft drink users and $n_2 = 15$ non users.

Results: She chose to do a Wilcoxon test because she was using small sample sizes. She designated the treatments X_1 for non-users and X_2 for high fructose soft drink users. After she obtained the double blind random samples, she realized that because she had unequal sample sizes, she had to use the Mann-Whitney approach to the Wilcoxon test[5]. That is, she had to rank the data both ways and use the lowest rank sum for W_R.

	Wilcoxon: Mann-Whitney Data Ranked Both Ways														Sum	Median
Fructose	189.4	191.6	126.4	197.2	153.8	224.6	106.8	241.8	156.7	151.7	104.1	197.4	183.8			183.8
X2 Rank +	20	21	8	23	12	27	3	28	13	11	2	24	19		211	
X2 Rank -	8	7	20	5	16	1	25	0	15	17	26	4	9		153	
No Fr.	125.6	209.0	147.5	173.9	195.9	122.2	87.6	173.6	174.1	133.4	157.4	176.2	122.2	223.3	110.1	157.4
X1 Rank +	7	25	10	16	22	5	1	15	17	9	14	18	6	26	4 195	
X1 Rank -	21	3	18	12	6	23	27	13	11	19	14	10	22	2	24 225	

Analysis: Using Equation 14.1,

$$Z_w = \frac{n_R(n_1 + n_2 + 1) - 2W_R}{\sqrt{\frac{n_1 n_2(n_1 + n_2 + 1)}{3}}},$$

where W_R is the smaller total of the two ranks, and n_R is its sample size ($n_1 = 13$),

$$Z_W = \frac{13(15 + 13 + 1) - 2 \times 153}{\sqrt{\frac{15 \times 13(15 + 13 + 1)}{3}}} = 1.63532.$$

Results and Discussion The cutoff value required to reject **H0** is $Z_{0.025} = 1.960$, but $Z_W = 1.63532$. So, she could not reject **H0**. Her professor warned her not to draw any conclusions from this procedure because her test was **not** statistically significant.

Exploring Effect size: She could not even publish an effect size because she did not get statistically significant evidence that an effect size exists. However, she wanted to unofficially explore. So, she calculated the **probability of superiority**, $\mathcal{P}_S$, using U, a value sometimes used in alternative versions of the Wilcoxon summed rank test (Grissom and Kim 2005),

$$U = W_R - \frac{n_R(n_R + 1)}{2} = 153 - \frac{13(13 + 1)}{2} = 62,$$

---

[5]For an idea of how to rank on a spreadsheet see Subsection 2.6.3

and using this she was able to estimate

$$\mathcal{P}_S = \frac{U}{n_A n_B} = \frac{62}{195} = 0.3179487,$$

which rounds to an effect size of $\mathcal{P}_S = 32\%$ additional probability (beyond 50%) that a score (an individual weight in this example) randomly drawn from one population would be greater than a score randomly drawn from the population with the smaller median. If the test had been significant, the population could be seen as a $18 : 82$ probability of a superior score rather than a $50 : 50$ probability of a superior score (in favor of X_2 for high fructose soft drink users).

If she had found statistical significance she could have used phi square as an estimate of the **percent of the variability** of the weights of male students that was associated with using high fructose soda,

$$\phi^2 = Z_W^2 / \mathbf{n} = (1.63532)^2 / 28 = 0.095510,$$

which rounds to 10%.

14.1.1.3 Subsequent power for the Wilcoxon test

Before returning to our student let's look at subsequent power for Wilcoxon's test. By rearranging Equation 14.3, it is possible to estimate the subsequent power, assuming there exists an a priori minimum reasonable difference (call it $\mathcal{P}_{S_{min}}$ rather than $\mathcal{D}_{min}$), the probability that a single randomly drawn individual (a score) from one population will be larger than if it were drawn from another population. That is the probability of coming across someone from a population that tended to have heavier individuals would be likely (more probable than $1/2$) than from the other population. Then using,

$$z = \mathcal{P}_{min} \sqrt{\mathbf{n} 12 c (1 - c)} - Z_{cutoff}, \quad c = 1 - n_R / \mathbf{n}, \tag{14.2}$$

where n_R is the sample size of the treatment with the smaller total of the ranks. It can be evaluated using appropriate tables, such as Appendix 4 or online sources via the sign rules and either Equation **7.13**,

$$\textit{iff } \hat{z} \geq 0, \text{ then } \hat{\mathbf{z}} = \hat{Z}_\beta, \text{ or}$$

or Equation **7.14**,

$$\textit{iff } \hat{z} < 0, \text{ then } |\hat{\mathbf{z}}| = \hat{Z}_{1-\beta}.$$

Example (continuing): Calculations that would be acceptable given a priori preparation:

The student's professor pointed out that because the test was not statistically significant, she didn't even know **if an effect existed**, or if any post hoc effect size estimate was merely a result of chance. However, he suggested that she do some data **exploration** of the power that would have been achieved given statistical significance, and also alternatives to $\mathcal{P}_S$ as an effect size.

Exploring Post Hoc Power: If she had planned for an appropriate power, a priori selecting not only α but also β and a reasonable minimum difference, $\mathcal{P}_{S_{min}}$, she could make a post hoc estimate of the power attained.[6] Purely for exploration, she decided to estimate the power of this

---

[6]Some authors point out that power estimates for nonsignificant tests are questionable, However they do show a rough estimate of the power in this one instance and for the sample size attained. But even a high power estimate must not be represented as supporting **H0**.

test using a reasonable minimum difference of $\mathcal{P}_{S_{min}} = 20\%$, which should have been a priori. That is, she calculated as though she was targeting a difference such that if she randomly chose a male student from whichever treatment happened to favor a greater weight, and compared him with a randomly chosen male student from the other treatment, he would be heavier in 20% more instances than if she selected him from the alternative treatment. Notice that this kind of reasonable minimum difference is not the usual one she would be likely to use for a Welch's (or Student's) t test, because she is looking for a difference in the percentage probability of outcomes rather than a difference in the average weights as per the t tests.

Analysis: She used Equation 14.2, $z = \mathcal{P}_{min}\sqrt{n12c(1-c)} - Z_{cutoff}$, $c = 1 - n_R/n$, to find

$$z = 0.20\sqrt{28 \times 12 \times (1 - 13/28) \times (13/28)} - 1.9600,$$

$$z = -0.1316518$$

where $c = 1 - n_{smaller}/n$.

Results and Discussion According to the subsequent inadequate power rule, Equation 7.14, z represents the Z value for the power, $1 - \beta$. Using online software, she found her estimate of the power, $1 - \beta = 0.44763$. So, her test appeared to have a power of only about a 45% likelihood of detecting a $30 : 70$ or a $70 : 30$ relationship, as contrasted with equality, a $50 : 50$ relationship, with respect to who was likely to outweigh who. Using two unequal sample sizes, $n_1 = 15$ and $n_2 = 13$, the test was not very sensitive even to this very large reasonable minimum difference for frequency, $\mathcal{P}_{S_{min}} = 20\%$.

However, even with equality, such small sample sizes as say, $n_1 = n_2 = 14$, would not have produced much more power than in this example which is weakened by an unbalanced sample size.

14.1.1.4 A priori sample size for Wilcoxon's two sample test

Noether (1987) estimated the sample size for a two sample Wilcoxon's test for $n = n_1 + n_2$ via

$$\widehat{n} = \frac{(Z_{cutoff} + Z_\beta)^2}{12c(1-c)\mathcal{P}^2_{S_{min}}}, \tag{14.3}$$

where $n = n_1 + n_2$, and $c = 1 - n_{smaller}/n$. So, if $n_1 = n_2$, then $c = 1/2$. The $_{cutoff}$ can be either α for one tail or $\alpha/2$ for two tails. Round $\widehat{n}$ **up** to an even number.

If you are interested in **odds ratios**, you can select the reasonable minimum difference to **target** a particular odds ratio, $O_R = \mathcal{O}_1/\mathcal{O}_2$, as a reasonable minimum difference using $\mathcal{P}^2_{S_{min}} = O_R/(1 + O_R)$ (Section 12.6.2.1).

Note that you do **not need a pilot study, or additional data**, to proceed, because the probability (in this case, c) implies the standard deviation via substitutions into, and rearrangement of, a normal approximation using the known characteristics of rank values.

14.1.1.5 Example: A Priori Sample Size for Wilcoxon

If you were planning to test the lifespan of spotted hyenas (*Crocuta crocuta*) versus the lifespan of striped hyenas (*Hyaena hyaena*) in the wild, then using the Wilcoxon two sample rank sum test, you could target a reasonable minimum difference, the added probability that a randomly chosen spotted hyena would have a significantly different lifespan from a striped hyena. For how many randomly chosen hyenas are you going to have to somehow find their lifespans?

Methods and Materials Assume you choose a reasonable minimum difference in the frequency of longer life of $\mathcal{P}_{S_{min}} = 20\%$. You choose $\alpha = 5\%$, because it is to be a two tail test, $\alpha/2 = 0.025$. You also choose the popular 80% power, even though it means that one time out of five you will not detect exactly $\mathcal{P}_{S_{min}}$ (but the probability of detection will increase if the $\mathcal{P}_S$ of a population that tends to have longer lived animals exceeds 70%). Thus $\beta = 1 - 0.80 = 0.20$. You use any of the many sources to find that $Z_{0.025} = 1.9600$ and that $Z_{0.20} = 0.8416$. Because you plan equal sample sizes, $c = 1/2$.

Analysis: Using Equation 14.3, $\widehat{n} = \dfrac{\left(Z_{(cutoff)} + Z_\beta\right)^2}{12c(1-c)\mathcal{P}_{S_{min}}^2}$, you find that

$$\widehat{n} = \frac{(1.9600 + 0.8416)^2}{12 \times 1/2 \times (1 - 1/2) \times 0.20^2},$$

$$\widehat{n} = 65.407,$$

which rounds up to a total of 66 samples.

Discussion: Ideally you should use equal sample sizes for each species, $n_1 = n_2 = 33$ animals of each species. This $\mathcal{P}_{S_{min}}$ is rather large, but large reasonable minimum differences certainly save on the cost in terms of sample size. Maybe you should consider some sort of stratified random sampling, that is, contriving to randomly sample equal numbers of each gender. That might best be done with a total of two even numbers, $34 + 34 = 68$ animals. And you might want to sample some extra animals in case some get lost and you cannot track them, or their collars, anymore. Anyway, have fun out there with the hyenas.

14.1.1.6 Wilcoxon, $n_1 \neq n_2$

If you are forced to have **one sample smaller** than the other due to a fixed maximum sample size, n_f (where f is for fixed), that will be less than your required more adjustable size, then before the test, you must a priori specify $\mathcal{P}_{S_{min}}$, and because they tend to be the minimum values acceptable to your peers, you must choose $\alpha \leq 5\%$ and $\beta \leq 20\%$.

- To get an **initial provisional estimate** using Noether's (1987) Equation 14.8,

$$\widehat{n} = \frac{\left(Z_{cutoff} + Z_\beta\right)^2}{12c\left(1 - c\right)\mathcal{P}_{S_{min}}^2} \text{ using } c = 1/2,$$

which, assuming equal sample sizes, is simply $\widehat{n} = 2n$ and has to be rounded **up to an even integer**.

- You are **not finished**; you must now calculate a properly adjusted estimate $\widehat{n}'$ using a more appropriate estimate of $c' = 1 - (n_{smaller}/\widehat{n})$ to correct for some of the power loss due to the fact that you started by estimating $\widehat{n} = 2n$ as though $n_1 = n_2$. So,

$$c' = 1 - n_f/\widehat{n} \text{ and}$$

$$\widehat{n}' = \frac{\left(Z_{cutoff} + Z_\beta\right)^2}{12c'\left(1 - c'\right)\mathcal{P}_{S_{min}}^2}.$$

Round **up**, and it might be prudent to iterate once. But **do not continue iterating** beyond once because the estimate of $\widehat{n}'$, the corrected sample size, will simply get larger and larger.

- Then for this two sample test with unequal sample sizes, the final adjusted estimate of the **adjustable sample size** is

$$\hat{n}'_{adj} = \hat{\mathbf{n}}' - n_f.$$

Note that the super script $'$ marks are simply to indicate results of adjustments.

14.2 The Wilcoxon Paired (or Single Sample) Signed Rank Test

The Wilcoxon signed rank test (Wilcoxon 1945, Siegal 1956) is a nonparametric counterpart of the Student's paired t test, and it is more robust for small samples than is the paired t test. However, even though it has less assumptions than the t test, it includes the assumption of symmetry, but to a lesser extent than a small sample paired t test. Any t test has an assumption that the differences between the values come from an equal interval scale. This disqualifies t tests, but not rank tests, from testing subjective ratings, IQ scores, and similar common scales. As an example of not meeting this assumption, Richard Lowrey (2018) refers to asking students to guesstimate probabilities out of their heads. The Wilcoxon test is **free of the equal interval scale assumption**. For equal interval scores the t test becomes more robust with sample size, and the Wilcoxon signed rank test does not, so you should consider using this Wilcoxon test for small sample sizes, $n_d < 30$ pairs (or if the skew is extreme, then $n_d < 60$ pairs), and the paired t test for n_d greater than that.

Some authors may say that losing power with already small samples is unacceptable, but the power lost by using a Wilcoxon test rather than Student's t test is generally less than 4.5% (assuming normality of the raw data), and a small sample paired t test that does not meet its assumptions, assuming it has any meaning at all, can easily lose more power than that.

The value, W, that we find by the following rank based procedure starts as a very platykurtic distribution, but when $n > 10$ it has sufficiently converged that it could be considered to be normally distributed. However, more precise probabilities are given in Appendix 9 (for $5 \leq n_d \leq 25$). It is more robust at smallish sample sizes than Student's t test, not requiring normality, but like the t test with small samples, it requires that the distribution be **symmetrical** (not skewed). J. Voraprateep observed that asymmetry caused a loss of power even with big sample sizes (2013). In light of this apparent persistent effect of asymmetry even with increasing sample size, it might be better to use a Student's paired t test instead and rely on the more direct (than in rank tests) effect of the CLT that will reduce the skew if the sample size is sufficiently large ($n_d > 30$ pairs or if the skew is extreme, then $n_d > 60$ pairs).

To do a Wilcoxon paired signed rank test start by forming a table with seven columns, one each for: x_H, x_K, $\mathbf{d} = x_A - x_B$, $|\mathbf{d}|$, rank$(|\mathbf{d}|)$, $+$rank$(|\mathbf{d}|)$, and $-$rank$(|\mathbf{d}|)$. Ties, $x_K = x_H$, also marked $\mathbf{d} = 0$, do not contribute to the ranking and the information is lost. However ties will require a correction because otherwise they cause the variance to be underestimated unless you adjust for them. So always try to collect the data to a greater number of places than seems reasonable in order to break ties. The operations for each column, starting with the two data columns, are illustrated in the following example (Subsubsection 14.2.0.1), which culminates with the summing of the $\pm$rank$(|\mathbf{d}|)$columns. The **greatest difference between the two sums**, $+$rank$(|\mathbf{d}|)$ and $-$rank$(|\mathbf{d}|)$ (regardless of sign) is W, and the expected value of W, given that there really is no difference, is

$$\mu_W = \frac{n_{\mathbf{d}}(n_{\mathbf{d}} + 1)}{4},$$

and whether there is a **difference or not**, the standard deviation of W is

$$\sigma_w = \sqrt{\frac{n_d \left(n_d - 1\right)\left(2n_d + 1\right)}{6}}. \tag{14.4}$$

Both are determined by the sample size.

Ties, identical values of $\pm rank\left(|d|\right)$ with the same sign that are equal to each other, require an **adjustment**, C_T, of the estimate of σ,

$$C_T = \sum_{all\,ties} C_i = \sum_{all\,ties} \frac{t_i^3 - t_i}{48},$$

where t is the number of such ties. If you have $+3.5, +3.5$ as final values of $+rank\left(|d|\right)$, that is $t = 2$ ties, or if you have $-3, -3, -3$ as final values $-rank\left(|d|\right)$, that is $t = 3$ ties. Each such group is a separate $_i$ in the summation. Then, you can estimate

$$Z_W = \frac{\left(W - \mu_W\right)}{\sigma_W - C_T}, \tag{14.5}$$

but if there are no ties, then $C_T = 0$.

For a total sample size, $n_d > 25$ **pairs**, refer the value, Z_W, to a table[7] of Z such as in Appendix 4 to challenge your a priori $\alpha/2$ cutoff.[8]

When $n_d \leq 25$ **pairs**

For $n_d \leq 25$ pairs, $T = \max(\sum rank\,|+d|\ or \sum rank\,|-d|)$, referring to the table in Appendix 9 as per Example 14.2.0.1 and its table (below).

14.2.0.1 Example: Pairing Revisited:

Let's look back at Example 5.4.0.1, but use the Wilcoxon signed rank test instead of a paired t test. .

The Creatures from the planet of two hands (revisited) Recall the group that did the nail
  polish experiment in Example 5.4.0.1. They formed a hypothesis pair,

H0: THERE IS A DIFFERENCE

HA: THERE IS NO DIFFERENCE,
  and since this is a two tail t test, $\alpha/2 = 2.5\%$

Method: As described in that example, each of the same lady's hands was to be given nail polish
  with either product H or product K at random. The product was presented in a double blind
  manner to the ladies, and even the time keeper did not know which product was which.

Analysis: To do the Wilcoxon signed rank test they created a table of times, $x_{i,j}$, and calculations:

---

[7]Some authors, very conservatively, use a continuity correction ± 0.5 in the denominator of Z_W; the sign is chosen so as
  to decrease the absolute value of Z_W regardless of direction. The effect is usually minimal.
[8]The copyright tables for Wilcoxon's paired signed rank test in Langley (1971) might be more exact for $n_d \leq 25$.

x_H	x_K	$d = x_K - x_H$	$|d|$	rank $|d|$	rank($|+d|$)	rank($|-d|$)
19.52	19.88	0.36	0.36	2	2	
18.55	21.95	3.40	3.40	12	12	
22.37	10.53	-11.84	11.84	18		18
19.37	10.53	-8.84	8.84	14		14
22.55	19.95	-2.60	2.60	10		10
17.31	19.39	2.08	2.08	8	8	
12.43	13.67	1.24	1.24	6	6	
10.19	10.11	-0.08	0.08	1		1
10.40	12.60	2.20	2.20	9	9	
21.22	15.18	-6.04	6.04	13		13
23.37	9.53	-13.84	13.84	20		20
19.34	18.46	-0.88	0.88	4		4
13.58	13.02	-0.56	0.56	3		3
11.31	9.39	-1.92	1.92	7		7
10.19	11.11	0.92	0.92	5	5	
22.34	10.46	-11.88	11.88	19		19
13.49	10.81	-2.68	2.68	11		11
12.25	22.25	10.00	10.00	16	16	
21.46	10.74	-10.72	10.72	17		17
13.49	22.81	9.32	9.32	15	15	
				T	$T_+ = 73$	$|T_-| = 137$

The researchers

used the first three columns to find the individual differences, $d_i = x_{K_i} - x_{H_i}$, shown in the third column. Then they worked across the remaining columns to find the absolute value of the difference, and finally reassigned the ranks to the appropriate columns, rank($|+d|$) and rank($|-d|$). Assume they summed each of the last two columns and carried the maximum sum (regardless of sign), $T_{max} = |T_-| = 137$, on as W to the 'large' sample calculation. They might handily have done some of this on a spreadsheet (see Subsubsection 2.6.3.1 to see how you might do this). Then suppose they went ahead as for $n_d \geq 25$ pairs and used Equation 14.4, $\sigma_w = \sqrt{\frac{n_d(n_d-1)(2n_d+1)}{6}}$, to get

$$\sigma_w = \sqrt{\frac{20\,(20-1)\,(40+1)}{6}} = 50.9575,$$

and then Equation 14.5, $Z_W = \frac{(W-\mu_W)\pm0.5}{\sigma_W}$ with no ties, to find

$$Z_W = \frac{137 - \frac{20(21)}{4} - 0.5}{50.9575} = 0.6181622.$$

Well, this falls short of $Z_{\alpha/2=0.025} = 1.960$, so the apparent difference does not seem to be significantly different.

But wait; this is **not a 'large' sample** because $n_d < 25$; they should have referred the $T_{min} = T_+ = 73$ to a table such as the one in Appendix 9 (which calls it T whereas other sources might call it W), where **it must not exceed** the given value, which is 52. The second column of the table in Appendix 9 is for two tailed tests with n overall $\alpha = 5\%$, so it gives the **maximum** statistically significant value for $T = 52$ for a cutoff of $\alpha/2 = 2.5\%$. So, $T_{min} = 73$ is **not**

statistically significant, even providing for small samples, and they know nothing new.

Discussion: If their sample size was $n_d > 25$, they most certainly should have used the large sample formula as in the mock error above. However, as is sometimes true, the sample size is $n_d \leq 25$ and they must use the table of calculations by Lee (2005) in Appendix 9. Of course the smaller the sample, the larger the difference must be, and the more limited the power will be. And, yes, the paired t test in Subsection 5.4.0.1 also failed to get significance for this same data.

Tied pairs ($d = 0$) would not count and that would mean they would calculate for one less pair for each tied pair. Differences that are equal will span two or more ranks, and these ranks should be averaged for a single rank that they will receive. That is, if the order (rank) positions 3, 4, and 5 are equal to each other, they should give them all a rank of 4; or if order positions 2 and 3 are equal to each other, they should give them both a rank of 2.5.

If you have statistical significance, you can report an alternative effect size as phi square, an estimate of the **fraction of the observed ordinal variance** among the pairs that is **due to treatments**,

$$\phi^2 = Z_W^2/n,$$

which is equivalent to $\phi^2 = \chi^2/n$, in this case where n is either the number of pairs, or for the one tail test (below), for the number of samples.

14.2.0.2 Wilcoxon's one sample test:

Wilcoxon's one sample test can be used as a substitute for the single value t test to use for small samples, and for when the scale of measurement is not on an equal interval scale. It is roughly analogous to a normal approximation, in which drawing a larger value is seen as a success.

By 'one sample' this book refers to one sample being compared to some standard, in this case a single value, x_0. This test is **not about medians**, it simply tests to see if there is a statistically significant difference in the ordinal distance from zero or from some other hypothesized x_o, which could be any value, positive, negative, or zero, instead of finding the differences between two columns of numbers. As the first step, we simply find the differences from a single value, x_0. An acceptable hypothesis pair would be

H0: THE POPULATION OF SCORES BEING SAMPLED ONLY DIFFERS FROM x_0 BY CHANCE ALONE

HA: THERE IS A DIFFERENCE BETWEEN THE SCORES AND x_0 THAT IS NOT ONLY DUE TO CHANCE ALONE.

Start by choosing an α, a β and a reasonable minimum difference, $\mathcal{P}_{S_{min}}$, your chosen percent above the 50% probability of drawing a value greater than x_0 from one of the treatments. Then do your best to estimate the appropriate sample size as discussed below. Next take a random sample and subtract the hypothesized x_0 from each of the $i = 1$ to n raw data values to get $d_i = x_i - x_0$. Finally calculate and evaluate your results just as for a paired difference Wilcoxon test (above).

14.2.1 Power, Wilcoxon's one sample and paired sample tests

14.2.1.1 A priori sample size for either single sample or paired sample Wilcoxon tests

To determine sample size n for either the single sample or the paired sample Wilcoxon test (with $c = 1/2$, meaning equal sample sizes), assuming you chose α or $\alpha/2$ for one or two tails respectively,

β, and a reasonable minimum difference, $\mathcal{P}_{S_{min}}$, I recommend another Noether's (1987) equation,

$$\hat{n} = \frac{(Z_{cutoff} + Z_\beta)^2}{3\mathcal{P}_{S_{min}}^2},\qquad(14.6)$$

where $\hat{n}$ can be either the number of pairs or the sample size itself, if there is only one treatment. The $\mathcal{P}_{S_{min}}$ is your a priori reasonable minimum difference, your chosen percent **above the expected** 50% probability for **H0: THERE IS NO DIFFERENCE**, of randomly drawing a larger value from one population than from another, or from a hypothesized population. There is agreement that it does work for large samples, and this is good news if your measures are not on an equal interval scale. However, there is not complete agreement in the literature as to whether this method is weak for small sample sizes. Maybe for small samples ($[n_{\mathbf{d}}$ or $n] < 30$ pairs of samples) you could compare this outcome with the sample size for a paired Student's t test (Subsection 7.6.1) multiplied by $1/0.95 = 1.05$ and then choose the larger sample size from the two methods estimate to be safe.

Post hoc power of either a single sample or a paired sample Wilcoxon signed rank test

If you have an a priori α, β, and $\mathcal{P}_{S_{min}}$, then you may run a post hoc power analysis by squaring, rearranging, and substituting z for Z_β and rearranging Equation 14.6, $\hat{n} = \frac{(Z_{cutoff}+Z_\beta)^2}{3\mathcal{P}_{S_{min}}^2}$ (itself a rearrangement of Noether's 1987 equation) to get

$$\hat{z} = \mathcal{P}_{S_{min}}\sqrt{3n} - Z_{cutoff},$$

where n is the sample size you **actually attained**. Then z can be evaluated using appropriate tables, such as Appendix 4 or online sources using the sign rules and either Equation 7.13,

$$iff\ \hat{z} \geq 0,\ then\ \hat{z} = \hat{Z}_\beta,$$

or Equation 7.14,

$$iff\ \hat{z} < 0,\ then\ |\hat{z}| = \hat{Z}_{1-\beta}.$$

14.2.2 Nonparametric test of spread

A useful variation on the ranking theme is *Siegel-Tukey ranking* of the combined data (sometimes called alternating counts). The smallest is ranked 1, the largest 2, the second smallest 3, the second largest 4, etc. (Lieges and Tukey 1960). The idea is that a statistically significant difference between these when separated back into their two categories will usually be obtained if the variabilities (the spreads) are not equal. That is, if it leads to a statistically significant outcome, it implies that $\sigma_2^2 \neq \sigma_1^2$. This, when combined into a Wilcoxon sum of ranks test (Section 14.1.1), is called a *Siegel-Tukey test* (Langley 1971). This approach is always superior to an F test for comparing two variabilities. It is handy, especially when the sample sizes are not huge. (Big data, national surveys etc., are a challenge for even a computer to rank after any fashion.)

You might be able to use this ranking approach with some other rank tests to address slightly different questions (Langley 1971, Lieges and Tukey 1960).

14.3 Kruskal-Wallis (One Way) Test

Kruskal and Wallis (1952) devised their nonparametric global rank based test to see if the treatments are all in the same population. Since the ANOVA is somewhat dependent on normality, those who must do global (in the sense of several treatments at once) tests using an experimental design like the one way ANOVA might be well advised to use a Kruskal-Wallis test instead. Also when the scale of measurement is not on an equal interval scale, you **cannot use the ANOVA**, but you can use the Kruskal-Wallis test. Examples that are not on an equal interval scale are customer ratings, subjective responses as on psychological surveys, and IQ, where half of the score, or of the rating, does not necessarily mean half as strong an effect. Since the underlying variance (the expected variance under **H0**) of ranks is already known, only the value of the denominator in Equation 8.14, $F_{1,2} = \left(\chi^2_{df_1}/df_1\right) / \left(\chi^2_{df_2}/df_2\right)$ is unknown; that is, the test only needs to estimate the equivalent of $\chi^2_{df_2}$ from the ranked sample data. This greatly reduces the potential for confounding. Although, like the ANOVA, the Kruskal Wallis test is somewhat affected by asymmetry, unlike the ANOVA it eliminates the more specific requirement of normality.

As always, to estimate a sample size, you must a priori specify an overall (family wide) $\alpha \leq$ 5%, and a targeted power of $(1 - \beta) \geq 80\%$ (or a $\beta \leq 20\%$) with respect to a judiciously chosen reasonable minimum difference, $\mathcal{D}_{min}$. Power analysis to estimate optimal sample size for attaining the desired power for global tests or for most nonparametric procedures is in a relatively primitive state. However, for the Kruskal-Wallis test, the power analysis procedure given in Subsection 14.4 should be an adequate approximation.

For $m \geq 3$ treatments, start the analysis by **combining** all the data and finding the ranks, $r_{i,j}$. (Ascending order is fine.) Alternatively, and **only** if you are interested in comparing variability, you could use Siegel-Tukey ranking as in Subsubsection 14.2.2). Then, sum each of the associated n_j rank values for each treatment j and square the totals to estimate

$$\mathsf{T}^2_j = \left(\sum_{i=1}^{i=n_j} \mathsf{r}_i\right)^2 \quad j = 1, 2, \ldots m,$$

the **square of each treatment rank total** for the $j = 1, \ldots, m$ treatments. Next use it to estimate

$$\mathsf{H} = \frac{12}{\mathbf{n}\,(\mathbf{n}+1)} \sum_{j=1}^{j=m} \frac{\mathsf{T}^2_j}{n_j} - 3\,(\mathbf{n}+1), \qquad df = m - 1, \tag{14.7}$$

where n_j is the size of sample j, m is the number of treatments, and $\mathbf{n} = nm$ (or $\sum_{i=1}^{i=m} n_i$) is the sum of all the sample sizes. You can **correct for ties** by dividing H by $1 - \frac{\sum_{i=1}^{i=g}\left(t_i^3 - t_i\right)}{\mathbf{n}^3 - \mathbf{n}}$ where g is the number of groupings of different tied ranks, and ti is the number of tied values within group i that are tied at a particular value.

For small numbers of treatments and small sample sizes, you can use the University of York values for H in Appendix 7 (Lee 2005). But such small sample sizes are unlikely to have much power unless you are targeting very large effect sizes.

14.3.0.1 Example: Worked Kruskal-Wallis Test

Using the fictitious Amadora data from Subsection 9.2.0.1, let's use a Kruskal-Wallis test as though the researchers neglected to properly do an appropriate a priori power analysis before the actual test. Assume they used the usual ordering (not the Siegel-Tukey, which might be difficult to interpret

C	Order C	D	Order D	E	Order E
28.74	89	17.69	71	11.92	35
18.32	73	16.91	66	16.00	62
11.30	32	20.39	83	12.13	38
12.93	43	6.49	8	19.39	76
15.42	54	12.30	42	6.02	6
12.26	41	15.22	53	9.62	21
11.51	33	18.85	75	10.77	28
9.80	23	29.07	90	17.60	69
20.32	82	15.54	57	6.00	5
25.49	88	25.26	87	13.00	45
14.38	51	9.36	19	17.91	72
17.61	70	19.39	77	5.71	4
12.10	36	8.03	12	7.35	10
10.75	27	12.20	40	9.67	22
8.56	15	10.73	26	6.86	9
12.96	44	17.37	67	3.77	2
7.50	11	19.45	78	3.15	1
21.71	85	8.41	14	9.13	18
11.11	30	15.58	58	9.86	24
16.00	61	21.84	86	17.41	68
12.16	39	15.95	60	16.15	65
13.95	49	16.06	64	3.80	3
13.68	47	20.24	81	12.10	37
18.47	74	15.47	56	13.76	48
8.77	16	11.84	34	14.02	50
6.09	7	19.60	79	10.27	25
8.21	13	19.66	80	8.99	17
15.43	55	20.95	84	13.12	46
11.00	29	14.80	52	16.05	63
11.23	31	15.70	59	9.54	20
Total	**1348**		**1758**		**989**
Median	**12.1300**		**15.5050**		**9.7650**
(Mean	13.9249		16.3450		11.0358)

Figure 14.1: Table for Kruskal-Wallis Amadora Perfume Trial
(The data was combined for ordering.)

in application); they simply wanted to see if they were different via a greater probability of higher scores.

Ordered omnibus odors The researchers had a locally contrived and carefully calibrated instrument something like a modified lie detector, the 'attention-o-meter', that measured in "attraction units." They formulated the following hypothesis pair:

H0: THERE IS NO DIFFERENCE AMONG THE TREATMENTS.

HA: THERE IS A DIFFERENCE AMONG THE TREATMENTS.

Criterion: $\alpha = 5\%$ (If this were a real test they would have wished to specify β and a **reasonable minimum difference** before getting this data. Later, following this first part, we will change our assumption and assume they did so and that $\beta = 10\%$).

Methods: They carefully scrubbed $n = 90$ women of the targeted age range and randomly assigned them to $m = 3$ equal size treatment groups of $n = 30$ each. The women were screened to be certain they had never worked for Amadora's or its competitors. This both eliminated certain types of bias and also reduced the possibility that they may have had some residual pheromone on them. Depending on which treatment group they fell into, they were anointed with treatments C, D, or E in a double blind random manner. They were then dressed identically and tested by being introduced in random order to the same randomly selected panel of men who were all monitored by the attention-o-meter that showed a collective level of attention. The data with appropriate combined ordering appears in Figure 14.1.

Analysis: Applying Equation 14.7 to the sums of $m = 3$ treatments from Table 14.1 in two steps with $df = 3 - 1$,

$$\mathsf{H} = \frac{12}{90\,(90 + 1)} \sum_{j=1}^{j=m} \frac{\mathsf{T}_j^2}{n_j} - 3\,(90 + 1) = \frac{1348^2}{30} + \frac{1758^2}{30} + \frac{909^2}{30} - 273,$$

$$\mathsf{H} = 7.04, \qquad df = 2.$$

Although the generally available tables of H such as Appendix 7 include experiments with $m = 3$ treatments, the researcher's sample sizes, n_i, exceeded anything in that table, so they appropriately found a chi square value, $\chi^2_{\alpha=0.05,\,df=2}$, from a table such as Appendix 5 to find that because $\mathsf{H} \geq 5.9915$, the result was **statistically significantly different**.

Results and Discussion: The result, $\mathsf{H} = 7.04$, showed the test to be statistically significant (as did the ANOVA in Subsection 9.2.0.1) at the 5% level with $2\,df$. Statistical significance in any global test, such as a Kruskal-Wallis Test, simply indicates that there is difference, not between what and what. In this example, the difference could be between an apparent attractant and an apparent repellent, or between any one of the treatments and just one of the other two. The complexity of the possible causes of statistical significance rapidly become greater as the number of comparisons becomes large. Using the Kruskal-Wallis test in this application was additionally appropriate because they did not know for **certain** that the metabolic processes measured by the attention-o-meter are on an **equal interval scale**. Unfortunately, although not a hanging offense, it would **not** really be proper for them to even report effect sizes unless their testing included multiple comparisons.

The best way: Parametric tests assume an equal interval scale; that is, they need a measurement scale upon which 10 attention units is really twice 5 attention units. Nonparametric tests such as the Kruskal-Wallis test, followed by subsequent comparisons between treatments using the Wilcoxon test (Subsection 14.1.1), are the **best** for many scales of measurement because they are likely to be valid, whether a scale is an **equal interval scale or not**. All they require is a tendency for their distribution to be sufficiently symmetrical, and that the values be arranged ordinarily. That is, 10 has to be less than 11, and 5 has to be less than 10, etc.

The Kruskal-Wallis test, like most non-parametric tests, rearranges and in some cases rephrases (as by ordering) the available information so that (given adequate sample size) an approach like a Pearson's chi square can assess whether something too unlikely to be due to chance alone has occurred. That is, the Kruskal-Wallis test in this example checks the outcome against the expected (due to chance) outcome of randomly drawing 3 samples with no treatment difference between samples. By choosing α, researchers define what is minimally too unlikely to be true in their opinion, $\alpha = 5\%$ being the maximum experiment-wide criterion that is acceptable to their peers.

Question 13 provides you with another example of the Kruskal-Wallis test and the chance to try this yourself.

> **Caution**: Wikipedia (https://en.wikipedia.org/wiki/Analysis_of_variance, 10/2020) warns that, "A common mistake is to use an ANOVA (or Kruskal–Wallis) for analysis using factors that are ordered groups as though they were treatments, e.g. in time sequence (changes over months), in disease severity (mild, moderate, severe), or in distance from a set point (10 km, 25 km, 50 km). Data in three or more ordered groups that are defined by the researcher should be analyzed by linear trend estimation."

14.3.0.2 Example: Continued Multiple Comparisons for Kruskal-Wallis

The observed statistical significance could have been due to the difference between one or both treatments and the control or between the treatments themselves. However, the researchers a priori chose to compare their outcomes with the control. Both because they a priori planned to use a control and because there were only two treatments to compare to it, this was not as problematic as it could have been without preplanning, or with more treatments. If they could be absolutely certain that there was no interaction, they could have used Sidak's correction (Subsection 10.1.3), but having the same men exposed to all the women (even at random) might conceivably result in some small interaction between reactions. So, they chose to use Bonferroni's correction (Subsection 10.1.2).

Correction for Multiple Comparisons: Using Bonferroni's correction, they applied Equation 10.3, $\alpha' = \alpha/k$, where $k = 2$, the number of comparisons they had a priori chosen to make.[9] Each comparison was a two tail test so their corrected cutoff for each comparison was

$$\frac{(\alpha/k)}{2} = \frac{\alpha'}{2} = \frac{(0.05/2)}{2} = 0.0125$$

or 1.25%, to whatever number of digits. (Anyone can choose to specify an exact criterion to any accuracy.)

The Comparisons: They then made the (two tail) comparisons using $\alpha'/2 = 0.0125$ for the cutoff and the Wilcoxon test (Equation 14.1),

$$Z_W = \frac{n_R(n_1 + n_2 + 1) - 2W_R}{\sqrt{\frac{n_1 n_2 (n_1 + n_2 + 1)}{3}}},$$

where W_R is the smallest total rank for the two treatments and n_R is the number of scores in W_R. From a suitable online calculator, the value of $Z_{\alpha'/2=0.0125} = 2.2414$, and it can also be found to an even higher accuracy using the qnorm() function in the R or S computer languages.[10] So for comparison of E to C,

$$Z_W = \frac{30(61) - 2(1348)}{\sqrt{\frac{900(61)}{3}}} = -6.401656,$$

and for D to C,

$$Z = \frac{30(61) - 2(909)}{\sqrt{\frac{900(61)}{3}}} = 0.08870655.$$

Discussion: So, with an experiment wide criterion of $\alpha = 0.05$, treatment E is statistically significantly different from the control, C, as a repellent (negative tail) and treatment D is not statistically significantly different from the control. Significance (but maybe not as marked) was also

---

[9] If the perfume researchers had **not a priori chosen** to make 2 particular comparisons, they would have to have used all $k = 3!/(2!(3-2)!) = 3$ possible corrections via either the Bonferroni correction or the Sidak correction. Because comparisons are always taken 2 at a time, this can be simplified to $k = m(m-1)/2$ for whatever number of comparisons, m.

[10] In the R and S languages, pnorm(Z) gives p, the area in one tail of a normal curve that is further away than Z standard deviations from the mean. Its opposite, qnorm(p), gives Z, the number of standard deviations that such an area is from the mean, $Z = 0$.

found by the ANOVA example using the same data (Subsection 10.2.0.1). It is possible for the ANOVA to be weaker than a nonparametric test, due to heteroscedasticity (which should have less effect on nonparametric tests) and/or lack of normality.[11]

Overall Effect Size Kruskal-Wallis: Because, for large samples, H is distributed as χ^2, it is possible to get an overall effect size for this test using phi,

$$\phi = \sqrt{H/n}\,,$$

where $0 < \phi < 1$, meaning it ranges from complete independence to absolute dependence. This is for the whole global test and it can also be presented as a percent. This mathematically approaches equality with R, the correlation coefficient, but except where you are interested in the concept that the things that your samples represent might vary together (especially in a linear fashion), equating to R might be inappropriate.[12] Squaring to estimate ϕ^2, the fraction of the variance that is due to treatment effects might make more sense.

14.4 Sample size and Kruskal-Wallis, Multiple Comparisons

You do **not need a pilot test**, **or extra data**, in order to estimate the sample size for a Kruskal-Wallis test **with preplanned multiple comparisons**, each of which will use a Wilcoxon test. That is, if you sum all the sample sizes for the treatments, using the sample sizes needed to use each treatment at least once in an individual comparison, you get the overall sample size for the Kruskal-Wallis test. That is, for the Kruskal-Wallis test, as for the ANOVA with equal sample sizes, you initially estimate the size of a single sample needed to allow for k a priori planned comparisons. Then you multiply this sample size estimate, $\hat{n}$, by m, the number of treatments.

For equal sample sizes, $n_1 = n_2 = \ldots = n_m$, where m is the number of treatments:

1. Use the Bonferroni correction (Subsection 10.1.2) or the Sidak correction (Subsection 10.1.3) to find the $cutoff'$, α' or $\alpha'/2$, and β' for each of the k multiple preplanned comparisons. (Generally you will be using $cutoff' = \alpha'/2$, but one way comparisons are possible if planned a priori).

2. Also specify the reasonable minimum difference, $\mathcal{P}_{S_{min}}$, and substitute all these choices into

---

[11]Russell Langley (1970) seems to have used a similar approach without directly mentioning Bonferroni, and he attributed his approach, in part, to Wilcoxon and Wilcox (1964).

[12]Those who think that the means and medians are shifted identically by whatever treatments that have an effect would do well to consider the means and medians from this sample where the individual treatment medians $\tilde{x}_C = 12.1300$, $\tilde{x}_D = 15.5050$, and $\tilde{x}_E = 9.7650$ (Table 14.1). The differences, $\tilde{x}_2 - \tilde{x}_1$, between the **medians** of the treatments and the control, C in original units of measure, 'attraction units', are

$$\mathcal{D}_{\tilde{D}-\tilde{C}} = 3.375, \text{ and } \mathcal{D}_{\tilde{E}-\tilde{C}} = -2.365,$$

(the second being the only significant comparison). The difference between the means is in original units of measure, but not identically shifted. The individual treatment means are $\bar{x}_C = 13.9249$, $\bar{x}_D = 16.3450$, and $\bar{x}_E = 11.0358$.

The differences, $\bar{x}_2 - \bar{x}_1$, between the **means** of the treatments and the control, C in original units of measure, 'attraction units' are,

$$D_{\bar{D}-\bar{C}} = 2.420, \text{ and } D_{\bar{E}-\bar{C}} = -2.8891,$$

(the second likewise being the only significant comparison).

Noether's (1987) Equation (14.3),

$$\widehat{2n} = \frac{(Z_{cutoff'} + Z_{\beta'})^2}{12c(1-c)\mathcal{P}^2_{S_{min}}}, \tag{14.8}$$

with $c = 1/2$ because this is for equal sample sizes; half of this $\widehat{2n}$ is the estimate of $\hat{n}$, the sample size of **each** treatment (assuming equal sample sizes, see Subsection 14.4.0.2. for unequal sample sizes). {This step will estimate adequate sample sizes for the individual Wilcoxon tests mentioned in step 4 below, for each multiple comparison that will follow this Kruskal-Wallis test.}

3. Now combine the results by multiplying by m, the number of treatments, to find the total sample size,

$$\widehat{\mathbf{n}} = m\hat{n},$$

for the **whole global test** with equal sample sizes. This total sample size, $\widehat{\mathbf{n}}$, allows you to estimate the cost of your planned experiment or observation in advance.

4. Finally obtain your data using $\hat{n}$ for each treatment, and analyze it with the Kruskal-Wallis test using the uncorrected α, followed by a Wilcoxon test for each comparison, using $\alpha'/2$ for two tails (or maybe α' for a priori chosen one tail comparisons).

If multiple comparisons are not planned a priori:

Regardless of how many multiple comparisons you want to run, if the multiple comparisons are **not planned** a priori, you must use a **Bonferroni** or **Sidak** correction for using $k = m(m-1)/2$ **as though** you were going to run **all possible** multiple comparisons; the Kruskal-Wallis test **must** be statistically significant before going on to the multiple comparisons; and you **must** test only for two tails.

14.4.0.1 Example, Sample Size for Kruskal-Wallis with $k = 5$ Comparisons and m Treatments:

Suppose you have an index of severity for tuberculosis (TB) and intend to use Wilcoxon tests for 5 multiple comparisons. So, you can easily use Noether's (1987) Equation 14.8 to estimate the total sample size $\widehat{2n} = \frac{\left(Z'_{cutoff'} + Z_{\beta'}\right)^2}{12c(1-c)\mathcal{P}^2_{S_{min}}}$ per comparison with **equal** sample sizes, $\hat{n}_1 = \hat{n}_2 = \widehat{2n}/2$. This does not need a pilot sample, extra data, or iteration. Because the sample sizes are equal (or provisionally equal) you use $c = 1/2$. (You could even write this as $\hat{n} = \frac{\left(Z'_{cutoff'} + Z_{\beta'}\right)^2}{3\mathcal{P}^2_{S_{min}}}$, given that $n_1 = n_2 = \ldots = n_m$.)

Methods and Materials: You wish to estimate a sample size to compare the severity of known infections on $m = 6$ pacific islands with an experiment wide criterion of $\alpha = 0.05$ with a power of $1 - \beta = 0.80$ for $\alpha = 5\%$ and $\beta = 20\%$ with a reasonable minimum difference of $\mathcal{P}_{S_{min}} = 20\%$, the probability that if you choose an individual with TB from one population their index will be larger than from the other. You will use a two tail Kruskal-Wallis test on the $m = 6$ samples, followed by Wilcoxon tests for the $k = 5$ multiple comparisons. For whichever of the tests, you hypothesize:

H0: THE POPULATIONS ARE ALIKE.

HA: THE POPULATIONS ARE DIFFERENT.

For big sample sizes, you can think of the minimum difference as the probability of superior scores equaling or exceeding

$$\frac{1}{2} \pm \mathcal{P}_{S_{min}},$$

that is, either a 70% : 30% or a 30% : 70% relationship, since this is a two tailed test. Your null hypothesis simply provides a reference point of no difference (a 50% : 50% relation). But in application, it is only a zero difference reference point to facilitate calculation.

The $\alpha/2$ and β must be corrected to compensate for inflation due to multiple comparisons. It is conservative, and handy, to use the Bonferroni correction for $k = 5$ comparisons; $\alpha'/2 = (\alpha/2)/5$ equals 0.005 and $\beta' = \beta/5$ which equals 0.04. The corresponding Z values are $Z_{\alpha'/2} = 2.5758$ and $Z_{\beta'} = 1.7507$.

Analysis: Applying Noether's equation,

$$\widehat{2n} = \frac{(2.5758 + 1.7507)^2}{12 \times 1/4 \times 0.04} = 155.9884,$$

which must be rounded **up to an even integer**, 156, so that we have $\hat{n} = 78$ in each sample. The **experiment wide** sample size would be $\widehat{\mathbf{n}} = m\hat{n}$, which comes out to $\widehat{\mathbf{n}} = 468$.

Discussion: The concept of the probability of selecting a superior item or individual from one population than from another is rather unfamiliar; it is not about averages. It makes sense if researchers were trying to discover if more patients would live longer after a certain vaccination than without it, and if the researchers also wanted to be certain that they were not living longer without it than with it. A greater average or median might really mean that a few are living a very long time while many others could be dying just a bit sooner. This idea of randomly selecting a superior score individual from one population than the other can be seen as a possibly less risky way of justifying a generalization.

14.4.0.2 Kruskal-Wallis Unequal Sample Sizes

When you intend to run a global test such as a Kruskal-Wallis it is only common sense to run multiple comparisons using Wilcoxon tests, but sometimes one or more of the samples have insufficient fixed maximum sizes, n_f. To estimate correct sample sizes, including for those treatments with one fixed and one adjustable sample size, it is necessary to:

- Start your sample size estimation as though you were going to be able to get equal sample sizes by estimating an adjustable sample size, $\hat{n}_{adj}$, that is **provisionally** for all the treatments with non-fixed sample sizes using Noether's Equation 14.8 with Bonferroni corrected α' and β' values for your k preplanned comparisons as in above Subsection 14.4 (gives the total needed for each comparison $\widehat{2n} = 2\hat{n}_{adj}$).

- Next it is necessary to adjust sample sizes of treatments that are adjustable and will be **included in comparisons** with treatments that have inconveniently fixed maximum sample sizes, meaning they will be insufficient to meet the need, $n_f \leq \hat{n}_{adj}$ (that is $(n_f + \hat{n}_{adj}) < \widehat{2n}$).

- Adjust these adjustable sample size estimates of the treatments in comparisons involving the **smallest fixed sample size**, n_f, **first** (then those involving the next smallest, n_f etc.), then use a Kifle and Destra based approach to restore the lost efficiency due to unequal sample sizes (also used in Subsection 7.11.2). This is calculated in two steps,

$$\widehat{2n'} = \widehat{2n}\left(2 - \frac{\bar{n}_h}{\bar{n}}\right),$$

where $\dddot{n}_h = \frac{2}{1/n_f + 1/n_{adj}}$ is the harmonic mean and $\bar{n} = \left(n_f + n_{adj}\right)/2$. Then

$$\hat{n}' = \widehat{2n'} - n_f$$

is the sample size for all the treatments that are planned to be compared with any sample that has a fixed sample size of n_f. Once you adjust a sample size to $\hat{n}'$, **do not adjust it again**.

- Choose the **next smallest** sample size, if any (let's also call it n_f), and repeat the last step until you have assigned adjusted sizes to all the treatments that are going to be compared with treatments having inconveniently non-adjustable sample sizes. (Never adjust downward.)

- Now, you have the m sample size estimates for the m treatments, so, sum all the treatment sample sizes to also get

$$\widehat{\mathbf{n}} = \sum_{j=1}^{j=m} \left(sample\ size\ estimates\right)_j,$$

the total sample size estimate for the **whole Kruskal-Wallis** test.

- Then run your experiment or make your observations using all these sample sizes. Then run the Kruskal-Wallis test, using the global overall α, and these sample sizes.

- If the Kruskal-Wallis test is significant, follow it with multiple comparisons, using Wilcoxon's tests with the Bonferroni corrected α'.

In application, this approach should be as good or better than using normal curve based sample size estimates under the same sample size constraints. If you have a high proportion of fixed sample sizes, or if the sample size of the control is fixed, or if the difference between the fixed sample sizes and the estimated sample sizes seems large, a final three place accuracy attained by rounding up the final result might be the best you can hope for when reporting effect sizes after statistical significance.

14.4.0.3 Example, Comparing 6 Locations for Health:

Assume you are burdened with two treatments out of $m = 6$ treatments where you cannot get enough samples to have all samples equal.

Methods and Materials: You have an index of severity for overall health and you want to compare people on $m = 6$ pacific islands, A, B, C, D, E, and F with an experiment wide criterion of $\alpha = 0.05$, a power of $1 - \beta = 0.80$ and a reasonable minimum difference of $\mathcal{P}_{S_{min}} = 0.20$. But due to bureaucratic weirdness within your agency you a priori are mandated to compare only $k = 4$ pairs of treatments (islands):

- The $k = 4$ preplanned comparisons are: | **AB** | **DF** | **CE** | **BF** |,

- but because of various circumstances B can have a **fixed** sample size no larger than $n_B = 35$ people,

- and F can have a **fixed** sample size no larger than $n_F = 25$ people.

Analysis: First you find an adjustable sample size for the global test as though all the sample sizes could be equal. The $\alpha/2$ and β must be corrected to compensate for inflation due to multiple comparisons. It is conservative, and handy, to use the Bonferroni correction for $k = 4$ comparisons, $\alpha'/2 = (\alpha/2)/4$ equals 0.00625 and $\beta' = \beta/4$ which equals 0.05. The corresponding Z values are $Z_{\alpha'/2} = 2.497706$ and $Z_{\beta'} = 1.644853$.

Applying Noether's equation $\widehat{2n} = \frac{\left(Z_{cutoff'} + Z_{\beta'}\right)^2}{12c(1-c)\mathcal{P}_{S_{min}}^2}$ with $c = 1/2$ you get an estimate of how many people are needed per comparison,

$$\widehat{2n} = \frac{(2.4977 + 1.6449)^2}{12 \times 1/4 \times 0.04} = 143.0095,$$

assuming equal sample sizes, and that must be rounded **up to an even integer**, 144. So that gives a provisional adjustable sample size $\hat{n}_{adj} = 72$ people for each island that does not have a fixed sample size. However, out of these $m = 6$ treatments, two of them have **fixed** maximum sample sizes, $n_B = 35$ people and $n_F = 25$ people, and they are both less than $\hat{n} = 72$.

DF Using a Kifle and Destra based approach to correct for the efficiency lost due to unequal sample sizes, select the **adjustable** samples to be compared with the smallest fixed sample size that is too small (inadequate), $n_F = 25$, which is going to be compared with island **D**. First calculate the harmonic mean $\dddot{n}_h = \frac{2}{1/n_F + 1/n_{adj}}$, which comes out $\dddot{n}_h = \frac{2}{1/25 + 1/72} = 37.1134$, and the mean $(n_F + n_{adj})/2$ to get $\bar{n} = (25 + 72)/2 = 48.5$. Then the Kifle and Destra approach gives the total sample size for comparing **D** to **F**,

$$\widehat{2n'} = \widehat{2n}\left(2 - \frac{\bar{n}_h}{\bar{n}}\right) = 143.0095\left(2 - \frac{37.1134}{48.5}\right) = 176.5846,$$

which rounds up to a final **total for this comparison**, $\widehat{2n'} = 177$. Then

$$\hat{n}' = \widehat{2n'} - n_F = 177 - 25 = 152.$$

This $\hat{n}'$ is the sample size for all the treatments (in this case island **D**) that are planned to be compared with any sample that has **this fixed sample size** $n_f = 25$ (which happens to be island **F**). So $n'_D = 152$. Once you adjust a sample size do not adjust it again.

AB Now there is a next smallest fixed inadequate sample size, $n_B = 35$, that you plan to compare with adjustable size treatment **A**. So, to use the Kifle and Destra based approach to correct the sample size for island **A**, first calculate the harmonic mean $\dddot{n}_h = \frac{2}{1/n_A + 1/n_{adj}}$, which comes out $\dddot{n}_h = \frac{2}{1/35 + 1/72} = 47.1028$, and the mean $(n_B + n_{adj})/2$ to get $\bar{n} = (35 + 72)/2 = 53.5$. Then the Kifle and Destra approach gives the total sample size for one comparison,

$$\widehat{2n'} = \widehat{2n}\left(2 - \frac{\bar{n}_h}{\bar{n}}\right) = 143.0095\left(2 - \frac{47.1028}{53.5}\right) = 160.1097$$

which rounds up to a final **total for this comparison** $\widehat{2n'} = 161$. Then

$$\hat{n}' = \widehat{2n'} - n_F = 161 - 35 = 126.$$

Then $\hat{n}'$ is the sample size for all the treatments (in this case island **A**) that are planned to be compared with any sample that has **this fixed sample size** of $n_f = 35$ (which happens to be island **B**).

There are no next smaller fixed sample sizes left that are to be compared with the adjustable ones, so let's look at a table of the resulting sample sizes:

Wilcoxon Test, $k = 4$, $m = 6$							Totals
Treatment:	**A**	**B**	**C**	**D**	**E**	**F**	
Provisional	$\widehat{72}$	35	$\widehat{72}$	$\widehat{72}$	$\widehat{72}$	25	**348**
Final Size $\hat{n}'$	126	35	72	152	72	25	**482**
Compared to	**B** **F**	**A**	**E**	**F**	**C**	**B** **D**	

Discussion: Well, this might have been an unlikely group of comparisons, but it demonstrates the rules affecting the sample size adjustments. If a sample size is increased to make one comparison work, it is going to be that big in all other comparisons involving it. Once again, **inequality costs**; we went from a grand total of $\hat{n} = 6 * 72 = 432$ people if the sample sizes could have been equal, to an adjusted final total of $\hat{n}' = 482$, meaning 50 more people were needed. This costly difference is due to the samples with fixed maximum sizes causing inequalities and loss of efficiency. Likewise, the inequality causes the power to vary over the comparisons.

BF However, when comparing the two fixed sample sized treatments that were too small, **B** and **F**, the power would be undesirably low. Equation 14.2, $z = \mathcal{P}_{S_{min}} \sqrt{\mathbf{n}12c(1-c)} - Z_{cutoff}$, $c = 1 - n_R/\mathbf{n}$ (so $c = 25/(25 + 35)$) and $Z_{\alpha'/2} = 2.497706$ to estimate the power for that comparison,

$$z = 0.20\sqrt{60 * 12(25/60)(1 - 25/60)} - 2.497706,$$

which comes out $z = 0.1480453$, and according to the positive z rule, $z = Z_\beta$. Referring to a suitable source of probabilities for Z gives $\beta = 0.4411535$ or about 44%, which means your power $1 - \beta$ is not so good, about 56%. Hopefully, you will not often have such pairs of awkwardly unequal sample sizes forced on you.

14.4.1 Wilcoxon's Test and Kruskal-Wallis vs Student's t and ANOVA

All tests can be seen as asking whether one population is different from another. However, you should be specifically aware of what different tests really ask. Most parametric tests, such as Student's t test, estimate the probability that a treatment sample comes from a population with a meaningfully different mean than a hypothesized mean, or than another sample's mean. However, most rank tests ask if the probability of drawing a larger value from one treatment population is greater than from another. If your high school has a few kids from billionaire families and the rest are from their employees, including their servants, and you are comparing that to other high schools, then t tests of family incomes will tend to find very different outcomes than will rank tests. It might be acceptable to conclude that rank tests involve differences between medians, but that concept only tends to be close to valid when the sample sizes are huge.

Both rank tests and parametric tests are intended to use individual data points to correctly affect the outcome of the test in one direction or another. Static-like background variation would have to be very strong in order to cause as much incorrect ordinal classification of individual points as it would on a continuous scale. So, although order statistics can be affected by differences in treatment variances, they are likely to be less affected by otherwise irrelevant variance differences than are parametric statistics. Order based tests are less affected by outliers because ordering reduces the magnitude of the effect of the outliers. So overall, some of the information overlooked by order statistics is irrelevant or even deleterious to making a correct decision.

In most cases it is best to use Wilcoxon tests after a Kruskal-Wallis Test. However, you could (a priori) consider using individual Student's t tests for the multiple comparisons following a Kruskal-Wallis test, or if your samples are unequal, you should use a Welch's t test instead of a Student's t test (Section 5.3). When sample sizes are large, Student's t test is rather robust, even when compared with non-parametric tests, and power usually requires large samples anyway. You would be able to estimate a total sample size $\hat{n}$ by combining planned multiple comparison sample sizes for t tests based on a reasonable minimum difference in original units of measure, as though they were going to be subsequent to a parametric omnibus test (e.g. Section 10.4). Then for the sample size for a rank based global test, such as the Kruskal-Wallis test, you can multiply this result $\hat{n}$ by $\pi/3 = 1.0472$ to get a sample size estimate.[13]

Unfortunately, nonparametric tests do not usually provide a way to get confidence interval estimates. Alternatively and especially for smallish samples ($n \leq 60$) you can get related information using boxplots and violin plots.

Multiple comparisons **appear** to lose a lot of degrees of freedom for each subsequent t test if the global test is not an ANOVA that provides a theoretically better estimate of the variance (MS_W) and more degrees of freedom per comparison (df_W). However, recall that, unless we can somehow guarantee that the raw data is quite close to normal, and the variance is homogeneous, the ANOVA will be weakened and possibly unable to detect a difference. It might even produce a totally erroneous MS_W, making the comparisons invalid. So, it is usually best in application to use the Kruskal-Wallis test, rather than the one way ANOVA, with multiple comparisons.

14.5 Friedman's Complete Block Test

Milton Friedman (1937) published a rank based test to avoid needing normality, implicit in the randomized complete block analysis of variance. It, like the Kruskal-Wallis test, assumes that the distribution from which the data arises is roughly **symmetrical**, but does not have to be as symmetrical as for the ANOVA. It is conceivable that the more complex an ANOVA is, the more compounded and confounded the deviations from the many assumptions of the ANOVA. So, Friedman's test is more robust, but extreme skew might still be a problem.[14] It, like the RCBD ANOVA, is a repeated measures test, meaning that it is used to determine if there is a statistically significant difference among the results across multiple occasions (blocks). Thus Friedman's test is a promising alternative to the randomized complete block design (RCBD) ANOVA. As for other rank tests, the data does not have to come from an equal-interval scale.

Historically, agricultural scientists tried to select the blocks for an RCBD ANOVA to be as similar as possible. No wonder they were doing so; one of the key assumptions, that of homogeneity of

---

[13]Mood (1954) concluded that the Wilcoxon test has $3/\pi = 95.493\%$ as much power with respect to sample size as the corresponding Student's t test. This implies that if you multiply the **sample size** you would need **for a Student's t test by** $\pi/3 = 1.0472$, you should be able to estimate the sample size for a Wilcoxon test. So, to estimate the sample size required for a global test based on the combined sample sizes found using Bonferroni corrected t tests (intended for subsequent multiple comparisons), you should at least get a **rough estimate** of what you would need for a non-parametric global test, such as the Kruskal-Wallis test. If there are very extreme outliers, you might try using MAD (Section 15.6.2).

[14]Friedman's, needs some symmetry

variance, is that the underlying variance, σ^2, is the same for the treatments as for the blocks. In application this refers to the variance of the residuals left after the partitioning out of variances said to be due to treatment effects and block effects, but this is a rather restrictive assumption. This might be exacerbated in situations like moderate to small sized islands where small distances can involve meaningful differences in soil and altitude. However, Friedman designed a test that is far **less restrictive** with regard to how you choose the blocks, and **less subject to confounding** due to differences between the blocks. In his original publication of this test, Friedman included fourteen drastically different blocks in a test of income level vs household expenditure. They were housing, household operation, food, clothing, furnishings and equipment, transportation, recreation, personal care, medical care, education, community welfare, vocation, gifts, and other. His treatments were seven income categories.

You could use a Montessori classroom block in the same test as a public inner city classroom block and an expensive private school classroom block. For medical studies you could use hospital wards all over the world as blocks. Since agricultural scientists now tend to do on farm experiments, and because farms are not necessarily adequately similar, they might be making a serious **mistake not to use Friedman's test instead** of an RCBD ANOVA. But, regardless of your area of interest, think carefully when laying out any such experimental design and be sure to consider exactly what is the question.

If your question is well thought out, this looks good until you consider that if someone supported by an interest group, or a fanatic, has control over the choice of blocks that are allowed to be drastically different, then the blocks, with a bit of background knowledge, could possibly be chosen according to type and number so as to force, or counter, statistical significance. It is best to either use blocks that fit into one group, or to be sure that every block is quite unrelated to the others. Whether we are considering blocks or their smaller cousins, pairs, we may have to be especially clear concerning what is the question. You should avoid grouping things that do not essentially demand to be grouped. If your data is spread across years, you have no choice but to block by year, but then you must qualify your conclusion with respect to what particular range of years (e.g. 1985-2015). Likewise, if you are applying a variety of skin tests to each of your patients and then somehow reading the results numerically, you will need to block, and you may have to specify how many males, females, and persons of each skin type were involved. Incidentally, in this case you could rank the results on each patient and put the rank values directly into a Friedman's test.

To use a Friedman's test, you must **a priori** specify an overall (family wide) criterion of $\alpha \leq 5\%$, however the accuracy tends to drop off as α gets smaller (Iman and Davenport 1980). Reading this paper suggests that the approach shown below is reasonably robust, especially for small samples, and that it is acceptably accurate at $\alpha = 5\%$, regardless of large or small samples. If, as you should, you have any interest in **either** an a priori sample size estimate, or a subsequent power estimate, you must use this pre chosen α, along with $\beta \leq 20\%$ or $(1 - \beta) \geq 80\%$, which, in turn, requires you to judiciously choose a reasonable minimum difference, $\mathcal{D}_{min}$. Conover (1980) presents the version used below, which uses an F test. Studies by Iman and Davenport (1980) demonstrate it to be clearly superior to the original calculations that were evaluated using χ^2.

For Friedman's test, the data should represent m_b blocks, each with a complete set of all m treatments, and for the approach shown here, Iman and Davenport (1980) suggest $m + m_b > 9$ and with more than $m > 2$ treatments and $m_b > 3$ blocks. The $m > 2$ requirement distinguishes a complete block from a paired test; only $m_b = 3$ blocks is, in effect, a sample size that is unlikely to have much chance of detecting anything, as would also be the problem with only $m + m_b = 9$.

Let's look at the calculations in steps: First rank $r = 1, \ldots, n_j$ within **each** block $_j$ (be sure to average any ties; ascending order is fine). Then sum each of the $i = 1, \ldots, m$ ranks to get

$$\mathcal{R}_j = \sum_{i=1}^{i=m} r_{i,j}$$

for each of the $i = 1, \ldots m$ treatments, where $r_{i,j}$ is the rank value of treatment i and block j. Square each R_j and sum those sums to get

$$\mathbf{T}^2 = \sum_{j=1}^{j=m_b} \mathcal{R}_j^2 = \sum_{j=1}^{j=m_b} \left(\sum_{i=1}^{i=m} \mathrm{r}_{i,j} \right)^2 .$$

Now we calculate two variables (calligraphic $\mathcal{A}$ and $\mathcal{B}$, don't confuse with α and β),

$$\mathcal{B} = \frac{1}{m_b} \mathbf{T}^2 .$$

Next, square and then sum every rank in the whole analysis table to get

$$A^2 = \sum_{i=1}^{i=m_b} \sum_{j=1}^{j=m} \mathrm{r}_{i,j}^2 .$$

But, *iff* **you have no ties**, you can more simply calculate it via

$$\left[A = \frac{m_b m (m+1)(2m+1)}{6} \right] ,$$

then you still estimate

$$\mathcal{B} = \frac{1}{m_b} \sum_{j=1}^{j=m_b} \mathcal{R}_j^2 .$$

Finally you can submit

$$\widehat{F} = \frac{(m_b - 1) \left[\mathcal{B} - m_b m (m-1)^2 / 4 \right]}{\mathcal{A} - \mathcal{B}} \tag{14.9}$$

to a table of Fisher's F with

$$df_1 = m - 1, \quad df_2 = (m-1)(m_b - 1).$$

In the hopefully rare instances that $\mathcal{A} = \mathcal{B}$, you are faced with division by zero, which is impossible to do. So, Conover (1980) indicates that in that case, you can estimate the $p\_value = (1/m!)^{m_b - 1}$ and compare it to your a priori α.

Quade's test (1979) is considered more powerful than Friedman's test for comparing up to five samples (Conover 1980), and can be adapted as a nonparametric ANCOVA, but there is controversy as to what restrictions exist regarding the kind of data that can contribute the ranks for Quade's test, so I am reluctant to recommend it yet, but some statistical packages provide it or can be adapted to do it.

14.5.0.1 Example: Friedman's Test of an RCBD Design

Let's see what Friedman's test can do with the data that was analyzed in Subsection 9.4.0.2.

RCBD via Friedman's test: Gollum's thesis research involved finding uses for vegetable waste, and he wanted to test them as fertilizers. He formulated a hypothesis pair,

H0: THERE IS NO TRUE DIFFERENCE AMONG THE TREATMENTS.

HA: THERE IS A TRUE DIFFERENCE AMONG THE TREATMENTS.

Criterion: $\alpha = 5\%$

Methods: He drew up a randomization plan (a map) of each of $m_B = 6$ experimental plots, randomizing the positions of a completed set of $m = 5$ treatments separately for each of these blocks. Gollum designated treatment A2 as the control. He also completely randomized the order in which the plots were planted, serviced, and harvested.

Analysis: For each block he assigned the appropriate rank values for the results in kilograms as shown in Table 9.2. He also found the sum of the squares of the treatment rank totals,

$$\mathbf{T}^2 = \sum_{j=1}^{j=m_B} \left(\sum_{i=1}^{i=m} \mathbf{r}_{i,j} \right)^2 = 1766,$$

shown in the left lower corner of the table.

Figure 14.2: Abomocrop with Friedman's Test

Treatm.	Block 1	rank	Block 2	rank	Block 3	rank	Block 4	rank	Block 5	rank	Block 6	rank	Sum	Sum^2
A1	52.16	5	57.89	4	58.78	4	79.57	5	80.06	5	46.93	4	27	729
A2	37.5	2	35.44	1	51.36	3	51.07	2	37.21	1	39.94	3	12	144
A3	35.09	1	66.57	5	80.53	5	64.66	4	48.41	3	37.83	2	20	400
A4	49.12	3	56.36	3	49.91	2	42.18	1	56.58	4	48.82	5	18	324
A5	50	4	49.79	2	49.43	1	53.38	3	37.51	2	32.33	1	13	169
													Sum	1766

He could have squared every rank in the body of the analysis table and added all the results together, and used $\mathcal{A} = \sum_{i=1}^{i=m_B} \sum_{j=1}^{j=m} (\mathbf{r}_{i,j})^2$ but, because there were no ties, Gollum used the easier to calculate

$$\mathcal{A} = \frac{m_B m (m+1)(2m+1)}{6},$$

which comes out with the same value, but only works **when there are no ties**, as is now the case,

$$\mathcal{A} = \frac{6 \times 5 (5+1)(2 \times 5 + 1)}{6} = 330.$$

Now he calculated the remaining steps to finally estimate F, and compare it to a table like Appendix 6:

$$\mathcal{B} = \frac{1}{m_B} \mathbf{T}^2,$$

$$\mathcal{B} = \frac{1}{6} 1766 = 294.3333.$$

$$\widehat{F} = \frac{(m_B - 1) \left[\mathcal{B}^2 - m_B m (m+1)^2 /4 \right]}{\mathcal{A} - \mathcal{B}}$$

$$df_1 = m - 1, \; df_2 = (m - 1)(m_B - 1)$$

$$\widehat{F} = \frac{(6 - 1)\left[294.3333 - 6 \times 5\,(5 + 1)^2 / 4\right]}{330 - 294.3333} = 3.411215$$

$$df_1 = 4, \quad df_2 = 20$$

From a suitable table, such as Appendix 6, he found that $F_{\alpha=0.05,\,df_1=4,\,df_2=20} = 2.87$, which is exceeded by $\widehat{F} = 3.41$ (conservatively rounded down), so the results are **statistically significant**.

Results and discussion: Both this F based Friedman's test and the RCBD ANOVA were statistically significant with this data (Subsection 9.4.0.2). Gollum, with prompting from Professor Snaggletooth, had a priori planned for multiple comparisons (something carelessly not mentioned to the service statistician who had OKed their plan anyway). So, they went on and examined their pre-planned comparisons with treatment A2 as a control (Subsubsection 14.5.0.3).

This version of Friedman's test is much more accurate than the χ^2, especially for any but very large samples, but it is not as conservative. So, it is only fair to round **down** your F estimate to three figures before comparing it to a table of F (of course you should round the table values **up**).

14.5.0.2 Overall effect size for Friedman's test

If you use Friedman's test as an omnibus test without multiple comparisons and their effect sizes you might need an overall effect size to meet editorial demands. You can calculate this if you revert to evaluating Friedman's test using the original (1937) equation,

$$\chi^2 = \frac{12\mathbf{T}^2}{m_b m\,(m + 1)} - 3m_b\,(m + 1).$$

Then you can estimate Cramer's phi squared, using Equation 12.8,

$$\psi_C^2 = \frac{\chi^2 / m m_b}{min(m, m_b) - 1},$$

which almost always ranges between 0 and 1. It can be expressed as a percent of the variation between the treatments that is due to nonrandom effects. It is an attempt to estimate phi, an estimator of the part of the variation that is due to the effect of the differences between the treatments, but with a correction for additional effects. This correction is not exact; it only gets us into the correct ballpark.

For an RCBD ANOVA, you could use 'variance' instead of 'variability' to describe Cramer's phi. However, when applied to Friedman's test, it refers to what could be called the variance of the ranks. It is somewhat (not perfectly) like you were comparing the variability for data that had been intensely corrected for outliers, as by a very aggressive use of MAD (Section 15.6.2). Some folks may struggle to fit some other effect size estimator from, say, Section 9.5 in a black box-like manner, but it will not be any better and it will be numerically inferior for Friedman's test.

14.5.0.3 Multiple comparisons for Friedman Test

Conover (1980 Second Edition) gives a more powerful method of doing multiple comparisons that is analogous to[15] the LSD in that it does not take the effect of alpha inflation into account (Section 10.1). Because Conover's form of Friedman's test is not particularly conservative, and because much of the calculation from it is carried over to these multiple comparisons, it is reasonable not to do any multiple comparisons unless it is at least significant at the 5% level. Although this method was not adequate as presented in Conover's 1980 printing without correction for alpha inflation, it can be very satisfactorily adapted by simply using the conservative Bonferroni corrected $\alpha' = \alpha/k$ where k is the number of comparisons you are going to make (Subsection 10.1.2). Whatever multiple comparison approach you use, **if specific multiple comparisons are not a priori** (planned in advance), you **must default** to expensively correcting as though you were testing every possible pairing (then $k = n(n-1)/2$).

So, for each of the $j = 1, 2, \ldots m$ treatments, you could sum the ranks across the $i = 1, 2, \ldots m_b$ blocks to get

$$\mathcal{R}_i = \sum_{j=1}^{j=m_b} \mathrm{r}_{i,j},$$

but you should have already done that when calculating Friedman's test itself. Likewise, you probably already have the $\mathcal{B} = \frac{1}{m_b} \sum_{j=1}^{j=m} \mathcal{R}_j^2$ and $\mathcal{A} = \sum_{i=1}^{i=m_b} \sum_{j=1}^{j=m} \left(\mathrm{r}_{i,j}\right)^2$ values (which could be found more easily *iff* **there are no** ties, using $\mathcal{A} = \frac{m_b m (m+1)(2m+1)}{6}$). Using Bonferroni's correction to correct the LSD-like method in Conover (1980), you can put together a decision rule that treatment 2 versus treatment 1, or that any other of the k preplanned comparisons, are statistically significant from each other if

$$|\mathcal{R}_2 - \mathcal{R}_1| > t_{1-\alpha'/2,\, df=(m_b-1)(m-1)} \sqrt{\frac{2m_b \left(\mathcal{A} - \mathcal{B}\right)}{(m_b - 1)(m - 1)}}. \tag{14.10}$$

Notice that the prime mark, ', is being used to indicate that the α or β have been corrected to prevent alpha inflation using either Sidak's correction, or Bonferroni's correction. It is usually preferable to use the more conservative Bonferroni correction. Because Conover's standard error of the ranks and the rank sums are somewhat independently determined, t is used rather than Z.

14.5.0.4 Example Friedman's Multiple Comparisons

Gollum was able to run his chosen multiple comparisons on the data used in Example 9.4.0.2 because he had preplanned to have multiple comparisons against a control. In this example, he adapted Friedman's multiple comparisons rule as suggested by Conover (1980), but with correction for alpha inflation.

Method and Materials: Gollum had preplanned to use treatment **A2 as the control**, so he only had to do $k = m - 1 = 4$ comparisons. (However, those specific 4 preplanned comparisons are the only ones he can make without defaulting to, and expensively adapting to all the $k = m(m-1)/2 = 10$ possible comparisons.) From the above Example 14.5.0.1, where he ran the complete Friedman's test, he already knew that $\mathcal{A} = 330$ and $\mathcal{B} = 294.333333$.

---

[15]The LSD, meaning *least significant difference*, was a method of applying uncorrected decision rules based on Student's t test without adjusting for alpha inflation. Like Cohen's d and 'small', 'medium', and 'large' it was a stumbling block that slowed the progress of research for decades.

Analysis: He used Bonferroni's correction to find the corrected cutoff for **each two tailed comparison**, $\alpha'/2 = (0.05/4)/2$, which gives a cutoff of 0.00625. Then, using $df = (6-1)(5-1) = 20$, he sought a suitable online calculator of t and found that

$$t_{\alpha'/2=0.00625,\, df=20} = 2.74437637.$$

Then with $R_j = \sum_{i=1}^{i=m_b} r_{i,j}$ and the inequality in Equation 14.10 as the decision rule,

$$|\mathcal{R}_2 - \mathcal{R}_1| > t_{1-\alpha'/2,\, df=(m_b-1)(m-1)} \sqrt{\frac{2m_b(\mathcal{A}-\mathcal{B})}{(m_b-1)(m-1)}};$$

for the pair by pair assessments he calculated

$$|\mathcal{R}_2 - \mathcal{R}_1| > 2.74437637 \sqrt{\frac{2 \times 6(330 - 294.333333)}{(6-1)(5-1)}},$$

which equals 12.6955. Referring back to the table in Example 10.6.0.1, he copied out the following ranks and their all important sums, $R_j = \sum_{i=1}^{i=m} r_{i,j}$, for each of the $j = 1, \ldots . m$ treatments:

Treatments	r_1	r_2	r_3	r_4	r_5	r_6	R_j	$|R_j - R_C|$	**Result**
A2=AC	2	1	3	2	1	3	**12**		
A1	5	4	4	5	5	4	**27**	15	Sig.
A3	1	5	5	4	3	2	**20**	8	NS
A4	3	3	2	1	4	5	**10**	2	NS
A5	4	2	1	3	2	1	**13**	1	NS

The difference between Treatment A1 and the control, Treatment A2, was the only comparison that was statistically significant. This agrees with the paired t tests following the RCBD ANOVA in Example 14.5.0.4.

If $\mathcal{A} = \mathcal{B}$ (as might very rarely occur), then you have statistical significance for Friedman's test, and if there are $m_b \geq 25$ blocks, you can use the Wilcoxon signed rank test with the Bonferroni correction for your multiple comparisons.

14.5.0.5 Only if m_b is largish: signed rank multiple comparisons

The Wilcoxon Matched-Pairs Signed-Rank test (Wilcoxon 1945, Segal 1956, Section 14.2) is another obvious nonparametric test that could, using Bonferroni's correction (or maybe Sidak's correction), be used for multiple comparisons following a Friedman's RCBD, **even if $\mathcal{A} = \mathcal{B}$**. However, the corrected fractional cutoff probabilities are not generally available in exact tables such as Appendix 9. So, Wilcoxon signed ranked multiple comparisons, using $Z_W = Z$ as the test evaluation, is only practical **for largish sample sizes**, n ≥ 25.

If $\mathcal{A} = \mathcal{B}$ and n < 25 there is little hope for multiple comparisons with useful power. **Sign tests might** work if the differences **are huge**, but they are the most insensitive tests in this book (Section 2.6). However, is it really true that $\mathcal{A} = \mathcal{B}$? What would happen if you calculated to more places?

Which block test?

If you were to decide you prefer testssuch as an RCBD ANOVA which is based on the assumption of normality, followed by paired t tests, you have to consider whether the added complexity of blocking is going to be able to give you an accurate estimate that is, a valid estimate, $MS_W = S^2$, of the overall variance, σ^2. An important consideration for choosing Friedman's test rather than the RCBD ANOVA is that Friedman's test allows large differences between the blocks, but the RCBD ANOVA might be increasingly risky as the difference between the blocks becomes larger. That is, an RCBD ANOVA might work well if you are using a group of plots in the same larger field, or wards in the same hospital, but Friedman's test is designed in such a way that you can use different farms for blocks or hospital wards in different countries as blocks.

If you do not wish to use a nonparametric method like the Wilcoxon Matched-Pairs Signed-Rank test for your comparisons, you can still use k Bonferroni corrected paired t tests, each with its own variance, after a Friedman's test, provided that k and α (hence α') are a priori. Because Friedman's test does not provide a MS_W, using paired t tests with it will provide less degrees of freedom than if you had run an ANOVA instead. However, ANOVAs are hypersensitive to non normality, so you gain nothing if MS_W does not turn out to be a valid estimate of σ^2.

14.5.0.6 Sample size for Friedman's Test

Here are two ways to estimate sample size for Friedman's test, neither of which is ideal for all applications.

As with the one way ANOVA, in some cases more so, the randomized complete block ANOVA (RCBD) tends to be sensitive to non normality and heteroscedasticity. However, it is sometimes suggested that you can estimate an acceptable sample size for nonparametric tests by using estimated sample sizes that you would use for the corresponding normal tests, in this case, an RCBD ANOVA. You can at least get a ballpark estimate for the **per comparison** sample size, $\widehat{n} = \widehat{2m_b}$, which will be the same for each of the individual pairs being compared, by using any of the approaches in Subsection 14.2.1.1 with a Bonferroni corrected $\alpha' = \alpha/k$ where k is the number of comparisons you plan to make. However, that will require a MS_W from a complete rehearsal or pilot test without the usual F test, or at least a variance estimate as from a reasonably large paired t test. Then, as for an RCBD ANOVA, you can estimate the total sample size for the test as

$$\widehat{\mathbf{n}} = \widehat{m_b}m,$$

where m is the number of treatments, and $\widehat{m_b}$ is the number of blocks, in effect, your **sample size per treatment** (half of $\widehat{2m_b}$).

The efficiency of Friedman's test compared to the ANOVA, assuming the assumptions of the ANOVA are met, is $\frac{0.955m_b}{m_b+1}$ (Prajapati, Dunne, and Armstrong 2010). So, if you base the sample size for a Friedman's test on an **estimation for an RCBD ANOVA or paired t test**, you should multiply the resulting sample size estimate by

$$\frac{m_b+1}{0.955m_b},$$

where m_b is the number of blocks. Round up until you have enough to have an **equal number in each block**; this will be conservative and will sometimes reduce any harmful effect of data loss. If feasible, adding an entire extra block to avoid data loss is also prudent for any complete block design.

The pilot data source should ideally include all the potential treatments. If a control is intended and you can only provide a pilot variance that is estimated as though a part of a paired t test, the control should be a member of the pair and the other treatment should be chosen to be an intended treatment that you think could be the most different from the control.

Here is a rough way of estimating sample sizes for tests like this which use Z or its equivalent for the final decision. It relies on you choosing a reasonable minimum percent (or fraction) of the variability, an unfamiliar concept at best. Let's start as though we were going to use Wilcoxon's signed rank test with only one comparison. You can choose a ϕ^2 (maybe as a percent) as the reasonable minimum amount of change that the treatment must cause in the variability of the data. Then using $Z = Z_W$, substitute $Z^2 = \chi^2_{1\,df}$, and starting with a slightly modified definition of phi,

$$\phi^2 = \chi^2/m_b\,,$$

substitute $\left(Z^2_{\alpha'/2} + Z^2_{\beta'}\right)$ for χ^2 and rearrange to get

$$\widehat{m_b} = \left(Z^2_{\alpha'/2} + Z^2_{\beta'}\right)/\phi^2.$$

Of course, $\widehat{m_b}$ estimates the number of pairs (hence blocks) you will need for a single comparison. Then the total sample size should be

$$\widehat{\mathbf{n}} = m\widehat{m_b}\,,$$

where m is the number of treatments. This **should work for** $\widehat{m_b} \geq 25$, which may seem like a rather large minimum number of blocks (such as plots, farms, or wards), but it is what we are stuck with.

14.6 Spearman's Rank Correlation Coefficient

Spearman (1904) was the first to suggest the use of ranks, the ordinal positions of the data in statistical analysis, rather than the data itself. **The mean of the ranks is not necessarily the mean of the raw data** and the **ranks center on the median**, $\tilde{x}$, of the combined raw data. The question is whether there is a tendency of two variables to vary together in the same direction, somewhat like Pearson's correlation coefficient (Section 11.3). However, whereas the Pearson's focuses on linear relationships, ***Spearman's rank correlation coefficient,*** ρ_s (the Greek letter rho-sub-s), measures the tendency the variables to be ***monotonically related***, meaning that the scores are always increasing together or decreasing together. When the ordinal relation of two samples is exactly monotonic, meaning the members of ranked data pairs follow the same trend, the Spearman's rank correlation coefficient, ρ_s, is ± 1. If the trend is toward the **same** direction (both increasing, or both decreasing), ρ_s approaches $+1$, but if the trend is toward the **opposite** way, one variable is increasing while the other variable is decreasing, then ρ_s approaches -1. Another way to say this is when the monotonic relation is perfect, Spearman's rank correlation coefficient has an absolute value of one. If there is absolutely no monotonic relation, then $\rho_s = 0$, but that does not necessarily mean no relationship, just no monotonic relationship (hence no linear relationship either).

If the square of Spearman's rank correlation coefficient, ρ_s^2, is expressed as a fraction or a percent, it may be thought of as a measure of the monotonic degree to which the variables are related. Being monotonically related is also required by Pearson's correlation, which differs by focusing only on linear, meaning straight line, relations. So, Spearman's rank correlation is intuitively more like what most people might picture when they ask if two things are correlated, and in application it is often much closer to the relation being sought than is Pearson's correlation coefficient (Section 11.3). Not all relations that are monotonic are particularly linear. However, relations that are linear are monotonic at least to the degree to which they are linear.

To calculate a Spearman's rank correlation, start by ranking the sampled values of A from the smallest to the largest, assigning a rank value to each, and repeat the process for the sampled values of B. Tied values are assigned a rank equal to the average of their ranks. Then, these rank values must be assigned back into the original sample pairs (x_i, y_i). This ranking transforms the (x_i, y_i) raw data points to $(r_{A,i}, r_{B,i})$ ranked points. Then you can estimate Spearman's rank correlation, ρ_s, using

$$\hat{\rho}_s = \frac{\sum_{i=1}^{i=n}\left(r_{A,i} - \bar{r}_A\right)\left(r_{B,i} - \bar{r}_B\right)}{\sqrt{\sum_{i=1}^{i=n}\left(r_{A,i} - \bar{r}_A\right)^2 \sum_{j=1}^{j=n}\left(r_{B,j} - \bar{r}_B\right)^2}}.$$ (14.11)

It looks like an ordinary Pearson's product moment correlation coefficient (Equation 11.7) with new symbols; what is happening here?[16] Well, neither association nor correlation is, in itself, dependent on normality. So this is a perfectly valid approach, even though normality would have been handy for estimating CIs, etc. Pearson's correlation tends to focus on the degree of linearity; Spearman's correlation tends to focus on relationships **between outcomes that might be somewhat curved** (curvilinear relations). The operational difference is that Spearman's Rank Correlation estimates the association in terms of the ranks drawn from the raw data rather than from the raw data itself. Otherwise the calculations of Spearman's and Pearson's are homologous. See Question 5 as both an exercise and as a worked example (answer at back of book).

14.6.1 The significance of Spearman's rank correlation

You cannot conclude that a Pearson's correlation coefficient is statistically significant if the residuals are not normally distributed. But they would rarely be normally distributed unless the points each had a sufficient sample size. Avoiding this restriction is yet another advantage of using Spearman's rank correlation. It only requires that A and B are **each from the same (their own) population**. Suppose you wished to test

H0: THERE IS NO CORRELATION BETWEEN A AND B

HA: THERE IS A MONOTONIC CORRELATION BETWEEN THEM.

Exact tables that result from evaluating all the possible outcomes, some for up to n $= 60$, are available in various texts and may be used for either one or two tail tests. However, the ρ_s values tend to rapidly converge towards a normal curve as the sample size grows large. That fact enables you to use

$$t = \hat{\rho}_s \sqrt{\frac{n-2}{(1-\rho_s^2)}} \qquad df = n-2$$ (14.12)

when the number of pairs (points), n, is > 30 (Choi, 1977, Fieller, Hartley, and Pearson, 1957). This can be compared against a table of t at the cutoff (α or $\alpha/2$) level because the slope of ρ_s, hence the tails, can go in either direction. Question 7 provides both an exercise and an example of this test.

14.6.2 A priori sample size estimate $\hat{n}$ for Spearman's rho

To a priori estimate how many pairs, n, you will need in order to see if a Spearman's correlation is statistically significant, first choose a reasonable minimum difference. Obviously, the reasonable

---

[16]Obviously, in the absence of ties, some terms such as $\sum_{i=1}^{i=n}\left(r_{A,i} - \bar{r}_A\right)^2$ and $\bar{r}$ are constant for a given sample size. Thus there are a variety of potential substitutions to make different formulas. So, it is possible to derive a widely published (e.g. Sokal and Rohlf 1969) and seemingly simpler approximation of the Spearman's rank correlation,

$$\hat{\rho}_s = 1 - \frac{6\Sigma(r_B - r_A)^2}{n\left(n^2-1\right)},$$

sometimes referred to as though it were Spearman's rho itself. It arises out of the fact that the mean and standard deviation is known for any set of ranks in the form $1, 2, \ldots, n$ (Section 14.1). But, this does **not work properly when tied ranks are present**, and they **usually are present**. Assuming there were almost no ties, this approach might have been somewhat justified in the days of hand calculation. With larger data sets and since the advent of computers to handle them via Equation 14.11, this alternative approach is now just a historical curiosity.

minimum difference has a strong relation to the power of your test. Even if you cannot do an a priori power analysis, you should a priori choose a credible reasonable minimum difference, a fraction, $0 < \rho^2_{s\_min} < 1$, of the observed variation resulting from a monotonic relation, that would be big enough to be of interest. Then use Equation 14.13, informally derived below.

Squaring Equation 14.12 (above) to get $t^2 = \rho_s^2 \frac{n-2}{1-\rho_s^2}$, then substituting $\left(t_{\hat{n}-2,\alpha/2} + t_{\hat{n}-2,\beta}\right)^2$ for t^2, and the reasonable minimum difference $\rho^2_{s\_min}$ for ρ_s^2, with more rearranging, gives

$$\frac{\left(1 - \rho^2_{s\_min}\right)}{\rho^2_{s\_min}} \left(t_{\hat{n}-2,\alpha/2} + t_{\hat{n}-2,\beta}\right)^2 = n - 2 .$$

Finally substituting $\hat{n}$ for n and adding 2 to each side gives

$$\hat{n} = \frac{\left(1 - \rho^2_{s\_min}\right)}{\rho^2_{s\_min}} \left(t_{\hat{n}-2,\alpha/2} + t_{\hat{n}-2,\beta}\right)^2 + 2 . \tag{14.13}$$

Because you do not know $\hat{n}$ for the initial estimate, you must **start** with $Z_{\alpha/2}$ and Z_β, and then **repeat** the estimation with $\left(t_{\hat{n}-2,\alpha/2} + t_{\hat{n}-2,\beta}\right)$ **until $\hat{n}$ stops changing**. For a one tail test, you will use α instead of $\alpha/2$. Question 9 provides an exercise, and its answer is an example of this estimation.

14.6.3 Subsequent power for Spearman's rho

For a subsequent power analysis of a Spearman's rank correlation, you must have a verifiable a priori reasonable minimum difference, $\rho^2_{s\_min}$, and a criterion, α. An informal derivation explanation starts with Equation 14.12, $t = \rho_s \sqrt{\frac{n-2}{1-\rho_s^2}}$. Then substituting $\left(t_{n-2,\alpha/2} + t_{n-2,\hat{\beta}}\right)$ for t gives

$$t_{n-2,\alpha/2} + t_{n-2,\hat{\beta}} = \sqrt{\frac{\rho_s^2 (n-2)}{1 - \rho_s^2}} .$$

Substituting $\rho^2_{s\_min}$ for ρ_s^2, and lowercase z for $t_{n-2,\hat{\beta}}$, and rearranging this gives

$$\hat{z} = \sqrt{\frac{\rho^2_{s\_min} (n-2)}{1 - \rho^2_{s\_min}}} - t_{n-2,\alpha/2} . \tag{14.14}$$

The lowercase $\hat{z}$ is used because it is possible for this equation to have a negative outcome, such that the interpretation of the outcome calls for the rules

$$iff \ \hat{z} < 0, \ then \ |\hat{z}| = \widehat{t}_{n-2,\,1-\beta} ,$$

or,

$$iff \ \hat{z} \geq 0, \ then \ \hat{z} = \widehat{t}_{n-2,\,\beta} .$$

After calculating, refer to a suitable table to get an estimate of the power, $1 - \beta$, or of β from which you can estimate the power. The explanation for these rules is given in Section 7.8. Mumby (2002) uses a Monte Carlo approach to the problem of power estimation for Spearman's rho.

Some folks prefer another nonparametric correlation for some applications by Kendall (1938, 1945; the latter for ties).

14.7 The Hypergeometric Distribution

Before going on, recall from Section 2.3 that the number of combinations of n things taken r at a time **without replacement** is $\binom{n}{r} = \frac{n!}{r!(n-r)!}$, the number of ways that you can select r things, without replacement, from n things, where order does not matter. Suppose this table $\begin{array}{|c|c|} \hline a & b \\ \hline c & d \\ \hline \end{array}$ has *George*, *Bill*, and *Jim* all included in the first column. Then $a + c = 3$, and if cell a has two of them in it, then a must include one of those $\binom{a+c}{a}$ possible pairs. That is, one of $\binom{3}{2} = 3!/(2!(3-2)!) = 3$ combinations: $\{George, Bill\}$, $\{George, Jim\}$ or $\{Jim, Bill\}$ is in cell a.

R.A. Fisher (1922) showed that if the combinations are drawn at random for each column and **H0** is true, the probability of getting any one particular set of values in a 2×2 contingency table is given by the ***hypergeometric distribution***, which may be defined by

$$p = \frac{\binom{a+b}{a}\binom{c+d}{c}}{\binom{n}{a+c}}, \tag{14.15}$$

where $\mathbf{n} = a + b + c + d$ is the total sample size.[17] Recall that $\binom{n}{r} = \frac{n!}{r!(n-r)!}$ The divisor in Equation 14.15, $\binom{n}{a+c}$ is the total number of ways you can draw $a + c$ things out of $\mathbf{n}$ things. It is the number of possible ways you might get the first column total. As for any 2×2 contingency table, if you know the sample size $\mathbf{n}$, and one row total and one column total, you only need to know one value in the interior cells to complete the table. This distribution tends to approach normality as the sample size becomes large. The hypergeometric distribution has a number of consequences of interest in statistics and in modeling. It is useful for small samples and, when computationally practical, for larger tests of independence. Fisher's Exact Test may help you to appreciate this distribution and how it might arise elsewhere.

14.8 Fisher's Exact Test

An ***exact test*** is a test in which no approximations of distributions are involved. Exact tests estimate the probabilities directly based on no more than an assumption of randomness, as opposed to ***approximate tests***, which estimate the probabilities based on the approximation of some particular distribution. Like most nonparametric tests, the power of exact tests is poorly understood, but their power is often assumed to be close to or equal to equivalent approximation tests such as the normal approximation or to Pearson's chi square, when applied to the same data.

Fisher's Exact Test (Fisher 1922) is a nonparametric test for categorical data that compares two or more different factors. It is especially useful for the analysis of 2×2 contingency tables where the expected values in **some cells are smaller than required for Pearson's** χ^2, or are otherwise unsuitable for that test. This is the primary reason that Fisher's exact test is included in this book. For instance, it would seem to be an ideal test for multiple comparisons where the comparisons include cells that are too small, $n < 10$, for a 2×2 Pearson's test. Its structure suggests a relation to the hypergeometric distribution, and that distribution might have some potential for someone to eventually work out a satisfactory power analysis specific to it. This test only depends on the assumption that the sampling is sufficiently random. Like all estimates involving randomly collected data, the estimate of the probabilities involved improve with sample size.

---

[17]Whereas the binomial distribution describes the number of successes for draws **with replacement**, the hypergeometric distribution is a discrete probability distribution that describes the number of successes in a sequence of $\mathbf{n}$ things drawn from a finite population **without replacement**.

So, the primary assumption of Fisher's exact test is that the data is random and that the factors are independent. Although the model intuitively seems to suggest that all the marginal values should be random values, some preplanned sample sizes, for instance, equal column totals or row totals, do not invalidate the test, and that can be quite handy. Indeed, I only recommend a **two tail** 2×2 Fisher's exact test for 2×2 tables which have either **two equal row totals** or which have **two equal column totals**.[18] Although a two way Fisher's exact test may be applied to two way tables, or to any $r \times c$ table (Mehta and Patel 1983), calculations involving large tables may be sufficiently complex as to become cumbersome, even for a computer. For relatively small sample sizes, it might be an old test whose time is dawning. However, Fisher's Exact test is most often used as a one tail test because of the asymmetry of the outcomes and due to other potentially valid controversies regarding its application as a two tail test. **Beware:** some major computer packages may merely use a Monte Carlo to approximate Fisher's exact test for large numbers of rows and columns.

Fisher's exact test is slightly hyper-conservative, meaning that when you reject, you are often just a little more certain than the p_value indicates. For sparse or unbalanced data, or small sample sizes, the exact (Fisher's) and asymptotic (Pearson's) p_values can be quite different, and may lead to opposite conclusions with regard to **H0** (Mehta, Patel, and Tsiatis 1984). In the unusual instance where an exact test gets significance and Pearson's χ^2 does not, **accept the exact test**. The slight conservativeness of the Fisher's test implies a slight loss of power, but published alternative tests have so far been inferior in other ways.

For 2×2 contingency tables, Fisher's exact test can generally be calculated by hand if necessary, and it can **even** be used when some cells involve values less than 5. So, let's focus on 2×2 tables. This involves calculating the number of ways the table can be modified by changing one cell to all its possible values while the marginal values are held constant. This is calculated by repeatedly using Equation 14.15. Although this process of obtaining rectangular tables and calculating the probabilities follows a hypergeometric distribution, the distribution is a characteristic of the process, not the data, so this exact test does not require the raw data to follow a hypergeometric distribution.

14.8.1 One tail 2×2 Fisher's exact test

For a one tail 2×2 exact test you must **a priori** hypothesize **which** one category, say Category A, has more effect than the other, say Category B.

In the examples here we will consider two categorical factors such as the two genders versus two categories, obese and not obese. Neither one factor nor the other have to be under your control; one of them may or may not be a treatment. The table will be identical to an analogous Pearson's chi square.

	Male	Female	Total
Obese	a	b	$a+b$
! Obese	c	d	$c+d$
Total	$a+c$	$b+d$	n

The important assumption is that the sampling is unbiased and random. You use Equation 14.15,

$$p = \frac{\binom{a+b}{a}\binom{c+d}{c}}{\binom{n}{a+c}},$$

[18]The two tail Fisher's exact test for tables greater than 2×2, like all sizes of two tail Fisher's exact tests, supposedly involves calculating all the tests as extreme or more extreme than the observed result. However, for larger tables, the exact interpretation of what that means is a matter of controversy, which makes its use impractical, unless that controversy can be resolved. Even for the 2×2 contingency tables, this difficulty is, at present, only avoided if both row marginal totals or both column marginal totals are the same.

to calculate the exact probabilities of all possible tables which have the same marginal values as this observed table. This involves changing one cell to all possible values that will not change the marginal values (see Example 14.8.1.1). Then you will **sum all the outcomes** that are **less than or equal to** the p_value_0 of the actually observed table, that is, you calculate

$$\Sigma(p\_value_{l>0} \leq p\_value_0) \, .$$

That is, you sum the probabilities of all the tables (including the observed table) that are as small as, or smaller than, the probability of your observed table. You then compare that sum to the criterion, say, $\alpha = 0.05$. You declare statistical significance if these summed probabilities are as small as or smaller than the criterion, α.

14.8.1.1 Example: One Tail Fisher's Exact Test

For working with Fisher's exact test, it helps to recall that Equation 2.3, $\binom{n}{r} = \frac{n!}{r!(n-r)!}$, defines combinations.

Obese cooks Your hypothesis pair for your hotel chain, where there seems to be a lot of absenteeism due to health problems involving overweight cooks, could be something like,

H0: THE PROPORTION OF FEMALE COOKS THAT ARE OBESE IS THE SAME AS THE PROPORTION OF OBESE MALE COOKS

HA: A GREATER PROPORTION OF THE FEMALE COOKS ARE OBESE,
with a criterion of $\alpha = 0.05$. Then you collect the data for a 2 × 2 table, which you will use to find the p_value_0, the likelihood of this one unique table occurring by chance alone. You have tried to attain random data. However, because of personnel policies, you cannot obtain a large sample.

Calculation: Assume you have drawn a random sample that can be represented by this table:

	male	female	Σ
obese	1	9	10
! obese	6	2	8
Σ	7	11	18

$p\_value_0 = 0.00879839$

The $p\_value_0 = \binom{10}{1}\binom{8}{6}/\binom{18}{7}$ is the likelihood that this **particular** table occurred by chance alone, given there was no difference.[19] But, you still need to find the various p_value_i probabilities associated with all the possible tables with the same marginal values. Choose a cell in the row or column with the **smallest marginal value**, (which is 7) so you can choose either cell a or c; in this example we will use cell a. Start by finding all such tables with the same marginal values by repeatedly **changing the value of the selected cell**. The selected cell can only be assigned values between zero and the smallest marginal value in its row or its column. You can only do this so many times before you run out of cell values which will produce the same marginal values. After each change you will have to adjust all the other three cells accordingly. After finding all the possible tables for these changes, various p_value_i can be calculated the same way as the original table, to get

---

[19]If you use the R language, factorial(x) returns $x!$. So, $\binom{8}{6}$ could be found using
`factorial(8)/(factorial(6)×factorial(8−6))`.

		Σ
2	8	10
5	3	8

Σ : 7 11 18

$p\_value_2 = 0.0791855$

		Σ
3	7	10
4	4	8

Σ : 7 11 18

$p\_value_3 = 0.263952$

		Σ
4	6	10
3	5	8

Σ : 7 11 18

$p\_value_4 = 0.369532$

		Σ
5	5	10
2	6	8

Σ : 7 11 18

$p\_value_5 = 0.221720$

		Σ
6	4	10
1	7	8

Σ : 7 11 18

$p\_value_6 = 0.0527904$

		Σ
7	3	10
0	8	8

Σ : 7 11 18

$p\_value_7 = 0.00377074$

		Σ
0	10	10
7	1	8

Σ : 7 11 18

$p\_value_8 = 0.0002513826$

Now sum all the outcomes that are **less than or equal** to the p_value_0 of the original observed table. That is, you sum the probabilities of all the observed tables with probabilities of occurring by chance alone that are **as small or smaller than** the probability from your observed table. So, in this instance, the result is

$$p\_value_{total} = \sum \left(p\_value_{l \geq 0} \leq p\_value_0\right) = p\_value_0 + p\_value_1 + p\_value_7 \,,$$

$$= 0.00879839 + 0.00377074 + 0.0002513826 = 0.01282051 \,.$$

You then compare that sum to the criterion, $\alpha = 0.05$. You declare statistical significance if this sum of the probabilities is as small or smaller than the criterion. So, this Fisher's exact test is **statistically significant** at the 5% probability level because it has a likelihood, p_value_{total}, of being observed by chance alone that is less than $\alpha = 0.05$.

Discussion: Although Fisher's exact test is widely accepted for such small samples, the chance of getting a Type II error is higher for small samples, meaning that a real difference might not always have a sufficient opportunity to be revealed.

Always remember to include the p_value from the observed table itself. As a check on your calculations, the sum of all the possible outcomes, including the first, should equal one.

CAUTION: Be certain the relation is in the correct direction; you can see that the proportion of obese female cooks is greater in the sample. You can only declare a one tail test significant if the observed proportion that your alternate hypothesis said was greater, was also greater in the data. Note that a statistical test only regards the population tested.

If you ill-advisedly attempt to use a spreadsheet, an error may occur at, say, the seventh or eighth figure, or maybe the third or fourth figure. (Excel has a particularly bad reputation for small probabilities. Keeling and Pavur 2012, McCullough and Wilson 2005, Yalta,2008).

14.8.2 Two tail 2×2 Fisher's exact test

When **either** the sums of two rows **or** the sums of two columns of a 2×2 table are the **same**, the distribution is sufficiently symmetrical, for two tail testing, that you may compare the final p_value

to $\alpha/2$ (Fisher 1922). Given that this condition is met, if you begin by using the marginal total of the first column as in Equation 14.15, and you reverse the roles, thus reversing the emphasis of the test, you would get the same result. Given that this condition is met, you can start calculating as though you wish to see if one category is significant to get half of the two tail p_{half_value}. That is, this symmetry will allow you to double the result of just doing it once, as though for one tail, and then double the resulting $\sum(p\_value_{l \geq 0} \leq p\_value_0)$, to compare it to the two tail total criterion, α. (That is, it will be significant if the underlying probability for either tail is less than $\alpha/2$. And if it is less than $\alpha/2$ for one tail, it will also be less than $\alpha/2$ for the other.) However, **when neither** the 2 row totals **nor** the 2 column marginal totals are the **same**, calculation of the two-tailed Fisher's exact test usually involves a great deal more calculation and is more suited to a computer (one possible reference is Glantz 1991).

14.8.2.1 Example: Two Tail 2×2 Fisher's Exact Test

Obese cooks (continued) Assume that instead of settling for unequal sample sizes, you somehow attain random data for **equal** numbers of males and females and because of the resulting **symmetry** you can choose a two tail hypothesis pair,

H0: THE PROPORTION OF OBESE MALE COOKS IS THE SAME AS THE PROPORTION OF OBESE FEMALE COOKS.

HA: THE PROPORTION OF ONE GENDER OF COOKS IS GREATER THAN THE OTHER.

Criterion: is still $\alpha = 0.05$, but because you have **split it** between two tails, you will have to **double** your result before comparing it to the criterion.

Calculation: Adequate randomness is assumed for the purposes of this example, and your table of original data comes out like this:

	male	female	Σ
obese	1	9	10
! obese	8	0	8
Σ	9	9	18

$p\_value_0 = 0.0002056767$

The $p\_value_0 = \binom{10}{1}\binom{8}{8}/\binom{18}{9}$ is the one tail probability that this particular table occurred by chance alone, given there was no difference (Equation 14.15). Then you need to calculate the probabilities of all the possible tables with the same marginal values. You can easily do this by repeatedly changing one of the values in the row or column with the smallest marginal value (either cell c or cell d). The rest of the cells can be adjusted accordingly to maintain the marginal values unchanged. You can only do this so many times before you run out of cell values which will preserve marginal values. That is, cell d can only assume values between zero and its smallest marginal value. So, for convenience, assume you choose cell d and change values from 0 up to and including 8:

		Σ
2	8	10
7	1	8
Σ 9	9	18

$p\_value_1 = 0.00740436$

		Σ
3	7	10
6	2	8
Σ 9	9	18

$p\_value_2 = 0.06910736$

		Σ
4	6	10
5	3	8
Σ 9	9	18

$p\_value_3 = 0.2418758$

		Σ
5	5	10
4	4	8
Σ 9	9	18

$p\_value_4 = 0.3628137$

	Σ					Σ		
6	4	10	$p\_value_5 = 0.2418758$		7	3	10	$p\_value_6 = 0.06910736$
3	5	8			2	6	8	
Σ 9	9	18			Σ 9	9	18	

	Σ					Σ		
8	2	10	$p\_value_7 = 0.00740436$		9	1	10	$p\_value_8 = 0.00020567671$
1	7	8			0	8	8	
Σ 9	9	18			Σ 9	9	18	

Only the p_values **as small or smaller than or equal to** p_value_0 are to be summed. The sum for one tail is

$$p\_value_{total/2} = \sum (p\_value_{l \geq 0} \leq p\_value_0) = p\_value_0 + p\_value_8 \,,$$

$$= 0.0002056767 + 0.00020567671 = 0.0004113534 \,.$$

Results and Discussion: But, this is a **two tail test**, so you **must double** the $p\_value_{total/2} = 0.0004113534$ to get the estimated sum for **both** tails: $p\_value_{total} = 0.0008227068$. This is less than $\alpha = 0.05$, so you can reject **H0**, meaning accept **HA**; the test is **statistically significant**. The check on the calculations comes out at exactly one, at over eight decimal places (don't forget to include the original p_value). You might feel uncomfortable with this sample not only because it is so small, but because it got only females who were obese. However, the calculations are not only based on the handful of numbers in the sample, but also upon the assumption that this is a completely random sample from an enormous number of possible samples.

14.8.3 A priori sample size estimate for Fisher's exact test

To estimate sample sizes for Fisher's Exact Test, your best hope so far is to conservatively apply an a priori calculation suggested for Pearson's χ^2, Section 13.2.1, and add a little, maybe 10%, of $\hat{n}$ and again round up.

Questions for Ch 14

Write the equations where appropriate, and then fill in the values. You should work all **odd numbered questions** and study their answers which are at the end of the book. The data tends to be fictitious.

1. Make a short table showing the ranks of the following numbers: 19, 21, 15, 18, 26, 18, 8, 26, 8, 13, 15, 27, 10, 21, 8, 14, 26, 12, 25, 19, 16, 5, 7, 30, 7 .

2. **Find the probability**, p, for this first Fisher's exact test table,

1	2	3
3	5	8
4	7	11

, occurring

by chance alone, as though you were starting to calculate all the tables for an exact test (Hint: Equation 14.15). If this were the first calculation in a Fisher's exact test **would you go on**, and **why**? Also, assuming you did this by hand and had lots of time, to **how many figures** can the answer be correctly represented?

3. Use a Wilcoxon summed rank test for a two tail test at $\alpha = 0.05$ to compare soil temperatures, using this fictitious data, allegedly from Northern Canada, July 4, 2020, in degrees centigrade, at the $\alpha = 0.05$ level.

Observ. A: 6.6, 6.5, 3.3, 4.0, 6.8, 4.5, 0.2, 2.5, 2.8 °C.
Observ. B: 3.9, 1.5, 5.0, 2.0, 7.8, 6.0, 5.5, 2.8, 5.1 °C.

4. To quote Spearman's original 1904 paper, "Suppose that it was desired to correlate acuteness of sight with that of hearing, and that for this purpose five persons were tested as to the greatest distance at which they could read and hear a standard alphabet and sound respectively. Suppose the results to be:

Person	Sight	Hearing
A	6 ft	6 ft
B	7	11
C	9 (median)	12
D	11	10 (median)
E	14	8

" ... "

Calculate the Spearman's rank correlation for this data. (You might want to do Question 5 first.)

5. **Find the Spearman's rank correlation coefficient** for the following set of data presented as (x_i, y_i) points,

(27, 58), (27, 64), (25, 77), (34, 74), (50, 75), (56, 76), (52, 73)

Also, what fraction, $\hat{\rho}_s^2$, of the observed variation do you estimate to have resulted from a monotonic relation? Because this is continuous data initially reported to two figures, try not to round until you find both answers, then round to three figure accuracy.

6. What would change about the ranking of a Wilcoxon summed rank test if $n_1 \neq n_2$?

7. Test the statistical significance of a Spearman's correlation in which Spearman's rho, $\hat{\rho}_s = 0.70$, and the sample size includes n = 36 points, (x_i, y_i), at the $\alpha = 5\%$ two tail level. (Recall that your hope of observing statistical significance when it represents the true situation is relative to the reasonable minimum difference and to the power, $1 - \beta$, but assume you have already provided for them and could validly proceed to run the test).

8. Can you use Student's t for IQ scores; why or why not?

9. Estimate the a priori sample size for a Spearman's correlation, given a power of $1 - \beta = 0.80$, a (two tail) criterion of 5%, and a reasonable minimum estimate of the contribution of observed monotonic variation of $\rho_{s\_min}^2 = 0.49$. (Although lesser correlations have many uses, naturalists and some agriculturalists tend not to accept any correlation, whether Spearman's or not, less than $\rho = 0.70$, which corresponds to a coefficient of determination of $\rho^2 = 49\%$ {almost 50%}.)

10. What might be superior to an F test for comparing variances via samples, especially when the sample size is not huge?

11. You are a statistician and your client is eager to preserve patient anonymity. The data is for 4 drastically different measurements (A through D) from each of 7 patients (Roman numerals I through VII) with severe birth defects. The client is looking to see if the patients (analyzed as though 'treatments') belong in different populations.

	A	B	C	D
I	1.2	16.7	0.001	90
II	0.9	22.5	0.012	120
III	2.0	30.0	0.011	75
IV	1.3	15.6	0.008	100
V	0.8	29.3	0.017	91
VI	1.9	22.1	0.013	85
VII	1.6	19.5	0.010	110

There is no way you could apply an ANOVA because of the obviously extreme heteroscedasticity among the measures ('blocks' A through D). You also suspect that some, or all, of the measurements are not on an equal interval scale. However, using the measurements as blocks, and considering the drastic combination of measurements successfully used in Friedman's 1937 paper, you agree to run a Friedman's test. So, do it with an $\alpha = 5\%$, show it, and report the results.

12. Assign Siegel-Tukey ranks to the following data: 2, 5, 1, 7, 9, 13, 41, 11 .

13. Mega Juiced had a soils and water problem in its vineyards, involving sodium, which was affecting the quality of their grapes. They decided upon an ambitious set of trials involving $m = 4$ different gypsum and irrigation treatments. They chose an $\alpha = 5\%$ criterion and made a hypothesis pair, **H0**: THERE IS NO DIFFERENCE IN SUGAR CONTENT vs **HA**: THERE IS A DIFFERENCE IN SUGAR CONTENT.

				Treatment				
	A	rank	B	rank	C	rank	D	rank
	254	38	222	28	225	32	228	33
	192	9	205	16.5	261	40	212	21
	166	1.5	197	10.5	214	24	183	4
	197	10.5	259	39	205	16.5	231	35
	166	1.5	200	14	201	15	187	6
	198	12	223	29	188	7	212	22
	224	30	177	3	219	26	237	36
	206	18.5	185	5	231	34	220	27
	199	13	219	25	190	8	208	20
	246	37	206	18.5	224	31	212	23
Total	2048	171	2093	188.5	2158	233.5	2130	227
Median	198.5		205.5		216.5		212	

They had a lot of individually irrigated vineyards to choose from; they were able to use $n = 10$ randomly selected areas for each treatment and obtained a total sample size of $\mathbf{n} = 40$ after a few years of these $m = 4$ regimes. They obtained the above sugar content data, in grams per liter (about ten times what the brewing industry would call Brix degrees). In this table, the ranking of combined data has already been done by repeatedly using the sort feature of a spreadsheet, to simplify the processing (see Subsection 2.6.3.1 for how to simplify the ranking process): There are 4 ties, and ties may weaken the test a little, so weighing to a higher precision may have been better, in that it might have eliminated them (Langley 1970). On the other hand, a few folks might see ties, especially the three ties with members in different treatments, as potential evidence against true difference, and thus a justification to permit the ties. Run a Kruskal-Wallis test and report the results. Beware, this enology data is **fictitious**. (It was also used in Chapter 9, Question 3.)

14. (You might wish to do Question 13 before this.) Mega Juiced was running trials to

attempt to compensate for a sodium problem. They involved $m = 4$ different gypsum and irrigation regimes.

Treatment							
A	rank	B	rank	C	rank	D	rank
651.36	13	914.71	18	1597.37	34	1239.68	27
556.86	7	631.50	9	2086.32	37	1348.93	32
883.08	17	282.69	1	1033.92	21	743.38	14
1044.10	22	834.14	16	2296.33	39	648.65	12
1018.78	20	519.96	5	2083.65	36	470.94	3
1192.80	25	1281.30	30	1471.64	33	1626.77	35
1326.06	31	638.97	10	2335.26	40	507.44	4
640.18	11	1004.70	19	2176.65	38	1156.94	23
373.54	2	770.48	15	1275.36	29	1264.44	28
558.17	8	529.08	6	1171.00	24	1217.50	26
Total 8244.93	156	7407.53	129	17527.5	331	10224.67	204

They chose an $\alpha = 5\%$ criterion and made a hypothesis pair, **H0**: THERE IS NO DIF-FERENCE IN YIELD vs **HA**: THERE IS A DIFFERENCE IN YIELD. They used $n = 10$ randomly selected areas for each treatment and obtained a total sample size of **n** $= 40$ after a few years of these regimes. They obtained this yield data, in kilograms of accept-able grapes per eighth hectare single treatment plot after a few years. The weights were reported to the nearest 10 grams (0.01 kg) and perhaps that is why there are no ties. Per-form a Kruskal-Wallis test to see if there are one or more significant differences between the $m = 4$ treatments. Do not bother with medians; they are optional and at best would only be rough estimates of the central measures under ranking. (This data was also used in a one way ANOVA in Question 4 of Chapter 9.)

15. You are a noxious weed eradication specialist and you want to compare two differ-ent anti-triffid[20] treatments applied a few months ago. But, your team discovers that the plants are alive and still blooming. Is there even a detectable difference in the vigor of this blooming? Your hypothesis pair is **H0**: THERE IS NO DIFFERENCE BETWEEN THE TREATMENTS vs **HA**: THERE IS A DIFFERENCE BETWEEN TREATMENT A AND TREAT-MENT B. Your technician has weighed 10 pairs of triffid blooms, each pair from a differ-ent location with a different altitude and climate. The data is in kilograms of blooms per plant:

Treatment $_A$	5.5	2.7	4.2	3.8	5.4	7.1	5.2	5.7	2.3	3.5	Kg.
Treatment $_B$	2.8	2.2	1.2	3.6	5.8	5.2	4.3	7.7	1.8	1.1	Kg.

Use a Wilcoxon signed rank test to challenge **H0** at the $\alpha = 5\%$ level, recalling that this is a two tail test, and report the results.

---

[20] John Wyndham 1951, The Day of the Triffids, Rosetta Books LLC, NY (novel)

15 DISTRIBUTIONS and TRANSFORMATIONS

If **sufficient sample sizes** are available, don't panic if sample distributions have some moderate deviations from normal (the ANOVA type tests being an exception). In application, there is usually sufficient sample size to compensate for lack of normality by using Student's t test or its variants if sufficient power for a truly reasonable minimum difference has been provided.

Researchers have been known to try all sorts of mathematical transformations to force their data to become normally distributed. When you transform data to attempt to get a normal distribution, your peers may reasonably ask what underlying reality justifies your conclusion that the transformed data relates to reality. Also some transformed data may have two or more sources of variation, one of which does not transform the same way as the other. So some people will question whether you could have possibly created heteroscedasticity, meaning a mixed variance condition invalidating the assumption that the variance, σ^2, is the same for all treatments. This criticism is a challenge to defend against when presenting the results of an ANOVA that used transformed data.[1] Most editors will allow you to use a transformation to meet the assumption of normality, especially if you can somehow meaningfully report your results (your ES) in terms of the converted units. For a graphical technique of getting some reassurance that your transformation might be working, see Section 11.10.

15.1 Exploratory Visual Data Analysis

This section presents a glimpse of *exploratory data analysis*, approaches to analyzing data sets to summarize their main characteristics, such as their distribution, in easy to understand form, without having to adhere to a particular statistical school of thought or having formulated a hypothesis. This subject was largely developed by John Tukey (1977) to encourage researchers to visually examine their data sets so as to formulate hypotheses that could be tested on new, or at least separate, data sets.

15.1.1 Boxplots

Confidence intervals generally depend on the normal distribution and refer to means, but *boxplots* are nonparametric and refer to ranges, medians, quartiles, etc. Boxplots, also written as two words, *box plots*, divide ranked data into four parts, each containing one fourth of the scores, with upper ends called *quartiles* (Figure 15.1). A boxplot has a box that starts at the first quartile and ends at the third quartile, and has a dot or line in it representing the median; then a vertical line, or *whisker*, at either end runs to the beginning and end of the ordered data that is not considered *outliers* (outliers are errors or just static, atypical values that lie so far from the rest of the sample data that researchers

---

[1]To check that the variances are equal to each other some researchers use Levene's test which is available on popular software suites, but it is not recommended because it is only valid under the assumption of normality, and it collapses as soon as one variable deviates even slightly from the normal distribution (Hayes and Cai, 2007, Rakotomalala, 2008). Derrick et. al. seem to disagree with this in respect to two sample comparisons (2018). (Recall that a two sample ANOVA is equivalent to a t test.) Another test, the Brown–Forsythe test, is more robust to non normality.

Figure 15.1: Typical Box Plot

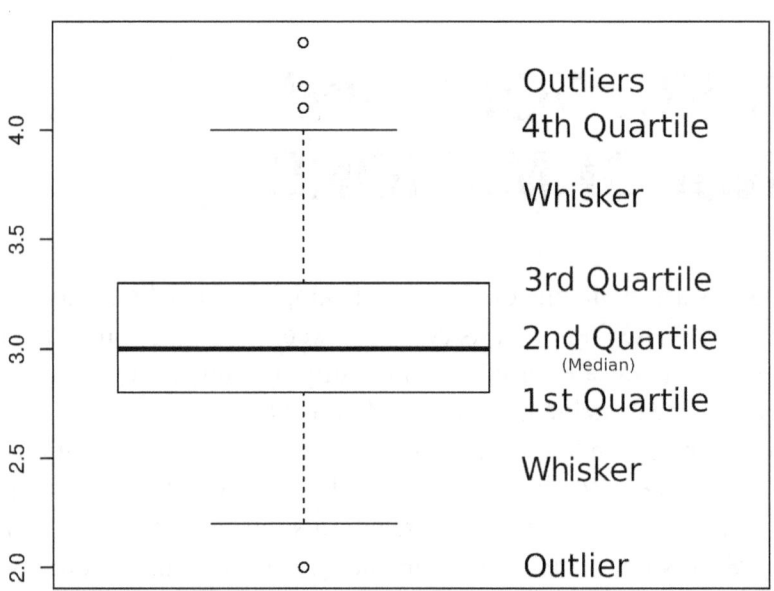

This is the combined, three species, petal widths of the classical iris data set (Fisher 1936). The **1st quartile** (*25th percentile*) is the bottom side of the box, the **2nd quartile** (*50th percentile*) is the median, the **3rd quartile** (*75th percentile*) is the top side of the box, and the **4th quartile** (*upper bound*) is the end of the top whisker (the *lower bound* is the end of the lower whisker). The vertical axis is marked off in original units of measure (cm). In application, the outliers are assumed to just be information static and **not** really part of the distribution.

tend to be suspicious as to whether they belong in the data). The whisker is intended to extend to the end of the range, but no further than a certain number, times the 'interquartile range' (the length of the box). So the *box* represents the **interquartile range**, the middle two quarters of the ranked data, and the whiskers represent the first quarter and the fourth quarter. Boxplots were developed by J. Tukey (1977).[2] They illustrate the distribution and are not sample size dependent, but greater sample sizes give greater accuracy. A popular definition of *outliers* is that they are observations that lie more than ± 1.5 times the interquartile range from the median. The multiplier of the interquartile range that determines the whisker length is therefore often 1.5, but this will vary **depending on the software**.

Below are some boxplots showing the effect of the CLT when taking larger and larger samples from the distribution of log_{10} transformed random numbers between 1 and 100 (Subsection 7.4). The rightmost boxplot ($n = 30$) is considered sufficiently normal for most applications (but extreme skew may require sample sizes as great as 80.

Boxplots provide information about the center, spread, and asymmetry of a population (Figure 15.3 below). The lower value side of the box is the first quartile, sometimes referred to as Q_1 and the higher side is Q_3, with the median being Q_2. The overall length of the box is therefore called the interquartile range, the **IQR**, the difference between the center two quartiles, $Q_1 = 25^{th} percentile$ and $Q_3 = 75^{th} percentile$ of the data (Section 15.6.1).[3] The extremes of the whiskers are either defined by you or your software, so if you do not pay attention your software will be choosing what you call data and what you call outliers.

However, there is considerable variation between statistical computer packages. For instance,

---

[2]Some of the ideas leading to Tukey's development of boxplots might have been based on the earlier work of Arthur Bowley.

[3]Another use of quartiles is the Tukey-Bowley **trimean** $T_M = \frac{Q_1 + 2Q_2 + Q_2}{4}$, a remarkably efficient and robust estimator of the population mean with moderately large samples (Tukey 1977).

Figure 15.2: Effect of Sample Size on Log Random Means

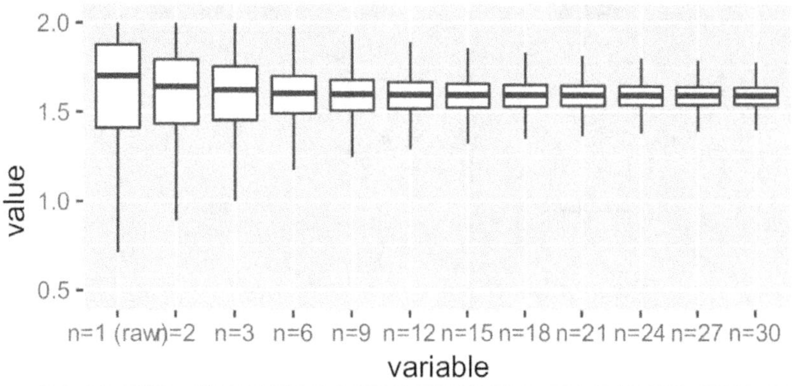

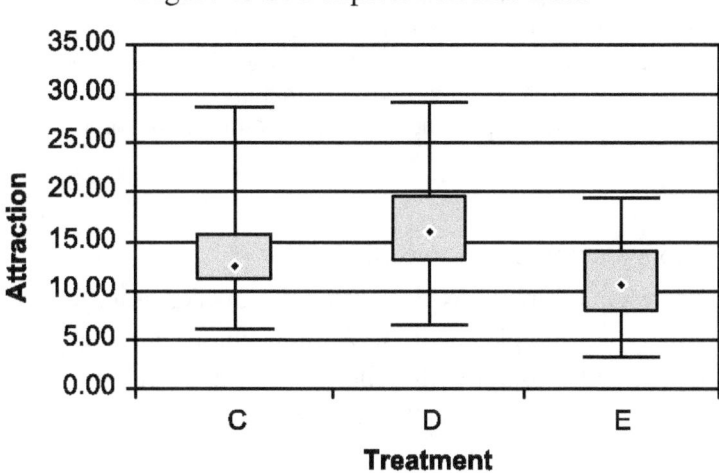

These plots correspond to Figure 7.2 .

Figure 15.3: Boxplots Perfume Data

Boxplots, also called ***box and whisker plots*** of data from Subsection 9.2.0.1. The diamond inside the box is the median and is not always at the center of the box. The top and bottom ends of the box, sometimes called ***hinges***, mark the first quartile and the third quartile. The horizontal lines, ***fences***, at the extremes of the whiskers, mark the minimum and maximum (no outliers here).

in a common R language package, maximum boxplot whisker lengths are optionally a user-selected number, times the interquartile range, and any data points, represented as dots, beyond the end of these are considered outliers. If there are tied outliers, the dots can be placed beside each other (Gonick and Smith 1993, p. 21). This effect is called ***jitter*** in some software. If you do not set the range variable, the whiskers often default to one and a half times the interquartile range from the box. The difference between a boxplot with whiskers that are a maximum of 1.5 interquartile distances long and a boxplot with whiskers that are a maximum of 3 interquartile distances long can be substantial. **If** the data were normal, values beyond a whisker length 3 times the interquartile range would only have 0.000001 (about 1 in a million) probability of occurring in a truly random draw. But values beyond whisker lengths that are only 1.5 times the interquartile range would have a 0.000373 probability (about 1 in 2681) of occurring by chance alone. Best be aware of how the maximum whisker lengths compare with interquartile range when you use any software to produce boxplots. Software that shows the outliers as separate points is best.

Statistical packages sometimes include the ability to make notched boxplots for which the nonoverlap of the notches allegedly indicates significance. This might work as a crude test for a **single** comparison of only two treatments but one difficulty is that the criterion for the non-overlap of these

Figure 15.4: Boxplots, Random Data

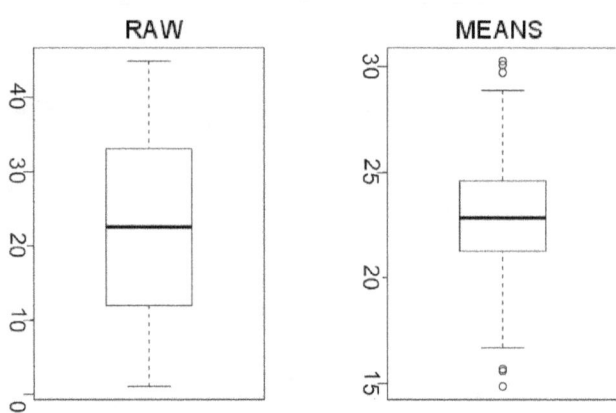

A raw data sample size of $n = 400$ raw scores is on the left, and an equal sized sample $m = 400$ of means (each of $n = 400$ raw scores) from the same population is on the right. There are a small number of outliers, ○, out of the sample of $m = 400$ means (not to the same scale, read the axis values). Considering the CLT and the law of large numbers, you might intuitively expect not to get any such outliers. However, chance can surprise us. The left boxplot represents a mere 400 scores, but the right boxplot represents $400 \times 400 = 160,000$ scores, so the probability of occasional rare random samples that are all large values and creating outliers is much greater. The darker outlier symbols represent overlapping outliers; in some software, a feature, 'jitter', would have avoided this.

notches, and its justification, are not always specified.

There are a number of ways to draw boxplots and boxplot-like graphics. A particularly interesting one suitable for relatively big samples is the ***violin plot*** that uses the width of the box to represent sample density and reveal structure found within the data (Hintze and Nelson 1998). Figure 15.5 is an example generated by using `GGPLOT2` in the R language. Violin plots are usually computer generated.

Figure 15.5: Violin Plots, Sable Fish Catch/Mo.

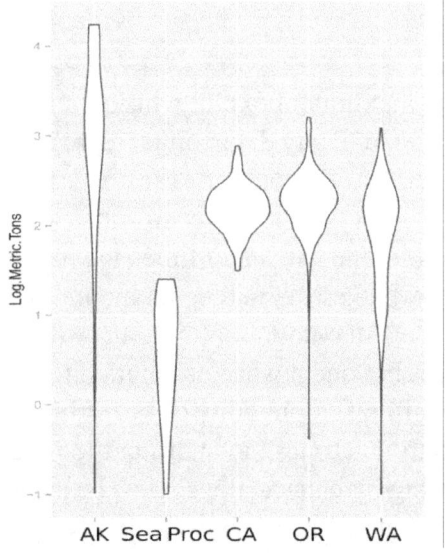

Next are some violin plots showing the effect of the CLT when taking the means of larger and larger samples from a flat ('truly random') distribution (Subsection 7.4):

Figure 15.6: Violin Plots of Effect Sample Size, n, on the Distribution of Means

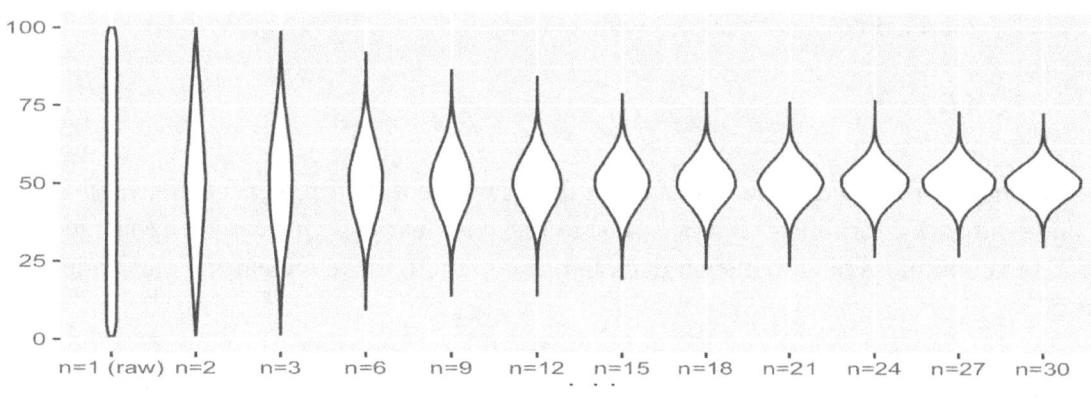

These plots correspond to Figure 7.2

The distribution in the rightmost plot ($n = 30$) is considered sufficiently normal for most applications.

15.1.2 Stem and leaf plots

Stem and leaf plots were also developed by John Tukey (1977).[4] They quickly summarize discrete data sets of $15 \leq n < 200$ while maintaining the individual data points. Like boxplots they are useful regardless of whether the data is assumed to fit a normal model or not. A stem and leaf plot retains the numbers themselves to show the same information as a histogram (but on its side), with little or no information loss. It also gives an idea of the shape of the distribution, the mode, the relative density of the data, and shows possible outliers. Unfortunately your audience needs to be familiar with the concept of stem and leaf plots in order to appreciate them. In addition to this text, they are presented in a very simple and acceptable manner in Gonick and Smith (1993). When presenting either stem and leaf plots or histograms, using between 5 and 20 subgroups (stems, or subsets, respectively) tends to maximize user friendliness.

15.1.2.1 Example: Stem and Leaf Plot

Stem and leaf plots are best understood by example:

A random sample: Let's start with a random sample of size, say, $n = 30$, from a population of random numbers:
-3 -14 22 -4 -6 -12 6 14 0 -5 16 -10 -2 24 1 -14 9 18 -12 -4 -3 24 11 -4 17 -15 12 1 13 -1
Order these from left to right in ascending order to get
-15 -14 -14 -12 -12 -10 -6 -5 -4 -4 -4 -3 -3 -2 -1 0 1 1 6 9 11 12 13 14 16 17 18 22 24 24 .
Examining these numbers, choose the highest numeral position in the data (tens place) to be the ***stems***, starting with -1 in the tens place and continuing all the way through to 2 still in the tens place, giving -1, -0, 1, 2 down the left side. Fill in the leaves on the right in ascending (absolute; regardless of ± sign) order, they are the numerals that are associated with each of the stem numerals on the left.

To illustrate: The stem and leaf plot will look like:

---

[4]Some of the work leading to Tukey's development of stem and leaf plots might have been based on earlier work by Arthur Bowley.

Stems	Leaves
-1	0 2 **2** 4 4 5
-0	1 2 3 3 4 4 4 5 6
0	0 1 1 6 9
1	1 2 3 **4** 6 7 8
2	2 4 4

The italic **bold 2** represents a value of -12 in the data, and the **bold 4** represents the value of 14 (italics and bold added for illustration only, just to make it clear how this works). You can also round and bin continuous data to integer data, but be certain to make it clear to your readers if you do so.

Discussion: This sample is drawn from a population that ranged from -15 to 24 inclusive and is totally random. A truly random population might be expected to have a uniform distribution (every value having the same relative frequency), however, chance has delivered a **sample** that is negatively skewed (has a lot of smaller numbers). Obviously a sample of $n = 30$ is not that free of the effects of chance. Random samples from random distributions will have random outcomes. Remember that the primary advantage of samples of size $n \geq 30$ is that the distribution of **means** of all such samples from most populations tend to approach being normally distributed. But this is not a sample of means, it is a sample that has a mean which could be calculated if you were interested.

15.2 Skew, Kurtosis, and Transformations

Skew and kurtosis are measures of distribution shape which formally address the manner in which some distributions are not normal. They are important when you are forced to work with small samples. It might seem ironic that you cannot validly test for them unless you have a very big sample of the data. But in some situations **a priori** data can come to the rescue. You might be able to check the distribution of the outcomes of hundreds of surgeries of a certain kind, but only have the capacity to perform a relatively small number of experimental surgeries. In that case you can look for skew and kurtosis in the old data before getting your sample (before performing the surgeries). Be aware that by examining data, you make it (the actually examined data) **ineligible** for subsequent inference.

When appropriate justification is available, a non-normal distribution might be transformed to see if there is a meaningful change toward normality. The graphical test at the beginning of Section 11.10, or tests of skew and kurtosis (Equation 15.1 and Equation 15.3), can be applied both before and after a transformation of data you really won't use for your conclusion. Except in very extreme instances, and for small sample sizes for which you have a great deal of understanding of the underlying distribution, you need to question whether transformations are appropriate. For instance, the Student's t test requires normality of the **distribution of the means, not the distribution of the data itself**. However, extremely asymmetrical or misshapen distributions may substantially increase the sample size necessary for the Central Limit Theorem, the CLT, to create a sufficient approach to normality in order for a t test to be valid. When power is taken into consideration for a Student's t test, a paired t test, or a Welch's t, the required sample sizes are usually enough that the CLT will compensate for substantial skew or kurtosis. In extreme situations where the data is too peculiar for the CLT to compensate, you might consider whether there is some intrinsic difficulty, such as inappropriate mixing of two or more very different populations. Could there be an experimental design difficulty? Are you asking the right question? However, in situations in which you must use data for which you do not have pre-existing information suggesting an appropriate transformation, and which defies even the CLT, maybe nonparametric approaches (Chapter 14) would be best for inferential tests or for simple correlation (e.g. Section 14.6).

In any case the post hoc exploration of the distribution of the data is often very informative. For instance, you might get additional insights for future hypotheses using boxplots (Subsection 15.1.1) or stem and leaf plots (Subsection 15.1.2).

15.2.1 Third moment skew

Skewness is a measure of the asymmetry of the probability distribution of a real-valued random variable. With regard to parametric tests, Tabachnick and Fidell (1996) point out that a non-normal kurtosis coefficient typically produces an underestimate of the variance of a variable, which, in turn, **increases** the Type I error rate (more false significance). Shlomo, Sawilowsky, and Blair used a Monte Carlo simulation to find that Student's t test failed to be robust primarily under distributions with extreme skew. Tests with unequal sample sizes, small sample sizes, and especially with only one tail, were most susceptible (1992). [5] Screening to remove outliers by using something like MAD (Section 15.6.2) may reduce this problem, but many authors recommend using one or another nonparametric test such as Wilcoxon's two sample test (Subsection 14.1.1) instead. However, once again, the difficulty might be in trying to use impractically small sample sizes. Ratcliffe examined sample sizes of up to 80 and concluded that "extreme non normality can as much as double the value of t at the 2.5% (one tail) probability level for small samples", but that increasing the sample sizes to $n = 50$ and $n = 80$ will for practical purposes remove the effect of moderate skewness and extreme skewness, respectively (1968). There do exist data sets that are too, for lack of a better word, unnatural, for any normal statistical procedure to handle without sample sizes approaching $n = 500$. An example of such data is medical costs which have a lot of zeros, a great deal of skew and many outliers, due to the odd relations between the government, the insurance companies, and the question of how much the individual patient is able to pay (skew might be a little over 8.0). Lumley, Diehr, Emerson, and Chen (2002) call attention to the challenge of medical cost data, and even a little about making it manageable, in their excellent review. Ironically, Student's t test, which Student originally intended for small samples, almost always works well for large enough samples in the presence of skew and kurtosis, but the tests for these characteristics are only effective for large samples. However, other than huge sample sizes for scrabbling hordes punctuated by capricious government mismanagement and the whims of billionaires and the insurance-government dance, it is quite rare to need sample sizes approaching half a thousand just to meet the assumptions of the t test. Attaining sufficient power to detect what you decide is sufficiently different to have meaning is an important matter, to be covered later.

If the data is so chaotic that you are unsure that the CLT can compensate, a nonparametric test might work.

Usually *skew* is estimated via the ***third standardized moment***,

$$\widehat{skew} = \frac{1}{n-1} \sum_{i=1}^{i=n} \left(\frac{x_i - \bar{x}}{S} \right)^3, \tag{15.1}$$

and this should be near 0 for the normal curve. For the normal approximation of the binomial,

$$skew = (p - q)pq \tag{15.2}$$

---

[5]The Sawilowsky and Blair paper seems to be, in part, a response to criticisms on the misuse of the normal curve by T. Micceri (1986, 1989).

(Lumley, Diehr, Emerson, and Chen 2002).

Positive skew indicates too much tail on the right and negative skew indicates too much tail on the left. The standard error of this skew estimate is $S_{\widehat{skew}} = \sqrt{6/n}$. It is popular to test to see if $skew > 2S_{\widehat{skew}}$, believed to mean the skew is too far to the right 'to be from a normal distribution', or to see if $skew < -2S_{\widehat{skew}}$, supposedly meaning that it is 'too far to the left to be from a normal distribution'. (I assume this is referring to t tests and to normal approximations.[6]) So we have a disguised Students t test requiring $n > 60$, that depends on an assumption of normality to test for non normality. However, the irony is that **almost no (if any)** raw data is from a normal distribution.

The real need is to know when and how to handle the deviations from normality, such as skew. Paul von Hippel (in Lovric, M., ed. 2010) gives the information you need to handle skew; he said, "Inferences grow more accurate as the sample size grows, with the required sample size depending on the amount of skew and the desired level of accuracy. A reliable rule is that, if you are using the normal or t distribution to calculate a nominal 95% confidence interval for the mean of a skewed variable, the interval will have at least **94%** coverage if the sample size is **at least** 25 **times** the absolute value of the (third-moment) skew"(Cochran 1977, Boos and Hughes-Oliver 2000) (**bold** highlighting added). For practical application this book accepts von Hippel's rule as adequate for using any common procedure involving the assumption of normality (except for the ANOVA).

As for kurtosis, being aware of third moment skew, with the associated rule, enables you to understand the printouts from a number of computer packages. Reporting skew and kurtosis might assist your colleagues in designing future experiments, even if you are reporting nonsignificant results. Indeed, studies specifically of the skew and kurtosis of certain populations might be very useful in themselves. Subsection 5.2.1.1 also discuses skew.

15.2.1.1 Imperfect skew rule

Many textbooks state that the mean is right of the median under right skew, and some state that the mode is then right of the median. Presumably this rule implies that the converse is true for unimodal continuous distributions with left skew. However, this rule should be sure to include the fact that unless the distribution is unimodal and continuous, these rules are **not** always true (von Hippel 2005).

15.2.2 Kurtosis

Kurtosis, $\mathcal{K}$, is a measure of the sharpness or flatness of the peak of the graph of a distribution which is assumed to produce a graph which is roughly symmetrical and ***unimodal*** (mound shaped). The graph of the distribution might be either too pointed or too flattened to be considered normal. The usual estimator is

$$\widehat{\mathcal{K}} = \frac{1}{n-1} \sum_{i=1}^{i=n} \left(\frac{x_i - \bar{x}}{S} \right)^4 - 3, \tag{15.3}$$

which would be about 0 **when** the distribution is **normal** (Tabachnick and Fidell 1996, Brown 1996). Its standard error estimate is about $S_{\widehat{\mathcal{K}}} = \sqrt{24/n}$, and it has been customary to assert that a $\widehat{\mathcal{K}}$ greater than $2S_{\widehat{\mathcal{K}}}$ means that the [raw] data is not normally distributed. Well, aside from the fact that this involves an approximation of a normal distribution based t test and requires samples greater than $n = 60$ to look for non normality, the vast majority of all raw data is non normal. Apparently, in application, a sample size that is adequate to detect a meaningful minimum difference and to adapt for skew will be adequate to adapt for kurtosis. However, this may need further study.[7]

---

[6] You might want to see the relevant footnote in Subsection 7.4

[7] You usually cannot assess kurtosis via a graph because varying the aspect ratio will vary the **apparent** peakedness or flatness.

Warning: the 'kurtosis value' given by some computer programs **might not subtract the 3** as in Equation 15.3. Some packages call this measure *excess kurtosis* if they mention the fact that it is 3 too high, if they mention it at all. **In that case** normality would have a 'kurtosis' equal to 3, so you might wish to subtract 3 from the value presented. Check the manual of the particular software package you happen to use.

Reporting kurtosis (even as a post hoc statistic) might also assist your colleagues in designing future experiments, even if you are reporting nonsignificant results.

15.2.2.1 Leptokurtosis

Leptokurtosis, $\mathcal{K} > 0$, is the condition in which the scores are more clustered about the mean than they would be in a normal distribution. The plot of a leptokurtotic distribution is more peaked than a normal distribution would be if plotted on the same scale.

15.2.2.2 Platykurtosis

Platykurtosis, $\mathcal{K} < 0$, is when the scores are more spread out than they would be in a normal distribution. The plot of a platykurtic distribution is flatter than a normal distribution would be if plotted on the same scale. Ratcliffe examined sample sizes of up to $n = 80$ and concluded that increasing the sample sizes to $n = 15$ and $n = 30$ will for practical purposes remove the effect of moderate flatness (platykurtosis), and extreme flatness, respectively (1968). Although Ratclffe did not cover both forms of kurtosis, it is unlikely that either will be a problem at sufficient sample sizes for practical application.

15.3 Transformations to Normality

Transformations which attempt to transform non-normal, especially highly skewed, distributions, to normal are easily available using various computer packages, but is transforming the raw data the proper thing to do? This question is not easy to answer and I cannot imagine any solution other than a priori examining a large sample from the population to be tested, but only if you can subsequently obtain new unexamined random data for the actual inferential test. Lumley, Diehr, Emerson, and Chen (2002) state, "Formal statistical tests for normality are especially undesirable as they will have low power in the small samples where the distribution matters and high power only in large samples where the distribution is unimportant."

Well, some data might be from a population so extremely skewed that you have no choice but to transform it. However, examining the distribution of the data before testing is only possible if you can subsequently obtain new unexamined random data for the actual inferential test. Then the transformation might be useful even if you are henceforth restricted to small samples. That could occur if the availability of the control is somewhat unlimited and the actual treatments are either too expensive or too difficult to obtain in large sample sizes. Transformation based on the distribution of a control might help with meeting the underlying normality assumption of a not very large sample Student's t test. Finding an apparently useful transformation for your data might give you a clue to the physiology or physics of whatever underlies what you are studying.

On the other hand the transformed results, **even the back-transformed data**, might provide **unintelligible results**, including effect sizes. With few exceptions, if you have analyzed using transformed data, it may not be appropriate to report your effect sizes in the original (pre-transformation) units of measure. For one thing, properly transforming back will almost always result in asymmetric confidence intervals. The most drastic and most trustworthy approach is to scrap the assumption of normality and use some nonparametric approach such as in Chapter 14. Nonparametric approaches do not waste as much information, in the sense of reduced sensitivity, as researchers often suspect, but they may limit the range of possible conclusions. Transformations that try to attain normality are acceptable if they sensibly relate to what you are considering.

It is a major step forward if you can find a transformation to a scale which meets the assumption of normality, and which also represents reality. When the scale of measurement is not on an equal interval scale, as is the case with customer ratings, and subjective responses as on psychological surveys and IQ evaluations, you cannot use such transformation and then feed it into a parametric test or correlation. However, when underlying causes are on an equal interval scale, such as various measures of energy, or some measurements such as the effects of sound energy on mammals, etc., it might be reasonable to transform the data using a ***log transformation***,

$$y_i = ln(x_i), \tag{15.4}$$

where you may have to shift your data by adding something to x so as to avoid $x = 0$. The ln refers to the log to the base $e = 2.71828\ldots$ which gives values which might be handy for various subsequent considerations in physics. However, some texts suggest using the more **familiar** log_{10}, meaning to the base 10. The difference is only equivalent to multiplication of the results of the transformation by a constant; $log_{10}(x) \approx 2.303 \times ln(x)$. Using log transformations to a higher base, say, log_{10}, shifts the curve more than do lower bases, say, $ln = log_e$ but does not change the shape of the transformed distribution and has no effect on the significance of inferential tests or of correlations.

The log transformation has many applications and it is often used in physical systems as well as in biological systems. Suppose you are researching the environmental impact of a machine which tends to radiate energy by impacting the air between 300 and 1000 times per second. Humans and other mammals generally perceive these audible vibrations similarly. At least over a certain range, as the energy increases exponentially, the perceived threshold at which the mammal detects the change will increase linearly. So, we have the decibel system of measurements where increases in energy input are transformed to decibels by using the log of the energy increase to the **base ten**. Even so, you may wish to verify that the transformed data, such as decibels, seems to be closer to normally distributed than the energy levels in Joules or whatever.[8]

Some researchers make such transformations merely because it happens to make the data look normal when plotted against a Gaussian distribution or when examined for skewness and kurtosis. Note that **if you transform the log transformed mean back, you get the geometric mean,**[9] not at all the mean, $\bar{x}$, in which you may be interested.

The pioneer work in the quantification of perception is usually attributed to Heinrich Weber (1846), who was especially famous for his researches into aural and cutaneous sensations. His law essentially says that the increase of stimulus necessary to produce an increase of sensation in any

---

[8] A **Joule** is the work done by passing an electric current of one ampere through a resistance of one ohm for one second.
There is a logarithmic relation between decibels and joules.

[9] If you multiply n sample values, x_i, together and take their n^{th} root, you get their geometric mean,

$$G_x = \sqrt[n]{x_1 \times x_2 \times \ldots \times x_{n-1} \times x_n}, \text{ also written } G = \left(\prod_{i}^{n} x_i\right)^{-1/n},$$

which also equals

$$G_x = ln^{-1}\left(\sum^n ln(x_i)/n\right).$$

Note that G_x is not generally equal to $\bar{x}$.

sense is not a fixed quantity but depends on the proportion which the increase bears to the immediately preceding stimulus. This is the principal generalization of that branch of scientific investigation which has come to be known as psycho-physics. However, Gustav Theodor Fechner expanded this to include the log transformation. The two are often combined as the Weber-Fechner law, which in reality is two laws seemingly conflated as one, and an extension of the Weber law. Gustav Theodor Fechner (1860) quantified this, stating that subjective sensation is proportional to the logarithm of the stimulus intensity. This relation holds reasonably for loud sounds, but not necessarily for soft ones. For decibels, this log is to the base ten, but transformations to other, smaller, bases might be more appropriate when other sensations are involved. Over a moderate range, the Weber–Fechner law holds for the human eye and stellar magnitudes, as they were first described on such a scale by the astronomer Hipparchus in about 150 BC, with 1 representing the brightest down to 6 representing the faintest. Now his scale has been extended beyond these limits; an increase in 5 magnitudes corresponds to a decrease in brightness by a factor of 100 (Bell Labs 1925). In both cases you need to consider the range of stimuli being observed. These relations often do not hold beyond certain extremes. In that case you may have to resort to nonparametric statistical approaches (Chapter 14).

There has been some interest in the Box-Cox family of power transformations (Box and Cox 1964). It is a modification of a transformation by Tukey (1957). It transforms your data from x_i values to the response,

$$y_i = \left(x_i^k - 1\right)/k,$$

where k is a value which may be determined by experimentation,[10] if you have sufficient data, and is often between $k = 0.05$ and $k = 1.00$. If $k = 0$, then the natural log of the data, $y_i = ln\left(x_i\right)$, is used instead of $y = \left(x_i^k - 1\right)/k$. Some of the values of this constant, k, may be available from various professional organizations or from literature. Handelsmanj (2002) includes his approach to finding k. The Box-Cox transformation is not very favorably reviewed by Sakia (1992). There is some overlap between potential Box-Cox and other transformations.

Stevens (1957, 1975) proposed a *stimulus power law*, a relationship between the strength of a physical stimulus and its perceived strength. The general form of the law is that the perceived strength, y, is the energy of the stimulus, x, to some power, c, times a constant, k which corresponds to the **power transformation**,

$$y_i = kx_i^c.$$

In the Stevens model, k is a proportionality constant that depends on the type of stimulation and c is often between 0.5 and 1.7, inclusive. That this transformation, when applied to stimulus response situations, necessarily reflects any reality other than line straightening, is contested by a number of researchers (e.g. Ellermeier and Faulhammer 2000, Luce 2002, Steingrimsson and Luce 2006, Zimmer 2005). Along with Zubin (1935), Stevens' power law and its opponents provide valuable points at which to enter a preliminary literature survey for psychology researchers potentially considering transformations. A citations index may help if you wish to know more about this. However, a purely *physical power law* may apply in some situations such as when you are studying an effect related to the distance from a point source of light, $c = -0.5$. In some situations, $k = 1$, making interpretation easier.

Suppose you are looking at the effect of the size of the area of the top surface of the bell on the survival of jellyfish washed in and out of a particular bay by tidal currents. Would the actual surface

---

[10]Actually the rule is that for $k = 0$, you use $x' = log(x)$. Also theoretical k may be outside the range mentioned here, but the range given seems to be the handiest for most applications of interest to researchers outside statistics itself. You may wish to know that Box and Cox (1964) used steps of 0.05 when trying out values. If you decide to do anything this elaborate, you should pay attention to the skew and kurtosis values as well as any graphical techniques. Outside advice may be particularly advisable as the proper techniques of obtaining k (sometimes written as lambda, λ) are still under development.

area, or its square root (which is proportional to either the diameter or the radius), be more representative of the potential for the jellyfish to be too damaged to survive? Some questions, maybe including this one about survival, are best answered using a simple *square root transformation*,

$$y_i = \sqrt{x_i}. \tag{15.5}$$

When the underlying data is continuous, or the transformed data to be tested is massive, the effectiveness of transformations in achieving a set of scores transformed to normality can be satisfactorily checked using the Cumulative Probability Plot program, or a Q-Q plot graphing function in a major computer stats program or in the R language. But, if the data is binomial or discrete, such a Q-Q plot type approach probably will not be suitable for assessing the effectiveness of the transformation for testing small or medium data sets (Section 11.10). Other graphical techniques may help you to subjectively assess the effectiveness in such instances. Recall that although data that has already been examined is ineligible for analysis, a random subsample can be so examined as long as it is not subsequently included in the analysis.

Jason Osborne (2002) has published a particularly readable paper on the log, inverse, and square root transformations. He is of the opinion that suitable variables for these transformations are becoming increasingly common in education and the social sciences, and need to be dealt with appropriately. One important point he makes is that transformations are sensitive to the magnitude (the size) of the numbers transformed. This is especially marked between 0 and ± 1 versus the rest of the number scale. **He recommends** that you find the minimum value in your combined data and add enough to every value to cause the data values to have a minimum of 1. Then, apply the appropriate transformation. Adding a fixed value has no effect on either the pre-transformation variance or standard deviation nor does it affect the difference between any two means. Inverse transformations and log transformations shift the curve higher (to the right if you graph the untransformed data on the horizontal) by compressing the higher values more than the lower ones. Log transformations to the base e have a less extreme effect than to the base 10. To shift in the other direction it is necessary to *reflect* the distribution by multiplying by -1 and then adding enough to bring its minimum back to where you want it. You usually want the pre-conversion minimum to be set at 1. Osborne (2002) said, "All three transformations will have the greatest effect if the distribution is anchored at 1.0, and as the minimum value of the distribution moves away from 1.0 the effectiveness of the transformation diminishes dramatically."

A good thing about these transformations is that they preserve the order, but **notice if they alter the direction**. However, they all result in some form of progressive change of spacing, also called a *curvilinear effect*. Be careful if you are using these transformations as black boxes. Your results will refer to the transformed data, not the original data. As with poorly thought out designs, poorly thought out (or un-thinking) transformations may result in answers for questions which you did not ask. For instance, a doubling of decibels is very different from a mere doubling of the physical energy they represent. Decibels is an appropriate measure for hearing, whereas the energy of sound is important for many uses, including sonar, ultrasound cleaners, etc. Often you can, at best, only say if things are statistically significantly 'different', regardless of what approach you use.

Suppose you had to compare the percent of male alcoholics in eastern states with western states. Percentages of counts are binomial in that either an object is included (success) or it is not. Percentages are also different from most other data, because when represented as real numbers, they usually are between 0 and 1 (a distribution truncated at 0 and 1).

The *arcsine transformation* is **only for data based on counts** (including percents, r_i/n_i. The formula is

$$y_i = \arcsin(\sqrt{r_i/n_i}), \tag{15.6}$$

where the part under the radical sign is equivalent to a percentage (expressed in its decimal form). Do be aware that such percentages **drawn from counts** may be distributed as binomial (Subsection 2.5.1) or categorical data (Section 12), and as such, might be analyzed in other ways. I may not be alone in seeing this transformation as having a bit of hocus-pocus about it. It usually reshapes the distribution to look normal, whether that relates to reality or not, because it tends to spread the values at both ends of the distribution compared with the central region. This in turn tends to homogenize apparent variances, making them work out to have similar values on the converted scale. These effects are due to the shape of the arcsine function, a function which returns the number (in radians) whose values are in the range 0 to 1. Numerous sources recommend that proportion data (including percentages) of this sort should be transformed with the arcsine function if the range of percentages is **greater than** 0.40 (Steel, Torrie and Dickey 1997). **But**, in what form would you report or postulate a meaningful effect size, how would you choose a sample size, how would you present a confidence interval your readers could accept, and would you lose power by artificially altering differences in apparent variances, some of which may be due to differences in the means?[11] Well, the arcsine transformation has a long history and a lot of outstanding workers have used it. At least it might be adequate to conclude that the populations were somehow different if the difference was statistically significant. But before using this transformation on your data, especially check to be certain that this transformation makes a big step toward meeting the assumption that the converted data actually approaches normality.

The *inverse transformation*

$$y_i = 1/x_i$$

works smoothly in a numerical sense if you first shift the distribution (by addition or subtraction) so that the minimum is ≥ 1. This transformation is not as common in application as some others, but is sometimes used for testing the relation between healthy activities and morbidity (as judged by certain indexes) or moribundity. For instance, there is an inverse relation between post diagnosis physical activity and colon cancer mortality (Meyerhardt, Giovannucci, Holmes, et al. 2006, Meyerhardt, Heseltine, Niedzwiezki, et al. 2006). Thus, in some contexts, the inverse transformation may better represent the information of interest than does the original data as collected.

Shifting population means by addition of a constant (often 1) to each data value does not change the variance or the shape of the distribution. And, it can allow you to take logs because it can be used to avoid the fact that $log(0)$ is undefined, and it similarly can allow the use of an inverse transformation even though $1/0$ is undefined.

15.3.0.1 Probit transformation

Probit conversion in which the data values are ranked, scored as percentiles and then given values equal to $5 +$ their corresponding Z score produces values that can be fed into processes designed for normally distributed data. This was first applied as *probit analysis* in insect toxicology by JD Finney (1990), but is sometimes used in other kinds of toxicology. This transformation is often used by big data analysts for data when it tends to form a *cumulative model*, a sigmoid curve-like distribution.

---

[11]A very widespread statistical test, the ANOVA (Section 9), assumes that there is only one underlying variance, $\sigma^2 \approx MS_W$. On the basis of this assumption it compares the calculated treatment variances to detect differences between them and the ideal. It is assumed that these differences are attributable to the effect of the means on the calculations of their variances.

15.4 Poisson Distribution

The *Poisson distribution* (Poisson 1837) is discrete and expresses the **probability of a given number**, r, of **independent** random events (sometimes called *arrivals*), occurring in a fixed interval of time (eg. von Bortkiewic 1898), **or** a fixed distance, **or** fixed instances of time and space. It is sometimes of interest in situations where the event is relatively uncommon (a relative term), but has a possibility of *clustering* (occurring a relatively large number of times in a given interval). With really rare events, it is difficult to discern whether occurrences are Poisson distributed (are independent chance events) or are not (are dependent events). The public most often witnesses the effects of Poisson distributions as seemingly counterintuitive clustering of rare events (seems like the clustering must have a cause, but in these cases the clustering is really by chance). Cot deaths seem to include clusters, but cot deaths are quite rare. Because rare events are difficult to observe it is difficult to tell whether cot deaths are Poisson distributed, or just not independent, or maybe some of both. The Poisson distribution contains 0, and you do **not have a Poisson distribution if zero is impossible**. On the other hand there can be more zeros (zero inflation) than predicted by the Poisson distribution. (There are models for both zero truncated and zero inflated distributions, but they are not covered here). The *law of rare events* says that the Poisson distribution can be applied to systems with a **large number of possible events**, **each occurrence of which is rare**. It is an important distribution which you may encounter and it is convenient to cover it here, because it may be mistaken for a binomial distribution, or occasionally for a normal distribution. It involves a value, λ (Greek lambda), which may be found, for instance, as the **average occurrence of an uncommon event** over a unit of time, or distance, or a unit of time and distance. So, $\widehat{\lambda} = r/n$, but in some designs, $n = 1$ because it refers to a single study unit of time, or distance, or time and space. Calling the number of occurrences (successes if you will) r, the distribution of λ may be expressed as a frequency function,

$$f(r, \lambda) = \frac{\lambda^r e^{-\lambda}}{r!},$$

which from a frequentist point of view might be written $p(r) = \frac{\lambda^r e^{-\lambda}}{r!}$. Either equation is unfamiliar looking but not unworkable. Of course $e = 2.71828183\ldots$ is the irrational number made famous as the basis of the natural logarithms[12] which themselves behave so well in higher mathematical applications, and have many additional applications and implications (Maor 2009). The Poisson distribution may occur in situations which are not all that unusual, as well as in situations that are extremely unusual. However for the distribution to be a Poisson distribution, then λ, even $\lambda > 1000$, should only be equal to a relatively small proportion of the population. The number of automobile accidents at a particular corner this year as compared to hundreds, or even thousands of identical intersections is often a cluster associated with the Poisson distribution of such events. In some situations, a square root transformation can be suitable to make a Poisson distribution more or less conform to the normal assumption of some tests (Section 15.3).

The parameter λ is **not only** the mean number of occurrences, it is **also** their variance. Thus, the number of observed occurrences[13] fluctuates about its mean, λ, with a standard deviation, $\sigma_r = \sqrt{\lambda}$. It is important to remember that the **means** and **standard deviations** of samples from a Poisson distribution will **tend to be correlated**. Observing such a correlation can be reason to suspect that the

---

[12]For math enthusiasts: The Poisson distribution may be derived from the binomial distribution using calculus with $\lim(p) \to 0$ and $n \to \infty$.

[13]In a *Poisson process*, the inter-event times (intervals) have a tendency to cluster because short inter-event times have a relatively high probability under the exponential distribution. Poisson processes are important to systems engineers who may wish to estimate how often their servers will be overloaded. They are also important when allotting emergency services, etc.

data is Poisson distributed.

A good approximation to the Poisson distribution is as follows: If λ is greater than about $\lambda = 10$, then use the normal approximation with a ***continuity correction*** (using $r_i + 0.5$ to adjust for the fact that the normal distribution is continuous, not discrete). If λ is sufficiently large, say $\lambda > 1000$, then use the normal approximation without a continuity correction (with mean λ and variance λ).

For comparing only two counts of successes, Sokal and Rohlf (1994) suggest a Poisson test that takes advantage of the fact that the variances equal the mean, giving the following comparative test:

$$Z_\lambda = \frac{r_2 - r_1}{\sqrt{r_1 + r_2}}, \tag{15.7}$$

where both values of r_i are the number of occurrences in a fixed unit of time or a fixed distance, or fixed instances of time and space. This could also be written as $Z_\lambda = \frac{\hat{\lambda}_2 - \hat{\lambda}_1}{\sqrt{\hat{\lambda}_1 + \hat{\lambda}_2}}$.

In a cohort study E might denote exposed, U might denote unexposed, and $\widehat{\lambda}_{E \text{ or } U} = r_{E \text{ or } U}/n_y$ is an estimate of the mean number of deaths per year. Some public health researchers and epidemiologists estimate n_y, the number of person years of follow up per exposure group, by using

$$\hat{n}_y = (Z_\alpha + Z_\beta)^2 \, \frac{\lambda_E + \lambda_U}{(\lambda_E - \lambda_U)^2},$$

where you **can substitute** $\mathcal{D}_{\lambda \, min}$ for $\lambda_E - \lambda_U$. If you have enough previous data that was not part of a testing process, you might be able to use this to plan a sample size (or time, etc.), $\hat{n}_y$, for designing a test of any pair of Poisson outcomes using the Poisson test (Equation 15.7). This equation might also be used to plan research involving blood cell counts, where an initial estimate λ_E could be derived from records of a clinical laboratory that is routinely doing counts.

The popular Wald CI is suitable for when the number of events counted is $n \geq 30$. It is estimated by

$$\widehat{\lambda} - Z_{\alpha/2}\sqrt{\widehat{\lambda}/n} \leq \lambda \leq \widehat{\lambda} + Z_{\alpha/2}\sqrt{\widehat{\lambda}/n},$$

where, of course, $\widehat{\lambda}$ is the estimated lambda. It seems to arise out of Abraham Wald's classic paper which gave rise to, or formalized, a variety of statistical procedures (Wald 1943).

Hopkins (2011) recommends using the square root of counts, provided you are **not close** to some upper bound in the counts. You should also consider square root of counts if you somehow know that the means and the standard deviations of counts are correlated. For instance, it is used for transforming

$$r_i = number\ of\ white\ blood\ cells\ observed$$

using a hemocytometer (generally agreed to be a Poisson distribution). When a Poisson distribution is not an established fact, you should check both the before and after transformation values to see which best fits the assumption of normality. Thus the square root is sometimes seen as transforming the Poisson distribution to approximate the normal distribution for testing (Section 15.4). Two variants of this transformation of small integers are

$$y_i = \sqrt{r_i + 0.5} \text{ and } y_i = \sqrt{r_i + 3/8}. \tag{15.8}$$

The first of these Equations 15.8 is most popular for some computer packages such as SAS. However the second, proposed by Anscombe (1948), may be preferable.

Subsection 11.9.1 discusses a so-called 'Poisson regression', a regression technique which, depending on the application, may, or may not involve the Poisson distribution. Guerriero, et al. (2009) propose a confidence interval for λ using samples containing at least $n = 15$ to $n = 20$ intervals.

His work, available both in Italian and in English, might be of particular interest to petroleum and geothermal engineers.

Sometimes you will see vague references to the Poisson distribution associated with Pearson's chi square. However, the real question might be to what degree Pearson's chi square is suitable for count data that is Poisson distributed.

Many details and specialized applications assuming the Poisson distribution are not covered in detail here. If you are working with such issues as the timing of arrivals, you may wish to find other sources for the Poisson distribution, such as the **Beta distribution** (the probability distribution of all the possible values of a probability, David Robinson 2017), and the **exponential distribution** (describes the times between events in a Poisson process). Gu et al. (2008) present a very complete discussion of testing and sample size estimation using two approaches to Poisson ratios. If you are interested in Bayesian statistics, you might consider looking up the subject of Markov chains with respect to arrivals and queuing.

15.5 Robustness and Measures of Efficiency

As you may recall, a robust estimator tends not to be affected by departures from an assumed model, such as normality (also see Subsection 5.2.1). Theoretically, the Student's t test assumes normality, but in application, due to the CLT, it is robust with respect to this requirement when the sample size is sufficiently large. Classically, 'large' is accepted to be $n \geq 30$, but some authors might advise $n \geq 50$ or even $n \geq 80$ unless you are confident that the distribution is not severely skewed (Subsection 15.2.1). As a rule you need samples of at least $25 \times skew$ for normal tests (Lovric, 2010). However, in extreme situations, this might be impractical, thus calling for a nonparametric approach such as Wilcoxon's test (Subsection 14.1.1). Researchers using nonparametric applications are naturally concerned about how they stack up against applications that depend on the normal distribution. Many tests, such as the ANOVA, are not robust when confronted with small sample sizes and deviations from the assumptions, or when confronted by extreme deviations from their model. Nonparametric tests are often more robust, especially for coping with the effects of outliers. As Lumley, Diehr, Emerson, and Chen (2002) state, "Formal statistical tests for normality are especially undesirable as they will have low power in the small samples where the distribution matters and high power only in large samples where the distribution is unimportant."

Efficiency is often measured as the comparative ratio between the number of samples needed for one approach (numerator) and the number of samples needed for some standard approach (denominator). The less efficient an estimator or test is, the larger the sample size it needs to achieve the same power as the standard to which it is being compared. A test with 50% efficiency requires twice as many samples as the standard to which it is being compared. The distribution itself affects the efficiency of a given estimator or test. In addition to having less assumptions to be robust about, many nonparametric tests are at least 90% (usually around 95%) as efficient as normality based tests. For many non-normal distributions, nonparametric tests based on rank may even have greater efficiency and therefore more power than normal distribution based tests. In such situations, the latter might not have any validity anyway.

15.6 Handling Outliers

Outliers are ridiculously large or ridiculously small data values which do not seem to belong in the population under investigation. They might be present due to any of many reasons. They could represent a gauge or instrument that sticks at the extremes of the range, or maybe the data is being collected on the edge of a chaotic phenomena, etc. Although they may represent a large or small measurement, the magnitude of that largeness or smallness might be difficult to pinpoint. The guy on the other end of that random telephone survey call may really be a 700 pound vegetarian who drinks 3 gallons of soft drinks a day, etc. Extreme values, whether outliers or not, can really mess up the data that you are trying to understand. A billionaire living in an otherwise average small town would severely distort the mean income (but have almost no effect on the median, hence the tendency to report medians as 'average' incomes. A rank based nonparametric approach might be able to handle the effects of health survey or wealth extremes, but it can only reduce, not eliminate, the effects of the sticking meter, or of electrical static that could favor erroneous conclusions.

So, you have to choose how to handle outliers. Are you going to minimize the degree to which they are different as by using ranks, or are you going to exclude them by some process of data cleaning? Maybe if they are few and your data set is large enough, it's best to leave them out. The problem with removal of outliers from your data is identifying what is really an outlier. Government and political data is often excessively cleaned, and of course, personal views may consciously or unconsciously affect what is considered an outlier.[14] Relevant publications regarding outliers and cleaning them from data include Hampel (1974), Rousseeuw and Croux (1993), and Hossjer, Rousseeuw and Croux (1996).

15.6.1 IQR when outliers are numerous

Percentiles are measures representing the cumulative frequency as a percent of the whole population, below which the observed value of a variable falls. *Quartiles* are three points, $Q_1 = 25^{th} percentile$, $Q_2 = 50^{th} percentile$, $Q_3 = 75^{th} percentile$. Together they divide a data set into four equal frequency groups, each group estimating a fourth of the distribution of the population.

The *IQR,* or *interquartile range*, which is the difference between quartiles Q_1 and Q_3, which enclose the middle half of the data is

$$IQR = Q_3 - Q_1.$$

Half of the distribution of the possible outcomes is estimated to lie between Q_1 and Q_3. If the distribution is normal this would mean that these two quartiles are each $Z_{0.25} = 0.67448$ standard deviation units from the mean (in the - and + direction respectively). So, if it were measured in numbers of standard deviation units, the $IQR = 2(0.67448)\sigma = 1.34986\sigma$. If you wish to estimate the standard deviation σ from the IQR under the assumption of normality, you must multiply the IQR times the reciprocal of 1.34986, which is 0.741312. So **assuming normality**, the IQR could be used to estimate the standard deviation,

$$S \approx 0.741312 \, IQR.$$

Unfortunately the IQR is **not very efficient** for estimating the standard deviation; it is only 33% efficient, even assuming normality. Provided the assumption of normality is valid, it would take 3 times as many samples to obtain the same power with regard to σ as you would need using

$$S = \sqrt{\sum (x - \bar{x}) / (n - 1)}.$$

Because the IQR is quite inefficient for estimating the standard deviation, it is **not suitable for estimating sample sizes**.

---

[14]How supposed outliers are removed from government data is not necessarily decided by the government as such, but by data collection technicians, and by government contractors. One could ask how the citizenry can run a democracy if they do not know the truth. The result of voting in ignorance is analogous to driving with the windshield blanked out.

However, for other applications, such as for providing a 50% CI-like estimate (the box of a box plot, Subsection 15.1.1), the IQR is very robust; whether using the assumption of normality or not, it is somewhat **robust** with respect to **skew**. It has a 25% **breakdown point**, meaning that it can handle **up to** 25% arbitrarily large observations (outliers), regardless of their sign, before giving an unacceptable large result.

Percentiles are used in many fields and are intuitively appealing. The IQR is also useful and is accepted for describing results; it and the median can be reported as indicators that are in some ways like confidence intervals, **but** note that a boxplot does that and more.

15.6.2 If frustrated, consider MAD

No, you don't have to walk on the dark side if you have a large number of outliers. In 1816, Carl Friedrich Gauss was the first known to consider **MAD**, the **median absolute deviation**, as a measure of variability for a univariate data set, $x_1, x_2, \ldots, x_n$ (Hampel 1974). MAD is the median of the absolute deviations from the sample's median, $\tilde{x}$,

$$MAD = median_{i=1}^{i=n} |x_i - \tilde{x}| . \tag{15.9}$$

This means that you find all the differences between the scores and the median regardless of sign, and then find the median (middle one) of those differences. This estimator may be seen as **focusing on** $3/4$ of the distribution. So, to estimate the variance for normally distributed data, MAD uses the fact that 0.67857 is the cumulative Z value for $3/4$ of the area under the normal curve and has a reciprocal of $1/0.67857 = 1.48262$.[15] So, MAD may be used to estimate the underlying standard deviation **even in the presence** of a lot of outliers;

$$S \approx 1.48262\, MAD . \tag{15.10}$$

This multiplier is about 1.5 for the normal curve or even for the uniform distribution (a distribution in which each value is equally likely). MAD is an extremely robust statistic (see Section 15.5) and it actually has a 50% breakdown point (Hampel, 1974), meaning that up to 50% of the data could be outliers without affecting the outcome. A 50% breakdown point is the largest possible.[16] With small to moderate data sets you can easily calculate it by hand. Unfortunately, even when the data is really from a normal distribution, it only has a 37% efficiency for estimating S. This means that if you use MAD to estimate sample size, you should **use** $1/0.37 = 2.8$ **times your initial estimate**. This is **not the best way** to find a standard deviation for sample size estimation, but it is not likely to underestimate it. Unfortunately, in order to make an a priori sample size estimate, you must use a different sample from the sample upon which you will do your test.

15.6.2.1 MAD screening for outliers

MAD is also **handy** for deciding which data points you feel could reasonably be declared to be outliers. If you feel you must drop some of the weirder data, a conservative approach would be to **discard all** x_i **for which**

$$\frac{|x_i - \tilde{x}|}{1.5\, MAD} > 3.0 . \tag{15.11}$$

---

[15]The multiplier 1.48262 for estimating σ from MAD is twice the multiplier 0.741312 for estimating σ from IQR.

[16]It may be of interest that the median estimator, $\tilde{x}$, has a 50% breakdown point, whereas the mean estimator, $\bar{x}$, has a 0% breakdown point.

That is, using $\frac{|x_i - \tilde{x}|}{1.5 MAD} = Z_p$ and **assuming normality**, $Z_{p \approx 0.001} = 3.0$. However, a lot of software defaults to $Z_p = 1.5$. The actual Z_P that you use is up to you, but p, the probability of Z_p happening by chance alone, must be very small in order to really distinguish an outlier. Screening data with MAD might make it more useful for getting an estimate of S for an a priori estimate of sample size when outliers appear to be present. This is especially true if you intend to use MAD with the same Z_p as part of the testing process.

If any data in your sample was edited out or ignored, as by MAD, you must report how you did it. For instance, you might report that MAD was used to remove outliers, amounting to 0.1% of the data.

MAD is far from perfect, but it is likely to be the best available screening process in popular computer packages. Remember that the procedures in this book have mostly depended on the Central Limit Theorem to work with means, because means of sufficiently large samples tend to be normally distributed. MAD is not about means, it is about the raw data, and data is rarely, if ever, normally distributed, and often not even relatively symmetrical. Unfortunately, MAD itself assumes symmetry (lack of skew), although using it does reduce skew a little and avoids the assumption of normality. Therefore you may wish to consider a discussion of more computationally intensive approaches in *"Alternatives to the Median Absolute Deviation"* by Peter J. Rouss and C. Christophe (1993). Depending on the program producing them, some boxplots show outliers too (Figure 15.1).

CAUTION: some software and many websites call the **mean average distance** 'MAD', but although it runs a great deal faster for huge data sets on computers, this is a less robust estimator that is **not ideal for defining outliers** for more moderate size data sets. To use the correct MAD in Python (Version 3+) use `scipy.stats.median_absolute_deviation(*args, **kwds)`. Also, many versions of the R language tend to provide the correct median_absolute_deviation for MAD, for instance, `mad(X, kwds)`, but in some languages you may have to write your own function.

Questions for Ch 15

1. Geothermal (volcanic steam) wells produce very corrosive superheated water which is flashed into steam for electric generators. You have records of the lifespan of 356 identical valves used on geothermal wells. You need to test the difference between otherwise identical valves made of two different **new** alloys (not the same as the 356). However, you can only obtain $n_i = n_2 = 12$ of each. Assuming the power were somehow sufficient, how might you check to see that the assumptions of a Student's t test has a reasonable chance of being valid?

2. Would you need to look for a suitable transformation of the data if you were using a Student's t test to examine 2 samples of $n = 500$ men's heights in Nairobi?

3. A particular large farm has had 3 calves with five legs born in the last decade. If you were to consider whether this farm is somehow different with respect to deformed cattle, what kind of data distribution do you think might be involved?

4. If you have data based on counts and the sample sizes ($n < 80$) for each count, what transformation might make the data normal? (The author is hesitant to use this or any transformation for any but relatively small sample tests, or maybe for modeling.)

5. Which is potentially useful for estimating σ^2 for an a priori sample size estimate if you cannot find a better way: MAD or IQR? How would you do it?

6. Assume you are somehow aware that the **means and variances** from the population under consideration are correlated. What transformation might you use? What does such a correlation suggest about the data? (No fair exploring the actual data you are to use.)

7. You are attempting to evaluate data obtained from the game port of an older computer, but every so often it leaps from reasonable readings to insanely different readings (game ports directly reading resistances often did that). What screening method would you use to pick out the crazy readings so that you could discard them?

8. Your statistics package gives you a boxplot with little dots beyond the ends of the whiskers. What are they?

9. Make a stem and leaf plot of the following data (also seen in Question 1, Ch.14):
   19, 21, 15, 18, 26, 18, 8, 26, 8, 13, 15, 27, 10, 21, 8, 14, 26, 12, 25, 19, 16, 5, 7, 30, 7.

CRITICALLY IMPORTANT POINTS

When I browse the internet, I see both postdoctoral fellows and established researchers asking how to test data they have already collected. Some of them assert impossible things such as claims that their counting data is "normally distributed." Then other 'experts' answer them and argue over how to analyze this poorly defined, preexamined data. But that is not the worst of it; they have collected this data without a plan that includes the targeted power, let alone how big a difference that they would consider to be a reasonable minimum difference (what might be seen as the effect size that they minimally intend to target). The careers of the researchers engaging in these nonsense exchanges are dependent on the publication of a sufficient number of peer reviewed papers, the validity of which is supposedly dependent on statistical conclusions.

They have statistical software packages at their disposal which provide bewildering universes of tests, models and procedures from which to choose. The printouts include the arcane, the archaic and the bewildering, in addition to what you need to know.[1] However, every such procedure needs an adequate knowledge of its requirements and peculiarities, as well as of basic statistics, as provided in this book. These folks would be better off with a clear understanding of the basics, and even then, they would need to keep their research as simple as possible so they can avoid getting involved with what are, to the uninitiated, black box models which may not be appropriate.

When you design your research, you must consider how you are going to analyze it. Simple approaches, maybe even avoiding global tests, are more understandable to your readers and thus a greater contribution to society. Exploiting the Central Limit Theorem by using simple parametric tests is usually the best strategy for largish samples, $n > 60$, as long as the data is on an equal interval scale.

Nonparametric tests are necessary for non-equal interval scales, such as Likert scales. They are also often better for small to medium samples on equal interval scales, $n < 80$, assuming the sample size provides enough power to find what you are looking for. At the very least, you should know most of the material in this book in sufficient detail so as to be able to refer back to it. If you are dependent on models and procedures that are specific to your field, these procedures may require specialized study for which this book provides a starting vocabulary and, hopefully, a solid foundation.

The following is merely a reminder of only a few salient points covered in this text:

- Collecting or examining the data that you intend to analyze, before carefully considering how you are going to statistically analyze it, is very poor science. The inferential method depends on demonstrating that a hypothesis is too unlikely to be true, by standards which must minimally meet the standards accepted by your peers. Examining the experimental results to look for a discovery or answer, or even just looking over the results, before running the statistical test, defeats that goal.

- The community of your peers has accepted the extravagant 5% standard of letting you declare statistical significance even though it means you take a one in twenty chance of being wrong each time you do an experiment. This extravagant 5% criterion is often forced upon us by costs and logistics, but using it as a minimum probability of getting a Type I error is scary, almost

---

[1]Whenever you get a printout from a computer analysis, it is always useful to note, for future designs for yourself and others, the standard deviation (or variance) and the skew, even if your results are otherwise disappointing.

457

CRITICALLY IMPORTANT POINTS

reckless. Using 5% as an absolute maximum cutoff is not a tyranny trying to deprive us of discovery.

- You need an appropriate sample size for any inferential test, or for estimating confidence intervals of a justifiable width. For all continuous variables, and some discrete variables, **tests based on the normal distribution** will generally require you to already have a good estimate of the standard deviation or variance. Even if you cannot have an a priori sample size estimate, you will always have to (a priori) have specified a reasonable minimum difference (as for a post hoc power estimation).

- Normally distributed data is almost nonexistent in application. Some tests, such as Student's t test, which were designed on the basis of the assumption of normality, attain normality in accordance with the CRT with large sample sizes; others such as the ANOVA are somewhat lacking in this regard. Although almost any test might be wretchedly weak at small sample sizes, nonparametric tests, such as the Wilcoxon test, will be more robust for small, $n < 30$, sample sizes. Student's t test is ironically superb for large samples even though originally created for small samples.

- Beware of: sample size estimation involving standardization (e.g. Cohen's d). Likewise, the still popular "large", "medium", and "small" (Cohen 1977) has been discredited for decades, including by the tacit admission of the author himself (Cohen 1988, p25).

- It is reasonable to expect a normal curve for the distribution of the means of adequately large random samples, from an equal interval scale, under the Central Limit Theorem, but not for the distribution of raw sample points. In application, the means of almost all random samples of $n \geq 30$ (or roughly $n \geq 80$ with **severe** skew) will, under the CLT, meet this requirement (the various forms of the ANOVA being an unfortunate exception). But that does not guarantee that the raw sample will be large enough to give you sufficient power (the probability of detecting your desired reasonable minimum difference).

- If it is practical to determine the skew before running a test, as by extra or historical data that will **not** be included in the test, then you can be more secure in your assumption of normality by using a sample size that is at least twenty five times the third moment skew. This will often be less than your estimated sample size, but it occasionally might be more.

- Global tests imply eventual pair by pair comparison. If you must do such multiple comparisons, they are not usually going to be practical unless you appropriately preplan them. And the most practical, least expensive plan is to designate an a priori control and compare to it.

- The sum, or mean, of the grades of one student is a nonrandom measure from its own population, the scores of that student. **It is not a sum, or weighted sum of randomly sampled scores** from the population of grades obtained by all the students. It is just a point estimate, a score. Therefore the distribution of individual student grades is not guaranteed, by the Central Limit Theorem, to approach being members of a normally distributed population.

There is what might be seen as an irony in the approach to parametric testing presented in this book. Most of the parametric procedures discussed here were originally introduced by their authors with the assumption that the data would be randomly sampled from normally distributed raw data, but raw data is almost never normally distributed. This is a very real problem for small samples, $n < 30$. Recall that the law of large numbers says that if you perform the same experiment a large number of times, the average of the results obtained should be close to an expected central measure, the true mean, and will tend to become closer as more trials are performed. The very questions you are attempting to answer assume a tendency for data to cluster around a central measure. Due to the mathematical phenomena described by the Central Limit Theorem, means, upon which parametric calculations depend, tend to approach being normally distributed as the sample size increases. Most discrete processes will involve, or can be contrived to involve, the binomial or multinomial distribution, but these distributions sufficiently approximate normal when the sample size is large enough. All this tends to produce sufficient robustness so that, in application, the parametric approaches presented in this book, the ANOVA tending to be an exception, are satisfactory for most large sample testing. An additional fact contributing to the robustness of these tests is that if you appropriately plan **for sufficient power**, the sample size will usually be larger than large, so the Central Limit Theorem will usually come to the rescue. Remember that 'large' is defined as $n \geq 30$, but in application, useful experimental design almost always requires $n \gg 30$. If you are forced to use small samples, then nonparametric tests might be best. The Kruskal-Wallis test and Friedman's test are generally more robust than the corresponding parametric tests at **small to moderate** sample sizes, but they are, to some extent, still adversely affected by asymmetry, and this is not remedied by larger sample sizes as it is for Student's t test. Friedman's test is also weakened by extreme differences in the variances of the treatments.

APPENDIXES

Unless stated otherwise, the tables presented here were derived using various software at various dates, so their formats vary. Some of the values in the tables of Appendixes 3, 5 and 6 were checked against NIST/SEMATECH e-Handbook of Statistical Methods, www.itl.nist.gov/div898/handbook (6/23/10), or by using the R language, generally using RStudio. . The intent of the extended part of all the tables is to be used in estimates of sample size and of power.

Interpolation is any process of estimating new data point pairs within the range of a discrete set of known ordered data, such as a table of Z and its corresponding p_values. One way to find a result $f(b)$ of function $f()$ for a value b in a table is to use the closest values, a and c, immediately above and below it, via the crude model, $\frac{f(b)-f(a)}{(b-a)} \approx \frac{f(c)-f(a)}{c-a}$. This allows you to approximate:

$$f(b) \approx \frac{f(c) - f(a)}{c - a} (b - a) - f(a) \ .$$

Be certain that you err to the conservative and remember that as the *df* increase, the probability tends to increase at a decreasing rate.

A rough definition of being conservative, as used in this book, is to always lean in the direction opposite to that which you desire. Especially round sample sizes up to avoid underestimating a sample size for an a priori power analysis. A sample that is slightly too big will only increase the power, but a sample that is slightly too small can ruin your work.

Appendix 1: TABLE FOR SIGN TEST OR 50% TEST ($p=1/2$)

	Minimum Successes for One Tail Significance				Minimum Successes for One Tail Significance			
	$\alpha = 0.5\%$	$\alpha = 1\%$	$\alpha = 2.5\%$	$\alpha = 5\%$	$\alpha = 0.5\%$	$\alpha = 1\%$	$\alpha = 2.5\%$	$\alpha = 5\%$
	Successes OR Failures, Two tail Significance				Successes OR Failures, Two tail Significance			
n	$\alpha = 1\%$	$\alpha = 2\%$	$\alpha = 5\%$	$\alpha = 10\%$	$\alpha = 1\%$	$\alpha = 2\%$	$\alpha = 5\%$	$\alpha = 10\%$
5				5				
6			6	6				
7		7	7	7				
8	8	8	8	7				
9	9	9	8	8				
10	10	10	9	9				
11	11	10	10	9				
12	11	11	10	10				
13	12	12	11	10				
14	13	12	12	11				
15	13	13	12	12				
16	14	14	13	12				
17	15	14	13	13				
18	15	15	14	13				
19	16	15	15	14				
20	17	16	15	15				
21	17	17	16	15				
22	18	17	17	16				
23	19	18	17	16				
24	19	19	18	17				
25	20	19	18	18				
26	20	20	19	18				
27	21	20	20	19				
28					22	21	20	19
29					22	22	21	20
30					23	22	21	20
31					24	23	22	21
32					24	24	23	22
33					25	24	23	22
34					25	25	24	23
35					26	25	24	23
36					27	26	25	24
37					27	27	25	24
38					28	27	26	25
39					28	28	27	26
40					29	28	27	26
41					30	29	28	27
42					30	29	28	27
43					31	30	29	28
44					31	31	29	28
45					32	31	30	29
46					33	32	31	30
47					33	32	31	30
48					34	33	32	31
49					35	34	32	31
50					35	34	33	32

Appendix 2: Sample Size for at Least 80% Power
50% TEST OR SIGN TEST (over all α = 5%)

TRUE p%	Sample Size		TRUE (1-p)% = q%
	One Tail α=5% n	Two Tail α=5% n	
70.0%	42	51*	30.0%
75.0%	26	33	25.0%
80.0%	18	20	20.0%
85.0%	13	15	15.0%
90.0%	11	12	10.0%
95.0%	8	9	5.0%
97.5%	5	6	2.5%
Both one tail:	α = 5%	α = 2.5%	

Table values are by iterating until stability attained.
*Above n = 51 use normal approximation approach.

The table here in Appendix 2 is only to indicate the minimum $\alpha = 5\%$ sample size at which Appendix 1 can be used with a power of at least 80%. A little consideration of these two tables together will demonstrate that the sign and 50% tests are not likely to be routinely useful for actual publication.

Small sample sizes intuitively make us feel uncomfortable, and there is some justification to feel uncomfortable. For instance, the law of large numbers says that if you perform the same experiment a large number of times, the average of the results obtained should be close to the expected value, and will tend to become closer as more trials are performed (Bernoulli,1713; Mlodinow 2008).

On the other hand, consider an instance that would be relevant even if your hypothesized $p \ll 1/2$, and the sample size was very small. Most of the readers have seen an ordinary coin land and stand on its edge, but it is so rare that they made little of it except as a novelty. Now, suppose I came to you and said, "This otherwise ordinary penny tends to land and stand on its edge more than most coins." So, naturally, you hypothesize that I am wrong, and try it three times and if you get even one instance of it landing and standing on edge, that would be strong evidence of a true tendency because it happened when you put it to the test, and the probability of getting a Type I error is almost infinitesimally small. So, if the hypothesis is a priori this event is consequential, but if it is only post hoc, we accept that we happened to observe but one unusual occurrence in tens, or hundreds, of thousands of trials. This **a priori prediction** followed by justifiable rejection of our initial belief is the key concept in all statistical inference.

Table refers to dark shaded area.

APPENDIX 3 One Tail Table of Student's t (Double the p for two tails.)

p df	20%	10%	5%	2.5%	1%	0.5%	0.25%	0.1%	p df	20%	10%	5%	2.5%	1%	0.5%	0.25%	0.1%
1	1.3764	3.0777	6.3138	12.7062	31.8205	63.6567	127.3213	318.3088	51	0.8487	1.2984	1.6753	2.0076	2.4017	2.6757	2.9343	3.2579
2	1.0607	1.8856	2.9200	4.3027	6.9646	9.9248	14.0890	22.3271	52	0.8486	1.2980	1.6747	2.0066	2.4002	2.6737	2.9318	3.2545
3	0.9785	1.6377	2.3534	3.1824	4.5407	5.8409	7.4533	10.2145	53	0.8485	1.2977	1.6741	2.0057	2.3988	2.6718	2.9293	3.2513
4	0.9410	1.5332	2.1318	2.7764	3.7469	4.6041	5.5976	7.1732	54	0.8483	1.2974	1.6736	2.0049	2.3974	2.6700	2.9270	3.2481
5	0.9195	1.4759	2.0150	2.5706	3.3649	4.0321	4.7733	5.8934	55	0.8482	1.2971	1.6730	2.0040	2.3961	2.6682	2.9247	3.2451
6	0.9057	1.4398	1.9432	2.4469	3.1427	3.7074	4.3168	5.2076	56	0.8481	1.2969	1.6725	2.0032	2.3948	2.6665	2.9225	3.2423
7	0.8960	1.4149	1.8946	2.3646	2.9980	3.4995	4.0293	4.7853	57	0.8480	1.2966	1.6720	2.0025	2.3936	2.6649	2.9204	3.2395
8	0.8889	1.3968	1.8595	2.3060	2.8965	3.3554	3.8325	4.5008	58	0.8479	1.2963	1.6716	2.0017	2.3924	2.6633	2.9184	3.2368
9	0.8834	1.3830	1.8331	2.2622	2.8214	3.2498	3.6897	4.2968	59	0.8478	1.2961	1.6711	2.0010	2.3912	2.6618	2.9164	3.2342
10	0.8791	1.3722	1.8125	2.2281	2.7638	3.1693	3.5814	4.1437	60	0.8477	1.2958	1.6706	2.0003	2.3901	2.6603	2.9146	3.2317
11	0.8755	1.3634	1.7959	2.2010	2.7181	3.1058	3.4966	4.0247	61	0.8476	1.2956	1.6702	1.9996	2.3890	2.6589	2.9127	3.2293
12	0.8726	1.3562	1.7823	2.1788	2.6810	3.0545	3.4284	3.9296	62	0.8475	1.2954	1.6698	1.9990	2.3880	2.6575	2.9110	3.2270
13	0.8702	1.3502	1.7709	2.1604	2.6503	3.0123	3.3725	3.8520	63	0.8474	1.2951	1.6694	1.9983	2.3870	2.6561	2.9093	3.2247
14	0.8681	1.3450	1.7613	2.1448	2.6245	2.9768	3.3257	3.7874	64	0.8473	1.2949	1.6690	1.9977	2.3860	2.6549	2.9076	3.2225
15	0.8662	1.3406	1.7531	2.1314	2.6025	2.9467	3.2860	3.7328	65	0.8472	1.2947	1.6686	1.9971	2.3851	2.6536	2.9060	3.2204
16	0.8647	1.3368	1.7459	2.1199	2.5835	2.9208	3.2520	3.6862	66	0.8471	1.2945	1.6683	1.9966	2.3842	2.6524	2.9045	3.2184
17	0.8633	1.3334	1.7396	2.1098	2.5669	2.8982	3.2224	3.6458	67	0.8470	1.2943	1.6679	1.9960	2.3833	2.6512	2.9030	3.2164
18	0.8620	1.3304	1.7341	2.1009	2.5524	2.8784	3.1966	3.6105	68	0.8469	1.2941	1.6676	1.9955	2.3824	2.6501	2.9015	3.2145
19	0.8610	1.3277	1.7291	2.0930	2.5395	2.8609	3.1737	3.5794	69	0.8469	1.2939	1.6672	1.9949	2.3816	2.6490	2.9001	3.2126
20	0.8600	1.3253	1.7247	2.0860	2.5280	2.8453	3.1534	3.5518	70	0.8468	1.2938	1.6669	1.9944	2.3808	2.6479	2.8987	3.2108
21	0.8591	1.3232	1.7207	2.0796	2.5176	2.8314	3.1352	3.5272	71	0.8467	1.2936	1.6666	1.9939	2.3800	2.6469	2.8974	3.2090
22	0.8583	1.3212	1.7171	2.0739	2.5083	2.8188	3.1188	3.5050	72	0.8466	1.2934	1.6663	1.9935	2.3793	2.6459	2.8961	3.2073
23	0.8575	1.3195	1.7139	2.0687	2.4999	2.8073	3.1040	3.4850	73	0.8466	1.2933	1.6660	1.9930	2.3785	2.6449	2.8949	3.2057
24	0.8569	1.3178	1.7109	2.0639	2.4922	2.7969	3.0905	3.4668	74	0.8465	1.2931	1.6657	1.9925	2.3778	2.6439	2.8936	3.2041
25	0.8562	1.3163	1.7081	2.0595	2.4851	2.7874	3.0782	3.4502	75	0.8464	1.2929	1.6654	1.9921	2.3771	2.6430	2.8924	3.2025
26	0.8557	1.3150	1.7056	2.0555	2.4786	2.7787	3.0669	3.4350	76	0.8464	1.2928	1.6652	1.9917	2.3764	2.6421	2.8913	3.2010
27	0.8551	1.3137	1.7033	2.0518	2.4727	2.7707	3.0565	3.4210	77	0.8463	1.2926	1.6649	1.9913	2.3758	2.6412	2.8902	3.1995
28	0.8546	1.3125	1.7011	2.0484	2.4671	2.7633	3.0469	3.4082	78	0.8463	1.2925	1.6646	1.9908	2.3751	2.6403	2.8891	3.1980
29	0.8542	1.3114	1.6991	2.0452	2.4620	2.7564	3.0380	3.3962	79	0.8462	1.2924	1.6644	1.9905	2.3745	2.6395	2.8880	3.1966
30	0.8538	1.3104	1.6973	2.0423	2.4573	2.7500	3.0298	3.3852	80	0.8461	1.2922	1.6641	1.9901	2.3739	2.6387	2.8870	3.1953
31	0.8534	1.3095	1.6955	2.0395	2.4528	2.7440	3.0221	3.3749	81	0.8461	1.2921	1.6639	1.9897	2.3733	2.6379	2.8860	3.1939
32	0.8530	1.3086	1.6939	2.0369	2.4487	2.7385	3.0149	3.3653	82	0.8460	1.2920	1.6636	1.9893	2.3727	2.6371	2.8850	3.1926
33	0.8526	1.3077	1.6924	2.0345	2.4448	2.7333	3.0082	3.3563	83	0.8460	1.2918	1.6634	1.9890	2.3721	2.6364	2.8840	3.1913
34	0.8523	1.3070	1.6909	2.0322	2.4411	2.7284	3.0020	3.3479	84	0.8459	1.2917	1.6632	1.9886	2.3716	2.6356	2.8831	3.1901
35	0.8520	1.3062	1.6896	2.0301	2.4377	2.7238	2.9960	3.3400	85	0.8459	1.2916	1.6630	1.9883	2.3710	2.6349	2.8822	3.1889
36	0.8517	1.3055	1.6883	2.0281	2.4345	2.7195	2.9905	3.3326	86	0.8458	1.2915	1.6628	1.9879	2.3705	2.6342	2.8813	3.1877
37	0.8514	1.3049	1.6871	2.0262	2.4314	2.7154	2.9852	3.3256	87	0.8458	1.2914	1.6626	1.9876	2.3700	2.6335	2.8804	3.1866
38	0.8512	1.3042	1.6860	2.0244	2.4286	2.7116	2.9803	3.3190	88	0.8457	1.2912	1.6624	1.9873	2.3695	2.6329	2.8795	3.1854
39	0.8509	1.3036	1.6849	2.0227	2.4258	2.7079	2.9756	3.3128	89	0.8457	1.2911	1.6622	1.9870	2.3690	2.6322	2.8787	3.1843
40	0.8507	1.3031	1.6839	2.0211	2.4233	2.7045	2.9712	3.3069	90	0.8456	1.2910	1.6620	1.9867	2.3685	2.6316	2.8779	3.1833
41	0.8505	1.3025	1.6829	2.0195	2.4208	2.7012	2.9670	3.3013	91	0.8456	1.2909	1.6618	1.9864	2.3680	2.6309	2.8771	3.1822
42	0.8503	1.3020	1.6820	2.0181	2.4185	2.6981	2.9630	3.2960	92	0.8455	1.2908	1.6616	1.9861	2.3676	2.6303	2.8763	3.1812
43	0.8501	1.3016	1.6811	2.0167	2.4163	2.6951	2.9592	3.2909	93	0.8455	1.2907	1.6614	1.9858	2.3671	2.6297	2.8755	3.1802
44	0.8499	1.3011	1.6802	2.0154	2.4141	2.6923	2.9555	3.2861	94	0.8455	1.2906	1.6612	1.9855	2.3667	2.6291	2.8748	3.1792
45	0.8497	1.3006	1.6794	2.0141	2.4121	2.6896	2.9521	3.2815	95	0.8454	1.2905	1.6611	1.9853	2.3662	2.6286	2.8741	3.1782
46	0.8495	1.3002	1.6787	2.0129	2.4102	2.6870	2.9488	3.2771	96	0.8454	1.2904	1.6609	1.9850	2.3658	2.6280	2.8734	3.1773
47	0.8493	1.2998	1.6779	2.0117	2.4083	2.6846	2.9456	3.2729	97	0.8453	1.2903	1.6607	1.9847	2.3654	2.6275	2.8727	3.1764
48	0.8492	1.2994	1.6772	2.0106	2.4066	2.6822	2.9426	3.2689	98	0.8453	1.2902	1.6606	1.9845	2.3650	2.6269	2.8720	3.1755
49	0.8490	1.2991	1.6766	2.0096	2.4049	2.6800	2.9397	3.2651	99	0.8453	1.2902	1.6604	1.9842	2.3646	2.6264	2.8713	3.1746
50	0.8489	1.2987	1.6759	2.0086	2.4033	2.6778	2.9370	3.2614	100	0.8452	1.2901	1.6602	1.9840	2.3642	2.6259	2.8707	3.1737

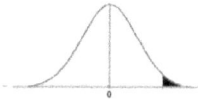 Table refers to dark shaded area.

df\ p	20%	10%	5%	2.5%	1%	0.5%	0.25%	0.1%	df\ p	20%	10%	5%	2.5%	1%	0.5%	0.25%	0.1%
101	0.8452	1.290	1.660	1.984	2.364	2.625	2.870	3.173	510	0.8423	1.283	1.648	1.965	2.334	2.586	2.819	3.106
102	0.8452	1.290	1.660	1.983	2.363	2.625	2.869	3.172	520	0.8423	1.283	1.648	1.965	2.334	2.585	2.819	3.106
104	0.8451	1.290	1.660	1.983	2.363	2.624	2.868	3.170	530	0.8423	1.283	1.648	1.964	2.333	2.585	2.819	3.106
106	0.8450	1.290	1.659	1.983	2.362	2.623	2.867	3.169	540	0.8423	1.283	1.648	1.964	2.333	2.585	2.819	3.105
108	0.8450	1.289	1.659	1.982	2.361	2.622	2.866	3.167	550	0.8423	1.283	1.648	1.964	2.333	2.585	2.818	3.105
110	0.8449	1.289	1.659	1.982	2.361	2.621	2.865	3.166	560	0.8423	1.283	1.648	1.964	2.333	2.585	2.818	3.105
115	0.8448	1.289	1.658	1.981	2.359	2.619	2.862	3.163	570	0.8423	1.283	1.648	1.964	2.333	2.584	2.818	3.105
120	0.8446	1.289	1.658	1.980	2.358	2.617	2.860	3.160	580	0.8422	1.283	1.647	1.964	2.333	2.584	2.818	3.104
125	0.8445	1.288	1.657	1.979	2.357	2.616	2.858	3.157	590	0.8422	1.283	1.647	1.964	2.333	2.584	2.818	3.104
130	0.8444	1.288	1.657	1.978	2.355	2.614	2.856	3.154	600	0.8422	1.283	1.647	1.964	2.333	2.584	2.817	3.104
135	0.8443	1.288	1.656	1.978	2.354	2.613	2.854	3.152	610	0.8422	1.283	1.647	1.964	2.332	2.584	2.817	3.104
140	0.8442	1.288	1.656	1.977	2.353	2.611	2.852	3.149	620	0.8422	1.283	1.647	1.964	2.332	2.584	2.817	3.103
145	0.8441	1.287	1.655	1.976	2.352	2.610	2.851	3.147	630	0.8422	1.283	1.647	1.964	2.332	2.584	2.817	3.103
150	0.8440	1.287	1.655	1.976	2.351	2.609	2.849	3.145	640	0.8422	1.283	1.647	1.964	2.332	2.584	2.817	3.103
155	0.8439	1.287	1.655	1.975	2.351	2.608	2.848	3.144	650	0.8422	1.283	1.647	1.964	2.332	2.583	2.817	3.103
160	0.8439	1.287	1.654	1.975	2.350	2.607	2.846	3.142	660	0.8422	1.283	1.647	1.964	2.332	2.583	2.817	3.103
165	0.8438	1.287	1.654	1.974	2.349	2.606	2.845	3.140	670	0.8422	1.283	1.647	1.964	2.332	2.583	2.816	3.102
170	0.8437	1.287	1.654	1.974	2.348	2.605	2.844	3.139	680	0.8422	1.283	1.647	1.963	2.332	2.583	2.816	3.102
175	0.8437	1.286	1.654	1.974	2.348	2.604	2.843	3.137	690	0.8421	1.283	1.647	1.963	2.332	2.583	2.816	3.102
180	0.8436	1.286	1.653	1.973	2.347	2.603	2.842	3.136	700	0.8421	1.283	1.647	1.963	2.332	2.583	2.816	3.102
190	0.8435	1.286	1.653	1.973	2.346	2.602	2.840	3.134	710	0.8421	1.283	1.647	1.963	2.332	2.583	2.816	3.102
200	0.8434	1.286	1.653	1.972	2.345	2.601	2.839	3.131	720	0.8421	1.283	1.647	1.963	2.332	2.583	2.816	3.102
210	0.8433	1.286	1.652	1.971	2.344	2.599	2.837	3.129	730	0.8421	1.283	1.647	1.963	2.331	2.583	2.816	3.101
220	0.8433	1.285	1.652	1.971	2.343	2.598	2.836	3.128	740	0.8421	1.283	1.647	1.963	2.331	2.582	2.815	3.101
230	0.8432	1.285	1.652	1.970	2.343	2.597	2.834	3.126	750	0.8421	1.283	1.647	1.963	2.331	2.582	2.815	3.101
240	0.8431	1.285	1.651	1.970	2.342	2.596	2.833	3.125	760	0.8421	1.283	1.647	1.963	2.331	2.582	2.815	3.101
250	0.8431	1.285	1.651	1.969	2.341	2.596	2.832	3.123	770	0.8421	1.283	1.647	1.963	2.331	2.582	2.815	3.101
260	0.8430	1.285	1.651	1.969	2.341	2.595	2.831	3.122	780	0.8421	1.283	1.647	1.963	2.331	2.582	2.815	3.101
270	0.8430	1.285	1.651	1.969	2.340	2.594	2.830	3.121	790	0.8421	1.283	1.647	1.963	2.331	2.582	2.815	3.101
280	0.8429	1.285	1.650	1.968	2.340	2.594	2.829	3.120	800	0.8421	1.283	1.647	1.963	2.331	2.582	2.815	3.100
290	0.8429	1.284	1.650	1.968	2.339	2.593	2.829	3.119	810	0.8421	1.283	1.647	1.963	2.331	2.582	2.815	3.100
300	0.8428	1.284	1.650	1.968	2.339	2.592	2.828	3.118	820	0.8421	1.283	1.647	1.963	2.331	2.582	2.815	3.100
310	0.8428	1.284	1.650	1.968	2.338	2.592	2.827	3.117	830	0.8421	1.283	1.647	1.963	2.331	2.582	2.815	3.100
320	0.8427	1.284	1.650	1.967	2.338	2.591	2.827	3.116	840	0.8420	1.283	1.647	1.963	2.331	2.582	2.814	3.100
330	0.8427	1.284	1.649	1.967	2.338	2.591	2.826	3.115	850	0.8420	1.283	1.647	1.963	2.331	2.582	2.814	3.100
340	0.8427	1.284	1.649	1.967	2.337	2.590	2.825	3.114	860	0.8420	1.283	1.647	1.963	2.331	2.582	2.814	3.100
350	0.8426	1.284	1.649	1.967	2.337	2.590	2.825	3.114	870	0.8420	1.283	1.647	1.963	2.331	2.581	2.814	3.100
360	0.8426	1.284	1.649	1.967	2.337	2.590	2.824	3.113	880	0.8420	1.283	1.647	1.963	2.331	2.581	2.814	3.100
370	0.8426	1.284	1.649	1.966	2.336	2.589	2.824	3.112	890	0.8420	1.283	1.647	1.963	2.331	2.581	2.814	3.099
380	0.8426	1.284	1.649	1.966	2.336	2.589	2.824	3.112	900	0.8420	1.282	1.647	1.963	2.330	2.581	2.814	3.099
390	0.8425	1.284	1.649	1.966	2.336	2.588	2.823	3.111	910	0.8420	1.282	1.647	1.963	2.330	2.581	2.814	3.099
400	0.8425	1.284	1.649	1.966	2.336	2.588	2.823	3.111	920	0.8420	1.282	1.647	1.963	2.330	2.581	2.814	3.099
410	0.8425	1.284	1.649	1.966	2.335	2.588	2.822	3.110	930	0.8420	1.282	1.646	1.963	2.330	2.581	2.814	3.099
420	0.8425	1.284	1.648	1.966	2.335	2.588	2.822	3.110	940	0.8420	1.282	1.646	1.962	2.330	2.581	2.814	3.099
430	0.8425	1.284	1.648	1.965	2.335	2.587	2.822	3.109	950	0.8420	1.282	1.646	1.962	2.330	2.581	2.814	3.099
440	0.8424	1.283	1.648	1.965	2.335	2.587	2.821	3.109	960	0.8420	1.282	1.646	1.962	2.330	2.581	2.814	3.099
450	0.8424	1.283	1.648	1.965	2.335	2.587	2.821	3.108	970	0.8420	1.282	1.646	1.962	2.330	2.581	2.813	3.099
460	0.8424	1.283	1.648	1.965	2.334	2.587	2.821	3.108	980	0.8420	1.282	1.646	1.962	2.330	2.581	2.813	3.099
470	0.8424	1.283	1.648	1.965	2.334	2.586	2.820	3.108	990	0.8420	1.282	1.646	1.962	2.330	2.581	2.813	3.098
480	0.8424	1.283	1.648	1.965	2.334	2.586	2.820	3.107	1000	0.8420	1.282	1.646	1.962	2.330	2.581	2.813	3.098
490	0.8424	1.283	1.648	1.965	2.334	2.586	2.820	3.107	2000	0.8418	1.282	1.646	1.961	2.328	2.578	2.810	3.094
500	0.8423	1.283	1.648	1.965	2.334	2.586	2.820	3.107	3000	0.8417	1.282	1.645	1.961	2.328	2.577	2.809	3.093
									4500	0.8417	1.282	1.645	1.960	2.327	2.577	2.808	3.092
									Infinity	0.8416	1.282	1.645	1.960	2.326	2.576	2.807	3.090

Looks at one tail at a time whether alpha, alpha/2, beta, beta/2, etc.

APPENDIXES

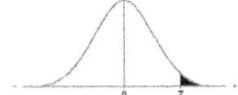

Table refers to dark shaded area.

Appendix 4 Z and its One Tail Probability

Z	p	Z	p	Z	p	Z	p	Z	p	Z	p	Z	p	Z	p
0.0000	0.50000	0.2550	0.39936	0.5100	0.30503	0.7650	0.22214	1.0200	0.15386	1.2750	0.10115	1.5300	0.06301	1.7850	0.03713
0.0050	0.49801	0.2600	0.39743	0.5150	0.30328	0.7700	0.22065	1.0250	0.15268	1.2800	0.10027	1.5350	0.06239	1.7900	0.03673
0.0100	0.49601	0.2650	0.39550	0.5200	0.30153	0.7750	0.21917	1.0300	0.15151	1.2850	0.09940	1.5400	0.06178	1.7950	0.03633
0.0150	0.49402	0.2700	0.39358	0.5250	0.29979	0.7800	0.21770	1.0350	0.15033	1.2900	0.09853	1.5450	0.06117	1.8000	0.03593
0.0200	0.49202	0.2750	0.39166	0.5300	0.29806	0.7850	0.21623	1.0400	0.14917	1.2950	0.09766	1.5500	0.06057	1.8050	0.03554
0.0250	0.49003	0.2800	0.38974	0.5350	0.29632	0.7900	0.21476	1.0450	0.14801	1.3000	0.09680	1.5550	0.05997	1.8100	0.03515
0.0300	0.48803	0.2850	0.38782	0.5400	0.29460	0.7950	0.21331	1.0500	0.14686	1.3050	0.09595	1.5600	0.05938	1.8150	0.03476
0.0350	0.48604	0.2900	0.38591	0.5450	0.29288	0.8000	0.21186	1.0550	0.14571	1.3100	0.09510	1.5650	0.05879	1.8200	0.03438
0.0400	0.48405	0.2950	0.38400	0.5500	0.29116	0.8050	0.21041	1.0600	0.14457	1.3150	0.09425	1.5700	0.05821	1.8250	0.03400
0.0450	0.48205	0.3000	0.38209	0.5550	0.28945	0.8100	0.20897	1.0650	0.14344	1.3200	0.09342	1.5750	0.05763	1.8300	0.03362
0.0500	0.48006	0.3050	0.38018	0.5600	0.28774	0.8150	0.20754	1.0700	0.14231	1.3250	0.09259	1.5800	0.05705	1.8350	0.03325
0.0550	0.47807	0.3100	0.37828	0.5650	0.28604	0.8200	0.20611	1.0750	0.14119	1.3300	0.09176	1.5850	0.05648	1.8400	0.03288
0.0600	0.47608	0.3150	0.37638	0.5700	0.28434	0.8250	0.20469	1.0800	0.14007	1.3350	0.09094	1.5900	0.05592	1.8450	0.03252
0.0650	0.47409	0.3200	0.37448	0.5750	0.28265	0.8300	0.20327	1.0850	0.13896	1.3400	0.09012	1.5950	0.05536	1.8500	0.03216
0.0700	0.47210	0.3250	0.37259	0.5800	0.28096	0.8350	0.20186	1.0900	0.13786	1.3450	0.08931	1.6000	0.05480	1.8550	0.03180
0.0750	0.47011	0.3300	0.37070	0.5850	0.27927	0.8400	0.20045	1.0950	0.13676	1.3500	0.08851	1.6050	0.05425	1.8600	0.03144
0.0800	0.46812	0.3350	0.36881	0.5900	0.27760	0.8450	0.19906	1.1000	0.13567	1.3550	0.08771	1.6100	0.05370	1.8650	0.03109
0.0850	0.46613	0.3400	0.36693	0.5950	0.27592	0.8500	0.19766	1.1050	0.13458	1.3600	0.08691	1.6150	0.05316	1.8700	0.03074
0.0900	0.46414	0.3450	0.36505	0.6000	0.27425	0.8550	0.19628	1.1100	0.13350	1.3650	0.08613	1.6200	0.05262	1.8750	0.03040
0.0950	0.46216	0.3500	0.36317	0.6050	0.27259	0.8600	0.19489	1.1150	0.13243	1.3700	0.08534	1.6250	0.05208	1.8800	0.03005
0.1000	0.46017	0.3550	0.36129	0.6100	0.27093	0.8650	0.19352	1.1200	0.13136	1.3750	0.08457	1.6300	0.05155	1.8850	0.02971
0.1050	0.45819	0.3600	0.35942	0.6150	0.26928	0.8700	0.19215	1.1250	0.13029	1.3800	0.08379	1.6350	0.05102	1.8900	0.02938
0.1100	0.45620	0.3650	0.35756	0.6200	0.26763	0.8750	0.19079	1.1300	0.12924	1.3850	0.08303	1.6400	0.05050	1.8950	0.02905
0.1150	0.45422	0.3700	0.35569	0.6250	0.26599	0.8800	0.18943	1.1350	0.12819	1.3900	0.08226	1.6450	0.04998	1.9000	0.02872
0.1200	0.45224	0.3750	0.35383	0.6300	0.26435	0.8850	0.18808	1.1400	0.12714	1.3950	0.08151	1.6500	0.04947	1.9050	0.02839
0.1250	0.45026	0.3800	0.35197	0.6350	0.26271	0.8900	0.18673	1.1450	0.12610	1.4000	0.08076	1.6550	0.04896	1.9100	0.02807
0.1300	0.44828	0.3850	0.35012	0.6400	0.26109	0.8950	0.18539	1.1500	0.12507	1.4050	0.08001	1.6600	0.04846	1.9150	0.02775
0.1350	0.44631	0.3900	0.34827	0.6450	0.25946	0.9000	0.18406	1.1550	0.12405	1.4100	0.07927	1.6650	0.04796	1.9200	0.02743
0.1400	0.44433	0.3950	0.34642	0.6500	0.25785	0.9050	0.18273	1.1600	0.12302	1.4150	0.07853	1.6700	0.04746	1.9250	0.02711
0.1450	0.44236	0.4000	0.34458	0.6550	0.25623	0.9100	0.18141	1.1650	0.12201	1.4200	0.07780	1.6750	0.04697	1.9300	0.02680
0.1500	0.44038	0.4050	0.34274	0.6600	0.25463	0.9150	0.18010	1.1700	0.12100	1.4250	0.07708	1.6800	0.04648	1.9350	0.02650
0.1550	0.43841	0.4100	0.34090	0.6650	0.25303	0.9200	0.17879	1.1750	0.12000	1.4300	0.07636	1.6850	0.04599	1.9400	0.02619
0.1600	0.43644	0.4150	0.33907	0.6700	0.25143	0.9250	0.17748	1.1800	0.11900	1.4350	0.07564	1.6900	0.04551	1.9450	0.02589
0.1650	0.43447	0.4200	0.33724	0.6750	0.24984	0.9300	0.17619	1.1850	0.11801	1.4400	0.07493	1.6950	0.04504	1.9500	0.02559
0.1700	0.43251	0.4250	0.33542	0.6800	0.24825	0.9350	0.17489	1.1900	0.11702	1.4450	0.07423	1.7000	0.04457	1.9550	0.02529
0.1750	0.43054	0.4300	0.33360	0.6850	0.24667	0.9400	0.17361	1.1950	0.11604	1.4500	0.07353	1.7050	0.04410	1.9600	0.02500
0.1800	0.42858	0.4350	0.33178	0.6900	0.24510	0.9450	0.17233	1.2000	0.11507	1.4550	0.07283	1.7100	0.04363	1.9650	0.02471
0.1850	0.42661	0.4400	0.32997	0.6950	0.24353	0.9500	0.17106	1.2050	0.11410	1.4600	0.07215	1.7150	0.04317	1.9700	0.02442
0.1900	0.42465	0.4450	0.32816	0.7000	0.24196	0.9550	0.16979	1.2100	0.11314	1.4650	0.07146	1.7200	0.04272	1.9750	0.02413
0.1950	0.42270	0.4500	0.32636	0.7050	0.24041	0.9600	0.16853	1.2150	0.11218	1.4700	0.07078	1.7250	0.04226	1.9800	0.02385
0.2000	0.42074	0.4550	0.32455	0.7100	0.23885	0.9650	0.16727	1.2200	0.11123	1.4750	0.07011	1.7300	0.04182	1.9850	0.02357
0.2050	0.41879	0.4600	0.32276	0.7150	0.23730	0.9700	0.16602	1.2250	0.11029	1.4800	0.06944	1.7350	0.04137	1.9900	0.02330
0.2100	0.41683	0.4650	0.32097	0.7200	0.23576	0.9750	0.16478	1.2300	0.10935	1.4850	0.06877	1.7400	0.04093	1.9950	0.02302
0.2150	0.41488	0.4700	0.31918	0.7250	0.23423	0.9800	0.16354	1.2350	0.10842	1.4900	0.06811	1.7450	0.04049	2.0000	0.02275
0.2200	0.41294	0.4750	0.31739	0.7300	0.23270	0.9850	0.16231	1.2400	0.10749	1.4950	0.06746	1.7500	0.04006	2.0050	0.02248
0.2250	0.41099	0.4800	0.31561	0.7350	0.23117	0.9900	0.16109	1.2450	0.10657	1.5000	0.06681	1.7550	0.03963	2.0100	0.02222
0.2300	0.40905	0.4850	0.31384	0.7400	0.22965	0.9950	0.15987	1.2500	0.10565	1.5050	0.06616	1.7600	0.03920	2.0150	0.02195
0.2350	0.40710	0.4900	0.31207	0.7450	0.22814	1.0000	0.15866	1.2550	0.10474	1.5100	0.06552	1.7650	0.03878	2.0200	0.02169
0.2400	0.40517	0.4950	0.31030	0.7500	0.22663	1.0050	0.15745	1.2600	0.10383	1.5150	0.06489	1.7700	0.03836	2.0250	0.02143
0.2450	0.40323	0.5000	0.30854	0.7550	0.22512	1.0100	0.15625	1.2650	0.10294	1.5200	0.06426	1.7750	0.03795	2.0300	0.02118
0.2500	0.40129	0.5050	0.30678	0.7600	0.22363	1.0150	0.15505	1.2700	0.10204	1.5250	0.06363	1.7800	0.03754	2.0350	0.02093

Z	p	Z	p	Z	p	Z	p	Z	p	Z	p	Z	p
2.2950	0.01087	2.5500	0.00539	2.8050	0.00252	3.0600	0.00111	3.3150	4.58E−04	3.5700	1.78E−04	3.8250	6.54E−05
2.3000	0.01072	2.5550	0.00531	2.8100	0.00248	3.0650	0.00109	3.3200	4.50E−04	3.5750	1.75E−04	3.8300	6.41E−05
2.3050	0.01058	2.5600	0.00523	2.8150	0.00244	3.0700	0.00107	3.3250	4.42E−04	3.5800	1.72E−04	3.8350	6.28E−05
2.3100	0.01044	2.5650	0.00516	2.8200	0.00240	3.0750	0.00105	3.3300	4.34E−04	3.5850	1.69E−04	3.8400	6.15E−05
2.3150	0.01031	2.5700	0.00508	2.8250	0.00236	3.0800	0.00104	3.3350	4.26E−04	3.5900	1.65E−04	3.8450	6.03E−05
2.3200	0.01017	2.5750	0.00501	2.8300	0.00233	3.0850	0.00102	3.3400	4.19E−04	3.5950	1.62E−04	3.8500	5.91E−05
2.3250	0.01004	2.5800	0.00494	2.8350	0.00229	3.0900	0.00100	3.3450	4.11E−04	3.6000	1.59E−04	3.8550	5.79E−05
2.3300	0.00990	2.5850	0.00487	2.8400	0.00226	3.0950	9.84E−04	3.3500	4.04E−04	3.6050	1.56E−04	3.8600	5.67E−05
2.3350	0.00977	2.5900	0.00480	2.8450	0.00222	3.1000	9.68E−04	3.3550	3.97E−04	3.6100	1.53E−04	3.8650	5.55E−05
2.3400	0.00964	2.5950	0.00473	2.8500	0.00219	3.1050	9.51E−04	3.3600	3.90E−04	3.6150	1.50E−04	3.8700	5.44E−05
2.3450	0.00951	2.6000	0.00466	2.8550	0.00215	3.1100	9.35E−04	3.3650	3.83E−04	3.6200	1.47E−04	3.8750	5.33E−05
2.3500	0.00939	2.6050	0.00459	2.8600	0.00212	3.1150	9.20E−04	3.3700	3.76E−04	3.6250	1.44E−04	3.8800	5.22E−05
2.3550	0.00926	2.6100	0.00453	2.8650	0.00209	3.1200	9.04E−04	3.3750	3.69E−04	3.6300	1.42E−04	3.8850	5.12E−05
2.3600	0.00914	2.6150	0.00446	2.8700	0.00205	3.1250	8.89E−04	3.3800	3.62E−04	3.6350	1.39E−04	3.8900	5.01E−05
2.3650	0.00902	2.6200	0.00440	2.8750	0.00202	3.1300	8.74E−04	3.3850	3.56E−04	3.6400	1.36E−04	3.8950	4.91E−05
2.3700	0.00889	2.6250	0.00433	2.8800	0.00199	3.1350	8.59E−04	3.3900	3.49E−04	3.6450	1.34E−04	3.9000	4.81E−05
2.3750	0.00877	2.6300	0.00427	2.8850	0.00196	3.1400	8.45E−04	3.3950	3.43E−04	3.6500	1.31E−04	3.9050	4.71E−05
2.3800	0.00866	2.6350	0.00421	2.8900	0.00193	3.1450	8.30E−04	3.4000	3.37E−04	3.6550	1.29E−04	3.9100	4.61E−05
2.3850	0.00854	2.6400	0.00415	2.8950	0.00190	3.1500	8.16E−04	3.4050	3.31E−04	3.6600	1.26E−04	3.9150	4.52E−05
2.3900	0.00842	2.6450	0.00408	2.9000	0.00187	3.1550	8.02E−04	3.4100	3.25E−04	3.6650	1.24E−04	3.9200	4.43E−05
2.3950	0.00831	2.6500	0.00402	2.9050	0.00184	3.1600	7.89E−04	3.4150	3.19E−04	3.6700	1.21E−04	3.9250	4.34E−05
2.4000	0.00820	2.6550	0.00397	2.9100	0.00181	3.1650	7.75E−04	3.4200	3.13E−04	3.6750	1.19E−04	3.9300	4.25E−05
2.4050	0.00809	2.6600	0.00391	2.9150	0.00178	3.1700	7.62E−04	3.4250	3.07E−04	3.6800	1.17E−04	3.9350	4.16E−05
2.4100	0.00798	2.6650	0.00385	2.9200	0.00175	3.1750	7.49E−04	3.4300	3.02E−04	3.6850	1.14E−04	3.9400	4.07E−05
2.4150	0.00787	2.6700	0.00379	2.9250	0.00172	3.1800	7.36E−04	3.4350	2.96E−04	3.6900	1.12E−04	3.9450	3.99E−05
2.4200	0.00776	2.6750	0.00374	2.9300	0.00169	3.1850	7.24E−04	3.4400	2.91E−04	3.6950	1.10E−04	3.9500	3.91E−05
2.4250	0.00765	2.6800	0.00368	2.9350	0.00167	3.1900	7.11E−04	3.4450	2.86E−04	3.7000	1.08E−04	3.9550	3.83E−05
2.4300	0.00755	2.6850	0.00363	2.9400	0.00164	3.1950	6.99E−04	3.4500	2.80E−04	3.7050	1.06E−04	3.9600	3.75E−05
2.4350	0.00745	2.6900	0.00357	2.9450	0.00161	3.2000	6.87E−04	3.4550	2.75E−04	3.7100	1.04E−04	3.9650	3.67E−05
2.4400	0.00734	2.6950	0.00352	2.9500	0.00159	3.2050	6.75E−04	3.4600	2.70E−04	3.7150	1.02E−04	3.9700	3.59E−05
2.4450	0.00724	2.7000	0.00347	2.9550	0.00156	3.2100	6.64E−04	3.4650	2.65E−04	3.7200	9.96E−05	3.9750	3.52E−05
2.4500	0.00714	2.7050	0.00342	2.9600	0.00154	3.2150	6.52E−04	3.4700	2.60E−04	3.7250	9.77E−05	3.9800	3.45E−05
2.4550	0.00704	2.7100	0.00336	2.9650	0.00151	3.2200	6.41E−04	3.4750	2.55E−04	3.7300	9.57E−05	3.9850	3.37E−05
2.4600	0.00695	2.7150	0.00331	2.9700	0.00149	3.2250	6.30E−04	3.4800	2.51E−04	3.7350	9.39E−05	3.9900	3.30E−05
2.4650	0.00685	2.7200	0.00326	2.9750	0.00146	3.2300	6.19E−04	3.4850	2.46E−04	3.7400	9.20E−05	3.9950	3.23E−05
2.4700	0.00676	2.7250	0.00322	2.9800	0.00144	3.2350	6.08E−04	3.4900	2.42E−04	3.7450	9.02E−05	4.0000	3.17E−05
2.4750	0.00666	2.7300	0.00317	2.9850	0.00142	3.2400	5.98E−04	3.4950	2.37E−04	3.7500	8.84E−05	4.0050	3.10E−05
2.4800	0.00657	2.7350	0.00312	2.9900	0.00139	3.2450	5.87E−04	3.5000	2.33E−04	3.7550	8.67E−05	4.0100	3.04E−05
2.4850	0.00648	2.7400	0.00307	2.9950	0.00137	3.2500	5.77E−04	3.5050	2.28E−04	3.7600	8.50E−05	4.0150	2.97E−05
2.4900	0.00639	2.7450	0.00303	3.0000	0.00135	3.2550	5.67E−04	3.5100	2.24E−04	3.7650	8.33E−05	4.0200	2.91E−05
2.4950	0.00630	2.7500	0.00298	3.0050	0.00133	3.2600	5.57E−04	3.5150	2.20E−04	3.7700	8.16E−05	4.0250	2.85E−05
2.5000	0.00621	2.7550	0.00293	3.0100	0.00131	3.2650	5.47E−04	3.5200	2.16E−04	3.7750	8.00E−05	4.0300	2.79E−05
2.5050	0.00612	2.7600	0.00289	3.0150	0.00128	3.2700	5.38E−04	3.5250	2.12E−04	3.7800	7.84E−05	4.0350	2.73E−05
2.5100	0.00604	2.7650	0.00285	3.0200	0.00126	3.2750	5.28E−04	3.5300	2.08E−04	3.7850	7.69E−05	4.0400	2.67E−05
2.5150	0.00595	2.7700	0.00280	3.0250	0.00124	3.2800	5.19E−04	3.5350	2.04E−04	3.7900	7.53E−05	4.0450	2.62E−05
2.5200	0.00587	2.7750	0.00276	3.0300	0.00122	3.2850	5.10E−04	3.5400	2.00E−04	3.7950	7.38E−05	4.0500	2.56E−05
2.5250	0.00578	2.7800	0.00272	3.0350	0.00120	3.2900	5.01E−04	3.5450	1.96E−04	3.8000	7.23E−05	4.0550	2.51E−05
2.5300	0.00570	2.7850	0.00268	3.0400	0.00118	3.2950	4.92E−04	3.5500	1.93E−04	3.8050	7.09E−05	4.0600	2.45E−05
2.5350	0.00562	2.7900	0.00264	3.0450	0.00116	3.3000	4.83E−04	3.5550	1.89E−04	3.8100	6.95E−05	4.0650	2.40E−05
2.5400	0.00554	2.7950	0.00259	3.0500	0.00114	3.3050	4.75E−04	3.5600	1.85E−04	3.8150	6.81E−05	4.0700	2.35E−05
2.5450	0.00546	2.8000	0.00256	3.0550	0.00113	3.3100	4.66E−04	3.5650	1.82E−04	3.8200	6.67E−05	4.0750	2.30E−05

APPENDIXES

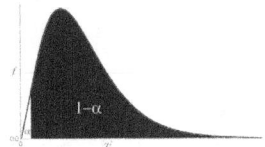

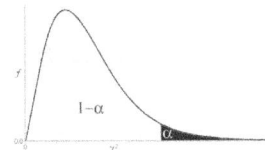

Appendix 5 Chi Square

p> df	0.995	0.99	0.9750	0.95	0.9	0.8	... df	0.2	0.1	0.05	0.025	0.01	0.005
1	3.927E−05	1.571E−04	9.821E−04	0.003932	0.01579	0.06418	1	1.6424	2.7055	3.8415	5.0239	6.6349	7.8794
2	1.003E−02	2.010E−02	5.064E−02	0.10259	0.21072	0.44629	2	3.2189	4.6052	5.9915	7.3778	9.2103	10.5966
3	0.07172	0.11483	0.21580	0.35185	0.58437	1.0052	3	4.6416	6.2514	7.8147	9.3484	11.3449	12.8382
4	0.20699	0.29711	0.48442	0.71072	1.0636	1.6488	4	5.9886	7.7794	9.4877	11.1433	13.2767	14.8603
5	0.41174	0.55430	0.83121	1.1455	1.6103	2.3425	5	7.2893	9.2364	11.0705	12.8325	15.0863	16.7496
6	0.67573	0.87209	1.2373	1.6354	2.2041	3.0701	6	8.5581	10.6446	12.5916	14.4494	16.8119	18.5476
7	0.98926	1.2390	1.6899	2.1673	2.8331	3.8223	7	9.8032	12.0170	14.0671	16.0128	18.4753	20.2777
8	1.3444	1.6465	2.1797	2.7326	3.4895	4.5936	8	11.0301	13.3616	15.5073	17.5345	20.0902	21.9550
9	1.7349	2.0879	2.7004	3.3251	4.1682	5.3801	9	12.2421	14.6837	16.9190	19.0228	21.6660	23.5894
10	2.1559	2.5582	3.2470	3.9403	4.8652	6.1791	10	13.4420	15.9872	18.3070	20.4832	23.2093	25.1882
11	2.6032	3.0535	3.8157	4.5748	5.5778	6.9887	11	14.6314	17.2750	19.6751	21.9200	24.7250	26.7568
12	3.0738	3.5706	4.4038	5.2260	6.3038	7.8073	12	15.8120	18.5493	21.0261	23.3367	26.2170	28.2995
13	3.5650	4.1069	5.0088	5.8919	7.0415	8.6339	13	16.9848	19.8119	22.3620	24.7356	27.6882	29.8195
14	4.0747	4.6604	5.6287	6.5706	7.7895	9.4673	14	18.1508	21.0641	23.6848	26.1189	29.1412	31.3193
15	4.6009	5.2293	6.2621	7.2609	8.5468	10.3070	15	19.3107	22.3071	24.9958	27.4884	30.5779	32.8013
16	5.1422	5.8122	6.9077	7.9616	9.3122	11.1521	16	20.4651	23.5418	26.2962	28.8454	31.9999	34.2672
17	5.6972	6.4078	7.5642	8.6718	10.0852	12.0023	17	21.6146	24.7690	27.5871	30.1910	33.4087	35.7185
18	6.2648	7.0149	8.2307	9.3905	10.8649	12.8570	18	22.7595	25.9894	28.8693	31.5264	34.8053	37.1565
19	6.8440	7.6327	8.9065	10.1170	11.6509	13.7158	19	23.9004	27.2036	30.1435	32.8523	36.1909	38.5823
20	7.4338	8.2604	9.5908	10.8508	12.4426	14.5784	20	25.0375	28.4120	31.4104	34.1696	37.5662	39.9968
21	8.0337	8.8972	10.2829	11.5913	13.2396	15.4446	21	26.1711	29.6151	32.6706	35.4789	38.9322	41.4011
22	8.6427	9.5425	10.9823	12.3380	14.0415	16.3140	22	27.3015	30.8133	33.9244	36.7807	40.2894	42.7957
23	9.2604	10.1957	11.6886	13.0905	14.8480	17.1865	23	28.4288	32.0069	35.1725	38.0756	41.6384	44.1813
24	9.8862	10.8564	12.4012	13.8484	15.6587	18.0618	24	29.5533	33.1962	36.4150	39.3641	42.9798	45.5585
25	10.5197	11.5240	13.1197	14.6114	16.4734	18.9398	25	30.6752	34.3816	37.6525	40.6465	44.3141	46.9279
26	11.1602	12.1981	13.8439	15.3792	17.2919	19.8202	26	31.7946	35.5632	38.8851	41.9232	45.6417	48.2899
27	11.8076	12.8785	14.5734	16.1514	18.1139	20.7030	27	32.9117	36.7412	40.1133	43.1945	46.9629	49.6449
28	12.4613	13.5647	15.3079	16.9279	18.9392	21.5880	28	34.0266	37.9159	41.3371	44.4608	48.2782	50.9934
29	13.1211	14.2565	16.0471	17.7084	19.7677	22.4751	29	35.1394	39.0875	42.5570	45.7223	49.5879	52.3356
30	13.7867	14.9535	16.7908	18.4927	20.5992	23.3641	30	36.2502	40.2560	43.7730	46.9792	50.8922	53.6720
31	14.4578	15.6555	17.5387	19.2806	21.4336	24.2551	31	37.3591	41.4217	44.9853	48.2319	52.1914	55.0027
32	15.1340	16.3622	18.2908	20.0719	22.2706	25.1478	32	38.4663	42.5847	46.1943	49.4804	53.4858	56.3281
33	15.8153	17.0735	19.0467	20.8665	23.1102	26.0422	33	39.5718	43.7452	47.3999	50.7251	54.7755	57.6484
34	16.5013	17.7891	19.8063	21.6643	23.9523	26.9383	34	40.6756	44.9032	48.6024	51.9660	56.0609	58.9639
35	17.1918	18.5089	20.5694	22.4650	24.7967	27.8359	35	41.7780	46.0588	49.8018	53.2033	57.3421	60.2748
36	17.8867	19.2327	21.3359	23.2686	25.6433	28.7350	36	42.8788	47.2122	50.9985	54.4373	58.6192	61.5812
37	18.5858	19.9602	22.1056	24.0749	26.4921	29.6355	37	43.9782	48.3634	52.1923	55.6680	59.8925	62.8833
38	19.2889	20.6914	22.8785	24.8839	27.3430	30.5373	38	45.0763	49.5126	53.3835	56.8955	61.1621	64.1814
39	19.9959	21.4262	23.6543	25.6954	28.1958	31.4405	39	46.1730	50.6598	54.5722	58.1201	62.4281	65.4756
40	20.7065	22.1643	24.4330	26.5093	29.0505	32.3450	40	47.2685	51.8051	55.7585	59.3417	63.6907	66.7660
41	21.4208	22.9056	25.2145	27.3256	29.9071	33.2506	41	48.3628	52.9485	56.9424	60.5606	64.9501	68.0527
42	22.1385	23.6501	25.9987	28.1440	30.7654	34.1574	42	49.4560	54.0902	58.1240	61.7768	66.2062	69.3360
43	22.8595	24.3976	26.7854	28.9647	31.6255	35.0653	43	50.5480	55.2302	59.3035	62.9904	67.4593	70.6159
44	23.5837	25.1480	27.5746	29.7875	32.4871	35.9743	44	51.6389	56.3685	60.4809	64.2015	68.7095	71.8926
45	24.3110	25.9013	28.3662	30.6123	33.3504	36.8844	45	52.7288	57.5053	61.6562	65.4102	69.9568	73.1661
46	25.0413	26.6572	29.1601	31.4390	34.2152	37.7955	46	53.8177	58.6405	62.8296	66.6165	71.2014	74.4365
47	25.7746	27.4158	29.9562	32.2676	35.0814	38.7075	47	54.9056	59.7743	64.0011	67.8206	72.4433	75.7041
48	26.5106	28.1770	30.7545	33.0981	35.9491	39.6205	48	55.9926	60.9066	65.1708	69.0226	73.6826	76.9688
49	27.2493	28.9406	31.5549	33.9303	36.8182	40.5344	49	57.0786	62.0375	66.3386	70.2224	74.9195	78.2307
50	27.9907	29.7067	32.3574	34.7643	37.6886	41.4492	50	58.1638	63.1671	67.5048	71.4202	76.1539	79.4900
51	28.7347	30.4750	33.1618	35.5999	38.5604	42.3649	51	59.2481	64.2954	68.6693	72.6160	77.3860	80.7467
52	29.4812	31.2457	33.9681	36.4371	39.4334	43.2814	52	60.3316	65.4224	69.8322	73.8099	78.6158	82.0008
53	30.2300	32.0185	34.7763	37.2759	40.3076	44.1987	53	61.4142	66.5482	70.9935	75.0019	79.8433	83.2526
54	30.9813	32.7934	35.5863	38.1162	41.1830	45.1167	54	62.4961	67.6728	72.1532	76.1920	81.0688	84.5019
55	31.7348	33.5705	36.3981	38.9580	42.0596	46.0356	55	63.5772	68.7962	73.3115	77.3805	82.2921	85.7490
56	32.4905	34.3495	37.2116	39.8013	42.9373	46.9552	56	64.6576	69.9185	74.4683	78.5672	83.5134	86.9938
57	33.2484	35.1305	38.0267	40.6459	43.8161	47.8755	57	65.7373	71.0397	75.6237	79.7522	84.7328	88.2364
58	34.0084	35.9135	38.8435	41.4920	44.6960	48.7965	58	66.8162	72.1598	76.7778	80.9356	85.9502	89.4769
59	34.7704	36.6982	39.6619	42.3393	45.5770	49.7182	59	67.8945	73.2789	77.9305	82.1174	87.1657	90.7153
60	35.5345	37.4849	40.4817	43.1880	46.4589	50.6406	60	68.9721	74.4370	79.0819	83.2977	88.3794	91.9517

At about df=6 the table progresses from left of the left figure, above, to right of the right figure (with skip between .8 and .2)

->

Appendix 5 Chi Square

p>	0.995	0.99	0.9750	0.95	0.9	0.8	...	0.2	0.1	0.05	0.025	0.01	0.005
df													
62	37.0684	39.0633	42.1260	44.8890	48.2257	52.4873		71.1253	76.6302	81.3810	85.6537	90.8015	94.4187
63	37.8382	39.8551	42.9503	45.7414	49.1105	53.4116		72.2010	77.7454	82.5287	86.8296	92.0100	95.6493
64	38.6098	40.6486	43.7760	46.5949	49.9963	54.3365		73.2761	78.8596	83.6753	88.0041	93.2169	96.8781
65	39.3831	41.4436	44.6030	47.4496	50.8829	55.2620		74.3506	79.9730	84.8206	89.1771	94.4221	98.1051
66	40.1582	42.2402	45.4314	48.3054	51.7705	56.1880		75.4245	81.0855	85.9649	90.3489	95.6257	99.3304
67	40.9350	43.0384	46.2610	49.1623	52.6588	57.1147		76.4978	82.1971	87.1081	91.5194	96.8278	100.554
68	41.7135	43.8380	47.0920	50.0202	53.5481	58.0418		77.5707	83.3079	88.2502	92.6885	98.0284	101.776
69	42.4935	44.6392	47.9242	50.8792	54.4381	58.9696		78.6429	84.4179	89.3912	93.8565	99.2275	102.996
70	43.2752	45.4417	48.7576	51.7393	55.3289	59.8978		79.7146	85.5270	90.5312	95.0232	100.425	104.215
71	44.0584	46.2457	49.5922	52.6003	56.2206	60.8266		80.7859	86.6354	91.6702	96.1887	101.621	105.432
72	44.8431	47.0510	50.4279	53.4623	57.1129	61.7558		81.8566	87.7430	92.8083	97.3531	102.816	106.648
73	45.6293	47.8577	51.2648	54.3253	58.0061	62.6856		82.9268	88.8499	93.9453	98.5163	104.010	107.862
74	46.4170	48.6657	52.1028	55.1892	58.9000	63.6158		83.9965	89.9560	95.0815	99.6783	105.202	109.074
75	47.2060	49.4750	52.9419	56.0541	59.7946	64.5466		85.0658	91.0615	96.2167	100.839	106.393	110.286
76	47.9965	50.2856	53.7821	56.9198	60.6899	65.4777		86.1346	92.1662	97.3510	101.999	107.583	111.495
77	48.7884	51.0974	54.6234	57.7864	61.5858	66.4094		87.2030	93.2702	98.4844	103.158	108.771	112.704
78	49.5816	51.9104	55.4656	58.6539	62.4825	67.3415		88.2709	94.3735	99.6169	104.316	109.958	113.911
79	50.3761	52.7247	56.3089	59.5223	63.3799	68.2740		89.3383	95.4762	100.749	105.473	111.144	115.117
80	51.1719	53.5401	57.1532	60.3915	64.2778	69.2069		90.4053	96.5782	101.879	106.629	112.329	116.321
81	51.9690	54.3566	57.9984	61.2615	65.1765	70.1403		91.4720	97.6796	103.010	107.783	113.512	117.524
82	52.7674	55.1743	58.8446	62.1323	66.0757	71.0741		92.5382	98.7803	104.139	108.937	114.695	118.726
83	53.5669	55.9931	59.6918	63.0039	66.9756	72.0083		93.6039	99.8805	105.267	110.090	115.876	119.927
84	54.3677	56.8130	60.5398	63.8763	67.8761	72.9429		94.6693	100.980	106.395	111.242	117.057	121.126
85	55.1696	57.6339	61.3888	64.7494	68.7772	73.8779		95.7343	102.079	107.522	112.393	118.236	122.325
86	55.9727	58.4559	62.2386	65.6233	69.6788	74.8132		96.7990	103.177	108.648	113.544	119.414	123.522
87	56.7769	59.2790	63.0894	66.4979	70.5810	75.7490		97.8632	104.275	109.773	114.693	120.591	124.718
88	57.5823	60.1030	63.9409	67.3732	71.4838	76.6851		98.9271	105.372	110.898	115.841	121.767	125.913
89	58.3888	60.9281	64.7934	68.2493	72.3872	77.6216		99.9906	106.469	112.022	116.989	122.942	127.106
90	59.1963	61.7541	65.6466	69.1260	73.2911	78.5584		101.054	107.565	113.145	118.136	124.116	128.299
91	60.0049	62.5811	66.5007	70.0035	74.1955	79.4956		102.117	108.661	114.268	119.282	125.289	129.491
92	60.8146	63.4090	67.3556	70.8816	75.1005	80.4332		103.179	109.756	115.390	120.427	126.462	130.681
93	61.6253	64.2379	68.2112	71.7603	76.0060	81.3711		104.241	110.850	116.511	121.571	127.633	131.871
94	62.4370	65.0677	69.0677	72.6398	76.9119	82.3093		105.303	111.944	117.632	122.715	128.803	133.059
95	63.2496	65.8984	69.9249	73.5198	77.8184	83.2478		106.364	113.038	118.752	123.858	129.973	134.247
96	64.0633	66.7299	70.7828	74.4005	78.7254	84.1867		107.425	114.131	119.871	125.000	131.141	135.433
97	64.8780	67.5624	71.6415	75.2819	79.6329	85.1259		108.486	115.223	120.990	126.141	132.309	136.619
98	65.6936	68.3957	72.5009	76.1638	80.5408	86.0654		109.547	116.315	122.108	127.282	133.476	137.803
99	66.5101	69.2299	73.3611	77.0463	81.4493	87.0052		110.607	117.407	123.225	128.422	134.642	138.987
100	67.3276	70.0649	74.2219	77.9295	82.3581	87.9453		111.667	118.498	124.342	129.561	135.807	140.169
105	71.4282	74.2520	78.5364	82.3537	86.9093	92.6504		116.962	123.947	129.918	135.247	141.620	146.070
110	75.5500	78.4583	82.8671	86.7916	91.4710	97.3624		122.250	129.385	135.480	140.917	147.414	151.948
120	83.8516	86.9233	91.5726	95.7046	100.624	106.806		132.806	140.233	146.567	152.211	158.950	163.648
130	92.2225	95.4510	100.331	104.662	109.811	116.272		143.340	151.045	157.610	163.453	170.423	175.278
140	100.655	104.034	109.137	113.659	119.029	125.758		153.854	161.827	168.613	174.648	181.840	186.847
150	109.142	112.668	117.985	122.692	128.275	135.263		164.349	172.581	179.581	185.800	193.208	198.360
160	117.679	121.346	126.870	131.756	137.546	144.783		174.828	183.311	190.516	196.915	204.530	209.824
170	126.261	130.064	135.790	140.849	146.839	154.319		185.293	194.017	201.423	207.995	215.812	221.242
180	134.884	138.820	144.741	149.969	156.153	163.868		195.743	204.704	212.304	219.044	227.056	232.620
190	143.545	147.610	153.721	159.113	165.485	173.430		206.182	215.371	223.160	230.064	238.266	243.959
200	152.241	156.432	162.728	168.279	174.835	183.003		216.609	226.021	233.994	241.058	249.445	255.264
225	174.116	178.609	185.348	191.281	198.278	206.980		242.631	252.578	260.992	268.438	277.269	283.390
250	196.161	200.939	208.098	214.392	221.806	231.013		268.599	279.050	287.882	295.689	304.940	311.346
250	196.161	200.939	208.098	214.392	221.806	231.013		268.599	279.050	287.882	295.689	304.940	311.346
200	152.241	156.432	162.728	168.279	174.835	183.003		216.609	226.021	233.994	241.058	249.445	255.264
400	330.903	337.155	346.482	354.641	364.207	376.022		423.590	436.649	447.632	457.305	468.724	476.606
500	422.303	429.388	439.936	449.147	459.926	473.210		526.401	540.930	553.127	563.852	576.493	585.207
750	653.997	662.852	676.003	687.452	700.814	717.225		782.386	800.043	814.822	827.785	843.029	853.514
1000	888.564	898.912	914.257	927.594	943.133	962.180		1037.43	1057.72	1074.68	1089.53	1106.97	1118.95
5000	4746.17	4770.31	4805.90	4836.66	4872.28	4915.65		5083.96	5128.58	5165.61	5197.88	5235.57	5261.34
10000	9639.48	9673.95	9724.72	9768.53	9819.19	9880.79		10118.8	10181.7	10233.7	10279.1	10331.9	10368.0

These compurer generated tables **might be inaccurate in the sixth place**. This sould be OK for testing.
However, derived results from them should be at least rounded to no more than five places.

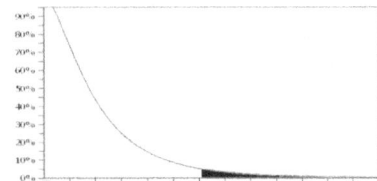

Graph of $df_1 = 4$, $df_2 = 11$ at 5%. But 1% and 99% tables are shown first. ($df1$ runs across, $df2$ goes down)

Appendix 6   1% Critical values of the F distribution

df1/df2	1	2	3	4	5	6	7	8	9	10	11	12	13	14	15
1	4052.18	4999.50	5403.35	5624.58	5763.65	5858.99	5928.36	5981.07	6022.47	6055.85	6083.32	6106.32	6125.86	6142.67	6157.28
2	98.5025	99.0000	99.1662	99.2494	99.2993	99.3326	99.3564	99.3742	99.3881	99.3992	99.4083	99.4159	99.4223	99.4278	99.4325
3	34.1162	9.55209	9.27663	9.11718	9.01346	8.94065	8.88674	8.84524	8.81230	8.78552	8.76333	8.74464	8.72868	8.71490	8.70287
4	21.1977	18.0000	16.6944	15.9770	15.5219	15.2069	14.9758	14.7989	14.6591	14.5459	14.4523	14.3736	14.3065	14.2486	14.1982
5	16.2582	13.2739	12.0600	11.3919	10.9670	10.6723	10.4555	10.2893	10.1578	10.0510	9.96265	9.88828	9.82481	9.77001	9.72222
6	13.7450	10.9248	9.77954	9.14830	8.74590	8.46613	8.26000	8.10165	7.97612	7.87412	7.78957	7.71833	7.65748	7.60490	7.55899
7	12.2464	9.54658	8.45129	7.84665	7.46044	7.19140	6.99283	6.84005	6.71875	6.62006	6.53817	6.46909	6.41003	6.35895	6.31433
8	11.2586	8.64911	7.59099	7.00608	6.63183	6.37068	6.17762	6.02887	5.91062	5.81429	5.73427	5.66672	5.60891	5.55887	5.51512
9	10.5614	8.02152	6.99192	6.42209	6.05694	5.80177	5.61287	5.46712	5.35113	5.25654	5.17789	5.11143	5.05451	5.00521	4.96208
10	10.0443	7.55943	6.55231	5.99434	5.63633	5.38581	5.20012	5.05669	4.94242	4.84915	4.77152	4.70587	4.64961	4.60083	4.55814
11	9.64603	7.20571	6.21673	5.66830	5.31601	5.06921	4.88607	4.74447	4.63154	4.53928	4.46244	4.39740	4.34162	4.29324	4.25087
12	9.33021	6.92661	5.95254	5.41195	5.06434	4.82057	4.63950	4.49937	4.38751	4.29605	4.21982	4.15526	4.09985	4.05176	4.00962
13	9.07381	6.70096	5.73938	5.20533	4.86162	4.62036	4.44100	4.30206	4.19108	4.10027	4.02452	3.96033	3.90520	3.85734	3.81537
14	8.86159	6.51488	5.56389	5.03538	4.69496	4.45582	4.27788	4.13995	4.02968	3.93940	3.86404	3.80014	3.74524	3.69754	3.65570
15	8.68312	6.35887	5.41696	4.89321	4.55561	4.31827	4.14155	4.00445	3.89479	3.80494	3.72990	3.66624	3.61151	3.56394	3.52219
16	8.53097	6.22624	5.29221	4.77258	4.43742	4.20163	4.02595	3.88957	3.78042	3.69093	3.61616	3.55269	3.49810	3.45063	3.40895
17	8.39974	6.11211	5.18500	4.66897	4.33594	4.10151	3.92672	3.79096	3.68224	3.59307	3.51851	3.45520	3.40072	3.35333	3.31169
18	8.28542	6.01290	5.09189	4.57904	4.24788	4.01464	3.84064	3.70542	3.59707	3.50816	3.43379	3.37061	3.31622	3.26888	3.22729
19	8.18495	5.92588	5.01029	4.50026	4.17077	3.93857	3.76527	3.63052	3.52250	3.43382	3.35960	3.29653	3.24221	3.19491	3.15334
20	8.09596	5.84893	4.93819	4.43069	4.10268	3.87143	3.69874	3.56441	3.45668	3.36819	3.29411	3.23112	3.17686	3.12960	3.08804
21	8.01660	5.78042	4.87405	4.36882	4.04214	3.81173	3.63959	3.50563	3.39815	3.30983	3.23587	3.17295	3.11874	3.07150	3.02995
22	7.94539	5.71902	4.81661	4.31343	3.98796	3.75830	3.58666	3.45303	3.34577	3.25761	3.18374	3.12089	3.06671	3.01949	2.97795
23	7.88113	5.66370	4.76488	4.26357	3.93919	3.71022	3.53902	3.40569	3.29863	3.21060	3.13682	3.07402	3.01987	2.97267	2.93112
24	7.82287	5.61359	4.71805	4.21845	3.89507	3.66672	3.49593	3.36287	3.25599	3.16807	3.09437	3.03161	2.97749	2.93029	2.88873
25	7.76980	5.56800	4.67546	4.17742	3.85496	3.62717	3.45675	3.32394	3.21722	3.12941	3.05577	2.99306	2.93895	2.89175	2.85019
26	7.72125	5.52633	4.63657	4.13996	3.81834	3.59108	3.42099	3.28840	3.18182	3.09411	3.02053	2.95785	2.90375	2.85655	2.81498
27	7.67668	5.48812	4.60091	4.10562	3.78477	3.55799	3.38822	3.25583	3.14939	3.06175	2.98823	2.92557	2.87149	2.82429	2.78270
28	7.63562	5.45294	4.56809	4.07403	3.75389	3.52756	3.35807	3.22587	3.11955	3.03199	2.95851	2.89588	2.84180	2.79460	2.75300
29	7.59766	5.42045	4.53779	4.04487	3.72540	3.49947	3.33025	3.19822	3.09201	3.00452	2.93108	2.86847	2.81440	2.76719	2.72558
30	7.56248	5.39035	4.50974	4.01788	3.69902	3.47348	3.30450	3.17262	3.06652	2.97909	2.90569	2.84310	2.78902	2.74181	2.70018
31	7.52977	5.36239	4.48369	3.99281	3.67453	3.44934	3.28059	3.14886	3.04285	2.95548	2.88211	2.81953	2.76546	2.71823	2.67659
32	7.49928	5.33634	4.45943	3.96948	3.65173	3.42688	3.25834	3.12675	3.02082	2.93351	2.86016	2.79759	2.74353	2.69629	2.65463
33	7.47080	5.31203	4.43679	3.94770	3.63046	3.40591	3.23757	3.10611	3.00026	2.91300	2.83968	2.77712	2.72305	2.67580	2.63413
34	7.44414	5.28928	4.41561	3.92733	3.61056	3.38631	3.21815	3.08681	2.98103	2.89381	2.82052	2.75797	2.70390	2.65664	2.61495
35	7.41912	5.26794	4.39575	3.90824	3.59191	3.36793	3.19995	3.06872	2.96301	2.87583	2.80256	2.74002	2.68594	2.63867	2.59697
36	7.39560	5.24789	4.37710	3.89031	3.57440	3.35068	3.18286	3.05173	2.94609	2.85895	2.78569	2.72315	2.66907	2.62179	2.58007
37	7.37344	5.22902	4.35954	3.87343	3.55792	3.33444	3.16677	3.03574	2.93016	2.84305	2.76982	2.70728	2.65320	2.60591	2.56417
38	7.35254	5.21122	4.34299	3.85752	3.54238	3.31913	3.15161	3.02067	2.91515	2.82807	2.75485	2.69232	2.63823	2.59093	2.54918
39	7.33279	5.19441	4.32736	3.84250	3.52771	3.30468	3.13730	3.00644	2.90097	2.81392	2.74072	2.67819	2.62410	2.57678	2.53501
40	7.31410	5.17851	4.31257	3.82829	3.51384	3.29101	3.12376	2.99298	2.88756	2.80055	2.72735	2.66483	2.61073	2.56340	2.52162
41	7.29638	5.16344	4.29856	3.81484	3.50070	3.27807	3.11093	2.98023	2.87486	2.78787	2.71469	2.65217	2.59806	2.55072	2.50892
42	7.27956	5.14914	4.28527	3.80207	3.48823	3.26579	3.09877	2.96814	2.86281	2.77585	2.70268	2.64016	2.58604	2.53869	2.49688
43	7.26358	5.13555	4.27265	3.78994	3.47640	3.25412	3.08722	2.95666	2.85137	2.76443	2.69127	2.62875	2.57463	2.52727	2.48544
44	7.24836	5.12263	4.26064	3.77841	3.46514	3.24303	3.07623	2.94574	2.84049	2.75357	2.68042	2.61790	2.56377	2.51640	2.47455
45	7.23387	5.11032	4.24921	3.76743	3.45442	3.23247	3.06577	2.93534	2.83013	2.74323	2.67008	2.60756	2.55343	2.50605	2.46418
46	7.22004	5.09858	4.23831	3.75696	3.44420	3.22240	3.05580	2.92543	2.82025	2.73337	2.66023	2.59771	2.54357	2.49617	2.45430
47	7.20684	5.08737	4.22790	3.74696	3.43444	3.21280	3.04628	2.91597	2.81082	2.72396	2.65083	2.58830	2.53416	2.48675	2.44486
48	7.19422	5.07666	4.21796	3.73742	3.42512	3.20362	3.03719	2.90693	2.80182	2.71497	2.64185	2.57932	2.52516	2.47775	2.43585
49	7.18214	5.06642	4.20845	3.72829	3.41621	3.19484	3.02849	2.89828	2.79320	2.70637	2.63325	2.57072	2.51656	2.46914	2.42722
50	7.17058	5.05661	4.19934	3.71955	3.40768	3.18643	3.02017	2.89001	2.78496	2.69814	2.62503	2.56250	2.50833	2.46089	2.41896

1% Critical values of the F distribution

df1/df2	1	2	3	4	5	6	7	8	9	10	11	12	13	14	15
51	7.15949	5.04721	4.19062	3.71117	3.39950	3.17838	3.01219	2.88208	2.77705	2.69025	2.61714	2.55461	2.50044	2.45299	2.41105
52	7.14885	5.03819	4.18225	3.70314	3.39167	3.17066	3.00454	2.87447	2.76948	2.68269	2.60958	2.54705	2.49287	2.44541	2.40345
53	7.13864	5.02954	4.17422	3.69543	3.38414	3.16325	2.99720	2.86718	2.76220	2.67543	2.60233	2.53979	2.48560	2.43813	2.39616
54	7.12882	5.02122	4.16650	3.68802	3.37691	3.15613	2.99015	2.86016	2.75521	2.66845	2.59535	2.53282	2.47862	2.43114	2.38916
55	7.11938	5.01322	4.15908	3.68090	3.36996	3.14928	2.98337	2.85342	2.74850	2.66174	2.58865	2.52611	2.47191	2.42442	2.38243
56	7.11029	5.00552	4.15194	3.67404	3.36328	3.14270	2.97685	2.84694	2.74203	2.65529	2.58220	2.51966	2.46545	2.41795	2.37595
57	7.10153	4.99811	4.14507	3.66745	3.35684	3.13636	2.97056	2.84069	2.73581	2.64908	2.57599	2.51344	2.45923	2.41172	2.36971
60	7.07711	4.97743	4.12589	3.64905	3.33888	3.11867	2.95305	2.82328	2.71845	2.63175	2.55867	2.49612	2.44188	2.39435	2.35230
65	7.04162	4.94741	4.09806	3.62235	3.31284	3.09302	2.92764	2.79802	2.69327	2.60661	2.53353	2.47097	2.41670	2.36912	2.32702
70	7.01140	4.92187	4.07440	3.59965	3.29069	3.07121	2.90603	2.77653	2.67186	2.58523	2.51216	2.44957	2.39528	2.34766	2.30552
75	6.98536	4.89988	4.05402	3.58011	3.27163	3.05244	2.88744	2.75804	2.65343	2.56682	2.49376	2.43116	2.37684	2.32919	2.28700
80	6.96269	4.88074	4.03630	3.56311	3.25505	3.03611	2.87127	2.74196	2.63740	2.55081	2.47775	2.41514	2.36079	2.31311	2.27088
85	6.94277	4.86393	4.02074	3.54819	3.24050	3.02178	2.85707	2.72785	2.62333	2.53676	2.46369	2.40107	2.34670	2.29899	2.25673
90	6.92514	4.84906	4.00697	3.53499	3.22763	3.00911	2.84451	2.71536	2.61088	2.52433	2.45126	2.38862	2.33423	2.28649	2.24420
95	6.90941	4.83580	3.99470	3.52323	3.21616	2.99781	2.83333	2.70424	2.59979	2.51325	2.44018	2.37753	2.32312	2.27536	2.23303
100	6.89530	4.82391	3.98370	3.51268	3.20587	2.98768	2.82330	2.69426	2.58984	2.50331	2.43024	2.36758	2.31316	2.26537	2.22301
110	6.87103	4.80346	3.96478	3.49456	3.18819	2.97028	2.80605	2.67712	2.57274	2.48623	2.41316	2.35048	2.29602	2.24819	2.20579
120	6.85089	4.78651	3.94910	3.47953	3.17355	2.95585	2.79176	2.66291	2.55857	2.47208	2.39900	2.33630	2.28181	2.23395	2.19150
130	6.83392	4.77223	3.93589	3.46688	3.16121	2.94371	2.77973	2.65094	2.54664	2.46015	2.38707	2.32436	2.26984	2.22195	2.17947
140	6.81942	4.76003	3.92462	3.45608	3.15068	2.93334	2.76946	2.64072	2.53645	2.44998	2.37689	2.31416	2.25962	2.21170	2.16919
150	6.80689	4.74949	3.91488	3.44675	3.14158	2.92438	2.76059	2.63190	2.52765	2.44118	2.36809	2.30535	2.25079	2.20284	2.16031
160	6.79596	4.74030	3.90638	3.43861	3.13365	2.91657	2.75285	2.62420	2.51998	2.43351	2.36042	2.29766	2.24309	2.19512	2.15255
170	6.78633	4.73220	3.89890	3.43144	3.12666	2.90969	2.74604	2.61743	2.51322	2.42676	2.35366	2.29090	2.23631	2.18832	2.14573
180	6.77779	4.72503	3.89227	3.42509	3.12047	2.90360	2.74000	2.61142	2.50723	2.42077	2.34767	2.28489	2.23029	2.18228	2.13968
190	6.77016	4.71861	3.88634	3.41942	3.11494	2.89815	2.73460	2.60606	2.50188	2.41543	2.34232	2.27954	2.22492	2.17690	2.13427
200	6.76330	4.71285	3.88102	3.41432	3.10997	2.89326	2.72976	2.60124	2.49707	2.41063	2.33751	2.27472	2.22009	2.17205	2.12942
210	6.75710	4.70765	3.87622	3.40972	3.10549	2.88885	2.72538	2.59688	2.49273	2.40629	2.33317	2.27037	2.21573	2.16768	2.12503
220	6.75148	4.70293	3.87185	3.40554	3.10142	2.88484	2.72141	2.59293	2.48879	2.40235	2.32923	2.26642	2.21178	2.16371	2.12104
230	6.74635	4.69862	3.86787	3.40173	3.09770	2.88118	2.71779	2.58933	2.48520	2.39876	2.32564	2.26283	2.20817	2.16009	2.11741
240	6.74165	4.69468	3.86423	3.39824	3.09430	2.87783	2.71447	2.58603	2.48191	2.39547	2.32235	2.25953	2.20486	2.15678	2.11408
250	6.73733	4.69105	3.86088	3.39504	3.09118	2.87476	2.71143	2.58300	2.47889	2.39245	2.31933	2.25650	2.20183	2.15373	2.11103
275	6.72793	4.68316	3.85359	3.38806	3.08438	2.86806	2.70479	2.57640	2.47230	2.38587	2.31274	2.24990	2.19521	2.14709	2.10437
300	6.72010	4.67659	3.84753	3.38225	3.07872	2.86249	2.69928	2.57092	2.46683	2.38040	2.30726	2.24442	2.18971	2.14157	2.09883
325	6.71349	4.67105	3.84241	3.37735	3.07394	2.85779	2.69462	2.56628	2.46221	2.37578	2.30264	2.23978	2.18506	2.13691	2.09415
350	6.70784	4.66630	3.83803	3.37316	3.06985	2.85376	2.69063	2.56231	2.45825	2.37182	2.29868	2.23582	2.18109	2.13292	2.09014
375	6.70294	4.66219	3.83423	3.36953	3.06632	2.85028	2.68718	2.55888	2.45482	2.36840	2.29525	2.23238	2.17764	2.12947	2.08667
400	6.69866	4.65860	3.83092	3.36635	3.06322	2.84724	2.68416	2.55588	2.45183	2.36541	2.29226	2.22938	2.17463	2.12645	2.08364
425	6.69488	4.65543	3.82800	3.36356	3.06050	2.84455	2.68150	2.55324	2.44919	2.36277	2.28962	2.22674	2.17198	2.12379	2.08097
450	6.69153	4.65262	3.82540	3.36107	3.05808	2.84217	2.67914	2.55089	2.44685	2.36043	2.28728	2.22439	2.16963	2.12142	2.07860
475	6.68853	4.65011	3.82308	3.35885	3.05591	2.84004	2.67703	2.54879	2.44476	2.35834	2.28518	2.22229	2.16752	2.11931	2.07648
500	6.68583	4.64785	3.82100	3.35685	3.05397	2.83812	2.67514	2.54690	2.44287	2.35646	2.28330	2.22040	2.16563	2.11741	2.07457
600	6.67730	4.64070	3.81440	3.35054	3.04781	2.83207	2.66913	2.54093	2.43692	2.35050	2.27734	2.21443	2.15964	2.11140	2.06853
700	6.67122	4.63560	3.80969	3.34604	3.04343	2.82775	2.66486	2.53668	2.43267	2.34626	2.27309	2.21017	2.15536	2.10711	2.06423
800	6.66667	4.63178	3.80617	3.34267	3.04014	2.82451	2.66165	2.53349	2.42949	2.34308	2.26990	2.20698	2.15216	2.10390	2.06100
900	6.66313	4.62881	3.80343	3.34005	3.03759	2.82200	2.65916	2.53101	2.42702	2.34061	2.26743	2.20450	2.14968	2.10140	2.05850
1000	6.66029	4.62644	3.80125	3.33795	3.03555	2.81999	2.65717	2.52903	2.42504	2.33863	2.26545	2.20252	2.14769	2.09941	2.05650
1100	6.65798	4.62450	3.79946	3.33624	3.03388	2.81835	2.65555	2.52741	2.42343	2.33702	2.26384	2.20090	2.14607	2.09778	2.05486
1200	6.65605	4.62289	3.79797	3.33481	3.03249	2.81698	2.65419	2.52606	2.42208	2.33567	2.26249	2.19955	2.14471	2.09642	2.05349
1300	6.65442	4.62152	3.79671	3.33361	3.03132	2.81583	2.65304	2.52492	2.42095	2.33454	2.26135	2.19841	2.14357	2.09527	2.05234
1400	6.65302	4.62035	3.79563	3.33258	3.03031	2.81483	2.65206	2.52395	2.41997	2.33356	2.26038	2.19743	2.14259	2.09429	2.05135

Appendix 6 (continued)

1% Critical values of the F distribution

df1/df2	16	17	18	19	20	21	22	23	24	25	26	27	28	29	30
1	6170.10	6181.43	6191.53	6200.58	6208.73	6216.12	6222.84	6228.99	6234.63	6239.83	6244.62	6249.07	6253.20	6257.05	6260.65
2	99.4367	99.4404	99.4436	99.4465	99.4492	99.4516	99.4537	99.4557	99.4575	99.4592	99.4607	99.4621	99.4635	99.4647	99.4658
3	26.8269	26.7867	26.7509	26.7188	26.6898	26.6635	26.6396	26.6176	26.5975	26.5790	26.5618	26.5460	26.5312	26.5174	26.5045
4	14.1539	14.1146	14.0795	14.0480	14.0196	13.9938	13.9703	13.9488	13.9291	13.9109	13.8940	13.8784	13.8639	13.8503	13.8377
5	9.68016	9.64287	9.60957	9.57966	9.55265	9.52812	9.50576	9.48529	9.46647	9.44912	9.43307	9.41818	9.40433	9.39141	9.37933
6	7.51857	7.48271	7.45066	7.42186	7.39583	7.37219	7.35063	7.33088	7.31272	7.29597	7.28047	7.26609	7.25270	7.24021	7.22853
7	6.27501	6.24010	6.20889	6.18082	6.15544	6.13238	6.11134	6.09205	6.07432	6.05795	6.04281	6.02874	6.01565	6.00344	5.99201
8	5.47655	5.44228	5.41163	5.38405	5.35909	5.33641	5.31571	5.29673	5.27926	5.26314	5.24822	5.23436	5.22145	5.20940	5.19813
9	4.92402	4.89019	4.85992	4.83266	4.80800	4.78556	4.76508	4.74629	4.72900	4.71303	4.69824	4.68450	4.67171	4.65976	4.64858
10	4.52045	4.48692	4.45691	4.42987	4.40539	4.38313	4.36278	4.34411	4.32693	4.31106	4.29635	4.28268	4.26995	4.25806	4.24693
11	4.21344	4.18013	4.15029	4.12340	4.09905	4.07688	4.05662	4.03803	4.02091	4.00509	3.99043	3.97680	3.96410	3.95224	3.94113
12	3.97237	3.93921	3.90950	3.88271	3.85843	3.83633	3.81612	3.79757	3.78049	3.76469	3.75005	3.73644	3.72375	3.71189	3.70079
13	3.77825	3.74520	3.71556	3.68884	3.66461	3.64254	3.62236	3.60383	3.58675	3.57096	3.55632	3.54271	3.53002	3.51815	3.50704
14	3.61868	3.58570	3.55611	3.52942	3.50522	3.48317	3.46300	3.44447	3.42739	3.41159	3.39694	3.38331	3.37060	3.35873	3.34760
15	3.48525	3.45231	3.42275	3.39608	3.37189	3.34984	3.32966	3.31112	3.29403	3.27822	3.26355	3.24990	3.23717	3.22526	3.21411
16	3.37205	3.33914	3.30960	3.28293	3.25874	3.23668	3.21649	3.19793	3.18081	3.16497	3.15028	3.13660	3.12385	3.11191	3.10073
17	3.27482	3.24193	3.21240	3.18573	3.16152	3.13944	3.11923	3.10064	3.08350	3.06764	3.05291	3.03921	3.02642	3.01445	3.00324
18	3.19043	3.15754	3.12801	3.10132	3.07710	3.05500	3.03476	3.01615	2.99897	2.98308	2.96832	2.95458	2.94176	2.92976	2.91852
19	3.11650	3.08361	3.05406	3.02736	3.00311	2.98098	2.96071	2.94207	2.92487	2.90894	2.89414	2.88037	2.86751	2.85548	2.84420
20	3.05120	3.01830	2.98873	2.96201	2.93774	2.91558	2.89528	2.87660	2.85936	2.84340	2.82857	2.81476	2.80187	2.78980	2.77848
21	2.99311	2.96019	2.93061	2.90386	2.87956	2.85737	2.83704	2.81832	2.80105	2.78505	2.77018	2.75634	2.74341	2.73131	2.71995
22	2.94109	2.90816	2.87855	2.85178	2.82745	2.80523	2.78486	2.76611	2.74880	2.73276	2.71786	2.70398	2.69102	2.67888	2.66749
23	2.89425	2.86130	2.83167	2.80487	2.78050	2.75825	2.73785	2.71907	2.70172	2.68565	2.67071	2.65679	2.64379	2.63161	2.62019
24	2.85185	2.81888	2.78923	2.76239	2.73800	2.71571	2.69527	2.67646	2.65907	2.64296	2.62799	2.61403	2.60100	2.58879	2.57733
25	2.81329	2.78030	2.75061	2.72375	2.69932	2.67701	2.65653	2.63768	2.62026	2.60411	2.58910	2.57511	2.56204	2.54980	2.53831
26	2.77807	2.74505	2.71534	2.68845	2.66399	2.64164	2.62113	2.60224	2.58479	2.56860	2.55356	2.53953	2.52643	2.51415	2.50262
27	2.74577	2.71273	2.68299	2.65607	2.63158	2.60919	2.58865	2.56973	2.55224	2.53602	2.52094	2.50688	2.49374	2.48143	2.46987
28	2.71605	2.68299	2.65322	2.62627	2.60174	2.57933	2.55875	2.53979	2.52227	2.50602	2.49090	2.47681	2.46364	2.45129	2.43970
29	2.68860	2.65552	2.62573	2.59874	2.57419	2.55174	2.53113	2.51214	2.49458	2.47830	2.46315	2.44902	2.43582	2.42344	2.41182
30	2.66319	2.63008	2.60026	2.57325	2.54866	2.52618	2.50553	2.48651	2.46892	2.45260	2.43742	2.42327	2.41003	2.39762	2.38597
31	2.63958	2.60645	2.57660	2.54956	2.52494	2.50243	2.48175	2.46270	2.44508	2.42873	2.41352	2.39933	2.38606	2.37362	2.36194
32	2.61760	2.58444	2.55457	2.52750	2.50285	2.48031	2.45960	2.44052	2.42286	2.40648	2.39124	2.37702	2.36372	2.35125	2.33954
33	2.59708	2.56390	2.53399	2.50690	2.48222	2.45965	2.43891	2.41980	2.40211	2.38570	2.37043	2.35618	2.34285	2.33035	2.31861
34	2.57788	2.54467	2.51474	2.48762	2.46292	2.44031	2.41955	2.40040	2.38269	2.36625	2.35095	2.33667	2.32331	2.31078	2.29902
35	2.55987	2.52665	2.49669	2.46954	2.44481	2.42218	2.40138	2.38221	2.36447	2.34800	2.33267	2.31836	2.30497	2.29242	2.28063
36	2.54296	2.50971	2.47973	2.45255	2.42779	2.40513	2.38431	2.36511	2.34734	2.33084	2.31548	2.30115	2.28773	2.27516	2.26333
37	2.52704	2.49376	2.46376	2.43656	2.41177	2.38909	2.36824	2.34901	2.33121	2.31468	2.29930	2.28494	2.27150	2.25889	2.24704
38	2.51202	2.47873	2.44870	2.42147	2.39666	2.37395	2.35307	2.33381	2.31599	2.29944	2.28403	2.26964	2.25617	2.24354	2.23167
39	2.49784	2.46452	2.43447	2.40722	2.38239	2.35965	2.33874	2.31946	2.30161	2.28503	2.26959	2.25518	2.24169	2.22904	2.21714
40	2.48442	2.45108	2.42101	2.39374	2.36888	2.34611	2.32518	2.30588	2.28800	2.27140	2.25593	2.24150	2.22798	2.21530	2.20338
41	2.47171	2.43835	2.40826	2.38096	2.35607	2.33329	2.31233	2.29300	2.27510	2.25847	2.24298	2.22852	2.21498	2.20228	2.19034
42	2.45965	2.42627	2.39615	2.36883	2.34392	2.32111	2.30014	2.28078	2.26285	2.24620	2.23069	2.21620	2.20264	2.18992	2.17795
43	2.44819	2.41479	2.38465	2.35731	2.33238	2.30954	2.28854	2.26916	2.25121	2.23454	2.21900	2.20450	2.19091	2.17816	2.16618
44	2.43729	2.40387	2.37371	2.34635	2.32139	2.29853	2.27751	2.25811	2.24013	2.22344	2.20788	2.19335	2.17975	2.16698	2.15497
45	2.42690	2.39347	2.36329	2.33590	2.31093	2.28805	2.26700	2.24757	2.22958	2.21286	2.19728	2.18273	2.16910	2.15631	2.14428
46	2.41700	2.38355	2.35335	2.32594	2.30094	2.27804	2.25698	2.23753	2.21951	2.20277	2.18717	2.17260	2.15895	2.14614	2.13409
47	2.40755	2.37408	2.34386	2.31643	2.29141	2.26849	2.24740	2.22793	2.20990	2.19314	2.17752	2.16292	2.14925	2.13642	2.12435
48	2.39852	2.36502	2.33479	2.30734	2.28230	2.25936	2.23825	2.21876	2.20071	2.18392	2.16828	2.15367	2.13998	2.12713	2.11504
49	2.38987	2.35637	2.32611	2.29865	2.27359	2.25063	2.22950	2.20999	2.19191	2.17511	2.15945	2.14481	2.13111	2.11824	2.10613
50	2.38160	2.34807	2.31780	2.29032	2.26524	2.24226	2.22111	2.20158	2.18349	2.16666	2.15098	2.13633	2.12261	2.10972	2.09759

Appendix 6 (continued)

df1/df2	16	17	18	19	20	21	22	23	24	25	26	27	28	29	30
51	2.37367	2.34013	2.30984	2.28234	2.25724	2.23424	2.21307	2.19352	2.17541	2.15857	2.14287	2.12820	2.11445	2.10155	2.08941
52	2.36606	2.33250	2.30220	2.27468	2.24957	2.22655	2.20536	2.18579	2.16766	2.15080	2.13508	2.12039	2.10663	2.09371	2.08155
53	2.35876	2.32519	2.29486	2.26733	2.24220	2.21916	2.19795	2.17837	2.16021	2.14334	2.12760	2.11290	2.09912	2.08618	2.07400
54	2.35174	2.31815	2.28781	2.26026	2.23511	2.21206	2.19084	2.17123	2.15306	2.13617	2.12041	2.10569	2.09189	2.07894	2.06674
55	2.34499	2.31139	2.28104	2.25347	2.22830	2.20523	2.18399	2.16437	2.14618	2.12927	2.11350	2.09876	2.08494	2.07197	2.05976
56	2.33850	2.30488	2.27451	2.24693	2.22174	2.19865	2.17740	2.15776	2.13956	2.12263	2.10684	2.09208	2.07825	2.06527	2.05304
57	2.33225	2.29861	2.26823	2.24063	2.21543	2.19232	2.17105	2.15140	2.13317	2.11623	2.10043	2.08566	2.07181	2.05880	2.04656
60	2.31480	2.28113	2.25070	2.22305	2.19781	2.17465	2.15334	2.13363	2.11536	2.09837	2.08252	2.06771	2.05381	2.04076	2.02848
65	2.28947	2.25573	2.22524	2.19752	2.17221	2.14898	2.12759	2.10782	2.08948	2.07242	2.05650	2.04161	2.02765	2.01453	2.00218
70	2.26791	2.23412	2.20356	2.17579	2.15041	2.12712	2.10567	2.08583	2.06743	2.05030	2.03432	2.01936	2.00534	1.99216	1.97975
75	2.24934	2.21550	2.18489	2.15706	2.13163	2.10828	2.08677	2.06687	2.04841	2.03123	2.01519	2.00018	1.98610	1.97286	1.96040
80	2.23318	2.19929	2.16864	2.14076	2.11527	2.09187	2.07031	2.05036	2.03185	2.01461	1.99852	1.98346	1.96933	1.95604	1.94353
85	2.21899	2.18506	2.15436	2.12643	2.10090	2.07746	2.05585	2.03585	2.01729	2.00000	1.98387	1.96876	1.95458	1.94125	1.92869
90	2.20643	2.17246	2.14172	2.11375	2.08818	2.06469	2.04304	2.02300	2.00439	1.98706	1.97088	1.95573	1.94151	1.92814	1.91554
95	2.19523	2.16122	2.13045	2.10244	2.07683	2.05330	2.03161	2.01153	1.99288	1.97552	1.95930	1.94411	1.92985	1.91644	1.90380
100	2.18518	2.15114	2.12033	2.09229	2.06665	2.04308	2.02135	2.00124	1.98256	1.96515	1.94890	1.93367	1.91938	1.90593	1.89325
110	2.16790	2.13381	2.10294	2.07484	2.04912	2.02550	2.00370	1.98352	1.96478	1.94731	1.93099	1.91570	1.90134	1.88783	1.87509
120	2.15357	2.11943	2.08851	2.06035	2.03459	2.01091	1.98906	1.96882	1.95002	1.93249	1.91612	1.90077	1.88636	1.87280	1.86001
130	2.14150	2.10731	2.07635	2.04814	2.02233	1.99860	1.97671	1.95642	1.93757	1.92000	1.90357	1.88818	1.87372	1.86011	1.84727
140	2.13118	2.09696	2.06595	2.03771	2.01186	1.98809	1.96615	1.94582	1.92693	1.90931	1.89285	1.87741	1.86291	1.84926	1.83638
150	2.12227	2.08801	2.05697	2.02870	2.00281	1.97900	1.95702	1.93666	1.91773	1.90008	1.88357	1.86810	1.85356	1.83987	1.82695
160	2.11449	2.08020	2.04914	2.02083	1.99491	1.97107	1.94905	1.92866	1.90969	1.89201	1.87547	1.85996	1.84539	1.83167	1.81872
170	2.10764	2.07333	2.04223	2.01390	1.98795	1.96408	1.94204	1.92161	1.90261	1.88490	1.86833	1.85280	1.83820	1.82444	1.81146
180	2.10157	2.06723	2.03611	2.00775	1.98177	1.95788	1.93581	1.91535	1.89633	1.87859	1.86199	1.84643	1.83181	1.81803	1.80502
190	2.09614	2.06178	2.03064	2.00226	1.97626	1.95234	1.93024	1.90976	1.89072	1.87295	1.85633	1.84075	1.82610	1.81229	1.79926
200	2.09126	2.05689	2.02573	1.99732	1.97130	1.94736	1.92524	1.90474	1.88567	1.86788	1.85124	1.83563	1.82096	1.80714	1.79408
210	2.08686	2.05247	2.02129	1.99286	1.96682	1.94286	1.92072	1.90020	1.88111	1.86330	1.84664	1.83101	1.81632	1.80247	1.78940
220	2.08286	2.04845	2.01725	1.98881	1.96275	1.93877	1.91662	1.89608	1.87697	1.85914	1.84246	1.82681	1.81210	1.79824	1.78515
230	2.07921	2.04479	2.01358	1.98512	1.95904	1.93504	1.91287	1.89232	1.87319	1.85535	1.83865	1.82298	1.80826	1.79437	1.78127
240	2.07587	2.04144	2.01021	1.98173	1.95564	1.93163	1.90945	1.88887	1.86973	1.85187	1.83516	1.81948	1.80473	1.79084	1.77771
250	2.07280	2.03835	2.00711	1.97862	1.95252	1.92849	1.90629	1.88570	1.86655	1.84867	1.83195	1.81625	1.80149	1.78758	1.77444
275	2.06612	2.03164	2.00037	1.97185	1.94571	1.92166	1.89942	1.87880	1.85962	1.84171	1.82495	1.80922	1.79443	1.78048	1.76731
300	2.06056	2.02605	1.99476	1.96621	1.94005	1.91597	1.89371	1.87306	1.85385	1.83591	1.81912	1.80337	1.78855	1.77458	1.76138
325	2.05586	2.02133	1.99002	1.96145	1.93527	1.91116	1.88888	1.86821	1.84897	1.83101	1.81420	1.79842	1.78358	1.76958	1.75636
350	2.05183	2.01729	1.98596	1.95737	1.93117	1.90704	1.88474	1.86405	1.84479	1.82681	1.80998	1.79418	1.77932	1.76530	1.75207
375	2.04835	2.01380	1.98245	1.95384	1.92762	1.90348	1.88116	1.86045	1.84118	1.82318	1.80633	1.79051	1.77563	1.76160	1.74834
400	2.04531	2.01074	1.97938	1.95076	1.92452	1.90036	1.87803	1.85731	1.83801	1.82000	1.80313	1.78730	1.77241	1.75836	1.74509
425	2.04263	2.00805	1.97667	1.94804	1.92179	1.89762	1.87527	1.85453	1.83522	1.81720	1.80032	1.78447	1.76956	1.75550	1.74221
450	2.04024	2.00565	1.97426	1.94562	1.91936	1.89518	1.87282	1.85207	1.83275	1.81471	1.79782	1.78196	1.76703	1.75296	1.73966
475	2.03811	2.00351	1.97211	1.94346	1.91719	1.89299	1.87062	1.84986	1.83053	1.81248	1.79558	1.77971	1.76477	1.75069	1.73738
500	2.03620	2.00159	1.97018	1.94152	1.91524	1.89103	1.86865	1.84788	1.82854	1.81048	1.79356	1.77769	1.76274	1.74865	1.73533
600	2.03013	1.99550	1.96406	1.93537	1.90906	1.88482	1.86241	1.84161	1.82223	1.80414	1.78719	1.77128	1.75631	1.74218	1.72883
700	2.02581	1.99116	1.95970	1.93098	1.90465	1.88039	1.85796	1.83713	1.81773	1.79962	1.78265	1.76671	1.75171	1.73756	1.72419
800	2.02257	1.98790	1.95643	1.92770	1.90135	1.87707	1.85462	1.83378	1.81436	1.79623	1.77924	1.76329	1.74827	1.73410	1.72071
900	2.02006	1.98537	1.95389	1.92515	1.89878	1.87449	1.85203	1.83117	1.81174	1.79359	1.77659	1.76063	1.74560	1.73141	1.71801
1000	2.01805	1.98335	1.95186	1.92310	1.89673	1.87243	1.84995	1.82908	1.80965	1.79149	1.77448	1.75850	1.74345	1.72926	1.71584
1100	2.01640	1.98170	1.95020	1.92144	1.89505	1.87074	1.84826	1.82738	1.80793	1.78976	1.77274	1.75676	1.74170	1.72750	1.71408
1200	2.01503	1.98032	1.94881	1.92004	1.89365	1.86934	1.84684	1.82596	1.80650	1.78833	1.77130	1.75531	1.74025	1.72604	1.71260
1300	2.01387	1.97916	1.94764	1.91887	1.89247	1.86815	1.84565	1.82476	1.80530	1.78711	1.77008	1.75408	1.73901	1.72480	1.71135
1400	2.01288	1.97816	1.94664	1.91786	1.89146	1.86713	1.84462	1.82373	1.80426	1.78607	1.76903	1.75303	1.73795	1.72373	1.71029

APPENDIXES

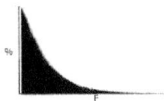

Graph of df_{1}=4 , df_{2}=11 at $\approx$ 95%; these follow 5%. **But** 1% and 99% tables are shown first.

Appendix 6 (continued, 99%)

99% Critical values of the F distribution

df1/df2	1	2	3	4	5	6	7	8	9	10	11	12	13	14	15
1	0.000247	0.010152	0.029312	0.047175	0.061508	0.072754	0.081657	0.088821	0.094684	0.099559	0.103670	0.107179	0.110207	0.112847	0.115166
2	0.000200	0.010101	0.032450	0.055556	0.075336	0.091535	0.104750	0.115619	0.124665	0.132285	0.138779	0.144371	0.149232	0.153495	0.157261
3	0.000185	0.010084	0.033948	0.059900	0.082919	0.102254	0.118325	0.131735	0.143022	0.152618	0.160856	0.167995	0.174235	0.179731	0.184605
4	0.000178	0.010076	0.034831	0.062590	0.087781	0.109310	0.127443	0.142733	0.155713	0.166824	0.176420	0.184776	0.192111	0.198595	0.204385
5	0.000174	0.010071	0.035414	0.064425	0.091182	0.114339	0.134040	0.150788	0.165100	0.177421	0.188111	0.197459	0.205693	0.212994	0.219509
6	0.000171	0.010067	0.035829	0.065760	0.093701	0.118118	0.139055	0.156969	0.172361	0.185673	0.197269	0.207444	0.216433	0.224426	0.231574
7	0.000169	0.010065	0.036138	0.066775	0.095643	0.121065	0.143004	0.161875	0.178162	0.192303	0.204663	0.215540	0.225175	0.233761	0.241456
8	0.000167	0.010063	0.036378	0.067573	0.097188	0.123432	0.146198	0.165869	0.182912	0.197758	0.210772	0.222254	0.232447	0.241549	0.249722
9	0.000166	0.010062	0.036569	0.068217	0.098447	0.125374	0.148837	0.169187	0.186876	0.202330	0.215911	0.227920	0.238602	0.248159	0.256753
10	0.000165	0.010060	0.036726	0.068748	0.099492	0.126998	0.151056	0.171990	0.190239	0.206222	0.220299	0.232772	0.243887	0.253846	0.262816
11	0.000164	0.010060	0.036856	0.069193	0.100375	0.128377	0.152948	0.174390	0.193129	0.209577	0.224093	0.236977	0.248477	0.258797	0.268104
12	0.000164	0.010059	0.036966	0.069572	0.101130	0.129562	0.154581	0.176469	0.195640	0.212501	0.227407	0.240659	0.252504	0.263148	0.272759
13	0.000163	0.010058	0.037060	0.069898	0.101783	0.130591	0.156005	0.178288	0.197843	0.215072	0.230329	0.243911	0.256069	0.267006	0.276892
14	0.000163	0.010058	0.037142	0.070182	0.102354	0.131494	0.157259	0.179893	0.199792	0.217352	0.232924	0.246806	0.259246	0.270450	0.280588
15	0.000162	0.010057	0.037213	0.070431	0.102857	0.132293	0.158370	0.181320	0.201528	0.219388	0.235246	0.249400	0.262098	0.273546	0.283914
16	0.000162	0.010057	0.037276	0.070652	0.103304	0.133004	0.159362	0.182597	0.203086	0.221217	0.237336	0.251739	0.264672	0.276344	0.286924
17	0.000162	0.010056	0.037332	0.070849	0.103704	0.133641	0.160254	0.183747	0.204491	0.222870	0.239227	0.253858	0.267009	0.278886	0.289661
18	0.000162	0.010056	0.037382	0.071025	0.104063	0.134216	0.161060	0.184787	0.205765	0.224371	0.240947	0.255787	0.269138	0.281206	0.292162
19	0.000161	0.010056	0.037427	0.071184	0.104388	0.134737	0.161791	0.185734	0.206925	0.225740	0.242518	0.257552	0.271088	0.283332	0.294457
20	0.000161	0.010055	0.037468	0.071329	0.104683	0.135211	0.162458	0.186599	0.207987	0.226994	0.243959	0.259173	0.272880	0.285289	0.296569
21	0.000161	0.010055	0.037504	0.071460	0.104952	0.135645	0.163069	0.187392	0.208962	0.228148	0.245286	0.260666	0.274533	0.287095	0.298521
22	0.000161	0.010055	0.037538	0.071580	0.105199	0.136043	0.163630	0.188122	0.209860	0.229212	0.246511	0.262046	0.276063	0.288767	0.300331
23	0.000161	0.010055	0.037569	0.071691	0.105426	0.136409	0.164148	0.188796	0.210691	0.230197	0.247645	0.263326	0.277483	0.290321	0.302012
24	0.000160	0.010055	0.037597	0.071792	0.105636	0.136748	0.164628	0.189420	0.211461	0.231111	0.248700	0.264516	0.278804	0.291767	0.303580
25	0.000160	0.010054	0.037624	0.071886	0.105830	0.137062	0.165072	0.190000	0.212178	0.231962	0.249682	0.265626	0.280036	0.293118	0.305044
26	0.000160	0.010054	0.037648	0.071973	0.106010	0.137354	0.165486	0.190541	0.212846	0.232756	0.250600	0.266663	0.281189	0.294383	0.306415
27	0.000160	0.010054	0.037671	0.072054	0.106178	0.137626	0.165872	0.191045	0.213470	0.233498	0.251458	0.267635	0.282270	0.295568	0.307702
28	0.000160	0.010054	0.037692	0.072130	0.106334	0.137880	0.166233	0.191518	0.214054	0.234195	0.252264	0.268547	0.283285	0.296683	0.308912
29	0.000160	0.010054	0.037711	0.072201	0.106480	0.138118	0.166571	0.191961	0.214603	0.234849	0.253021	0.269404	0.284240	0.297732	0.310052
30	0.000160	0.010054	0.037729	0.072267	0.106617	0.138341	0.166889	0.192377	0.215119	0.235464	0.253734	0.270213	0.285141	0.298722	0.311128
31	0.000160	0.010054	0.037747	0.072329	0.106746	0.138550	0.167188	0.192769	0.215606	0.236044	0.254407	0.270976	0.285991	0.299657	0.312145
32	0.000160	0.010053	0.037763	0.072387	0.106867	0.138748	0.167469	0.193138	0.216065	0.236592	0.255043	0.271697	0.286796	0.300542	0.313108
33	0.000159	0.010053	0.037778	0.072442	0.106982	0.138934	0.167735	0.193488	0.216499	0.237111	0.255644	0.272380	0.287558	0.301381	0.314021
34	0.000159	0.010053	0.037792	0.072493	0.107090	0.139111	0.167987	0.193818	0.216909	0.237602	0.256214	0.273028	0.288281	0.302178	0.314889
35	0.000159	0.010053	0.037806	0.072542	0.107192	0.139277	0.168225	0.194131	0.217299	0.238068	0.256756	0.273643	0.288968	0.302934	0.315713
36	0.000159	0.010053	0.037818	0.072589	0.107288	0.139435	0.168451	0.194429	0.217669	0.238510	0.257270	0.274228	0.289622	0.303655	0.316498
37	0.000159	0.010053	0.037831	0.072633	0.107380	0.139585	0.168666	0.194711	0.218021	0.238931	0.257760	0.274784	0.290244	0.304341	0.317246
38	0.000159	0.010053	0.037842	0.072674	0.107467	0.139728	0.168870	0.194980	0.218355	0.239332	0.258226	0.275315	0.290838	0.304996	0.317960
39	0.000159	0.010053	0.037853	0.072714	0.107550	0.139864	0.169064	0.195236	0.218674	0.239715	0.258671	0.275822	0.291404	0.305621	0.318642
40	0.000159	0.010053	0.037863	0.072752	0.107629	0.139993	0.169249	0.195480	0.218979	0.240080	0.259096	0.276306	0.291946	0.306219	0.319294
41	0.000159	0.010053	0.037873	0.072788	0.107704	0.140116	0.169426	0.195713	0.219270	0.240428	0.259502	0.276768	0.292464	0.306791	0.319919
42	0.000159	0.010053	0.037883	0.072822	0.107776	0.140234	0.169595	0.195936	0.219548	0.240762	0.259891	0.277211	0.292960	0.307338	0.320517
43	0.000159	0.010053	0.037892	0.072855	0.107844	0.140347	0.169756	0.196149	0.219814	0.241081	0.260263	0.277636	0.293435	0.307864	0.321091
44	0.000159	0.010053	0.037900	0.072886	0.107910	0.140454	0.169911	0.196354	0.220069	0.241388	0.260620	0.278043	0.293892	0.308368	0.321641
45	0.000159	0.010053	0.037908	0.072916	0.107973	0.140557	0.170059	0.196549	0.220314	0.241681	0.260963	0.278433	0.294329	0.308852	0.322170
46	0.000159	0.010053	0.037916	0.072945	0.108033	0.140656	0.170201	0.196737	0.220548	0.241963	0.261292	0.278809	0.294750	0.309317	0.322679
47	0.000159	0.010052	0.037924	0.072972	0.108091	0.140751	0.170338	0.196917	0.220774	0.242234	0.261608	0.279170	0.295155	0.309764	0.323168
48	0.000159	0.010052	0.037931	0.072999	0.108146	0.140843	0.170469	0.197091	0.220991	0.242495	0.261912	0.279517	0.295544	0.310195	0.323639
49	0.000159	0.010052	0.037938	0.073024	0.108200	0.140930	0.170595	0.197257	0.221199	0.242745	0.262205	0.279851	0.295919	0.310610	0.324093
50	0.000159	0.010052	0.037945	0.073048	0.108251	0.141014	0.170716	0.197418	0.221400	0.242987	0.262487	0.280173	0.296281	0.311010	0.324531

Appendix 6 (continued)

df1 /df2	1	2	3	4	5	6	7	8	9	10	11	12	13	14	15
51	0.0001586	0.010052	0.037951	0.073072	0.108300	0.141095	0.170833	0.197572	0.221593	0.243219	0.262759	0.280484	0.296629	0.311396	0.324953
52	0.0001586	0.010052	0.037957	0.073094	0.108347	0.141174	0.170945	0.197721	0.221780	0.243444	0.263021	0.280783	0.296966	0.311769	0.325361
53	0.0001586	0.010052	0.037963	0.073116	0.108393	0.141249	0.171054	0.197865	0.221960	0.243660	0.263274	0.281072	0.297291	0.312128	0.325755
54	0.0001585	0.010052	0.037969	0.073137	0.108437	0.141322	0.171158	0.198004	0.222133	0.243869	0.263518	0.281352	0.297605	0.312476	0.326135
55	0.0001585	0.010052	0.037974	0.073157	0.108480	0.141392	0.171259	0.198137	0.222301	0.244071	0.263754	0.281622	0.297908	0.312812	0.326504
56	0.0001585	0.010052	0.037980	0.073177	0.108521	0.141459	0.171357	0.198267	0.222463	0.244267	0.263983	0.281883	0.298202	0.313138	0.326860
57	0.0001585	0.010052	0.037985	0.073195	0.108561	0.141525	0.171451	0.198392	0.222620	0.244455	0.264204	0.282136	0.298486	0.313452	0.327205
60	0.0001584	0.010052	0.037999	0.073248	0.108672	0.141709	0.171716	0.198743	0.223061	0.244987	0.264826	0.282848	0.299286	0.314339	0.328177
65	0.0001583	0.010052	0.038020	0.073326	0.108835	0.141978	0.172106	0.199260	0.223710	0.245770	0.265743	0.283898	0.300468	0.315650	0.329615
70	0.0001582	0.010052	0.038038	0.073392	0.108976	0.142211	0.172442	0.199707	0.224271	0.246447	0.266537	0.284808	0.301492	0.316788	0.330863
75	0.0001581	0.010052	0.038054	0.073450	0.109098	0.142414	0.172735	0.200097	0.224761	0.247039	0.267231	0.285604	0.302389	0.317785	0.331958
80	0.0001581	0.010052	0.038068	0.073501	0.109206	0.142592	0.172992	0.200439	0.225192	0.247560	0.267843	0.286306	0.303181	0.318665	0.332925
85	0.0001580	0.010052	0.038080	0.073546	0.109301	0.142749	0.173221	0.200743	0.225575	0.248023	0.268387	0.286931	0.303885	0.319448	0.333786
90	0.0001580	0.010051	0.038091	0.073586	0.109386	0.142890	0.173424	0.201015	0.225917	0.248437	0.268873	0.287489	0.304516	0.320149	0.334557
95	0.0001579	0.010051	0.038101	0.073622	0.109462	0.143016	0.173607	0.201259	0.226224	0.248809	0.269310	0.287992	0.305083	0.320781	0.335252
100	0.0001579	0.010051	0.038109	0.073654	0.109530	0.143130	0.173772	0.201479	0.226502	0.249145	0.269706	0.288447	0.305597	0.321353	0.335881
110	0.0001578	0.010051	0.038125	0.073710	0.109649	0.143327	0.174059	0.201861	0.226984	0.249730	0.270393	0.289238	0.306491	0.322349	0.336978
120	0.0001577	0.010051	0.038137	0.073757	0.109748	0.143492	0.174298	0.202181	0.227388	0.250220	0.270971	0.289902	0.307242	0.323186	0.337900
130	0.0001577	0.010051	0.038148	0.073796	0.109832	0.143632	0.174502	0.202454	0.227732	0.250637	0.271462	0.290468	0.307883	0.323900	0.338687
140	0.0001576	0.010051	0.038157	0.073830	0.109905	0.143753	0.174677	0.202688	0.228028	0.250997	0.271886	0.290956	0.308435	0.324516	0.339367
150	0.0001576	0.010051	0.038165	0.073860	0.109968	0.143857	0.174830	0.202892	0.228285	0.251310	0.272255	0.291381	0.308916	0.325053	0.339959
160	0.0001576	0.010051	0.038172	0.073886	0.110023	0.143949	0.174963	0.203070	0.228511	0.251584	0.272578	0.291755	0.309339	0.325526	0.340480
170	0.0001576	0.010051	0.038178	0.073909	0.110071	0.144030	0.175081	0.203229	0.228711	0.251827	0.272865	0.292086	0.309714	0.325944	0.340942
180	0.0001575	0.010051	0.038184	0.073929	0.110115	0.144103	0.175187	0.203369	0.228890	0.252044	0.273121	0.292381	0.310048	0.326318	0.341354
190	0.0001575	0.010051	0.038189	0.073947	0.110153	0.144167	0.175281	0.203496	0.229050	0.252239	0.273351	0.292645	0.310348	0.326653	0.341725
200	0.0001575	0.010051	0.038193	0.073963	0.110188	0.144226	0.175366	0.203610	0.229194	0.252414	0.273558	0.292884	0.310619	0.326956	0.342059
210	0.0001575	0.010051	0.038197	0.073978	0.110220	0.144278	0.175443	0.203713	0.229325	0.252573	0.273746	0.293101	0.310865	0.327231	0.342363
220	0.0001574	0.010051	0.038201	0.073992	0.110249	0.144327	0.175513	0.203807	0.229444	0.252718	0.273917	0.293299	0.311089	0.327481	0.342639
230	0.0001574	0.010051	0.038204	0.074004	0.110275	0.144371	0.175577	0.203893	0.229552	0.252850	0.274073	0.293479	0.311294	0.327710	0.342892
240	0.0001574	0.010051	0.038207	0.074015	0.110299	0.144411	0.175636	0.203972	0.229652	0.252972	0.274217	0.293645	0.311482	0.327921	0.343125
250	0.0001574	0.010051	0.038210	0.074026	0.110322	0.144448	0.175690	0.204044	0.229744	0.253084	0.274349	0.293798	0.311656	0.328115	0.343340
275	0.0001574	0.010051	0.038216	0.074049	0.110370	0.144529	0.175808	0.204203	0.229946	0.253329	0.274639	0.294133	0.312036	0.328539	0.343809
300	0.0001573	0.010051	0.038221	0.074068	0.110411	0.144597	0.175907	0.204336	0.230114	0.253534	0.274881	0.294413	0.312353	0.328895	0.344202
325	0.0001573	0.010051	0.038225	0.074084	0.110445	0.144654	0.175991	0.204448	0.230256	0.253707	0.275086	0.294650	0.312623	0.329196	0.344536
350	0.0001573	0.010051	0.038229	0.074097	0.110474	0.144703	0.176063	0.204544	0.230379	0.253857	0.275263	0.294854	0.312854	0.329456	0.344822
375	0.0001573	0.010051	0.038232	0.074109	0.110500	0.144746	0.176125	0.204628	0.230485	0.253986	0.275416	0.295031	0.313055	0.329681	0.345072
400	0.0001573	0.010051	0.038235	0.074120	0.110522	0.144784	0.176180	0.204702	0.230578	0.254100	0.275550	0.295186	0.313232	0.329878	0.345290
425	0.0001573	0.010051	0.038237	0.074129	0.110542	0.144817	0.176228	0.204766	0.230660	0.254200	0.275669	0.295324	0.313388	0.330053	0.345483
450	0.0001573	0.010051	0.038240	0.074137	0.110560	0.144846	0.176271	0.204824	0.230734	0.254289	0.275774	0.295446	0.313527	0.330208	0.345655
475	0.0001573	0.010051	0.038242	0.074145	0.110575	0.144872	0.176310	0.204876	0.230799	0.254369	0.275869	0.295555	0.313651	0.330348	0.345809
500	0.0001572	0.010051	0.038243	0.074151	0.110590	0.144896	0.176344	0.204922	0.230858	0.254441	0.275954	0.295654	0.313763	0.330473	0.345948
600	0.0001572	0.010051	0.038249	0.074172	0.110634	0.144971	0.176454	0.205070	0.231045	0.254669	0.276225	0.295967	0.314119	0.330872	0.346389
700	0.0001572	0.010050	0.038253	0.074187	0.110666	0.145025	0.176533	0.205175	0.231179	0.254833	0.276418	0.296191	0.314373	0.331157	0.346706
800	0.0001572	0.010050	0.038256	0.074198	0.110691	0.145065	0.176591	0.205255	0.231280	0.254956	0.276564	0.296359	0.314565	0.331372	0.346944
900	0.0001572	0.010050	0.038258	0.074207	0.110709	0.145097	0.176637	0.205316	0.231359	0.255051	0.276677	0.296491	0.314714	0.331539	0.347129
1000	0.0001572	0.010050	0.038260	0.074214	0.110724	0.145122	0.176674	0.205366	0.231421	0.255128	0.276768	0.296596	0.314834	0.331673	0.347277
1100	0.0001572	0.010050	0.038262	0.074220	0.110737	0.145142	0.176704	0.205406	0.231473	0.255191	0.276842	0.296682	0.314932	0.331783	0.347399
1200	0.0001572	0.010050	0.038263	0.074225	0.110747	0.145159	0.176729	0.205440	0.231516	0.255243	0.276904	0.296754	0.315013	0.331875	0.347500
1300	0.0001571	0.010050	0.038264	0.074229	0.110755	0.145174	0.176751	0.205468	0.231552	0.255288	0.276957	0.296815	0.315083	0.331952	0.347586
1400	0.0001571	0.010050	0.038265	0.074232	0.110763	0.145186	0.176769	0.205493	0.231583	0.255326	0.277002	0.296867	0.315142	0.332019	0.347660

Appendix 6 (continued)

							99%	Critical values of the F distribution						
df1 /df2	16	17	18	19	20	21	22	23	24	25	26	27	28	29	30
1	0.117220	0.119051	0.120694	0.122176	0.123518	0.124741	0.125859	0.126885	0.127830	0.128703	0.129513	0.130265	0.130965	0.131619	0.132232
2	0.160611	0.163610	0.166309	0.168751	0.170971	0.172998	0.174855	0.176563	0.178139	0.179598	0.180952	0.182212	0.183387	0.184487	0.185517
3	0.188957	0.192864	0.196391	0.199589	0.202503	0.205168	0.207615	0.209869	0.211952	0.213882	0.215677	0.217348	0.218910	0.220371	0.221742
4	0.209530	0.214180	0.218387	0.222209	0.225698	0.228895	0.231834	0.234545	0.237054	0.239382	0.241548	0.243568	0.245457	0.247227	0.248888
5	0.225356	0.230631	0.235411	0.239764	0.243743	0.247393	0.250755	0.253859	0.256735	0.259406	0.261894	0.264217	0.266390	0.268428	0.270342
6	0.238003	0.243813	0.249089	0.253899	0.258303	0.262348	0.266078	0.269526	0.272724	0.275697	0.278468	0.281058	0.283482	0.285757	0.287896
7	0.248389	0.254666	0.260373	0.265585	0.270362	0.274756	0.278811	0.282564	0.286047	0.289289	0.292313	0.295140	0.297790	0.300278	0.302618
8	0.257098	0.263785	0.269875	0.275442	0.280551	0.285255	0.289600	0.293626	0.297365	0.300848	0.304099	0.307142	0.309994	0.312674	0.315197
9	0.264521	0.271574	0.278004	0.283889	0.289295	0.294278	0.298885	0.303156	0.307127	0.310828	0.314285	0.317522	0.320559	0.323414	0.326103
10	0.270934	0.278314	0.285050	0.291221	0.296896	0.302130	0.306974	0.311468	0.315650	0.319549	0.323195	0.326610	0.329816	0.332831	0.335673
11	0.276537	0.284211	0.291223	0.297654	0.303572	0.309036	0.314096	0.318794	0.323168	0.327250	0.331068	0.334646	0.338008	0.341171	0.344152
12	0.281477	0.289419	0.296682	0.303350	0.309490	0.315164	0.320421	0.325306	0.329857	0.334107	0.338084	0.341813	0.345318	0.348618	0.351729
13	0.285870	0.294055	0.301548	0.308432	0.314776	0.320643	0.326082	0.331140	0.335854	0.340258	0.344382	0.348252	0.351889	0.355316	0.358548
14	0.289802	0.298211	0.305915	0.312997	0.319530	0.325574	0.331182	0.336398	0.341264	0.345812	0.350072	0.354072	0.357833	0.361378	0.364723
15	0.293346	0.301960	0.309858	0.317124	0.323830	0.330038	0.335802	0.341167	0.346173	0.350854	0.355242	0.359363	0.363240	0.366895	0.370346
16	0.296556	0.305360	0.313437	0.320873	0.327740	0.334101	0.340010	0.345512	0.350649	0.355456	0.359962	0.364196	0.368182	0.371940	0.375490
17	0.299479	0.308458	0.316702	0.324295	0.331312	0.337816	0.343860	0.349491	0.354750	0.359673	0.364291	0.368632	0.372719	0.376574	0.380217
18	0.302151	0.311294	0.319693	0.327433	0.334590	0.341226	0.347397	0.353148	0.358522	0.363555	0.368278	0.372718	0.376900	0.380847	0.384577
19	0.304606	0.313900	0.322443	0.330321	0.337609	0.344369	0.350658	0.356523	0.362005	0.367140	0.371961	0.376496	0.380768	0.384801	0.388614
20	0.306867	0.316304	0.324982	0.332988	0.340398	0.347276	0.353676	0.359647	0.365230	0.370463	0.375377	0.380000	0.384358	0.388472	0.392363
21	0.308959	0.318528	0.327332	0.335460	0.342985	0.349972	0.356477	0.362548	0.368227	0.373552	0.378553	0.383260	0.387698	0.391890	0.395855
22	0.310898	0.320592	0.329516	0.337756	0.345390	0.352481	0.359085	0.365250	0.371020	0.376431	0.381515	0.386301	0.390816	0.395081	0.399116
23	0.312703	0.322514	0.331549	0.339897	0.347632	0.354821	0.361518	0.367773	0.373628	0.379121	0.384284	0.389146	0.393733	0.398067	0.402170
24	0.314385	0.324307	0.333447	0.341896	0.349728	0.357009	0.363795	0.370135	0.376071	0.381642	0.386879	0.391813	0.396469	0.400869	0.405035
25	0.315958	0.325984	0.335224	0.343768	0.351692	0.359060	0.365930	0.372350	0.378363	0.384008	0.389316	0.394318	0.399040	0.403503	0.407730
26	0.317432	0.327556	0.336891	0.345525	0.353535	0.360987	0.367936	0.374433	0.380519	0.386234	0.391610	0.396677	0.401461	0.405984	0.410269
27	0.318816	0.329033	0.338457	0.347178	0.355270	0.362800	0.369825	0.376394	0.382551	0.388332	0.393773	0.398902	0.403745	0.408326	0.412666
28	0.320118	0.330424	0.339933	0.348734	0.356905	0.364510	0.371607	0.378245	0.384468	0.390314	0.395816	0.401004	0.405904	0.410540	0.414933
29	0.321346	0.331735	0.341325	0.350204	0.358448	0.366125	0.373291	0.379995	0.386281	0.392188	0.397749	0.402993	0.407948	0.412637	0.417080
30	0.322504	0.332974	0.342640	0.351593	0.359908	0.367653	0.374884	0.381651	0.387999	0.393964	0.399581	0.404879	0.409886	0.414625	0.419117
31	0.323600	0.334146	0.343885	0.352908	0.361291	0.369101	0.376395	0.383222	0.389627	0.395648	0.401319	0.406670	0.411727	0.416514	0.421052
32	0.324639	0.335256	0.345065	0.354155	0.362603	0.370475	0.377829	0.384714	0.391175	0.397249	0.402971	0.408372	0.413477	0.418310	0.422893
33	0.325623	0.336310	0.346186	0.355340	0.363850	0.371781	0.379192	0.386132	0.392646	0.398772	0.404544	0.409992	0.415143	0.420021	0.424647
34	0.326559	0.337312	0.347251	0.356467	0.365035	0.373024	0.380490	0.387483	0.394048	0.400223	0.406042	0.411536	0.416731	0.421652	0.426320
35	0.327449	0.338265	0.348265	0.357539	0.366165	0.374208	0.381726	0.388770	0.395384	0.401606	0.407471	0.413009	0.418247	0.423209	0.427917
36	0.328296	0.339172	0.349231	0.358562	0.367242	0.375337	0.382906	0.389999	0.396660	0.402927	0.408836	0.414417	0.419696	0.424697	0.429443
37	0.329104	0.340038	0.350153	0.359538	0.368270	0.376415	0.384033	0.391173	0.397879	0.404190	0.410141	0.415763	0.421081	0.426121	0.430904
38	0.329876	0.340865	0.351034	0.360470	0.369252	0.377446	0.385110	0.392295	0.399045	0.405399	0.411391	0.417051	0.422408	0.427485	0.432303
39	0.330613	0.341656	0.351876	0.361362	0.370192	0.378433	0.386142	0.393370	0.400162	0.406556	0.412587	0.418286	0.423679	0.428792	0.433645
40	0.331318	0.342412	0.352681	0.362216	0.371092	0.379378	0.387130	0.394400	0.401232	0.407666	0.413735	0.419470	0.424899	0.430045	0.434932
41	0.331993	0.343136	0.353453	0.363034	0.371955	0.380283	0.388078	0.395388	0.402259	0.408731	0.414836	0.420607	0.426070	0.431250	0.436168
42	0.332640	0.343831	0.354194	0.363819	0.372783	0.381153	0.388987	0.396337	0.403246	0.409753	0.415894	0.421699	0.427195	0.432407	0.437357
43	0.333261	0.344497	0.354904	0.364572	0.373577	0.381988	0.389861	0.397248	0.404193	0.410736	0.416911	0.422749	0.428277	0.433520	0.438500
44	0.333857	0.345137	0.355587	0.365296	0.374341	0.382790	0.390701	0.398124	0.405105	0.411682	0.417890	0.423759	0.429319	0.434592	0.439601
45	0.334429	0.345752	0.356243	0.365992	0.375076	0.383562	0.391509	0.398968	0.405982	0.412592	0.418832	0.424733	0.430322	0.435624	0.440661
46	0.334980	0.346344	0.356874	0.366661	0.375783	0.384305	0.392287	0.399780	0.406827	0.413469	0.419740	0.425670	0.431289	0.436619	0.441684
47	0.335510	0.346913	0.357482	0.367306	0.376464	0.385021	0.393037	0.400562	0.407642	0.414314	0.420615	0.426575	0.432221	0.437579	0.442670
48	0.336020	0.347462	0.358068	0.367928	0.377120	0.385712	0.393760	0.401317	0.408427	0.415130	0.421460	0.427448	0.433121	0.438506	0.443623
49	0.336512	0.347990	0.358632	0.368527	0.377753	0.386377	0.394458	0.402046	0.409186	0.415917	0.422275	0.428290	0.433991	0.439401	0.444543
50	0.336987	0.348500	0.359177	0.369106	0.378364	0.387020	0.395131	0.402749	0.409918	0.416678	0.423063	0.429105	0.434831	0.440266	0.445432

Appendix 6 (continued)

							99%	Critical values of the F distribution						
df1 /df2	16	17	18	19	20	21	22	23	24	25	26	27	28	29	30
51	0.337444	0.348993	0.359703	0.369664	0.378955	0.387641	0.395782	0.403429	0.410626	0.417413	0.423824	0.429892	0.435643	0.441102	0.446292
52	0.337887	0.349468	0.360211	0.370204	0.379525	0.388241	0.396411	0.404086	0.411310	0.418124	0.424561	0.430654	0.436429	0.441912	0.447125
53	0.338314	0.349928	0.360702	0.370725	0.380076	0.388822	0.397020	0.404722	0.411972	0.418811	0.425274	0.431391	0.437190	0.442696	0.447930
54	0.338727	0.350372	0.361177	0.371230	0.380610	0.389383	0.397608	0.405337	0.412613	0.419477	0.425964	0.432105	0.437926	0.443455	0.448711
55	0.339126	0.350802	0.361636	0.371718	0.381126	0.389927	0.398178	0.405933	0.413234	0.420122	0.426633	0.432796	0.438640	0.444190	0.449468
56	0.339513	0.351218	0.362081	0.372191	0.381626	0.390453	0.398731	0.406510	0.413836	0.420748	0.427281	0.433467	0.439333	0.444904	0.450202
57	0.339887	0.351621	0.362512	0.372649	0.382111	0.390964	0.399266	0.407070	0.414419	0.421354	0.427910	0.434117	0.440004	0.445596	0.450914
60	0.340942	0.352758	0.363728	0.373942	0.383478	0.392404	0.400778	0.408650	0.416067	0.423067	0.429686	0.435956	0.441903	0.447553	0.452929
65	0.342504	0.354440	0.365528	0.375858	0.385507	0.394542	0.403021	0.410998	0.418515	0.425613	0.432328	0.438691	0.444729	0.450467	0.455929
70	0.343861	0.355904	0.367095	0.377526	0.387273	0.396404	0.404978	0.413046	0.420652	0.427838	0.434637	0.441082	0.447201	0.453018	0.458556
75	0.345051	0.357188	0.368471	0.378991	0.388826	0.398043	0.406700	0.414849	0.422535	0.429797	0.436672	0.443191	0.449381	0.455269	0.460875
80	0.346104	0.358324	0.369689	0.380289	0.390202	0.399495	0.408227	0.416448	0.424205	0.431538	0.438480	0.445065	0.451320	0.457270	0.462939
85	0.347041	0.359336	0.370774	0.381446	0.391430	0.400791	0.409590	0.417877	0.425698	0.433093	0.440097	0.446741	0.453055	0.459062	0.464786
90	0.347881	0.360243	0.371748	0.382485	0.392532	0.401955	0.410815	0.419161	0.427040	0.434492	0.441551	0.448250	0.454616	0.460675	0.466450
95	0.348638	0.361062	0.372626	0.383422	0.393526	0.403006	0.411921	0.420321	0.428253	0.435756	0.442866	0.449614	0.456029	0.462136	0.467957
100	0.349324	0.361803	0.373423	0.384272	0.394429	0.403960	0.412925	0.421375	0.429355	0.436905	0.444061	0.450855	0.457314	0.463464	0.469327
110	0.350520	0.363096	0.374812	0.385755	0.396004	0.405626	0.414679	0.423216	0.431281	0.438915	0.446152	0.453026	0.459563	0.465790	0.471728
120	0.351526	0.364186	0.375982	0.387005	0.397333	0.407031	0.416160	0.424771	0.432909	0.440614	0.447921	0.454863	0.461467	0.467760	0.473763
130	0.352385	0.365116	0.376982	0.388074	0.398469	0.408234	0.417428	0.426102	0.434303	0.442069	0.449437	0.456438	0.463100	0.469449	0.475508
140	0.353127	0.365919	0.377846	0.388998	0.399451	0.409274	0.418525	0.427255	0.435510	0.443330	0.450750	0.457803	0.464516	0.470915	0.477023
150	0.353774	0.366620	0.378600	0.389804	0.400309	0.410183	0.419483	0.428263	0.436566	0.444433	0.451900	0.458998	0.465756	0.472199	0.478350
160	0.354344	0.367237	0.379265	0.390515	0.401065	0.410984	0.420328	0.429151	0.437497	0.445406	0.452914	0.460052	0.466850	0.473332	0.479522
170	0.354849	0.367785	0.379854	0.391145	0.401737	0.411695	0.421079	0.429940	0.438324	0.446270	0.453815	0.460990	0.467823	0.474340	0.480564
180	0.355299	0.368273	0.380380	0.391708	0.402336	0.412330	0.421750	0.430646	0.439064	0.447044	0.454622	0.461829	0.468694	0.475243	0.481498
190	0.355704	0.368712	0.380853	0.392215	0.402875	0.412902	0.422353	0.431280	0.439729	0.447740	0.455348	0.462584	0.469479	0.476056	0.482339
200	0.356070	0.369109	0.381280	0.392672	0.403363	0.413419	0.422899	0.431854	0.440331	0.448370	0.456004	0.463268	0.470189	0.476792	0.483100
210	0.356402	0.369469	0.381668	0.393088	0.403805	0.413888	0.423394	0.432376	0.440879	0.448942	0.456602	0.463890	0.470834	0.477461	0.483793
220	0.356705	0.369798	0.382022	0.393467	0.404209	0.414316	0.423847	0.432852	0.441378	0.449464	0.457147	0.464457	0.471424	0.478072	0.484425
230	0.356982	0.370098	0.382346	0.393814	0.404579	0.414709	0.424261	0.433289	0.441836	0.449943	0.457647	0.464978	0.471964	0.478633	0.485005
240	0.357236	0.370375	0.382644	0.394133	0.404919	0.415069	0.424642	0.433690	0.442257	0.450384	0.458107	0.465456	0.472462	0.479149	0.485540
250	0.357471	0.370630	0.382919	0.394427	0.405233	0.415402	0.424994	0.434060	0.442646	0.450791	0.458531	0.465899	0.472921	0.479625	0.486033
275	0.357985	0.371188	0.383520	0.395072	0.405920	0.416132	0.425765	0.434872	0.443498	0.451683	0.459462	0.466868	0.473929	0.480671	0.487115
300	0.358415	0.371655	0.384024	0.395612	0.406496	0.416743	0.426411	0.435552	0.444212	0.452431	0.460243	0.467682	0.474775	0.481548	0.488024
325	0.358781	0.372052	0.384452	0.396071	0.406985	0.417262	0.426960	0.436131	0.444820	0.453067	0.460908	0.468374	0.475495	0.482295	0.488798
350	0.359095	0.372393	0.384821	0.396466	0.407407	0.417709	0.427433	0.436629	0.445343	0.453615	0.461480	0.468971	0.476115	0.482939	0.489465
375	0.359368	0.372690	0.385141	0.396809	0.407773	0.418098	0.427844	0.437063	0.445798	0.454092	0.461978	0.469490	0.476655	0.483499	0.490045
400	0.359607	0.372950	0.385421	0.397110	0.408094	0.418439	0.428205	0.437443	0.446198	0.454510	0.462416	0.469946	0.477129	0.483992	0.490556
425	0.359819	0.373180	0.385670	0.397377	0.408378	0.418741	0.428525	0.437780	0.446552	0.454881	0.462803	0.470350	0.477549	0.484427	0.491007
450	0.360007	0.373385	0.385891	0.397614	0.408631	0.419010	0.428809	0.438080	0.446867	0.455211	0.463148	0.470709	0.477924	0.484816	0.491410
475	0.360176	0.373569	0.386089	0.397827	0.408858	0.419252	0.429064	0.438349	0.447150	0.455507	0.463458	0.471032	0.478259	0.485165	0.491771
500	0.360329	0.373734	0.386268	0.398018	0.409063	0.419469	0.429295	0.438591	0.447405	0.455775	0.463737	0.471323	0.478562	0.485479	0.492098
600	0.360812	0.374260	0.386836	0.398628	0.409714	0.420160	0.430026	0.439363	0.448215	0.456624	0.464625	0.472250	0.479526	0.486481	0.493136
700	0.361159	0.374637	0.387243	0.399065	0.410180	0.420656	0.430551	0.439917	0.448798	0.457235	0.465263	0.472915	0.480219	0.487200	0.493882
800	0.361420	0.374921	0.387549	0.399394	0.410532	0.421030	0.430947	0.440334	0.449236	0.457694	0.465744	0.473417	0.480741	0.487743	0.494444
900	0.361623	0.375142	0.387788	0.399650	0.410806	0.421321	0.431255	0.440659	0.449578	0.458053	0.466119	0.473808	0.481149	0.488166	0.494883
1000	0.361786	0.375319	0.387980	0.399856	0.411025	0.421554	0.431502	0.440920	0.449853	0.458341	0.466420	0.474122	0.481475	0.488506	0.495236
1100	0.361920	0.375464	0.388137	0.400024	0.411205	0.421746	0.431705	0.441134	0.450077	0.458577	0.466667	0.474379	0.481743	0.488784	0.495525
1200	0.362031	0.375586	0.388267	0.400165	0.411355	0.421906	0.431874	0.441312	0.450265	0.458773	0.466873	0.474594	0.481967	0.489017	0.495766
1300	0.362125	0.375688	0.388378	0.400284	0.411483	0.422041	0.432017	0.441463	0.450424	0.458940	0.467047	0.474776	0.482157	0.489214	0.495970
1400	0.362206	0.375776	0.388473	0.400386	0.411592	0.422157	0.432140	0.441593	0.450561	0.459083	0.467197	0.474933	0.482320	0.489383	0.496146

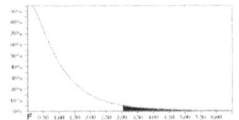

Graph of $df_1 = 4$, $df_2 = 11$ at approx 5%

Appendix 6 (continued 5%)

5% Critical values of the F distribution

df1/df2	1	2	3	4	5	6	7	8	9	10	11	12	13	14	15
1	161.448	199.500	215.707	224.583	230.162	233.986	236.768	238.883	240.543	241.882	242.983	243.906	244.690	245.364	245.950
2	18.5128	19.0000	19.1643	19.2468	19.2964	19.3295	19.3532	19.3710	19.3848	19.3959	19.4050	19.4125	19.4189	19.4244	19.4291
3	10.1280	9.55209	9.27663	9.11718	9.01346	8.94065	8.88674	8.84524	8.81230	8.78552	8.76333	8.74464	8.72868	8.71490	8.70287
4	7.70865	6.94427	6.59138	6.38823	6.25606	6.16313	6.09421	6.04104	5.99878	5.96437	5.93581	5.91173	5.89114	5.87335	5.85781
5	6.60789	5.78614	5.40945	5.19217	5.05033	4.95029	4.87587	4.81832	4.77247	4.73506	4.70397	4.67770	4.65523	4.63577	4.61876
6	5.98738	5.14325	4.75706	4.53368	4.38737	4.28387	4.20666	4.14680	4.09902	4.05996	4.02744	3.99994	3.97636	3.95593	3.93806
7	5.59145	4.73741	4.34683	4.12031	3.97152	3.86597	3.78704	3.72573	3.67667	3.63652	3.60304	3.57468	3.55034	3.52923	3.51074
8	5.31766	4.45897	4.06618	3.83785	3.68750	3.58058	3.50046	3.43810	3.38813	3.34716	3.31295	3.28394	3.25902	3.23738	3.21841
9	5.11736	4.25649	3.86255	3.63309	3.48166	3.37375	3.29275	3.22958	3.17889	3.13728	3.10249	3.07295	3.04755	3.02547	3.00610
10	4.96460	4.10282	3.70826	3.47805	3.32583	3.21717	3.13546	3.07166	3.02038	2.97824	2.94296	2.91298	2.88717	2.86473	2.84502
11	4.84434	3.98230	3.58743	3.35669	3.20387	3.09461	3.01233	2.94799	2.89622	2.85362	2.81793	2.78757	2.76142	2.73865	2.71864
12	4.74723	3.88529	3.49029	3.25917	3.10588	2.99612	2.91336	2.84857	2.79638	2.75339	2.71733	2.68664	2.66018	2.63712	2.61685
13	4.66719	3.80557	3.41053	3.17912	3.02544	2.91527	2.83210	2.76691	2.71436	2.67102	2.63465	2.60366	2.57693	2.55362	2.53311
14	4.60011	3.73889	3.34389	3.11225	2.95825	2.84773	2.76420	2.69867	2.64579	2.60216	2.56550	2.53424	2.50726	2.48373	2.46300
15	4.54308	3.68232	3.28738	3.05557	2.90129	2.79046	2.70663	2.64080	2.58763	2.54372	2.50681	2.47531	2.44811	2.42436	2.40345
16	4.49400	3.63372	3.23887	3.00692	2.85241	2.74131	2.65720	2.59110	2.53767	2.49351	2.45637	2.42466	2.39725	2.37332	2.35222
17	4.45132	3.59153	3.19678	2.96471	2.81000	2.69866	2.61430	2.54796	2.49429	2.44992	2.41256	2.38065	2.35306	2.32895	2.30769
18	4.41387	3.55456	3.15991	2.92774	2.77285	2.66130	2.57672	2.51016	2.45628	2.41170	2.37416	2.34207	2.31430	2.29003	2.26862
19	4.38075	3.52189	3.12735	2.89511	2.74006	2.62832	2.54353	2.47677	2.42270	2.37793	2.34021	2.30795	2.28003	2.25561	2.23406
20	4.35124	3.49283	3.09839	2.86608	2.71089	2.59898	2.51401	2.44706	2.39281	2.34788	2.30999	2.27758	2.24951	2.22496	2.20327
21	4.32479	3.46680	3.07247	2.84010	2.68478	2.57271	2.48758	2.42046	2.36605	2.32095	2.28292	2.25036	2.22216	2.19747	2.17567
22	4.30095	3.44336	3.04912	2.81671	2.66127	2.54906	2.46377	2.39650	2.34194	2.29670	2.25852	2.22583	2.19750	2.17269	2.15078
23	4.27934	3.42213	3.02800	2.79554	2.64000	2.52766	2.44223	2.37481	2.32011	2.27473	2.23642	2.20361	2.17516	2.15024	2.12822
24	4.25968	3.40283	3.00879	2.77629	2.62065	2.50819	2.42263	2.35508	2.30024	2.25474	2.21631	2.18338	2.15482	2.12980	2.10767
25	4.24170	3.38519	2.99124	2.75871	2.60299	2.49041	2.40473	2.33706	2.28210	2.23647	2.19793	2.16489	2.13623	2.11111	2.08889
26	4.22520	3.36902	2.97515	2.74259	2.58679	2.47411	2.38831	2.32053	2.26545	2.21972	2.18107	2.14793	2.11917	2.09395	2.07164
27	4.21001	3.35413	2.96035	2.72777	2.57189	2.45911	2.37321	2.30531	2.25013	2.20429	2.16554	2.13230	2.10345	2.07815	2.05575
28	4.19597	3.34039	2.94669	2.71408	2.55813	2.44526	2.35926	2.29126	2.23598	2.19004	2.15120	2.11787	2.08893	2.06354	2.04107
29	4.18296	3.32765	2.93403	2.70140	2.54539	2.43243	2.34634	2.27825	2.22287	2.17684	2.13791	2.10449	2.07547	2.05000	2.02746
30	4.17088	3.31583	2.92228	2.68963	2.53355	2.42052	2.33434	2.26616	2.21070	2.16458	2.12556	2.09206	2.06296	2.03742	2.01480
31	4.15962	3.30482	2.91133	2.67867	2.52254	2.40943	2.32317	2.25491	2.19936	2.15316	2.11405	2.08048	2.05131	2.02569	2.00301
32	4.14910	3.29454	2.90112	2.66844	2.51225	2.39908	2.31274	2.24440	2.18877	2.14249	2.10331	2.06966	2.04042	2.01474	1.99199
33	4.13925	3.28492	2.89156	2.65887	2.50264	2.38939	2.30298	2.23456	2.17886	2.13250	2.09325	2.05954	2.03023	2.00448	1.98167
34	4.13002	3.27590	2.88260	2.64989	2.49362	2.38031	2.29383	2.22534	2.16956	2.12314	2.08382	2.05004	2.02066	1.99486	1.97199
35	4.12134	3.26742	2.87419	2.64147	2.48514	2.37178	2.28524	2.21668	2.16083	2.11434	2.07496	2.04111	2.01167	1.98581	1.96288
36	4.11317	3.25945	2.86627	2.63353	2.47717	2.36375	2.27714	2.20852	2.15261	2.10605	2.06661	2.03270	2.00321	1.97729	1.95431
37	4.10546	3.25192	2.85880	2.62605	2.46965	2.35618	2.26951	2.20083	2.14485	2.09824	2.05873	2.02477	1.99522	1.96925	1.94622
38	4.09817	3.24482	2.85174	2.61899	2.46255	2.34903	2.26230	2.19356	2.13753	2.09086	2.05129	2.01728	1.98767	1.96165	1.93857
39	4.09128	3.23810	2.84507	2.61231	2.45583	2.34226	2.25549	2.18668	2.13060	2.08387	2.04425	2.01018	1.98053	1.95445	1.93133
40	4.08475	3.23173	2.83875	2.60597	2.44947	2.33585	2.24902	2.18017	2.12403	2.07725	2.03758	2.00346	1.97376	1.94764	1.92446
41	4.07855	3.22568	2.83275	2.59997	2.44343	2.32977	2.24289	2.17399	2.11780	2.07096	2.03125	1.99708	1.96733	1.94116	1.91795
42	4.07265	3.21994	2.82705	2.59426	2.43769	2.32399	2.23707	2.16812	2.11187	2.06499	2.02523	1.99101	1.96122	1.93501	1.91175
43	4.06705	3.21448	2.82163	2.58884	2.43224	2.31850	2.23153	2.16253	2.10624	2.05931	2.01950	1.98524	1.95540	1.92915	1.90585
44	4.06171	3.20928	2.81647	2.58367	2.42704	2.31326	2.22625	2.15721	2.10087	2.05390	2.01405	1.97974	1.94986	1.92357	1.90024
45	4.05661	3.20432	2.81154	2.57874	2.42209	2.30827	2.22122	2.15213	2.09576	2.04874	2.00884	1.97450	1.94458	1.91825	1.89487
46	4.05175	3.19958	2.80684	2.57404	2.41736	2.30351	2.21642	2.14729	2.09087	2.04381	2.00387	1.96949	1.93953	1.91316	1.88975
47	4.04710	3.19506	2.80236	2.56954	2.41284	2.29896	2.21183	2.14266	2.08620	2.03910	1.99912	1.96470	1.93471	1.90830	1.88486
48	4.04265	3.19073	2.79806	2.56524	2.40851	2.29460	2.20744	2.13823	2.08173	2.03459	1.99458	1.96012	1.93009	1.90365	1.88017
49	4.03839	3.18658	2.79395	2.56112	2.40438	2.29043	2.20323	2.13399	2.07745	2.03028	1.99023	1.95573	1.92567	1.89920	1.87569
50	4.03431	3.18261	2.79001	2.55718	2.40041	2.28644	2.19920	2.12992	2.07335	2.02614	1.98606	1.95153	1.92143	1.89493	1.87138

Appendix 6 (continued)

df1/df2	16	17	18	19	20	21	22	23	24	25	26	27	28	29	30
1	246.464	246.918	247.323	247.686	248.013	248.309	248.579	248.826	249.052	249.260	249.453	249.631	249.797	249.951	250.095
2	19.4333	19.4370	19.4402	19.4431	19.4458	19.4481	19.4503	19.4523	19.4541	19.4558	19.4573	19.4587	19.4600	19.4613	19.4624
3	8.69229	8.68290	8.67452	8.66699	8.66019	8.65402	8.64839	8.64324	8.63850	8.63414	8.63010	8.62635	8.62286	8.61961	8.61658
4	5.84412	5.83197	5.82112	5.81136	5.80254	5.79453	5.78723	5.78054	5.77439	5.76872	5.76347	5.75859	5.75406	5.74983	5.74588
5	4.60376	4.59044	4.57853	4.56782	4.55813	4.54933	4.54129	4.53393	4.52715	4.52090	4.51512	4.50974	4.50474	4.50008	4.49571
6	3.92228	3.90826	3.89571	3.88441	3.87419	3.86489	3.85640	3.84862	3.84146	3.83484	3.82872	3.82303	3.81774	3.81279	3.80816
7	3.49441	3.47988	3.46686	3.45514	3.44452	3.43487	3.42604	3.41795	3.41049	3.40361	3.39723	3.39131	3.38579	3.38063	3.37581
8	3.20163	3.18670	3.17332	3.16125	3.15032	3.14037	3.13128	3.12293	3.11524	3.10813	3.10155	3.09543	3.08972	3.08440	3.07941
9	2.98897	2.97370	2.96000	2.94765	2.93646	2.92626	2.91693	2.90837	2.90047	2.89318	2.88641	2.88012	2.87426	2.86878	2.86365
10	2.82757	2.81201	2.79805	2.78545	2.77402	2.76360	2.75407	2.74532	2.73725	2.72978	2.72286	2.71642	2.71042	2.70481	2.69955
11	2.70091	2.68510	2.67090	2.65808	2.64645	2.63584	2.62613	2.61720	2.60897	2.60136	2.59430	2.58772	2.58159	2.57586	2.57049
12	2.59888	2.58284	2.56843	2.55541	2.54359	2.53281	2.52293	2.51386	2.50548	2.49773	2.49054	2.48384	2.47760	2.47176	2.46628
13	2.51492	2.49867	2.48407	2.47087	2.45888	2.44794	2.43792	2.42870	2.42020	2.41232	2.40501	2.39820	2.39185	2.38591	2.38033
14	2.44461	2.42818	2.41340	2.40004	2.38790	2.37681	2.36665	2.35731	2.34868	2.34069	2.33327	2.32636	2.31991	2.31387	2.30821
15	2.38488	2.36827	2.35333	2.33982	2.32754	2.31632	2.30603	2.29657	2.28783	2.27973	2.27221	2.26520	2.25866	2.25253	2.24679
16	2.33348	2.31672	2.30164	2.28798	2.27557	2.26423	2.25383	2.24425	2.23541	2.22721	2.21959	2.21250	2.20587	2.19966	2.19384
17	2.28880	2.27189	2.25667	2.24289	2.23035	2.21890	2.20839	2.19871	2.18977	2.18148	2.17377	2.16659	2.15988	2.15360	2.14771
18	2.24959	2.23255	2.21720	2.20330	2.19065	2.17909	2.16847	2.15870	2.14966	2.14129	2.13350	2.12624	2.11946	2.11311	2.10714
19	2.21490	2.19773	2.18226	2.16825	2.15550	2.14383	2.13313	2.12326	2.11414	2.10569	2.09782	2.09049	2.08363	2.07721	2.07119
20	2.18398	2.16670	2.15112	2.13701	2.12416	2.11240	2.10160	2.09165	2.08245	2.07392	2.06598	2.05858	2.05166	2.04517	2.03909
21	2.15626	2.13887	2.12319	2.10898	2.09603	2.08419	2.07331	2.06328	2.05400	2.04540	2.03739	2.02992	2.02294	2.01639	2.01025
22	2.13126	2.11377	2.09799	2.08369	2.07066	2.05873	2.04777	2.03767	2.02832	2.01964	2.01157	2.00404	1.99700	1.99040	1.98420
23	2.10860	2.09101	2.07515	2.06075	2.04764	2.03563	2.02460	2.01442	2.00501	1.99627	1.98814	1.98055	1.97345	1.96679	1.96054
24	2.08796	2.07028	2.05433	2.03986	2.02666	2.01458	2.00348	1.99324	1.98376	1.97496	1.96677	1.95912	1.95197	1.94526	1.93896
25	2.06909	2.05132	2.03529	2.02074	2.00747	1.99532	1.98415	1.97385	1.96431	1.95545	1.94720	1.93950	1.93229	1.92554	1.91919
26	2.05176	2.03391	2.01780	2.00318	1.98984	1.97763	1.96639	1.95603	1.94643	1.93751	1.92921	1.92146	1.91421	1.90740	1.90101
27	2.03579	2.01787	2.00169	1.98699	1.97359	1.96131	1.95002	1.93959	1.92994	1.92097	1.91262	1.90482	1.89752	1.89067	1.88424
28	2.02103	2.00304	1.98678	1.97203	1.95856	1.94622	1.93487	1.92439	1.91469	1.90567	1.89727	1.88942	1.88208	1.87519	1.86871
29	2.00735	1.98928	1.97297	1.95815	1.94462	1.93222	1.92082	1.91029	1.90053	1.89147	1.88302	1.87513	1.86774	1.86081	1.85429
30	1.99462	1.97650	1.96012	1.94524	1.93165	1.91920	1.90775	1.89716	1.88736	1.87825	1.86976	1.86183	1.85440	1.84743	1.84087
31	1.98276	1.96457	1.94813	1.93320	1.91956	1.90706	1.89555	1.88492	1.87507	1.86592	1.85739	1.84941	1.84195	1.83494	1.82834
32	1.97168	1.95343	1.93694	1.92195	1.90826	1.89571	1.88415	1.87348	1.86358	1.85439	1.84581	1.83780	1.83030	1.82325	1.81662
33	1.96130	1.94300	1.92645	1.91141	1.89767	1.88507	1.87347	1.86275	1.85281	1.84358	1.83496	1.82692	1.81938	1.81230	1.80564
34	1.95157	1.93321	1.91660	1.90151	1.88773	1.87508	1.86344	1.85268	1.84270	1.83342	1.82478	1.81669	1.80912	1.80200	1.79531
35	1.94241	1.92400	1.90735	1.89221	1.87838	1.86569	1.85400	1.84320	1.83318	1.82387	1.81519	1.80707	1.79946	1.79231	1.78559
36	1.93378	1.91532	1.89862	1.88344	1.86956	1.85683	1.84510	1.83427	1.82421	1.81486	1.80615	1.79799	1.79035	1.78318	1.77642
37	1.92564	1.90713	1.89039	1.87516	1.86124	1.84847	1.83671	1.82583	1.81574	1.80636	1.79761	1.78942	1.78175	1.77455	1.76776
38	1.91794	1.89939	1.88260	1.86733	1.85338	1.84057	1.82876	1.81785	1.80773	1.79831	1.78953	1.78131	1.77361	1.76638	1.75957
39	1.91066	1.89206	1.87523	1.85992	1.84593	1.83308	1.82124	1.81029	1.80014	1.79069	1.78187	1.77363	1.76590	1.75864	1.75180
40	1.90375	1.88511	1.86824	1.85289	1.83886	1.82598	1.81410	1.80312	1.79294	1.78346	1.77461	1.76634	1.75858	1.75129	1.74443
41	1.89719	1.87851	1.86160	1.84622	1.83215	1.81923	1.80733	1.79631	1.78610	1.77659	1.76771	1.75941	1.75163	1.74431	1.73743
42	1.89096	1.87224	1.85529	1.83987	1.82577	1.81282	1.80088	1.78984	1.77959	1.77005	1.76115	1.75282	1.74501	1.73767	1.73076
43	1.88502	1.86627	1.84928	1.83383	1.81969	1.80671	1.79474	1.78367	1.77339	1.76383	1.75490	1.74654	1.73871	1.73135	1.72441
44	1.87936	1.86057	1.84356	1.82807	1.81390	1.80089	1.78889	1.77779	1.76748	1.75789	1.74893	1.74056	1.73270	1.72531	1.71835
45	1.87397	1.85514	1.83809	1.82257	1.80837	1.79533	1.78330	1.77217	1.76184	1.75222	1.74324	1.73484	1.72696	1.71955	1.71257
46	1.86881	1.84995	1.83287	1.81732	1.80309	1.79002	1.77796	1.76681	1.75645	1.74680	1.73780	1.72938	1.72147	1.71404	1.70704
47	1.86388	1.84499	1.82788	1.81230	1.79804	1.78494	1.77286	1.76167	1.75129	1.74162	1.73260	1.72415	1.71622	1.70877	1.70175
48	1.85917	1.84025	1.82310	1.80749	1.79320	1.78008	1.76797	1.75676	1.74635	1.73666	1.72761	1.71914	1.71119	1.70372	1.69668
49	1.85465	1.83570	1.81852	1.80288	1.78857	1.77542	1.76328	1.75205	1.74162	1.73191	1.72283	1.71434	1.70637	1.69888	1.69182
50	1.85031	1.83133	1.81413	1.79846	1.78412	1.77095	1.75879	1.74753	1.73708	1.72734	1.71825	1.70974	1.70175	1.69424	1.68716

APPENDIXES

5% Critical values of the F distribution

df1/df2	1	2	3	4	5	6	7	8	9	10	11	12	13	14	15
51	4.03039	3.17880	2.78623	2.55340	2.39660	2.28260	2.19534	2.12602	2.06942	2.02217	1.98205	1.94749	1.91736	1.89083	1.86725
52	4.02663	3.17514	2.78260	2.54976	2.39295	2.27892	2.19163	2.12228	2.06564	2.01836	1.97821	1.94362	1.91346	1.88689	1.86329
53	4.02302	3.17163	2.77911	2.54627	2.38944	2.27539	2.18806	2.11868	2.06201	2.01470	1.97452	1.93989	1.90970	1.88311	1.85948
54	4.01954	3.16825	2.77576	2.54292	2.38607	2.27199	2.18463	2.11522	2.05852	2.01118	1.97097	1.93631	1.90609	1.87947	1.85581
55	4.01620	3.16499	2.77254	2.53969	2.38282	2.26872	2.18133	2.11189	2.05516	2.00779	1.96755	1.93286	1.90261	1.87596	1.85228
56	4.01297	3.16186	2.76943	2.53658	2.37970	2.26557	2.17816	2.10869	2.05193	2.00453	1.96425	1.92954	1.89926	1.87259	1.84888
57	4.00987	3.15884	2.76644	2.53358	2.37668	2.26253	2.17509	2.10560	2.04881	2.00138	1.96108	1.92634	1.89604	1.86934	1.84560
60	4.00119	3.15041	2.75808	2.52522	2.36827	2.25405	2.16654	2.09697	2.04010	1.99259	1.95221	1.91740	1.88702	1.86024	1.83644
65	3.98856	3.13814	2.74592	2.51304	2.35603	2.24172	2.15410	2.08441	2.02742	1.97980	1.93930	1.90437	1.87388	1.84700	1.82309
70	3.97778	3.12768	2.73554	2.50266	2.34559	2.23119	2.14348	2.07369	2.01660	1.96887	1.92828	1.89325	1.86266	1.83568	1.81168
75	3.96847	3.11864	2.72659	2.49370	2.33658	2.22211	2.13431	2.06444	2.00726	1.95945	1.91876	1.88364	1.85297	1.82591	1.80182
80	3.96035	3.11077	2.71878	2.48588	2.32872	2.21419	2.12632	2.05637	1.99911	1.95122	1.91046	1.87526	1.84451	1.81738	1.79322
85	3.95321	3.10384	2.71192	2.47902	2.32181	2.20723	2.11930	2.04928	1.99195	1.94398	1.90315	1.86789	1.83707	1.80987	1.78565
90	3.94688	3.09770	2.70584	2.47293	2.31569	2.20106	2.11307	2.04299	1.98559	1.93757	1.89667	1.86134	1.83047	1.80321	1.77893
95	3.94122	3.09222	2.70041	2.46749	2.31022	2.19555	2.10751	2.03737	1.97992	1.93184	1.89088	1.85550	1.82457	1.79726	1.77292
100	3.93614	3.08730	2.69553	2.46261	2.30532	2.19060	2.10251	2.03233	1.97483	1.92669	1.88569	1.85026	1.81927	1.79191	1.76753
110	3.92739	3.07882	2.68714	2.45421	2.29687	2.18208	2.09391	2.02364	1.96605	1.91783	1.87673	1.84121	1.81014	1.78269	1.75823
120	3.92012	3.07178	2.68017	2.44724	2.28985	2.17501	2.08677	2.01643	1.95876	1.91046	1.86929	1.83370	1.80255	1.77503	1.75050
130	3.91399	3.06584	2.67429	2.44135	2.28393	2.16904	2.08074	2.01034	1.95261	1.90424	1.86301	1.82735	1.79614	1.76856	1.74397
140	3.90874	3.06076	2.66926	2.43632	2.27887	2.16393	2.07559	2.00513	1.94735	1.89893	1.85763	1.82192	1.79066	1.76302	1.73838
150	3.90420	3.05637	2.66491	2.43197	2.27449	2.15952	2.07113	2.00062	1.94280	1.89432	1.85299	1.81722	1.78591	1.75823	1.73354
160	3.90024	3.05253	2.66111	2.42816	2.27067	2.15566	2.06724	1.99669	1.93882	1.89031	1.84892	1.81312	1.78177	1.75404	1.72931
170	3.89674	3.04915	2.65776	2.42482	2.26730	2.15226	2.06381	1.99322	1.93532	1.88676	1.84534	1.80950	1.77811	1.75035	1.72558
180	3.89364	3.04615	2.65479	2.42184	2.26431	2.14925	2.06076	1.99015	1.93220	1.88362	1.84216	1.80629	1.77487	1.74707	1.72227
190	3.89087	3.04347	2.65214	2.41919	2.26164	2.14655	2.05804	1.98740	1.92942	1.88081	1.83932	1.80342	1.77197	1.74414	1.71931
200	3.88837	3.04106	2.64975	2.41680	2.25924	2.14413	2.05559	1.98492	1.92693	1.87828	1.83677	1.80084	1.76936	1.74151	1.71665
210	3.88612	3.03888	2.64760	2.41464	2.25707	2.14194	2.05338	1.98269	1.92467	1.87600	1.83446	1.79850	1.76700	1.73912	1.71424
220	3.88407	3.03690	2.64564	2.41268	2.25509	2.13995	2.05137	1.98066	1.92261	1.87392	1.83236	1.79638	1.76486	1.73696	1.71205
230	3.88221	3.03509	2.64385	2.41089	2.25330	2.13814	2.04954	1.97881	1.92074	1.87203	1.83045	1.79445	1.76290	1.73498	1.71006
240	3.88050	3.03344	2.64221	2.40926	2.25165	2.13648	2.04786	1.97711	1.91903	1.87030	1.82870	1.79267	1.76111	1.73317	1.70823
250	3.87892	3.03192	2.64071	2.40775	2.25013	2.13495	2.04632	1.97555	1.91745	1.86870	1.82708	1.79104	1.75946	1.73151	1.70655
275	3.87550	3.02860	2.63743	2.40447	2.24683	2.13162	2.04296	1.97215	1.91401	1.86522	1.82357	1.78749	1.75587	1.72788	1.70288
300	3.87264	3.02585	2.63470	2.40174	2.24409	2.12885	2.04016	1.96932	1.91115	1.86233	1.82064	1.78453	1.75288	1.72486	1.69983
325	3.87023	3.02352	2.63239	2.39943	2.24177	2.12651	2.03779	1.96693	1.90873	1.85989	1.81817	1.78203	1.75035	1.72230	1.69725
350	3.86816	3.02152	2.63042	2.39746	2.23978	2.12450	2.03577	1.96488	1.90666	1.85779	1.81605	1.77989	1.74819	1.72011	1.69504
375	3.86638	3.01979	2.62871	2.39574	2.23806	2.12277	2.03401	1.96311	1.90487	1.85598	1.81422	1.77803	1.74631	1.71822	1.69312
400	3.86481	3.01828	2.62721	2.39425	2.23655	2.12125	2.03248	1.96156	1.90330	1.85439	1.81261	1.77641	1.74467	1.71656	1.69144
425	3.86343	3.01695	2.62590	2.39293	2.23522	2.11991	2.03113	1.96019	1.90192	1.85299	1.81120	1.77498	1.74322	1.71510	1.68997
450	3.86221	3.01576	2.62472	2.39176	2.23405	2.11872	2.02993	1.95898	1.90069	1.85175	1.80994	1.77371	1.74194	1.71380	1.68865
475	3.86111	3.01471	2.62368	2.39071	2.23299	2.11766	2.02885	1.95789	1.89959	1.85064	1.80882	1.77257	1.74079	1.71263	1.68748
500	3.86012	3.01375	2.62273	2.38977	2.23204	2.11670	2.02788	1.95691	1.89860	1.84964	1.80780	1.77155	1.73975	1.71159	1.68642
600	3.85700	3.01074	2.61975	2.38678	2.22904	2.11367	2.02483	1.95382	1.89547	1.84647	1.80460	1.76831	1.73648	1.70828	1.68308
700	3.85478	3.00859	2.61763	2.38466	2.22690	2.11151	2.02264	1.95161	1.89324	1.84421	1.80232	1.76600	1.73414	1.70592	1.68069
800	3.85311	3.00698	2.61603	2.38306	2.22530	2.10990	2.02101	1.94996	1.89157	1.84252	1.80061	1.76427	1.73239	1.70415	1.67890
900	3.85181	3.00573	2.61479	2.38182	2.22405	2.10864	2.01974	1.94867	1.89027	1.84121	1.79928	1.76292	1.73103	1.70277	1.67751
1000	3.85077	3.00472	2.61380	2.38083	2.22305	2.10763	2.01872	1.94765	1.88923	1.84015	1.79821	1.76185	1.72994	1.70167	1.67639
1100	3.84993	3.00391	2.61299	2.38002	2.22224	2.10681	2.01789	1.94681	1.88838	1.83929	1.79734	1.76097	1.72905	1.70077	1.67548
1200	3.84922	3.00322	2.61232	2.37935	2.22156	2.10612	2.01720	1.94610	1.88767	1.83858	1.79662	1.76023	1.72831	1.70002	1.67472
1300	3.84862	3.00265	2.61175	2.37877	2.22098	2.10554	2.01661	1.94551	1.88707	1.83797	1.79600	1.75961	1.72768	1.69938	1.67408
1400	3.84811	3.00215	2.61126	2.37828	2.22049	2.10505	2.01611	1.94500	1.88655	1.83745	1.79548	1.75908	1.72714	1.69884	1.67353

Appendix 6 (continued)

df1/df2	16	17	18	19	20	21	22	23	24	25	26	27	28	29	30
51	1.84616	1.82715	1.80992	1.79422	1.77986	1.76666	1.75447	1.74320	1.73272	1.72296	1.71385	1.70532	1.69731	1.68978	1.68268
52	1.84216	1.82313	1.80587	1.79015	1.77576	1.76253	1.75033	1.73903	1.72853	1.71875	1.70962	1.70107	1.69304	1.68549	1.67837
53	1.83832	1.81926	1.80198	1.78623	1.77182	1.75857	1.74634	1.73502	1.72450	1.71470	1.70555	1.69698	1.68893	1.68137	1.67423
54	1.83463	1.81554	1.79824	1.78247	1.76803	1.75476	1.74251	1.73116	1.72063	1.71081	1.70164	1.69305	1.68498	1.67740	1.67025
55	1.83107	1.81196	1.79463	1.77884	1.76438	1.75108	1.73881	1.72745	1.71689	1.70706	1.69786	1.68926	1.68118	1.67357	1.66641
56	1.82765	1.80851	1.79116	1.77534	1.76086	1.74755	1.73526	1.72387	1.71329	1.70344	1.69423	1.68560	1.67751	1.66989	1.66271
57	1.82435	1.80518	1.78781	1.77197	1.75747	1.74413	1.73182	1.72042	1.70982	1.69995	1.69072	1.68208	1.67397	1.66633	1.65914
60	1.81511	1.79589	1.77845	1.76255	1.74798	1.73459	1.72222	1.71077	1.70012	1.69019	1.68091	1.67222	1.66406	1.65638	1.64914
65	1.80166	1.78233	1.76480	1.74881	1.73415	1.72067	1.70822	1.69668	1.68595	1.67595	1.66660	1.65783	1.64960	1.64185	1.63454
70	1.79017	1.77075	1.75313	1.73706	1.72233	1.70877	1.69624	1.68463	1.67383	1.66376	1.65434	1.64551	1.63722	1.62941	1.62204
75	1.78023	1.76074	1.74304	1.72690	1.71210	1.69847	1.68588	1.67420	1.66334	1.65321	1.64373	1.63484	1.62649	1.61863	1.61121
80	1.77156	1.75200	1.73424	1.71803	1.70316	1.68947	1.67682	1.66509	1.65417	1.64398	1.63445	1.62551	1.61711	1.60920	1.60173
85	1.76392	1.74430	1.72648	1.71021	1.69529	1.68154	1.66884	1.65705	1.64608	1.63585	1.62627	1.61729	1.60884	1.60088	1.59337
90	1.75714	1.73747	1.71959	1.70327	1.68830	1.67450	1.66175	1.64992	1.63890	1.62862	1.61900	1.60997	1.60149	1.59349	1.58594
95	1.75109	1.73136	1.71344	1.69707	1.68205	1.66821	1.65542	1.64354	1.63248	1.62216	1.61250	1.60344	1.59491	1.58687	1.57929
100	1.74565	1.72588	1.70791	1.69150	1.67643	1.66256	1.64972	1.63780	1.62671	1.61635	1.60665	1.59755	1.58899	1.58092	1.57330
110	1.73627	1.71641	1.69837	1.68188	1.66674	1.65279	1.63988	1.62790	1.61674	1.60631	1.59655	1.58739	1.57877	1.57064	1.56296
120	1.72846	1.70854	1.69043	1.67388	1.65868	1.64467	1.63170	1.61966	1.60844	1.59796	1.58814	1.57892	1.57025	1.56207	1.55434
130	1.72187	1.70189	1.68373	1.66712	1.65186	1.63780	1.62478	1.61268	1.60142	1.59089	1.58102	1.57176	1.56304	1.55482	1.54705
140	1.71623	1.69620	1.67798	1.66133	1.64603	1.63191	1.61885	1.60671	1.59540	1.58483	1.57492	1.56562	1.55686	1.54860	1.54079
150	1.71135	1.69127	1.67301	1.65631	1.64097	1.62682	1.61371	1.60154	1.59019	1.57958	1.56964	1.56030	1.55151	1.54321	1.53537
160	1.70708	1.68697	1.66867	1.65193	1.63655	1.62236	1.60922	1.59701	1.58563	1.57499	1.56501	1.55564	1.54682	1.53849	1.53062
170	1.70332	1.68317	1.66484	1.64807	1.63265	1.61843	1.60526	1.59302	1.58161	1.57094	1.56093	1.55153	1.54268	1.53433	1.52643
180	1.69997	1.67980	1.66143	1.64463	1.62919	1.61494	1.60174	1.58947	1.57803	1.56733	1.55731	1.54788	1.53900	1.53063	1.52270
190	1.69698	1.67678	1.65839	1.64156	1.62609	1.61182	1.59859	1.58630	1.57483	1.56411	1.55406	1.54461	1.53571	1.52731	1.51937
200	1.69430	1.67407	1.65565	1.63880	1.62331	1.60901	1.59576	1.58344	1.57196	1.56121	1.55114	1.54167	1.53275	1.52433	1.51637
210	1.69187	1.67161	1.65317	1.63630	1.62079	1.60647	1.59320	1.58086	1.56935	1.55859	1.54850	1.53901	1.53007	1.52163	1.51365
220	1.68966	1.66938	1.65092	1.63403	1.61850	1.60416	1.59087	1.57851	1.56699	1.55620	1.54609	1.53659	1.52763	1.51918	1.51118
230	1.68764	1.66735	1.64887	1.63196	1.61640	1.60205	1.58874	1.57637	1.56482	1.55403	1.54390	1.53438	1.52541	1.51694	1.50892
240	1.68580	1.66548	1.64699	1.63006	1.61449	1.60011	1.58679	1.57440	1.56284	1.55203	1.54189	1.53235	1.52337	1.51488	1.50685
250	1.68410	1.66377	1.64526	1.62831	1.61273	1.59834	1.58500	1.57259	1.56102	1.55019	1.54004	1.53049	1.52149	1.51299	1.50495
275	1.68039	1.66003	1.64148	1.62450	1.60888	1.59446	1.58109	1.56865	1.55705	1.54619	1.53600	1.52642	1.51739	1.50886	1.50079
300	1.67731	1.65691	1.63834	1.62133	1.60568	1.59123	1.57783	1.56536	1.55373	1.54285	1.53264	1.52303	1.51398	1.50542	1.49733
325	1.67470	1.65428	1.63568	1.61864	1.60297	1.58849	1.57507	1.56258	1.55093	1.54002	1.52979	1.52016	1.51109	1.50251	1.49440
350	1.67247	1.65202	1.63340	1.61634	1.60065	1.58615	1.57271	1.56020	1.54853	1.53760	1.52735	1.51770	1.50861	1.50001	1.49188
375	1.67053	1.65007	1.63142	1.61435	1.59864	1.58412	1.57066	1.55814	1.54645	1.53550	1.52523	1.51557	1.50646	1.49785	1.48970
400	1.66884	1.64836	1.62970	1.61261	1.59688	1.58235	1.56887	1.55633	1.54462	1.53367	1.52338	1.51370	1.50458	1.49596	1.48779
425	1.66735	1.64685	1.62817	1.61107	1.59533	1.58078	1.56729	1.55474	1.54302	1.53205	1.52175	1.51206	1.50292	1.49428	1.48611
450	1.66602	1.64551	1.62682	1.60970	1.59395	1.57939	1.56589	1.55332	1.54159	1.53060	1.52029	1.51059	1.50144	1.49280	1.48461
475	1.66483	1.64431	1.62561	1.60848	1.59271	1.57814	1.56463	1.55205	1.54031	1.52932	1.51899	1.50928	1.50012	1.49147	1.48327
500	1.66376	1.64323	1.62452	1.60738	1.59160	1.57702	1.56350	1.55091	1.53916	1.52815	1.51782	1.50810	1.49894	1.49027	1.48206
600	1.66038	1.63982	1.62107	1.60390	1.58809	1.57347	1.55992	1.54730	1.53552	1.52448	1.51412	1.50437	1.49517	1.48648	1.47824
700	1.65797	1.63738	1.61861	1.60141	1.58557	1.57094	1.55736	1.54472	1.53291	1.52186	1.51147	1.50170	1.49248	1.48377	1.47551
800	1.65616	1.63555	1.61676	1.59954	1.58369	1.56904	1.55544	1.54278	1.53096	1.51989	1.50949	1.49970	1.49046	1.48173	1.47346
900	1.65475	1.63413	1.61533	1.59809	1.58223	1.56756	1.55395	1.54128	1.52944	1.51836	1.50794	1.49814	1.48889	1.48015	1.47187
1000	1.65363	1.63299	1.61418	1.59693	1.58106	1.56638	1.55276	1.54007	1.52823	1.51713	1.50671	1.49690	1.48764	1.47889	1.47059
1100	1.65271	1.63206	1.61324	1.59598	1.58010	1.56541	1.55178	1.53909	1.52723	1.51613	1.50570	1.49588	1.48661	1.47785	1.46955
1200	1.65194	1.63129	1.61246	1.59519	1.57930	1.56461	1.55097	1.53827	1.52640	1.51529	1.50485	1.49503	1.48575	1.47699	1.46868
1300	1.65129	1.63063	1.61179	1.59453	1.57862	1.56392	1.55028	1.53757	1.52570	1.51459	1.50414	1.49431	1.48503	1.47626	1.46794
1400	1.65074	1.63007	1.61123	1.59395	1.57805	1.56334	1.54969	1.53698	1.52510	1.51398	1.50353	1.49369	1.48441	1.47563	1.46731

Graph of $df_1 = 4$, $df_2 = 11$ at approx 95%

Appendix 6 (continued 95%)

95% Critical values of the F distribution

df1 / df2	1	2	3	4	5	6	7	8	9	10	11	12	13	14	15
1	0.006194	0.054017	0.09874	0.12972	0.15133	0.16702	0.17884	0.18805	0.19541	0.20143	0.20643	0.21065	0.21426	0.21739	0.22012
2	0.005013	0.052632	0.10469	0.14400	0.17283	0.19443	0.21109	0.22427	0.23494	0.24373	0.25111	0.25738	0.26277	0.26746	0.27157
3	0.004636	0.052180	0.10780	0.15171	0.18486	0.21021	0.23005	0.24593	0.25890	0.26967	0.27875	0.28651	0.29321	0.29905	0.30419
4	0.004453	0.051957	0.10968	0.15654	0.19260	0.22057	0.24270	0.26056	0.27525	0.28752	0.29791	0.30683	0.31455	0.32131	0.32727
5	0.004345	0.051823	0.11095	0.15985	0.19801	0.22793	0.25179	0.27119	0.28722	0.30068	0.31212	0.32197	0.33053	0.33804	0.34467
6	0.004274	0.051734	0.11185	0.16226	0.20201	0.23343	0.25867	0.27928	0.29641	0.31083	0.32314	0.33376	0.34302	0.35116	0.35836
7	0.004224	0.051671	0.11253	0.16409	0.20509	0.23772	0.26406	0.28568	0.30370	0.31893	0.33197	0.34325	0.35310	0.36177	0.36946
8	0.004186	0.051624	0.11306	0.16553	0.20754	0.24115	0.26840	0.29086	0.30964	0.32556	0.33921	0.35105	0.36141	0.37055	0.37867
9	0.004157	0.051587	0.11348	0.16670	0.20954	0.24396	0.27198	0.29515	0.31457	0.33108	0.34528	0.35761	0.36841	0.37796	0.38645
10	0.004134	0.051557	0.11382	0.16766	0.21119	0.24631	0.27499	0.29876	0.31875	0.33577	0.35043	0.36319	0.37439	0.38430	0.39313
11	0.004116	0.051533	0.11411	0.16847	0.21259	0.24830	0.27754	0.30185	0.32232	0.33979	0.35487	0.36801	0.37956	0.38979	0.39891
12	0.004100	0.051513	0.11436	0.16916	0.21378	0.25000	0.27975	0.30451	0.32542	0.34329	0.35874	0.37221	0.38407	0.39460	0.40399
13	0.004087	0.051496	0.11456	0.16975	0.21481	0.25149	0.28166	0.30684	0.32813	0.34636	0.36213	0.37591	0.38806	0.39884	0.40848
14	0.004076	0.051482	0.11475	0.17026	0.21571	0.25278	0.28335	0.30889	0.33053	0.34907	0.36514	0.37920	0.39160	0.40262	0.41248
15	0.004066	0.051469	0.11490	0.17071	0.21651	0.25393	0.28484	0.31071	0.33266	0.35149	0.36783	0.38214	0.39477	0.40601	0.41607
16	0.004057	0.051458	0.11504	0.17111	0.21721	0.25495	0.28617	0.31234	0.33456	0.35366	0.37024	0.38478	0.39763	0.40906	0.41931
17	0.004050	0.051448	0.11517	0.17147	0.21784	0.25587	0.28737	0.31380	0.33628	0.35562	0.37243	0.38717	0.40021	0.41183	0.42225
18	0.004043	0.051440	0.11528	0.17179	0.21841	0.25669	0.28845	0.31513	0.33784	0.35739	0.37441	0.38934	0.40257	0.41435	0.42493
19	0.004037	0.051432	0.11538	0.17208	0.21892	0.25744	0.28942	0.31633	0.33925	0.35901	0.37621	0.39133	0.40472	0.41666	0.42738
20	0.004032	0.051425	0.11547	0.17234	0.21939	0.25812	0.29032	0.31743	0.34055	0.36049	0.37787	0.39315	0.40669	0.41878	0.42964
21	0.004027	0.051419	0.11555	0.17258	0.21981	0.25874	0.29113	0.31843	0.34173	0.36185	0.37939	0.39482	0.40851	0.42073	0.43172
22	0.004023	0.051413	0.11563	0.17279	0.22020	0.25931	0.29188	0.31936	0.34283	0.36310	0.38079	0.39636	0.41019	0.42254	0.43365
23	0.004019	0.051408	0.11570	0.17299	0.22056	0.25983	0.29257	0.32021	0.34384	0.36426	0.38209	0.39780	0.41174	0.42421	0.43543
24	0.004015	0.051403	0.11576	0.17318	0.22089	0.26032	0.29321	0.32100	0.34477	0.36533	0.38329	0.39912	0.41319	0.42577	0.43710
25	0.004012	0.051399	0.11582	0.17335	0.22119	0.26077	0.29381	0.32174	0.34564	0.36633	0.38441	0.40036	0.41454	0.42722	0.43865
26	0.004009	0.051395	0.11587	0.17351	0.22148	0.26118	0.29436	0.32242	0.34645	0.36726	0.38546	0.40152	0.41580	0.42858	0.44010
27	0.004006	0.051391	0.11592	0.17365	0.22174	0.26157	0.29487	0.32306	0.34721	0.36813	0.38644	0.40260	0.41698	0.42986	0.44146
28	0.004003	0.051387	0.11597	0.17379	0.22199	0.26194	0.29535	0.32365	0.34792	0.36895	0.38736	0.40362	0.41809	0.43105	0.44274
29	0.004001	0.051384	0.11601	0.17392	0.22222	0.26228	0.29580	0.32421	0.34858	0.36971	0.38822	0.40457	0.41913	0.43218	0.44394
30	0.003998	0.051381	0.11606	0.17404	0.22243	0.26259	0.29623	0.32474	0.34920	0.37043	0.38903	0.40547	0.42011	0.43324	0.44508
31	0.003996	0.051378	0.11609	0.17415	0.22264	0.26289	0.29662	0.32523	0.34979	0.37111	0.38980	0.40632	0.42104	0.43424	0.44615
32	0.003994	0.051376	0.11613	0.17426	0.22283	0.26318	0.29700	0.32570	0.35035	0.37175	0.39052	0.40712	0.42191	0.43518	0.44717
33	0.003992	0.051373	0.11616	0.17436	0.22301	0.26344	0.29735	0.32614	0.35087	0.37236	0.39120	0.40787	0.42274	0.43608	0.44813
34	0.003991	0.051371	0.11620	0.17445	0.22318	0.26369	0.29769	0.32655	0.35137	0.37293	0.39185	0.40859	0.42352	0.43693	0.44904
35	0.003989	0.051369	0.11623	0.17454	0.22334	0.26393	0.29800	0.32695	0.35184	0.37347	0.39246	0.40927	0.42427	0.43774	0.44991
36	0.003987	0.051366	0.11625	0.17462	0.22349	0.26416	0.29830	0.32732	0.35229	0.37399	0.39305	0.40992	0.42498	0.43851	0.45073
37	0.003986	0.051364	0.11628	0.17470	0.22363	0.26437	0.29859	0.32768	0.35271	0.37448	0.39360	0.41054	0.42566	0.43924	0.45152
38	0.003984	0.051363	0.11631	0.17478	0.22377	0.26457	0.29886	0.32802	0.35311	0.37495	0.39413	0.41113	0.42630	0.43994	0.45226
39	0.003983	0.051361	0.11633	0.17485	0.22390	0.26477	0.29912	0.32834	0.35350	0.37539	0.39463	0.41169	0.42691	0.44060	0.45298
40	0.003982	0.051359	0.11635	0.17492	0.22402	0.26495	0.29936	0.32865	0.35387	0.37582	0.39512	0.41222	0.42750	0.44124	0.45366
41	0.003981	0.051358	0.11638	0.17498	0.22414	0.26513	0.29960	0.32894	0.35422	0.37623	0.39558	0.41273	0.42806	0.44185	0.45432
42	0.003979	0.051356	0.11640	0.17504	0.22426	0.26529	0.29982	0.32922	0.35455	0.37661	0.39602	0.41322	0.42860	0.44243	0.45494
43	0.003978	0.051355	0.11642	0.17510	0.22436	0.26545	0.30004	0.32949	0.35487	0.37699	0.39644	0.41369	0.42911	0.44299	0.45554
44	0.003977	0.051353	0.11644	0.17516	0.22447	0.26561	0.30024	0.32975	0.35518	0.37734	0.39684	0.41414	0.42961	0.44352	0.45612
45	0.003976	0.051352	0.11646	0.17521	0.22457	0.26576	0.30044	0.32999	0.35547	0.37768	0.39723	0.41457	0.43008	0.44404	0.45667
46	0.003975	0.051351	0.11647	0.17527	0.22466	0.26590	0.30063	0.33023	0.35576	0.37801	0.39760	0.41499	0.43053	0.44453	0.45721
47	0.003974	0.051349	0.11649	0.17532	0.22475	0.26603	0.30081	0.33045	0.35603	0.37833	0.39796	0.41538	0.43097	0.44501	0.45772
48	0.003973	0.051348	0.11651	0.17536	0.22484	0.26616	0.30098	0.33067	0.35629	0.37863	0.39830	0.41577	0.43139	0.44546	0.45821
49	0.003973	0.051347	0.11652	0.17541	0.22492	0.26629	0.30115	0.33088	0.35654	0.37892	0.39863	0.41614	0.43180	0.44590	0.45868
50	0.003972	0.051346	0.11654	0.17545	0.22500	0.26641	0.30131	0.33108	0.35678	0.37920	0.39895	0.41649	0.43219	0.44633	0.45914

Appendix 6 (continued)

95% Critical values of the F distribution

df1 /df2	1	2	3	4	5	6	7	8	9	10	11	12	13	14	15
51	0.003971	0.05134	0.11655	0.17550	0.22508	0.26652	0.30146	0.33128	0.35701	0.37947	0.39926	0.41683	0.43256	0.44674	0.45958
52	0.003970	0.05134	0.11656	0.17554	0.22515	0.26663	0.30161	0.33146	0.35724	0.37973	0.39955	0.41716	0.43293	0.44713	0.46001
53	0.003970	0.05134	0.11658	0.17558	0.22523	0.26674	0.30176	0.33164	0.35745	0.37998	0.39984	0.41748	0.43328	0.44751	0.46042
54	0.003969	0.05134	0.11659	0.17561	0.22529	0.26684	0.30189	0.33182	0.35766	0.38023	0.40011	0.41779	0.43362	0.44788	0.46081
55	0.003968	0.05134	0.11660	0.17565	0.22536	0.26694	0.30203	0.33199	0.35786	0.38046	0.40038	0.41809	0.43394	0.44824	0.46120
56	0.003968	0.05134	0.11662	0.17569	0.22543	0.26704	0.30216	0.33215	0.35806	0.38069	0.40064	0.41837	0.43426	0.44858	0.46157
57	0.003967	0.05134	0.11663	0.17572	0.22549	0.26713	0.30228	0.33230	0.35825	0.38091	0.40089	0.41865	0.43457	0.44891	0.46193
60	0.003965	0.05134	0.11666	0.17582	0.22566	0.26739	0.30263	0.33275	0.35878	0.38152	0.40159	0.41943	0.43543	0.44985	0.46294
65	0.003963	0.05133	0.11671	0.17596	0.22592	0.26778	0.30315	0.33339	0.35956	0.38243	0.40262	0.42059	0.43670	0.45124	0.46444
70	0.003960	0.05133	0.11675	0.17608	0.22614	0.26811	0.30359	0.33395	0.36023	0.38322	0.40351	0.42159	0.43780	0.45244	0.46573
75	0.003959	0.05133	0.11678	0.17618	0.22633	0.26840	0.30398	0.33444	0.36082	0.38390	0.40429	0.42246	0.43876	0.45349	0.46687
80	0.003957	0.05133	0.11681	0.17627	0.22650	0.26865	0.30432	0.33487	0.36133	0.38450	0.40498	0.42323	0.43962	0.45442	0.46787
85	0.003955	0.05132	0.11684	0.17636	0.22665	0.26887	0.30462	0.33525	0.36179	0.38504	0.40559	0.42392	0.44037	0.45524	0.46877
90	0.003954	0.05132	0.11686	0.17643	0.22678	0.26907	0.30489	0.33559	0.36220	0.38552	0.40614	0.42453	0.44105	0.45598	0.46957
95	0.003953	0.05132	0.11689	0.17649	0.22690	0.26925	0.30514	0.33590	0.36257	0.38595	0.40663	0.42508	0.44166	0.45665	0.47028
100	0.003952	0.05132	0.11691	0.17655	0.22701	0.26942	0.30535	0.33617	0.36290	0.38634	0.40707	0.42558	0.44221	0.45725	0.47094
110	0.003950	0.05132	0.11694	0.17665	0.22720	0.26970	0.30573	0.33665	0.36348	0.38701	0.40785	0.42644	0.44316	0.45830	0.47207
120	0.003949	0.05132	0.11697	0.17674	0.22735	0.26993	0.30605	0.33705	0.36396	0.38758	0.40849	0.42717	0.44397	0.45918	0.47302
130	0.003947	0.05131	0.11699	0.17681	0.22748	0.27013	0.30632	0.33739	0.36438	0.38806	0.40904	0.42779	0.44465	0.45993	0.47384
140	0.003946	0.05131	0.11701	0.17687	0.22760	0.27030	0.30655	0.33768	0.36473	0.38847	0.40952	0.42832	0.44524	0.46057	0.47454
150	0.003945	0.05131	0.11703	0.17692	0.22770	0.27045	0.30675	0.33794	0.36504	0.38884	0.40993	0.42878	0.44576	0.46114	0.47515
160	0.003944	0.05131	0.11705	0.17697	0.22778	0.27058	0.30693	0.33816	0.36531	0.38915	0.41029	0.42919	0.44621	0.46163	0.47569
170	0.003944	0.05131	0.11706	0.17701	0.22786	0.27069	0.30708	0.33836	0.36555	0.38943	0.41061	0.42955	0.44661	0.46207	0.47616
180	0.003943	0.05131	0.11707	0.17705	0.22793	0.27080	0.30722	0.33853	0.36576	0.38968	0.41090	0.42988	0.44697	0.46246	0.47659
190	0.003943	0.05131	0.11708	0.17708	0.22799	0.27089	0.30735	0.33869	0.36595	0.38991	0.41116	0.43016	0.44729	0.46281	0.47697
200	0.003942	0.05131	0.11709	0.17711	0.22804	0.27097	0.30746	0.33883	0.36612	0.39011	0.41139	0.43043	0.44758	0.46313	0.47732
210	0.003942	0.05131	0.11710	0.17714	0.22809	0.27105	0.30756	0.33896	0.36628	0.39029	0.41160	0.43066	0.44784	0.46342	0.47763
220	0.003941	0.05131	0.11711	0.17716	0.22814	0.27111	0.30765	0.33908	0.36642	0.39046	0.41179	0.43088	0.44808	0.46368	0.47791
230	0.003941	0.05130	0.11712	0.17719	0.22818	0.27118	0.30774	0.33919	0.36655	0.39061	0.41197	0.43107	0.44830	0.46392	0.47817
240	0.003940	0.05130	0.11712	0.17721	0.22822	0.27123	0.30782	0.33928	0.36667	0.39075	0.41213	0.43126	0.44850	0.46414	0.47841
250	0.003940	0.05130	0.11713	0.17723	0.22825	0.27129	0.30789	0.33938	0.36678	0.39088	0.41227	0.43142	0.44868	0.46434	0.47863
275	0.003939	0.05130	0.11714	0.17727	0.22833	0.27140	0.30804	0.33957	0.36702	0.39116	0.41260	0.43179	0.44909	0.46479	0.47912
300	0.003939	0.05130	0.11716	0.17730	0.22839	0.27150	0.30817	0.33974	0.36722	0.39140	0.41287	0.43209	0.44943	0.46516	0.47952
325	0.003938	0.05130	0.11717	0.17733	0.22845	0.27158	0.30828	0.33988	0.36739	0.39160	0.41310	0.43235	0.44971	0.46548	0.47987
350	0.003938	0.05130	0.11717	0.17735	0.22849	0.27165	0.30838	0.34000	0.36754	0.39177	0.41330	0.43257	0.44996	0.46575	0.48016
375	0.003937	0.05130	0.11718	0.17738	0.22853	0.27171	0.30846	0.34010	0.36766	0.39192	0.41347	0.43277	0.45018	0.46598	0.48042
400	0.003937	0.05130	0.11719	0.17740	0.22857	0.27176	0.30853	0.34020	0.36777	0.39205	0.41362	0.43293	0.45036	0.46619	0.48064
425	0.003937	0.05130	0.11719	0.17741	0.22860	0.27181	0.30860	0.34028	0.36787	0.39217	0.41375	0.43308	0.45053	0.46637	0.48084
450	0.003937	0.05130	0.11720	0.17743	0.22863	0.27185	0.30865	0.34035	0.36796	0.39227	0.41387	0.43322	0.45068	0.46653	0.48102
475	0.003936	0.05130	0.11720	0.17744	0.22865	0.27189	0.30870	0.34041	0.36804	0.39236	0.41397	0.43334	0.45081	0.46668	0.48117
500	0.003936	0.05130	0.11721	0.17745	0.22867	0.27192	0.30875	0.34047	0.36811	0.39244	0.41407	0.43344	0.45093	0.46681	0.48132
600	0.003935	0.05130	0.11722	0.17749	0.22874	0.27203	0.30889	0.34065	0.36833	0.39271	0.41437	0.43378	0.45131	0.46723	0.48177
700	0.003935	0.05130	0.11723	0.17752	0.22879	0.27210	0.30900	0.34079	0.36849	0.39289	0.41459	0.43403	0.45158	0.46752	0.48210
800	0.003935	0.05130	0.11723	0.17754	0.22883	0.27216	0.30908	0.34088	0.36861	0.39304	0.41475	0.43421	0.45178	0.46775	0.48234
900	0.003934	0.05130	0.11724	0.17755	0.22886	0.27221	0.30914	0.34096	0.36871	0.39315	0.41487	0.43435	0.45194	0.46792	0.48253
1000	0.003934	0.05130	0.11724	0.17757	0.22888	0.27224	0.30918	0.34102	0.36878	0.39323	0.41498	0.43447	0.45207	0.46806	0.48268
1100	0.003934	0.05130	0.11725	0.17758	0.22890	0.27227	0.30922	0.34107	0.36884	0.39331	0.41506	0.43456	0.45217	0.46818	0.48281
1200	0.003934	0.05130	0.11725	0.17759	0.22892	0.27230	0.30926	0.34112	0.36889	0.39337	0.41513	0.43464	0.45226	0.46827	0.48291
1300	0.003934	0.05130	0.11725	0.17759	0.22893	0.27232	0.30928	0.34115	0.36894	0.39342	0.41519	0.43471	0.45233	0.46835	1.00000
1400	0.003934	0.05130	0.11725	0.17760	0.22894	0.27233	0.30931	0.34118	0.36897	0.39346	0.41524	0.43476	0.45240	0.46842	0.48307

APPENDIXES

							95% Critical values of the F distribution							
df1 /df2	16	17	18	19	20	21	22	23	24	25	26	27	28	29	30
1	0.22252	0.22465	0.22656	0.22827	0.22982	0.23122	0.23251	0.23368	0.23476	0.23575	0.23668	0.23753	0.23832	0.23906	0.23976
2	0.27520	0.27843	0.28133	0.28394	0.28630	0.28845	0.29041	0.29222	0.29387	0.29540	0.29682	0.29814	0.29937	0.30051	0.30158
3	0.30875	0.31282	0.31646	0.31976	0.32275	0.32547	0.32796	0.33025	0.33236	0.33431	0.33612	0.33780	0.33936	0.34083	0.34220
4	0.33257	0.33730	0.34156	0.34541	0.34891	0.35210	0.35502	0.35771	0.36019	0.36249	0.36462	0.36660	0.36845	0.37018	0.37180
5	0.35058	0.35587	0.36064	0.36496	0.36888	0.37247	0.37576	0.37879	0.38158	0.38417	0.38658	0.38882	0.39091	0.39287	0.39470
6	0.36479	0.37055	0.37576	0.38047	0.38477	0.38869	0.39230	0.39562	0.39869	0.40154	0.40419	0.40665	0.40895	0.41111	0.41313
7	0.37634	0.38251	0.38809	0.39315	0.39777	0.40200	0.40588	0.40946	0.41277	0.41585	0.41871	0.42137	0.42386	0.42620	0.42839
8	0.38594	0.39247	0.39838	0.40375	0.40865	0.41314	0.41727	0.42109	0.42461	0.42789	0.43094	0.43378	0.43644	0.43893	0.44127
9	0.39406	0.40092	0.40712	0.41276	0.41792	0.42265	0.42700	0.43101	0.43474	0.43819	0.44141	0.44442	0.44723	0.44987	0.45235
10	0.40104	0.40818	0.41464	0.42053	0.42592	0.43086	0.43541	0.43961	0.44351	0.44713	0.45051	0.45366	0.45661	0.45938	0.46198
11	0.40710	0.41450	0.42120	0.42731	0.43290	0.43804	0.44277	0.44714	0.45120	0.45497	0.45849	0.46178	0.46486	0.46775	0.47046
12	0.41243	0.42005	0.42697	0.43328	0.43906	0.44437	0.44927	0.45380	0.45801	0.46192	0.46557	0.46898	0.47217	0.47517	0.47800
13	0.41714	0.42498	0.43210	0.43859	0.44454	0.45001	0.45506	0.45974	0.46408	0.46811	0.47188	0.47541	0.47871	0.48182	0.48474
14	0.42135	0.42938	0.43667	0.44334	0.44945	0.45507	0.46026	0.46506	0.46953	0.47369	0.47757	0.48120	0.48460	0.48780	0.49082
15	0.42513	0.43333	0.44080	0.44761	0.45387	0.45963	0.46495	0.46988	0.47446	0.47872	0.48271	0.48644	0.48994	0.49323	0.49633
16	0.42854	0.43691	0.44453	0.45149	0.45788	0.46377	0.46921	0.47425	0.47894	0.48330	0.48739	0.49121	0.49480	0.49817	0.50135
17	0.43164	0.44016	0.44792	0.45502	0.46153	0.46754	0.47309	0.47824	0.48303	0.48749	0.49166	0.49557	0.49924	0.50269	0.50595
18	0.43447	0.44313	0.45102	0.45824	0.46487	0.47099	0.47665	0.48189	0.48678	0.49133	0.49559	0.49958	0.50333	0.50685	0.51017
19	0.43707	0.44585	0.45387	0.46120	0.46794	0.47416	0.47992	0.48526	0.49023	0.49487	0.49921	0.50327	0.50709	0.51069	0.51408
20	0.43945	0.44836	0.45649	0.46393	0.47078	0.47709	0.48294	0.48837	0.49342	0.49814	0.50255	0.50669	0.51058	0.51424	0.51769
21	0.44165	0.45067	0.45891	0.46645	0.47340	0.47980	0.48574	0.49125	0.49638	0.50117	0.50566	0.50986	0.51382	0.51754	0.52105
22	0.44369	0.45282	0.46115	0.46880	0.47583	0.48232	0.48834	0.49392	0.49913	0.50399	0.50855	0.51282	0.51683	0.52061	0.52418
23	0.44558	0.45481	0.46324	0.47097	0.47809	0.48467	0.49076	0.49642	0.50170	0.50663	0.51124	0.51557	0.51964	0.52348	0.52710
24	0.44735	0.45667	0.46519	0.47300	0.48020	0.48685	0.49302	0.49875	0.50409	0.50909	0.51376	0.51815	0.52228	0.52617	0.52984
25	0.44899	0.45840	0.46701	0.47490	0.48218	0.48890	0.49514	0.50093	0.50634	0.51139	0.51613	0.52057	0.52475	0.52869	0.53241
26	0.45053	0.46003	0.46871	0.47668	0.48403	0.49082	0.49712	0.50298	0.50845	0.51356	0.51835	0.52284	0.52707	0.53106	0.53483
27	0.45198	0.46155	0.47031	0.47836	0.48577	0.49263	0.49899	0.50491	0.51043	0.51560	0.52044	0.52498	0.52926	0.53330	0.53711
28	0.45334	0.46299	0.47182	0.47993	0.48741	0.49433	0.50075	0.50673	0.51230	0.51752	0.52241	0.52700	0.53133	0.53541	0.53926
29	0.45462	0.46434	0.47324	0.48141	0.48896	0.49593	0.50241	0.50844	0.51407	0.51934	0.52427	0.52891	0.53328	0.53740	0.54129
30	0.45582	0.46561	0.47458	0.48282	0.49042	0.49745	0.50398	0.51006	0.51574	0.52105	0.52604	0.53072	0.53513	0.53929	0.54322
31	0.45696	0.46682	0.47584	0.48414	0.49180	0.49889	0.50547	0.51160	0.51732	0.52268	0.52771	0.53243	0.53688	0.54108	0.54505
32	0.45804	0.46796	0.47704	0.48540	0.49311	0.50025	0.50688	0.51306	0.51883	0.52423	0.52930	0.53406	0.53855	0.54278	0.54679
33	0.45906	0.46904	0.47818	0.48659	0.49435	0.50154	0.50822	0.51444	0.52025	0.52570	0.53080	0.53561	0.54013	0.54440	0.54844
34	0.46003	0.47007	0.47926	0.48772	0.49553	0.50277	0.50949	0.51576	0.52161	0.52709	0.53224	0.53708	0.54164	0.54595	0.55002
35	0.46096	0.47104	0.48029	0.48880	0.49666	0.50394	0.51070	0.51701	0.52291	0.52843	0.53361	0.53848	0.54308	0.54742	0.55152
36	0.46184	0.47197	0.48127	0.48982	0.49773	0.50505	0.51186	0.51821	0.52414	0.52970	0.53491	0.53982	0.54445	0.54882	0.55295
37	0.46267	0.47286	0.48220	0.49080	0.49875	0.50611	0.51296	0.51935	0.52532	0.53091	0.53616	0.54110	0.54576	0.55016	0.55433
38	0.46347	0.47371	0.48309	0.49174	0.49972	0.50713	0.51402	0.52044	0.52644	0.53207	0.53736	0.54233	0.54702	0.55145	0.55564
39	0.46423	0.47451	0.48394	0.49263	0.50066	0.50810	0.51503	0.52148	0.52752	0.53318	0.53850	0.54350	0.54822	0.55268	0.55690
40	0.46496	0.47529	0.48476	0.49348	0.50155	0.50903	0.51599	0.52248	0.52855	0.53424	0.53959	0.54462	0.54937	0.55386	0.55810
41	0.46566	0.47603	0.48554	0.49430	0.50241	0.50992	0.51692	0.52344	0.52954	0.53526	0.54064	0.54570	0.55048	0.55499	0.55926
42	0.46633	0.47673	0.48628	0.49509	0.50323	0.51078	0.51780	0.52436	0.53049	0.53624	0.54165	0.54674	0.55154	0.55607	0.56037
43	0.46697	0.47741	0.48700	0.49584	0.50401	0.51160	0.51866	0.52524	0.53140	0.53718	0.54262	0.54773	0.55256	0.55712	0.56144
44	0.46758	0.47807	0.48769	0.49656	0.50477	0.51238	0.51947	0.52609	0.53228	0.53809	0.54355	0.54869	0.55354	0.55812	0.56247
45	0.46818	0.47869	0.48835	0.49726	0.50549	0.51314	0.52026	0.52691	0.53312	0.53896	0.54444	0.54961	0.55448	0.55909	0.56345
46	0.46874	0.47930	0.48899	0.49792	0.50619	0.51387	0.52102	0.52769	0.53394	0.53980	0.54530	0.55049	0.55539	0.56002	0.56441
47	0.46929	0.47987	0.48960	0.49857	0.50687	0.51457	0.52175	0.52845	0.53472	0.54060	0.54613	0.55135	0.55627	0.56092	0.56533
48	0.46981	0.48043	0.49019	0.49919	0.50752	0.51525	0.52245	0.52918	0.53547	0.54138	0.54694	0.55217	0.55711	0.56179	0.56621
49	0.47032	0.48097	0.49076	0.49978	0.50814	0.51590	0.52313	0.52988	0.53620	0.54213	0.54771	0.55297	0.55793	0.56262	0.56707
50	0.47081	0.48149	0.49130	0.50036	0.50874	0.51653	0.52378	0.53056	0.53690	0.54286	0.54846	0.55373	0.55872	0.56343	0.56790

Appendix 6 (continued)

df1/df2	16	17	18	19	20	21	22	23	24	25	26	27	28	29	30
						95% Critical values of the F distribution									
51	0.47128	0.48199	0.49183	0.50092	0.50933	0.51714	0.52441	0.53121	0.53758	0.54356	0.54918	0.55448	0.55948	0.56421	0.56870
52	0.47174	0.48247	0.49234	0.50145	0.50989	0.51772	0.52503	0.53185	0.53823	0.54423	0.54987	0.55519	0.56022	0.56497	0.56947
53	0.47218	0.48294	0.49284	0.50197	0.51043	0.51829	0.52562	0.53246	0.53887	0.54489	0.55055	0.55589	0.56093	0.56570	0.57022
54	0.47260	0.48339	0.49331	0.50247	0.51096	0.51884	0.52619	0.53305	0.53948	0.54552	0.55120	0.55656	0.56162	0.56641	0.57094
55	0.47301	0.48383	0.49378	0.50296	0.51147	0.51937	0.52674	0.53362	0.54008	0.54613	0.55183	0.55721	0.56229	0.56709	0.57165
56	0.47341	0.48425	0.49422	0.50343	0.51196	0.51989	0.52727	0.53418	0.54065	0.54673	0.55245	0.55784	0.56293	0.56776	0.57233
57	0.47379	0.48466	0.49465	0.50388	0.51243	0.52038	0.52779	0.53472	0.54121	0.54730	0.55304	0.55845	0.56356	0.56840	0.57299
60	0.47488	0.48581	0.49587	0.50517	0.51378	0.52179	0.52926	0.53624	0.54278	0.54893	0.55472	0.56018	0.56534	0.57022	0.57485
65	0.47648	0.48752	0.49768	0.50707	0.51577	0.52387	0.53143	0.53849	0.54512	0.55134	0.55721	0.56274	0.56797	0.57293	0.57763
70	0.47787	0.48900	0.49925	0.50872	0.51751	0.52568	0.53331	0.54045	0.54715	0.55345	0.55938	0.56498	0.57028	0.57529	0.58005
75	0.47909	0.49029	0.50062	0.51017	0.51903	0.52728	0.53497	0.54218	0.54894	0.55530	0.56129	0.56695	0.57231	0.57738	0.58219
80	0.48016	0.49144	0.50184	0.51145	0.52038	0.52869	0.53645	0.54371	0.55053	0.55694	0.56299	0.56870	0.57411	0.57923	0.58409
85	0.48112	0.49246	0.50292	0.51259	0.52158	0.52994	0.53776	0.54508	0.55195	0.55841	0.56451	0.57027	0.57572	0.58088	0.58579
90	0.48198	0.49338	0.50389	0.51362	0.52265	0.53107	0.53893	0.54630	0.55322	0.55973	0.56587	0.57167	0.57716	0.58237	0.58732
95	0.48275	0.49420	0.50477	0.51454	0.52363	0.53209	0.54000	0.54741	0.55437	0.56092	0.56710	0.57294	0.57847	0.58372	0.58870
100	0.48345	0.49495	0.50556	0.51538	0.52451	0.53301	0.54096	0.54841	0.55541	0.56200	0.56822	0.57410	0.57966	0.58494	0.58996
110	0.48467	0.49625	0.50694	0.51684	0.52604	0.53462	0.54264	0.55017	0.55723	0.56389	0.57017	0.57611	0.58174	0.58708	0.59216
120	0.48570	0.49735	0.50810	0.51807	0.52734	0.53598	0.54406	0.55164	0.55877	0.56548	0.57182	0.57782	0.58350	0.58889	0.59402
130	0.48657	0.49828	0.50910	0.51912	0.52844	0.53714	0.54528	0.55291	0.56009	0.56685	0.57324	0.57928	0.58500	0.59044	0.59561
140	0.48733	0.49909	0.50996	0.52003	0.52940	0.53814	0.54633	0.55400	0.56123	0.56803	0.57446	0.58054	0.58631	0.59179	0.59700
150	0.48799	0.49979	0.51070	0.52082	0.53023	0.53902	0.54724	0.55496	0.56222	0.56906	0.57553	0.58165	0.58745	0.59296	0.59821
160	0.48856	0.50041	0.51136	0.52152	0.53097	0.53979	0.54805	0.55580	0.56310	0.56997	0.57647	0.58262	0.58846	0.59400	0.59927
170	0.48908	0.50096	0.51195	0.52214	0.53162	0.54048	0.54877	0.55655	0.56388	0.57078	0.57731	0.58349	0.58935	0.59492	0.60022
180	0.48954	0.50145	0.51247	0.52269	0.53220	0.54109	0.54941	0.55722	0.56457	0.57151	0.57806	0.58427	0.59015	0.59575	0.60107
190	0.48995	0.50189	0.51294	0.52319	0.53273	0.54164	0.54998	0.55782	0.56520	0.57216	0.57873	0.58496	0.59087	0.59649	0.60184
200	0.49032	0.50229	0.51336	0.52363	0.53320	0.54214	0.55050	0.55836	0.56576	0.57274	0.57934	0.58559	0.59153	0.59716	0.60253
210	0.49065	0.50265	0.51374	0.52404	0.53363	0.54259	0.55098	0.55886	0.56628	0.57328	0.57990	0.58617	0.59212	0.59777	0.60316
220	0.49096	0.50298	0.51409	0.52441	0.53402	0.54300	0.55141	0.55931	0.56675	0.57377	0.58040	0.58669	0.59266	0.59833	0.60373
230	0.49124	0.50328	0.51442	0.52475	0.53438	0.54338	0.55180	0.55972	0.56718	0.57421	0.58087	0.58717	0.59316	0.59884	0.60426
240	0.49150	0.50356	0.51471	0.52506	0.53471	0.54372	0.55217	0.56010	0.56757	0.57462	0.58129	0.58761	0.59361	0.59931	0.60475
250	0.49174	0.50381	0.51498	0.52535	0.53501	0.54404	0.55250	0.56045	0.56794	0.57500	0.58169	0.58802	0.59403	0.59975	0.60519
275	0.49226	0.50437	0.51558	0.52598	0.53568	0.54474	0.55324	0.56122	0.56874	0.57584	0.58255	0.58891	0.59496	0.60070	0.60618
300	0.49270	0.50484	0.51608	0.52651	0.53624	0.54533	0.55385	0.56186	0.56941	0.57653	0.58327	0.58966	0.59573	0.60150	0.60700
325	0.49307	0.50524	0.51650	0.52696	0.53671	0.54583	0.55437	0.56241	0.56998	0.57712	0.58389	0.59030	0.59639	0.60218	0.60770
350	0.49339	0.50558	0.51686	0.52735	0.53712	0.54626	0.55482	0.56288	0.57047	0.57763	0.58442	0.59085	0.59696	0.60277	0.60831
375	0.49366	0.50587	0.51718	0.52768	0.53747	0.54663	0.55521	0.56329	0.57089	0.57808	0.58488	0.59133	0.59745	0.60328	0.60883
400	0.49390	0.50613	0.51746	0.52798	0.53778	0.54696	0.55556	0.56365	0.57127	0.57847	0.58528	0.59175	0.59788	0.60373	0.60929
425	0.49412	0.50636	0.51770	0.52824	0.53806	0.54725	0.55586	0.56396	0.57160	0.57881	0.58564	0.59212	0.59827	0.60412	0.60970
450	0.49431	0.50657	0.51792	0.52847	0.53830	0.54750	0.55613	0.56425	0.57189	0.57912	0.58596	0.59245	0.59861	0.60448	0.61007
475	0.49448	0.50675	0.51811	0.52868	0.53852	0.54773	0.55637	0.56450	0.57216	0.57940	0.58625	0.59274	0.59892	0.60479	0.61039
500	0.49464	0.50692	0.51829	0.52886	0.53872	0.54794	0.55659	0.56473	0.57240	0.57964	0.58650	0.59301	0.59919	0.60508	0.61069
600	0.49512	0.50744	0.51885	0.52946	0.53935	0.54860	0.55729	0.56546	0.57316	0.58043	0.58732	0.59386	0.60007	0.60599	0.61163
700	0.49548	0.50782	0.51925	0.52988	0.53980	0.54908	0.55779	0.56598	0.57370	0.58100	0.58791	0.59447	0.60071	0.60664	0.61230
800	0.49574	0.50810	0.51956	0.53020	0.54014	0.54944	0.55816	0.56637	0.57411	0.58143	0.58836	0.59493	0.60118	0.60713	0.61281
900	0.49595	0.50832	0.51979	0.53045	0.54040	0.54972	0.55845	0.56668	0.57443	0.58176	0.58870	0.59529	0.60155	0.60752	0.61321
1000	0.49611	0.50850	0.51998	0.53066	0.54062	0.54994	0.55869	0.56692	0.57469	0.58203	0.58898	0.59558	0.60185	0.60783	0.61352
1100	0.49624	0.50865	0.52013	0.53082	0.54079	0.55012	0.55888	0.56712	0.57490	0.58225	0.58921	0.59582	0.60210	0.60808	0.61378
1200	0.49636	0.50877	0.52026	0.53096	0.54093	0.55027	0.55904	0.56729	0.57507	0.58243	0.58940	0.59601	0.60230	0.60829	0.61400
1300	0.49645	0.50887	0.52037	0.53107	0.54106	0.55040	0.55918	0.56743	0.57522	0.58258	0.58956	0.59618	0.60247	0.60847	0.61419
1400	0.49653	0.50896	0.52047	0.53117	0.54116	0.55051	0.55929	0.56756	0.57535	0.58272	0.58970	0.59632	0.60262	0.60862	0.61435

APPENDIXES

Appendix 7 Kruskal-Wallis H {Courtesy of Peter M Lee, and University of York (2005)}

Upper Critical Values for the Kruskal-Wallis Test

The critical values give significance levels as close as possible to, but not exceeding α. The actual levels of significance are in brackets. When the tabled sample sizes, n, are exceeded use,

$$H_\alpha = \chi^2_\alpha, \quad df = m - 1.$$

Treatment Sizes, n			Nominal size α			
			0.10	0.05	0.025	0.01
2	2	2	4.571 (.06667)	- -	- -	- -
3	2	1	4.286 (.10000)	- -	- -	- -
3	2	2	4.500 (.06667)	4.714 (.04762)	- -	- -
3	3	1	4.571 (.10000)	5.143 (.04286)	- -	- -
3	3	2	4.556 (.10000)	5.361 (.03214)	5.556 (.02500)	- -
3	3	3	4.622 (.10000)	5.600 (.05000)	5.956 (.02500)	7.200 (.00357)
4	2	1	4.500 (.07619)	- -	- -	- -
4	2	2	4.458 (.10000)	5.333 (.03333)	5.500 (.02381)	- -
4	3	1	4.056 (.09286)	5.208 (.05000)	5.833 (.02143)	- -
4	3	2	4.511 (.09841)	5.444 (.04603)	6.000 (.02381)	6.444 (.00794)
4	3	3	4.709 (.09238)	5.791 (.04571)	6.155 (.02476)	6.745 (.01000)
4	4	1	4.167 (.08254)	4.967 (.04762)	6.167 (.02222)	6.667 (.00952)
4	4	2	4.555 (.09778)	5.455 (.04571)	6.327 (.02413)	7.036 (.00571)
4	4	3	4.545 (.09905)	5.598 (.04866)	6.394 (.02476)	7.144 (.00970)
4	4	4	4.654 (.09662)	5.692 (.04866)	6.615 (.02424)	7.654 (.00762)
5	2	1	4.200 (.09524)	5.000 (.04762)	- -	- -
5	2	2	4.373 (.08995)	5.160 (.03439)	6.000 (.01852)	6.533 (.00794)
5	3	1	4.018 (.09524)	4.960 (.04762)	6.044 (.01984)	- -
5	3	2	4.651 (.09127)	5.251 (.04921)	6.004 (.02460)	6.909 (.00873)
5	3	3	4.533 (.09697)	5.648 (.04892)	6.315 (.02121)	7.079 (.00866)
5	4	1	3.987 (.09841)	4.985 (.04444)	5.858 (.02381)	6.955 (.00794)
5	4	2	4.541 (.09841)	5.273 (.04877)	6.068 (.02482)	7.205 (.00895)
5	4	3	4.549 (.09892)	5.656 (.04863)	6.410 (.02496)	7.445 (.00974)
5	4	4	4.668 (.09817)	5.657 (.04906)	6.673 (.02429)	7.760 (.00946)
5	5	1	4.109 (.08586)	5.127 (.04618)	6.000 (.02165)	7.309 (.00938)
5	5	2	4.623 (.09704)	5.338 (.04726)	6.346 (.02489)	7.338 (.00962)
5	5	3	4.545 (.09965)	5.705 (.04612)	6.549 (.02436)	7.578 (.00968)
5	5	4	4.523 (.09935)	5.666 (.04931)	6.760 (.02490)	7.823 (.00978)
5	5	5	4.560 (.09952)	5.780 (.04878)	6.740 (.02475)	8.000 (.00946)
6	1	1	- -	- -	- -	- -
6	2	1	4.200 (.09524)	4.822 (.04762)	5.600 (.02381)	- -
6	2	2	4.545 (.08889)	5.345 (.03810)	5.745 (.02063)	6.655 (.00794)
6	3	1	3.909 (.09524)	4.855 (.05000)	5.945 (.02143)	6.873 (.00714)
6	3	2	4.682 (.08528)	5.348 (.04632)	6.136 (.02294)	6.970 (.00909)
6	3	3	4.590 (.09773)	5.615 (.04968)	6.436 (.02229)	7.410 (.00779)
6	4	1	4.038 (.09437)	4.947 (.04675)	5.856 (.02424)	7.106 (.00866)
6	4	2	4.494 (.09986)	5.340 (.04906)	6.186 (.02453)	7.340 (.00967)
6	4	3	4.604 (.09997)	5.610 (.04862)	6.538 (.02498)	7.500 (.00966)
6	4	4	4.595 (.09847)	5.681 (.04881)	6.667 (.02495)	7.795 (.00990)
6	5	1	4.128 (.09271)	4.990 (.04726)	5.951 (.02453)	7.182 (.00974)
6	5	2	4.596 (.09807)	5.338 (.04729)	6.196 (.02481)	7.376 (.00982)
6	5	3	4.535 (.09932)	5.602 (.04956)	6.667 (.02452)	7.590 (.00999)
6	5	4	4.522 (.09974)	5.661 (.04991)	6.750 (.02473)	7.936 (.00998)
6	5	5	4.547 (.09835)	5.729 (.04973)	6.788 (.02484)	8.028 (.00988)
6	6	1	4.000 (.09774)	4.945 (.04779)	5.923 (.02381)	7.121 (.00932)
6	6	2	4.438 (.09824)	5.410 (.04993)	6.210 (.02443)	7.467 (.00982)
6	6	3	4.558 (.09948)	5.625 (.04999)	6.725 (.02462)	7.725 (.00985)
6	6	4	4.548 (.09982)	5.724 (.04950)	6.812 (.02458)	8.000 (.00998)
6	6	5	4.542 (.09987)	5.765 (.04993)	6.848 (.02489)	8.124 (.00990)
6	6	6	4.643 (.09874)	5.801 (.04905)	6.889 (.02493)	8.222 (.00994)
7	7	7	4.594 (.09933)	5.819 (.04911)	6.954 (.02446)	8.378 (.00992)
8	8	8	4.595 (.09933)	5.805 (.04973)	6.995 (.02485)	8.465 (.00991)

Do not use more than 5% per tail ever.

Appendix 8 {Courtesy of Peter M Lee, and University of York (2005)}
(Mann-Whitney-Wilcoxon test aka Wilcoxon rank-sum test)

Critical values for the Mann-Whitney rank-sum test

The pairs of values below are approximate critical values of R for two-tailed tests at levels $P = 0.10$ (upper value) and $P = 0.05$ (lower value).

(Use relevant $P = 0.10$ entry for one-tailed test at level 0.05).

		\multicolumn{7}{c}{larger sample size, n_2}						
		4	5	6	7	8	9	10
smaller sample	4	12,24	13,27	14,30	15,33	16,36	17,39	18,42
size n_1		11,25	12,28	12,32	13,35	14,38	15,41	16,44
	5		19,36	20,40	22,43	23,47	25,50	26,54
			18,37	19,41	20,45	21,49	22,53	24,56
	6			28,50	30,54	32,58	33,63	35,67
				26,52	28,56	29,61	31,65	33,69
	7				39,66	41,71	43,76	46,80
					37,68	39,73	41,78	43,83
	8					52,84	54,90	57,95
						49,87	51,93	54,98
	9						66,105	69,111
							63,108	66,114
	10							83,127
								79,131

Less than is significant. Also do **not** use more than 5% per tail ever.

(Notice that these European tables use commas for decimal points.)

Appendix 9 {Courtesy of Peter M Lee, and University of York (2005)}

Critical Values for T in the Wilcoxon Signed Mixed-Pairs Signed-Rank Test

The values below are the approximate critical values of T for two-tailed tests at level P. For a signicant result, the calculated T must be less than or equal to the tabulated value.

(Values of P are halved for one-tailed tests using R_- and R_+.)

n	$P = 0.10$	$P = 0.05$
5	2	-
6	2	0
7	3	2
8	5	3
9	8	5
10	10	8
11	14	10
12	17	13
13	21	17
14	26	21
15	30	25
16	36	29
17	41	34
18	47	40
19	53	46
20	60	52
21	67	58
22	75	65
23	83	73
24	91	81
25	100	89

The n here is for the number or pairs (n_d in the text). **Reject if R_W is less than shown**. These tables are for overall two tail significance. Do not ever use more than 5% per tail.

ANSWERS to ODD NUMBERED QUESTIONS

(Comments in parentheses are just that, comments, not expected to be part of the answers.)

Answers to Odd Chapter 1 Questions

1. Although a hearsay rumor about this fiendish activity already existed, Sue only had an **observation of one**. It does not confirm universal beheadings. These stories arise from the small number of people whose dogs have not only **bitten** someone but have displayed **certain kinds of uncharacteristic behavior** when quarantined at the pound. If she had asked enough people, she actually would have found that if the dogs behaved normally at the pound, as most do, they were returned. Otherwise they are humanely killed and then beheaded for lab examination of the brain to confirm that they have rabies. (However, when I was four, the adult neighbors actually hid their dog when it bit a small child because they simply attributed this certainty of beheading to the nastiness of authority in general. Fortunately, in this case neither the dog nor the child died. The neighbors might not have known that unless someone examined the dog to see that it was not showing rabies behavior, the public health authorities might need to require the child to undergo the then very painful rabies intraperitoneal injection process.)

3. This approach involves adjusting the degree of belief associated with preexisting information. So, it is primarily **Bayesian**; it contains subjective material, even though based on a sample of professional experience. (Its limitations are the possibility of bias, of opportunism, of uncontrolled coincidence, or of tricks of human perception. However, it is often all you have and in similar situations it has tended to come close to its goal. Thus it, like so many formal Bayesian approaches, is valuable because you cannot build a sample of identical buildings to get a proper sample before building the first. Advanced management and social science students should especially consider getting training in Bayesian statistics as well as frequentist statistics. However, whichever you use, you must make what you did and how you did it very clear whenever you go to press, or to your stockholders.)

5. As Mother Nature knows, the number of the sea lions in your area with the parasite is $N_A = 650$. The total size of the sea lion population is $N = 2000$. Dividing the absolute frequency by total number in the population gives $f = N_A/N = 650/2000 = 0.325$ meaning 32.5% **is, by definition**, the true **frequency**. Assuming that you are going to sample one sea lion truly at random, she knows you have a $P = 32.5\%$ chance of it being parasitized. However, unlike Mother Nature, your **sample relative frequency** is subject to chance, so you are stuck with estimating

$$\hat{f} = r_{parasitized}/n_{sampled} = 10/50 = 0.20,$$

which is only an **estimate** of the probability $\hat{p} = \hat{f} = 20\%$. The estimated probability and the true probability are not the same; it is to be expected that the **estimate $\hat{p}$ based on a sample will differ** from P; especially with less than huge samples, they can differ by quite a bit, due to chance alone. (A potential challenge to randomization in this sort of situation is that parasitized animals often act differently, so they may have a different probability of being captured for examination. This is also true for insects, etc.)

7. Drilling operations are well aware that their history of successes and failures involve chance. From a frequentist view it is possible to equate the percentages given as estimates of probability taken from actual counts. Therefore the answer may be found using the addition law as given by Equation 1.5 for probability. So, the probability of adequate steam is $P_S = 0.20$ and the probability of being below budget is $P_B = 0.51$ and the wells that are both have a probability of $P_{S \cap B} = 0.13$. Referring to the addition law, the probability of a well being either or both is $P_{S \cup B} = P_S + P_B - P_{S \cap B}$, which works out to $P_{S \cup B} = 0.20 + 0.51 - 0.13 = \mathbf{0.58}$, **or** you can say **58%**.

9. H.Y. Pothetical, MAg, has a test that has a probability of $P_{B|A} = 80\%$ of detecting a variety that is A, resistant to chicksucker, and it has a probability, $P_{B|!A} = 5\%$, that a randomly selected variety will produce a false positive. Past experience indicates that about $P_A = 1\%$ of new chicory varieties found by plant explorers are really resistant to chicksucker. What is the probability that a variety that the test claims is resistant will really be resistant? Well, $P_A + P_{!A} = 100\%$ provides the final input, $P_{!A} = 99\%$, needed to use Bayes' theorem,

$$P_{A|B} = \frac{P_{B|A} P_A}{P_{B|A} P_A + P_{B|!A} P_{!A}}$$

so $P_{A|B} = \frac{0.80 \times 0.01}{0.80 \times 0.01 + 0.05 \times 0.99}$ gives $P_{A|B} \approx 0.13913$; **only about 14% of the positive tests** will be truly positive.

11. The probability of such an organism being in the sample is $P_A = 0.0050$, so $P_{!A} = 0.9950$ because it is always true that $P_A + P_{!A} = 1$. The probability of a true positive is $P_{B|A} = 0.90$ of detecting a suitable organism, if present, that produces antibiotic effects with regard to any of the several pathogens that are being targeted. The probability of a false positive is $P_{B|!A} = 0.08$. According to Bayes' theorem, the probability that a sample with a positive test really contains such an organism is

$$P_{A|B} = \frac{P_{B|A} P_A}{P_{B|A} P_A + P_{B|!A} P_{!A}} .$$

So, plugging in the numbers,

$$P_{A|B} = \frac{0.90 \times 0.0050}{0.90 \times 0.0050 + 0.08 \times 0.9950} = 0.0535077 ,$$

which estimates that about **5.35%** of the positives will be **true** positives. The rest, 94.65%, will be false positives.

(Retesting might help but the retest might contain individuals which will falsely test positive regardless of how many tests you run. However, the number of chance false positives will tend to decrease at each retest. Assuming the tests characteristics are not altered by individuals which will falsely test positive regardless of how many tests you run, then

$$P_{A|B} = \frac{0.90 \times 0.05358}{0.90 \times 0.05358 + 0.08 \times 0.9465} = 0.388750 ,$$

so you can expect as many as 38.9% true positives on the first retest. Well, that is a lot better. At some point you might wish to change the testing protocol.)

Answers to Odd Chapter 2 Questions

The following answers are more detailed than is generally required for homework.

1. Tempting one or more of the subjects was an **experiment** in the purest sense of the word because the researchers did more than just observe, they created a situation, and that is an experiment. Experiments that have an effect on the subjects are discouraged in field anthropology, so anthropologists might say that it was an 'abomination'. It may have given a hint about the subjects' society, in that they have among them someone who would steal from strangers, or hates them, or that a bad egg in the clan was dominant, or that the group was absolutely desperate, or ... Experiments often give undeniable results, as in this case, but that does not mean that the experimenter knows **what the results imply**.

3. Scientists often start the calculations about the data by pretending for the sake of estimation that the null hypothesis is true and try to find if there is enough evidence to conclude that it is sufficiently unlikely that it must be rejected. This is merely a temporary starting point upon which to base their calculations. It does not necessarily represent reality. So: **No**, you **cannot accept** a hypothesis solely because you cannot prove that it is wrong. In other words, you can **never accept** the null hypothesis.

5. You use the binomial theorem to find the probability of randomly drawing $r = 4$ successes (statistically speaking) out of $n = 12$ trials, but with respect to the given $p = 1/2$. Recall that multiplying $(1/2)^r$ times $(1/2)^{n-r}$ gives you $(1/2)^n$. You multiply this result times the relevant number of combinations to get a variant of the binomial equation, Equation 2.5, so

$$P_4 = \binom{n}{r} \left(\frac{1}{2}\right)^n = \frac{n!}{r!\,(n-r)!} \left(\frac{1}{2}\right)^n = \frac{12!}{4!\,(12-4)!} \left(\frac{1}{2}\right)^n$$

$$= \frac{12 \times 11 \times 10 \times 9}{4!} \left(\frac{1}{2}\right)^{12} = \frac{11880}{24} \frac{1}{4096} = 0.12085$$

or **about 12%**.

7. Your hypothesis pair is **H0**: THERE ARE THE SAME NUMBER OF EACH GENDER. and **HA**: THERE ARE MORE FEMALES, and you are using the $\alpha = 0.05$ criterion. You found $r = 35$ females out of $n = 50$ plants. So, using the 50% test, you could calculate a cumulative binomial probability, adding up the probability of all outcomes this large or larger, using

$$p\_value = \sum_{i=r}^{i=n} \binom{n}{i} \left(\frac{1}{2}\right)^i \left(\frac{1}{2}\right)^{n-i} \text{ or } = \left(\frac{1}{2}\right)^n \sum_{i=r}^{i=n} \binom{n}{i} = \left(\frac{1}{2}\right)^{50} \sum_{i=35}^{i=50} \binom{n}{i}$$

(neither would be fun to work out by hand). However, you could not validly stop with the p_value here because you are using a hypothesis **pair**, and that necessitates a Neyman-Pearson test that either gives you a definite yes, there is sufficient evidence to reject **H0**, or gives you the result that there is insufficient evidence to reject **H0**. Nonrejection of

H0 in no way supports **H0**. So, you could only compare your observed p_value to α, only declaring statistical significance if $p\_value \leq \alpha$. The table in Appendix 1 has done all this painful and highly subject to rounding calculation for you and provides the minimum number of successes necessary to reject **H0** and declare statistical significance. For one tail, given $n = 50$, this table indicates that for any value of $r \geq 32$, this test is **statistically significant at** $\alpha = 0.05$. That is, the $p\_value_{given\,r} \leq 0.05$, your prechosen criterion; how much less (or more) is irrelevant because that has its own limitations resulting from chance, when used this way. You therefore accept **HA** and conclude that there are more female plants than male plants. If chance has led you to be wrong in declaring statistical significance, you have gotten a Type I error, rejecting **H0** due to chance alone when it is, in fact, true. (A solution to less than all the quadrats having the targeted weed is to take the map and random number table with you and keep randomly selecting quadrats until you have examined 50 specimens. Incidentally, using RStudio, I got a $p\_value = 0.00330$.)

9. This is a two tail test, meaning that it simultaneously tests whether something is either greater than or less than hypothesized. Because the total criterion is $\alpha = 0.05$, you must use half of it, $\alpha/2 = 0.025$, to evaluate each direction. That is, for the tables provided, r is counted in the direction in which it is largest. **Choosing between the tails counts as part of the analysis**. Now when you look at the table in Appendix 1 you look for the value, r, that is closest to statistically significant at $\alpha/2 = 0.025$ for $n = 50$ and that is $r_{\alpha/2} \geq 33$; this is larger than for r_α one tail. (Because you used a hypothesis pair rather than one hypothesis, you have committed to the Neyman-Pearson model, so you must also specify a criterion, α.) Because $r = 35$, which is even less probable than $\alpha/2$, the test is **statistically significant**. That means that by your a priori commitment to a criterion of your choice, **H0** is rejected because its likelihood is **less than** the cutoff, $\alpha/2$. Therefore you accept **HA**: THE NUMBER OF FEMALES IS NOT THE SAME AS THE NUMBER OF MALES. (Note that at the same sample size as in Question 7, $n = 50$, it takes as large or larger numbers of successes, r, to get two tail statistical significance as contrasted with one tail statistical significance.)

11. Because the results are paired vertically, by days in this case, you simply mark the pairs as to whether the difference is + or - to do a one tail **sign test**; in accordance with the alternative hypothesis, greater than A is a success (+):

$Site A:$	5.5	7.3	2.7	6.2	4.2	0.1	3.8	5.4	2.9	7.1	1.1	5.2	5.7	2.3	3.5
$Site B:$	2.8	2.2	9.4	8.2	1.2	7.1	3.6	5.8	5.2	4.3	7.7	0.8	9.5	0.9	1.1
$Sign$	+	+	−	−	+	−	+	−	−	+	−	+	−	+	+

There are $r = 8$ successes out of the sample of size $n = 15$ valid trials (the ones without ties) with the criterion previously selected to be $\alpha = 5\%$. So, you use the table in Appendix 1 to find that it would require at least $r_\alpha = 12$ successes to be statistically significant at this 5% level. Because $r < r_\alpha$ is more likely than what you have chosen as a reasonable minimum probability, you should conclude that **you cannot reject H0**, and that the results are inconclusive. Alternatively, you could obtain a p_value using $p\_value = \left(\frac{1}{2}\right)^n \sum_{i=r}^{i=n} \binom{n}{i}$. That calculation would be a little cumbersome, but not too difficult. Comparing that to $\alpha = 5\%$ would result in the same conclusion.

13. Either a sign test or a 50% test challenges whether **H0**: $p = p_0 = 1/2$ is true and you can choose to take, at the very most, an $\alpha = 5\%$ probability of getting a Type I error. However, the first part of this question is about β, the chance of a Type II error when the truth is really that $p = p_A = p_0 + \mathcal{D}$. First we need to find β, the probability of being so

unlucky as to get $r_\alpha < 5$ successes out of a total of $n = 5$ random trials by chance alone **when** $p_A = 80\%$. To find β, the probability of not detecting statistical significance when it is really true that $p = p_A = 0.80$, you could use a version of Equation 2.6,

$$\beta = \sum_{i=0}^{i=r_\alpha-1} \binom{n}{i} (p_A)^i (q_A)^{n-i},$$

where $q_A = 1 - p_A$, to calculate the cumulative probability of all outcomes **that are less than** r_α. This equation sums the probabilities of successes from 0 successes to $r_\alpha - 1$ successes (up to but not including r_α), the largest number of successes that is not as large as $r_a = 5$ (r_α is the minimum number of successes to get statistical significance). Recall that $0! = 1$ and $x^0 = 1$ for any x. Your calculations, whether by some sort of electronic device or by hand, would produce something like

$$\beta = \sum_{i=0}^{i=4} \binom{5}{i} \frac{5!}{i!\,(5-i)!} (0.80)^i (0.20)^{5-i},$$

$$\beta = \frac{120}{1\,(120)}(1)(0.00032) + \frac{120}{1\,(24)}(0.80)(0.0016) + \frac{120}{2\,(6)}(0.64)(0.008)$$

$$+ \frac{120}{6\,(2)}(0.512)(0.04) + \frac{120}{24\,(1)}(0.4096)(0.20) = 0.67232,$$

or about a 67% **probability**, β, **of getting a Type II erro**r, failing to reject **H0** when it is, in fact, false. So, even though $p_A = 80\%$ is very large, the probability of detecting even this large a difference, $\mathcal{D} = 0.30$, above $p_0 = 0.50$, using a sample of size $n = 5$, the

power is $1 - \beta = 1 - 0.67232$ or about 33%.

That is, the power can be seen as the probability of **not** getting a Type II error, given that the difference you choose actually exists. Clearly the **power** of a sign test or a 50% test (for instance $1 - \beta = 33\%$ with even a very big $p_A = 80\%$) to detect a difference this large or larger with a sample size as small as $n = 5$ does not provide enough power to be useful. You used $r_\alpha - 1$ to calculate β because you were interested in finding the total probability of getting any number of successes that was 1 **or more less than the minimally statistically significant** r_α. This is because you did not wish to include r_α because it would be counted as statistically significant.

15. Remember this refers to true randomness without flimflam. On the first draw there would be 4 chances in 52 of getting a queen. So the probability would be $4/52$. Then the probability of getting another of the 3 queens from the remaining pack of 51 cards by chance alone would be $3/51$. But if that happened there will be only 2 queens left in the remaining 50 cards, so the probability of drawing another queen would be $2/50$. Now only 1 of the queens would be left in the remaining cards and the probability of getting it would be $1/49$. Having found this by counting and applying the multiplication law repeatedly, the probability that your roommate would get these four cards in only four random draws turns out to be

$$p_{4\,queens} = \frac{4}{52} \times \frac{3}{51} \times \frac{2}{50} \times \frac{1}{49} = 0.00000369.$$

There is another, more formal, way to approach this problem. You can see that this is just the probability of getting $r = 4$ things out of $n = 52$ things, at random without regard to order $(1/[number\ of\ ways])$. You can get the same answer using the reciprocal of the combinations of $n = 52$ things drawn $r = 4$ at a time,

$$1/{}^nC_r = 1/\left(\frac{52!}{4!\,(52-4)!}\right) = 1/(\frac{52 \times 51 \times 50 \times 49}{4 \times 3 \times 2 \times 1}) = 24/6497400 = 0.00000369\,,$$

also written $1/\binom{n}{r} = 3.69^{-6}$. (This called is a combination and it resembles a permutation except that the first $r! = 4!$ in the denominator is there because there would be $4 \times 3 \times 2 \times 1$ ways to order, that is arrange, the cards. But we are not interested in what order the roommate draws cards. The $(52 - 4)! = 48!$ canceled out, top and bottom.)

17. You can arrange n meetings $n!$ ways. So there are

$$7! = 7 \times 6 \times 5 \times 4 \times 3 \times 2 \times 1 = 5040$$

ways to meet with them one at a time. If you intend to choose the order at random you only have a probability of $p = 1/5040 = 0.000198$ of getting it right. Better head for the bomb shelter!

19. You have ${}^5P_3 = \frac{5!}{(5-3)!} = 5 \times 4 \times 3 = 60$ possible ordered ways (you canceled out $(2)! = (2 \times 1)$ in the numerator and the denominator) to cut 3 wires out of 5 wires in the correct order. So, you have a $p = 1/60$ or $\approx 1.67\%$ probability of moving on to the next situation of survival-rather-improbable.

21. Since the teams are identical, then the order is not important and you can choose the 9 countries out of 12 countries $\binom{12}{9} = \frac{12!}{9!(12-9)!} = \frac{12 \times 11 \times 10}{3!} = 2 \times 11 \times 10 = 220$ different ways. So, you have a $p = 1/220 = 0.004545$ probability of getting the optimal selection of which 9 countries get teams, when randomly choosing from the 12 countries.

23. This is simply a question of what is the chance of getting $r = 5$ white flowering plants out of ten plants when $p = 0.25$ for each randomly chosen seedling. Under random selection, the binomial probability is

$$\binom{10}{5}\left(\frac{1}{4}\right)^5\left(\frac{3}{4}\right)^{(10-5)} = \frac{10!}{5!\,(10-5)!}\left(\frac{1}{4}\right)^5\left(\frac{3}{4}\right)^{(10-5)}$$

$$= [(10 \times 9 \times 8 \times 7 \times 6)/(5 \times 4 \times 3 \times 2)]\,(0.000977)\,(0.237) = 0.008342743$$

or about 0.8%. (There were two instances of $5!$ in the denominator and only one could cancel out against the numerator.)

25. This is simply the binomial problem of the probability of getting $r = 2$ successes, in this case arbitrarily designated girls, out of $n = 4$ live births, assuming $p = 1/2$ for each birth. So the probability of 2 children of each gender is

$$P = \binom{4}{2}(1/2)^4 = \frac{4!}{2!\,(4-2)!}\,(1/16) = \frac{4 \times 3}{2 \times 1 \times 16} = 3/8,$$

which equals 37.5%.

27. The probability of 2 to 4 of their offspring surviving to age four is

$$B(7,\,0.75,\,2,\,4) = \sum_{i=2}^{i=4}\binom{7}{i}0.75^i 0.25^{(7-i)} = 0.2422\,,$$

about 24%. (Where there is no social security, surviving offspring are a source of support in old age. This along with the low infant survival rate is one of the many reasons that developing world folks have a lot of babies. However, with social security maybe they would stop at a lower number. This makes a pragmatist wonder if social security might be a good investment, at least in environmental terms. Also note that the idea that the third world and our ancestors led a 'simpler life' is a bad joke, about as funny as a bag of dead babies.)

Answers to Odd Chapter 3 Questions

1. They appear not to be aware that zero is merely a reference point. Regardless of their power, statistical tests become less sensitive as the difference from zero becomes small, never getting close to zero difference. Experiments and observations should target a value (a reasonable minimum difference) that the user can accept as meaningfully different from the zero reference point.

3. Even though 0 difference is only a reference point, the p_value is the likelihood that your results occurred by chance alone if your **H0** hypothesis was perfectly true (there was really 0 difference). However, α is a probability (a single point p_value) that you have a priori agreed will be the cutoff below or equal to which you will reject **H0** as false.

5. NHST originally meant Null Hypothesis Significance Testing. However, because of the mistaken idea that the tests challenge absolute zero differences, it is also used derogatorily to mean Nil Hypothesis Significance Testing. Both words have similar meaning, but the word nil tends to imply non existent, while null may classically have been intended to imply lack of a detectable effect (a meaningful difference).

7. The $1 - \beta$ is the **power**, the probability of detecting the (a priori chosen) reasonable minimum difference. (It is interesting that the power is the complement of β, but that is not the meaning that is most relevant to the question.)

9. Delta, δ, is the unknowable exact true difference between two means or probabilities. In this book $\mathcal{D}$, or D, symbolizes the observed difference between samples, and $\mathcal{D}_{min}$ is the **reasonable minimum difference** that you a priori select. (Note that in this book, $\mathcal{D}$ is calligraphic and that D and D_{min} are italic).

11. A small true difference, δ, makes it more difficult to achieve your desired power, and a large true difference makes it easier to achieve your desired power (in terms of sample size, n).

13. Check the literature to see what your peers see as a minimally meaningful difference, or in business, make the reasonable minimum difference only marginally larger than or equal to (**not** smaller than) added cost.

15. In application, either discrete or continuous paired data would be suitable for a sign test, and counts or other discrete data would be suitable for a 50% test.

17. No, you cannot accept **H0**, you are only challenging **H0**. Even the failure of a very powerful experiment to detect a difference from **H0** would not justify accepting **H0** (reporting that there was no difference).

19. You are justified in doing a subsequent (post hoc) power analysis if you have no control over the sample size or if you lose data, as by patients dropping out of a medical

trial. You should **always** do a beforehand (a priori) sample size power analysis to estimate what the sample size should be. Even if you cannot control the sample size, an a priori estimate tells you when to stop or when to go on.

21. You must choose the **reasonable minimum difference** (your target) in advance whether you can control the sample size or not. Otherwise you would not be able to estimate your attained power to discover if the reasonable minimum difference actually exists, or is exceeded.

23. In $\beta = \sum_{i=0}^{i=r_{cutoff}-1} \binom{n}{i} (p_A)^i (q_A)^{n-i}$, you use $r_{cutoff} - 1$ in the summation because you are summing all the probabilities that are **less than** (not equal to) the r value that is the next integer below r_{cutoff}, the value (number of successes) at which statistical significance would be declared. The probability, p_A, is the value hypothesized by **H0** (usually by implication) plus the reasonable minimum difference, $p_A = p_0 + \mathcal{D}_{min}$.

25. Put simply, this sort of difficulty arises when you combine discrete numbers with continuous numbers. In this case we are attempting to map the results, binomial counts, onto a continuous probability scale. The outcome is inexact with overlapping, etc.

Answers to Odd Chapter 4 Questions

1. Try to round only **after** you get a final answer that is ready to report.

3. The following $n = 6$ measurements are in cubic centimeters:

$$4.5cc \ 1.2cc \ 3.7cc \ 1.3cc \ 4.3cc \ 1.8cc$$

To hand calculate a sample variance estimate and the standard deviation estimate, you need an estimate of the **sample mean**:
$\bar{x} = (4.5 + 1.2 + 3.7 + 1.3 + 4.3 + 1.8)/6 = 2.80\,cc$. Now for the variance you need to sum the squared differences, $\sum (x - \bar{x})^2 =$
$(4.5 - 2.8)^2 + (1.2 - 2.8)^2 + (3.7 - 2.8)^2 + (1.3 - 2.8)^2 + (4.3 - 2.8)^2 + (1.8 - 2.8)^2$
$= (1.7)^2 + (-1.6)^2 + (0.9)^2 + (-1.5)^2 + (1.5)^2 + (1.0)^2 ,$

so $\sum (x - \bar{x})^2 = 2.89 + 2.56 + 0.81 + 2.25 + 2.25 + 1.00 = 11.76\,cc^2$ is the sample sum of squares, and the sample **variance** is

$$S^2 = \frac{\sum (x - \bar{x})^2}{n - 1} = 11.76/(6 - 1) = 2.3520\,cc^2,$$

or, simply 2.35, as it will ultimately be reported. But a personal record to more figures might be helpful if someone wants to continue calculating, something like the standard error of the mean, or the sample size for a future experiment. The units of measure are often left out for variance because measures like cc^2 seem to be somewhere in the fourth dimension. To estimate **the sample standard deviation**, take the square root of the unrounded variance estimate,

$$S = \sqrt{S^2} = \sqrt{2.3520cc^2} = 1.53362\,cc,$$

and this can conveniently be reported in the original units of measure. Using the customary rounding for most journals, it would be reported as 1.53 cc. It might be preferable rounded to 1.5336 cc as a reference for possible future sample size calculations, but not for publication. Note that the standard deviation is in the original units of measure. (The 'unrounded' values are from a calculator, so they will tend to contain rounding, at least in the last place.)

5. You should have used $n - 1 = 5$ degrees of freedom, where $n = 6$ is the sample size, in Question 3. This is because you used up a degree of freedom estimating $\bar{x}$. Every time you use an additional estimate such as a sample mean in calculating a variance, you must reduce the degrees of freedom by one.

7. The Central Limit Theorem (CLT) states that given that a distribution has a finite mean, μ, and standard deviation, σ, then as the sample size, n, becomes larger, the distribution of the sample means will approach a normal distribution with mean, $\mu_{\bar{x}} = \mu$ and standard error of the means, $\sigma_{\bar{x}} = \sigma/\sqrt{n}$ even when the underlying frequency distribution is not normal. (This theorem makes most of the rest of this book work.)

9. The variance of the data set in Question 3 is $S^2 = 2.3520\,\text{cc}^2$ (you will need a value that is as unrounded as possible). The standard error of the mean can be calculated from its square root, that is, from the **standard deviation**, $S = \sqrt{S^2} = 1.53362\,\text{cc}$. Then via the CLT, the **standard error of the mean**,

$$S_{\bar{x}} = S/\sqrt{n} = 1.53362\,\text{cc}/\sqrt{6} = 0.626099\text{cc},$$

approximately, and you might round this to 5 significant figures for some applications. However, the error deviation of the mean may also be calculated **directly from the variance** using $S_{\bar{x}} = \sqrt{S^2/n} = \sqrt{2.3520/6} = 0.626099$. (There might be a small difference due to rounding; your calculator might produce different values in, say, the 6th decimal place. And, of course any statistics you actually report in literature are customarily only to one more place than the data.)

11. As an exercise in the calculation process, pretend you could use a 'Z test' for **H0**: THERE IS NO DIFFERENCE vs **HA**: μ IS GREATER THAN $\mu_0 = 5.0$ METERS, with a criterion, $\alpha = 5\%$, given the following sample in meters:
5.5m, 5.9m, 5.0m, 5.1m, 5.3m, 5.4m, 5.1m, 5.5m, 5.8m, 5.4m. You first calculate the mean,
$\sum x = 5.5 + 5.9 + 5.0 + 5.1 + 5.3 + 5.4 + 5.1 + 5.5 + 5.8 + 5.4 = 54\,\text{m}$, so the mean $\bar{x} = \sum x/n = 54/10 = 5.40\,\text{m}$. Then you need $\sum (x - \bar{x})^2$, the sum of squared deviations from the (sample) mean, which equals $(5.5 - 5.4)^2 + (5.9 - 5.4)^2 + (5.0 - 5.4)^2 + (5.1 - 5.4)^2 + (5.1 - 5.4)^2 + (5.5 - 5.4)^2 + (5.3 - 5.4)^2 + (5.4 - 5.4)^2 + (5.4 - 5.4)^2 + (5.8 - 5.4)^2$, which sums to $\sum (x - \bar{x})^2 = 0.01 + 0.25 + 0.16 + 0.09 + 0.01 + 0.0 + 0.09 + 0.01 + 0.0 + 0.16 = 78\text{m}^2$. The variance $S^2 = \sum (x - \bar{x})^2 / (n - 1) = 0.78\text{m}^2/9 = 0.08666667\text{m}^2$. So, the standard error of the mean could be calculated either of two equivalent ways, $S_{\bar{x}} = S/\sqrt{n}$ where $S = \sqrt{S^2}$, or $S_{\bar{x}} = \sqrt{S^2/n}$. The student may do it either way; the answer will be **the same**. For convenience let's use $S_{\bar{x}} = \sqrt{S^2/n} = \sqrt{0.08666667\text{m}^2/10} = 0.09309493\,\text{m}$. Finally you have the information you need for an (under sized) one tail-one-sample 'Z test',

$$Z = \frac{\bar{x} - \mu_0}{S_{\bar{x}}} = \frac{(5.40 - 5.00)\,\text{m}}{0.093095\text{m}} = 4.2967,$$

where the meters/meters have canceled out. So the difference between the sample mean and the true mean is 4.2967 standard errors of the mean ($Z = 4.2967$). You check this outcome against the minimum Z value required for statistical significance, which turns out to be $Z_{\alpha=0.5} = 1.645$, and discover that your result, $Z = 4.2967$, is far enough into the upper tail (sufficiently unlikely) as to justify rejecting **H0**. So, you reject **H0** and you **accept HA**. (This is simply to demonstrate the calculation and you would not use this particular test on anything but a very large sample; stay tuned.)

13. A standardized z_i value can be the difference between a single data value, x_i, and the mean divided by the **sample standard deviation**, meaning the divisor is the standard deviation of the **raw data**. However, a Z value is always a difference between means, divided by the **standard error of the means**.

15. The **mean** is $\Sigma x / n$ where $\Sigma x = 1 + 2 + 3 + 4 + 5 = 15$. So $\bar{x} = 15/5 = 3$. Next, to estimate the **variance** you need to sum the squared differences from the mean,

$$\Sigma (x - \mu)^2 = (1 - 3)^2 + (2 - 3)^2 + (3 - 3)^2 + (4 - 3)^2 + (5 - 3)^2$$

$$\Sigma (x - \mu)^2 = 4 + 1 + 0 + 1 + 4 = 10,$$

$$S^2 = \Sigma (x - \mu)^2 / (n - 1) = 10/4 = 2.5.$$

The divisor is the **degrees of freedom**, in this case, $df = n - 1 = 4$, to find the variance, The other, very important, name for the variance is the **mean square error**. The **standard deviation** is $\sqrt{S^2} = \sqrt{2.5} = 1.58114$ (in the same units as the mean and the data). Incidentally these were counts, so the input numbers are perfect and the amount of rounding is at your discretion.

(Another more classical approach to start the calculation is to create a table, like in a spreadsheet,

x	$x - \bar{x}$	$(x - \bar{x})^2$
1	-2	4
2	-1	1
3	0	0
4	1	1
5	2	4
Σ 15	$0\checkmark$	10

$\bar{x} = 15/5 = 3$

$S^2 = \Sigma (x - \mu)^2 / df = 10/ (5 - 1) = 2.5$, the variance, also called the mean square error.

$\sqrt{S^2} = \sqrt{2.5} = 1.58114$, the standard deviation.)

17. Using Equation 4.3 to find the pooled variance:

The means are $\bar{x}_1 = \Sigma x_{i,1}/n_1 = 4.2000 \, \text{kg}$ and $\bar{x}_2 = \Sigma x_{i,2}/n_2 = 4.6533 \, \text{kg}$.

The **sum of squares** for population X_1 is $\sum_{1, i=1}^{i=n_1} (x_{1,i} - \bar{x}_1)^2 = 63.0200 \text{kg}^2$ and

the **sum of squares** for population X_2 is $\sum_{2, i=1}^{j=n_2} (x_{2,j} - \bar{x}_2)^2 = 139.6573 \text{kg}^2$.

So, the **pooled variance** is $S_{1,2}^2 = \frac{\sum_{j=1}^{j=n_1} (x_{1,j} - \bar{x}_1)^2 + \sum_{j=1}^{j=n_2} (x_{2,j} - \bar{x}_2)^2}{n_1 + n_2 - 2} \text{kg}^2$, which means that

$$S_{1,2}^2 = \frac{63.0200 + 139.6573}{15 + 15 - 2} \, \text{kg}^2 = 7.238476190 \, \text{kg}^2.$$

To estimate any variance, you avoid rounding (at least minimize it as much as possible) until after taking the square root, but this variance can eventually be reported as 7.24.

However, do not rush to round statistical calculations until you are finished. It might then be reasonable to round this and report $\bar{x}_1 = 4.20\,\text{kg}$ and $\bar{x}_2 = 4.65\,\text{kg}$. The **degrees of freedom** are $df = n_1 - n_2 - 2$ equals 28 because you **used up 2 degrees of freedom estimating two different values of the mean**. The pooled **standard deviation** is the square root of the pooled variance[1], $S_{1,2} = \sqrt{7.23847619\,\text{kg}^2} = 2.69044163\,\text{kg}$, which will ultimately be reported as $S_{1,2} = 2.69\,\text{kg}$. The means are $\bar{x}_1 = 4.2000\,\text{kg}$ and $\bar{x}_2 = 4.6533\,\text{kg}$ which will be reported as $\bar{x}_1 = 4.20\,\text{kg}$ and $\bar{x}_2 = 4.65\,\text{kg}$, respectively. Note that the outcomes are to be reported only **to one more figure than the input data**. The input is not integers, and rounding as a last step before reporting is necessary. In general, it would be best to keep such values to at least five figures in your personal archive records as long as you do not take the extra figures too seriously.

19. No, 'standard error' never refers to the population of raw data. It most often refers to the population of **all possible means** of the raw data population. 'Standard error' can refer to the standard deviation of any **statistic**, but not raw data.

21. The Neyman-Pearson approach compares the p_value to α and only permits publication of 'yes it did exceed α' (statistical significance), or 'no it did not exceed α' (no conclusion regarding **H0**); other information regarding the magnitude of the p_value does not in any way qualify for publication with this approach. Fisher's approach lets the researcher report the p_value as evidence against **H0**; the actual acceptance or rejection is up to the researcher (it does not use α for decision making).

23. Fisher intended the researcher to use the p_value to provide evidence against a hypothesis regarding the actual population in nature. His approach assumes that sometimes the researcher can not reject **H0**, because the risk of being wrong is too large. (This might be seen as a stepping stone to the presently popular Neyman-Pearson approach which differs by making yes-no decisions. But, Fisher disliked this idea.)

25. Even Fisher (who hated them) acknowledged that Neyman and Pearson's approach was excellent for quality control. So, the Neyman-Pearson approach is the answer. (But Neyman and Pearson felt that it can also be used in most instances where Fisher's approach might also be acceptable though different.)

27. A confidence interval is a range of values that has a $1 - \alpha$ probability of including some particular statistic. For instance, a 95% confidence interval has a $1 - \alpha = 0.95$ probability of including the statistic of interest (more in Chapter 6).

Answers to Odd Chapter 5 Questions

1. Student's t test adjusts for the fact that you do not know the variance (thus the standard deviation). Especially for inference based on the normal distribution, you need to adjust for the estimation of the data's variability unless the degrees of freedom is big (much larger than 'large', $n = 30$).

3. The hypothesis pair,
H0: THE TREATMENTS HAVE THE SAME EFFECT
HA: TREATMENT $_2$ HAS MORE EFFECT THAN TREATMENT $_1$
tells you that this is a one tail test. That is, you need to find out if the result of Treatment $_2$ is really **greater than** the result of Treatment $_1$ at $\alpha = 1\%$, using a two sample, one tail

---

[1] Recall that units of measure square along with their numerical values.

Student's t test,

$$t = \frac{\bar{x}_2 - \bar{x}_1}{S_{\bar{x}_1, \bar{x}_2}} \qquad df = 2n - 2.$$

You can use $df = 2n - 2$ for $n_1 = n_2$ rather than $df = n_1 + n_2 - 2$. You have estimated the pooled sample variance $S_{1,2}^2 = 7.23847619\,\text{kg}^2$, and the means $\bar{x}_1 = 4.200000\,\text{kg}$ and $\bar{x}_2 = 4.653333\,\text{kg}$. The pooled standard deviation, $S_{1,2} = \sqrt{7.23847619\,\text{kg}^2} = 2.69044163\,\text{kg}$, and the pooled standard error of the means, $S_{\bar{x}_1, \bar{x}_2} = S_{1,2}\sqrt{(2/n)}$ is $S_{\bar{x}_1, \bar{x}_2} = 2.69044163\sqrt{(2/15)}\,\text{kg} = 0.9824104\,\text{kg}$. So,

$$t = \frac{(4.653333 - 4.200000)\,\text{kg}}{0.9824104\,\text{kg}} = \frac{0.45333333333}{0.9824104} = 0.4614497 \text{ and } df = 28.$$

Use a table of t such as Appendix 3 to find that $t_{1\%, 28} = 2.467$. The t value from the test is less than $t_{1\%, 28}$, so the two samples are **not** statistically significantly different; the test tells you **nothing about H0**.

(You can rarely find anything using such small samples with an $\alpha = 1\%$ unless the difference between the treatments is huge. A researcher might a priori take a $1/20$ chance of getting a Type I error and use $\alpha = 5\%$, but there might be nothing meaningful to detect anyway.)

5. A paired t test does reduce the degrees of freedom and that could make it harder in the sense of needing a greater sample size to detect a meaningful difference. But if you pair correctly, it greatly reduces the variance and eliminates some of the unrelated interference. **In that case** statistical significance is **easier** to attain even with less degrees of freedom. Such pairing is great for experiments where each worker gets both treatments alternately, such as before and after modifying working conditions on a shop by shop basis, etc. So, pairing **almost always** makes it easier to detect meaningful differences in spite of the loss of degrees of freedom.

7. Almost all authors would agree that it would be unacceptable to use the binomial approximation **when either** $np < 5$ **or** $nq < 5$. (You should be aware that $q = 1 - p$, so when one is large the other is small. Such rules help to achieve a reasonable approximation of the normal curve. As will become obvious in some following chapters, these rules by no means guarantee a sufficient power to detect as small a difference as you are likely to deem meaningful.)

9. The **standard error of the proportion** (the probability) of a binomial distribution defined by $n = 72$ and $p = 0.42$ is

$$\sigma_p = \sqrt{pq/n} = \sqrt{p\,(1-p)\,/n} = \sqrt{0.42\,(0.58)\,/72},$$

so

$$\sigma_p = 0.058166.$$

(Note this defines the standard deviation σ_p, not its estimate, S_p, which would be calculated the same way, but with $\hat{p}$ and $\hat{q}$. Three figures, the input plus one, $\sigma_p = 0.0582$, would be adequate for reporting, but do not round before the square root calculation or before subtraction.)

11. The granting agency has based the funding on the assumption that the snails are common in 45% of the rice paddy areas within your very large developing world tropical rice growing region. So, according to the granting agency, $p_0 = 0.45$. Before obtaining

the data, you must choose a convincing criterion, so you choose $\alpha = 1\%$ and form a hypothesis pair,

H0: THERE IS NO DIFFERENCE FROM 45%.

HA: THERE IS MORE THAN 45%.

Your crews randomly sample $n = 360$ rice paddy watersheds and find the snails in $r = 252$ of them. So, your sample estimate is $\hat{p} = 252/360 = 0.70$ (to any number of decimal places). You analyze this one tail one sample test using Equation 5.11, $Z = \frac{r/n - p_0}{\sqrt{p_0 q_0 / n}}$, and by substitution you get

$$Z = \frac{0.70 - 0.45}{\sqrt{(0.45 \times 0.55)/360}} = 9.53463.$$

You then go to a table of t and find that the cutoff for a one tail 1% criterion is $Z_{\alpha=1\%,} = 2.3263$. Because $Z = 9.6346$ clearly exceeds the cutoff, your test is statistically significant and you reject **H0** and accept **HA**. That is, you have sufficient evidence to make a good case for getting your funding increased.

(This parasite is no joke and is what most worried me when fording streams that received rice paddy drainage in the tropics. It also occurs in South America, such as in the Amazon river. But this data is fictitious and is only for illustration.)

13. You **do not** need to correct for continuity for big samples, say, $n > 60$.

15. Cohen's $d = \frac{\bar{x}_2 - \bar{x}_1}{\sigma}$ is an attempt to create an ES via standardizing measures on different scales, or measures of different things, so that their effect sizes might be compared. But it breaks down when you compare effect sizes. You cannot know σ, so you must estimate it, and that introduces confounding, and dependence on the degrees of freedom. The only 'usefulness' is to be aware that Cohen's d is misleading with respect to sample size and inference as well as for comparing effect sizes, so Cohen's d is **not** useful.

Answers to Odd Chapter 6 Questions

1. Although the **sample** mean, $\bar{x}$ (or other **statistic** of interest), is **often marked** at the center of a CI, the location of the **true** mean, μ (or other **parameter** of interest), is probabilistic and remains unknown. You can only estimate the **likelihood** that the CI includes it, not where the true mean μ actually is. The parameter of interest, for instance, μ, can really be anywhere on the real number line. However the confidence interval attempts to trap it. So, presuming random sampling, etc. for, say, a 99% CI, you can expect to be 99% confident that it is **somewhere** inside the CI.

3. It is conservative to use $Z_{0.025} \approx 2$ for the CI if $n \geq 60$. The standard error of the mean, $S_{\bar{x}} = \sqrt{35.5/100} = 0.595819$. So, a conservative (meaning more likely to be wide enough) 95% **right or left** whisker length can be calculated as $2(0.59582) = 1.19164$, which is really only accurate to three figures because the estimated central measure $\bar{x} = 5.20$ is assumed to be to three figures. So, round the whisker length up to three figures, 1.20, and report the 95% CI as $4.00 < \mu < 6.40$. It could be sketched as a line between 4.00 and 6.40 with a mark at 5.20 to represent the point estimate of the sample mean. (When the mean, or other central measure is of fixed length, round the whisker length up before either adding or subtracting.)

5. To find the 95% confidence interval you need to calculate the mean, and standard error of the mean. You can do this by first summing the data and dividing the result by $n = 15$ somewhere out of the way to get an estimate of the mean, $\bar{x} = 4.20000$ kg, which you will ultimately report as 4.20 kg. Now, although you could combine some column operations, let's describe this as though on a ledger ruled sheet. The next column is the first column minus the mean, the 'errors'. If everything is done correctly this column will sum to zero as a check. Now for a column which will be important to some other procedures as well as estimating a simple standard error. It will contain the squares of the differences, and its total will be the sum of squares of error, $\Sigma (x - \bar{x})^2 = 63.02000$ kg^2, for this sample, but often it will be a very large value. Dividing this by $n - 1 = 14$ gives the variance as $S^2 = 4.5014285714$kg^2 (or to even more figures). The estimated standard deviation (the sample standard deviation) is the square root of this $S = \sqrt{S} = 2.121657034355$ kg. It would be best if either the sample standard deviation or the variance were reported to maybe five figures, because they would be useful to researchers wishing to estimate sample sizes for future research. Maybe your journal will insist you report them to one more figure than the data, if at all, but that is not the question here. The standard error of the mean can be estimated from the sample standard deviation, $S_{\bar{x}} = S/\sqrt{n} = 0.54780949$ kg, give or take some figures. However, you can skip the calculation of the sample standard deviation and estimate the standard error of the mean directly from the variance estimate, $S_{\bar{x}} = \sqrt{S^2/n} = 0.54780949$ kg. You need a source that gives the probabilities associated with t to at least four decimal places and to look up $t_{0.025, \, 14df} = 2.1448$ (which also equals $|t_{0.975, \, 14df}|$). Now you can estimate the 95% CI, using

$$\bar{x} - t_{\alpha/2, \, n-1} S_{\bar{x}} \leq \mu \leq \bar{x} + t_{\alpha/2, \, n-1} S_{\bar{x}} \, .$$

With substitution of the values,

$$4.2000 - 2.1448 \, (0.54781) \leq \mu \leq 4.2000 + 2.1448 \, (0.54781) \text{ killograms,}$$

this comes out to
$$3.02506 \text{ kg} \leq \mu \leq 5.3749 \text{ kg} \, ,$$

which should be appropriately rounded (down on the left and up on the right) in light of the accuracy of the the data. This is

$$3.02 \text{ kg} \leq \mu \leq 5.38 \text{ kg} \, ,$$

the CI to one more figure than the data. Note that the rounding is always up to the right and down to the left. Always try to keep lots of figures around until this last step, reporting the outcome. (Some software, such as a spreadsheet, produces a small rounding error in the third decimal place for this calculation.)

7. You want the 99% confidence interval. There is a total of $\mathbf{n} = 2n = 30$ and it is necessary to estimate two means, so there are 28 degrees of freedom. The variances are $S_1^2 = 4.501429$, and $S_2^2 = 9.975524$. So the pooled standard error of the means is

$$S_{x_1, x_2} = \sqrt{\frac{S_1^2 \, (n_1 - 1) + S_2^2 \, (n_2 - 1)}{n_1 + n_2 - 2}}$$

$$= \sqrt{\frac{4.501429\,(14) + 9.975524\,(14)}{28}} = 2.690442\,\text{kg},$$

then

$$S_{\bar{x}_1, \bar{x}_2} = S_{x_1, x_2}\sqrt{2/15} = 0.9824104\,\text{kg}.$$

From a table such as Appendix 3 you can find that $t_{\alpha/2=0.005\%,\,28} = 2.76326246$. The difference between the **estimated** means, the effect size (which estimates delta δ), is $D = \bar{x}_2 - \bar{x}_1 = 4.65333 - 4.2000 = 0.45333$. The CI is found by using t in Equation 6.6,

$$D - t_{\alpha/2.\,df} S_{\bar{x}_1, \bar{x}_2} \le \mu \le D + t_{\alpha/2.\,df} S_{\bar{x}_1, \bar{x}_2},$$

which is estimated as

$$0.453333\,\text{kg} - 2.76326246\,(0.9824104\,\text{kg}) \le \mu \le 0.453333\,\text{kg} + 2.76326246\,(0.9824104\,\text{kg})$$

and that equals

$$-2.261325\,\text{kg} \le \mu \le 3.167991\,\text{kg},$$

which when rounded is

$$-2.27\,\text{kg} \le \mu \le 3.17\,\text{kg}.$$

Notice how rounding down on the left made the negative more negative. No, you cannot infer anything about statistical significance from a CI. (CIs and tests are like different universes.)

9. Your sample is $n = 250$ salmon and $r = 155$ have sea lice. The 99% confidence interval for the proportion of this fish population that have sea lice can be estimated using $\hat{p} - Z_{\alpha/2}\sqrt{\hat{p}\hat{q}/n} \le p \le \hat{p} + Z_{\alpha/2}\sqrt{\hat{p}\hat{q}/n}$. Starting with $\hat{p} = r/n = 0.6200$ so that $\hat{q} = 1 - \hat{p} = 0.3800$, and looking up $Z_{0.005} = 2.575829$, then

$$0.6200 - 2.57583\sqrt{0.6200 \times 0.3800/250} \le p \le 0.6200 + 2.57583\sqrt{0.6200 \times 0.3800/250},$$

$$0.54093 \le p \le 0.69907.$$

Because the input is all integers, you can report endless numbers of figures, but the accuracy of your source of Z actually limits the accuracy. However, in practice you might only be interested in the nearest tenth of a percent, $54.0\% \le p \le 70.9\%$. That seems pretty wide for a lot of work and expenditure, but research is expensive.

11. You take a random sample of $n = 250$ out of the tens of thousands of new inductees and your very accurate test indicates that $r = 88$ inductees in this sample have never been exposed to chickenpox, Herpes zoster. The 95% CI for the number of soldiers per company which have never been exposed to this common viral infection can be estimated using Equation 6.10, $r - Z_{\alpha/2}\sqrt{n\hat{p}\hat{q}} \le \mu \le r - Z_{\alpha/2}\sqrt{n\hat{p}\hat{q}}$ where $\hat{p} = r/n = 0.352$ and $\hat{q} = 1 - p = 0.648$ and you decide to conservatively use $Z_{0.025} \approx 2$. By substitution,

$$88 - 2\sqrt{250 \times 0.352 \times 0.648} \le \mu \le 88 + \sqrt{250 \times 0.352 \times 0.648}$$

$$72.89715 \le \mu \le 103.10265.$$

Which appropriately (that means with the least unjustified optimism) rounds to:

$$72 \leq \mu \leq 104.$$

Obviously you should round (down on the left and up on the right) to whole numbers.

13. If you were to only compare two CIs the primary difficulty would be that overlap does not mean statistical non significance, and a lack of overlap requires a larger difference than does an actual test.

Answers to Odd Chapter 7 Questions

1. Assuming that you have two distributions, you can, in effect, increase the separation between means on the Z scale by increasing the sample size (increase n).

3. The word 'reduce' indicates a single direction, hence one tail, and because you are testing against a hypothesized mean, μ_0, it will be a one sample test (one sample to find $\bar{x}$). The value of μ_0 will not come into the estimate, but the reasonable minimum difference, $D_{min} = 2.000$ mmhg, will. Starting with a good estimate of the standard deviation of the blood pressures, $S = 14.530$ mmhg and Equation 7.3, $\hat{n} = \frac{\left(Z_a + Z_\beta\right)^2 S^2}{D_{min}^2}$ (its variant $\hat{n} = \left(\frac{(Z_\alpha + Z_\beta) S}{D_{min}}\right)^2$ is also acceptable). You can find that $Z_{0.05} = 1.64485$ and $Z_{0.80} = 0.84162$ using tables like in Appendix 4. Then $\hat{n} = \frac{(1.6449 + 0.8416)^2 (14.530)^2}{2.0000^2}$ works out to be $\hat{n} = 326.324$, which would round up to 327, but that sample size would be too small to use Z in your test. So, appropriately rounding down (because this is not the last step) instead, you have a tentative estimate of $\hat{n} = 326$, giving a provisional $df = 326 - 1 = 325$. Referring to tables and using $t_{\alpha=5\%, df=325} = 1.6514$, and $t_{\beta=20\%, df=325} = 0.8432$ with Equation 7.4, $\hat{n} = \frac{\left(t_{a, \hat{n}-1} + t_{\beta, \hat{n}-1}\right)^2 S^2}{D_{min}^2}$,

$$\hat{n} = \frac{(1.6514 + 0.8432)^2 (14.530)^2}{4.0000},$$

which is 328.4529. Rounding up gives $\hat{n} = 329$. Consideration of a table of t **suggests** that results would not change at this level of precision if you iterated again, so this might be the sample size you want, but let's check: $t_{\alpha=5\%, df=\hat{n}-1=328} = 1.649513$, and $t_{\beta=5\%, df=\hat{n}-1=3285} = 0.8432$, so

$$\hat{n} = \frac{(1.649513 + 0.8427185)^2 (14.530)^2}{4.0000} = 327.8295,$$

which clearly rounds to $\hat{n} = 328$, a change of one, which in some situations could be a meaningful saving as opposed to leaping to conclusions. This is the amount we used to find the input values, and trying again with $df = 327$ does not change the result, so $\hat{n} = 328$ is the **final** estimate.

5. The most popular power is 80%, meaning 4 chances out of 5 of detecting the reasonable minimum difference you have chosen in advance. It is also the **minimum power** that is going to be taken at all seriously by granting agencies and professors. Higher power costs more, and can be, but is not always, greatly more expensive.

7. **In effect**, the t test reduces the overlap by examining the populations of the sample means rather than the populations of the raw data. Two populations with different means will overlap less in terms of their **sample** means rather than in terms of their raw scores. (The larger the sample size for estimated means, the less the conceptual populations of means will overlap. The means show less variability than the raw data points and thus less spread, meaning variability. You might want to see Subsection 7.4.)

9. This may seem to be a repeat of Question 3, but it is not. You wish to see if a particular diet will **change** the average blood pressure by $D_{min} = 2.000$ mmhg, using a criterion $\alpha = 5\%$ and $\beta = 20\%$. The standard deviation is $S = 14.530$ mmhg. Notice that the difference from Question 3 is that this one is about 'change', suggesting either direction. Therefore this question is about a two tail one sample test and you must use $Z_{\alpha/2} = 1.9600$ and, if necessary $t_{\alpha/2, df}$ (rather than $_\alpha$); β is not affected by the number of tails so you still use $Z_{\beta=0.20} = 0.8416$ (and, if necessary $t_{\beta, df}$). So, $\hat{n} = \frac{\left(Z_{a/2}+Z_\beta\right)^2 S^2}{D_{min}^2}$ (its variant $\hat{n} = \left(\frac{\left(Z_{\alpha/2}+Z_\beta\right)S}{D_{min}}\right)^2$ is also acceptable), the estimate $\hat{n} = \frac{(1.9600+0.8416)^2(14.530)^2}{2.0000^2}$ works out to be $\hat{n} = 414.27$. That would round up to $\hat{n} = 415$ if it were the final estimate, but it isn't. It is too small to justify testing with Z. So, appropriately rounding down, you have a tentative estimate of $\hat{n} = 414$,

giving a provisional $df = 414-1 = 413$. So referring to tables and using $t_{\alpha/2=2.5\%, df=413} = 1.9657$, and $t_{\beta=20\%, df=413} = 0.842$ in Equation 7.6, $\hat{n} = \frac{\left(t_{a/2, \hat{n}-1}+t_{\beta, \hat{n}-1}\right)^2 S^2}{D_{min}^2}$

$$\hat{n} = \frac{(1.9657 + 0.8425)^2 (14.530)^2}{4.0000},$$

which is $\hat{n} = 416.22$. Rounding up gives a **final** estimate of $\hat{n} = 417$. Another iteration will **not** change the sample size by a whole integer, so this is the sample size you want. (Notice that two tails **require larger samples** than one tail.)

11. When your estimated sample size exceeds 5%, meaning $1/20$, of your total population, you must apply a finite sample size correction (Subsection 7.12). You have estimated that you need to sample a total of $\hat{n} = 178$ out of a total population with only $N = 870$ adult subjects, which is a lot more than 5% of 870. You improve your estimate and usually reduce your cost in terms of sample size by applying a finite correction, using Equation 5.15,

$$\hat{n}' = \frac{\hat{n}}{1 + \frac{\hat{n}}{N}} = \frac{178}{1 + \frac{178}{870}} = 147.767 \text{ subjects,}$$

which is **rounded up** to **148 subjects**. (178 has already been rounded up as part of the original sample size estimation.)

13. The agency was challenging **H0** $:\mu_0 \leq 1.0$ ppb with $\alpha = 5\%$ and they were willing to accept 0.50 ppb as a reasonable minimum difference in a **one tail** test. From $n = 25$ random air quality samples they found an airborne benzene level of $\bar{x} = 2.6$ ppb. The historical standard deviation estimate was $S = 1.3$ ppb. Because the sample size was small they used t to evaluate it. So the **subsequent power** of their test can be found using Equation 7.11, $\hat{z} = \left|\frac{D_{min}}{S\sqrt{n}}\right| - t_{\alpha, df=n-1}$, and after looking up $t_{0.05, 24} = 1.710882$ they

would have gotten

$$\hat{z} = \left| \frac{0.50}{1.3/\sqrt{25}} \right| - 1.710882 = 0.2121949 \,,$$

give or take a little in the seventh decimal place. Then according to the $+\hat{z}$ rule, Equation 7.14, $|\hat{z}| = \hat{t}_{\beta, 24}$. So, they could read the $\beta = 0.5831$ from a $p >$ table of cumulative probabilities of t. So, the power, $1 - \beta = 0.4169$ or about 41.6% (always round power **down** to whatever accuracy makes sense). Notice they had less than a coin flipping chance of detecting a reasonable minimum difference equal to 0.50 ppb. Yes, there is a **difficulty** regarding assumptions here; the **sample size**, $n = 25$, is **too small** to just assume normality. (This example is an improvement on some actual government reports which not only used such sample sizes, but which failed to put much effort into randomization. Sample sizes this small, and vague population definitions (how do they define 'central city'?), often characterize the information upon which the legislators and regulators of nations act. The hypothetical mean, μ_0, and the actual $\bar{x}$ that they found are irrelevant to the subsequent power except for the fact that the use of a hypothetical μ_0 instead of an additional sample mean indicates a **one sample test**.)

15. The agronomy team examined the effect of two trace element additives on a **large sample size** total of $n = n_1 + n_2 = 2000$ rice paddies in a small region of Southeast Asia. This is obviously more than adequate for the CLT to cut in. (However, the results were **far from statistically significant**.) Equal numbers of paddies with each treatment were sampled randomly, so there were $n = 1000$ paddies per treatment. They could have used t for greater accuracy, but looking at the tables at such high degrees of freedom, they decided a final three figure accuracy would be possible with Z. The sample means are irrelevant to this estimation of the power, but the historical standard deviation for the earlier even larger data they had was $S = 1187$ kg/ha. They had chosen an a priori reasonable minimum difference of $D_{min} = 50$ kg/ha, in favor of **either** treatment, so this is a **two tail test**. They sought to find the power, given the cutoff of $\alpha/2 = 2.5\%$ in each tail. So, they estimated, using Equation 7.17, $\hat{z} = \left| \frac{D_{min}}{S\sqrt{2/n}} \right| - Z_{0.025}$,

$$\hat{z} = \left| \frac{50 \, \text{kg/ha}}{1187 \, \text{kg/ha} \, \sqrt{2/1000}} \right| - 1.9600 = -1.0181 \,.$$

Because the result is negative, the minus $\hat{z}$ rule, Equation 7.14, applies, so $|\hat{z}| = \hat{Z}_{1-\beta} = 1.0181$. Referring to tables or software, they found that the power was only about $1 - \beta = 15.4\%$. What went wrong here? The difficulty resulted from choosing a D_{min} that was so vastly **smaller than the standard deviation** (even of the standard error of the mean).

(Suppose they had chosen $D_{min} = 200$ kg/ha. Then, $\hat{z} = \left| \frac{200 \, \text{kg/ha}}{1187 \, \text{kg/ha} \, \sqrt{2/1000}} \right| - 1.9600 =$ 1.8076 . Because $\hat{z} > 0$, the positive $\hat{z}$ rule, Equation 7.13, $\hat{z} = Z_{\beta}$, so, from either tables or software, $\beta = 0.03533$. Thus, the power would equal $1 - \beta = 0.96466$, 96.4%. But, they did not have a difference that big. A bigger sample might have revealed something, but they simply do not know.) Always plan ahead; the validity of your outcome, and the practicality of what you seek, both depend on the standards you choose **before** the experiment.

17. They used a criterion of $\alpha = 1\%$ and formed a hypothesis pair, **H0**: THERE IS NO DIFFERENCE VS **HA**: THE FRENCH PRODUCT IS HEAVIER The latter, **HA**, indicates a two sample **one tail test**. Their product costs the retailers 0.02 francs more, equals about 2% more than their competitor's product. Assuming both gave you the same amount of candy for the same amount of money before they weighed a random sample of size $n_1 = 250$ of their own product, and only $n_2 = 172$ of the competitor's product, the alleged weight of each was 100 grams. So, $0.0200 \ldots \times 100$ grams $= 2.00 \ldots$ grams, which is the acceptable **reasonable minimum difference**, D_{min}. It is obvious because it is the point where the added weight equals the added cost. (In a sense, tightly linked economic D_{min} values can sometimes be considered to be pre-existing.) The pooled variance for the jawbreakers was $S_{1,2}^2 = 1.6210$. Then $n_1 + n_2 - 2 = 420\,df$, and because $420\,df < 1001\,df$ needed for even three figures using Z, they used t. However, the sample size is unequal, $n_1 \neq n_2$, so they had to use $\frac{n_1+n_2}{n_1 n_2} = 1/n_1 + 1/n_2$ under the radical sign,

$$\hat{z} = \left| \frac{D_{min}}{\sqrt{S^2 \left(\frac{n_1+n_2}{n_1 n_2} \right)}} \right| - t_{cutoff,\, n_1+n_2-2}\,.$$

They consulted a suitable reference and found that $t_{\alpha,\, n_1+n_2-2} = t_{0.01,\, 420} = 2.33525$, and substituting, they got

$$\hat{z} = \left| \frac{2.0000}{\sqrt{1.6210 \left(\frac{250+172}{250 \times 172} \right)}} \right| - 2.33525 = 13.5216\,.$$

This is a positive result, so by the rule about positive outcomes, $\hat{z} = \hat{t}_{\beta.\,420}$, and referring to either tables or to software, results in $\beta = 0.000005$. So, $(1 - \beta) > 0.999995$, a **power** approaching, but not quite reaching, 100% of detecting a difference as large as D_{min}.

19. The U.S National Center for Health Statistics, NCHS, typically does not allow the reporting of an estimate if the relative standard error, RSE, exceeds 30%. Suppose you suspect that the Sweaty Archipelago has a tuberculosis morbidity of 1% ($p = 0.01$), but you want to check this by sampling. To find a reasonable minimum sample size, you can start with

$$n_{min} = \frac{\sqrt{pq}}{0.30p},$$

and plug in the one value you need to start, $p = 0.01$,

$$n_{min} = \frac{\sqrt{0.01 \times 0.99}}{0.30 \times 0.01} = 33.16625\,,$$

which rounds up to a random sample size of **at least** $n = 34$, which would be acceptable if the morbidity was as high as 1%. (However, this is a minimum; if p turned out to be less than 1% you would need a larger sample size just to meet this bureaucratic requirement.)

21. Mega Binary suggests challenging the equality of the number of fish of each gender using a normal approximation with criterion $\alpha = 1\%$ and $n = 1024$ in hopes of detecting a reasonable minimum difference of $\mathcal{D}_{min} = 0.16$. The null hypothesis is **H0**: THERE IS NO DIFFERENCE and the alternative hypothesis is **HA**: SURVIVAL TO ADULTHOOD

IS ASSOCIATED WITH GENDER. **HA** indicates that a two tailed test is called for. Mega's plan would use Equation 7.26, $\hat{z} = \dfrac{|\mathcal{D}_{min}|}{\sqrt{(p_0q_0+p_1q_1)/n}} - Z_{cutoff}$, with $p_0 = 1/2$. We get $p_0 = 1/2$ because the population is $50 : 50$, males:females under **H0**, and we get $p_1 = 1/2 + \mathcal{D}_{min}$. (Notice that this sums two estimates of the variance for pooling, so the denominator under the radical sign is times n rather than $2/n$.) Because this is a two tail question, $\alpha/2 = 0.005$ must be used, and from a dependable program, Mega finds that $Z_{\alpha/2=0.005} = 2.575829$. So, by substitution,

$$\hat{z} = \frac{0.16}{\sqrt{(0.5 \times 0.5) + (0.66 \times 0.34)/1024}} - 2.57583 = 4.857743 .$$

This is positive, so by the rule, Equation 7.13, $\hat{z} = \hat{Z}_\beta$ and referring again to software, Mega finds that β is almost zero to a lot of decimal places. That is, the power would approach 100%. What happened here? Well, $\mathcal{D}_{min} = 0.16$ is a rather **huge** reasonable minimum difference in a binomial situation. (Obviously, $\mathcal{D}_{min}$ needs to be reduced. Maybe you would like to try estimating the power $\mathcal{D}_{min} = 0.10$ as a practice estimate that is **not** required. Hopefully you will rarely have to segue into an experimental plan by starting with something like Mega's proposal.)

23. You have estimated that each of the two samples for your test needs to have a size of $\hat{n} = 86$, but the maximum available sample size for one of the samples is $n_f = 58$. The provisional **total** sample size, $\hat{\mathbf{n}} = 2\hat{n} = 172$ and the provisional **adjustable** sample size is $\hat{\mathbf{n}} - n_f = 114$. Recall that, in testing one sample against another, $efficiency\_loss = 1 - \bar{n}_h/\bar{n}$ as in Section 7.11.2 where $\bar{n} = \hat{\mathbf{n}}/2 = \hat{n}$ is the mean and

$$\bar{n}_h = 2/\left(1/\left(\hat{\mathbf{n}} - n_f\right) + 1/n_f\right) = 86$$

is the harmonic mean. Plugging the values in, the mean, $\bar{n} = 172/2 = 86$ and the harmonic mean is $\bar{n}_h = 2/(1/114 + 1/58) = 76.8837$. (You do not really need to calculate the $efficiency\_lost = 1 - \frac{76.8837}{86} = 0.106003$.) So, the **adjusted total sample size** is

$$\hat{\mathbf{n}}' = \hat{\mathbf{n}}\left(2 - \frac{\bar{n}_h}{\bar{n}}\right) = 172\,(1.106003) = 190.233 ,$$

which rounds up to $\hat{\mathbf{n}}' = 191$. The final size of the member of the pair that is adjustable is

$$\hat{x}_{j,k} = \frac{\left(m_A \sum_{j=1}^{j=m_B} x_{j,k} + m_B \sum_{k=1}^{k=m_A} x_{j,k}\right) - \sum_{h=1}^{h=m_B} \sum_{l=1}^{l=m_A} x_{h,l}}{(m_A - 1)(m_B - 1)}$$

$$\hat{n}' = \hat{\mathbf{n}}' - n_f = 191 - 58 = 133 .$$

Once again, inequality can be expensive.

25. The advantage of an a priori power analysis is that it estimates the sample size you need to properly **center** the confidence interval.

Answers to Odd Chapter 8 Questions

1. The hypothesized $\sigma_0 = 6.552$ mph and the sample had $S = 6.945$ mph with $\alpha = 0.05$ and $n = 150$, so you can apply Equation 8.1, $\chi^2 = \frac{(n-1)S^2}{\sigma_0^2}$ with $n - 1 = 149\,df$. However,

because **HA** refers to any difference (greater or smaller), the criterion is divided between two tails, greater variation and less variation. So you are looking at $\alpha/2 = 0.025$ in each tail. But the tables are just for more, $p >$, variation so you look up $\chi^2_{\alpha=0.025,\,149} = 184.69$ for more variation, and $\chi^2_{\alpha=0.975,\,149} = 117.98$ for less variation. (You might notice that the one for more variation is greater than the $df - 1$ and the one for less variation is less than the $df - 1$ because the theoretical expected χ^2 would equal the $df = n - 1$ under the null hypothesis). The logic of the test is more obvious if the equation is written as $\chi^2 = (n - 1)\frac{S^2}{\sigma^2}$). Substituting into the equation,

$$\chi^2 = \frac{(149)\,(6.945\,\text{mph})^2}{(6.552\,\text{mph})^2} = 149\frac{48.2330}{42.9287} = 167.41\,,$$

which is greater than the df, but it is not as large or larger than $\chi^2_{\alpha=0.025,\,149} = 184.687$. So the evidence from the sample is **not statistically significantly different** (in either direction).

The 95% **confidence interval** of the variance is found by using Equation 8.7,

$$\frac{(n - 1)\,S^2}{\chi^2_{\alpha/2,\,n-1}} \leq \sigma^2 \leq \frac{(n - 1)\,S^2}{\chi^2_{1-\alpha/2,\,n-1}}\,,$$

and substituting gives

$$\frac{(149)\,48.2330\,\text{mph}^2}{184.687} \leq \sigma^2 \leq \frac{(149)\,48.2330\,\text{mph}^2}{117.098}\,.$$

When rounded down on the left and up on the right, this is

$$38.9129\,\text{mph}^2 \leq \sigma^2 \leq 61.3736\,\text{mph}^2.$$

BEWARE: Some tables and software will read from the opposite end of the distribution, so that the divisors above must be reversed. Remember to always put the largest divisor on the left.

(Would you really tell your board of directors or the public about an estimate in terms of something called mph^2. This is not in original units of measure, and how is the layperson supposed to picture a squared mile per hour? That does not mean that there are not appropriate audiences for such considerations, but who is your intended audience? Question 2 relates to this issue. It is best to save the unrounded results of these calculations for Question 2.)

3. You wish to perform an a priori sample size estimate with a reasonable minimum difference of $D_{\sigma\_min} = 0.3\,\text{mph}$ in Question 1 where you have a hypothesized $\sigma_0 = 6.552$ mph and your hypothesis pair indicates a two tail test. You choose $\alpha = 0.05$, so you use $\alpha/2 = 0.025$, and because you have chosen a power of $1 - \beta = 0.90$, you use $\beta = 0.10$. You then use tables or software to find $Z_{\alpha/2} = 1.9600$ and $Z_\beta = 1.2816$. Substituting into Equation 8.11, $\hat{n} = \frac{(Z_{\alpha/2}+Z_\beta)^2 \sigma^2}{4D_\sigma^2}$, you calculate

$$\hat{n} = \frac{(1.9600 + 1.2816)^2\,(6.552)^2}{4(0.3)^2} = 1253.04\,,$$

which should be rounded up to $\hat{n} = 1254$ cars. (In subsequent problems, be aware of whether you are being given the hypothesized standard deviation σ_0, or its square, the hypothesized variance, σ_0^2. They might occur in the same problem in any order.)

5. To calculate an ES for Question 1 as a percent change in the variance due to the treatment, recall that $S = 6.945$ mph and $\sigma_0 = 6.552$ mph and $n = 150$. Then calculate phi square $\phi^2 = \chi^2/n$ as a percent which has great usefulness and meaning to some workers despite the fact that the original units of measure have canceled out. Substituting into $\chi^2 = \frac{(n-1)S^2}{\sigma_0^2}$,

$$\chi^2 = \frac{(149)\,(6.945\,\text{mph})^2}{(6.552\,\text{mph})^2} = 149\frac{48.2330}{42.9287} = 167.41\,.$$

Then substituting into $\phi^2 = \chi^2/n$,

$$\phi^2 = 167.41/150 = 1.11607\,,$$

estimating about 112% or about 1.12 times as much more variance than σ_0^2. (If the test had been statistically significant, one could also say that the effect size was an increase in the variance of about +12%, but statements about percentages can be misunderstood by the public.)

7. Because this is a two tail test (either more variation, or less variation), you can find the subsequent power using Equation 8.13, $\hat{z} = \frac{|D_{min}|}{S/2\sqrt{n}} - Z_{\alpha/2}$. This equation is justified because the standard deviation of the standard deviation (its standard error) is $S_S \approx S/2\sqrt{n}$ and the distribution of S_S quickly approaches normal as the sample size gets large. You are interested in $1 - \beta$, the power, in this case relative to a reasonable minimum difference of $D_{\sigma\_min} = 0.3$ mph. Question 1 gives the official sample standard deviation, $\sigma_0 = 6.552$ mph, and it is the a priori value from which you must work. The sample size, $n = 150$ was also given, and you can look up or remember that $Z_{\alpha/2=0.025} = 1.9600$. So, by substitution,

$$\hat{z} = \frac{|0.3\,\text{mph}|}{6.552\,\text{mph}/2\sqrt{150}} - 1.9600 = -0.8384\,.$$

Apply the negative $\hat{z}$ rule, Equation 7.14, that a negative result ($-\hat{z}$) applies directly to the power. So, looking up $\widehat{Z}_{1-\beta} = -1.16694$, you find $1 - \beta = 0.200903$ or about 20%. That's not much chance of detecting a difference in the standard deviation as small as $D_{\sigma\_min} = 0.3$ mph.

9. This is an F test, $F = S_1^2/S_2^2$ (Equation 8.15). The a priori hypothesis pair is **HO**: THE VARIANCES ARE THE SAME vs **HA**: THE VARIANCES ARE DIFFERENT. The criterion is $\alpha = 0.05$, but the hypothesis is about difference, indicating two tails, so the cutoff is $\alpha/2$ in each tail. For the fruits grown under organic conditions, $S_A^2 = 926.56\,\text{g}^2$ with a sample size of $n = 50$, versus when grown with commercial pesticides, $S_B^2 = 815.18\,\text{g}^2$ with a sample size of $n = 100$ fruits. It is handy, as part of the analysis, to choose the largest variance and put it on top. As the first step (only after everything else), determining **which is larger** is part of this two tail F calculation. Designate the largest unknown true variance σ_1^2 and the other σ_2^2 and put the S^2 sample estimate corresponding to σ_1^2 on top. Then you find that

$$F = S_{1=A}^2/S_{2=B}^2 = 926.56\,\text{g}^2/815.18\,\text{g}^2 = 1.1366$$

with $df_1 = 50 - 1 = 49$ and $df_2 = 100 - 1 = 99$. (The cutoffs for the two tails are different because the F distribution itself is not symmetrical. But choosing the largest avoids this problem). The result, F, is evaluated in an obviously two tail manner using $\alpha/2 = 0.025$ **in each tail**. Use outside tables or software[2] (this book does not have F tables for 2.5%) to show that $F_{0.025,49,99} = 1.5974$ is the cutoff for the upper tail, and $1.1366 < 1.5974$, so the test is **not** statistically significant. (Unfortunately it is difficult to control or assess the power for this test. So, this is not the best use of the F test. A better approach might be to use the Siegel-Tukey ranking as suggested in Section 14.2.2.)

Answers Odd Chapter 9 Questions

1. A randomized complete block ANOVA assumes that the underlying variance, other than the effect of the factors themselves, is the same, σ^2, throughout the experiment. This forces us to design or choose the blocks to be as alike as possible. It greatly limits the versatility of RCBD ANOVAs. Failure to achieve sufficient likeness among blocks will lead to confounding (inclusion of inseparable variance components affecting the results). (The nonparametric Friedman's test avoids some of this problem by allowing a much greater diversity among the blocks.)

3. Mega Juiced Inc. decided to use an ANOVA and a set of trials involving $m = 4$ gypsum and irrigation soil treatments. They chose an $\alpha = 5\%$ criterion and made a hypothesis pair, **H0**: THERE IS NO DIFFERENCE IN SUGAR CONTENT vs **HA**: THERE IS A DIFFERENCE IN SUGAR CONTENT, and they used $n = 10$ randomly selected plots for each treatment, to obtain a grand total of $\mathbf{n} = 40$. They sampled after a few years of these regimes and obtained the sugar content data in grams per liter. The means were:

gr/l	A	B	C	D	$\bar{\bar{x}}$
Mean	204.8	209.3	215.8	213.0	**210.725**

For even better understanding you might wish to refer back to the table in the original question. There were $\mathbf{n} - 1 = 40 - 1 = 39\ df_T$ degrees of freedom. They made a table of the sums of squares,

Squared Deviations from Grand Mean

1872.725625	127.1256	203.7756	298.4256
350.625625	32.77562	2527.576	1.625625
2000.325625	188.3756	10.72563	768.6756
188.375625	2330.476	32.77562	411.0756
2000.325625	115.0256	94.57562	562.8756
161.925625	150.6756	516.4256	1.625625
176.225625	1137.376	68.47563	690.3756
22.325625	661.7756	411.0756	86.02563
137.475625	68.47563	429.5256	7.425625
1244.325625	22.32562	176.2256	1.625625 **TSS**
8154.65625	4834.406	4471.156	2829.756 20289.98

Sq Deviations Treatment Mean fr Grand Mean   **SSA**
Among Treatments   35.1056   2.0306   25.7556   5.1756   680.68 .

Note that SSA was $n = 10$ times the summed squares of the $m = 4$ deviations of the means from the grand mean. Their ANOVA table corresponded to Table 9.1,

---

[2]I found an F to p calculator at David Lane's site, http://davidmlane.com/hyperstat/F_table.html and iterated to find the value used here. Tables of $F_{0.025}$ are not easy to find.

Sources of Variation	df	SS	MS	F
Among treatments	3	680.675	226.8917	$F = 0.41654$
Within	36	19609.300	544.7028	
Total	39	20289.975		

They sought to avoid rounding, however the SS values that take part in a subtraction can be made to have the same number of decimal places as the one with the least (unrounded) decimal places. The MS should be as unrounded as reasonably possible before finding F. They went to a table of F with $df_1 = n - m = 36 = df_W$ and $df_2 = m - 1 = 3 = df_A$ to find the cutoff, $F_{\alpha, df_1, df_2} = F_{0.05, 3, 36} = 2.8663$. But the result, rounded to four decimal places, $F = MS_A/MS_W = 0.4165$ was a lot less than the cutoff $F_{\alpha=0.05, df_1=3, df_2=36} = 2.8663$, so that the test is **not significantly different**. The ANOVA tests for a combined difference between means which would increase the variance. So it uses a one tail F test for **more** variance. They used $F_{0.05, df_1, df_2}$ for an overall $\alpha = 5\%$. (The lack of statistical significance means that such an outcome is reasonably likely to occur by chance alone. So, there is no reason to continue to comparisons between individual treatments using this data. F should finally be reported to four or five decimal places. The $MST = 520.2558$ is not reported here, because it is not used in any of the calculations. These **fictitious** numbers were really drawn from normal distributions, differing only in their finite means but having the same finite common variance. Maybe the sample was too small to be able to detect the difference or maybe chance just happened to obscure the difference. Otherwise this describes the design of an on-farm experiment, a popular approach to agricultural research. However, it is difficult to sufficiently match farms so as to have approximately the same underlying variance, σ^2, and it is hard to adequately motivate and control the farmer-cooperators.)

5. This consists of $m_A = 4$ variants of the candy bar recipe run at random order in $m_B = 8$ almost identical factories (blocks). The hypothesis pair is **H0**: THERE IS NO DIFFERENCE vs **HA**: THERE IS A DIFFERENCE, with a criterion of $\alpha = 0.01$. The resulting sums of squares are shown in the following table. **Replace the $* * *$ marks** with the proper values from calculation and the preceding information. Is the test significant?

Sources Variation	df	SS	MS	F
Among Treatments	$* * *$	29304.17	$* * *$	$* * *$
Among Blocks	$* * *$	36216.73	$* * *$	(*optional*)
Within	$* * *$	$* * *$	$* * *$	
Total	$* * *$	107519.40		

Avoid rounding sums of squares, SS, as much as possible, but try to work with the same number of decimal places where subtraction is involved. The MS should be represented to at least six figures if possible, and F should generally be to four (or five) decimal places.

Sources Variation	df	SS	MS	F
Among treatments	$m_A - 1 = 3$	29304.17	9768.06	4.8842
(due to) Blocks	$m_B - 1 = 7$	36216.73	5173.82	2.5870
Within	$3 \times 7 = 21$	41998.50	1999.93	
Total	$n - 1 = m_A \times m_B - 1 = 31$	107519.40		

Going to a table or suitable software, $F_{0.01, 3, 21} = 3.0725$, so the result for **Among treatments**, $F = 4.8842$, is **statistically significant** at the 1% level of statistical significance. Some authors would incorrectly say that it is 'highly statistically significant', but that would violate the model.

Note: some software, for instance, the default in the R suite, would give $F_{0.01, 7, 21} = 0.1631$ and $F_{0.99, 21, 7} = 3.6396$ because it is looking at the area under the left tail. **When in doubt** look up the value for both α and $1-\alpha$, to **use the larger F value for the upper tail**.

The effect of **blocking** is $F = 2.5870$, and $F_{0.05, 7, 21} = 2.4205$, so the effect of blocking reached statistical significance at the $\alpha = 5\%$ level. However, even if the effect of blocking is not statistically significant, you have **no evidence** that the block effect was not beneficial. Using an RCBD is generally beneficial.

7. Given the following RCBD data, which is missing a value in a cell, you first need to sum the corresponding row and column data as well as the grand total, all as though $* * * = 0$.

Treatments	Block 1	Block 2	Block 3	Unadjusted Sums
A1	52.16	57.89	58.78	
A2	37.50	$0'$	51.36	**88.86**
A2	35.09	66.57	80.53	
A4	49.12	56.36	49.91	
Unadjusted Sums	173.87	**180.82**	240.58	**595.02**

The $0'$ is just temporary as are the sums of the row and column that include it. To fill in the missing $* * *$ data value, multiply the sum of the row by the number of cells in that row and multiply the sum of the column by the number of cells in the column while **temporarily setting** the missing cell equal to 0 for each. Then subtract the grand total and divide the result by **one less than** the number of cells in the row, times **one less than** the number of cells in the column, as in Equation 9.6,

$$\hat{x}_{j,k} = \frac{\left(m_A \sum_{j=1}^{j=m_B} x_{j,k} + m_B \sum_{k=1}^{k=m_A} x_{j,k}\right) - \sum_{h=1}^{h=m_B} \sum_{l=1}^{l=m_A} x_{h.l}}{(m_A - 1)(m_B - 1)}$$

$$= \frac{(4 \times 88.86 + 3 \times 180.82) - 595.02}{(4 - 1) \times (3 - 1)} = 50.48 = x_{A2, B2}$$

The complete table (**adjusted values** are bold) with marginal sums is:

Treatments	Block 1	Block 2	Block 3	Sums
A1	52.16	57.89	58.78	168.83
A2	37.50	**50.48**	51.36	**139.34**
A2	35.09	66.57	80.53	182.19
A4	49.12	56.36	49.91	155.39
Sums	173.87	**231.30**	240.58	**645.75**

Then you need to adjust the degrees of freedom of the MS_w of this RCBD, using $df'_W = (df_T - 1) - df_A - df_B$. That is, you have lost a degree of freedom because one cell is only a substitute. So, $df_T = 10$ and $df'_W = 10 - 3 - 2 = 5$. Now the ANOVA can go on with the reduced degrees of freedom. Of course, the adjusted grand mean ($\bar{\bar{x}} = 645.75/12$), and other means, etc. will also reflect this adjustment.

9. The data represents the computer memory space used per student, in kilobytes, from a sample of $m = 3$ tutorials with $n = 21$ students in each. Your ANOVA has a $MS_W = 600\,\text{KB}^2$, with $df_w = 60$, so the sample size is not quite large enough to allow you to use $Z_{\alpha/2 = 0.025} = 1.9600$, so you need to use $t_{0.025, 60} = 2.00029782$. You **cannot** honestly round down to $Z \approx 2$, so, assuming the data is accurate to at least three places, you must round up to $t \approx 2.001$, no more than four places. The sample means are $\bar{x}_C = 96.10\text{KB}$, $\bar{x}_A = 76.60\text{KB}$, and $\bar{x}_B = 104.3\text{KB}$, with equal sample sizes, $n = 21$. Using

Equation 9.10, $\bar{x} - 2.001\sqrt{MS_W/n} \le \mu \le \bar{x} + 2.001\sqrt{MS_W/n}$, where the whisker length is $2.001\sqrt{600\,\text{KB}^2/21} = 10.6958$. So the CIs for the three treatment means rounded to three figures (four are OK) are:

$$96.1\text{KB} - 10.7\text{KB} \le \mu_C \le 96.1\text{KB} + 10.7\text{KB}, \quad \therefore \; \mathbf{85.4\text{KB}} \le \mu_C \le \mathbf{106.8\text{KB}},$$

$$76.6\text{KB} - 10.7\text{KB} \le \mu_A \le 76.6\text{KB} + 10.7\,\text{KB}, \quad \therefore \; \mathbf{65.9\text{KB}} \le \mu_A \le \mathbf{87.3\text{KB}},$$

$$104.3\text{KB} - 10.7\text{KB} \le \mu_B \le 104.3\text{KB} + 10.7\text{KB}, \quad \therefore \; \mathbf{93.6\text{KB}} \le \mu_B \le \mathbf{115.0\text{KB}}.$$

11. From a sample of $m = 3$ tutorials with equal sample sizes, $n = 21$ students each, your ANOVA has a $MS_W = 600\,\text{KB}^2$, with $df_w = 60$. The control mean $\bar{x}_C = 96.1\text{KB}$, and the effect sizes, ES, are the differences between it and the treatment means, $\bar{x}_A = 76.6\text{KB}$ and $\bar{x}_B = 104.3\text{KB}$. **So**, $ES_A = 76.60\,k - 96.10\text{KB} = \mathbf{-19.50\text{KB}}$, and $ES_B = 104.3\text{KB} - \mathbf{96.10\text{KB}} = \mathbf{8.20\text{KB}}$. The sample size is not quite large enough to allow you to use $Z_{\alpha/2=0.025} = 1.9600$, so you need to use $t_{0.025,\,60} = 2.00029782$. You cannot honestly round down to $Z \approx 2$, so, assuming the data is accurate to at least three places, you must round up to $t \approx 2.001$, no more than four places. The CI formula is

$$(\bar{x}_i - \bar{x}_C) - 2.001\sqrt{MS_W\,(2/n)} \le \mu_i - \mu_C \le (\bar{x}_i - \bar{x}_C) + 2.001\sqrt{MS_W\,(2/n)},$$

the equal sample size version of Equation 9.11, where the whisker length is

$$2.001\sqrt{MS_W\,(2/n)} = 2.001\sqrt{600\,\text{KB}^2\,(2/21)} = 15.13\text{KB}.$$

So the 95% CIs for the ES values (rounded to 3 places due some loss in subtraction) are:

$$-19.50\text{KB} - 15.13\text{KB} \le \mu_A - \mu_C \le -19.50\text{KB} + 15.13\text{KB}, \quad \therefore \; \mathbf{-34.8\text{KB}} \le ES_A \le \mathbf{-4.37\text{KB}},$$

$$8.20\text{KB} - 15.12\text{KB} \le \mu_B - \mu_C \le 8.20\text{KB} + 15.12\text{KB}, \quad \therefore \; -6.92\text{KB} \le ES_B \le \mathbf{23.4\text{KB}}.$$

Answers Odd Chapter 10 Questions

1. Because I have been foolish enough to leave the decision to do multiple comparisons to the last minute, I must compare the $k_{max} = m!/2!\,(m-2)!$ pairings of **all** $m = 3$ treatments with each other. But with only three treatments I need not despair. By substitution, I work this out to be only $k_{max} = (3 \times 2 \times 1)/(2 \times 1)(1) = 3$ comparisons. Having only a few treatments tends to be convenient, but with increased numbers of treatments, the number of comparisons can quickly become impractical. However, with only $k = 3$ comparisons, I can comfortably use an $\alpha' = 1\%$ criterion for each comparison, because as Equation 10.1 shows, $\alpha_{strict} = 1 - (1 - \alpha')^k$, and that works out to $\alpha_{strict} = 1 - (1 - 0.01)^3$ and $\alpha_{strict} = 1 - 0.99^3 = 0.029701$ or about 3%, which is $< 5\%$. (Now try Question 2, if it surprises you, you might have the correct answer.)

3. Notice that when calculating a corrected cutoff, α', to achieve a certain α_{strict}, you generally use α in place of α_{strict} because the context is obvious. To calculate a **Sidak corrected** $\alpha'/2$ needed for an overall $\alpha = 5\%$, where you **a priori** plan to make $k = 8$ independent two tail comparisons, use $\alpha'/2 = \left(1 - (1-\alpha)^{1/k}\right)/2$. So, using substitution

into Equation 10.4, divided by 2 for two tail testing,

$$\alpha'/2 = \left(1 - (1 - 0.05)^{1/8}\right)/2 = 0.006391/2 = 0.0031956\,.$$

Yep, that is all the chance you can take of getting a Type I error for each tail of each comparison! The corresponding t cutoff for $16\,df$ is $t_{(\alpha'/2),\,16'} = 3.1352$. Many researchers consider $t > 3$ or $Z > 3$ to be so unlikely as to be impossible in application. (Experiments with this many treatments and a global $\alpha = 5\%$ are common, but are almost never practical.) There is a special assumption of the Sidak correction, which is that the comparisons are **independent**.

5. To find the efficiency, REEB, for a randomized complete block that has $MS_W = 107.4285$ and $MS_B = 241.2854$, with $m_B = 6$ blocks, and $m_A = 5$ treatments, use Equation 10.14, $REEB = \frac{(m_B-1)MS_B + m_B(m_A-1)MS_W}{(m_A m_B - 1)MS_W}$, to get

$$REEB = \frac{(6 - 1)\,241.2854 + 6\,(5 - 1)\,107.4285}{(30 - 1)\,107.4285}\,,$$

which equals 1.214829 and could be rounded a little and called 121.5%, estimating that this RCBD is 1.215 times as efficient as a one way ANOVA.

7. The number of comparisons is $k = 2$, with an a priori choice of a criterion of $\alpha = 5\%$ but with power, $1 - \beta = 95\%$, and a reasonable minimum difference of $D_{min} = 3.00$ attraction units and an estimate of $S^2 = MS_W = 24.5$ with $87\,df$. They will reasonably plan for all sample sizes to be equal. They will first use Bonferroni's correction (Section 10.1.2) to find the per-treatment $\alpha'/2 = (\alpha/k)/2 = 0.0125$ and $\beta' = \beta/k = 0.025$. Then they must use the corresponding t values, $t_{\alpha'/2=0.0125,\,df=87} = 2.28086$ and $t_{\beta'=0.0250,\,df=87} = 1.98761$ in Equation 10.6 to get

$$\hat{n}' = \frac{2\left(t_{\alpha'/2,\,df_W} + t_{\beta'\,df_W}\right)^2 MS_W}{D_{min}^2} = \frac{2\,(2.28086 + 1.98761)^2\,24.5}{3^2} = 99.1969\,.$$

So, appropriately rounding **up**, $\hat{n}' = 100$. Now because there are $m = 3$ treatments, Equation 10.7 estimates a **total sample size** of $\widehat{n}' = m\hat{n}' = 300$ is necessary. Note that there is **no need for iteration** because the estimate of $S = \sqrt{MS_w}$ is the same with the same number of degrees of freedom, df_W.

9. They found three p_values, $p\_value_1 = -0.42200$, $p\_value_2 = 0.05300$, and $p\_value_3 = 0.15771$. From a suitable source they found that $Z_{-0.42200} = -0.196778$, $Z_{0.05300} = 1.616436$, and that $Z_{0.15771} = 1.00390$. They combined the results, using Stouffer's Z, Equation 10.15, $\widehat{Z} = \sum_{i=1}^{i=k} Z_i/\sqrt{k}$,

$$\widehat{Z}_{p\_value} = (Z_{p\_value1} + Z_{p\_value2} + Z_{p\_value3})/\sqrt{k}\,,$$

where k is the number of experiments. Then by substitution,

$$\widehat{Z}_{p\_value} = (-0.196779 + 1.616436 + 1.003914)/\sqrt{3} = 1.399249\,,$$

which is positive, favoring the marigold powder. Referring to a suitable source of values of Z, they find that their estimated $\widehat{p\_value} = 0.0809$, approximately 8%, which is above the minimum $\alpha = 5\%$ that could <u>suggest</u> statistical significance. (Various notations are acceptable. Generally this sort of calculation is purely exploratory.)

11. Alpha inflation is often a major problem with $k > 10$ outcomes. Extremely tiny probabilities ($\alpha/2 < 0.0025$) per comparison would be needed to assure that all k comparisons could be statistically significant. This would require very large reasonable minimum differences (assuming such could be 'reasonable') or huge sample sizes (and possible hypersensitivity for at least some of the comparisons).

Answers to Odd Chapter 11 Questions

1. It is **not** necessary for the residuals to be normally distributed in order for you to make valid correlation or regression estimates. (But it is necessary if you want to use a t test to see if the results are statistically significant.)

3. Recall that the **explanatory variable does not** have to be normally distributed for a regression, even for testing it. So, let's treat the % sign as though it were a unit of measure (per 100). Find $\hat{b} = \frac{\sum(x-\bar{x})(y-\bar{y})}{\sum(x-\bar{x})^2}$, into which the values from the suitable totals in the table, Figure 11.13 can be substituted to estimate a slope of

$$\hat{b} = \frac{-\$212506.751\%}{1554.885093\%^2} = -\$136.670389/\%\,,$$

which, in this case, estimates an annual per capita loss due to the percent of students not finishing high school. This is a reasonable description of at least some of the downward trend in the scatter plot. To report such a number, you would have to consider the accuracy of the original data. In this case you would round to no more than five figures to report a slope of $-\$136.67$ per % of students not finishing high school, but, **before rounding**, there are still calculations to go. For the whole estimation equation, $a = \bar{y} - b\bar{x}$, you still need to estimate the intercept, a, using $\bar{x} = 23.71577\%$ and $\bar{y} = \$9144.180$ from the table, and equation 11.2, $\hat{a} = \bar{y} - \hat{b}\bar{x}$,

$$a = \$9144.180 - (-\$136.670389/\%) \times 23.71577\% = \$12385.4238\,,$$

with the % canceling out. Now, via the linear equation, $\hat{y} = \hat{a} + \hat{b}\bar{x}$, the average personal income of an individual state, given that the percent of students without high school, $x_i\%$, is known, can be estimated to be

$$\$\hat{y}_i = \$12385.4238 - \$136.670389\, x_i \left(\frac{\%}{\%}\right)\,,$$

and the % cancels out. When conveniently rounded and simplified,

$$\$\hat{y}_i = \$\left(12385 - 136.67x_i\right)\,.$$

Because one or both variables were to only **four figure** accuracy, you should round the final results, $\hat{y}_i$, to **five figure** accuracy. (The $\hat{y}$ might not be fully rounded, in the sense that final zeros can be included for easy reading so long as you say something like, '**rounded to the nearest ten cents**'; **then** $\bar{y} = \$9144.20$ would be accurate to only five figures.) It will be necessary to explore this result further in some of the following questions using your un-rounded information during calculation.

(Note the negative value of $\hat{b}$ indicates that there is a downward slope. Also note that the $\%^2$ is treated as a squared unit of measure, not as an instruction. (**When percentages are very small**, say mostly $< 10\%$, you might consider a log transformation of the dependent variable (Section 11.9.1). To justify having used a log transformation you should be able to demonstrate some correlation between the residuals and x. Aside from the effect of chance and of possible non-linear effects in this question, there remains the question of cause-effect. Is lack of a high school diploma causing the effect, or is history and poverty effecting both the literacy rate and the poverty together? Such questions cannot simply be answered by statistics; it is only a tool for examining, estimating, and describing. they often helps you to discover and observe effects, but it, especially regression and correlation, cannot just hand you an answer in terms of causes.)

5. For this F test you need to construct a table, something like the one below. From Question 1, the SS for Regression is 2.010635, with $df = m - 1$, where m is the number of variables (always $m = 2$ in this kind of simple regression). Of course you will need to fill in actual values from Question 1, Figure 11.13, and the table-like fragment in Question 3. The remaining sum of squares for residuals is found by subtraction. The MS values are found by dividing the SS by their respective df. $F = MSR/MS_\epsilon$ is from another division with $df_1 = 1$ and $df_2 = 48$.

Sources	SS	df	MS	F
Regression	29042163.36	1	29042163.36	26.1990
Residuals, ϵ	53208956.02	48	1108519.92	
Total	82251119.38	49	(not needed)	

Referring to a suitable source, $F_{0.05, 1, 48} = 4.0384$, and this result exceeds it, so the result is statistically significant at the 5% level. Because this is an F test **with $df_{1=1}$**, you could use a t test; that is, you can simply use the square root of $F_{\alpha,1,df_2} = t_{\alpha/2,df_2}$ (note α vs $\alpha/2$),

$$t = \sqrt{F} = \sqrt{MSR/MS_\epsilon} = \sqrt{SSR}/S_\epsilon = 5.1185,$$

which exceeds $t_{\alpha/2, df_\epsilon=48} = 2.0107$.

The F test does **not** directly distinguish between tails. Therefore neither can the derived t test, in which t is the **square root** of an F value with one degree of freedom in the numerator. (One tail Student's t tests can be feasible, but not by this approach; the hypothesis and criterion must be stated **before** you observe or test the particular direction (+ or -) of slope.)

7. It has two tails in this instance because this hypothesis pair, when used directly with a simple t test, indicates either direction. It uses $t = \frac{\hat{b}-b_0}{S_b}$, where $b_0 = \$100/\%$, but b_0 can be any slope you want to test (even $b = \$0/\%$). This explanation uses $\$/\%$ as original units of measure. However, that can be tricky and it is not essential to your answers of this kind of question, especially because they cancel out in finding t. First you need to look at Question 3 to estimate $S_\epsilon^2 = \$^2 53208956.02/48 = \$^2 1108519.92$.

Then substituting this and information from the original table,

$$S_{\mathbf{b}} = S_\epsilon / \sqrt{\sum (x - \bar{x})^2}$$

$$= \sqrt{\$^2 1108519.92} / \sqrt{1554.885093\,\%^2} = \$26.7006972/\%,$$

or for reporting, after all calculation is finished, $S_{\mathbf{b}} = \$26.7007/\%$. This used $\sqrt{MS_\epsilon} = S_\epsilon$, where MS_ϵ was found as in Question 1. **Or**, you could alternatively calculate all of the $\hat{y}_i$ and then the squared differences from y_i, finally summing them to get $\sum (y - \hat{y})^2$ for Equation 11.11,

$$S_{\mathbf{b}} = \sqrt{\sum (y - \hat{y})^2 / (n - 2)} / \sqrt{\sum (x - \bar{x})^2}.$$

Substitution from the table in Question 1 gives the same value. So, using $\hat{\mathbf{b}} = \$136.670389/\%$ before rounding from Question 1,

$$t = \frac{\hat{\mathbf{b}} - \mathbf{b_0}}{S_{\mathbf{b}}} = \frac{\$\,(136.670389 - 100)\,/\%}{\$26.7006972/\%} = 1.37338695$$

and 1.3733 is smaller than $t_{0.025,\,48} = 2.3139$ from a table such as Appendix 3. So, the difference is not statistically significantly different. The hypothesis pair and choosing $\mathbf{b_0}$ must be made **before** the data is obtained. (The sign of $\hat{\mathbf{b}} - \mathbf{b_0}$ determines which tail, $+$, for the right tail and $-$ for the left tail.)

The $RMSE = S_\epsilon = \sum (y - \hat{y})^2 / (n - 1)$ might qualify as an effect size if one is required. The numerator is the square of the differences from the estimated line. For this regression,

$$RSME = 1554.885093/48 = 32.39344\,.$$

The RMSE (also called RMSD) the root mean square error or the root mean square deviation, estimates the S_ϵ, the residual mean square. (RMSE is sensitive to outliers but is still popular and something similar is also used in model building. So, you may need this in your vocabulary.)

9. From the calculations in Question 1, the slope $\mathbf{b} = \$136.670389/\%$. Question 5 shows two ways to calculate $S_{\mathbf{b}} = \$26.4268365/\%$ (with units of measure shown) and $t_{0.025,\,48} = 2.313899$ from an online source or a table. Then the CI is rather straight-forward, $\hat{\mathbf{b}} - t_{\alpha/2,\,df} S_{\mathbf{b}} \le \mathbf{b} \le \hat{\mathbf{b}} + t_{\alpha/2,\,df} S_{\mathbf{b}}$,

$$\$136.670389/\% - 2.313899 \times \$26.7006972/\% \le \mathbf{b} \le \$136.670389 + 2.313899 \times \$26.7006972/\%$$

$$\$75/\% \le \mathbf{b} \le \$198/\%,$$

to the nearest dollar, where $/\%$ reads as "**per percent**." (This CI could have been adjusted to five figures because the input data was mostly around four figures. But the question was about the nearest dollar.)

From Question 1, note that any t adjusts for sample size. This adjustment can often be useful all the way up to over $n = 1000$, as when estimating sample sizes for tests. From Question 1 the unrounded intercept $\mathbf{a} = \$12385.4238$ and from Equation 11.13, $S_\alpha =$

$S_\epsilon \sqrt{\frac{1}{n} + \frac{\bar{x}^2}{\sum(x-\bar{x})^2}}$, where, from Question 5, $S_\epsilon = \sqrt{\$^2 1108519.92} = \$1052.86273$. So,

$$S_\mathbf{a} = \$1052.86273 \sqrt{\frac{1}{50} + \frac{23.71577^2}{1554.88509}} = \$197.681605 .$$

Continuing with $\hat{\mathbf{a}} - t_{\alpha/2,\,df} S_\mathbf{a} \le \mathbf{a} \le \hat{\mathbf{a}} + t_{\alpha/2,\,df} S_\mathbf{a}$,

$$\$12385.4238 - 2.313899 \times \$197.681605 \le \mathbf{a} \le \$12385.4238 + 2.313899 \times 197.681605$$

which is

$$\$11928 \le \mathbf{a} \le \$12843$$

to the nearest dollar and after appropriate rounding (down on the left and up to the right).

11. Pearson's product moment correlation coefficient for the data in Question 1 is found using values already calculated in above odd questions, along with Equation 11.7,

$$R = \frac{\sum(x-\bar{x})(y-\bar{y})}{\sqrt{\sum(x-\bar{x})^2}\sqrt{\sum(y-\bar{y})^2}}$$

$$R = \frac{-\$212506.751\%}{\sqrt{1554.8850093\%^2}\sqrt{\$^2 82251119.380}} = -0.5942274 ,$$

where the + means sloping up, and - means sloping down (as in this case). These signs can be seen as being caused by the denominator always being positive (as is assumed for square roots) while the numerator is not squared. Notice that the units of measure, $\$/\%$, cancel out. However, its square, the coefficient of determination,

$$R^2 = 0.3531062 ,$$

estimates that about 35% of the variation in the response variable is due to a linear relation.

13. An estimated regression line normally hinges on the means at $(\bar{x}, \bar{y})$. However when you force it through the origin, you force it to hinge on $(0,0)$ **as though** $(\bar{x}, \bar{y}) = (0,0)$, meaning it must pass through the origin, and therefore $\mathbf{a} = 0$. To calculate a regression that forces the line from the data in example Subsection 11.4.0.1 through the origin, $(x, y) = (0, 0)$, use Equation 11.6, $\hat{\mathbf{b}} = \frac{\sum_{i=1}^{i=n} x_i y_i}{\sum_{i=1}^{i=n} x_i^2}$. The following illustrates this calculation of $\hat{\mathbf{b}}$:

$$\hat{\mathbf{b}} = \frac{241582384}{190336163} = 1.269241$$

	Cost vs Size, **b|a=0**		
Size x	Cost y	x*x	x*y
2590	3906	6708100	10116540
5311	7000	28206721	37177000
1371	5288	1879641	7249848
4486	4820	20124196	21622520
6964	7312	48497296	50920768
2199	4956	4835601	10898244
2774	5288	7695076	14668912
4542	6704	20629764	30449568
5860	4590	34339600	26897400
2558	4956	6543364	12677448
3298	5732	10876804	18904136
Sum	**=**	190336163	241582384

Sum xy / **Sum** x² = $\hat{\mathbf{b}}$ = **1.2692406**

The result is $\hat{y}_i = 1.269241 x_i$ which is $\hat{y}_i = \mathbf{a} + \hat{\mathbf{b}} x_i$ where $\mathbf{a} = 0$. (This estimate of the slope, $\hat{\mathbf{b}} = 1.2692$, is a lot steeper than the estimate, $\hat{\mathbf{b}} = 0.35088$, from the original example. Of course, you would report to a total of no more than five meaningful figures for publication because the recorded accuracy of the data was only to four figures. (Although foolish in this example, forcing through zero might be appropriate if you were crudely calibrating a meter which should logically go to $y = 0$ when $x = 0$. For such a calibration you might want to compare the slopes, when forced and not forced, anyway. In such an application, they should be close to equal. Also, this technique works better if about half your measurements are negative and the rest are positive.)

15. The environmental scientist who is considering the thickness, y mm, of California Condor egg shells over the x decades since DDT was prohibited, has to use a criterion, $\alpha = 5\%$, and a power, $1 - \beta = 80\%$, to detect a linear relation as strong as, or stronger than, $R_{min}^2 = 0.50$ (as a reasonable minimum difference) to estimate the sample size for a Student's t test of correlation. To do this using $\beta = 1 - (1 - \beta) = 0.20$ and $\alpha = 0.05$, it is first necessary to look up $Z_{\beta=0.20} = 0.841621$ and $Z_{\alpha/2=0.025} = 1.959964$. Summing these gives $Z_{\alpha/2} + Z_{\beta} = 2.801585$, and to get a preliminary size estimate, substituting these for t in Equation 11.19, $\hat{n} = \frac{\left(t_{\alpha/2,\,n-2} + t_{\beta,\,n-2}\right)^2 \left(1 - R_{min}^2\right)}{R_{min}^2} + 2$,

$$\hat{n} = \frac{\left(Z_{\alpha/2,} + Z_{\beta}\right)^2 \left(1 - R_{min}^2\right)}{R_{min}^2} + 2 = \frac{(2.801585)^2 \times (1 - 0.5)}{0.5} + 2$$

$$= \frac{7.84888 \times 0.50}{0.50} + 2 = 9.84888 \,.$$

But t values for very small sample sizes are larger than are Z values, which are only suitable for huge sample sizes. So, rounding this initial estimate gives $\hat{n} = 10$ for the first iteration, using $t_{\alpha/2=0.025,\,\hat{n}-2=8} = 2.306004$ and $t_{\beta=0.20,\,\hat{n}-2=8} = 0.888892$. Then by summing, $t_{\alpha/2} + t_{\beta} = 3.194896$, so the next estimate is

$$\hat{n} = \frac{(3.194896)^2 \times 0.5}{0.5} + 2 = \frac{10.207360 \times 0.5}{0.5} + 2 = 12.207360 \,,$$

which could round up to $\hat{n} = 13$ shells if this were the final estimate. This is equal to the initial estimate, so it is necessary to iterate again, using this sample size estimate rounded the usual way to $n = 12$, meaning $df = 10$, and substituting $t_{\alpha=0.025,\,\hat{n}-2=10} = 3.107197$

and $t_{\beta=0.20,\,\hat{n}-2=10} = 0.879058$, to find

$$\hat{n} = \frac{3.145561^2 \times 0.5}{0.5} + 2 == \frac{9.654673 \times 0.5}{0.5} + 2 = 11.65467\,,$$

which rounds to $\hat{n} = 12$ shells and the iteration process has obviously reached a stable estimate, so stop.

Answers to Odd Chapter 12 Questions

1. The hypothesis pair is **H0:** THE POPULATION REPRESENTED BY THIS SAMPLE CONFORMS TO THE DISTRIBUTION OF AD HOC CUSTOMER PREFERENCE vs **HA:** THE POPULATION SAMPLED DOES NOT CONFORM TO THE AD HOC CUSTOMER PREFERENCE SURVEY, with $\alpha = 5\%$. I counted the number of each surveyed color in the 'random sample' of **n** = 381 candies and produced the following table of expected, E, and observed, O:

	Yellow	Red	Green	Blue	Orange	Total
E	0.16×381 $= 60.96$	0.25×381 $= 95.25$	0.20×381 $= 76.20$	0.31×381 $= 118.11$	0.08×381 $= 30.48$	381
O	41	51	67	116	106	381

Applying Equation 12.1, $P\chi^2 = \sum_{j=1}^{j=k} \frac{(O_j - E_j)^2}{E_j}$, Pearson's chi square,

$$O\chi^2 = \frac{(41 - 60.96)^2}{60.96} + \frac{(51 - 95.25)^2}{95.25} + \frac{(67 - 76.20)^2}{76.20} + \frac{(116 - 118.11)^2}{118.11} + \frac{(106 - 30.48)^2}{30.48}$$

gave

$$P\chi^2 = 215.3562\,.$$

I was only interested in **differences of proportion** resulting in **more variation** than the distribution of interest. So I placed the whole rejection region, $\alpha = 0.05$, in the upper tail of the chi square distribution. This test of fit has $5 - 1 = 4\,df$, and a p >table, such as Appendix 5, gave $\chi^2_{\alpha=0.05,\,df=4} = 9.4877$. The estimated chi square is greater than the cutoff (is less likely); so I **rejected H0**, and accepted **HA**. That is, I found the distribution of the sampled population (OK, whatever that bag represented) was statistically significantly different from the distribution suggested by ad hoc customer preference.

Because of the lack of randomness when sampling one bag, this example is simply an exercise and **not a truly random** sampling process. The website survey only surveyed people who would visit such a website and then opt to enter their preference. Should M and M have changed their color distribution? They could consider a more careful random sample of the shoppers and change the distribution in the bags to match. Notice that this a goodness of fit test; it does not need multiple comparisons.

3. Assuming that your editor will accept a whole row of effect sizes as meeting the ES requirement, probabilities might be more useful effect sizes in application. Reporting $\mathcal{D}$, the percent of the individual difference between the observed probability, p_O, and the expected probability, p_E, for each category, might be seen as a more detailed, as well as more informative, way of presenting effect sizes. That is, ideally, you should be able to present estimates of the difference in relative frequency, which is, under the frequentist approach, an estimate of the **difference in probability**, $\mathcal{D} = \hat{p}_O - p_E$:

color	yellow	red	green	blue	orange
$\widehat{p}_{Oi} = O_i/\mathbf{n}$	0.11	0.13	0.18	0.30	0.28
$p_{Ei} = E_i/\mathbf{n}$	0.16	0.25	0.20	0.31	0.08
$\mathcal{D}$	-5%	-12%	-2%	-1%	20%

(In a sense, frequency could be seen as the original unit of measure. Note that this is not a global ES, but rather a category by category ES. In Question 5 you will find a CI for each difference, $\mathcal{D}_i$, as decimal probability values, using Equation 6.10.)

5. The $1 - \alpha = 95\%$ confidence intervals for the effect sizes, with respect to the probability differences in Question 1, can be based on $\mathcal{D}$, the standard error of the difference. Since $\mathcal{D}$ is a probability, its CI is found by using $S_\mathcal{D}$ the same as you would use $S_{\hat{p}}$ for finding a CI for any probability. This gives the estimates of the differences and their standard errors in the following table:

	yellow	red	green	blue	orange
$\mathcal{D} = (O - E)/\mathbf{n}$	-0.05	-0.12	-0.02	-0.01	0.20
$S_\mathcal{D} = \sqrt{\hat{p}(1-\hat{p})/\mathbf{n}}$	0.0112	0.0166	0.00717	0.00510	0.0205

Notice that each probability (likelihood) estimate is based on a subsample of the total sample size, $\mathbf{n} = 381$. Use the **absolute** estimated probability difference, $|\mathcal{D}|$, for **calculating the bimodal variance**, $S_{\hat{p}}$. Create the CIs for them using Equation 12.13 to get $\delta_p = \mathcal{D} \pm Z_{\alpha/2}S_\mathcal{D}$ (you can use $Z_{\alpha/2=0.05} \approx 2$). Then, **for yellow**, the CI is

$$-0.05 - 2 \times 0.0112 \le \delta_p \le -0.05 + 2 \times 0.0112,$$

which as a percent **difference from** the expected, E, works out to

$$-7.24\% \le \delta_p \le -2.76\%$$

For red the CI is $-15.3\% \le \delta_p \le -8.68\%$.

For green the CI is $-3.43\% \le \delta_p \le -0.566\%$.

For blue the CI is $-2.02\% \le \delta_p \le 0.020\%$.

For orange the CI is $15.9\% \le \delta_p \le 24.1\%$.

The expected in every cell was indeed greater than 10, but the normal approximation works best when p is close to $1/2$. So, a **source of concern** here is that, for some of the CIs, the probabilities involved are very small for a normal approximation. Although this might not invalidate approximations with very large samples, it is a good policy whenever $\widehat{p}_i < 10\%$, to round your results as much as your intended practical application will allow. In this example, rounding to two figures might be better than three. (Maybe reporting CIs for difference from the expected is a bit of overkill for reporting this particular goodness of fit experiment, but it demonstrates a possible technique.)

7. The risk ratio can either be written as

$$R_R = P(exposed \rightarrow outcome)/P(not\_exposed \rightarrow outcome),$$

or $R_R = \frac{a/(a+b)}{c/(c+d)}$. So, for workers exposed to such a spill, the risk ratio can be estimated as follows:

	Illness	No Illness			Illness	No Illness
Exposed	a	b	$=$	Exposed	70	$1190 - 70 = 1120$
Not Exposed	c	d		Not Exposed	15	$3150 - 15 = 3135$

Using Equation 12.16, $R_R = \frac{a/(a+b)}{c/(c+d)}$,

$$R_R = \frac{70/(1120 + 70)}{15/(3135 + 15)} = 12.35294,$$

which is **much** greater than one. This estimates that there is over twelve times more risk for workers exposed to such a spill as compared to workers who are not. It tends to support the idea that the illness is more likely to happen as a result of the exposure. (Question 8 relates to the fact that risk ratio tends to be approximated by the odds ratio when the number of illness cases in the unexposed is $< 10\%$.)

9. **Matthew's correlation coefficient** is used to measure the degree of machine learning at each step. In this set of trials (this experiment) you randomly provide **1000** stationary (negative) objects and **100** moving (positive) objects. However, at this point in your program and test cycle it only attacks **78** moving objects, and it also attacks **120** stationary objects.

	Machine acts				Positive	Negative	
		Positive	Negative	=			
Reality	Positive	TP	FN		Positive	78	22	**100**
	Negative	FP	TN		Negative	120	880	**1000**
						198	**902**	

Using Equation 12.17, $M_{CC} = \frac{TP \times TN - FP \times FN}{\sqrt{(TP+FP)(TP+FN)(TN+FP)(TN+FN)}}$,

$$M_{CC} = \frac{66000}{\sqrt{(198)(100)(1000)(902)}} = 0.4939.$$

A value near $M_{CC} = +1$ is close to complete agreement, and a value near $M_{CC} = -1$ is close to total disagreement. This outcome indicates a partial agreement, and you hope to increase it towards complete agreement.

11. This table is an adaptation of data you might find in the records of a largish rural hospital after the first wave of COVID 19, for the deaths of people age 65 and older vs for people under 65, out of everyone who was hospitalized for COVID 19. The hypothesis pair is **H0**: THERE IS NO DIFFERENCE IN THE NUMBER OF HOSPITALIZED COVID 19 PATIENTS OVER 65 YEARS OLD DYING THAN FOR YOUNGER PATIENTS vs **HA**: THERE IS A DIFFERENCE with $\alpha = 5\%$. To use Pearson's chi square test to see if the difference is statistically significant, it is handy to put the expected values into the 2×2 cells, using equation $E_{i,j} = (n_{\mathbf{r}_i} \times n_{\mathbf{c}_i})/\mathbf{n}$:

COVID 19 Deaths Age $\geq$ 65 vs $<$ 65			
Age	Lived	Died	Total
65 and older	124	50	**174**
E	144.695	29.305	
Under 65	192	14	**206**
E	171.305	34.695	
Total	**316**	**64**	**380**

So for the first cell in the first row, $E_{1,1} = (\mathbf{174} \times \mathbf{316})/\mathbf{380}$ which equals 144.6947 and can be reasonably rounded to 144.695 as shown. (Just check that everything adds up after reasonable rounding.) You can work this equation as often as needed or you can simply subtract to fill out the rest of the row, $E_{1,2} = \mathbf{174} - 144.695 = 29.305$, and likewise for the remaining cell in the first column, $E_{2,1} = \mathbf{316} - 144.695 = 171.305$. You could check all this by filling in the last E cell using the formula $E_{2,2} = (\mathbf{64} \times \mathbf{206})/\mathbf{380} = 34.69474$, which can be rounded to 34.695, which corresponds to what you would get by subtraction and will make every row and column sum to the **marginal** totals. Now you can test to see if there is a statistically significant difference, using the shortcut formula $P\chi^2 = (O - E)^2 \sum_{all} \frac{1}{E}$ and any handy cell, say $O_{1,2} - E_{1,2} = 20.7$, so that $P\chi^2 = (20.7)^2 \left(\frac{1}{144.695} + \frac{1}{29.305} + \frac{1}{171.305} + \frac{1}{34.695}\right)$ which equals 32.41892. From a table of chi square, $\chi^2_{0.05,1df} = 3.84146$, so because the $P\chi^2$ value is bigger, the difference is **statistically significant**.

(Notice that even relatively young people had a sufficiently unpleasant experience as to land them in the hospital and this does not record how many had continuing health problems.)

13. When applied to 2×2 tables, phi square, $\phi^2 = \chi^2/\mathbf{n}$, is equivalent to the coefficient of determination, R^2, and in this case it estimates the proportion of the variance in the combined data that is due to the (mostly linear) effect of age equal to or over 65 vs younger, as

$$\phi^2 = 32.41892/\mathbf{380} = 0.085312,$$

or about 8.5%; the rest might be seen as being due to chance. (However, you will see a very big increase in this value for age 85 and older versus younger in Question 12.)

Answers to Odd Chapter 13 Questions

1. You wish to recalculate the example in Sub-subsection 13.2.1.2, to find the estimated sample size $\widehat{n}$ for each city, to test if there is a reasonable minimum difference in the probability of a party of one ordering more than two beers, still using $\alpha = 5\%$ and $\beta = 10\%$, **but** with $\mathcal{D}_{min} = 0.05000$. Recall that your previous year combined receipts for the two locales suggest $p_0 = 64\%$ of all such customers in a party of one buying more than two drinks. Your hypothesis pair is **H0**: THERE IS NO ASSOCIATION BETWEEN CITY AND THE NUMBER OF REGULAR CUSTOMERS BUYING MORE THAN TWO GLASSES vs **HA**: THERE IS AN ASSOCIATION (a lack of homogeneity). From tables or software you find that $Z_{0.0250} = 1.9600$ and $Z_{0.1000} = 1.2816$. This is clearly a two sample test, so use Equation 7.10.2, $\hat{n} = \left(Z_{\alpha/2} + Z_{\beta}\right)^2 (p_0 q_0 + p_1 q_1)/\mathcal{D}_{min}^2$, where $p_1 = 0.6400 + 0.5000 = 0.6900$ and $Z_{0.0250} = 1.9600$ and $Z_{0.1000} = 1.2816$, to make your estimate. Then

$$\hat{n} = (1.9600 + 1.2816)^2 (0.6400 \times 0.3600 + 0.6900 \times 0.3100)/0.05000^2 = 1867.477$$

for **each pub**, all calculated to four or more figures. This rounds up to a sample size of 1868. This estimated sample is much **larger** than the estimate in the example, **due to the smaller** reasonable minimum difference, $\mathcal{D}_{min}$.

3. Mendel crossed F1 hybrid (Yy) pea plants which had yellow seeds with pure recessive (yy) plants which had green seeds. In accord with his theory, he expected a $p_0 = 1/2$ of the offspring having each color of seeds. Assume that before he crossed these, that he hypothesized **H0**: THE NUMBER OF PLANTS IN EACH CATEGORY WILL BE THE SAME vs **HA**: THE NUMBERS WILL NOT BE THE SAME. He chose $\alpha = 5\%$, $1 - \beta = 80\%$, meaning $\beta = 20\%$, and a reasonable minimum difference of $\mathcal{D}_{min} = 2\%$ so that $p_1 = 0.50000 + 0.020000 = 0.520000$. To estimate the sample size, he sought a highly accurate

table to find that $Z_{\alpha=0.025} = 1.95996$ and $Z_{\beta=0.20} = 0.84162$ and used Equation 7.24,

$$\hat{n} = \frac{\left(Z_{\alpha/2}\sqrt{p_0\left(1-p_0\right)} + Z_\beta\sqrt{p_1\left(1-p_1\right)}\right)^2}{\mathcal{D}_{min}^2} \quad \text{to estimate}$$

$$\frac{\left(1.95996\sqrt{0.50000 * 0.50000} + 0.84162\sqrt{0.520000 * 0.480000}\right)^2}{0.02^2} = 3501.133,$$

which rounds **up** to 3502.

5. The hypothesized probability is specified to be $p_0 = 0.60$, with $c = 8$ species (treatments), and there are $k = 7$ comparisons against the control. So, it will be necessary to calculate $\hat{n}'$, the number of adult aphids to be sampled per plant species against cotton, the control, using an experiment wide $\alpha = 0.05$ and power, $1 - \beta = 0.80$, with a reasonable minimum difference, $\mathcal{D}_{min} = p - p_0 = 0.15$. Using the Sidak correction,

$$(\alpha/2)' = 1 - \sqrt[k]{1 - \alpha/2} = 1 - \sqrt[7]{0.975} = 0.0036103$$

$$\beta' = 1 - \sqrt[k]{1 - \beta} = 1 - \sqrt[7]{0.800} = 0.031375$$

Then looking at tables or software, $Z_{(\alpha/2)'} = 2.686495$ and $Z_{\beta'} = 1.860959$. Now to estimate the sample, use Equation 7.10.2, $\hat{n} = \left(Z_{cutoff} + Z_\beta\right)^2 \left(p_0 q_0 + p_1 q_1\right)/\mathcal{D}_{min}^2$, where $p_A = p_0 + \mathcal{D}_{min} = 0.75$. All this together estimates

$$\hat{n}' = (2.686495 + 1.860959)^2 (0.6000 (0.4000) + 0.7500 (0.2500)) / 0.1500^2.$$

So, $\hat{n}' = 392.9074$, which rounds up to $\hat{n}' = 393$, which is the number required for **each** column (treatment), and the total number of aphids needed is $\widehat{\mathbf{n}}' = c\hat{n}' = 8 \times 393 = 3144$. Sample processing by leaning over a microscope for this experiment would indeed be a pain in the neck. (Maybe use a pattern recognition AI?)

(Note that with such large samples, the Sidak correction produces a slightly smaller estimate than would the more conservative Bonferroni correction. Sidak's correction is permissible because there won't be any interaction between the mothers on different plants in different cages.)

7. Assume I am planning to do the same experiment as Question 5, but with an expensive and hard to get species in the genus <u>Sedum</u> as the **control**. But because I have reason to doubt that I can depend on getting more than $n_f = 100$ parthenogenetic female aphids from this species, I need to adjust the sample size of any column that is compared with it, and that means all the columns. My initial estimate of the per treatment column (plant species) sample size is $\hat{n}' = 220$. I would estimate the unadjusted total sample size to be $\widehat{\mathbf{n}}' = 2\hat{n}' = 440$ per comparison. So, an initial estimate of the sample size for the adjustable member of each pair would be $\widehat{\mathbf{n}}' - n_f = 440 - 100 = 340$. The mean sample size is still $\bar{n} = \left(n_f + \hat{n}'\right)/2 = 220$ and the harmonic mean is

$$\bar{n}_h = \frac{2}{1/n_f + 1/\left(\widehat{\mathbf{n}}' - n_f\right)} = \frac{2}{1/100 + 1/340} = 154.5455.$$

(Do not round much, if any, yet.) In application it is not really necessary to estimate the efficiency lost due to our particular method of sample size estimation. So, I go on to adjust the total sample size for that comparison via Equation 13.2,

$$\widehat{2n''} = 2\hat{n}'\left(2 - \frac{\bar{n}_h}{\bar{n}}\right) = 440\left(2 - \frac{154.5455}{220}\right) = 570.909,$$

which rounds **up** to 571 aphids per comparison. That means that every sample except the control will have a sample size of $\widehat{2n''} - n_f = 571 - 100 = 471$ aphids from every plant **but** the control. So, the total sample size estimate, including the control, is

$$\hat{n}'' = 100 + 7 \times 471 = 3397 \text{ aphids.}$$

This is 1506 more aphids than even in Question 5. (A fixed maximum control size can be very costly. Also, Pearson's chi square with huge sample sizes is suspected to be hyper-sensitive.)

(In reality I might stop to reconsider this experimental design in any case. Such a rapidly reproducing organism would require me to carefully consider timing, etc. No one is going to be able to examine that many aphids under a microscope in even a day or two. Then there is the question of what to do when my technician gets tired of peering at aphids and joins the Foreign Legion. Actually if I were obliged to attempt this I might follow my technician's example.)

Answers to Odd Chapter 14 Questions

1. Start with the raw data, 19, 21, 15, 18, 26, 18, 8, 26, 8, 13, 15, 27, 10, 21, 8, 14, 26, 12, 25, 19, 16, 5, 7, 30, 7, and then order it (sort by ascending value). Assign an order position, or order number, its *rank*. Subsubsection 2.6.3.1 might help. However there are ties, so you must then average the order of the position values for the ties. Here is a (folded) table with the **ranks in the enclosed boxes**:

sorted	5	7	7	8	8	8	10	12	13	14	15	15	16
order	1	2	3	4	5	6	7	8	9	10	11	12	13
finished	1	2.5	2.5	5	5	5	7	8	9	10	11.5	11.5	13

18	18	19	19	21	21	25	26	26	26	27	30
14	15	16	17	18	19	20	21	22	23	24	25
14.5	14.5	16.5	16.5	18.5	18.5	20	22	22	22	24	25

(This data has inconveniently many ties. If the data was not restricted to integers, and if it could have been recorded to some decimal places instead of just nearest integers, it might have been possible to avoid many of the ties, which tend to weaken the results.)

3. Using a Wilcoxon summed rank test for a two tail test at $\alpha = 0.05$ to compare soil temperatures, start with a hypothesis pair, **H0**: THESE TEMPERATURES ARE THE SAME vs **HA**: THEY ARE DIFFERENT. Then rank all the data together and redistribute the rank values to their respective locales as a table:

Soil Temperature, °C, Wilcoxon Test										Sum
Locale A:	6.6	6.5	3.3	4.0	6.8	4.5	0.2	2.5	2.8	
Rank	16	15	7	9	17	10	1	4	5	84
Locale B:	3.9	1.5	5.0	2.0	7.8	6.0	5.5	2.8	5.1	
Rank	8	2	11	3	18	14	13	6	12	87

Since $n_1 = n_2$, you only need to rank the data one way (ascending). The lowest rank sum, W_R, is what you want in this test. Using Equation 14.1,

$$Z_W = \frac{n_R(n_1 + n_2 + 1) - 2W_R}{\sqrt{\frac{n_1 n_2 (n_1 + n_2 + 1)}{3}}},$$

where $W_R = 84$ is the smaller total of the two ranks, and its sample size is $n_R = 9$, so

$$Z_W = \frac{9(9 + 9 + 1) - 2 \times 84}{\sqrt{\frac{9 \times 9(9 + 9 + 1)}{3}}} = 0.1324532,$$

which appears not to be statistically significant ($Z_{\alpha/2=0.025} = 1.959964$). However, at such a small sample size, using this formula is the wrong approach; **instead**, a special table such as Appendix 8 should be used to evaluate W_R directly. Appendix 8 shows that $W_R > 63.108$ would **not** be statistically significant at the two tail overall 5% level. So, the two sample locales are **not statistically significantly** different, so do not reject **H0**. But that does not validate **H0** either. (Notice that these European tables use commas for decimal points.)

5. The Spearman's rank correlation coefficient is the correlation coefficient of individually ranked members of paired data sets. So, for

(27,58), (27,64), (25,77), (34,74), (50,75), (56,76), (52,73),

it is found by first ordering x_i and y_i independently, assigning ranks, and then assigning the **order values**, r_i, back to their respective pairs, and proceeding as per this table:

(x,y)	r_x	$(r_x - \bar{r}_x)$	$(r_x - \bar{r}_x)^2$	r_y	$(r_y - \bar{r}_y)$	$(r_y - \bar{r}_y)^2$	$(r_x - \bar{r}_x)(r_y - \bar{r}_y)$
(27,58)	2.5	-1.5	2.25	1	-3	9	4.5
(27,64)	2.5	-1.5	2.25	2	-2	4	3
(25,77)	1	-3	9	7	3	9	-9
(34,74)	4	0	0	4	0	0	0
(50,75)	5	1	1	5	1	1	1
(56,76)	7	3	9	6	2	4	6
(52,73)	6	2	4	3	-1	1	-2
$\sum$		0	27.5		0	28	3.5
Mean	4			4			

Then to estimate Spearman's rho, ρ_s, the amount of correlation, apply Equation 14.11,

$$\hat{\rho}_s = \frac{\sum_{i=1}^{i=n}(r_{x,i} - \bar{r}_x)(r_{y,i} - \bar{r}_y)}{\sqrt{\sum_{i=1}^{i=n}(r_{x,i} - \bar{r}_x)^2 \sum_{j=1}^{j=n}(r_{y,i} - \bar{r}_y)^2}}$$

$$= \frac{3.5}{\sqrt{27.5 \times 28}} = 0.1261312,$$

which rounds to $\hat{\rho}_s = 0.126$. (That is not very much of a correlation, whether non-linear or not.) The square of this gives the fraction $\hat{\rho}_s^2 = 0.0159091$, which rounds to estimate that $\hat{\rho}_s^2 = 0.0159$, or 1.59%, of the observed variation resulted from a monotonic relation, the tendency to vary either way.

7. To test the statistical significance of a Spearman's correlation in which $\hat{\rho}_s = 0.70$ and $n = 36$ points, (x_i, y_i), at the prechosen $\alpha = 5\%$ level, you need to use Equation 14.12, $t = \rho_s \sqrt{\frac{n-2}{1-\rho_s^2}}$ with $n - 2$ degrees of freedom. This works out to

$$t = 0.7\sqrt{\frac{34}{1 - 0.49}} = 5.715476\,,$$

which is greater than $t_{0.025,\,34} = 2.0322$ and therefore **statistically significant**. (This works, not because the raw data is normally distributed, **it is not**, but because of the tendency of ranks to converge to normality.)

9. To estimate the a priori sample size for a Spearman's correlation, given a power of 80%, a (two tail total) criterion of 5%, and a reasonable minimum estimate of the contribution to observed variation of $\rho_{s\_min}^2 = 0.49$ (a Spearman's correlation estimate of $\rho_s = 0.70$), you can use Equation 14.13, $\hat{n} = \frac{(1-\rho_{s\_min}^2)}{\rho_{s\_min}^2}\left(t_{\hat{n}-2,\alpha/2} + t_{\hat{n}-2,\beta}\right)^2 + 2$. However, because you do not have a beginning provisional $\hat{n}$, you must start with $Z_{0.025} = 1.9600$ and $Z_{0.200} = 0.8416$. So,

$$\hat{n} = \frac{(0.51)}{0.49}\left(1.9600 + 0.8416\right)^2 + 2 = 10.16933\,,$$

which would round up to $\hat{n} = 11$ sample pairs, but because we are not on the last iteration it rounds the usual way to 10. Now you can try this as a trial sample size for a provisional $df = 10 - 2$ to use t, with $t_{0.025,\,8} = 2.306002$ and $t_{0.200,\,9} = 0.8834039$, to estimate that

$$\hat{n} = \frac{(0.51)}{0.49}\left(2.26216 + .888860\right)^2 + 2 = 12.3305\,,$$

which rounds up to $\hat{n} = 13$ if this were the last iteration; because that is a noticeable change, we need to check if it has stopped changing. To do that, try it rounded the usual arithmetic way, to 12, then using $t_{0.025,\,10} = 2.22814$ and $t_{0.200,\,10} = 0.879058$ gives

$$\hat{n} = \frac{(0.51)}{0.49}\left(2.22814 + 0.87906\right)^2 + 2 = 12.04876\,,$$

and that still rounds up to $\hat{n} = 13$, the **final answer**. (That might not be sufficient for the requirement of normality to be met, because the aid of the CLT will be minimal. If you can afford it, you would be well advised to try for a higher power or a smaller criterion, or both, to try to attain an a priori plan with $n \geq 30$ pairs.)

11. The data is for $m_b = 4$ drastically different measurements (A through D) from each of $m = 7$ patients (Roman numerals I through VII) with severe birth defects. The client wants to challenge whether the patients fit into the same population, so you a priori choose an $\alpha = 5\%$ and a hypothesis pair, something like **H0**: THE PATIENTS FIT INTO THE SAME POPULATION VS **HA**: THEY DON'T. Using the measurements (A through D) as

	A	rank	B	rank	C	rank	D	rank	Sum	Sum^2
I	1.2	3	16.7	2	0.001	1	90	3	9	81
II	0.9	2	22.5	5	0.012	5	120	7	19	361
III	2.0	7	30.0	7	0.011	4	75	1	19	361
IV	1.3	4	15.6	1	0.008	2	100	5	12	144
V	0.8	1	29.3	6	0.017	7	91	4	18	324
VI	1.9	6	22.1	4	0.013	6	85	2	18	324
VII	1.6	5	19.5	3	0.010	3	110	6	17	289
										1884

blocks, and considering the drastic combination of measurements successfully used in Friedman's 1937 paper, you decide to use a Friedman's test. However, you use the improved estimation method, Equation 14.9. First you find the sum of the squares of the treatment rank totals,

$$\mathbf{T}^2 = \sum_{j=1}^{j=m_B} \left(\sum_{i=1}^{i=m} \mathbf{r}_{i,j} \right)^2 = 1884 \,,$$

shown in the right lower corner of the above table. Because there no ties you can easily find

$$A = \frac{m_b m \,(m+1)\,(2m+1)}{6} \,, \text{ so}$$

$$A = \frac{4 \times 7 \times (7+1) \times (2 \times 7 + 1)}{6} = 560 \,.$$

(If there were ties you would have had to sum the squares of all the individual ranks of the data to get A.) Next,

$$B = \mathbf{T}^2 / m_b = \frac{1}{4} 1884 = 471 \,, \text{ and}$$

$$\widehat{F} = \frac{(m_b - 1) \left[B - m_b m \,(m+1)^2 /4 \right]}{A - B} \,, \text{ so}$$

$$\widehat{F} = \frac{(4-1) \left[471 - 4 \times 7 \times (7+1)^2 /4 \right]}{560 - 471} = 0.7752809 \,.$$

Referring to a table of Fisher's F (such as Appendix 6) with $df_1 = m - 1 = 6$ and $df_2 = (m-1)(m_b - 1) = 18$, you find that $F_{\alpha=0.05, 6, 18} = 3.8957$. Your $\widehat{F}$ was smaller, so the test is not statistically significantly different and the client does not know anything new. (You find you have to repeatedly explain to the client that a **lack** of statistical significance is **not** evidence that the patients are in the same population.)

13. Mega Juiced had a soils and water quality problem in its vineyards, involving sodium, which was affecting the quality of their grapes. They decided upon an ambitious set of trials involving $m = 4$ different gypsum and irrigation treatments. They chose an $\alpha = 5\%$ criterion and made a hypothesis pair, **H0**: THERE IS NO DIFFERENCE IN SUGAR CONTENT vs **HA**: THERE IS A DIFFERENCE IN SUGAR CONTENT. They had a lot of individually irrigated vineyards to choose from and they were able to use $n = 10$ randomly selected areas for each treatment and obtained a total sample size of $\mathbf{n} = 40$, after a few years of these regimes. They obtained the following sugar content data, in grams per liter. The ranking has already been done to simplify your processing:

				Treatment				
A	**rank**	B	**rank**	C	**rank**	D	rank
254	**38**	222	**28**	225	**32**	228	**33**
192	**9**	205	**16.5**	261	**40**	212	**21**
166	**1.5**	197	**10.5**	214	**24**	183	**4**
197	**10.5**	259	**39**	205	**16.5**	231	**35**
166	**1.5**	200	**14**	201	**15**	187	**6**
198	**12**	223	**29**	188	**7**	212	**22**
224	**30**	177	**3**	219	**26**	237	**36**
206	**18.5**	185	**5**	231	**34**	220	**27**
199	**13**	219	**25**	190	**8**	208	**20**
246	**37**	206	**18.5**	224	**31**	212	**23**
Total	2048	**171**	2093	**188.5**	2158	**233.5**	2130	**227**
T^2	-	28900	-	35532.25	-	54522.25	-	51529
Median	198.5		205.5		216.5		212	

From the hypothesis pair, it is obvious that this is a two tail test, but we are going to evaluate it against a chi square value that is for two tails total anyway. So we start the Kruskal-Wallis by applying Equation 14.7 to the sums of the ranks of the $m = 4$ treatments, in two steps with $df = 4 - 1$,

$$\sum_{j=1}^{j=m} \frac{T^2_j}{n_j} = \frac{28900}{10} + \frac{35532.25}{10} + \frac{54522.25}{10} + \frac{51529}{10} = 17048.35,$$

$$H = \frac{12}{n(n+1)} \sum_{j=1}^{j=4} \frac{T^2_j}{n_j} - 3(n+1)$$

$$H = \frac{12}{40(41)} \times 17048.35 - 3(41) = 1.744$$

Compare this H to $\chi^2_{\alpha=0.05, df=3} = 7.815$, and you can see that H is too small, and therefore **not** statistically significant, and Mega Juiced knows nothing new. (This agrees with the ANOVA, Question 3, Chapter 9.)

(I have included the median in this particular table because, if Mega Juiced was interested in a central measure, this would have been closest to what the test sees. It does not fully take account of the more extreme values, and therefore has little, often no, relation to means. In general you can ignore the medians, as in Question 14, because although they are closer to the point, they are not exactly the focus of these rank tests.)

15. You want to compare two different anti-triffid treatments applied a few months ago. But, you discover the plants are alive and still blooming. Is there even a detectable difference in the vigor of this blooming? Your hypothesis pair is **H0**: THERE IS NO DIFFERENCE BETWEEN THE TREATMENTS vs **HA**: THERE IS A DIFFERENCE BETWEEN TREATMENT A AND TREATMENT B. Your technician has weighed 10 pairs of triffid blooms, each pair from a different location with a different altitude and climate. The data is in kilograms of blooms per plant:

Treatment A	5.5	2.7	4.2	3.8	5.4	7.1	5.2	5.7	2.3	3.5	Kg.
Treatment B	2.8	2.2	1.2	3.6	5.8	5.2	4.3	7.7	1.8	1.1	Kg.

Use a Wilcoxon signed rank test to challenge **H0** at the $\alpha = 5\%$ level, recalling that this is a two tail test.

Analysis: Create a table to facilitate this test,

x_A	x_B	$d = x_A - x_B$	$\lvert d \rvert$	rank $\lvert d \rvert$	rank($\lvert +d \rvert$)	rank($\lvert -d \rvert$)
5.5	2.8	2.7	2.7	9	9	
2.7	2.2	0.5	0.5	3.5	3.5	
4.2	1.2	4.0	4.0	10	10	
3.8	3.6	0.2	0.2	1	1	
5.4	5.8	-0.4	0.4	2		-2
7.1	5.2	1.9	1.9	6	6	
5.2	4.3	0.9	0.9	5	5	
5.7	7.7	-2.0	2.0	7		-7
2.3	1.8	0.5	0.5	3.5	3.5	
3.5	1.1	2.4	2.4	8	8	
				T	$T_+ = 46$	$T_- = 9$

This is **not** a 'large' sample size; when, as in this instance, $n \leq 25$, you should refer to a table such as the one in Appendix 9. The second column is for two tails at $\alpha = 0.05$ ($\alpha/2 = 0.25$ for each tail). It says that, given $n = 10$ samples, T_{min} is statistically significant if $T_{min} \leq 8$. So, since $T_{min} = 9$, it is **not** statistically significant, and you know nothing new with respect to the populations represented by the bloomin' triffids.

(However, do not forget that **for $n > 25$** there exists the very good Equation 14.5 for the estimation of the probability of getting a Type I error. It finds Z_W, an approximation of Z that works even at $n = 25$, because the ranks tend to converge towards normality rather rapidly.)

Answers to Odd Chapter 15 Questions

1. You could a priori use the historical population data of the records of the lifespan of the 356 otherwise identical valves made of the alloy presently in use, to see if the distribution is roughly normal, either using the graphical technique in Section 11.10, and/or you could test it for skew (and maybe kurtosis too). If the historical data fails any of these tests you can use it to look for ways to transform it to normally distributed, but the transformations should make some sort of chemical or physical sense. Student's t test heavily depends on the means, and actual raw data is rarely normal. So, because you will have fixed size samples of $n = 12$, why not explore how the means of this size behave? You could have a computer take, say, $m = 250$ random samples, with replacement, of size $n = 12$ from the 356 valve lifespans that are not actually in the study and examine the distribution of their means, $\bar{x}_i$, using a Q-Q plot and maybe calculating the skew. You still cannot be absolutely certain that the distributions of valve lifespans are similar with respect to normality across various alloys, but it is a reasonable guess. (However, maybe a Wilcoxon test would usually be more appropriate for such small samples.)

3. Although rare in any one place, a lot of calves are born with five legs each decade in the world. (Most die almost immediately.) Notice that the number of calves born in the world each decade is enormous. The Poisson distribution involves a very big number of possible outcomes that include a rare event that is of interest, such as calves with five legs, or even two headed calves. Poisson distributions actually tend to have clusters that could sometime include $r = 3$ per decade on one farm due to chance alone. (But, there

is also the possibility that there is a genetic bias in that herd, or a pollution causing such deformities. This is a hard matter to study and would probably involve a survey over many years, even a world-wide study.)

5. Use MAD; IQR would be a disaster. First find the median, $\tilde{x}$, of a random subset of the data that will not be analyzed, then $MAD = median_{i=1}^{i=n} |x_i - \tilde{x}|$. Discard each value, x_i, for which $\frac{|x_i - \tilde{x}|}{1.5\,MAD} > 3.0$, because it can be considered an outlier, and outliers distort estimates of the variance. Then the estimate of the σ^2 for the actual underlying population could be found the usual way, $S^2 = \frac{\sum_{i=1}^{i=n}(x_i - \bar{x})^2}{n-1}$, where the x_i are only from the MAD cleaned data. This information can then be used to find the sample size, $\hat{n}$. Finally use MAD to screen a random sample of size $\hat{n}$ from the remaining data and **use it** in the test.

7. MAD would be good.

9. The stems are 0, 1, 2 and 3 for the powers of ten. (But, depending on the data, they could be the number of hundreds or whatever.) So,

Stems	leaves
0	5 7 7 8 8 8
1	0 2 3 4 5 5 6 8 8 9 9
2	1 1 5 6 6 6 7
3	0

is the stem and leaf plot for 19, 21, 15, 18, 26, 18, 8, 26, 8, 13, 15, 27, 10, 21, 8, 14, 26, 12, 25, 19, 16, 5, 7, 30, 7. The data from which it was drawn happened to be random, so it could be truncated at 5 and 30, only there is no way to know if that is the case. If it were the case, such truncation would affect even what could appear on the beginning stem (0), and there would be less possibilities for the ending stem. Data from truncated samples can be distorted at each extreme by the truncation. Truncation might be why the distribution looks skewed. Making this diagram would have been easier if the data was rearranged from smallest to largest,
5, 7, 7, 8 , 8, 8, 10, 12, 13, 14, 15, 15, 16, 18, 18, 19, 19, 21, 21, 25, 26, 26, 26, 27, 30,
as might be done with the sort feature of a spreadsheet (Subsubsection 2.6.3.1).

BIBLIOGRAPHY

Acock, Alan C. and Gordon R. Staving 1979. A measure of association for nonparametric statistics. Social Forces, University of North Carolina Press.

A'Court, Christine; Stevens, Richard; Heneghan, Carl 2012. Against all odds? Improving the understanding of risk reporting. Br. J. of General Practice, Volume 62, Number 596(4) e220-e223.

Agresti, Alan and Brent A. Coull 1998. Approximate is better than "exact" for interval estimation of binomial proportions. The Am. Statistician Vol. 52, No. 2, pp. 119-126. (May)

Aitken, A.C. 1935. On least squares and linear combinations of observations, Proc. of the Royal Soc. of Edinburgh, vol. 55, pp. 42–48.

Altman D.G., JM. Bland 1995. Absence of evidence is not evidence of absence. BMJ

American Psychological Association 2001. Publication manual of the American Psychological Association (5th ed.). Washington, DC.

Almiron, M.G., B. Lopes, Alyson L. C. O., A. C. Medeiros, and A.C. Frery 2010, On the numerical accuracy of spreadsheets. J. Statistical Software, Vol, 34, No. 4.

Anderson, Gabriella, Haiko Sprott, and Bjorn R Olsen 2013. Non-confirmatory or "negative" results are not worthless. (Opinion) The Scientist (January 15, 2013)

Anscombe, F. J. 1948. The validity of comparative experiments. J. of the Royal Statistical Soc.. Series A (General) 111 (3): 181–211.

Aschwanden, Christie (2015-11-24). Not even scientists can easily explain p-values. FiveThirtyEight. Archived from the original on 25 September 2019. Retrieved 11 October 2019.
https://fivethirtyeight.com/contributors/christie-aschwanden/

Bagui, Sikha and Subhash Bagui 2005. JMASM17: An algorithm and code for computing exact critical values for Friedman's nonparametric ANOVA. Journal of Modern Applied Statistical Methods Copyright © 2005 JMASM, Inc. May, 2005, Vol. 4, No.1, 312-318

Barker, L., Luman, E., McCauley, M., Chu, S. 2002. Assessing equivalence: An alternative to the use of difference tests for measuring disparities in vaccination coverage. Am. J. of Epidemiology

Barnes, J.W. 1994. Statistical Analysis for Engineers and Scientists. McGraw-Hill, New-York.

Barnette J. Jackson and J E McLean 1999 Empirically based criteria for determining meaningful effect size paper , Annual Meeting of the Mid-South Educational Research Association Point Clear, Alabama November 19, 1999

Barnette, J. Jackson, and J E McLean 2001. Distribution characteristics and applications in evaluation of standardized effect sizes in the independent t test. Paper presented to the American Educational Research Association April 11, 2001, Seattel, WA

Barnette, J. Jackson and J E McLean, 2002. The need to abolish the arbitrary effect size standards. paper presented at The Annual Meeting of the Am., Educational Research Association, April 4, 2002, New Orleans LA

Barnette, J. Jackson and J E McLean 2006. Confidence intervals of common effect sizes: What are they good for? Eval. Inst. San Francisco: CA, Annual Meeting of the Am. Educational Research Association.
posted at www.eval.org/summerinstitute/06SIHandouts/SI06.Barnette. TR2.Online.pdf

Barnette, J. Jackson, and J E Mclean 2002. Shedding light on eta squared and omega squared relationships with the standardized effect size. Paper presented to the American Educational Research Association April 2002, New Orleams, LA.

Bayes, Thomas 1763. An essay towards solving a Problem in the Doctrine of Chances. Philosophical Transactions of the Royal Soc. of London 53: 370–418.

Bell Labs 1925. Transmission Circuits for Telephonic Communication.

Benjamin, Daniel J, et al 2020. Redefine statistical significance.PsyArXiv. July 22. (signed by seventy-two scientists)

Bernoulli, Jakob 1714. *Ars Conjectandi: Usum & Applicationem Praecedentis Doctrinae in Civilibus*. Moralibus & Oeconomicis, 1713, Chapter 4, (Translated into English by Oscar Sheynin)

Bienaymé I.J. 1853. *Considérations àl'appui de la découverte de Laplace*. Comptes Rendus de l'Académie des Sciences 37: 309–324.

Blair, R. C., and Higgins, J. J. 1980. A comparison of power of Wlicoxon's rank- sum statistic to that of Student's t statistic under various nonnormal distributions. Journal of Educational Statistics, 5, 309-335.

Blanca, María J, Rafael Alarcón, Jaume Arnau, Roser Bono, Rebecca Bendayan 2017. Non-normal data: Is ANOVA still a valid option? Psicothema 2017, Vol. 29, No. 4, 552-557 (45 citations)

Blanca, María J, Rafael Alarcón, Jaume Arnau, Roser Bono & Rebecca Bendayan 2018, Effect of variance ratio on ANOVA robustness: Might 1.5 be the limit? Behavior Research Methods volume 50, pages937–962 (34 Citations)

Bluman, A.G. 2005. Elementary Statistics. McGraw-Hill.

Bonferroni, C.E. 1935. *Il calcolo delle assicurazioni su gruppi di teste*. In Studi in Onore del Professore Salvatore Ortu Carboni. Rome: Italy, pp. 13-60.

Bonferroni, C.E. 1936. *Teoria statistica delle classi e calcolo delle probabilità*. Pubblicazioni del R Istituto Superiore di Scienze Economiche e Commerciali di Firenze 8, 3-62.

Boos, D.D., and Hughes-Oliver, J.M. 2000. How large does n have to be for Z and t intervals? The Am. Statistician 54:(2) 121-128.

Box G.E.P and D.R. Cox 1964. An analysis of transformations. J R Stat Soc Ser B. 26:211–246.

Box, G.E.P., J.S. Hunter and W.G. Hunter 1978. Statistics for Experimenters: An Introduction to Design, Data Analysis, and Model Building. Wiley.

Box, G.E.P,, J.S. Hunter and W.G. Hunter 2005. Statistics for experimenters, Innovation, and Discovery (2nd Edition). Wiley.

Bradley, J. V. 1978. Robustness? British Journal of Mathematical and Statistical Psychology, 31, 144-152.

Bridge, P. K., and Sawilowsky, S. S. 1999.. Increasing physician's awareness of the impact of statistical tests on research outcomes: Investigating the comparative power of the Wilcoxon Rank-Sum test and independent samples t test to violations from normality. Journal of Clinical Epidemiology, 52, 229-235.

Brown, J.D. 1996. Testing in language programs. Upper Saddle River, NJ: Prentice Hall.

Brown, Morton B., Forsythe, Alan B. 1974. Robust tests for the equality of variances. Journal of the American Statistical Association. 69: 364–367.

Calin-Jageman, Robert J., and Geoff Cumming 2019 Estimation for Better Inference in Neuroscience. eNeuro. 2019 Jul-Aug; 6(4) (Prepublished online 2019 https://www.ncbi.nlm.nih.gov/pmc/articles/PMC6709206)

Camilli, G., and Hopkins, K. D. 1978. Applicability of chi-square to 2×2 contingency tables with small expected cell frequencies. Psychological Bulletin, 85(1), 163-167.

Chalmers, Iain , 2012.The need for reports of new research to begin with up-to-date analyses of what is already known. Scientific Symposium including 4th EQUATOR Annual Lecture. EQUATOR Network and the German Cochrane Centre Freiburg, Germany.

Chen, Z. 2011. Is the weighted z-test the best method for combining probabilities from independent tests? J. of Evolutionary Biology 24: 926-930.

Choi, S.C. 1977. Tests of equality of dependent correlation coefficients. Biometrika 64 (3): 645–647.

Clarke M, Chalmers I. Discussion sections in reports of controlled trials published in general medical journals: islands in search of continents? JAMA. 1998;280:280-2.

Cochran, W.G. 1977. Sampling Techniques. New York: Wiley.

Cochran, G.G. and G.M. Cox 1950. Experimental Designs 2nd ed. John Wiley and Sons. London.

Cohen, J, 1960. A coefficient of agreement for nominal scales. Educational and Psychological Measurement 20 (1): 37–46.

Cohen, J. 1962, The statistical power of abnormal-social psychological research; A review. J. of Abnormal Psychology, **65**,145-153.

Cohen, J. 1977. Statistical Power Analysis for the Behavioral Sciences, Revised Ed. Academic Press. N.Y.

Cohen, J. 1988. Statistical Power Analysis for the Behavioral Sciences (2nd ed.). Hillsdale, NJ: Erlbaum.

Conover, W J 1980. Practical Nonparametric Statistics (2nd ed). New York: Wiley. (Differs from the 1971 printing of the 2nd ed.)

Cornfield, J. 1951. A method for estimating comparative rates from clinical data. applications to cancer of the lung, breast, and cervix. J. of the National Cancer Institute 11:1269–1275.

Cramer, H. 1946. Mathematical Methods of Statistics. Princeton: Princeton University Press, p282.

Cumming, G. 2012. Understanding The New Statistics: Effect Sizes, Confidence Intervals, and Meta-Analysis. New York: Routledge.

Daly L.E. 1998 Confidence limits made easy: interval estimation using a substitution method. Am. J. of Epidemiology 147:783-790

Davenport, C.B. 1904. Statistical Methods with Special Reference to Biological Variation, 2nd ed. Uohn Wiley and Sons NY.

Davenport, E. and El-Sanhury, N. 1991. Phi/phimax: Review and synthesis. Educational and Psychological Measurement, 51:821–828.

Delacre, M., Lakens, D., and Leys, C. (2017, February 17). Why psychologists should by default use Welch's t-test Instead of Student's t-test (in press for the International Review of Social Psychology). https://doi.org/10.31219/osf.io/sbp6k

de Moivre, A. 1733. *Approximatio ad Summam Terminorum Binomii $(a + b)^n$ in Seriem expansi*. (English trans.: The Doctrine of Chances … , 2nd ed. (London, England: H. Woodfall, 1738), pp. 235–243) https://books.google.com/books?id=PII_AAAAcAAJ&pg=PA235#v=onepage&q&f=false

Derrick, B; Ruck, A; Toher, D; White, P 2018. Tests for equality of variances between two samples which contain both paired observations and independent observations. Journal of Applied Quantitative Methods. 13 (2): 36–47.

Dumas-Mallet,Estelle, Katherine S. Button, Thomas Boraud, Francois Gonon, and Marcus R. Munafò 2017. Low statistical power in biomedical science: a review of three human research domains. R Soc Open Sci. 2017 Feb; 4(2): 160254. Published online 2017 at
https://www.ncbi.nlm.nih.gov/pmc/articles/PMC5367316/

Dunn, O.J. 1961. Multiple comparisons among means. J. of the Am. Statistical Association, 56:52-64.

Eisenhart, C. 1947. The assumptions underlying the analysis of variance. Biometrics 3:1-21.

Ellermeier, W. and G. Faulhammer 2000. Empirical evaluation of axioms fundamental to Stevens's ratio-scaling approach: I. Loudness production, Perception and Psychophysics 62 (8): 1505–1511.

Erceg-Hurn, D. M., and Mirosevich, V. M. 2008. Modern robust statistical methods: an easy way to maximize the accuracy and power of your research. American Psychologist 63: (7), 591.

Everitt, B.S. 2003. The Cambridge Dictionary of Statistics, 2nd ed. New York : Cambridge University Press.

Fay, Michael P. and Michael A 2010. Wilcoxon-Mann-Whitney or t-test? On assumptions for hypothesis tests and multiple interpretations of decision rules. PMC Stat Surv. 4: 1–39 National Center for Biotechnology Information, U.S. National Library of Medicine.

Fechner G.T. 1860. *Elemente der Psychophysik*. Leipzig: Breiftkopf and Härtel (an English "introduction" to this is in Wozniak, R. H. 1999. Classics in Psychology, 1855-1914: Historical Essays.)

Fieller, E.C.; Hartley, HO.; Pearson, E. S. 1957. Tests for rank correlation coefficients. I. Biometrika 44:470–481.

Finney, D.J. 1990. Probit Analysis, a Statistical Treatment of the Sigmoid Response Curve. Cambridge UP, Cambridge UK. 253pp

Fisher, R.A. 1922. On the interpretation of $\chi 2$ from contingency tables, and the calculation of. P. J. of the Royal Statistical Soc. 85(1):87–94.

Fisher, R.A. 1925. Applications of "Student's" distribution. Metron 5:90–104.

Fisher, R.A. 1925. Statistical Methods for Research Workers. Oliver and Boyd, Edinburgh.

Fisher, R.A. 1935. The logic of inductive inference (with discussion). J.Roy.Statistici. Soc. 98:39-54.

Fisher, R. A. 1936. The use of multiple measurements in taxonomic problems. Annals of Eugenics. 7 (2): 179–188

Fisher, R.A. 1941. Statistical Methods for Research Workers, Oliver and Boyd, Edinburgh.

Fisher, R.A. 1948. Questions and answers #14. The Am. Statistician 2 (5):30–31.

Fisher, R.A. 1953. The Design of Experiments. Edinburgh, Oliver and Boyd.

Fisher, R.A. 1956. Statistical Methods and Scientific Inference, Edinburgh, Oliver and Boyd.

Fisher, R.A. and Yates, Frank 1963. Statistical Tables for Biological, Agricultural, and Medical Research. 6th ed. New York, Hafner Pub. Co. 146 pp.

Fleischer, A.I. 1980. Confidence intervals for correlation ratios. Educational and Psychological Measurement, 40:659–670

Fox, N. and Mathers N. 1997. Empowering research: statistical power in general practice research. Fam Pract 14:324–29

Freedman, K.B., S. Back and J. Bernstein 2001. Sample size and statistical power of randomized, controlled trials in orthopaedics. J Bone Joint Surg. [Br] 2001;83-B:397-402.

Freund, J.E. and R.E. Walpole 1980. Mathematical Statistics. 3rd. ed. Prentice Hall, London.

Friedman, Milton 1937. The use of ranks to avoid the assumption of normality implicit in the analysis of variance. J. of the Am. Statistical Association (December 1937) 32 (200):675–701. (Also see errata referring to top p 695: Journal of the American Statistical Association Volume 34, 1939 - Issue 205, p 109)

Gauss, Carl Friedrich 1816. *Bestimmung der Genauigkeit der Beobachtungen* (tr. Gr: Determination of the accuracy of the observations). Zeitschrift für Astronomie und verwandt Wissenschaften 1:187–197.

Geary, R.C. 1947. Testing for normality. Biometrika, vol. 34:209 - 242.

Gigerenzer, Gerd, Zeno Swijtink, Theodore Porter, Lorraine Daston, John Beatty and Lorenz Kruger 1989. The Empire of Chance: How Probability Changed Science and Everyday Life. Cambridge University Press.

Gill, Jeff 1999. The insignificance of null hypothesis significance Testing. Political Research Quarterly, 52:(3)647-674.

Glantz, S.A. 1991. Primer of Biostatistics (3rd Ed). New York: McGraw-Hill. pp 144-148.

Glass, G.V., McGaw, B. and Smith, M.L. 1981. Meta-Analysis in Social Research. London: Sage.

Gleick, James 1988. Chaos: Making a New Science. Penguin 352 pp.

Gonick, L. and Smith, W. 1993. The Cartoon Guide to Statistics. New York: Harper Perennial.

Goodman, Steven N. 1999. Toward evidence-based medical statistics. 1: The p value fallacy. Annals of Internal Medicine.130:(12)995-1004.

Grissom, R.J. 1994. Probability of the superior outcome of one treatment over another Journal of Applied Psychology. 79:314-316.

Grissom R.J., and Kim J.J. 2005. Effect Sizes for Research. Mahwah, New Jersey: Lawrence Erlbaum Associates.

Grissom, R. J. and Kim, J.J. 2012. Effect Sizes for Research: Univariate and Multivariate Applications. (2nd ed.). New York, NY: Taylor and Francis

Grunkemeier, Gary L. and Ruyun Jin 2007. Power and Sample Size: How Many Patients Do I Need? Ann Thorac Surg. 83:1934-1939.

Gu, Kangxia; Hon Keung; Tony Ng; Man Lai Tang; and William R. Schucany 2008, Testing the ratio of two poisson rates, Biom J. 50(2):283-98.

Guerriero, V.A. Iannace, S. Mazzoli, M. Parente, S. Vitale, M. Giorgioni 2009. Quantifying uncertainties in multi-scale studies of fractured reservoir analogues: Implemented statistical analysis of scan line data from carbonate rocks. J. of Structural Geology (Elsevier)

Guilford, J. 1936. Psychometric Methods. New York: McGraw–Hill Book Company, Inc.

Hair, J.F., Jr.Anderson, R.E. Tatham, R.L. and Black, W.C. 1998. Mutivariate Data Analysis, New Jersey, Prentice Hall.

Hamilton, Basil L. 1973. An empirical investigation of the effects of heterogeneous regression slopes in analysis of covariance. Educational and Psychological Measurement October 1977. 37:(3)701-712.

Hampel, F.R. 1974. The influence curve and its role in robust estimation. J. Amer. Statist. Assoc. 69:383-393.

Handelsmanj, David J 2002. Optimal power transformations for analysis of sperm concentration and other semen variables. J. of Andrology. 23:(5)629–634

Hampel, F.R. 1974. The Influence Curve and its Role in Robust Estimation. J. of the Am. Statistical Association, 69:383-393.

Harlow, L.L., Mulaik, S. A., and Steiger, J. H. (Eds.) 1997, What if there were no significance tests. Mahwah, NJ: Lawrence Erlbaum Associates.

Hart, Marilyn and Robert Hart 2002. Statistical Process Control for Health Care. Pacific Grove, California: Duxbury Press.

Harremoës, P. and Tusnády, G. 2012. Information divergence is more chi squared distributed than the chi squared statistic. Proc. ISIT 2012. pp. 538–543.

Hayes, A. F., and Cai, L. 2007. Further evaluating the conditional decision rule for comparing two independent means. British Journal of Mathematical and Statistical Psychology, 60:(2), 217–244.

Herrnstein, Richard J. and Charles Murray, 1994. Bell Curve: Intelligence and Class Structure in Am. Life. Free Press Paperbacks N.Y. N.Y. (a rather extreme popularization, but not without some merit)

Hintze, Jerry L. and Ray D. Nelson. 1998. Violin plots: A box plot-density trace synergism. The Am. Statistician 52(2):181-84.

Hoening, John M. and Dennis M. Heisey 2001 The Abuse of Power: The Pervasive Fallacy of Power Calculations for Data Analysis. The American Statistician. 55:(1)19-24.[in reference to attempting to accept **H0**]

Hoey, J. 2012. The two-way likelihood ratio (G) test and comparison to two-way chi-squared test. arXiv:1206.4881

Holcomb W.L. Jr. Chaiworapongsa T. Luke DA. Burgdorf KD. 2001. An odd measure of risk: use and misuse of the odds ratio. Obstetrics and Gynaecology. 98(4): 685-8.

Holm, S. 1979. A simple sequentially rejective multiple test procedure. Scandinavian J. of Statistics 6:(2)65–70.

Hopkins W.G. 2004. Square-root and arcsine-root transformation. In: A New View of Statistics.
(Accessed 29 Nov.l 2011 at www.sportsci.org/resource/stats/index.html)

Hosseini, Davood Seyed 2000. Determination of the mean daily stool weight, frequency of defecation and bowel transit time: Assessment of 1000 Healthy Subjects. Archives of Iranian Med. Volume 3(4) (online journal at) http://www.ams.ac.ir/aim/0034/asl0034.html

Hossjer, Ola, Peter J. Rousseeuw and Christophe Croux 1996, Asymptotics of an Estimator of a robust spread functional. Statistica Sinica 6(1996)375-388.

Hubbard, Raymond and R. Murray Lindsay 2008. Why p values are not a useful measure of evidence in statistical significance testing. Theory Psychology. 18:(1)69-88

Huff, Derrell 1954. How to Lie with Statistics. WW Norton and Co. NY

Hunter, William G. 1981 The practice of statistics: The real world is an idea whose time has come. The Am. Statistician, 35:(2)72-76.

Inman, Hanry, F. 1994. Karl Pearson and R. A. Fisher on statistical tests: A 1935 exchange from nature. The Am. Statistician. 48(1)2-11.

Iman, R.L., and Davenport, J.M. 1980. Approximation of the critical region of the Friedman Statistic. Communications in Statistics A9(6) 571-595.

Ioannidis John P. A. 2005 Why most published research findings are false. PLOS medicine 2(8): e124

Isserlis, L. 1918. On the value of a mean as calculated from a sample. J. of the Royal Statistical Soc. 81:(1)75-81

Jedlicka, Peter 2017. Revisiting the quantum brain hypothesis: Toward quantum (neuro) biology? Front Mol Neurosci. 2017; 10: 366.

Jensen, A.R. 1970. IQ's of identical twins reared apart. Behavjour Genetics. 1:(2)133-146,.

Julious, S.A. 2004, Using confidence intervals around individual means to assess statistical significance between two means. Pharmaceut. Statist. 3: 217-222.

Keats, Jonathon 2013. 20 Things You didn't know about failure. Discover, September 2013.

Kellie B. and Robert J. Pavur 2012. Statistical accuracy of spreadsheet software pages. Statistical Computing Software Reviews265-273 I Received 01 Apr 2009, Published online: 24 Jan 2012
https://www.tandfonline.com/doi/abs/10.1198/tas.2011.09076

Kemmler G, M Hummer, C Widschwendter, and WW Fleischhacker 2005. Dropout rates in placebo-controlled and active-control clinical trials of antipsychotic drugs: a meta-analysis. Arch Gen Psychiatry. 62(12):1305-12.

Kendall, M.G. 1938. A new measure of rank correlation, Biometrika 30:81–93. https://dx.doi.org/10.1093/biomet/30.1-2.81.

Kendall, M.G. 1945. The treatment of ties in rank problems. Biometrika 33:239–251. ttps://dx.doi.org/10.1093/biomet/33.3.239.

Kendall, M.G., A. Stuart 1973. The Advanced Theory of Statistics, Volume 2: Inference and Relationship. Griffin.Kendall.

Kifle,Temesgen, and Isaac Hailemariam Desta 2012 Gender differences in domains of job satisfaction: Evidence from doctoral graduates from Australian universities. Economic Analysis and Policy, Vol. 42:(3)319-338.

Kilem, Gwet 2002. Inter-rater reliability: Dependency on trait prevalence and marginal homogeneity. Statistical Methods for Inter-Rater Reliability Assessment 2:1–10.

Klein R.J., Proctor SE, Boudreault MA, Turczyn KM. 2002. Healthy People 2010 criteria for data suppression. Statistical Notes, no 24. Hyattsville, Maryland: National Center for Health Statistics.

Knezevic, Andrea 2008. StatNews #73. Statistical Consulting Unit Cornell University, Ithaca, NY

Kraemer HC, and Thiemann S 1987. How Many Subjects? Statistical Power Analysis in Research; Sage Publications.

Kruskal, W. and Wallis, W.A. 1952 Use of ranks in one-criterion variance analysis. J. of the Am. Statistical Association 47:(260)583–621.

Lancaster G.A, Dodd S, Williamson PR 2004. Design and analysis of pilot studies: recommendations for good practice. J Eval Clin Pract. 10:307-12.

Lancaster H.O. 1949. The combination of probabilities arising from date in discrete distributions. Biometrica 30:370-382.

Lane, David 1999. Hyperstat, Second Edition. Atomic Dog Publishing, Houston, TX.

Lane, David, M. (PI) 2011.Online statistics education: A multimedia course of study. Rice University (viewed at http://onlinestatbook.com
Also see: http://davidmlane.com/hyperstat/A19196.html).

Langley, Russell 1971. Practical Statistics Simply Explained. Dover, NY

Laplace, Pierre Simon 1812. *Théorie analytique des probabilités*. Paris: Courcier. Reprinted as Oeuvres completes de Laplace7, 1878-1912. Paris: Gauthier-Villars.

Larsen, R.J. and M.L Marx 1986. An Introduction to Mathematical Statistics and Its Applications. Prentice Hall, Englewood Cliffs, NY.

Lee, Peter M 2005. Upper Critical Values for the Kruskal-Wallis Test, University of York.
https://www.york.ac.uk/depts/maths/tables/kruskalwallis.pdf

Leech, Nancy L. and Onwuegbuzie, Anthony J. 2002. A call for greater use of nonparametric statistics. Paper presented at the Annual Meeting of the Mid-South Educational Research Association, Chattanooga, TN, November 6-8, 2002. (Contains 52 references.)
https://files.eric.ed.gov/fulltext/ED471346.pdf;

Leland Wilkinson and the Task Force on Statistical Inference APA Board of Scientific Affairs 1999. Statistical methods in psychology journals guidelines and explanations Leland Wilkinson and the Task Force on Statistical Inference APA Board of Scientific Affairs August 1999. Am. Psychologist. 54:(8)594-604

Lenth, R.V. 2001. Some practical guidelines for effective sample size determination. Am. Statistician. 55:187–193.

Lewicki, Pawel and Thomas Hill 2007. Statistics, Methods and Applications. StatSoft, Tulsa, OK.

Liebetrau, Albert M. 1983. Measures Of Association. Series: Quantitative Applications in the Social Sciences Series, #32 SAGE Publications

Lieges, Sidney 1956. Nonparametric Statistics for the Behavioural Sciences. McGraw-Hill, Kogakusha, LTD, New Delhi.

Lieges, Sidney and Castellan, N. J. 1988. Nonparametric Statistics for the Behavioural Sciences. McGraw-Hill, New York.

Lieges, Sidney and John W. Tukey 1960. A nonparametric sum of ranks procedure for relative spread in unpaired samples. Journal of the American Statistical Association. 55:(291)429-445.

Lindstrom. E.W. 1918. Cornell Experiment Station Memoir 13 (Cited from Snedecor and Cochran 1967)

Lipsey, Mark 1990. Design Sensitivity: Statistical Power for Experimental Research; Sage Publications.

Lipsey, Mark W., Kelly Puzio, Cathy Yun, Michael A. Hebert, Kasia Steinka-Fry, Mikel Cole, Megan Roberts, Karen S. Anthony, Matthew D. Busick 2012. Translating the Statistical Representation of the Effects of Education Interventions into More Readily Interpretable Forms. (NCSER 2013-3000). Washington, D.C: U.S. Government Printing Office, 2012.

Liu, Mozhu 2012. Power and Sample Size for some Chi-Square Goodness of Fit Tests. (Thesis) Department of Mathematics Emporia State Univ. Emporia, Kansas.

Lohr, Sharon L. 1999. Sampling: Design and Analysis. Pacific Grove, California: Duxbury Press.

Lovric, M. (Ed) 2010. International Encyclopedia of Statistical Science. New York: Springer, New York. London.

Derrick, B; Ruck, A; Toher, D; White, P 2018. Tests for equality of variances between two samples which contain both paired observations and independent observations. Journal of Applied Quantitative Methods. 13 (2): 36–47.

Lowry, Richard 2012. Concepts and applications of inferential statistics. VassarStats: Web Site for Statistical Computation (http://vassarstats.net).

Lowry, Richard 2018. Concepts and Applications of Inferential Statistics. (http://vassarstats.net/textbook/ch12a.html).

Luce, R.D. 2002, A psychophysical theory of intensity proportions, joint presentations, and matches. Psychological Review 109:(3)520–532.

Lumley, Thomas, Paula Diehr, Scott Emerson, and Lu Chen 2002. The importance of the normality assumption in large public health data sets, Annual Review of Public Health 23:151-169

Macdonald, Ranald R. 2004, Statistical inference and Aristotle's rhetoric, British Journal of Mathematical and Statistical Psychology. 57, 193–203

Mahalanobis, Prasanta Chandra 1936. On the generalized distance in statistics. Proc. of the National Institute of Sciences of India. 2:(1)49–55.

Mann, Henry B.; Whitney, Donald R. 1947. On a test of whether one of two random variables is stochastically larger than the other. Annals of Mathematical Statistics. 18:(1)50–60.

Maor, Eli 2009. *e*, the Story of a Number. Princeton University Press, 227 pp.

Matthews, B.W 1975. Comparison of the predicted and observed secondary structure of T4 phage lysozyme. Biochim. Biophys. Acta. 405:442–451

McCullough, B.D. and Wilson, Berry 2005. On the accuracy of statistical procedures in Microsoft Excel 2003. Computational Statistics and Data Analysis 49:(4)1244–1252.

McCartney, K., and Rosenthal, R. 2000. Effect size, practical importance, and social policy for children. Child Development , 71:(1)173-180.

McDonald, J.H. 2009. Handbook of Biological Statistics (2nd ed.). Sparky House Publishing, Baltimore, Maryland. (**careful**: check against other sources).

McHugh, Mary L 2012. Interrater reliability: the kappa statistic. Biochemia Medica 2012;22(3):276-82

Mead R. 1988. The design of experiments. Cambridge, New York: Cambridge University Press. 620 p.

Mehta C.R., Patel N.R. 1983. A network algorithm for performing Fisher's exact test in r Xc contingency tables. J. of the Am. Statistical

Mélard, Guy 1914-15 On the accuracy of statistical procedures in Microsoft Excel 2010. Computational Statistics 29(5)

Mendel, J.G. 1866. *Versuche über Pflanzenhybriden Verhandlungen des naturforschenden Vereines*. in Brünn, Bd. IV für das Jahr, 1865 Abhandlungen:3–47. [Translated by: Druery, C.T and William Bateson 1901. "Experiments in plant hybridization". J. of the Royal Horticultural Soc. 26:1–32. [Retrieved 2009-10-09. Association 78:(382)427–434.]

Mehta C.R., Patel N.R. 1983. A Network Algorithm for Performing Fisher's Exact Test in r Xc Contingency Tables. J. of the Am. Statistical Association 78 (382):427–434.

Mehta, C R; Patel, Nitin R; Tsiatis, Anastasios A 1984. Exact significance testing to establish treatment equivalence with ordered categorical data. Biometrics 40:(3)819–825.

Meng, Xiao-Li, Statistics: Your chance for happiness (or misery), Amstat News, September, 2009, p. 43

Meyerhardt J.A., Heseltine D, Niedzwiezki D., et al. 2006. Impact of physical activity on cancer occurrence and survival in patients with stage III colrectal cancer: findings from CALGB 89803 J Clin Oncol 24:(33)3535-3541.

Meyerhardt J.A., Giovannucci EL, Holmes MD, et. al. 2006. Physical activity and survival after colrectal cancer diagnosis. J Clin Oncol 24:(22) 3527-3434.

Micceri, T. 1986 (November). A futile search for that statistical chi mera of normality. Paper presented at the 31st Annual Convention of the Florida Educational Research Association, Tampa.

Micceri, T. 1989. The unicorn, the normal curve, and other improbable creatures. Psychological Bulletin, 105:156-166.

Michael F.W. Festing 2010. Isogenic.info. On line at http://www.isogenic.info/html/resource_equation.html (site intended to minimize cost in terms of animal sacrifice in research)

Mlodinow, Leonard. 2008. The drunkard's walk: How randomness rules our lives. New York: Pantheon. (may note p50)

Mood, A M 1954. On the asymptotic efficiency of certain nonparametric two-sample tests. The Annals of Mathematical Statistics Vol. 25, No. 3 (Sep.), pp. 514-522. P.J.

Mumby P.J. 2002. 'Statistical power of non-parametric tests: a quick guide for designing sampling strategies. Mar Pollut Bull. 2002 Jan;44(1):85-7.

Nanna, M. J., and Sawilowsky , S. S. 1998. Analysis of Likert scale data in disability and medical rehabilitation evaluation. Psychometric Methods, 3, 55-67.

Neyman, J. 1937. Outline of a theory of statistical estimation based on the classical theory of probability. Philosophical Transactions of the Royal Soc. of London A, 236:333–380.

Neyman, Jerzy and Egon Pearson 1933. On the problem of the most efficient tests of statistical hypotheses. Philosophical Transactions of the Royal Soc. of London. Series A, Containing Papers of a Mathematical or Physical Character 231:289–337.

NIST 2013. NIST/SEMATECH e-Handbook of Statistical Methods (years cited in text), http://www.itl.nist.gov/div898//handbook/eda/section3/eda3665.htm

Noether, Gottfried E. 1987. Sample size determination for some common nonparametric tests. J. of the Am. Statistical Association. 82:398. Theory and Methods.

Noether, Gottfried E. 1976. Wilcoxon confidence intervals for location parameters in the discrete case. J. of the Am. Statistical Association. 82:184-188.

Nosek, Brian A., Aarts, Alexander A., Anderson, Joanna E. and Kappes, Heather Barry 2015. Estimating the reproducibility of psychological science. Science, 349 (6251). aac4716-aac4716.

Olejnik, Stephen and James Algina 2003. Generalized Eta and Omega Squared Statistics: Measures of effect size for some common research designs. Psychological Methods. 8:(4)434–447.

Owuegbuzie, A.J., and J.R. Lavine 2003. Without supporting evidence where would measures of substantive importance lead? J. Modern Applied Statistical Methods 2:133-151.

Osborne, Jason 2002. Notes on the use of data transformations. Practical Assessment, Research and Evaluation, 8:(6). http://pareonline.net/getvn.asp?v=8&n=6

Paterson T. A., Harms P. D., Steel P., and Credé M. 2016. An assessment of the magnitude of effect sizes: Evidence from 30 years of meta-analysis in management. J. of Leadership and Organizational Studies 23:(1)66-81.

Pearson, Karl; Lee, Alice; Bramley-Moore, Lesley 1899. Genetic (reproductive) selection: Inheritance of fertility in man, and of fecundity in thoroughbred racehorses. Philosophical Transactions of the Royal Society A. 192: 257–330.

Pearson, Karl 1900. On the criterion that a given system of deviations from the probable in the case of a correlated system of variables is such that it can be reasonably supposed to have arisen from random sampling. Philosophical Magazine, Series 5 50:(302)157–175.

Pearson, Karl 1904. On the theory of contingency and its relation to association and normal correlation, part of the Drapers' Company Research Memoirs Biometric Series I.

Pearson, Karl 1914-31. Tables for Statisticians and Biometricians. Cambridge, University Press, Cambridge, Eng.

Pearson, Karl 1930. The Life, Letters and Labors of Francis Galton, Cambridge University Press.

Pearson, Karl 1935. Statistica tests. Nature, 136:296-297.

Pickover, C.A. 1990. Computers Pattern Chaos and Beauty. St. Martin's Press, New York. 394pp.

Poisson, S.D. 1837. *Probabilité des jugements en matière criminelle et en matière civile, précédées des règles générales du calcul des probabilitiés.* Bachelier, Paris, France

Prajapati, Bhavna, Mark Dunne and Richard Armstrong 2010. Sample size estimation and statistical power analyses. Ot PeerReviewed (clinical) 16/07/10. 10pp.

Purcell, E.M. 1977 Life at low Reynolds number. Lyman Laboratory, Harvard University, Cambridge, Mass. Am. J. of Physics. 45:3-11. (http://jilawww.colorado.edu/perkinsgroup/Purcell_life_at_low_reynolds_number.pdf)

Quade, Dana. 1979. Using Weighted Rankings in the Analysis of Complete Blocks with Additive Block Effects. In: Journal of the American Statistical Association. 74 (367) / 1979, pp. 680-682, doi : 10.1080 / 01621459.1979.10481670

Rakotomalala, R. 2008. *Comparaison de populations.* Tests Non Paramétriques, Université Lumiere Lyon, 2. http://www.academia.edu/download/ 44989200/Comp_Pop_Tests_Nonparametriques. pdf.

Ramsey, Philip H. 1980. Exact Type 1 Error rates for robustness of Student's t test with unequal variances. J Educational and Behavioral Statistics. 5:(4)337-349.

Ramsey, Philip H. and Patricia P. Ramsey 1988. Evaluating the normal approximation to the binomial test. J. of educational and behavioral statistics. 13:(2)173-182.

Rand Corporation 1955. A Million Random figures with 100,000 Normal Deviates. The Free Press, New York.

Ratcliffe JF. 1968. The effect on the t distribution of non-normality in the sampled population. Appl. Stat. 17:42–48.

Rasch, D., Kubinger, K. D., and Moder, K. 2011. The two-sample t test: pre-testing its assumptions does not pay off. Statistical Papers 52:(1), 219–231.

Rein M.G.J. Houbenb, Paco M.J. Welsingb, Gerhard A. Zielhuisa 2005. An investigation of clinical studies suggests those with multiple objectives should have at least 90% power for each endpoint. 59:(1)1-6. (posted on line January 2006. http://www.jclinepi.com/article/S0895-4356%2805%2900264-7/fulltext)

Reinhart, Alex 2015. Statistics Done Wrong The Woefully Complete Guide by 176 pp. No Starch Press, San Francisco

Rice, William R. and Seven D. Gaines 1989. One-way analysis of variance with unequal variances (data analysis/Behrens-Fisher problem/biostatistics). Proc. Nol. A(ad. S(i. USA 86: 8183-8184.

Robinson, David 2017. What is the intuition behind beta distribution? {Cross Validated} https://stats.stackexchange.com/q/47782 (version: 2017-04-08)

Rodger, R.S. 1973. Confidence intervals for multiple comparisons and the misuse of the bonferroni inequality. Br. J. of Mathematical and Statistical Psychology Br. J. of Mathematical and Statistical Psychology. 26(1)58–60.

Rosenthal, R., Rosnow, R. L., and Rubin, D.B. 2000.Contrasts and effect sizes in behavioral research: A correlational approach. New York: Cambridge University Press.

Roush, W.B. and P.R. Tozer, 2004. The power of tests for bioequivalence in feed experiments with poultry. J ANIM SCI 82:13 suppl E110-E118

Rouss, Peter J, and Christophe C. 1993. Alternatives to the median absolute deviation. J. Amer. Statist Assoc. 88:(424)1373-1283 (Theory and Methods).

Rousseeuw, P. J. and Croux, C. 1993. Alternatives to the median absolute deviation. J. Amer. Statist. Assoc. 88:1273-1283

Ruxton, G. D. 2006. The unequal variance t-test is an underused alternative to Student's t-test and the Mann-Whitney U test. Behavioral Ecology. 17:(4), 688-690.

Sakia, R.M. 1992. The Box-Cox transformation technique: a review. The Statistician. 41:169-178,

Sedlmier, P and G. Gigerenzer 1989, Do studies of statistical power have an effect on power of studies? Psychological Bulletin, 105:309-316.

Sawilowsky, Shlomo S.; Blair, R. Clifford 1992. A more realistic look at the robustness and Type II error properties of the t test to departures from population normality. Psychological Bulletin. Vol 111:(2)352-360.

Shapiro, Samuel 1990. How to Test Normality and Other Distribution Assumptions. Milwaukee, WI: Am. Soc. for Quality.

Shieh, Gwowen and Show-Li Jan, 2004. The effectiveness of randomized complete block design. Statistica Neerlandica Vol. 58, nr. 1, pp. 111–124.

Shtaynberger, Jonathan and Haim Bar 2013. Equivalence testing StatNews #85 Cornell U. Statistical Consulting News.

Sidak, Z. 1967. Rectangular confidence regions for the means of multivariate normal distributions. J. of the Am. Statistical Association, 62:626-633.

Siddiqui, Kamran 2013. Heuristics for sample size determination in multivariate statistical techniques. World Applied Sciences Journal 27 (2): 285-287, 2013

Sim, J. and C.C. Wright 2005. The kappa statistic in reliability studies: Use, interpretation, and sample size requirements. Physical Therapy 85:257–268.

Sinclair J.C. Bracken M.B. 1994 Clinically useful measures of effect in binary analyses of randomized trials. J. of Clinical Epidemiology. 47:(8)881-9.

Skovlund D. and Fenstad G.U. 2001. Should we always choose a nonparametric test when comparing two apparently nonnormal distributions? J. Clin. Epidemiol. 54:86–92

Smith, Martha K. 2012. Summer Statistics Institute Course (May 20 - 23, 2013): Common mistakes in using statistics: Spotting them and avoiding them. http://www.ma.utexas.edu/users/mks/statmistakes/StatisticsMistakes.html

Smith, Zachary R. and Craig S. Wells 2006. Central Limit Theorem and Sample Size. Northeastern Educational Research Association, Kerhonkson, New York, 22pp.

Snedecor, G.W. and W.G. Cochran 1967. Statistical Methods, 6th Edition. Iowa State Univ. Press. Ames, IO.

Spearman, C. 1904. The proof and measurement of association between two things. Amer. J. Psychol, 15:72–101

Sokal R.R and Rohlf FJ 1969. Biometry: The Principles and Practice of Statistics in Biological Research. Oxford: W.H. Freeman

Sokal, R.R. and F.J. Rohlf. 2012. Biometry: the Principles and Practice of Statistics in Biological Research. 4th edition. W. H. Freeman and Co.: New York. 937 pp.

StatSoft, Inc. 2011. Electronic Statistics Textbook. Tulsa, OK: StatSoft. http://www.statsoft.com/textbook (You may also get a hard copy from StatSoft via Amazon.com)

Steel R.G.D , J.H. Torrie and D.A. Dickey 1997. Principles and Procedures of Statistics, 3rd Edition. McGraw-Hill.

Steiger, James H. 2004. Beyond the F test: Effect size confidence intervals and tests of close fit in the analysis of variance and contrast analysis. Psychological Methods 9:164-182.

Steingrimsson, R. and Luce, R.D. 2006. Empirical evaluation of a model of global psychophysical judgments: III. A form for the psychophysical function and intensity filtering. J. of Mathematical Psychology 50:(1)15–29.

Steven R, M.Dd Cummings and D.G Grady 2001. Designing Clinical Research, Second Edition.

Stevens, S.S. 1957. On the psychophysical law. Psychological Review 64:(3)153–181.

Stevens, S.S. 1975. Geraldine Stevens, editor. Psychophysics: introduction to its perceptual, neural, and social prospects. Thransaction Books, New Brunswick

Stigler, S.M. 1989. Francis Galton's Account of the Invention of Correlation. Statistical Science 4:(2)73–79.

Stouffer, S.A., E.A. Suchman, L.C. DeVinney, S.A. Star, R.M. Jr. Williams 1949. The Am. Soldier, Vol.1: Adjustment During Army Life. Princeton University Press, Princeton.

Student 1908. The probable error of a mean. Biometrika 6:(1)1–25.

Tabachnick, B.G., and L.S. Fidell 1996. Using multivariate statistics (3rd ed.). New York: Harper Collins.

Taeger, D., Yi Sun and Kurt Straif 1998. On the use, misuse and interpretation of odds ratios. BMJ 1998:316:989.

Taleb, Nassim Nicholas 2007. The Black Swan: The Impact of the Highly Improbable. New York:Random House

Tchebichef, P. 1867. *Des valeurs moyennes*. J. de mathématiques pures et appliquées. 2:(12)177–184.

Theil, Henri 1961. Economic Forecasts and Policy. Holland, Amsterdam: North

Thompson, Bruce 1994. The concept of statistical significance testing. ERIC/AE EDO-TM-94-1 Measurement Update 4:(1)5-6.

Tornetta, Paul III, Heather Locher, and Mohit Bhandari 2000. Type II Error Rates (Beta Errors) of Randomized Trials in Orthopaedic Trauma. Paper #28, 8:44 am. Orthopaedic Trauma Assn, 2000. Annual Meeting, Session V, Friday October 13, 2000.

Trout AT, Kaufmann T.J., Kallmes DF. 2007. No significant difference… Says who? AJNR Am J Neuroradiol. 28:195–197.

Tsang, R., Colley, L., and Lynd, L. D. 2009. Inadequate statistical power to detect clinically significant differences in adverse event rates in randomized controlled trials. Journal of Clinical Epidemiology. 62 (6): 609–616

Tukey, J.W. 1957. The comparative anatomy of transformations. Annals of Mathematical Statistics, 28:602-632.

Tukey, J.W. 1977. Exploratory Data Analysis. Addison-Wesley.

Tuttle, Robert J., Brian Oliver, and Ray McGinnis 2007. Cumulative Probability Plot 3.0. Boeing and U.S. Department of Energy (provided for free free download courtesy of Boing and DOE on http://www.radprocalculator.com).

Valliant, Richard , Alan H. Dorfman, Richard M. Royall 2000. Finite Population Sampling and Inference: A Prediction Approach John Wiley and Sons (Series in Survey Methodology)

van der Vaart, H.R. 1950. Some remarks on the power of Wilcoxon's test for the problem of two samples. Proc. Koninklijke Nederlandse Akadamie van Wetenschappen. 53:507-520.

van Belle, Gerald and Steven P Millard 1998. STRUTS: Statistical Rules of Thumb. Univ. of WA, Seattle, W.A. 14pp. (lots of good information, but the rules of thumb may underestimate required sample sizes.)

Vankov I, Bowers J, and Munafò MR. 2014. On the persistence of low power in psychological science. Q J. Exp. Psychol. (Colchester) 67, 1037-1040.

Veldman, Donald J. and McNemar, Quinn 2007 In Defense of the Chi-Square Continuity Correction. Us Department of Health Education and Welfare Document ED 069 688. 4pp. (also presented at the American Psychological Assn.)

Velleman, Paul F. 2008 "Cited by: Truth, Damn Truth, and Statistics"Cornell University J. of Statistics Education Vol. 16, No. 2

Verbeks, Geert; E. Lesaffre, and B Spiessens 2001. The practical use of different strategies to handle dropout in longitudinal studies. Drug Information J. 35:419–434.

Vickers, Andrew J. 2006. How to randomize. J Soc Integr Oncol. 4:(4)194–198.

Viera, Anthony J. 2008. Odds ratios and risk ratios: What's the difference and why does it matter? Southern Medical J.: July 2008 - Volume 101:(7)730-734.

von Bortkiewicz, Ladislaus 1898. *Das Gesetz der kleinen Zahlen* [The law of small numbers]. BG Teubner, Germany: Teubner.

von Hippel, Paul T. 2005. Mean, median, and skew: Correcting a Textbook Rule. J. of Statistics Education Vol. 13, No. 2
www.amstat.org/publications/jse/v13n2/vonhippel.html

von Hippel, Paul T. 2010. Skewness, Ohio State University, USA *in* Lovric, M. (Ed) 2010. International Encyclopedia of Statistical Science. New York: Springer.
https://lbj.utexas.edu/sites/default/files/file/news/Skew.pdf

Voraprateep, J. 2013. Robustness of Wilcoxon Signed-Rank Test Against the Assumption of Symmetry. (Thesis) University of Birmingham.

Wald, Abraham 1943. Tests of statistical hypotheses concerning several parameters When the number of observations is large. Transactions of the Am. Mathematical Soc., 54:426-482

Wang, Hansheng , Bin Chen, and Shein-Chung Chow 2002. Sample size determination based on rank tests in clinical trials. Hansheng. J. of Biopharmaceutical Statistics. 13:(4)735–751, 2003 735

Wasserstein, Ronald L.; Lazar, Nicole A. (7 March 2016). The ASA's statement on p-values: context, process, and purpose. The American Statistician. 70 (2): 129–133.

Weber, E.H. 1846. *Der Tastsinn und das Gemeingefühl.* In: Wagner R, editor. Handwörterbuch der Physiologie mit Rücksicht auf physiologische Pathologie Vol. 3, pp.481–588 Braunschweig: Vieweg (Available free for Kindle, in German, from Amazon.com).

Whitlock, M.C. 2005. Combining probability from independent tests: the weighted Z-method is superior to Fisher's approach. J. of Evolutionary Biology, 18:1368–1373.

Wilcox, Rand R. 2006. Graphical methods for assessing effect size: Some alternatives to Cohen's d. Journal of Experimental Education, 74 (4), 353–367

Wilcoxon, Frank 1945. Individual comparisons by ranking methods. Biometrics Bulletin 1 (6): 80–83.

Wilcoxon, Frank and Roberta A. Wilcox 1964. Some Rapid Approximate Statistical Procedures, edition revised. Lederle Laboratories. 60 pp

Wolfe R, and J Hanley 2002. If we're so different, why do we keep overlapping? When 1 plus 1 doesn't make 2. CMAJ 166:(1)65–66.

Wright, Gerald C. Jr. 1976. Linear models for evaluating relationships. Am. J. of Political Science 20:349-373.

Yalta, A.T. 2008. On the accuracy of statistical distributions in Microsoft Excel 2007 Computational Statistics and Data Analysis 52:(10)4579-4586

Yates, F 1934 Contingency table involving small numbers and the χ^2 test. Supplement to the J. of the Royal Statistical Soc. 1:(2)217–235.

Yule, G. U. 1903. Notes on the theory of association of attributes in statistics. Biometrika. 2 (2): 121–134.

Zimmer, K. 2005. Examining the validity of numerical ratios in loudness fractionation. Perception and Psychophysics 67:569–579.

Zimmerman, Donald W. 2011. Inheritance of properties of normal and non-normal distributions after transformation of scores to ranks. Psicológica 32:65-85.

Zimmerman D.W. and Zumbo DW. 1992. Para metric alternatives to the student t test under violation of normality and homogeneity of variance. Percept. Motor. Skill. 74:835–44.

Zumbo, B. D., and Coulombe, D. 1997. Investigation of the robust rank-order test for non-normal populations with unequal variances: The case of reaction time. Canadian Journal of Experimental Psychology/Revue Canadienne de Psychologie Expérimentale 51:(2), 139.

Index